AF684150

ELECTROKINETIC AND COLLOID TRANSPORT PHENOMENA

ELECTROKINETIC AND COLLOID TRANSPORT PHENOMENA

JACOB H. MASLIYAH
University of Alberta

SUBIR BHATTACHARJEE
University of Alberta

WILEY-INTERSCIENCE

A JOHN WILEY & SONS, INC., PUBLICATION

Copyright © 2006 by John Wiley & Sons, Inc. All rights reserved.

Published by John Wiley & Sons, Inc., Hoboken, New Jersey
Published simultaneously in Canada

No part of this publication may be reproduced, stored in a retrieval system, or transmitted in any form or by any means, electronic, mechanical, photocopying, recording, scanning, or otherwise, except as permitted under Section 107 or 108 of the 1976 United States Copyright Act, without either the prior written permission of the Publisher, or authorization through payment of the appropriate per-copy fee to the Copyright Clearance Center, Inc., 222 Rosewood Drive, Danvers, MA 01923, (978) 750-8400, fax (978) 750-4470, or on the web at www.copyright.com. Requests to the Publisher for permission should be addressed to the Permissions Department, John Wiley & Sons, Inc., 111 River Street, Hoboken, NJ 07030, (201) 748-6011, fax (201) 748-6008, or online at http://www.wiley.com/go/permission.

Limit of Liability/Disclaimer of Warranty: While the publisher and authors have used their best efforts in preparing this book, they make no representations or warranties with respect to the accuracy or completeness of the contents of this book and specifically disclaim any implied warranties of merchantability or fitness for a particular purpose. No warranty may be created or extended by sales representatives or written sales materials. The advice and strategies contained herein may not be suitable for your situation. You should consult with a professional where appropriate. Neither the publisher nor the authors shall be liable for any loss of profit or any other commercial damages, including but not limited to special, incidental, consequential, or other damages.

For general information on our other products and services or for technical support, please contact our Customer Care Department within the United States at (800) 762-2974, outside the United States at (317) 572-3993 or fax (317) 572-4002.

Wiley also publishes its books in a variety of electronic formats. Some content that appears in print, may not be available in electronic formats. For more information about Wiley products, visit our web site at www.wiley.com.

Library of Congress Cataloging-in-Publication Data is available.

Electrokinetic and Colloid Transport Phenomena
Masliyah, Jacob H. and Bhattacharjee, Subir

ISBN 13: 978-0-471-78882-9
ISBN 10: 0-471-78882-1

Printed in the United States of America

10 9 8 7 6 5 4 3 2 1

530.417
M379e

Chemistry

This work is dedicated to our parents,
Heskel Haim & Naima Masliyah,
and Sambhu Nath & Swarna Bhattacharjee.

—

For our families, Odette, Tamara, Ruth, & Daniel,
and Kamaljit & Arya.

—

A tribute to our mentors

CONTENTS

PREFACE

Electrokinetics is a subject that has been at the core of numerous fundamental advancements in the field of colloid science for over a century. Electrokinetics is a self-contained body of science that has led to spectacular applications in separations, characterization of surface properties, manipulation of colloidal materials, and facilitation of fluid transport in microchannels. For instance, electrophoresis is one of the common techniques for separation of biological macromolecules (such as proteins). Reverse osmosis is nowadays almost a household name in the context of water filtration and purification. Streaming potential or electrophoretic mobility measurements are extremely common for evaluation of the surface potentials of charged surfaces or colloidal particles in a variety of industrial applications. The above examples provide a few glimpses of the diverse applications of electrokinetic transport phenomena. Concurrently, the theoretical knowledge base underlying electrokinetics has undergone significant improvement over the past several decades, providing a complete and unified picture of the basic physical and chemical phenomena associated with these transport processes.

The subject of electrokinetics is also fraught with considerable enigma and misconceptions, perhaps leading to a general apathy toward it by a large body of scientists and researchers. There are even notions that the theoretical treatment in this area is often incomplete, and sometimes inaccurate. A plausible reason behind this ambivalent approach to this subject is the lack of a coordinated understanding of fluid dynamics, mass transport, and electrostatics—the three pillars on which the subject stands. Very rarely do we provide students in a single discipline of science and engineering a complete course covering all three aspects of electrokinetic transport in sufficient detail. Often such lack of coordination has indeed led to some misconceptions and inaccurate models. However, it is perhaps unfair to consider this highly evolved and extraordinarily beautiful subject incomplete. There has been a resurgence of interest in this fairly mature subject since the advent of microfluidics during the last decade. In light of this renewed interest in this subject, it is perhaps pertinent to systematically reevaluate the theoretical underpinnings of electrokinetic transport phenomena.

The present book can be considered as an outgrowth of an earlier book titled *Electrokinetic Transport Phenomena*, published in 1994. The guiding principle of the earlier book was to provide a detailed description of the fundamental electrochemical transport processes underlying electrokinetic and colloid transport phenomena. Particular emphasis was laid toward rendering the mathematical developments sufficiently tractable such that a graduate or a senior undergraduate student approaching the subject for the first time can easily interpret the theories. We follow this style in the present book as well.

The present book aims to provide a fundamental perspective of electrokinetic and colloid transport processes, emphasizing the coupling between electrical, mass transport, and fluid dynamical aspects of these complex transport phenomena. In the first five chapters, the principles of electrostatics are described, providing a comprehensive coverage of basic electrostatics *in vacuo* and perfect dielectrics, as well as electric double layer phenomena in electrolyte solutions. Following this, Chapter six elaborates the fundamental aspects of mass transport phenomena in electrolyte systems. Chapters seven through ten describe the four basic electrokinetic phenomena, namely, electroosmosis, streaming potential, electrophoresis, and sedimentation potential. In Chapter eleven, we present the Derjaguin–Landau–Verwey–Overbeek (DLVO) model of colloidal interactions, which forms the basis of subsequent discussion on two aspects of colloid transport phenomena, namely, coagulation and particle deposition, in Chapters twelve and thirteen, respectively. Chapter fourteen presents a self-contained description of numerical approaches for solving electrokinetic transport problems. Finally, in Chapter fifteen, we provide several examples of application of electrokinetic and colloid transport phenomena in different disciplines of science and engineering.

The book has been a product of several years of teaching graduate and senior undergraduate students in the subjects of colloidal phenomena, colloid transport, electrokinetics, and microfluidics. The courses were offered independently by both of the authors at various institutions in Chemical Engineering, Materials Engineering, and Civil and Environmental Engineering, as well as Mechanical Engineering. The audience for these courses virtually always consisted of a diverse array of graduate students or senior undergraduate students from different branches of engineering (Chemical, Civil, Environmental, Electrical, and Mechanical), physics, and chemistry. The courses took titles ranging from Colloidal Phenomena, Colloidal Phenomena in Aquatic Systems, Colloidal Hydrodynamics, Colloidal and Electrokinetic Phenomena, to Microfluidics and Electrokinetics.

The subject of electrokinetics is generally regarded as highly mathematical in nature. Most treatments of this subject rely extensively on vector and tensor notations. Experience gleaned from our teaching this course leads us to believe that using concise vector and tensor notations often makes the subject appear fairly complex to a graduate student being initiated to this area. Consequently, our approach has been to use the "exploded" forms of the governing equations. This might initially give the reader the impression that the book contains an unusually large number of equations and formulas. However, our teaching experience shows that the detailed derivations of the mathematical models (where we rarely skimp on intermediate steps) can make the process of learning and applying the mathematical principles fairly simple for a student with a modest mathematical background. The style of the mathematics used in the book should also render writing computer codes on the basis of these formulae fairly straightforward. Although the primary objective of the book is to serve as a graduate-level textbook, we feel that it can also serve as a comprehensive and updated reference material on the theoretical aspects of electrokinetic and colloid transport phenomena relevant to a wide range of practicing engineers and scientists.

The writing of the book led us through a significant learning process. In this respect, we are grateful to Dr. Emiliy Zholkovskij for clarifying numerous questions, doubts, and misconceptions, and generally illuminating us with a clear and unambiguous theoretical and physical picture of the subject matter. His contributions toward writing the material on hydrodynamic dispersion in Chapter 8 are gratefully acknowledged. Finally, without his meticulous proofreading of the mathematical formulae and the derivations, the book would have been incomplete. We are grateful to Mr. Glen Thomas for his assistance with the proofreading of the manuscript. We would also like to acknowledge many of our graduate student researchers, whose contributions are featured in many parts of the book. The financial support of Natural Sciences and Engineering Research Council (NSERC), Canada Research Chairs program (CRC), Canada Foundation for Innovation (CFI), Alberta Science and Research Authority, Alberta Energy Research Institute, Alberta Ingenuity Fund, and several other funding sources toward our research has been the catalyst that encouraged the learning process leading to this book. Our sincere thanks to Dr. Arza Seidel from John Wiley & Sons for her patience in dealing with our interminable revisions and her generous help with the final production of the manuscript. Finally, we are thankful to our families and colleagues for understanding our preoccupation with this book, and foregoing their demands on our time.

We hope that this book serves its intended purpose. We will be rewarded if the material helps make the subject of electrokinetics readily accessible to a broad range of scientific professionals. Finally, virtually no work that is human could be complete and error-free. Despite our careful revisions, there may be errors and omissions that have crept into the manuscript. We will be grateful to be informed of any errors and omissions in this work so that we can improve the book in subsequent editions.

JACOB H. MASLIYAH
SUBIR BHATTACHARJEE

Edmonton, Alberta
August, 2005

COPYRIGHT ACKNOWLEDGMENTS

Table 2.1: Reproduced by permission of the Royal Society of Chemistry.

Table 2.2: Copyright (1989), with permission from Cambridge University Press.

Table 3.4: Reprinted from Intermolecular and Surface Forces, Israelachvili, J.N., Copyright (1985), with permission from Elsevier.

Table 6.2: Copyright (2002) from CRC Handbook of Chemistry and Physics by Lide, D.R. (Ed.). Reproduced by permission of Routledge/Taylor & Francis Group, LLC.

Table 6.4: Reprinted from Atkinson, G., Electrochemical Information, in American Institute of Physics Handbook, 3rd ed., Gray, D.E. (Ed.), Copyright (1972), with permission from The McGraw-Hill Companies.

Figure 8.20: Copyright (1975), with permission from Elsevier.

Figure 9.10: Reproduced by permission of The Royal Society of Chemistry.

Figure 9.11: Reproduced by permission of The Royal Society of Chemistry.

Figure 9.17: Copyright (1989), with permission from Elsevier.

Figure 9.18: Copyright (1989), with permission from Elsevier.

Figure 9.19: Copyright (1989), with permission from Elsevier.

Table 11.5: Copyright (1989), with permission from Cambridge University Press.

Figure 11.6: Reprinted from Introduction to Colloid and Surface Chemistry, Shaw, D.J., Copyright (1980), with permission from Elsevier.

Figure 11.7: Copyright (1989), with permission from Cambridge University Press.

Figure 11.8: Reprinted from Intermolecular and Surface Forces, Israelachvili, J.N., Copyright (1985), with permission from Elsevier.

Figure 11.9: Copyright (1978), with permission from The Royal Society of Chemistry.

Figure 11.10: Copyright (1981), with permission from Elsevier.

Figure 12.6: Copyright (1989), with permission from Cambridge University Press.

Table 12.2: Copyright (1989), with permission from Cambridge University Press.

Table 12.3: Reproduced with permission. Copyright ©1977 AIChE.

Figure 12.12: Reprinted from Colloidal Hydrodynamics, van de Ven, T.G.M., Copyright (1989), with permission from Elsevier.

Figure 12.13: Copyright (1983), with permission from Cambridge University Press.

Figure 12.14: Copyright (1981), with permission from Elsevier.

Figure 12.17: Copyright (1983), with permission from Cambridge University Press.

Figure 12.18: Reproduced with permission. Copyright ©1977 AIChE.

Figure 12.20: Copyright (1985), with permission from Elsevier.

Figure 13.5: Copyright (1969), with permission from Cambridge University Press.

Table 13.1: Copyright (1973), with permission from Elsevier.

Figure 13.9: Copyright (1983), with kind permission of Springer Science and Business Media.

Figure 13.10: Copyright (1983), with kind permission of Springer Science and Business Media.

Figure 13.11: Reproduced with permission. Copyright ©1974 AIChE.

Figure 13.12: Reproduced with permission. Copyright ©1974 AIChE.

Figure 13.13: Reproduced with permission. Copyright ©1974 AIChE.

Figure 13.14: Reproduced with permission. Copyright ©1974 AIChE.

Figure 13.15: Reproduced with permission. Copyright ©1974 AIChE.

Figure 13.16: Reproduced with permission. Copyright ©1974 AIChE.

Figure 13.20: Copyright (1973), with permission from Elsevier.

Figure 13.21: Copyright (1973), with permission from Elsevier.

Figure 13.23: Copyright (1973), with permission from Elsevier.

Figure 13.24: Copyright (1973), with permission from Elsevier.

Figure 13.25: Copyright (1973), with permission from Elsevier.

Figure 13.26: Copyright (1973), with permission from Elsevier.

Figure 13.31: Copyright (1973), with permission from Elsevier.

Figure 13.32: Copyright (1973), with permission from Elsevier.

Figure 13.33: Copyright (1973), with permission from Elsevier.

Figure 14.4: Reprinted with permission. Copyright (2003) American Chemical Society.

Figure 14.6: Reprinted with permission. Copyright (2003) American Chemical Society.

Figure 14.15: Copyright (2005), with permission from Elsevier.

Figure 15.1: Reprinted with permission. Copyright (1972) American Chemical Society.

Figure 15.2: Reprinted with permission. Copyright (1972) American Chemical Society.

Figure 15.3: Copyright (1987), with permission from Elsevier.

Figure 15.6: Copyright (1988), with permission from Elsevier.

Figure 15.7: Copyright (1988), with permission from Elsevier.

Figure 15.12: Copyright (1972), with permission from Elsevier.

Figure 15.14: Reproduced by permission of The Royal Society of Chemistry.

Figure 15.15: Copyright (1972), with permission from Elsevier.

Figure 15.18: Copyright (1992), with permission from Elsevier.

Figure 15.22: Reprinted with permission. Copyright (2003) American Chemical Society.

Figure 15.23: Reprinted with permission. Copyright (2003) American Chemical Society.

Figure 15.25: Reproduced with permission. Copyright ©2004 AIChE.

Figure 15.26: Copyright (2002), with permission from Elsevier.

Figure 15.30: Reprinted with permission. Copyright (2003) American Chemical Society.

Figure 15.31: Reprinted with permission. Copyright (2003) American Chemical Society.

Figure 15.32: Reprinted with permission. Copyright (2003) American Chemical Society.

Figure 15.33: Reprinted with permission. Copyright (2003) American Chemical Society.

Figure 15.35: Copyright (2005), with permission from Elsevier.

Figure 15.36: Copyright (2005), with permission from Elsevier.

Figure 15.37: Copyright (2005), with permission from Elsevier.

Figure 15.38: Copyright (2005), with permission from Elsevier.

Figure 15.45: Copyright (2005), with permission from Elsevier.

CHAPTER 1

MATHEMATICAL PRELIMINARIES

Treatment of electrokinetic transport phenomena requires understanding of fluid mechanics, colloidal phenomena, and the interaction of charged particles, surfaces, and electrolytes with an external electrical field. Accordingly, dealing with electrokinetic transport processes requires familiarity with the units and dimensions of fundamental quantities from a diverse range of subjects. In this chapter, we outline the pertinent units and dimensions of the fundamental quantities encountered in electrokinetic transport processes.

Historically, the centimeter-gram-second (cgs) system of units was widely used in most colloid science and electrokinetics literature. However, with the popularity of the *Système Internationale d'Unités* (the SI system), most of the modern treatment of these subjects are based on SI units. Accordingly, most of the topics covered in this book are based on the SI system. To facilitate the conversion of other units into the SI system, the first few sections of this Chapter are devoted to definitions of the fundamental units and dimensions, description of the derived units in the SI system, values of the commonly encountered physical constants in various units, and conversion factors for different quantities from SI to non-SI units. The latter half of the Chapter outlines some of the mathematical fundamentals required to develop the theoretical treatments in the rest of the book, including a short primer of series functions, vector, and tensor operations.

Electrokinetic and Colloid Transport Phenomena, by Jacob H. Masliyah and Subir Bhattacharjee
Copyright © 2006 John Wiley & Sons, Inc.

1.1 UNITS

The fundamental quantities required in electrokinetic transport analysis are shown in Table 1.1, along with their SI units and symbols.

These fundamental quantities can be combined to yield different derived quantities, the units of which are combinations of the fundamental units. Table 1.2 provides some of the commonly used derived quantities and their SI units.

To provide a facile transition of the basic dimensions over large ranges, it is often convenient to use scale factors for the basic units. This is particularly important in terms of the length scales used to define the dimensions of extremely small colloidal particles. For instance, it is convenient to express particle sizes in terms of nanometer (nm) or micrometer (μm) instead of meter (m). Similarly, the colloidal forces are conveniently expressed in terms of nano-newtons (nN) or pico-newtons (pN) rather than newtons (N). Table 1.3 provides the commonly used scale factors for the basic units.

TABLE 1.1. Fundamental Quantities Used in Electrokinetic Transport Analysis, their SI Units and Symbols.

Quantity	Name of SI Unit	Symbol
Mass	kilogram	kg
Length	meter	m
Time	second	s
Temperature	Kelvin	K
Quantity of mass	mole	mol
Electric current	Ampere	A

TABLE 1.2. Derived Quantities and their SI Units.

Quantity	SI Unit Name	Symbol	Definition
Force	Newton	N	kg m s^{-2}
Pressure	Pascal	Pa	$\text{N m}^{-2} = \text{kg m}^{-1}\text{s}^{-2}$
Energy	Joule	J	$\text{N m} = \text{kg m}^2\text{s}^{-2}$
Power	Watt	W	$\text{J s}^{-1} = \text{kg m}^2\text{s}^{-3}$
Electric charge	Coulomb	C	A s
Electric potential	Volt	V	$\text{J C}^{-1} = \text{kg m}^2\text{s}^{-3}\text{A}^{-1}$
Electric resistance	Ohm	Ω	$\text{V A}^{-1} = \text{kg m}^2\text{s}^{-3}\text{A}^{-2}$
Electric conductance	Siemens	S	$\text{A V}^{-1} = \text{kg}^{-1}\text{m}^{-2}\text{s}^3\text{A}^2$
Electric capacitance	Farad	F	$\text{C V}^{-1} = \text{kg}^{-1}\text{m}^{-2}\text{s}^4\text{A}^2$
Frequency	Hertz	Hz	s^{-1}
Magnetic inductance	Henry	H	$\text{J A}^{-2} = \text{kg m}^2\text{s}^{-2}\text{A}^{-2}$
Dynamic viscosity			$\text{Pa s} = \text{N s m}^{-2} = \text{kg m}^{-1}\text{s}^{-1}$
Material density			kg m^{-3}

Source: Adapted from Russel *et al.* (1989) and Probstein (2003).

TABLE 1.3. Scale Factors for the Basic Units.

Factor	Prefix	Symbol	Factor	Prefix	Symbol
10^{-1}	deci	d	10	deca	da
10^{-2}	centi	c	10^{2}	hecto	h
10^{-3}	milli	m	10^{3}	kilo	k
10^{-6}	micro	μ	10^{6}	mega	M
10^{-9}	nano	n	10^{9}	giga	G
10^{-12}	pico	p	10^{12}	tera	T
10^{-15}	femto	f	10^{15}	peta	P
10^{-18}	atto	a	10^{18}	exa	E

1.2 PHYSICAL CONSTANTS AND CONVERSION FACTORS

The commonly used physical constants and their values in SI units are listed in Table 1.4. The use of non-SI units for various quantities is still common in electrokinetics literature. Conversion factors between SI and other units are provided for some of these quantities in Table 1.5.

TABLE 1.4. Common Physical Constants and their Values in SI Units (Lide, 2001).

Quantity	Symbol	Value	SI Units
Avogadro number	N_A	6.022×10^{23}	mol^{-1}
Boltzmann constant	k_B	1.381×10^{-23}	$J\,K^{-1}$
Elementary charge	e	1.602×10^{-19}	C
Faraday constant	$\mathcal{F}$	9.649×10^{4}	$C\,mol^{-1}$
Magnetic permeability of vacuum	μ_0	1.2566×10^{-7}	NA^{-2}, $NC^{-2}s^{-2}$, or Hm^{-1}
Universal gas constant	R	8.314	$JK^{-1}mol^{-1}$
Permittivity of vacuum	ϵ_0	8.854×10^{-12}	$CV^{-1}m^{-1}$, $C^2N^{-1}m^{-2}$, or Fm^{-1}
Planck constant	h	6.626×10^{-34}	J s
Speed of light in vacuum	c	2.9979×10^{8}	$m\,s^{-1}$
Standard gravitational acceleration	g	9.8066	$m\,s^{-2}$
Standard atmospheric pressure (at sea level and 288.16 K)	p_0	1.01325×10^{5}	Pa
Zero of Celsius scale	T_0	273.15	K
1 liter	L	1.0000028×10^{-3}	m^3
k_BT/e at 298.16 K		25.69×10^{-3}	V
1 molar solution	M	1.0	mol/dm^3 or $kmol/m^3$

Source: Adapted from Hiemenz (1986), Russel *et al.* (1989), and Probstein (2003).

TABLE 1.5. Conversion Factors for Non-SI Units.

Unit	Abbreviation	Value
atmosphere	atm	101325 Pa (definition)
torr	torr	133.322 Pa $= 1/760$ atm
atomic mass unit	amu	1.6605×10^{-27} kg
bar	bar	1×10^5 Pa
electron volt	eV	1.6022×10^{-19} J
poise	P	0.1 kg m^{-1} s^{-1}
liter	L	1×10^{-3} $m^3 = 1$ dm^3
Angstrom	Å	1×10^{-10} m
Debye	D	3.3356×10^{-30} C m
calorie	cal	4.184 J (definition)
inch	in	0.0254 m (definition)
pound	lbm	0.4536 kg

1.3 FREQUENTLY USED FUNCTIONS

Here we list some of the common series expansions and functions used frequently in this book. Excellent compilation of mathematical formulae is given by Jeffrey (1995).

$$\exp(x) = 1 + x + \frac{x^2}{2!} + \frac{x^3}{3!} + \cdots \qquad -\infty < x < \infty$$

$$\ln(1+x) = x - \frac{x^2}{2} + \frac{x^3}{3} - \frac{x^4}{4} + \cdots = \sum_{k=1}^{\infty} (-1)^{k+1} \frac{x^k}{k} \qquad -1 < x \leq 1$$

$$\ln(1-x) = -\left(x + \frac{x^2}{2} + \frac{x^3}{3} + \frac{x^4}{4} + \cdots \right) = -\sum_{k=1}^{\infty} \frac{x^k}{k} \qquad -1 \leq x < 1$$

$$\sinh(x) = \frac{1}{2}[\exp(x) - \exp(-x)] = x + \frac{x^3}{3!} + \frac{x^5}{5!} + \frac{x^7}{7!} + \cdots \qquad -\infty < x < \infty$$

$$\cosh(x) = \frac{1}{2}[\exp(x) + \exp(-x)] = 1 + \frac{x^2}{2!} + \frac{x^4}{4!} + \frac{x^6}{6!} + \cdots \qquad -\infty < x < \infty$$

$$\tanh(x) = \frac{\sinh(x)}{\cosh(x)} = x - \frac{x^3}{3} + \frac{2x^5}{15} - \frac{17x^7}{315} + \cdots \qquad |x| < \pi/2$$

$$\cosh^2(x) = 1 + \sinh^2(x)$$

$$\sinh(x) \cong \cosh(x) \quad x \to \infty$$

$$\tanh(x) \longrightarrow 1 \quad x \to \infty$$

$$\sinh(x/2) = \pm\sqrt{\left(\frac{\cosh(x) - 1}{2}\right)} \quad [+ \text{ if } x > 0 \text{ and } - \text{ if } x < 0]$$

$$\sinh(x) = 2\sinh(x/2)\cosh(x/2)$$

$$\sinh^2(x) = \frac{1}{2}[\cosh(2x) - 1]$$

$$\sinh(2x) = 2\sinh(x)\cosh(x)$$

$$\cosh(x/2) = \left(\frac{1+\cosh(x)}{2}\right)^{1/2}$$

$$\cosh(x) = \cosh^2(x/2) + \sinh^2(x/2)$$

$$\cosh^2(x) = \frac{1}{2}[1 + \cosh(2x)]$$

$$\cosh(2x) = 2\cosh^2(x) - 1$$

$$\tanh(x/2) = \frac{\sinh(x)}{1+\cosh(x)} = \frac{\cosh(x)-1}{\sinh(x)} = \frac{1-\exp(-x)}{1+\exp(-x)}$$

$$1 - \tanh(x/2) = \frac{2\exp(-x)}{1+\exp(-x)}$$

$$\tanh(x) = \frac{2\tanh(x/2)}{1+\tanh^2(x/2)} = \frac{2\coth(x/2)}{\operatorname{csch}^2(x/2)+2}$$

$$\tanh^2(x) = \frac{\cosh(2x)-1}{1+\cosh(2x)}$$

$$\tanh(2x) = \frac{2\tanh(x)}{1+\tanh^2(x)}$$

Some commonly used integrals are provided next. These being indefinite integrals, one should remember to add an integration constant to each result.

$$\int \frac{dx}{\sinh^2(x)} = -\coth(x)$$

$$\int \frac{dx}{\cosh^2(x)} = \tanh(x)$$

$$\int \frac{dx}{1+\cosh(x)} = \tanh(x/2)$$

$$\int \frac{dx}{1-\cosh(x)} = \coth(x/2)$$

$$\int \tanh(kx)dx = \frac{1}{k}\ln[\cosh(kx)]$$

$$\int \tanh^2(kx)dx = x - \frac{1}{k}\tanh(kx)$$

$$\int \coth(kx)dx = \frac{1}{k}\ln|\sinh(kx)|$$

1.4 VECTOR OPERATIONS

A scalar quantity is defined by a single real number. Temperature and mass are good examples of scalar quantities. A vector quantity is defined by a magnitude and a direction. Velocity of a projectile is a vector quantity. The magnitude of a vector **u** is given by $|\mathbf{u}|$ or simply u.

Addition and subtraction of two vectors, **u** and **v**, are illustrated in Figure 1.1.

Multiplication of a vector **u** by a scalar quantity s results in changing the magnitude of the vector to $s|\mathbf{u}|$ or simply su. The vector direction remains same.

A vector can be multiplied with another vector in several ways. Scalar or dot product of two vectors, **u** and **v**, is given by

$$\mathbf{u} \cdot \mathbf{v} = \mathbf{v} \cdot \mathbf{u} = uv \cos \phi$$

where ϕ is the angle formed between the vectors **u** and **v**. Here, u and v are the magnitudes of the vectors **u** and **v**, respectively. The scalar product rules are

Commutative:

$$\mathbf{u} \cdot \mathbf{v} = \mathbf{v} \cdot \mathbf{u}$$

Not associative:

$$(\mathbf{u} \cdot \mathbf{v})\, \mathbf{w} \neq \mathbf{u}\, (\mathbf{v} \cdot \mathbf{w})$$

Distributive:

$$\mathbf{u} \cdot (\mathbf{v} + \mathbf{w}) = \mathbf{u} \cdot \mathbf{v} + \mathbf{u} \cdot \mathbf{w}$$

Vector product or cross product of two vectors **u** and **v** is given by another vector defined by

$$\mathbf{u} \times \mathbf{v} = uv \sin \phi \, \mathbf{n}$$

where ϕ is the angle between the two vectors and **n** is a vector of unit length (magnitude) normal to both the vectors **u** and **v** in the sense in which a right-handed screw

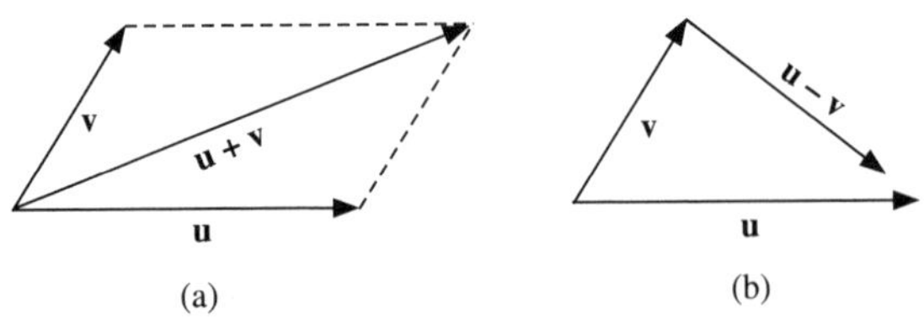

Figure 1.1. (a) Addition and (b) subtraction of two vectors.

would advance if rotated from **u** to **v**. A convenient form of a cross product is given by

$$\mathbf{u} \times \mathbf{v} = \begin{pmatrix} \mathbf{i}_1 & \mathbf{i}_2 & \mathbf{i}_3 \\ u_1 & u_2 & u_3 \\ v_1 & v_2 & v_3 \end{pmatrix}$$

where $\mathbf{u} = u_1\mathbf{i}_1 + u_2\mathbf{i}_2 + u_3\mathbf{i}_3$ and $\mathbf{v} = v_1\mathbf{i}_1 + v_2\mathbf{i}_2 + v_3\mathbf{i}_3$. Here, $\mathbf{i}_1$, $\mathbf{i}_2$, and $\mathbf{i}_3$ are orthogonal unit vectors, with u_1, u_2, and u_3 being the magnitudes of vector **u** in the $\mathbf{i}_1$, $\mathbf{i}_2$, and $\mathbf{i}_3$ directions, respectively. Similarly, v_1, v_2, and v_3 are the magnitudes of the vector **v** in the $\mathbf{i}_1$, $\mathbf{i}_2$, and $\mathbf{i}_3$ directions, respectively. The magnitude of a vector **u** is given by

$$|\mathbf{u}| = u = \sqrt{\sum_{i=1}^{3} u_i^2}$$

One can also multiply two vectors to obtain a tensor or a dyadic product. The dyadic product of two vectors **u** and **v** is given by **uv**. We will discuss dyadic products in the next section.

A compilation of useful vector identities and vector operations is given by Bird *et al.* (2002).

Some commonly used differential vector operations in different orthogonal coordinate systems are given below. Here, ψ is used for a scalar and **u** is used for a vector.

Cartesian Coordinates (x, y, z)

The orthogonal curvilinear coordinates, in the case of Cartesian coordinates, are defined by the unit vectors $\mathbf{i}_x$, $\mathbf{i}_y$, and $\mathbf{i}_z$ directed along the x, y, and z coordinates, respectively, and the vector **u** is given by

$$\mathbf{u} = u_x\mathbf{i}_x + u_y\mathbf{i}_y + u_z\mathbf{i}_z$$

The differential operator ∇ is given by

$$\nabla = \frac{\partial}{\partial x}\mathbf{i}_x + \frac{\partial}{\partial y}\mathbf{i}_y + \frac{\partial}{\partial z}\mathbf{i}_z$$

Some useful differential operations are given by

$$\nabla\psi = \frac{\partial\psi}{\partial x}\mathbf{i}_x + \frac{\partial\psi}{\partial y}\mathbf{i}_y + \frac{\partial\psi}{\partial z}\mathbf{i}_z$$

$$\nabla^2\psi = \frac{\partial^2\psi}{\partial x^2} + \frac{\partial^2\psi}{\partial y^2} + \frac{\partial^2\psi}{\partial z^2}$$

$$\nabla\cdot\mathbf{u} = \frac{\partial u_x}{\partial x} + \frac{\partial u_y}{\partial y} + \frac{\partial u_z}{\partial z}$$

$$\nabla \times \mathbf{u} = \begin{pmatrix} \mathbf{i}_x & \mathbf{i}_y & \mathbf{i}_z \\ \partial/\partial x & \partial/\partial y & \partial/\partial z \\ u_x & u_y & u_z \end{pmatrix}$$

$$= \left(\frac{\partial u_z}{\partial y} - \frac{\partial u_y}{\partial z}\right)\mathbf{i}_x + \left(\frac{\partial u_x}{\partial z} - \frac{\partial u_z}{\partial x}\right)\mathbf{i}_y + \left(\frac{\partial u_y}{\partial x} - \frac{\partial u_x}{\partial y}\right)\mathbf{i}_z$$

Normally, $\nabla\psi$ is referred to as the gradient of the scalar ψ, $\nabla \cdot \mathbf{u}$ is known as the divergence of the vector $\mathbf{u}$, and $\nabla \times \mathbf{u}$ is known as the curl of vector $\mathbf{u}$. Here, ∇^2 is a Laplacian operator.

Cylindrical Coordinates (r, θ, z)

The cylindrical orthogonal coordinate system is defined by the three orthogonal unit vectors given by $\mathbf{i}_r$, $\mathbf{i}_\theta$, and $\mathbf{i}_z$ acting along the r, θ, and z directions, respectively. A vector $\mathbf{u}$ is given by

$$\mathbf{u} = u_r\mathbf{i}_r + u_\theta\mathbf{i}_\theta + u_z\mathbf{i}_z$$

The differential operator ∇ is given by

$$\nabla = \frac{\partial}{\partial r}\mathbf{i}_r + \frac{1}{r}\frac{\partial}{\partial \theta}\mathbf{i}_\theta + \frac{\partial}{\partial z}\mathbf{i}_z$$

Some useful differential operations are given by

$$\nabla\psi = \frac{\partial \psi}{\partial r}\mathbf{i}_r + \frac{1}{r}\frac{\partial \psi}{\partial \theta}\mathbf{i}_\theta + \frac{\partial \psi}{\partial z}\mathbf{i}_z$$

$$\nabla^2\psi = \frac{\partial^2 \psi}{\partial r^2} + \frac{1}{r}\frac{\partial \psi}{\partial r} + \frac{1}{r^2}\frac{\partial^2 \psi}{\partial \theta^2} + \frac{\partial^2 \psi}{\partial z^2}$$

$$\nabla \cdot \mathbf{u} = \frac{1}{r}\frac{\partial}{\partial r}(ru_r) + \frac{1}{r}\frac{\partial}{\partial \theta}u_\theta + \frac{\partial}{\partial z}u_z$$

$$\nabla \times \mathbf{u} = \frac{1}{r}\begin{pmatrix} \mathbf{i}_r & r\,\mathbf{i}_\theta & \mathbf{i}_z \\ \partial/\partial r & \partial/\partial \theta & \partial/\partial z \\ u_r & ru_\theta & u_z \end{pmatrix}$$

$$= \left(\frac{1}{r}\frac{\partial u_z}{\partial \theta} - \frac{\partial u_\theta}{\partial z}\right)\mathbf{i}_r + \left(\frac{\partial u_r}{\partial z} - \frac{\partial u_z}{\partial r}\right)\mathbf{i}_\theta + \left(\frac{1}{r}\frac{\partial}{\partial r}(ru_\theta) - \frac{1}{r}\frac{\partial u_r}{\partial \theta}\right)\mathbf{i}_z$$

Spherical Coordinates (r, θ, ϕ)

The spherical orthogonal coordinate system is defined by the three orthogonal unit vectors given by $\mathbf{i}_r$, $\mathbf{i}_\theta$, and $\mathbf{i}_\phi$ acting along the r, θ, and ϕ directions, respectively. A vector $\mathbf{u}$ is given by

$$\mathbf{u} = u_r\mathbf{i}_r + u_\theta\mathbf{i}_\theta + u_\phi\mathbf{i}_\phi$$

The differential operator ∇ is given by

$$\nabla = \frac{\partial}{\partial r}\mathbf{i}_r + \frac{1}{r}\frac{\partial}{\partial \theta}\mathbf{i}_\theta + \frac{1}{r\sin\theta}\frac{\partial}{\partial \phi}\mathbf{i}_\phi$$

Some useful differential operations are given by

$$\nabla\psi = \frac{\partial\psi}{\partial r}\mathbf{i}_r + \frac{1}{r}\frac{\partial\psi}{\partial\theta}\mathbf{i}_\theta + \frac{1}{r\sin\theta}\frac{\partial\psi}{\partial\phi}\mathbf{i}_\phi$$

$$\nabla^2\psi = \frac{1}{r^2}\frac{\partial}{\partial r}\left(r^2\frac{\partial\psi}{\partial r}\right) + \frac{1}{r^2\sin\theta}\frac{\partial}{\partial\theta}\left(\sin\theta\frac{\partial\psi}{\partial\theta}\right) + \frac{1}{r^2\sin^2\theta}\frac{\partial^2\psi}{\partial\phi^2}$$

$$\nabla\cdot\mathbf{u} = \frac{1}{r^2}\frac{\partial}{\partial r}(r^2u_r) + \frac{1}{r\sin\theta}\frac{\partial}{\partial\theta}(u_\theta\sin\theta) + \frac{1}{r\sin\theta}\frac{\partial u_\phi}{\partial\phi}$$

$$\begin{aligned}\nabla\times\mathbf{u} &= \frac{1}{r^2\sin\theta}\begin{pmatrix}\mathbf{i}_r & r\,\mathbf{i}_\theta & r\sin\theta\,\mathbf{i}_\phi\\ \partial/\partial r & \partial/\partial\theta & \partial/\partial\phi\\ u_r & ru_\theta & r\sin\theta u_\phi\end{pmatrix}\\ &= \frac{1}{r\sin\theta}\left[\frac{\partial}{\partial\theta}(\sin\theta u_\phi) - \frac{\partial u_\theta}{\partial\phi}\right]\mathbf{i}_r\\ &\quad + \frac{1}{r}\left[\frac{1}{\sin\theta}\frac{\partial u_r}{\partial\phi} - \frac{\partial}{\partial r}(ru_\phi)\right]\mathbf{i}_\theta\\ &\quad + \frac{1}{r}\left[\frac{\partial}{\partial r}(ru_\theta) - \frac{\partial u_r}{\partial\theta}\right]\mathbf{i}_\phi\end{aligned}$$

In any orthogonal curvilinear system of coordinates, the cross product of a gradient, *i.e.*, $\nabla\psi$, is zero. In other words

$$\nabla\times\nabla\psi = 0$$

Also, for a vector, **u**, one can write

$$\nabla\cdot(\nabla\times\mathbf{u}) = 0$$

1.5 TENSOR OPERATIONS

A second order tensor is a quantity that has nine components that are associated with three orthogonal directions and normal planes. The components of a tensor quantity $\bar{\bar{T}}$ is given by

$$\bar{\bar{T}} = \begin{pmatrix}T_{11} & T_{12} & T_{13}\\ T_{21} & T_{22} & T_{23}\\ T_{31} & T_{32} & T_{33}\end{pmatrix}$$

For $\bar{\bar{T}}$ to be a stress tensor, each term T_{ij} represents the stress on the ith plane in the j-direction. In general, $T_{ij} \neq T_{ji}$ unless the second order tensor is *symmetric*, where $T_{ij} = T_{ji}$.

A second order tensor can also be given as

$$\bar{\bar{T}} = \sum_{i=1}^{3}\sum_{j=1}^{3}\mathbf{i}_i\mathbf{i}_j T_{ij}$$

and

$$\bar{\bar{T}} = \mathbf{i}_1\mathbf{i}_1 T_{11} + \mathbf{i}_1\mathbf{i}_2 T_{12} + \mathbf{i}_1\mathbf{i}_3 T_{13} + \mathbf{i}_2\mathbf{i}_1 T_{21} + \mathbf{i}_2\mathbf{i}_2 T_{22} + \mathbf{i}_2\mathbf{i}_3 T_{23} + \mathbf{i}_3\mathbf{i}_1 T_{31} + \mathbf{i}_3\mathbf{i}_2 T_{32} + \mathbf{i}_3\mathbf{i}_3 T_{33}$$

where $\mathbf{i}_1$, $\mathbf{i}_2$, and $\mathbf{i}_3$ are unit vectors. Here, $\mathbf{i}_i\mathbf{i}_j$ is called the unit dyad, which follows certain rules when associated with a vector operation such as:

$$\mathbf{i}_i\mathbf{i}_j \cdot \mathbf{i}_k = \mathbf{i}_i(\mathbf{i}_j \cdot \mathbf{i}_k) = \mathbf{i}_i\delta_{jk}$$

$$\mathbf{i}_i \cdot \mathbf{i}_j\mathbf{i}_k = (\mathbf{i}_i \cdot \mathbf{i}_j)\mathbf{i}_k = \mathbf{i}_k\delta_{ij}$$

where δ_{ij} is the Kronecker delta defined as

$$\delta_{ij} = 1 \quad \text{if} \quad i = j$$
$$\delta_{ij} = 0 \quad \text{if} \quad i \neq j$$

When the components of a second order tensor are formed from components of two vectors **u** and **v**, the resulting product is called a dyadic product of **u** and **v**, given by **uv**, where

$$\mathbf{uv} = \sum_{i=1}^{3}\sum_{j=1}^{3} \mathbf{i}_i\mathbf{i}_j u_i v_j$$

A unit tensor is defined as

$$\bar{\bar{I}} = \sum_i \sum_j \mathbf{i}_i\mathbf{i}_j \delta_{ij}$$

or

$$\bar{\bar{I}} = \begin{pmatrix} 1 & 0 & 0 \\ 0 & 1 & 0 \\ 0 & 0 & 1 \end{pmatrix}$$

The magnitude of a tensor is given by

$$|\bar{\bar{T}}| = \sqrt{\frac{1}{2}\sum_i \sum_j T_{ij}^2}$$

The addition of tensors $\bar{\bar{T}}$ and $\bar{\bar{A}}$ or dyadic products simply follows

$$\bar{\bar{T}} + \bar{\bar{A}} = \sum_i \sum_j (T_{ij} + A_{ij})$$

The multiplication of a tensor or a dyadic product by a scalar gives

$$s\bar{\bar{T}} = \sum_i \sum_j \mathbf{i}_i \mathbf{i}_j (sT_{ij})$$

The vector product (or dot product) of a tensor with a vector is given by

$$(\bar{\bar{T}} \cdot \mathbf{u}) = \sum_i \mathbf{i}_i \left\{ \sum_j T_{ij} u_j \right\}$$

The vector product (or dot product) of a vector with a tensor is given by

$$(\mathbf{u} \cdot \bar{\bar{T}}) = \sum_i \mathbf{i}_i \left\{ \sum_j u_j T_{ji} \right\}$$

In general $\bar{\bar{T}} \cdot \mathbf{u} \neq \mathbf{u} \cdot \bar{\bar{T}}$ unless the tensor $\bar{\bar{T}}$ is symmetric where $T_{ij} = T_{ji}$. In expanded form one can write for $\bar{\bar{T}} \cdot \mathbf{u}$ in Cartesian coordinates

$$\begin{aligned} \bar{\bar{T}} \cdot \mathbf{u} = {} & (T_{xx} u_x + T_{xy} u_y + T_{xz} u_z)\mathbf{i}_x \\ & + (T_{yx} u_x + T_{yy} u_y + T_{yz} u_z)\mathbf{i}_y \\ & + (T_{zx} u_x + T_{zy} u_y + T_{zz} u_z)\mathbf{i}_z \end{aligned}$$

In other orthogonal coordinates, one can simply replace (x, y, z) by the respective new coordinates *e.g.*, with (r, θ, ϕ) for a spherical coordinate system.

1.6 VECTOR AND TENSOR INTEGRAL THEOREMS

1.6.1 The Divergence and Gradient Theorems

If a volume V is enclosed by a surface S, then

$$\int_V (\nabla \cdot \mathbf{u}) dV = \oint_S (\mathbf{u} \cdot \mathbf{n}) dS$$

where $\mathbf{n}$ is the outwardly directed unit normal vector. As $\mathbf{u} \cdot \mathbf{n} = \mathbf{n} \cdot \mathbf{u}$, one can write

$$\int_V (\nabla \cdot \mathbf{u}) dV = \oint_S (\mathbf{n} \cdot \mathbf{u}) dS$$

Two related theorems for scalars, ψ, and tensors, $\bar{\bar{T}}$, can be written as

$$\int_V \nabla \psi dV = \oint_S \mathbf{n} \psi dS$$

and

$$\int_V (\nabla \cdot \overline{\overline{T}}) dV = \oint_S (\mathbf{n} \cdot \overline{\overline{T}}) dS$$

Clearly, the tensor $\overline{\overline{T}}$ can be replaced by a dyadic product.

1.6.2 The Stokes Theorem

If a surface S is bounded by a closed curve C, then

$$\int_S \mathbf{n} \cdot (\nabla \times \mathbf{u}) dS = \oint_C (\mathbf{t} \cdot \mathbf{u}) dC$$

where $\mathbf{t}$ is a unit tangential vector in the direction of integration along path C, $\mathbf{n}$ is the unit normal vector to the surface S in the direction that a right-hand screw would move if its head were twisted in the direction of integration along contour C (Bird *et al.* 2002).

1.7 REFERENCES

Bird, R. B., Stewart, W. E., and Lightfoot, E. N., *Transport Phenomena*, John Wiley, New York, (2002).

Hiemenz, P. C., *Principles of Colloid and Surface Chemistry*, 2nd ed., Marcel Dekker, New York, (1986).

Jeffrey, A., *Handbook of Mathematical Formulas and Integrals*, Academic Press, San Diego, (1995).

Lide, D. R., Editor in Chief, *Handbook of Chemistry and Physics*, 82nd ed., CRC Press, Cleveland, (2001).

Probstein, R. F., *Physicochemical Hydrodynamics, An Introduction*, 2nd ed., Wiley Interscience, New York, (2003).

Russel, W. B., Saville, D. A., and Schowalter, W. R., *Colloidal Dispersions*, Cambridge University Press, Cambridge, (1989).

CHAPTER 2

COLLOIDAL SYSTEMS

The subject of electrokinetic transport is intricately related to the field of colloid and interface science, and hence, a brief primer of colloidal systems and colloidal phenomena is a prerequisite for a sound understanding of electrokinetic processes.

2.1 THE COLLOIDAL STATE

The term colloid originates from the Greek word '$\kappa o\lambda\lambda\alpha$' – meaning glue. In the 19th century, Thomas Graham (1805–1869) coined the terms colloid and crystalloid to classify two types of matter. While colloidal particles form a dispersion or suspension, crystalloids form a homogeneous solution when dissolved in a solvent. Colloidal dispersions are distinct from true (homogeneous) solutions in several ways. In a true solution, the solute is supposed to have lost its identity (consider dissolution of a salt in water: the salt dissociates into its constituent ions, and apparently undergoes a change in property). Colloidal particles, however, retain their identity in a suspension. Therefore, a colloidal suspension is considered a heterogeneous system.

The above classification between colloidal dispersions and homogeneous solutions may appear to be ludicrous to a modern physical chemist. We now consider the particulate nature of matter as a universal truth. In this scenario, the classification of colloids from crystalloids (solutes or ions) has become inconsequential, or even redundant. Hence, we generally resort to the simple size based demarcation between colloidal particles and smaller ionic or molecular solutes.

Electrokinetic and Colloid Transport Phenomena, by Jacob H. Masliyah and Subir Bhattacharjee
Copyright © 2006 John Wiley & Sons, Inc.

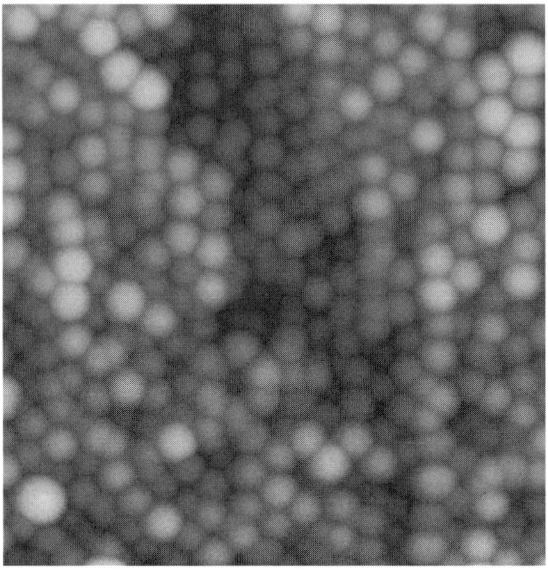

Figure 2.1. Spherical polystyrene latex particles observed using an atomic force microscope. The individual particles have a diameter of *ca.* 140 nm.

A colloidal dispersion is loosely defined as a multi-phase system, in which a *discrete* phase (the dispersed phase) is suspended in a continuous medium called the dispersant. The use of the term "discrete" is crucial in the above definition, since this word imposes a restriction on the size of the dispersed phase relative to the molecules of the dispersant. For instance, if we consider an aqueous dispersion of 25 nanometer (1 nm = 10^{-9} m) diameter silica particles, the silica particles will be nearly 100 times larger than the water molecules (which have an approximate diameter of 0.276 nm). In this case, although the water molecules are discrete themselves, they are so much smaller than the silica particle, that they will appear as part of a continuous medium relative to the silica particle. In contrast, if we add a salt (say NaCl) to the dispersion, the sodium and chloride ions, with hydrated diameters approximately 0.4–0.5 nm, will be of the same size range as the solvent. In this case, the ions will also appear to be part of a continuum relative to the silica particles. Thus, in order to have a colloidal system, the suspended or dispersed medium should have a size that is approximately one order of magnitude larger than the solvent molecules. In keeping with the scope of the above definition, colloidal particles are usually defined as entities having a size range of 1 nm to about 10 micro-meter (1 μm = 10^{-6} m). Figure 2.1 depicts spherical polystyrene latex particles of about 140 nm in diameter. Some examples of colloidal systems are outlined in Table 2.1, and typical particle size for some colloidal systems are given in Figure 2.2.

TABLE 2.1. Some Typical Colloidal Systems.

Examples	Class	Phase	
		Dispersed	Continuous
Disperse Systems			
Fog, mist, smoke, aerosol sprays	Liquid aerosol or aerosol of liquid particles[a]	Liquid	Gas
Industrial smokes	Solid aerosol or aerosol of solid particles[a]	Solid	Gas
Milk, butter, mayonnaise, asphalt, pharmaceutical creams	Emulsions	Liquid	Liquid
Inorganic colloids (gold, silver iodide, sulphur, metallic hydroxides, *etc.*), paints[b]	Sols or colloidal suspensions	Solid	Liquid
Clay slurries, toothpaste, mud, polymer lattices	When very concentrated called a paste	Solid	Liquid
Opal, pearl, stained glass, pigmented plastics	Solid suspension or dispersion	Solid	Solid
Froths, foams	Foam[c]	Gas	Liquid
Meerschaum, expanded plastics	Solid foam	Gas	Solid
Microporous oxides, silica gel, porous glass, microporous carbons, zeolites	Xerogels[d]		
Macromolecular Colloids			
Jellies, glue	Gels	Macromolecules	Solvent
Association Colloids			
Soap/water, detergent/water, dye solutions	—	Micelles	Solvent
Biocolloids			
Blood		Corpuscles	Serum
Bone		Hydroxyapatite	Collagen
Muscle, cell membrane		Protein structures, thin lecithin films, *etc.*	
		Coexisting Phases	
Three Phase Colloidal Systems (Multiple Colloids)			
Oil bearing rock	Porous rock	Oil	Water
Capillary condensed vapors	Porous solid	Liquid	Vapor
Frost heaving	Porous rock or soil	Ice	Water
Mineral flotation	Mineral	Water	Air bubbles or oil droplets
Double emulsions	Oil	Aqueous phase	Water

Source: Adapted from Everett (1988).

[a]Preferred nomenclature according to IUPAC recommendations.

[b]Many modern paints are more complex, containing both dispersed pigment and emulsion droplets.

[c]In a foam, it is usually the thickness of the film of dispersion medium which is of colloidal dimensions, although the dispersed phase may also be finely divided.

[d]In some cases both phases are continuous, forming interpenetrating networks both of which have colloidal dimensions.

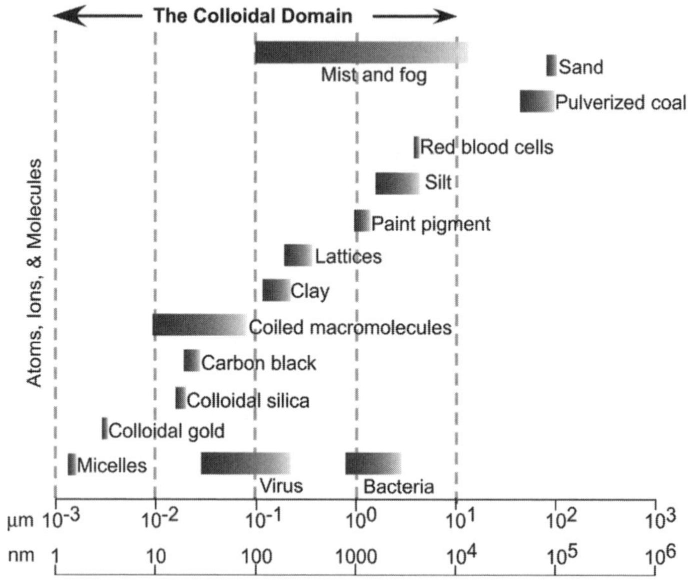

Figure 2.2. Typical particle size ranges in the colloidal domain.

2.2 COLLOIDAL PHENOMENA

Colloidal phenomena are concerned with small particles or systems where the ratio of surface area to volume is very large. For spherical particles, the surface area to volume ratio varies as $6/d$, where d is the particle diameter. Naturally, as the particle becomes smaller, the surface area per unit volume of the particle increases. Thus, in colloidal systems, in addition to the standard body forces (forces that act over the entire volume of a body) encountered in macroscopic objects, surface forces become important. The surface forces are typically engendered by the interactions occurring at the interfaces between the dispersed phase and the dispersing medium. These surface forces are often dominant in colloidal systems, leading to unique behaviors of colloidal dispersions, which are collectively termed as colloidal phenomena. By and large, the factors which contribute most to the overall nature of a colloidal system are:

- Particle size and shape
- Surface properties, both chemical and physical
- Continuous phase chemical and physical properties
- Particle–particle interactions
- Particle–continuous phase interactions

The various forces that enter into the interactions are:

- Electric repulsive or attractive force
- Attractive or repulsive London–van der Waals force

- Brownian force
- Viscous force
- Inertial force
- Gravitational force
- Steric (size-exclusion) forces
- Surface tension

The electric force between two particles can be derived from Coulomb's law and it is of the order $\epsilon\psi_s^2$, where ϵ is the dielectric permittivity of the continuous phase and ψ_s is the surface electric potential.

The London–van der Waals forces on the atomic scale yield the force between macroscopic bodies known as the dispersion force which is of the order A/a. Here A is the Hamaker constant, which is a function of the properties of both the dispersed and continuous phases. The characteristic length of the particle is given by a.

The thermal energy of the molecules manifests itself as a Brownian force of the order $k_B T/a$, where k_B is the Boltzmann constant and T is the absolute temperature.

The fluid (continuous phase) viscosity gives rise to a viscous force of the order $\mu U a$ where μ is the viscosity of the continuous phase and U is the particle velocity through the continuous phase. Due to the movement of the bulk particles, an inertial force comes into play which is of the order $\rho a^2 U^2$, where ρ is the continuous medium density.

The gravitational body force on a particle is of the order $a^3 g \Delta\rho$ where g is the acceleration due to gravity and $\Delta\rho$ is the density difference between the particle and the continuous phase.

Surface tension arises from the interaction between the two phases represented by the particle and the continuous medium. This force is of the order γa, where γ is the surface (interfacial) tension of the particle in the medium.

Table 2.2 gives the order of magnitude values for the relative significance of the force cited above (Russel *et al.*, 1989). For the particular values chosen, it is clear that the ratio of the inertial to viscous forces is not important and that the ratio of repulsive electric force to Brownian force is fairly high. This would indicate a stable colloidal system. However, it should be recognized that the electrical and attractive forces between particles are greatly influenced by the separation distance between the particles. Consequently, the analysis cannot be considered as being complete.

Table 2.2 shows that the ratio of various forces contains the particle dimension, which can vary from 10^{-8} to 10^{-5} m, and consequently, can affect the nature of the interactions. As the surface to volume ratio is of the order $1/a$, it becomes clear that for very small particles, the surface to volume ratio can be significantly large and a high percentage of the molecules in such a particle will lie within or close to the region of inhomogeneity associated with the particle/medium interface. These molecules have properties different from those in the bulk phases more distant from the interface. It is no longer possible to describe a colloidal system (dispersed and continuous phases) simply in terms of the sum of contributions from molecules in the bulk phases, calculated as if both phases had the same properties as they would have

TABLE 2.2. Magnitudes of the Characteristic Force: $T = 300$ K, $a = 1$ μm, $\mu = 10^{-3}$ Pa s, $U = 1 \times 10^{-6}$ m/s, $\rho = 10^3$ kg/m^3, $\Delta\rho/\rho = 10^{-2}$, $g = 10$ m/s^2, $A = 10^{-20}$ J, $\psi_s = 0.05$ V, $\epsilon = 8.85 \times 10^{-10}$ C/Vm, $\gamma = 0.1$ N/m, $k_B = 1.381 \times 10^{-23}$ J/K.

$\dfrac{\text{electrical force}}{\text{Brownian force}}$	$\dfrac{\epsilon\psi_s^2}{k_BT/a}$	$\approx 10^3$
$\dfrac{\text{attractive force}}{\text{Brownian force}}$	$\dfrac{A/a}{k_BT/a}$	≈ 1
$\dfrac{\text{Brownian force}}{\text{viscous force}}$	$\dfrac{k_BT/a}{\mu U a}$	≈ 1
$\dfrac{\text{gravitational force}}{\text{viscous force}}$	$\dfrac{a^3 g\Delta\rho}{\mu U a}$	$\approx 10^{-1}$
$\dfrac{\text{inertial force}}{\text{viscous force}}$	$\dfrac{\rho a^2 U^2}{\mu U a}$	$\approx 10^{-6}$
$\dfrac{\text{surface tension}}{\text{gravitational force}}$	$\dfrac{\gamma a}{a^3 g\Delta\rho}$	$\approx 10^9$

Source: Adapted from Russel *et al.* (1989).

in the individual bulk state. A significant and dominating contribution comes from the molecules residing at the interface. This is why surface chemistry plays an important role in colloid science and why colloidal properties become important as the particle decreases in size, say below 10 μm (Everett, 1988).

The following example demonstrates how the ratio of the number of the "surface molecules" to the total molecules increases with the decreasing ratio of particle size to the molecule size.

Consider a particle in the shape of a cube having a side length of L. Assume that the particle material is made up of molecules having the shape of a cube whose side length is S. The total number of molecules N_t contained in the particles is given by

$$N_t = \left(\frac{L}{S}\right)^3 \tag{2.1}$$

The number of surface particles is given by

$$N_s = 2\left(\frac{L}{S}\right)^2 + 2\left(\frac{L}{S} - 2\right)\left(\frac{L}{S}\right) + 2\left(\frac{L}{S} - 2\right)^2 \quad \text{for } L/S \geq 2 \tag{2.2}$$

Figure 2.3 shows the subdivision of the particle. The ratio of the number of surface molecules to the total number of molecules is then given by

$$\frac{N_s}{N_t} = 6\left(\frac{S}{L}\right) - 12\left(\frac{S}{L}\right)^2 + 8\left(\frac{S}{L}\right)^3 \tag{2.3}$$

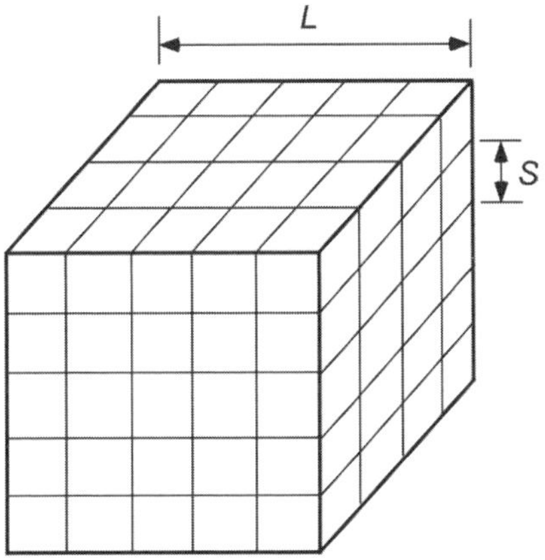

Figure 2.3. Subdivision of a cube by its molecules.

For the case of $S/L \ll 1$, Eq. (2.3) gives

$$\frac{N_s}{N_t} = 6\left(\frac{S}{L}\right) \tag{2.4}$$

For $L = 10$ nm and $S = 0.3$ nm, Eq. (2.3) gives $N_s/N_t = 0.17$. This indicates that 17% of the molecules reside on the particle surface, *i.e.*, on the interface of the particle and the medium. Such a high percentage would indicate that the properties of the molecules are important in describing the behavior of a colloidal particle. Figure 2.4 gives the variation of N_s/N_t with S/L.

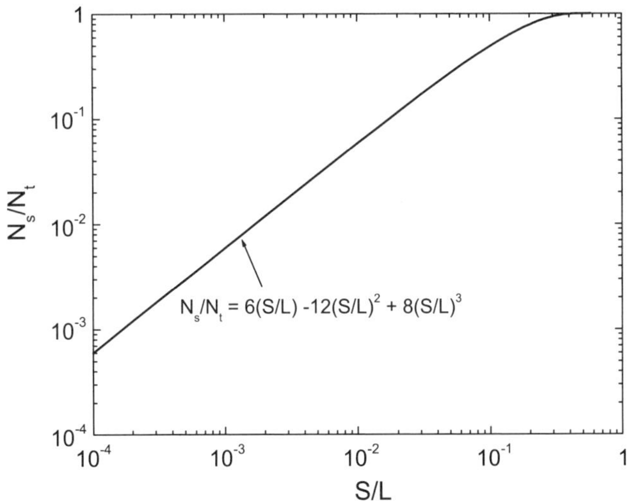

Figure 2.4. Ratio of surface to total molecules as a function of size ratio.

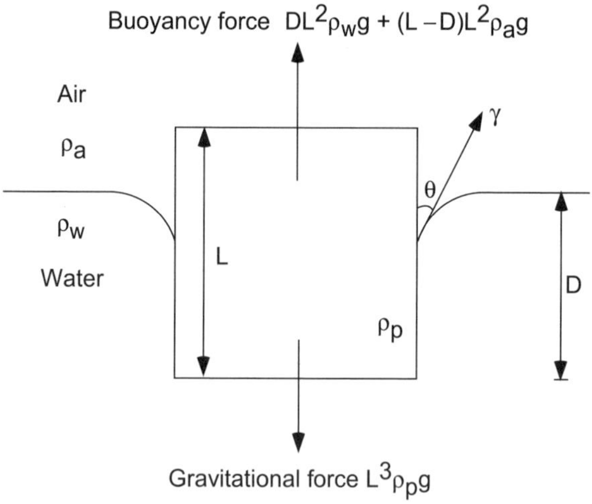

Figure 2.5. A particle straddling an air–water interface.

Another example is that of a particle straddling an interface. It illustrates the significance of the particle size. Consider a cube of side length L and density ρ_p straddling the interface of air and water. The water density is ρ_w and the air density is ρ_a. Figure 2.5 shows such a particle.

Assuming a downward force is positive, the forces acting on the cube are given by

$$\begin{aligned} \text{Surface Tension:} \quad & -4L\gamma\cos\theta \\ \text{Gravitation:} \quad & L^3\rho_p g \\ \text{Buoyancy:} \quad & -DL^2\rho_w g - (L-D)L^2\rho_a g \end{aligned}$$

At equilibrium, the sum of all the forces is zero, hence

$$L^3\rho_p g - 4L\gamma\cos\theta - DL^2\rho_w g - (L-D)L^2\rho_a g = 0 \tag{2.5}$$

where D is the submerged length and θ is the static contact angle which is related to the surface tensions of the particle-air, particle-water, and water-air. As $\rho_a/\rho_w \ll 1$, Eq. (2.5) leads to

$$\frac{D}{L} = \frac{\rho_p}{\rho_w} - \frac{4\gamma\cos\theta}{L^2\rho_w g} \tag{2.6}$$

and

$$\frac{D}{L} = \frac{\rho_p}{\rho_w} - \frac{4\cos\theta}{\text{Bo}} \tag{2.7}$$

where

$$\frac{\rho_w g L^2}{\gamma} = \frac{\text{gravitational force}}{\text{surface tension force}} = \text{Bo}$$

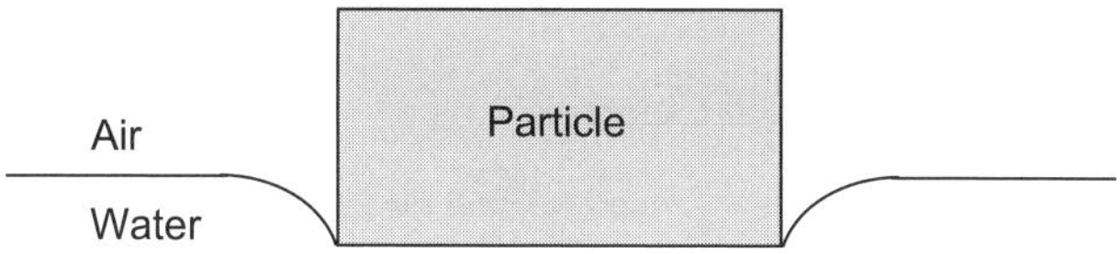

Figure 2.6. A floating particle at an air–water interface having a small Bond number.

The dimensionless group Bo is known as the Bond number. For given physical properties ρ_p, ρ_w, γ, and $\cos\theta$, and constant g, the Bond number $\to 0$ as the particle dimension $L \to 0$.

For $0 < D/L < 1$, the particle floats on the surface. The (D/L) value is a balance between (ρ_p/ρ_w) and $4\cos\theta/\text{Bo}$. For a large particle, Bo is large and Eq. (2.7) leads to $D/L = \rho_p/\rho_w > 1$. This is the situation where the particle sinks into the water. This criterion is only related to the density ratio. For the case of a very small particle, *i.e.*, $L \to 0$, and $\text{Bo} \to 0$, Eq. (2.7) gives

$$\frac{D}{L} \approx -\frac{4\cos\theta}{\text{Bo}} \tag{2.8}$$

A negative (D/L) value signifies that a small particle will always float on the water surface irrespective of the density ratio. This is shown in Figure 2.6. Consequently, it is clear that the particle size has a major role in determining its behavior.

2.3 STABILIZATION OF COLLOIDS

Colloidal particles can be stabilized against coagulation (or flocculation) by electrostatic repulsion due to the presence of ions near their surfaces or by steric effects arising from polymer chains being attached to the surface of particles. The properties of a colloidal system, *i.e.*, its rheology, shear stability, and stability to added electrolyte or polymers are much affected by the nature of the stabilizing mechanism (Walbridge, 1987).

Colloidal particles can be stabilized by the electrostatic forces that arise as a result of the charged particle surface and the presence of an associated diffuse atmosphere (double layer) of counterions. Surface charge can arise from different causes, *e.g.*, surface ionization and physically adsorbed ionized surfactant groups. Electrostatic stabilization plays a dominant role in aqueous systems. Electrostatically stabilized particles can be flocculated by the addition of an electrolyte and by shear. Figure 2.7 shows the effect of electrolyte addition on the total interaction potential energy ϕ of a colloidal particle. A positive value of the derivative of the interaction potential energy with respect to the separation distance indicates an attractive force. Figure 2.7 shows that increasing the electrolyte concentration changes the shape of the interaction potential energy leading to weak repulsive force and a possible attractive force at a large separation distance between the particles.

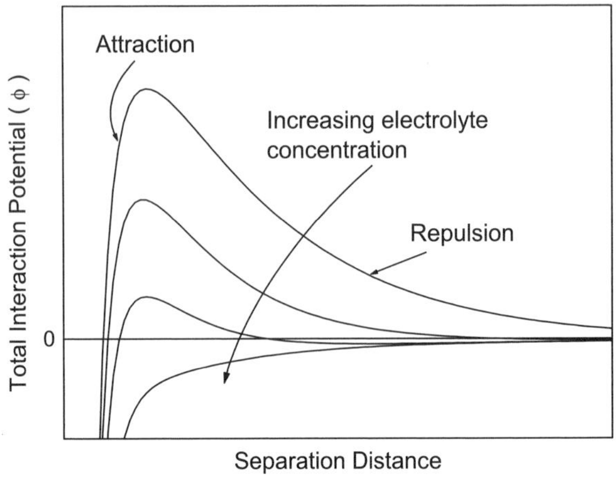

Figure 2.7. Variation of the total potential at different electrolyte concentrations for charge stabilized particles.

A colloidal dispersion can be stabilized in either an aqueous or non-aqueous continuous phase by solvated polymeric moieties adsorbed on the colloidal particle surface. The polymeric chains attached to the surface can be regarded as a barrier around each particle, preventing their close approach to each other. For sterically stabilized colloidal systems, the continuous phase (medium) must be a good solvent to the attached polymer with nearly complete surface coverage as shown in Figure 2.8(a). The case of a poor solvent is depicted in Figure 2.8(b).

The total potential energy for sterically stabilized particles is shown in Figure 2.9. Sterically stabilized colloidal particles are usually very stable over a wide range of particle sizes and shear rates, and at high dispersed phase concentrations. They can

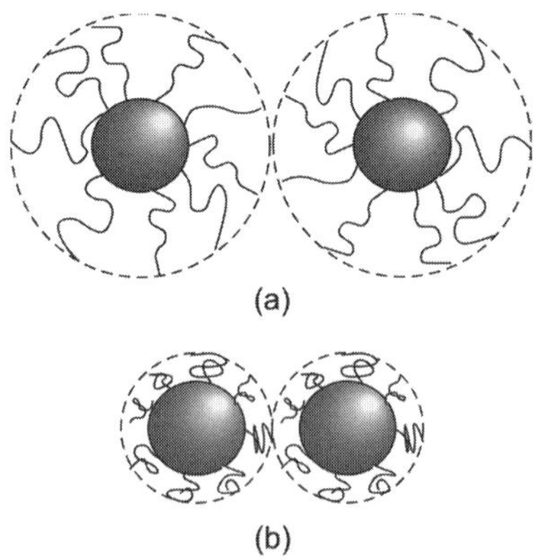

Figure 2.8. (a) Highly sterically stabilized colloidal particle (polymer is in a good solvent); (b) poorly sterically stabilized colloidal particles (polymer is in poor solvent).

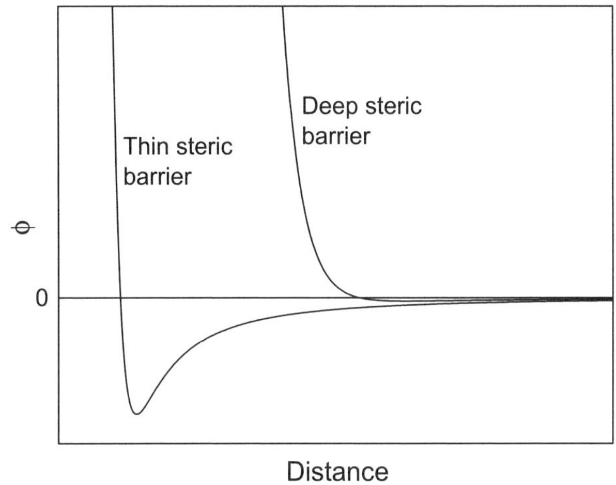

Figure 2.9. Variation of the total interaction potential energy for sterically stabilized particles.

be flocculated by changing the solvency of the continuous phase or by desorbing the attached polymeric moieties. In practice, molecular weights above 1000 are desirable (Walbridge, 1987).

2.4 PREPARATION OF COLLOIDAL SYSTEMS

There are two fundamentally different ways in which colloidal dispersions can be made: either by breaking down (splitting) bulk coarse matter to colloidal dimensions or by building up molecular aggregates to colloidal size. The first method is referred to as the dispersion method and the second is referred to as the condensation or nucleation method.

2.4.1 Dispersion Methods

Energy changes are associated with the breakup of a coarse particle. The free energy change associated with creation of new surface area, A, is given by

$$\Delta G = \gamma A \tag{2.9}$$

where γ is the solid material surface tension and ΔG is the change in Gibbs free energy due to the creation of new surface area. ΔG is then the work required to create the new surface area and it is directly related to γ and A (Everett, 1988).

The diameter of particles obtained in the grinding or milling of a solid, either in a mortar or in a ball mill, is about 1 to 5 μm. The size of the colloidal particles can be reduced by grinding the solids in a liquid having a low surface tension. A further decrease in size can be achieved by the addition of a surface-active electrolyte to the grinding medium. These electrolytes, such as soaps or long chain alkylaryl-sulphonates, are adsorbed on the particle surfaces and hence tend to stabilize the dispersion of the colloidal particles. Particle sizes of the order of 0.1 μm can be achieved by wet milling (Jirgensons and Straumanis, 1962).

Formation of emulsions by the "breaking down" of one liquid in the presence of another can be achieved either by simple stirring or by applying high shear to the two liquid phases. Colloidal mills or high speed homogenizers are commonly used to prepare a liquid-in-liquid dispersion, *i.e.*, an emulsion. The success of an emulsification process depends, to a large extent, on the interfacial tension between the two liquid phases and on the stabilizing force present. Emulsifying agents, *i.e.*, agents that lower interfacial tension, are usually required for successful emulsification.

In addition to the mechanical approach used in breaking up solids and liquids, ultrasonic waves can be used for the breaking up of solids in the emulsification process. The velocity of sound v is related to the wavelength λ and the frequency f by

$$v = f\lambda \tag{2.10}$$

Ultrasonic vibrations can be easily generated in the range of 25 kHz–2 MHz with commercially available equipment. An oscillating quartz disk is able to produce very high intensity ultrasonic vibrations of up to 10^5 W/m^2. This intensity is about 10^{10} times larger than the intensity produced by a loud radio. Such an energy level can disintegrate coarse particles and form emulsions (Jirgensons and Straumanis, 1962). It is interesting to note that with proper particle size and vibration frequency, the reverse process of solids agglomeration or de-emulsification can occur.

2.4.2 Condensation Methods

Colloidal particles can be made by combining small molecules into larger units. This can be achieved either by a decrease in solubility or through chemical methods. The chart in Figure 2.10 illustrates the types of condensation methods used in making of a colloidal system (or sol).

A simple way to prepare a colloidal system is to pour a true molecular solution of a solute into another liquid in which the solute is practically insoluble. A sulphur sol can be prepared in this way by decreasing the sulphur solubility due to a change in the solvent. Sulphur is dissolved in alcohol and then the solution is diluted with

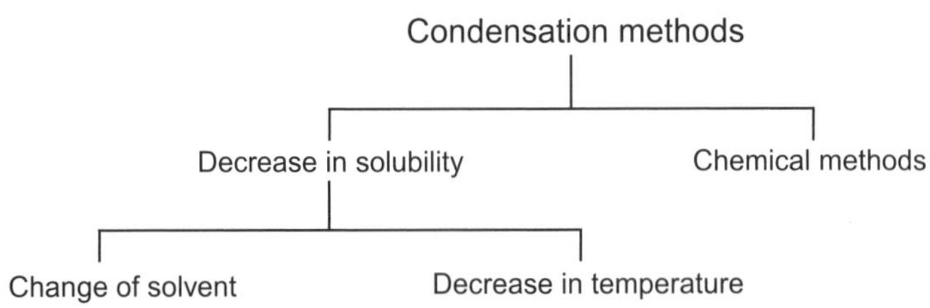

Figure 2.10. Types of condensation methods used for preparation of colloidal systems (sols).

water in which the sulphur is much less soluble. Various hydrosols of resins and fats are prepared similarly.

Instead of reducing the solubility by changing the solvent, one can decrease the temperature. A sol of ice in pentane can be prepared by cooling pentane that contains traces of water.

Chemical methods provide excellent means for preparing colloidal systems. The basic idea is to perform a chemical reaction in which an insoluble or practically insoluble substance is formed so that the solid remains dispersed as small particles. For example, sols of hydroxides can be prepared by hydrolysis. Specifically, colloidal ferric hydroxide is prepared by the hydrolysis of ferric chloride at the boiling point:

$$FeCl_3 + 3H_2O \rightarrow Fe(OH)_3 + 3HCl$$

Methods of preparation and characterization of monodisperse metal hydrous oxide sols, *i.e.*, colloidal dispersions, are given by Matijevic (1976). Monodisperse silica spheres can be prepared using techniques advanced by Stöber *et al.* (1968), van Blaaderen and Vrij (1993), and Nyffenegger *et al.* (1993). Some recipes for preparation of some simple sols are given below (Everett, 1988).

Gold Sol: Add 1 cm^3 of a 1% solution of gold chloride ($HAuCl_4 \cdot 3H_2O$) to 100 cm^3 of distilled water, bring to boil and add 2.5 cm^3 of 1% sodium citrate solution. Keep the solution just boiling. After a few minutes observe the appearance of a blue coloration, followed by the formation of a ruby-red gold sol.

Sulphur Sol: Rapidly mix 4 mM (millimolar, 1 mM = 0.001 mol dm^{-3}) sodium thiosulphate solution and 4 mM hydrochloric acid. The mixture becomes cloudy after a few minutes and then develops to an opaque white dispersion of colloidal sulphur.

Silver Bromide Sol: Mix equal amounts of 20 mM sodium bromide solution to 18 mM silver nitrate solution. A colloidal dispersion of silver bromide is formed immediately. A silver iodide sol may be prepared in a similar manner.

Ferric Hydroxide Sol: Add 2 cm^3 of a 30% solution of ferric chloride slowly, with stirring, to 500 cm^3 of boiling distilled water. A clear reddish-brown dispersion of ferric hydroxide is formed.

Polymeric Dispersions: Polymeric dispersions can be prepared by emulsion polymerization, dispersion polymerization, and suspension polymerization (Walbridge, 1987).

In *emulsion polymerization*, a monomer is emulsified in a non-solvent (usually water) in the presence of a surfactant. A water-soluble initiator is added. Particles of the polymer form and grow in the aqueous continuous phase. Production of the particles proceeds until the monomer is used up.

In *dispersion polymerization*, monomer, initiator, stabilizer, and solvent initially form a homogeneous solution. Polymer precipitates when the solubility limit is exceeded. Polymer particles continue to grow until the monomer is used up.

In *suspension polymerization*, the monomer is emulsified in the medium using a surfactant. The initiator is often dissolved within the monomer droplet. The droplet is gradually converted into the insoluble particles, but no new particles are formed. The size distribution of the dispersion becomes that of the original monomer droplets.

In most of the processes involving preparation of a colloidal system, the formation of the sols passes through nucleation and growth stages. Control of such stages would determine the size of the colloidal particles and their size distribution.

2.5 PURIFICATION OF SOLS

In the previous section, we dealt with methods of making colloidal systems of solids or of immiscible liquids in either an aqueous or non-aqueous continuous phase. Prepared sols of solid particles that are insoluble in water or in a solvent usually contain some contamination having low molecular weight. These contaminants can be removed by various methods such as dialysis, ultrafiltration, and electrodialysis.

Dialysis, in its simplest form, involves placing the sol in a container having one end covered with a semi-permeable membrane. The membrane side of the container is placed in water (or solvent) as shown in Figure 2.11. The membrane is permeable to the solvent and the other small molecular weight impurities, but impermeable to the colloidal particles. Diffusion of the impurities through the membrane leads to the eventual depletion of the impurities within the prepared sols (Overbeek, 1952).

Ultrafiltration is a pressure driven membrane separation process. The solvent and low molecular weight impurities pass through the membrane but the membrane rejects particulate matter. Ultrafiltration is similar to common filtration processes except that the membrane pores are very small. The applied pressure is usually in the range of 200–1000 kPa. Ultrafiltration is not a method of purification but rather of concentration. Upon the rejection of the solvent, together with the associated impurities, fresh solvent is added to the sol and ultrafiltration is once again applied to the sol. The addition of the fresh solvent acts as a washing medium and makes it possible to obtain a sol with little impurities.

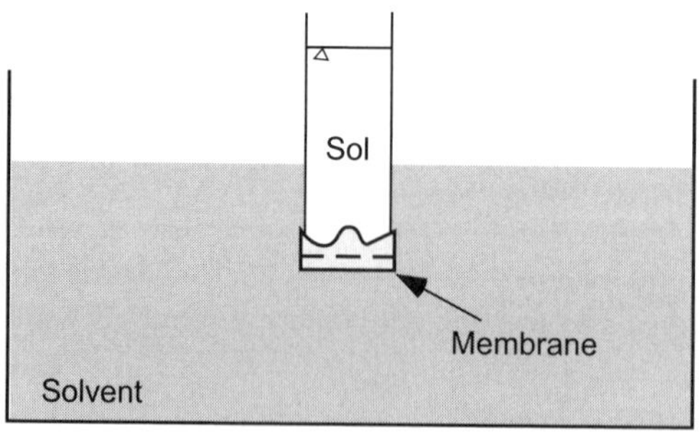

Figure 2.11. A simple dialysis cell.

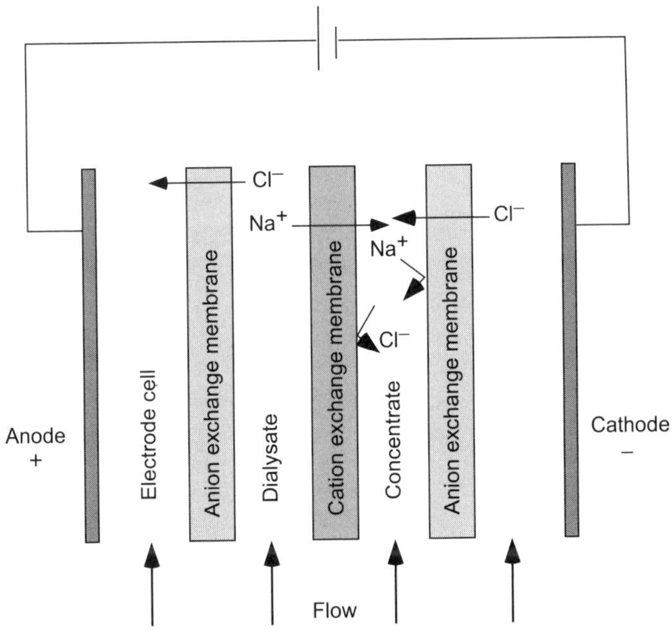

Figure 2.12. Electrodialysis cell pair. Sol purification from Na^+ and Cl^- ions. Adapted from Probstein (2003).

Electrodialysis is a membrane process in which dissolved ions are removed from an aqueous solution through membranes under the driving force of a dc electric field. Electrodialysis membranes are ion-exchange membranes (Probstein, 2003).

An electrodialysis cell has alternating anion and cation exchange membranes. The sol to be purified is made to flow between the membranes. An anion exchange membrane allows the passage of the anions but prevents passage of the cations. Conversely, a cation exchange membrane allows the cations to pass, while retaining the anions. In this manner, it is possible to concentrate the undesirable electrolyte and to deplete the sol from the electrolyte (*e.g.*, NaCl). This is shown in Figure 2.12. The chamber containing the electrolyte-free sol is called the dialysate channel and the chamber containing the electrolyte-rich sol is called the concentrate channel.

2.6 A HISTORICAL SUMMARY

As stated in the beginning of this chapter, Thomas Graham classified colloids and crystalloids as two types of matter. His classification was based on diffusion experiments through membranes. Colloidal systems were observed and studied for at least two centuries prior to Graham's studies, since about the seventeenth century, when alchemists produced sols by treating gold chloride solutions with reducing agents.

Observation of the behavior of colloidal particles in a suspension dates back to the works of Robert Brown (1773–1858). He studied the erratic motion of particles in pollen grains suspended in water under a microscope in 1827. A detailed description of his experiments and a recreation of the original experiments is given by Ford (1992). The molecular basis of Brownian motion was settled by Perrin (1870–1942) in his book "Brownian Motion and Molecular Reality" (1910). The theory of Brownian motion was presented by Einstein (1879–1955). His papers on Brownian movement, published between 1905 and 1911, have been translated into English and the compilation is available as a book edited with notes by Fürth (1926). Shortly after Einstein's first work on Brownian motion, Langevin (1872–1946) formulated a theory in which the minute fluctuations in the position of a particle were due explicitly to a random force (1908). Langevin's approach proved to have great utility in describing molecular fluctuations in colloidal systems, including non-equilibrium thermodynamics.

The early experimental observations on dynamics of colloidal systems were intricately related to the development of optical microscopy. Following Faraday's discovery that small particles could be detected by focussing light rays into a cone, Zsigmondy and Siedentopf invented the ultramicroscope in 1903. Studies with this instrument probed the nature of the erratic Brownian motion of individual particles.

The discovery that naturally occurring colloidal particles were charged was made by Reuss in 1809, who observed the motion of clay particles in an electric field. Schulz (1882) and Hardy (1900) elucidated the role of added electrolytes in suppressing the effects of charge and promoting coagulation. Their work provided strong evidence that stability of aqueous dispersions derived from electrostatic repulsion. Smoluchowski (1872–1917) provided the celebrated formula relating the surface electric potential [more appropriately, the zeta (ζ) potential] to the electrophoretic mobility (1903). A significant amount of the early theoretical developments in electrokinetic transport phenomena can be attributed to Smoluchowski and Helmholtz (1821–1894).

Gouy (1910) and Chapman (1913) presented the theory for the screening of surface charge by the diffuse layer of counterions, thereby relating the thickness of the diffuse layer to the ionic strength of the solution. Smoluchowski (1917) deduced expressions for the rate of formation of small aggregates by Brownian and shear-induced collisions between particles, which related the flocculation rates to the screening of the electrostatic repulsion with excess electrolyte.

Although the theoretical interpretation of colloid stabilization employing electrostatic repulsion was becoming more formal in the early twentieth century, the nature of the attractive force that causes destabilization of colloids and subsequent flocculation was not explored formally until the works of de Boer (1936) and Hamaker (1937). They developed a theory to obtain the attractive van der Waals interactions between colloidal particles by pairwise summation of the fundamental atomistic (Lennard–Jones) interaction potentials. This led to the representation of the total interparticle interaction potential as the sum of the attractive van der Waals and the repulsive electrostatic interactions. The first formal theory of colloidal stability was presented independently by Derjaguin and Landau (1941) in the Soviet Union, and Verwey and Overbeek (1948) in the Netherlands, who used the total interparticle interaction as a linear combination of the van der Waals and electrostatic interactions between a pair of particle. The DLVO theory, named after these four scientists, forms a basis of modern theoretical colloid science, and remains a cornerstone for theoretical elucidation of colloid transport phenomena.

2.7 ELECTROKINETIC PHENOMENA IN MODERN COLLOID SCIENCE

Electrokinetic transport phenomena have been an important aspect of colloid science since the early twentieth century. Despite the fact that the first observations of electrokinetic phenomena date back to the studies conducted by Reuss in 1809, the first theoretical developments of electrokinetic transport are attributed to Helmholtz (1879) and Smoluchowski (1903). The development of the subject following these early works has been summarized by Dukhin and Derjaguin (1974). Considerable attention has since been conferred on four key electrokinetic phenomena, namely, streaming potential, electroosmotic flow, electrophoresis, and sedimentation of charged suspensions. Several books provide details of these developments (Russel *et al.*, 1989; Probstein, 2003), although attention has mostly been devoted to dilute suspensions and single charged particles. The last two decades of the past century have seen a steady improvement in the understanding of these phenomena in concentrated and confined systems. In this book, we will discuss these developments in considerable detail.

Electrokinetic processes involving colloidal systems have received heightened attention over the past decade, particularly due to the rapid developments associated with biological separations, micro-electromechanical systems (MEMS), and nano-scale separation processes. The renewed interest in electrokinetic phenomena primarily stems from the requirement of moving fluids in confinements (or channels) that are extremely narrow. Clearly, fluid handling devices, particularly pumps, operating in the macroscopic regime are incapable of handling the requirements of the microscopic transport realm, the single biggest challenge being the enormous pressure gradients necessary to drive a fluid through a capillary of microscopic radius. Consequently, there has been an immense effort toward exploration of alternative means of fluid transport in microscopic channels. Electrokinetic transport is perhaps the most suitable candidate for providing the motive force to the fluids in microscopic domains without requiring enormous pressure gradients.

Electrophoresis is one of the major processes utilized in biological separations. Biocolloids can be manipulated in a small environment by applying an electric field. This principle is utilized to separate proteins of different sizes or charges, DNA, and cells. Electrokinetic phenomena like electrophoresis and streaming potential are routinely employed to estimate the charge of colloidal particles in different media. Electrical forces are used to cut, transport, mix, separate, and manipulate microscopic liquid droplets in microfluidic chips. Processes like electrowetting (EW) or electrowetting on a dielectric (EWOD) are being used more frequently for moving fluids (Cho *et al.*, 2003). There is a tremendous surge in application of alternating current (ac) electrokinetic phenomena like dielectrophoresis as separation methods for cells and colloidal particles (Hughes, 2003). In all these modern applications, we need to understand colloidal and electrokinetic transport phenomena in a new light. Unlike the historical perspective of colloid science, which dealt with these phenomena to understand colloidal behavior in bulk systems, modern colloid science has to deal with overall system sizes that are often comparable to the colloidal particle dimensions envisioned by the early colloid scientists. This necessitates a re-evaluation of many traditional theories and concepts of colloid science and electrokinetics.

This book aims to bridge the gap between the traditional understanding of electrokinetic transport phenomena in colloidal systems and the modern approaches evolving from the necessity of interpreting electrokinetic and colloid transport behavior in extremely small domains. It provides a detailed evaluation of the fundamental concepts underlying electrokinetic transport processes, while pointing out the modifications in traditional theories needed to address the modern challenges posed by micro-scale transport devices. Accordingly, the book builds up on the basic construct of electrokinetic transport processes in Chapters 3 to 6 by providing a detailed background on electrostatics in vacuum and pure dielectrics, electrical double layers in the presence of free charge, and basic transport equations for charged electrolyte solutions and colloidal suspensions in presence of an external electric field. Following a brief overview of the four key electrokinetic phenomena in Chapter 7, detailed analyses of these four major electrokinetic phenomena, namely, electroosmosis, streaming potential, electrophoresis, and sedimentation potential, are given in Chapters 8 to 10. Following this, we introduce the van der Waals forces and the DLVO theory of colloid stability in Chapter 11. Chapters 12 and 13 present the two important aspects of colloid transport phenomena, namely, coagulation and deposition, respectively. In Chapter 14, detailed numerical approaches for modeling electrokinetic transport phenomena in micro-scale processes are described. In the final chapter of the book, we present some practical applications of the theoretical principles outlined and developed in the book, including membrane filtration processes, environmental cleanup, extraction of bitumen, and microfluidic applications.

2.8 NOMENCLATURE

a	particle radius, m
Bo	Bond number, dimensionless
D	immersed depth in water, m
d	diameter of sphere, m
f	frequency of sound, s^{-1}
g	acceleration due to gravity, m/s^2
ΔG	free energy, J
k_B	Boltzmann constant, J/K
L	cube side length, m
N_s	total number of surface molecules
N_t	total number of inner molecules
S	molecule side length, m
T	absolute temperature, K

U particle velocity, m/s
v speed of sound, m/s

Greek Symbols

γ interfacial tension, N/m
ϵ dielectric permittivity of medium, C/Vm
ϕ total interaction potential energy, J
λ sound wave length, m
μ fluid viscosity, Pa.s
θ angle related to contact angle $= \pi -$ contact angle
ρ density, kg/m^3
ψ_s surface electric potential, V

2.9 REFERENCES

Chapman, D. L., A contribution to the theory of electroencapillarity, *Phil. Mag.*, **25**, 475–481, (1913).

Cho, S. K., Moon, H. J., and Kim, C. J., Creating, transporting, cutting, and merging liquid droplets by electrowetting-based actuation for digital microfluidic circuits, *J. Microelectromech. Systems*, **12**, 70–80, (2003).

de Boer, J. H., The influence of van der Waals forces and primary bonds on binding energy, strength and orientation, with special reference to some artificial resins, *Trans. Faraday Soc.*, **32**, 10–38, (1936).

Derjaguin, B. V., and Landau, L. D., Theory of the stability of strongly charged lyophobic colloids and the adhesion of strongly charged particles in solutions of electrolytes, *Acta Physicochim. URSS*, **14**, 633–662, (1941).

Dukhin, S. S., and Derjaguin, B. V., Electrokinetic Phenomena, in *Surface and Colloid Science*, vol. 7, E. Matijevic (Ed.), Wiley, (1974).

Einstein, A., *Investigations on the Theory of the Brownian Movement*, R. Fürth (Ed.), Dutton, NY, (1926) (also a Dover publication, 1956).

Everett, D. H., *Basic Principles of Colloid Science*, Royal Society of Chemistry, London, (1988).

Ford, B. J., Brownian movement in clarkia pollen: A reprise of the first observations, *The Microscope*, **40**, 235–241, (1992).

Gouy, G., Sur la constitution de la electrique a la surface d'un electrolyte, *J. Phys. Radium*, **9**, 457–468, (1910).

Hamaker, H. C., London–van der Waals attraction between spherical particles, *Physica*, **4**, 1058–1072, (1937).

Hardy, W. B., A preliminary investigation of the conditions which determine the stability of irreversible hydrosols, *Proc. Roy. Soc. Lond*, **66**, 110–125, (1900).

Helmholtz, H. V., Studien uber elctrische grenschichten, *Ann. der Physik und Chimie*, **7**, 337–387, (1879).

Hughes, M. P., *Nanoelectromechanics in Engineering and Biology*, CRC Press, Boca Raton, (2003).

Jirgensons, B., and Straumanis, M. E., *A Short Textbook of Colloid Chemistry*, McMillan Co., London, (1962).

Langevin, P., Theory of Brownian motion, *C. R. Acad. Sci.*, **146**, 530–533, (1908).

Matijevic, E., Preparation and characterization of monodisperse metal hydrous oxide sols, *Prog. Colloid Polymer Sci.*, **61**, 24–25, (1976).

Nyffenegger, R., Quellet, C., and Ricka, J., Synthesis of fluorescent, monodisperse, colloidal silica particles, *J. Colloid Interface Sci.*, **159**, 150-157, (1993).

Overbeek, J. Th. G., Phenomenology of lyophobic systems, Chapter II, in *Colloid Science*, H. R. Kruyt (Ed.), Elsevier, Amsterdam, (1952).

Perrin, J., *Brownian Motion and Molecular Reality*, Taylor and Francis, London, (1910).

Probstein, R. F., *Physicochemical Hydrodynamics, An Introduction*, 2nd ed., Wiley Interscience, New York, (2003).

Russel, W. B., Saville, D. A., and Schowalter, W. R., *Colloidal Dispersions*, Cambridge University Press, Cambridge, (1989).

Schulz, H., Schwefelarsen im wasseriger losung, *J. Prakt. Chem.*, **25**, 431–452, (1882).

Smoluchowski, M. von, Contribution a la theorie de l'endosmose electrique et de quelques phenomenes correlatifs, *Bull. International de l'Academie des Sciences de Cracovie*, **8**, 182–200, (1903).

Smoluchowski, M. von, Versuch einer mathematischen Theorie der Koagulationkinetik kollider losungen, *Z. Phys. Chem.*, **92**, 129–168, (1917).

Stöber, W., Fink, A., and Bohm, E., Controlled growth of monodisperse silica spheres in micron size range, *J. Colloid Interface Sci.*, **26**, 62–69, (1968).

van Blaaderen, A., and Vrij, A., Synthesis and characterization of monodisperse colloidal organo-silica spheres, *J. Colloid Interface Sci.*, **156**, 1–18, (1993).

Verwey, E. J. W., and Overbeek, J. Th. G., *Theory of Stability of Lyophobic Colloids*, Elsevier, Amsterdam, (1948).

Walbridge, D. J., Preparation of liquid-solid dispersions, Chapter 2, in *Solid/Liquid Dispersions*, Tadros, Th. F. (Ed.), Academic Press, London, (1987).

CHAPTER 3

ELECTROSTATICS

It was pointed out earlier that most surfaces attain a surface charge when immersed in an electrolyte solution. The electrostatic forces arising from the surface charge are essential for stabilizing colloidal suspensions and they play a central role in biological systems and industrial processes.

In order to appreciate and better understand the role of charged surfaces immersed in an electrolyte solution where free electric charges are present, we will deal in this chapter with basic electrostatics in free space, *i.e.*, in vacuum, and in materials called dielectrics in the absence of free charge. In other words, we will deal with physical situations where there is no electric current.

3.1 BASIC ELECTROSTATICS IN FREE SPACE

3.1.1 Fundamental Principles of Electrostatics

One can state that there are four fundamental principles (or laws) of electrostatics (Feynman *et al.* 1964; Slater and Frank, 1969; Eyges, 1980; Griffiths, 1989).

The *first principle* is that of charge conservation. It simply states that the total charge on an isolated body cannot be changed. In other words, for an isolated body, $\sum Q_i =$ constant, where Q_i is the ith charge contained in the body.

The *second law* is that charge is "quantized". This means that the total charge on a body is an integral multiple of a fundamental charge. Here, the magnitude of the charge carried by an electron (1.602×10^{-19} Coulombs) is the fundamental elementary

Electrokinetic and Colloid Transport Phenomena, by Jacob H. Masliyah and Subir Bhattacharjee
Copyright © 2006 John Wiley & Sons, Inc.

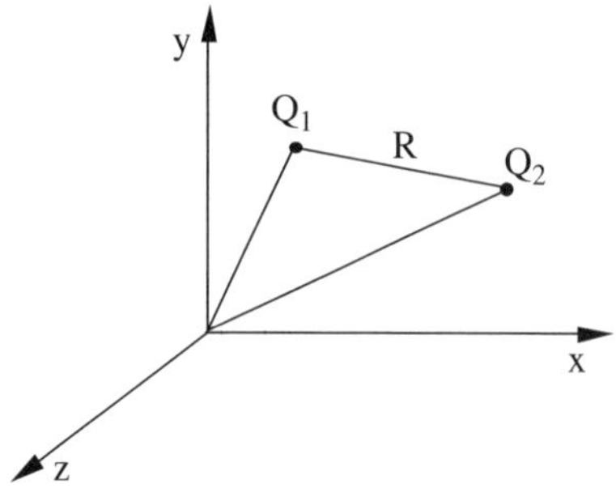

Figure 3.1. Two point charges separated by distance R in vacuum.

charge, often denoted by the symbol e. Consequently, a charge is given by Ne where N is an integer number and e is the fundamental charge.

The *third principle* is Coulomb's law that describes the force between two point charges. Consider two stationary point charges of magnitude Q_1 and Q_2 in free space separated by a distance R as shown in Figure 3.1. Coulomb's law states that the mutual force between these two charges, F, is given by

$$F = \frac{Q_1 Q_2}{4\pi \epsilon_o R^2} \tag{3.1}$$

The force is in Newtons (N), the charge is in Coulombs (C), and the separation distance is in meters (m). Here, ϵ_o is the permittivity of free space, *i.e.*, vacuum, its value in SI units being 8.854×10^{-12} C/Vm or C^2/Nm^2, where V is an abbreviation for Volts, the SI unit for electric potential.

The significance of Eq. (3.1) is that the force is proportional to the product of the charges and it is inversely proportional to the square of the separation distance between the charges. As force is a vector, Eq. (3.1) representing the mutual force between the two charges needs more clarification. Accordingly, Coulomb's law can be formally written in a vectorial form as

$$\mathbf{F}_{12} = \frac{Q_1 Q_2 \mathbf{R}_{12}}{4\pi \epsilon_o |\mathbf{R}_{12}|^3} \tag{3.2}$$

Here $\mathbf{F}_{12}$ is the force exerted by charge Q_1 on charge Q_2, or analogously, the force felt by charge Q_2 due to the charge Q_1. The force acts along a straight line joining the two point charges in the direction of charge Q_1 to charge Q_2. This direction is represented by the direction of the vector $\mathbf{R}_{12}$, which is often termed as the separation (or displacement) vector for charges Q_1 to Q_2. The position vectors of the two charges Q_1 and Q_2 are given as $\mathbf{r}_1$ and $\mathbf{r}_2$, respectively. The separation vector $\mathbf{R}_{12}$ is related to the position vectors as

$$\mathbf{R}_{12} = \mathbf{r}_2 - \mathbf{r}_1 \tag{3.3}$$

and its magnitude provides the linear distance between the two charges

$$|\mathbf{R}_{12}| = R \tag{3.4}$$

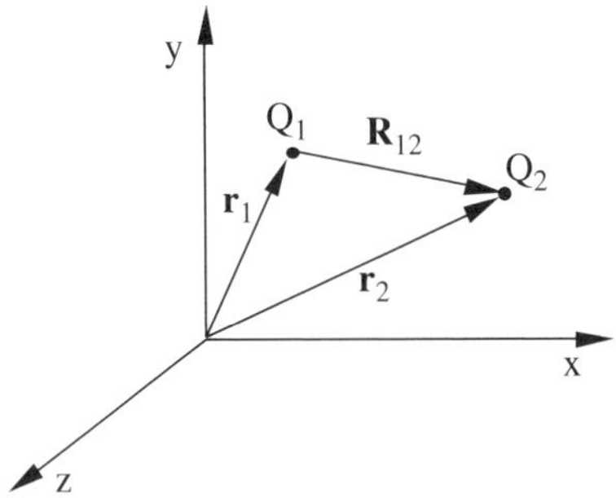

Figure 3.2. Position and displacement vectors for two point charges.

The position and separation vectors are shown in Figure 3.2.

Denoting the separation vector as

$$\mathbf{R}_{12} = |\mathbf{R}_{12}|\,\mathbf{i}_{12} \tag{3.5}$$

where $\mathbf{i}_{12}$ is a unit vector in the direction of the separation vector $\mathbf{R}_{12}$ directed from point charge 1 to point charge 2, Eq. (3.2) can be written as

$$\mathbf{F}_{12} = \frac{Q_1 Q_2 \mathbf{i}_{12}}{4\pi\epsilon_o |\mathbf{R}_{12}|^2} \tag{3.6}$$

Equations (3.2) and (3.6) are identical.

By definition, Eq. (3.6) provides that

$$\mathbf{F}_{12} = -\mathbf{F}_{21} \tag{3.7}$$

or

$$\mathbf{F}_{21} = \frac{Q_1 Q_2 \mathbf{i}_{21}}{4\pi\epsilon_o\, |\mathbf{R}_{12}|^2} \tag{3.8}$$

where the unit vector $\mathbf{i}_{21}$ is directed from point charge Q_2 to point charge Q_1.

It should be noted that apart from the direction of the separation vector $\mathbf{R}_{12}$, the sign of the product $Q_1 Q_2$ also dictates the direction of the force. The electrostatic forces for the cases of $Q_1 Q_2 > 0$ and $Q_1 Q_2 < 0$ are shown in Figures 3.3(a) and 3.3(b), respectively. The case $Q_1 Q_2 > 0$ represents charges of the same sign, whereas the case $Q_1 Q_2 < 0$ represents charges with opposite signs. For the case of $Q_1 Q_2 > 0$ the forces are repulsive, while they are attractive for $Q_1 Q_2 < 0$.

The *fourth principle* of electrostatics is that of superposition. It states that the force between any two charges is unaffected by the presence of other charges. In other words, Coulomb's law between any two point charges is not affected by the presence of any other charges. This law may sound trivial, but it forms the cornerstone of electrostatics. The use of superposition principle is not confined to electrostatics as it is also applicable to Newtonian mechanics.

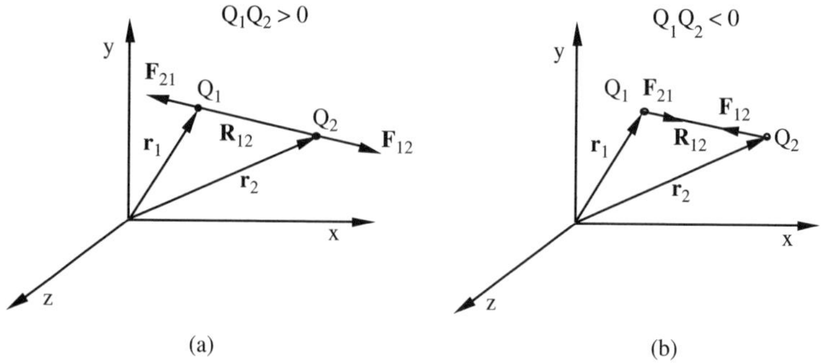

Figure 3.3. (a) Repulsive forces for $Q_1Q_2 > 0$ case. (b) Attractive forces for $Q_1Q_2 < 0$ case.

3.1.2 Electric Field Strength

Consider many point charges $Q_1, Q_2, \ldots, Q_n$ at distances $R_{1A}, R_{2A}, \ldots, R_{nA}$, respectively, from a positive test charge that we designate as Q_A. The geometry is shown in Figure 3.4. Utilizing the superposition principle, the total force (in a vectorial form) exerted by the point charges $i = 1, 2, \ldots, n$ on the test point charge Q_A is given by

$$\begin{aligned} \mathbf{F} &= \mathbf{F}_{1A} + \mathbf{F}_{2A} + \cdots + \mathbf{F}_{nA} \\ &= \frac{1}{4\pi\epsilon_o}\left[\frac{Q_1 Q_A \mathbf{i}_{1A}}{|\mathbf{R}_{1A}|^2} + \frac{Q_2 Q_A \mathbf{i}_{2A}}{|\mathbf{R}_{2A}|^2} + \cdots + \frac{Q_n Q_A \mathbf{i}_{nA}}{|\mathbf{R}_{nA}|^2}\right] \end{aligned} \tag{3.9}$$

where $\mathbf{i}_{iA}$ is the unit vector along point charges Q_i and Q_A. Here, $\mathbf{R}_{iA}$ represents the displacement vector between point charges Q_i and Q_A, such that

$$\mathbf{R}_{iA} = \mathbf{r}_A - \mathbf{r}_i \tag{3.10}$$

where $\mathbf{r}_A$ and $\mathbf{r}_i$ represent the position vectors of the point charges Q_A and Q_i, respectively. Equation (3.9) can be rewritten in a concise form as

$$\mathbf{F} = \frac{Q_A}{4\pi\epsilon_o}\sum_{i=1}^{n}\frac{Q_i}{|\mathbf{R}_{iA}|^2}\mathbf{i}_{iA} \tag{3.11}$$

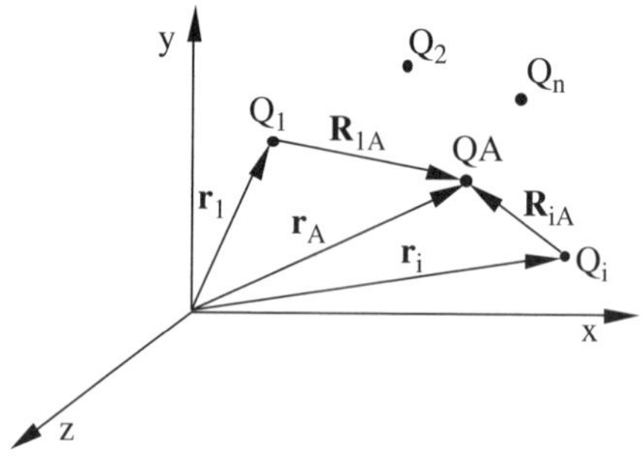

Figure 3.4. Point charges in free space.

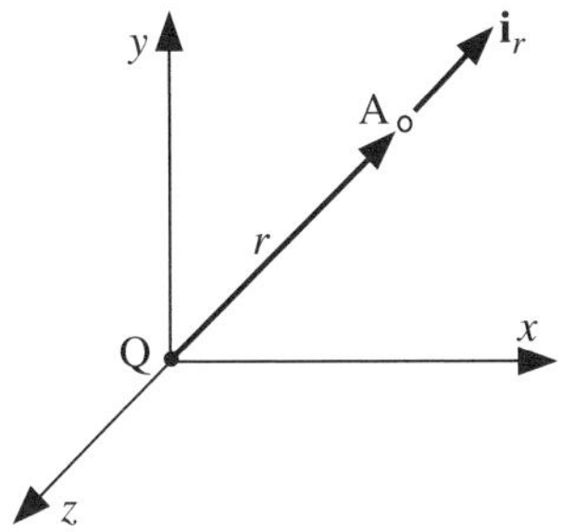

Figure 3.5. Electric field strength at position A due to a point charge, Q, located at the origin.

Equation (3.11) leads us to the definition of the electric field strength. Formally, the electric field strength, **E**, is defined as the force per unit positive test charge. In mathematical notation,

$$\mathbf{E} = \frac{\mathbf{F}}{Q_A} \tag{3.12}$$

where **F** is the electric force on a test point charge Q_A. Following Eq. (3.12), the electric field strength at the location of the test charge Q_A is obtained by rewriting Eq. (3.11) as

$$\mathbf{E} = \frac{\mathbf{F}}{Q_A} = \frac{1}{4\pi\epsilon_o}\sum_{i=1}^{n}\frac{Q_i}{|\mathbf{R}_{iA}|^2}\mathbf{i}_{iA} \tag{3.13}$$

The electric field strength, **E**, is due to the point charges Q_i; it acts at the location of the test charge Q_A and does not depend on the magnitude of the charge Q_A. It has units of V/m.

As a special case of Eq. (3.13), the electric field strength at a location r due to a point charge, Q, located at the origin is given by

$$\mathbf{E} = \frac{Q}{4\pi\epsilon_o}\frac{\mathbf{i}_r}{r^2} \tag{3.14}$$

where $\mathbf{i}_r$ is a unit vector in the radial (r) direction of the spherical coordinate system. The geometry of a point charge at the origin is shown in Figure 3.5.

As the electric field is a vector quantity, one needs to consider its direction as well. As was observed in the case of force, the direction of the electric field is dictated by the direction of the unit separation vector as well as the sign of the charge. To illustrate this, we consider the special case of a point charge Q at the origin, given by Eq. (3.14). When this charge is negative, the field appears to converge to the point charge (the vectors point toward the charge Q), while if Q is positive, the field appears to diverge from the point charge (the vectors point away from the charge Q).

One can generalize Eq. (3.13) by replacing the discrete point charges with a continuum bearing infinitesimal charges pertaining to a line, surface, and volume.

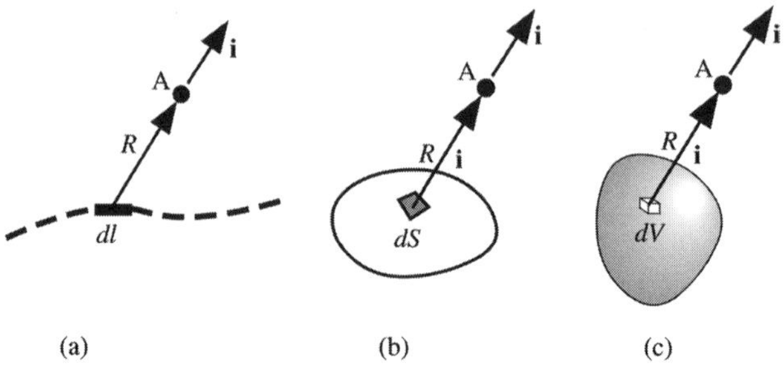

Figure 3.6. Geometry of a line, surface, and volume bearing distributed charges.

The relevant electric field strength at a location A becomes:
For a line charge

$$\mathbf{E} = \frac{1}{4\pi\epsilon_o}\int_{line}\frac{\mathbf{i}}{R^2}\lambda\, dl \tag{3.15}$$

For a surface charge

$$\mathbf{E} = \frac{1}{4\pi\epsilon_o}\int_{surface}\frac{\mathbf{i}}{R^2}q_s\, dS \tag{3.16}$$

For a volume charge

$$\mathbf{E} = \frac{1}{4\pi\epsilon_o}\int_{volume}\frac{\mathbf{i}}{R^2}\rho\, dV \tag{3.17}$$

where the unit vector **i** is directed from the charge bearing element to point A and R is the distance between them as shown in Figure 3.6.

We define:
λ is the line charge density, C/m.
q_s is the surface charge density, C/m^2.
ρ is the volume charge density, C/m^3.

Here, l, S, and V represent the length, surface, and volume, respectively, over which the charge is distributed.

EXAMPLE 3.1

Electric field strength due to two point charges in free space. Evaluate the electric field strength for two point charges of Q_1 and Q_2 in vacuum. The charges are located on the x-axis and separated by a distance $2b$ as shown in Figure 3.7.

Solution Consider point A located at (r, θ) of a spherical coordinate system. The electric field strength at point A is given by Eq. (3.13) as

$$\mathbf{E} = \frac{1}{4\pi\epsilon_o}\left[\frac{Q_1}{R_1^2}\mathbf{i}_{1A} + \frac{Q_2}{R_2^2}\mathbf{i}_{2A}\right] \tag{3.18}$$

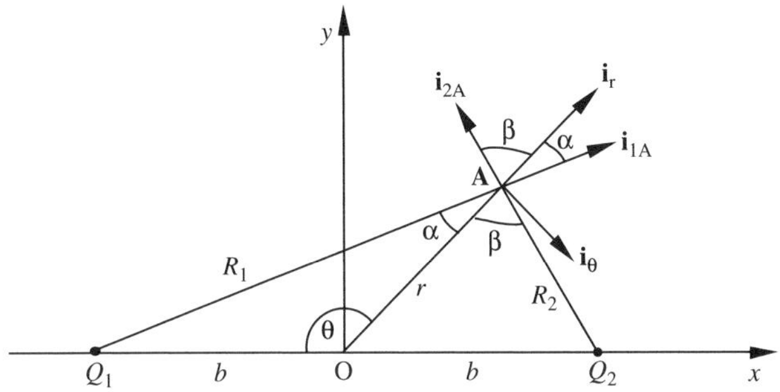

Figure 3.7. Two point charges Q_1 and Q_2 equally spaced from the origin.

As the electric field strength is a vector, we need to evaluate the components of **E** in r and θ directions. The components are given by E_r and E_θ, respectively, where

$$\mathbf{E} = E_r\mathbf{i}_r + E_\theta\mathbf{i}_\theta \tag{3.19}$$

Here, $\mathbf{i}_r$ and $\mathbf{i}_\theta$ are the unit vectors in r and θ directions, respectively.

Let us evaluate the r-component of the electric field strength, E_r. One can use Eq. (3.18) together with Eq. (3.19) to give

$$E_r\mathbf{i}_r + E_\theta\mathbf{i}_\theta = \frac{1}{4\pi\epsilon_o}\left[\frac{Q_1}{R_1^2}\mathbf{i}_{1A} + \frac{Q_2}{R_2^2}\mathbf{i}_{2A}\right] \tag{3.20}$$

Performing dot product on the terms in Eq. (3.20) with $\mathbf{i}_r$ gives

$$E_r = \frac{1}{4\pi\epsilon_o}\left[\frac{Q_1}{R_1^2}\mathbf{i}_{1A}\cdot\mathbf{i}_r + \frac{Q_2}{R_2^2}\mathbf{i}_{2A}\cdot\mathbf{i}_r\right] \tag{3.21}$$

From the geometry of Figure 3.7, one can write

$$\mathbf{i}_{1A}\cdot\mathbf{i}_r = \cos\alpha \tag{3.22}$$

$$\mathbf{i}_{2A}\cdot\mathbf{i}_r = \cos\beta \tag{3.23}$$

$$\sin\alpha = \frac{b}{R_1}\sin\theta \tag{3.24}$$

$$\sin\beta = \frac{b}{R_2}\sin(\pi - \theta) \tag{3.25}$$

$$R_1^2 = r^2 + b^2 - 2rb\cos\theta \tag{3.26}$$

$$R_2^2 = r^2 + b^2 - 2rb\cos(\pi - \theta) \tag{3.27}$$

Making use of Eqs. (3.22)–(3.27), Eq. (3.21) becomes

$$E_r = \frac{1}{4\pi\epsilon_o}\left[\frac{Q_1\sqrt{R_1^2 - b^2\sin^2\theta}}{R_1^3} + \frac{Q_2\sqrt{R_2^2 - b^2\sin^2\theta}}{R_2^3}\right] \tag{3.28}$$

where R_1 and R_2 are provided by Eqs. (3.26) and (3.27), respectively.

The θ-component, E_θ, can be evaluated by performing dot product of the terms in Eq. (3.20) by $\mathbf{i}_\theta$ and recognizing that

$$\mathbf{i}_{1A}\cdot\mathbf{i}_\theta = \cos\left(\frac{\pi}{2} - \alpha\right) = \sin\alpha \tag{3.29}$$

$$\mathbf{i}_{2A}\cdot\mathbf{i}_\theta = \cos\left(\frac{\pi}{2} + \beta\right) = -\sin\beta \tag{3.30}$$

and using the sine theorem

$$\sin\beta = \frac{b}{R_2}\sin\theta \tag{3.31}$$

to give

$$E_\theta = \frac{b\sin\theta}{4\pi\epsilon_o}\left[\frac{Q_1}{R_1^3} - \frac{Q_2}{R_2^3}\right] \tag{3.32}$$

The magnitude of $\mathbf{E}$ is given by

$$E = \sqrt{E_r^2 + E_\theta^2} \tag{3.33}$$

Equations (3.28) and (3.32) provide the components of the electric field strength, $\mathbf{E}$, at point A.

Let us consider the special case of $Q_1 = +Q$ and $Q_2 = -Q$. Let the separation distance $2b$ be very small, *i.e.*, $R_1 \gg b$, $R_2 \gg b$, and $r \gg b$. Neglecting the small terms of the order of (b/r), we can write from Eqs. (3.26) and (3.27)

$$R_1 \simeq R_2 \simeq r$$

and

$$(R_2 - R_1) \simeq 2b\cos\theta$$

The r-component of $\mathbf{E}$ as given by Eq. (3.28) for this case gives

$$E_r = \frac{Q}{4\pi\epsilon_o}\left[\frac{1}{R_1^2} - \frac{1}{R_2^2}\right] = \frac{Q}{4\pi\epsilon_o}\left[\frac{(R_2 - R_1)(R_2 + R_1)}{R_1^2 R_2^2}\right]$$

or

$$E_r = \frac{2(2b)Q\cos\theta}{4\pi\epsilon_o r^3} \tag{3.34}$$

The θ-component of **E** as provided by Eq. (3.32) gives

$$E_\theta = \frac{Qb \sin\theta}{4\pi\epsilon_o}\left[\frac{1}{R_1^3} + \frac{1}{R_2^3}\right] = \frac{Q(2b)\sin\theta}{4\pi\epsilon_o r^3} \tag{3.35}$$

Making use of Eq. (3.19), the electric field strength is given by

$$\mathbf{E} = \frac{Q(2b)}{4\pi\epsilon_o r^3}\left[2\cos\theta\,\mathbf{i}_r + \sin\theta\,\mathbf{i}_\theta\right] \tag{3.36}$$

This is the field strength for two point charges $+Q$ and $-Q$ located on the x-axis and separated by a distance $2b$.

A system of two point charges of $+Q$ and $-Q$ separated by a very small distance of $2b$ constitutes what is called a point-like dipole. It is interesting to note that two charges of opposite signs at very small separations produce an electric field. Equation (3.36) represents the electric field strength generated by a dipole in free space. We will discuss dipoles at a later stage as they are important in non-conducting materials, *i.e.*, dielectric materials. ■

EXAMPLE 3.2

Electric field strength in free space due to a charged spherical shell. Evaluate the electric field strength at a distance z from the center of a spherical surface having radius a placed in vacuum. The surface charge density of the spherical shell is q_s (C/m^2). The geometry is shown in Figure 3.8.

Solution Let us start with Eq. (3.16) which describes the electric field strength at a location in space due to a surface charge distribution

$$\mathbf{E} = \frac{1}{4\pi\epsilon_o}\int_S \frac{q_s \mathbf{i}}{R^2}\,dS \tag{3.37}$$

Let point A, located at a distance z from the origin, represent the position where the electric field strength is to be evaluated. For the problem at hand, the unit vector $\mathbf{i}_r$ is along OB, unit vector $\mathbf{i}$ is along BA, and $\mathbf{i}_z$ is the unit vector along the z-coordinate. From the geometry of Figure 3.8, one can write

$$\mathbf{i}\cdot\mathbf{i}_z = \cos\alpha \tag{3.38}$$

$$R^2 = a^2 + z^2 - 2az\cos\theta \tag{3.39}$$

$$\cos\alpha = \frac{(z - a\cos\theta)}{(a^2 + z^2 - 2az\cos\theta)^{1/2}} \tag{3.40}$$

and

$$dS = a^2 \sin\theta\, d\phi\, d\theta \tag{3.41}$$

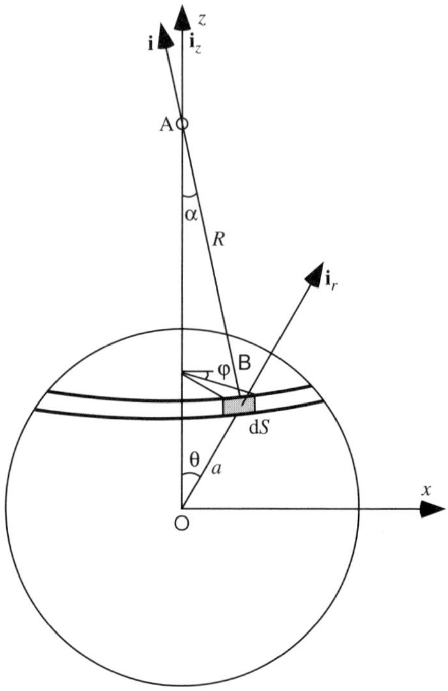

Figure 3.8. Spherical surface with a specified surface charge density q_s.

Recognizing that $E_z = \mathbf{E} \cdot \mathbf{i}_z$ and making use of Eqs. (3.38)–(3.41), Eq. (3.37) becomes

$$E_z = \frac{q_s}{4\pi\epsilon_o} \int_0^{\pi} \int_0^{2\pi} \frac{a^2(z - a\cos\theta)\sin\theta\, d\phi\, d\theta}{(a^2 + z^2 - 2az\cos\theta)^{3/2}} \tag{3.42}$$

Integration over ϕ leads to

$$E_z = \frac{a^2 q_s}{2\epsilon_o} \int_0^{\pi} \frac{(z - a\cos\theta)\sin\theta\, d\theta}{(a^2 + z^2 - 2az\cos\theta)^{3/2}} \tag{3.43}$$

Making the substitution of $\cos\theta = t$, Eq. (3.43) can then be integrated to yield

$$E_z = \frac{a^2 q_s}{\epsilon_o z^2} \qquad \text{for} \quad |z| > a \tag{3.44}$$

outside the spherical shell, and

$$E_z = 0 \qquad \text{for} \ |z| < a \tag{3.45}$$

inside the spherical shell.

The outer electric field strength decays with the square of the distance from the center and it is zero inside the spherical shell. ■

3.1.3 The Gauss Law

The Gauss law relates the electric field strength flux through a closed surface to the enclosed charge. To derive the Gauss law, let us consider a point charge Q located in some arbitrary volume, V, bounded by a surface S as shown in Figure 3.9.

Let $\mathbf{n}$ be a unit outward normal vector to the bounding surface. The electric field strength at the element of surface dS due to the charge Q is given by

$$\mathbf{E} = \frac{Q\,\mathbf{i}}{4\pi\epsilon_o R^2} \tag{3.46}$$

where the unit vector $\mathbf{i}$ is directed from the point charge to the surface element dS and R is the distance between them.

Performing dot product with $\mathbf{n}\,dS$ and integrating over the bounding surface S, Eq. (3.46) gives

$$\oint_S (\mathbf{E}\cdot\mathbf{n})\,dS = \oint_S \frac{(\mathbf{i}\cdot\mathbf{n})Q}{4\pi\epsilon_o R^2}dS \tag{3.47}$$

Recognizing that the element of solid angle $d\Omega$ is given by $(\mathbf{i}\cdot\mathbf{n})dS/R^2$, Eq. (3.47) becomes

$$\oint_S (\mathbf{E}\cdot\mathbf{n})\,dS = \frac{1}{4\pi\epsilon_o}\oint_0^{4\pi} Q d\Omega \tag{3.48}$$

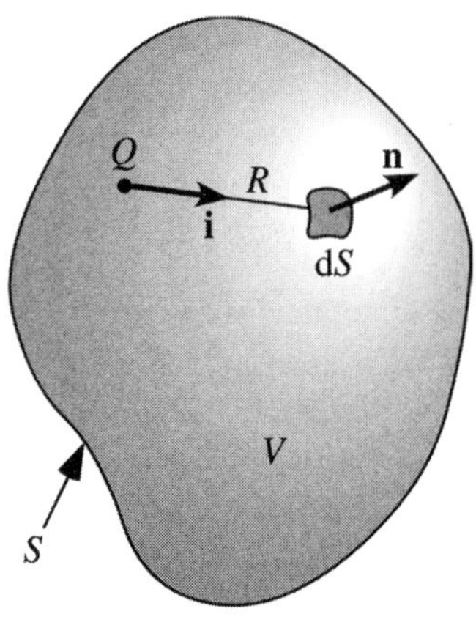

Figure 3.9. Surface enclosing a point charge Q.

Upon integration, Eq. (3.48) gives

$$\oint_S (\mathbf{E} \cdot \mathbf{n})\, dS = \frac{Q}{\epsilon_o} \tag{3.49}$$

Equation (3.49) states that the electric field strength flux through the closed surface S is proportional to the charge enclosed by the surface. Such a relationship is independent of the surface shape or the position of the charge within the surface enclosure. Making use of the superposition principle, one can generalize Eq. (3.49) so that the point charge Q is the sum of all point charges within the surface enclosure S.

Equation (3.49) is the integral form of the Gauss law or theorem. It simply states that if there is no net charge within the enclosure surface S, then the electric field flux is zero. Based on the Gauss law one should expect a null electric flux arising from, say, a piece of polymer where its net charge is zero.

The differential form of the Gauss law can be derived quite readily using the divergence theorem, which states

$$\int_V (\nabla \cdot \mathbf{E})\, dV = \int_S (\mathbf{E} \cdot \mathbf{n})\, dS \tag{3.50}$$

Here, volume V is enclosed by surface S. One can express the charge density ρ within the volume V in terms of the total charge Q as

$$Q = \int_V \rho\, dV \tag{3.51}$$

From Eqs. (3.49) to (3.51), one can write

$$\int_V (\nabla \cdot \mathbf{E})\, dV = \frac{1}{\epsilon_o} \int_S \rho\, dV \tag{3.52}$$

As volume V is arbitrary, Eq. (3.52) reduces to

$$\nabla \cdot \mathbf{E} = \frac{\rho}{\epsilon_o} \tag{3.53}$$

This is the differential expression of the Gauss law.

3.1.4 Electric Potential

It is of interest to investigate certain properties of the electric field strength, $\mathbf{E}$, that would facilitate the solution of problems involving point and distributed charges. Without loss of generality, let us consider a single point charge located at the origin of the system of coordinates.

Utilizing Eq. (3.14), one can write

$$\mathbf{E} = \frac{Q\mathbf{r}}{4\pi\epsilon_o r^3} \tag{3.54}$$

where $\mathbf{r} = r\mathbf{i}_r$, following Eq. (3.5).

A useful vector identity is given by

$$\nabla \times (f\mathbf{G}) = f(\nabla \times \mathbf{G}) + \nabla f \times \mathbf{G} \tag{3.55}$$

where f is a scalar function and $\mathbf{G}$ is a vector function. We may represent Eq. (3.54) as a product of a scalar and a vector function by writing

$$f = \frac{Q}{4\pi\epsilon_o r^3} \tag{3.56}$$

and

$$\mathbf{G} = \mathbf{r} \tag{3.57}$$

Taking the curl ($\nabla\times$) of Eq. (3.54) and utilizing the vector identity, Eq. (3.55), we obtain

$$\nabla \times \mathbf{E} = \left(\frac{Q}{4\pi\epsilon_o r^3}\right)(\nabla \times \mathbf{r}) + \nabla\left(\frac{Q}{4\pi\epsilon_o r^3}\right) \times \mathbf{r} \tag{3.58}$$

Performing the differentiation in Eq. (3.58) provides

$$\nabla \times \mathbf{E} = \left(\frac{Q}{4\pi\epsilon_o r^3}\right)(\nabla \times \mathbf{r}) - \left(\frac{3Q}{4\pi\epsilon_o r^4}\right)\nabla r \times \mathbf{r} \tag{3.59}$$

Recognizing that

$$\nabla = \frac{\partial}{\partial x}\mathbf{i}_x + \frac{\partial}{\partial y}\mathbf{i}_y + \frac{\partial}{\partial z}\mathbf{i}_z\,, \tag{3.60}$$

$$\mathbf{r} = x\mathbf{i}_x + y\mathbf{i}_y + z\mathbf{i}_z \tag{3.61}$$

and utilizing the definition of a curl, one can show that

$$\nabla \times \mathbf{r} = 0 \tag{3.62}$$

and

$$\nabla r \times \mathbf{r} = 0 \tag{3.63}$$

These identities, when substituted in Eq. (3.59), yield

$$\nabla \times \mathbf{E} = 0 \tag{3.64}$$

Equation (3.64) states that the curl of the electric field strength is zero.

Such a property would lend the name irrotational field to the electric field strength **E** and would allow it to possess certain properties that are useful in electrostatics. Some of these properties are (Griffiths, 1989):

(a) $\int_1^2 \mathbf{E} \cdot d\mathbf{t}$ is independent of path for given end points 1 and 2.

(b) $\oint \mathbf{E} \cdot d\mathbf{t} = 0$ for any closed path. This is a "conservative" property.

(c) $\mathbf{E} = -\nabla\psi$ where ψ is some scalar function.

Here **t** is a unit vector along a line path traced on a surface.

The property under (c) is of much interest. It implies that the electric field strength can be calculated as a gradient of a single scalar function, the potential function, which we will designate as ψ. We write this fundamental relationship between the irrotational field and the scalar potential as

$$\mathbf{E} = -\nabla\psi \tag{3.65}$$

By using the scalar function ψ, one can reduce a vectorial problem to a scalar one. Consequently, in many cases it is easier to work with the electric potential, ψ, rather than with the electric field strength, **E**.

Let us find some physical meaning to the electric potential, ψ. The total work, W_{12}, required to move a point test charge Q a finite distance between two points 1 and 2 in an electric field **E**, as shown in Figure 3.10, is given by

$$W_{12} = -\int_1^2 Q\mathbf{E} \cdot d\mathbf{t} \tag{3.66}$$

where we note that $Q\mathbf{E}$ is the force according to Eq. (3.12) and **t** is the path along which the charge Q travels. From Eqs. (3.65) and (3.66), one can write

$$W_{12} = \int_1^2 Q\nabla\psi \cdot d\mathbf{t} \tag{3.67}$$

and obtain

$$W_{12} = Q(\psi_2 - \psi_1) \tag{3.68}$$

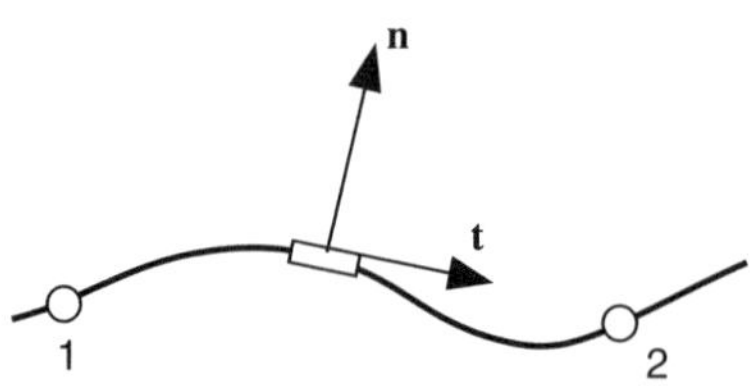

Figure 3.10. A point charge moving from location 1 to 2 along path **t** whose normal is **n**.

Equation (3.68) shows that the difference in the electric potential between two points is the work done to move a unit test (and point) charge in an electric field, **E**. This equation also indicates that the work is independent of the path taken between points 1 and 2. Such a property is characteristic of a conservative field.

Let us now derive an expression for the electric potential due to a point charge located at the origin. Making use of the electric field strength expression for a point charge given by Eq. (3.14) and utilizing the relationship between **E** and ψ from Eq. (3.65) one can write

$$\nabla\psi = -\frac{Q\,\mathbf{i}_r}{4\pi\epsilon_o r^2} \tag{3.69}$$

As the variation of ψ is in the radial direction only, Eq. (3.69) simplifies to

$$\frac{d\psi}{dr} = -\frac{Q}{4\pi\epsilon_o r^2} \tag{3.70}$$

Integration of Eq. (3.70) provides

$$\psi = \psi_1 + \frac{Q}{4\pi\epsilon_o}\left(\frac{1}{r} - \frac{1}{r_1}\right) \tag{3.71}$$

where $\psi = \psi_1$ at $r = r_1$. Here, we can think of ψ_1 being a reference value for the electric potential at a radial distance r_1. By taking the reference location at infinity, and letting ψ_1 to be zero at infinity, Eq. (3.71) gives the electric potential for a point charge located at the origin.

$$\psi = \frac{Q}{4\pi\epsilon_o r} \tag{3.72}$$

As indicated by Eq. (3.72), the potential due to a point charge decays with r rather than r^2 as was the case for the electric field strength.

A generalization for the electric potential at location A due to a point charge Q_1, shown in Figure 3.11, is given by

$$\psi = \frac{Q_1}{4\pi\epsilon_o|\mathbf{r}_A - \mathbf{r}_1|} \tag{3.73}$$

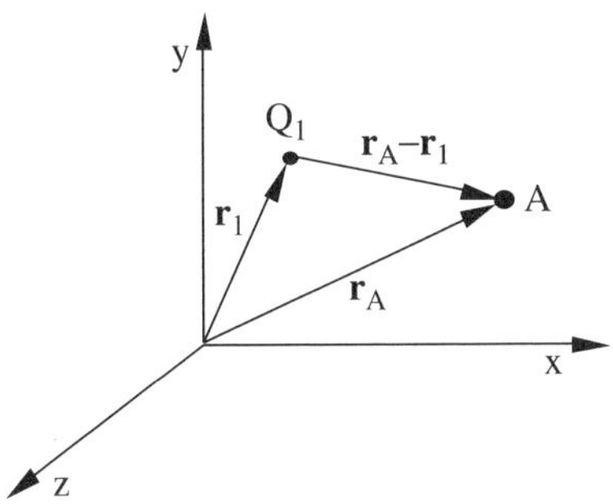

Figure 3.11. Electric potential at location A due to a point charge Q_1 located at $\mathbf{r}_1$.

$|\mathbf{r}_A - \mathbf{r}_1|$ represents the distance between the point charge Q_1 and location A. By letting $|\mathbf{r}_A - \mathbf{r}_1| = R_1$, one obtains

$$\psi = \frac{Q_1}{4\pi\epsilon_o R_1} \tag{3.74}$$

where R_1 is the separation distance between point charge Q_1 and point A.

By invoking the superposition principle, one can write an expression for the electric potential at location A due to the presence of many point charges.

$$\psi = \frac{1}{4\pi\epsilon_o}\sum_{i=1}^{n}\frac{Q_i}{|\mathbf{r}_A - \mathbf{r}_i|} = \frac{1}{4\pi\epsilon_o}\sum_{i=1}^{n}\frac{Q_i}{R_i} \tag{3.75}$$

where $R_i = |\mathbf{r}_A - \mathbf{r}_i|$ represents the separation distance between point charge Q_i and location A and $\mathbf{r}_i$ is the position vector of charge Q_i. The total number of point charges is n.

Similar to the case of the electric field strength, one can write expressions for the electric potential at location A due to a line, surface, and volume carrying a charge distribution. The relevant electric potential at location A in space are expressed as:
For a line charge

$$\psi = \frac{1}{4\pi\epsilon_o}\int_{line}\frac{\lambda}{R}\,dl \tag{3.76}$$

For a surface charge

$$\psi = \frac{1}{4\pi\epsilon_o}\int_{surface}\frac{q_s}{R}\,dS \tag{3.77}$$

For a volume charge

$$\psi = \frac{1}{4\pi\epsilon_o}\int_{volume}\frac{\rho}{R}\,dV \tag{3.78}$$

where λ, q_s, and ρ are the line, surface, and volume charge densities, respectively, as was defined previously for the electric field strength. Here, R is the distance between the charge carrying element and location A.

The differential form of the electric potential can be easily derived by combining Eqs. (3.53) and (3.65) to give

$$\nabla\cdot(\nabla\psi) = -\frac{\rho}{\epsilon_o} \tag{3.79}$$

or,

$$\nabla^2\psi = -\frac{\rho}{\epsilon_o} \tag{3.80}$$

Equation (3.80) is known as Poisson's equation. In cases where the electric charge density within a body is absent, *i.e.*, $\rho = 0$, The Poisson equation (3.80) becomes

$$\nabla^2 \psi = 0 \tag{3.81}$$

Equation (3.81) is known as Laplace's equation.

Solution of either the Poisson or Laplace equations is subject to appropriate boundary conditions. Discussion on the boundary conditions will be given in Section 3.5.

EXAMPLE 3.3

Electric potential in free space due to a charged spherical shell. Let us solve Example 3.2 by evaluating the electric potential at point z due to the placement of a charged spherical surface having a surface charge density of q_s (C/m^2).

Solution From Eq. (3.77) we can write,

$$\psi(z) = \frac{1}{4\pi\epsilon_o}\int_S \frac{q_s}{R}\, dS \tag{3.82}$$

Making use of Figure 3.8, one obtains

$$R^2 = a^2 + z^2 - 2az\cos\theta \qquad \text{(cosine law)} \tag{3.83}$$

$$dS = a^2 \sin\theta \; d\phi \; d\theta \tag{3.84}$$

From Eqs. (3.83) and (3.84), Eq. (3.82) becomes

$$\psi(z) = \frac{q_s}{4\pi\epsilon_o}\int_0^{\pi}\int_0^{2\pi} \frac{a^2 \sin\theta \; d\phi \; d\theta}{(a^2 + z^2 - 2az\cos\theta)^{1/2}} \tag{3.85}$$

Upon integration, we obtain

$$\psi(z) = \frac{a^2 q_s}{\epsilon_o z} \qquad \text{outside the surface} \tag{3.86}$$

$$\psi(z) = \frac{a q_s}{\epsilon_o} \qquad \text{inside the surface} \tag{3.87}$$

Let the total charge, Q_s, be related to the surface charge density, q_s, as

$$Q_s = 4\pi a^2 q_s \tag{3.88}$$

Equation (3.86) then gives

$$\psi(z) = \frac{Q_s}{4\pi\epsilon_o z} \qquad \text{for } z \geq a. \tag{3.89}$$

At the shell's surface $z = a$, one obtains

$$\psi_s = \frac{Q_s}{4\pi\epsilon_o a} \tag{3.90}$$

Equation (3.90) relates the total charge to the surface potential at the spherical shell. In terms of surface charge density distribution, we obtain

$$q_s = \frac{\epsilon_o \psi_s}{a} \tag{3.91}$$

Clearly, one would have obtained the same results by making use of the solution of the field **E** obtained in Example 3.2. ■

3.2 SUMMARY OF ELECTROSTATIC EQUATIONS IN FREE SPACE

The pertinent governing electrostatic equations in free space are given below in their integral and differential forms. The symbols were previously defined and they are also provided in the nomenclature at the end of this chapter.

3.2.1 Integral Form

The relevant electrostatic equations in free space are listed in Tables 3.1 and 3.2. In Table 3.1, $\mathbf{i}_{1A}$ is the unit vector along point charge Q_1 to point A, $\mathbf{R}_{iA}$ is the displacement vector defined as $\mathbf{r}_A - \mathbf{r}_i$, $\mathbf{r}_A$ is the position vector of location A, and $\mathbf{r}_i$ is the position vector of point charge i.

3.2.2 Differential Form

The relevant electrostatic equations in free space in their differential form are provided in Table 3.3, where $\mathbf{E} = -\nabla\psi$ and $\nabla \times \mathbf{E} = 0$.

In order to use the differential form of the electrostatic equations, we need to define the boundary conditions pertinent to a given physical situation. A discussion on the

TABLE 3.1. Integral Forms of the Electrostatic Equations in Free Space for Discrete Point Charges.

	Electric Field Strength, **E**	Electric Potential, ψ
Point charge Q_1 at origin	$\dfrac{Q_1 \mathbf{i}_r}{4\pi\epsilon_o r^2}$	$\dfrac{Q_1}{4\pi\epsilon_o r}$
Point charge Q_1 at $\mathbf{r}_1$	$\dfrac{Q_1 \mathbf{i}_{1A}}{4\pi\epsilon_o \lvert\mathbf{R}_{1A}\rvert^2}$	$\dfrac{Q_1}{4\pi\epsilon_o \lvert\mathbf{R}_{1A}\rvert}$
Multiple point charges	$\dfrac{1}{4\pi\epsilon_o}\sum \dfrac{Q_i \mathbf{i}_{iA}}{\lvert\mathbf{R}_{iA}\rvert^2}$	$\dfrac{1}{4\pi\epsilon_o}\sum \dfrac{Q_i}{\lvert\mathbf{R}_{iA}\rvert}$

TABLE 3.2. Integral Forms of the Electrostatic Equations in Free Space for Distributed Charges.

	Electric Field Strength, **E**	Electric Potential, ψ
Line charge	$\frac{1}{4\pi\epsilon_o}\int\frac{\mathbf{i}\lambda}{R^2}dl$	$\frac{1}{4\pi\epsilon_o}\int\frac{\lambda}{R}dl$
Surface charge	$\frac{1}{4\pi\epsilon_o}\int\frac{\mathbf{i}q_s}{R^2}dS$	$\frac{1}{4\pi\epsilon_o}\int\frac{q_s}{R}dS$
Volume charge	$\frac{1}{4\pi\epsilon_o}\int\frac{\mathbf{i}\rho}{R^2}dV$	$\frac{1}{4\pi\epsilon_o}\int\frac{\rho}{R}dV$

TABLE 3.3. Differential Forms of the Electrostatic Equations in Free Space.

Electric field strength, **E**	$\nabla\cdot\mathbf{E}=\frac{\rho}{\epsilon_o}$
Electric potential, ψ	$\nabla^2\psi=-\frac{\rho}{\epsilon_o}$

boundary conditions will follow at a later stage after we have discussed electrostatics in real media, *i.e.*, in presence of free charge.

The electrostatic equations derived so far are applicable to charges in free space, *i.e.*, vacuum. When free space is replaced by a real medium, the electrostatic equations for free space need to be modified. First, we shall discuss briefly the electrical classification of materials and then show how to modify the electrostatic equations in real media.

3.3 ELECTROSTATIC CLASSIFICATION OF MATERIALS

Interaction of electrical potentials and fields with physical materials gives rise to different effects depending on how the material responds to the electrical parameters. From the perspective of classical electrostatics, it is possible to broadly differentiate matter into two classes, namely, *conductors* and *dielectrics* or *insulators*.

Conductors are materials that have a large supply of *free charge*, which can freely move from one region of the material to another under the influence of an externally imposed electric potential difference across the material. Free charge can exist in different forms in different materials. For instance, in conducting metals (such as copper or gold), the free charge carriers are the mobile electrons. In aqueous electrolyte solutions (for instance, a solution of sodium chloride in water), the dissociated ions are the free charge carriers. When an electrical potential difference is applied across such a conducting material, the free charges will move to the regions of different potentials depending on the type of charge they carry. For instance, in a metallic conductor, the mobile electrons (negative charge) will be transported to the positive potential. If the metallic conductor is also connected to a source of electrons (negative potential), then electrons will move into the conductor to fill in the voids left by the depleting electrons. Such a process will set up an electric current (flow of electrons or charge) through the conducting material. This process is called electrical conduction. The extent of current flow under an applied potential difference across a conducting material is governed by the electric conductivity of the material. The inverse of the electric conductivity is known as the electric resistivity. Low conductivity (or high resistivity) signifies a poor conductor or an *insulator*. A poor conductor either does not possess sufficient amount of free charges, or is unable to transport the free charges easily. The voltage-current characteristics of conducting materials are related by Ohm's law

$$\Delta V = \frac{I}{K} = IR$$

where ΔV is the applied potential difference (Volt) across the material, I is the current (Ampere) flowing through the material, K is the electric conductance of the material (Ampere/Volt = Siemens), and $R = K^{-1}$ is the electric resistance of the material, (Volt/Ampere = Ohm).

Dielectrics, on the other hand, are materials that have a capacity of storing charge across their surfaces when a potential difference is applied across such a material. Ideally, dielectrics are materials with no conductance or infinite resistance, implying that these materials do not have free or mobile charges. Consequently, there will be no direct current flow through perfect dielectric materials. When a potential difference is set up across such dielectric materials, the material tends to develop opposite charges on its two surfaces. The mechanism of accumulation of charges at the two ends of a dielectric is called *polarization*.

Dielectric materials, although devoid of free or mobile charges, may contain bound charge carriers which can move or reposition in an atomic or molecular scale. Such repositioning allows dielectric molecules or atoms to act as electric dipoles. These dipoles can orient themselves under the influence of an electric field. The alignment of dipoles under the influence of an external electric field is termed as electric polarization. Let us consider a dielectric material placed in an external electric field. To simplify the discussion, let us assume that the dielectric material is made up of "perfect" neutral spherical atoms. What is meant by a "perfect" neutral spherical atom is that there is a positively charged core, *i.e.*, a nucleus, and a negatively charged electron *cloud* surrounding the nucleus. Collectively, the nucleus carries a charge $+Q$ and the electrons a charge of $-Q$, although both types of charges are centered at the same location. For such a "perfect" atom, its net charge is zero, and there is no dipole moment in absence of an external electric field. However, when an external electric field is present, the atom can acquire a dipole moment. It should be recalled from Example 3.1, that a pair of point charges $+Q$ and $-Q$ *separated* by a small distance is called an electric dipole and such a charge configuration would normally produce an electric field.

Upon the placement of such a dielectric material in an external electric field, there will be a charge separation within each atom as shown in Figure 3.12(a). The electrons within an atom are shifted slightly from their original position with respect to the nucleus. Such a shift in position would transform each atom into a dipole,

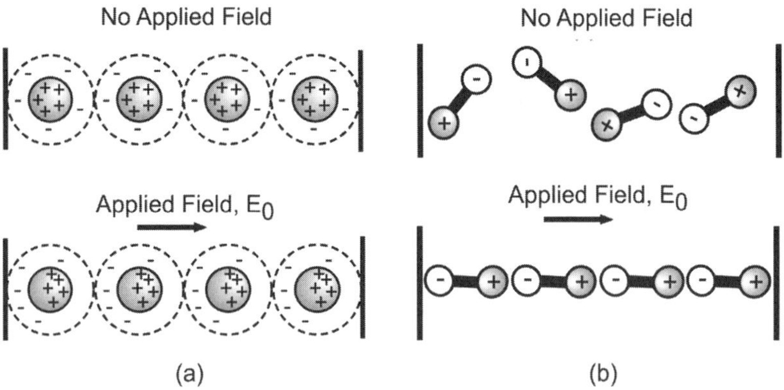

Figure 3.12. Polarization of a dielectric material in presence of an electric field. (a) Deformational polarization of neutral spherical atoms. (b) Orientational polarization of a permanent dipole molecule. In both cases, there is no polarization charge in absence of an external electric field. In presence of an external electric field, the polarization charge is set up due to the two mechanisms shown.

i.e., two charges of $+Q$ and $-Q$ separated by a finite but small distance. We refer to these atoms as being polarized. The mechanism of polarization of neutral atoms in the above manner is often termed as "deformational polarization". What is of interest is that the induced charge separation within the atoms manifests itself as a "polarization" charge at the extremities of the dielectric material. Such a polarization charge is called "bound charge". In other words, when a dielectric material is placed in an external electric field, bound charges (or polarization charges) appear at the surface of the dielectric medium. In electrostatics, the appearance of bound charges is of interest. Development of bound charges is not limited to dielectrics made up of neutral atoms (*i.e.*, non-polarized atoms or molecules). Bound charges are also manifested when dielectrics made up of molecules that are already polarized (*i.e.*, they are dipoles in absence of any external electric field, better known as permanent dipoles) are subjected to an external field. In this case, the bound charge appearance would be due to the rotation of the molecules as well as to charge separation. The polarization caused by re-orientation of a molecule with a permanent dipole in an externally imposed electric field is often referred to as "orientational polarization", and the mechanism is depicted in Figure 3.12(b). Some examples of dielectric materials are glass, silk, poly-(*methyl methacrylate*), water, ethyl alcohol, and air.

The ability of a dielectric to accumulate charges at its outer surfaces is termed as its capacitance. The capacitance is related to the charge accumulated across the dielectric's surfaces and the applied potential difference by the expression

$$\Delta V = \frac{Q}{C} \tag{3.92}$$

where ΔV is the potential difference across the dielectric (Volt), Q is the total accumulated charge across the dielectric (Coulomb), and C is the capacitance

(Coulomb/Volt = Farad). The above equation may be considered as analogous to Ohm's law for conducting materials, with the conductance replaced by capacitance and the current replaced by the accumulated charge.

Let us consider the phenomenon of charging of a capacitor, and assess how different dielectric materials influence the charging process. For this, we will refer to Figure 3.13. Figure 3.13(a) depicts two parallel plates being held at different potentials V_1 and V_2, which are separated by a distance d. The intervening medium is vacuum. In this case, applying the Gauss law, it can be shown that the electric field E_0 between the plates will be related to the charge on the plates, Q, through

$$E_0 = \frac{Q}{\epsilon_o A} \tag{3.93}$$

Here, Q is the charge on one of the plates, ϵ_o is the permittivity of vacuum, and A is the cross-section area of the plate, implying that the charge density on the plate is $q = Q/A$. Note that in this type of parallel plate capacitors, if $V_1 > V_2$, then the plate on the left will acquire a charge of $+Q$ and the plate on the right will acquire a charge $-Q$. We also note that the field can be expressed as $E_0 = \Delta V/d$, where $\Delta V = V_1 - V_2$. Therefore, Eq. (3.93) can be written as

$$\Delta V = \frac{Q}{\epsilon_o A/d} \tag{3.94}$$

Comparing Eqs. (3.92) and (3.94) it may be noted that $\epsilon_o A/d$ is the capacitance of the vacuum between the plates.

Consider now that a dielectric medium is placed between the two plates instead of the vacuum as shown in Figure 3.13(b). The medium will be polarized in presence of

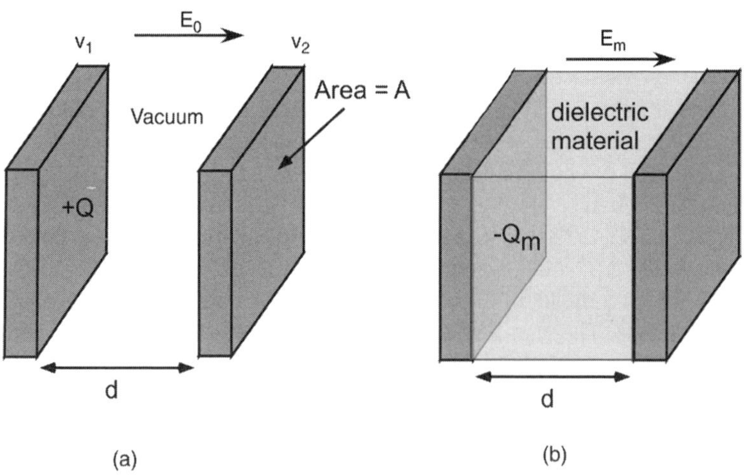

Figure 3.13. Modification of electric field by a dielectric material. (a) Two charged parallel plates in vacuum. (b) Same plates with a dielectric material in between.

the applied electric field, and as a consequence, the surface of the medium adjacent to the left plate will acquire a negative charge $-Q_m$, while the surface of the medium adjacent to the right hand side plate will acquire a positive charge $+Q_m$. These charges will oppose the original charges of the plates, and the magnitude of the net charge will be $Q - Q_m$ at the plate-medium interface. In this case, the effective electric field, E_m, between the plates through the dielectric medium will be

$$E_m = \frac{(Q - Q_m)}{\epsilon_o A} \tag{3.95}$$

The effective field is an outcome of the counteracting field due to the polarized medium opposing the original electric field in vacuum, E_0. Writing the field as $E_m = \Delta V/d$, Eq. (3.95) can be rearranged as

$$\Delta V = \frac{(Q - Q_m)}{\epsilon_o \dfrac{A}{d}} = \frac{Q}{\epsilon_o \dfrac{Q}{(Q - Q_m)} \dfrac{A}{d}} = \frac{Q}{\epsilon_o \epsilon_r \dfrac{A}{d}} \tag{3.96}$$

where

$$\epsilon_r = \frac{Q}{(Q - Q_m)}$$

The dimensionless parameter ϵ_r is termed as the relative permittivity of the medium, or the dielectric constant of the medium, and signifies how effectively the medium polarizes in an applied electric field compared to vacuum. If the medium polarizes very poorly, then $Q_m \to 0$ and $\epsilon_r \to 1$. However, if the medium is highly polarizable, the polarized charge Q_m is significant, and hence, the dielectric constant of the medium becomes a large number. The capacitance of the dielectric material is expressed as

$$C = \epsilon_o \epsilon_r \frac{A}{d} \tag{3.97}$$

Comparing Eqs. (3.94) and (3.96), it becomes clear that when the same potential difference ΔV is applied between the parallel plates, the charge accumulation across the plates when the intervening medium is a dielectric material is increased by a factor ϵ_r compared to the charge accumulation in vacuum.

The dielectric constant, ϵ_r of a dielectric material is a positive quantity with a minimum value of 1 corresponding to vacuum. The dielectric constant is often quite large for highly polarizable materials. For instance, water has a dielectric constant of 78.54 at 25 °C.

When describing conducting materials, we briefly mentioned that aqueous electrolyte solutions can act as conductors. Electrolyte solutions are peculiar in that they contain mobile charges in the form of ions suspended in a dielectric solvent. A simple example is the case of water containing dissolved common salt, NaCl. Upon dissolution, the salt dissociates into sodium, Na^+, and chloride, Cl^-, ions, which can freely move around in the water. Such an electrolyte solution can be considered as an *ionic* conductor. By placing two electrodes connected to a battery in an aqueous solution containing salts, an electric current flow can be established between the two electrodes. The current is due to the movement of the salt ions. In the case of sodium chloride salt, it is the movement of the Na^+ and Cl^- ions that gives rise to the electric current between the electrodes. The mechanism of current flow in ionic conductors is quite different from that in metallic or electronic conductors. Ionic conductors play a major role in biological functions and in many aspects of electrostatics, particularly electrokinetic phenomena. This notion of an ionic conductor, or simply a dielectric material containing free ions, is a prerequisite for dealing with electrostatic double layers, which will be discussed in Chapter 5.

3.4 BASIC ELECTROSTATICS IN DIELECTRICS

The basic electrostatic equations in free space were discussed in Section 3.1. We need now to extend our discussion to include dielectric materials. In this section we will derive the relevant electrostatic equations for a dielectric medium.

The molecules of a dielectric material constitute dipoles in the presence of an electric field. A dipole comprises two equal and opposite charges, $+Q$ and $-Q$, separated by a distance d. A *dipole moment* is defined as $Q\mathbf{d}$, a vector quantity, where $\mathbf{d}$ is the vector orientation between the two charges as shown in Figure 3.14 (a vector describing the average orientation of the dipole and the charge separation distance). For bulk phases, the *polarization dipole moment* or *polarization field strength* $\mathbf{P}$ is given by

$$\mathbf{P} = NQ\mathbf{d} \tag{3.98}$$

where N is the number of dipoles per unit volume. For most dielectric materials, when the electric field is not too strong, the polarization is directly proportional to the applied field, and one can write

$$\mathbf{P} = \chi\epsilon_o\mathbf{E} \tag{3.99}$$

A material that obeys the proportionality expressed by Eq. (3.99) is called a linearly polarized dielectric material or simply a linear dielectric. Here χ is the electric susceptibility of the dielectric medium and it is dimensionless. Through complex averaging arguments, one can write

$$\nabla \cdot \mathbf{P} = -\rho_p \tag{3.100}$$

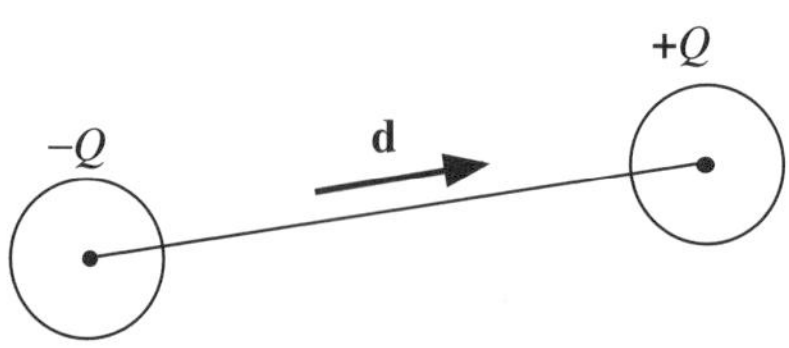

Figure 3.14. Two charges separated by a distance d.

where ρ_p is the volumetric polarization (or bound) charge density. Equation (3.100) simply states that the polarization field $\mathbf{P}$ is related to the polarization charge.

Within a dielectric material, the total volumetric charge density is made up of two types of charge densities, a polarization and a free charge density

$$\rho = \rho_p + \rho_f \tag{3.101}$$

For lack of better terminology, ρ_f is termed as free volumetric charge density. This "free" charge can consist of ions in an electrolyte solution, electrons on a conductor, or ions embedded in the dielectric material. In other words, the term free charge, ρ_f, constitutes any charge that is not the result of polarization. Combining the definition of total charge density provided by Eq. (3.101) with the Gauss law, Eq. (3.53), gives

$$\nabla \cdot \mathbf{E} = \frac{1}{\epsilon_o}(\rho_p + \rho_f) \tag{3.102}$$

Substituting for the polarization charge, ρ_p from Eq. (3.100), the divergence of the electric field becomes

$$\nabla \cdot \mathbf{E} = \frac{1}{\epsilon_o}(-\nabla \cdot \mathbf{P} + \rho_f) \tag{3.103}$$

leading to

$$\nabla \cdot (\epsilon_o \mathbf{E} + \mathbf{P}) = \rho_f \tag{3.104}$$

The term $(\epsilon_o \mathbf{E} + \mathbf{P})$ is usually designated as the electric displacement vector, $\mathbf{D}$. Equation (3.104) can then be written in the format of the Gauss law as

$$\nabla \cdot \mathbf{D} = \rho_f \tag{3.105}$$

Utilizing the divergence theorem, Eq. (3.105) can be written in the integral form as

$$\int_S \mathbf{D} \cdot \mathbf{n}\, dS = Q_f \tag{3.106}$$

where Q_f is the total free charge enclosed by surface S. Equations (3.105) and (3.106) are part of the Maxwell set of electrostatic equations for dielectrics.

We will now make use of the equation relating the polarization dipole moment to the electric field as given by Eq. (3.99). Combining Eqs. (3.99) and (3.104) leads to

$$\nabla \cdot [\epsilon_o(1 + \chi)\mathbf{E}] = \rho_f \tag{3.107}$$

Let

$$\epsilon = \epsilon_o(1 + \chi) \tag{3.108}$$

We will call ϵ the permittivity of the material. Combining Eqs. (3.105), (3.107), and (3.108) gives

$$\mathbf{D} = \epsilon \mathbf{E} \tag{3.109}$$

In free space, there is no material to be polarized, and therefore χ is zero. Here, the permittivity simply reduces to ϵ_o, which is the permittivity of free space. Consequently, the definition of ϵ provided by Eq. (3.108) is consistent.

Combining Eqs. (3.107) and (3.108) leads to

$$\nabla \cdot (\epsilon \mathbf{E}) = \rho_f \tag{3.110}$$

For constant permittivity, ϵ, Eq. (3.110) gives

$$\epsilon \nabla \cdot \mathbf{E} = \rho_f \tag{3.111}$$

Equation (3.111) is Maxwell's equation for a dielectric material.

If one is to define $(1 + \chi)$ as

$$\epsilon_r = (1 + \chi) \tag{3.112}$$

then the equivalent equations to Eqs. (3.110) and (3.111) become, respectively,

$$\epsilon_o \nabla \cdot (\epsilon_r \mathbf{E}) = \rho_f \tag{3.113}$$

and for ϵ_r being constant

$$\epsilon_o \epsilon_r \nabla \cdot \mathbf{E} = \rho_f \tag{3.114}$$

The term ϵ_r is called the dielectric constant or relative permittivity of the dielectric material and is a dimensionless number. The use of $(\epsilon_o \epsilon_r)$ instead of ϵ is convenient as ϵ_r for different dielectric materials varies within a numerical range that is easily remembered. The minimum value of ϵ_r is unity for vacuum. Its values range from near unity for most gases, 2–4 for oil crudes, 15–35 for alcohols, and about 80 for water. The use of relative permittivity (or dielectric constant) is akin to the use of the specific gravity (or relative density) of substances using the density of water as a reference. If one is to strictly adhere to SI convention, then the use of a material permittivity ϵ is more appropriate. In some literature the symbols ϵ_r and ϵ are used interchangeably.

Making use of the potential relationship given by Eq. (3.65), Eq. (3.110) can be written as

$$\nabla \cdot (\epsilon \nabla \psi) = -\rho_f \tag{3.115}$$

For constant permittivity, one can write

$$\nabla^2 \psi = -\frac{\rho_f}{\epsilon} \tag{3.116}$$

Equation (3.116) represents the Poisson's equation for the electric potential distribution in a dielectric material. In situations where there is no free charge, one can set ρ_f to zero, and Eq. (3.116) becomes

$$\nabla^2 \psi = 0 \tag{3.117}$$

Equation (3.117) is referred to as Laplace's equation, which describes the electric potential distribution in a material having no free charge.

EXAMPLE 3.4

Electric Potential Distribution. Obtain an expression for the electric potential distribution for a point charge Q in a dielectric in the absence of free charge.

Solution In the absence of free charge, Laplace's equation is applicable, where

$$\nabla^2 \psi = 0 \tag{3.118}$$

Since there is no θ- and ϕ-dependence in a spherical coordinate system, Eq. (3.118) becomes

$$\frac{1}{r^2}\frac{d}{dr}\left(r^2 \frac{d\psi}{dr}\right) = 0 \tag{3.119}$$

the solution of which is

$$\frac{d\psi}{dr} = \frac{A_1}{r^2} \tag{3.120}$$

and

$$\psi = -\frac{A_1}{r} + A_2 \tag{3.121}$$

where A_1 and A_2 are integration constants. The constant A_2 can be evaluated using the condition that at large distances, *i.e.*, as $r \to \infty$, the effect of the point charge is negligible and consequently, $\psi \to 0$. This condition implies that $A_2 = 0$.

In order to evaluate A_1 we need to make use of the charge Q. Consider Eq. (3.111) where

$$\epsilon \nabla \cdot \mathbf{E} = \rho_f$$

Consider a spherical shell enclosing the point charge. Applying the divergence theorem yields

$$\int_V (\nabla \cdot \mathbf{E})\, dV = \int_S (\mathbf{E} \cdot \mathbf{i}_r)\, dS = \int_V \frac{\rho_f}{\epsilon} dV \tag{3.122}$$

Since $\mathbf{E} \cdot \mathbf{i}_r = E_r = -d\psi/dr$, the above equation becomes

$$-\int_S \frac{d\psi}{dr} dS = -\int_S \left(\frac{A_1}{r^2}\right) dS = \frac{Q}{\epsilon} \tag{3.123}$$

where the term $\int_V \rho_f dV$ represents the charge Q. Note that the charge Q is the *free* charge in the spherical volume. As the surface element $dS = r^2 d\phi$, Eq. (3.123) becomes

$$-\int_0^{4\pi} A_1\, d\phi = \frac{Q}{\epsilon} \tag{3.124}$$

leading to

$$A_1 = -\frac{Q}{4\pi\epsilon}$$

Therefore, the potential is given as

$$\psi = \frac{Q}{4\pi\epsilon r} \tag{3.125}$$

The above equation for the potential due to a point charge in a dielectric is similar to that given in free space, Eq. (3.72) with ϵ_o being replaced by ϵ.

If one replaces the point charge by a spherical shell of radius a having a uniform surface charge distribution of q_s that is equivalent to a total free surface charge of Q_s, then the equivalent to Eq. (3.125) becomes

$$\psi_s = \frac{Q_s}{4\pi\epsilon a} \tag{3.126}$$

The above equation can be compared to the case of free space as provided by Eq. (3.90).

The analysis conducted in this example and in the previous Example 3.3, clearly shows how the surface potential and surface charge density can be related to each other. It should be noted that Eq. (3.126) is also an expression given to a capacitor, where $4\pi\epsilon a$ is known as the capacity of the spherical shell

$$\frac{Q_s}{\psi_s} = 4\pi\epsilon a = C \tag{3.127}$$

where C is the capacitance. ■

Now, let us return to the force exerted by two free point charges Q_1 and Q_2 in a dielectric medium. Making use of Eq. (3.1) the equivalent expression for Coulomb's law in a dielectric is given by

$$F = \frac{Q_1 Q_2}{4\pi\epsilon R^2} \tag{3.128}$$

Allowing ourselves to write $\epsilon = \epsilon_o\epsilon_r$, Eq. (3.128) becomes

$$F = \frac{Q_1 Q_2}{4\pi\epsilon_o\epsilon_r R^2} \tag{3.129}$$

where R is the distance separating the two point charges. Comparing the force between the two point charges in free space as was given by Eq. (3.1) and that in a dielectric medium of permittivity $\epsilon_o\epsilon_r$, as given by Eq. (3.129), it is clear that the force between the two charges is reduced by a factor ϵ_r due to the presence of the dielectric material in which the point charges are located.

One can think of the dielectric constant or relative permeability, ϵ_r, in terms of a ratio which is given by

$$\epsilon_r = \frac{\text{Force between two charges in free space}}{\text{Force between the same charges, at the same separation, in a dielectric medium}}$$

ϵ_r is a dimensionless quantity with $\epsilon_r > 1$.

Typical values of the dielectric constant are provided in Table 3.4. For gases at atmospheric pressure, $\epsilon_r \approx 1$, and ϵ_r increases with pressure. For example, air at 1 atm has an ϵ_r value of about 1.0006 whereas at 100 atm, $\epsilon_r = 1.055$. The dielectric constant for water is about 80. The dielectric constant is a function of temperature and for the case of water ϵ_r variation with temperature is provided in Table 3.5.

TABLE 3.4. Dielectric Constants* ϵ_r of Some Common Liquids and Solids at 25 °C.

Compound		ϵ_r	Compound	ϵ_r
Hydrogen Bonding			*Polymers*	
Methyl formamide	$HCONHCH_3$	182.4	Nylon	3.7–4.2
Formamide	$HCONH_2$	109.5	Fluorocarbons	2.1–3.6
Hydrogen fluoride	HF	84.0	Polycarbonate	3.0
	(at 0 °C)		Polystyrene	2.4
Water	H_2O	78.5	PTFE	2.0
Heavy water	D_2O	77.9		
Formic acid	HCOOH	58.5		
	(at 16 °C)		*Glasses*	
Ethylene glycol	$C_2H_4(OH)_2$	40.7	Fused quartz SiO_2	3.8
Methanol	CH_3OH	32.6	Soda glass	7.0
Ethanol	C_2H_5OH	24.3	Borosilicate glass	4.5
n-Propanol	C_3H_7OH	20.2		
Ammonia	NH_3	16.9	*Crystalline Solids*	
Acetic acid	CH_3COOH	6.2	Diamond (carbon)	5.7
			Quartz SiO_2	4.5
Non-Hydrogen Bonding			Mica	5.4–7.0
Acetone	$(CH_3)_2CO$	20.7	Sodium chloride	6.0
Chloroform	$CHCl_3$	4.8	Alumina Al_2O_3	8.5
Benzene	C_6H_6	2.3		
Carbon tetrachloride	CCl_4	2.2	*Miscellaneous*	
Cyclohexane	C_6H_{12}	2.0	Paraffin (liq.)	2.2
Dodecane	$C_{12}H_{26}$	2.0	Paraffin wax (solid)	2.2
Hexane	C_6H_{14}	1.9	Silicone oil	2.8
			Helium (liq. at 2-3 K)	1.055
			Water (liq. at 0 °C)	87.9
			Water (ice at 0 °C)	91.6–106.4
			Air (dry)	1.00054

*Taken from Israelachvili (1985).

TABLE 3.5. Variation of the Dielectric Constant of Water with Temperature.

Temperature, °C	ϵ_r, Literature*	ϵ_r, Curve Fit†
0	87.90	87.9
10	83.96	84.0
20	80.20	80.2
25	78.54	78.5
30	76.60	76.6
40	73.17	73.2
50	69.88	69.9
60	66.73	66.7
70	63.73	63.7
80	60.86	60.8
90	58.12	58.1
100	55.51	55.5

*Adapted from Archer and Wang (1990).

†Fitting equation: $\epsilon_r = 87.86 - 0.3963T + 7.306 \times 10^{-4}T^2$, where T is in °C.

In summary, Eq. (3.116) is referred to as Poisson's equation that describes the electric potential in a dielectric. It is one of the fundamental equations to be used for the evaluation of the potential ψ in an electrolyte solution (*e.g.*, NaCl in water). Equation (3.116) is therefore the relevant equation to be used in solving problems involving electrostatics in dielectric materials. A remarkable characteristic of Eq. (3.116) is that the charge density is the "free" charge density in the material. It is not the bound or polarization charge density, which is induced through the influence of an electric field. In an electrolyte solution, say, Na_2SO_4 in water, the free charges are Na^+ and SO_4^{2-} ions that are free to move in water.

In order to solve Poisson's equation, we need to define appropriate boundary conditions. In the following section, we discuss the boundary conditions associated with dielectric materials.

3.5 BOUNDARY CONDITIONS FOR ELECTROSTATIC EQUATIONS

Poisson's equation, given by Eq. (3.116), describes the electric potential distribution in a dielectric material in the presence of free charge for a space independent permittivity ϵ. In the absence of free charge density within the interior of the material under consideration, Poisson's equation becomes Laplace's equation, Eq. (3.117). In order to evaluate the electric field strength for a given physical problem, boundary conditions must be specified in order to solve the relevant differential equations. We shall present here the applicable boundary conditions at an interface between two dielectric materials having permittivities of ϵ_1 and ϵ_2. If one of the materials is free space, then ϵ_o will be used instead.

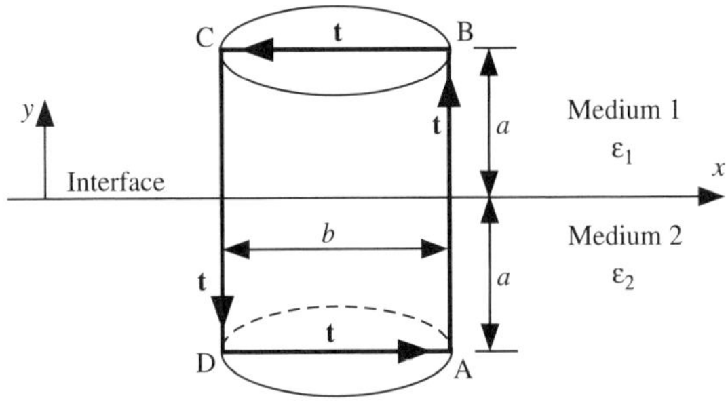

Figure 3.15. Cylindrical surface element straddling an interface separating two dielectrics.

For convenience, let us consider a cylindrical volume whose surface is S and is spanned by the closed curve C, denoted by ABCD. The tangent to C is $\mathbf{t}$. The cylindrical volume straddles the interface of the two dielectric materials. We seek boundary conditions at the interface. The geometry is depicted in Figure 3.15.

Recognizing that the electric field strength is conservative, one can write *cf.*, Eq. (3.64),

$$\nabla \times \mathbf{E} = 0 \tag{3.130}$$

Making use of Stokes theorem, one can write

$$\int_S (\nabla \times \mathbf{E}) \cdot \mathbf{n}\, dS = \int_c \mathbf{E} \cdot d\mathbf{t} \tag{3.131}$$

where $\mathbf{n}$ is an outward unit normal vector to the surface S and $\mathbf{t}$ is an element vector along a contour C spanning the surface S.

Combining Eqs. (3.130) and (3.131) leads to

$$\int_{ABCD} \mathbf{E} \cdot d\mathbf{t} = 0 \tag{3.132}$$

Carrying out the integration along ABCD with AB and DC parallel to the y-axis and CB and DA parallel to the x-axis, one can write

$$E_{2y}a + E_{1y}a - E_{1x}b - E_{1y}a - E_{2y}a + E_{2x}b = 0 \tag{3.133}$$

leading to

$$E_{1x} = E_{2x} \tag{3.134}$$

where E_{1x} and E_{2x} are the electric field strengths in x-direction for media 1 and 2, respectively. Similarly, E_{1y} and E_{2y} are the electric field strengths in y-direction for media 1 and 2, respectively.

Recognizing that Eq. (3.65) gives

$$E_{1x} = -\frac{\partial \psi_1}{\partial x} \tag{3.135-a}$$

$$E_{2x} = -\frac{\partial \psi_2}{\partial x} \tag{3.135-b}$$

and using Eq. (3.134), one obtains

$$\frac{\partial \psi_1}{\partial x} = \frac{\partial \psi_2}{\partial x} \tag{3.136}$$

Upon integration, Eq. (3.136) gives

$$\psi_1 = \psi_2 + \text{Constant} \tag{3.137}$$

Equation (3.137) indicates that the potential at the interface between the two media will differ by at most a constant. Referring back to Eq. (3.68), we note that the potential difference is related to the work required in moving a unit charge. Now, consider the scenario when the cylinder in Figure 3.15 is shrunk by reducing the lengths AB and DC, and take the limit when the two lines BC and AD coincide (corresponding to $a \to 0$). This is analogous to folding the cylindrical volume into the interface such that the two circular faces of the cylinder coincide. We note that for this limiting situation, there will be no work associated with moving a charge across the interface from medium 1 to 2 (as there is no finite length over which a charge is moved). Accordingly, the constant in Eq. (3.137) vanishes across an interface, and we obtain

$$\psi_1 = \psi_2 \tag{3.138}$$

Equation (3.138) states that the electric potential at the interface is the same whether one approaches the interface from medium 1 or medium 2. In other words, there is continuity of electric potential.

As Poisson's equation is a second order differential equation, it requires a second boundary condition. We will now proceed to establish the second boundary condition. Poisson's equation is given by

$$\epsilon \nabla \cdot \mathbf{E} = \rho_f \tag{3.139}$$

Integrating Eq. (3.139) over a volume V and applying the divergence theorem on the volume V enclosed by a surface S, we obtain

$$\int_V \epsilon(\nabla \cdot \mathbf{E})\, dV = \int_V \rho_f \, dV = \int_S (\epsilon \mathbf{E} \cdot \mathbf{n})\, dS$$

or,

$$\int_S (\epsilon \mathbf{E} \cdot \mathbf{n})\, dS = \int_V \rho_f \, dV \tag{3.140}$$

The unit vector $\mathbf{n}$ is outward normal to the surface S.

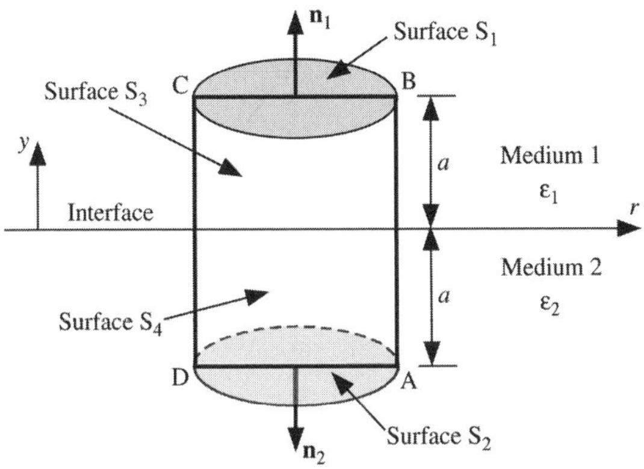

Figure 3.16. Cylindrical volume element showing the bounding surfaces S_1, S_2, S_3, and S_4. Here, the unit normal vectors $\mathbf{n}_1$ and $\mathbf{n}_2$ point toward the positive and negative y directions, respectively. Both $\mathbf{n}_3$ and $\mathbf{n}_4$ (not shown) point in the radial direction.

Consider an element volume at the interface of two dielectric media as shown in Figure 3.16. Applying Eq. (3.140) to the surface elements leads to

$$\epsilon_1 E_{1y} S_1 - \epsilon_2 E_{2y} S_2 + \epsilon_1 E_{1r} S_3 + \epsilon_2 E_{2r} S_4 = \int_V \rho_f \, dV \tag{3.141}$$

Subscripts 1 and 2 refer to the media 1 and 2, respectively, and subscripts y and r refer to the cylindrical coordinates. Setting $S_1 = S_2 = S$, and letting the depth 'a' approach zero, we get $S_3 = S_4 = 0$, and Eq. (3.141) becomes

$$\epsilon_1 E_{1y} - \epsilon_2 E_{2y} = \frac{1}{S} \int_V \rho_f \, dV = q_{sf} \tag{3.142}$$

We note that the term $\int_V \rho_f \, dV/S$ represents the free charge per unit area at the interface, or the free surface charge density at the interface, q_{sf}. One can write Eq. (3.142) in terms of the electric potential, ψ, as

$$-\epsilon_1 \frac{\partial \psi_1}{\partial y} + \epsilon_2 \frac{\partial \psi_2}{\partial y} = q_{sf} \tag{3.143}$$

In the absence of any free charge at the interface, *i.e.*, $q_{sf} = 0$, Eq. (3.143) becomes

$$\epsilon_1 \frac{\partial \psi_1}{\partial y} = \epsilon_2 \frac{\partial \psi_2}{\partial y} \tag{3.144}$$

In other words, for a zero surface charge density, the normal component, $D_n = \mathbf{D} \cdot \mathbf{n}$, of the electric displacement vector, $\mathbf{D}$, given by Eqs. (3.105) and (3.109) is continuous

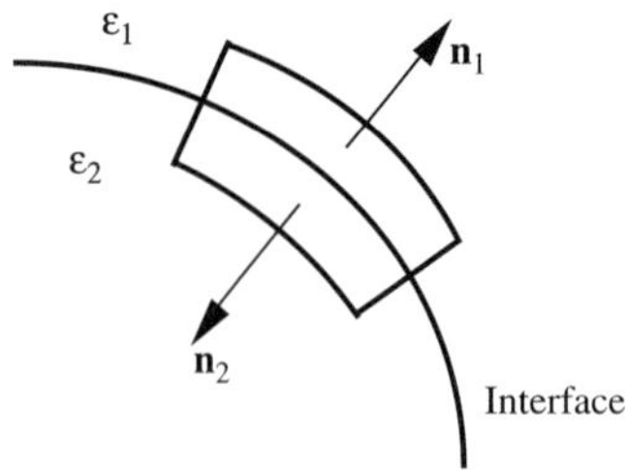

Figure 3.17. Volume element straddling an arbitrary interface where "a" of Figure 3.16 approaches zero.

across an interface. However, since the dielectric permittivities of the two media separated by the interface are different ($\epsilon_1 \neq \epsilon_2$), the electric field, **E**, is discontinuous across the interface.

Let us now derive a different form of Eq. (3.143) in terms of unit outward normal vectors applicable to interfaces of any arbitrary shape. Applying Eq. (3.140) to the element of Figure 3.17 and allowing $a \to 0$, one can write

$$[\epsilon \mathbf{E} \cdot \mathbf{n}]_1 + [\epsilon \mathbf{E} \cdot \mathbf{n}]_2 = \frac{1}{S} \int_V \rho_f \, dV = q_{sf} \tag{3.145}$$

where $\mathbf{n}_1$ and $\mathbf{n}_2$ are the outward unit normal vectors acting on S_1 and S_2, respectively, as shown in Figure 3.16.

It is more convenient to express Eq. (3.145) in terms of the normal derivatives of the potential, which is given by

$$-\mathbf{E} \cdot \mathbf{n} = \nabla\psi \cdot \mathbf{n} = \frac{\partial \psi}{\partial n}$$

This renders Eq. (3.145) as

$$\epsilon_1 \frac{\partial \psi_1}{\partial n_1} + \epsilon_2 \frac{\partial \psi_2}{\partial n_2} = -q_{sf} \tag{3.146}$$

Here n_1 and n_2 are the *coordinates* normal to a surface in directions $\mathbf{n}_1$ and $\mathbf{n}_2$, respectively. For example, in Cartesian coordinates, n_1 can represent the x-coordinate and $\mathbf{n}_1$ becomes $\mathbf{i}_x$.

Equations (3.138) and (3.146) are the necessary boundary conditions at an interface separating two media having different permittivities. They apply at any surface, flat or curved, and whether there is a charge or not. These boundary conditions are simply a statement of the Gauss law and the fact that the electric field is irrotational *i.e.*, $\nabla \times \mathbf{E} = 0$.

It should be pointed out that the free surface charge density at the interface is not an acquired state due to the imposition of the external electric field as is the case for the polarization charge density. It is a surface charge that already *exists* prior to the introduction of the external field. For instance, when a rubber ball is rubbed against a woolen cloth, a surface charge is acquired by the ball. In the process of applying the boundary conditions, this acquired "free surface charge" must be included in the analysis for the boundary conditions when the rubber ball is in an external electric field. This in spite of the fact that there might not be any free charge within the interior of the rubber ball; a case that would allow the use of Laplace's equation to describe the potential inside the rubber ball.

We have already discussed the relevant electrostatic equations for a dielectric medium. We will now briefly discuss polarization bound charges. As mentioned earlier, when a dielectric material is placed in an electric field, a polarization charge is developed at the interface. To derive the necessary boundary conditions involving polarization charges, we will utilize the derivations previously applied to the free charges.

Combining Eqs. (3.101), (3.102), and (3.110), and assuming space-independent permittivity inside the dielectric material, we obtain

$$(\epsilon - \epsilon_o)\nabla \cdot \mathbf{E} = -\rho_p \tag{3.147}$$

Application of the divergence theorem over volume V enclosed by a surface S leads to

$$\int_S (\epsilon - \epsilon_o)(\mathbf{E} \cdot \mathbf{n})\, dS = -\int_V \rho_p\, dV \tag{3.148}$$

By analogy to the analysis employed for Eqs. (3.140), (3.145), and (3.146), one can write

$$(\epsilon_1 - \epsilon_o)\frac{\partial \psi_1}{\partial n_1} + (\epsilon_2 - \epsilon_o)\frac{\partial \psi_2}{\partial n_2} = q_{sp} \tag{3.149}$$

where q_{sp} is the polarization surface charge density defined as

$$q_{sp} = \frac{1}{S}\int_V \rho_p\, dV \qquad (\text{C/m}^2) \tag{3.150}$$

Equation (3.149) can be used to evaluate surface polarization charge density at an interface between two dielectrics.

In terms of the total charge density, one can write

$$\epsilon_o \nabla \cdot \mathbf{E} = \rho \tag{3.151}$$

leading to

$$-\epsilon_o\left[\frac{\partial \psi_1}{\partial n_1} + \frac{\partial \psi_2}{\partial n_2}\right] = q_s \tag{3.152}$$

where q_s is the total charge density at an interface.

In this section, the relevant boundary conditions were derived together with expressions for the free, polarization, and total charge densities at an interface.

When a body is subjected to an electric field, stresses develop within the body as well as at its outer boundary. When the body is elastic, the electric stresses can deform it. In the next section, we will develop expressions for the electrostatic stresses caused by an electric field.

3.6 MAXWELL STRESS FOR A LINEAR DIELECTRIC

It was pointed out that Poisson's equation, Eq. (3.116), describes the electric potential in a dielectric subject to boundary conditions associated with a given physical situation. In the case of an interface, *i.e.*, a boundary separating two media having different permittivities, the appropriate boundary conditions were discussed. It remains now to evaluate the force exerted on a dielectric body in an electric field.

The Korteweig–Helmholtz electric force per unit volume, **f**, for an incompressible fluid is given as

$$\mathbf{f} = \rho_f \mathbf{E} - \frac{1}{2} E^2 \nabla \epsilon \tag{3.153}$$

This expression is obtained by neglecting a third contribution to the volumetric force density called electrostriction. The electrostriction term is given by

$$\nabla \left(\frac{1}{2} E^2 \rho \frac{\partial \epsilon}{\partial \rho} \right)$$

where ρ is the fluid material density. For further details on the formulation of the electrical body force, see Saville (1997). In all problems discussed in the present book, the strictional term can be omitted.

The term $\rho_f \mathbf{E}$ in Eq. (3.153) represents the body force due to the interaction of the free charges in the fluid with the electric field. The last term in Eq. (3.153) accounts for the inhomogeneity in the permittivity of the medium. Recognizing that for a linear dielectric material $\mathbf{D} = \epsilon \mathbf{E}$, $\nabla \cdot \mathbf{D} = \rho_f$, and after some vector manipulation, one can show that the electric force per unit volume is given by

$$\mathbf{f} = \nabla \cdot \left[\epsilon \mathbf{E}\mathbf{E} - \frac{1}{2} \epsilon \mathbf{E} \cdot \mathbf{E} \bar{\bar{I}} \right] \tag{3.154}$$

The term within the square brackets in Eq. (3.154) is tensorial, and is generally referred to as the Maxwell stress tensor, $\bar{\bar{T}}$, in a rest frame. Thus, the force per unit volume in a given dielectric is

$$\mathbf{f} = \nabla \cdot \bar{\bar{T}} \tag{3.155}$$

where

$$\bar{\bar{T}} = \epsilon \mathbf{E}\mathbf{E} - \frac{1}{2} \epsilon \mathbf{E} \cdot \mathbf{E} \bar{\bar{I}} \tag{3.156}$$

In Eqs. (3.154) and (3.156), the second order unit tensor, $\overline{\overline{I}}$, is defined as

$$\overline{\overline{I}} = \begin{pmatrix} 1 & 0 & 0 \\ 0 & 1 & 0 \\ 0 & 0 & 1 \end{pmatrix} \tag{3.157}$$

and the term **EE** is a dyadic product, *i.e.*, a second order tensor, defined as

$$\mathbf{EE} = \begin{pmatrix} E_1^2 & E_1 E_2 & E_1 E_3 \\ E_2 E_1 & E_2^2 & E_2 E_3 \\ E_3 E_1 & E_3 E_2 & E_3^2 \end{pmatrix} \tag{3.158}$$

where E_1, E_2, and E_3 are the components of the vector **E** that represents the field strength.

The force arising from Maxwell electric stress on a body of volume V enclosed by surface S is given by

$$\mathbf{F} = \int_V \mathbf{f}\, dV = \int_V (\nabla \cdot \overline{\overline{T}}) dV \tag{3.159}$$

Utilizing the divergence theorem, Eq. (3.159) gives

$$\mathbf{F} = \int_V (\nabla \cdot \overline{\overline{T}}) dV = \int_S (\mathbf{n} \cdot \overline{\overline{T}}) dS$$

Recognizing that the Maxwell stress tensor $\overline{\overline{T}}$ is symmetric,[1] the force on the body due to Maxwell stress can be written as

$$\mathbf{F} = \int_S (\overline{\overline{T}} \cdot \mathbf{n}) dS \tag{3.160}$$

where **n** is the unit outward normal vector to the surface S enclosing the volume V, as shown in Figure 3.18. From a physical point of view, one can view the stress tensor as that quantity which when dotted with the outward normal **n**, gives the stress vector acting on a surface whose outward normal is **n** (Whitaker, 1968).

Using Eqs. (3.156) and (3.160), the force acting on a body due to Maxwell stresses becomes

$$\mathbf{F} = \int_S \left[\epsilon \mathbf{EE} - \frac{1}{2} \epsilon \mathbf{E} \cdot \mathbf{E} \overline{\overline{I}} \right] \cdot \mathbf{n}\, dS \tag{3.161}$$

Once the electric field strength, **E**, is known, the force on a body in an electric field can be evaluated through the use of Eq. (3.161). The force acting on the body is

[1]For the general case, $\overline{\overline{T}} \cdot \mathbf{n} \neq \mathbf{n} \cdot \overline{\overline{T}}$, unless the tensor $\overline{\overline{T}}$ is symmetric.

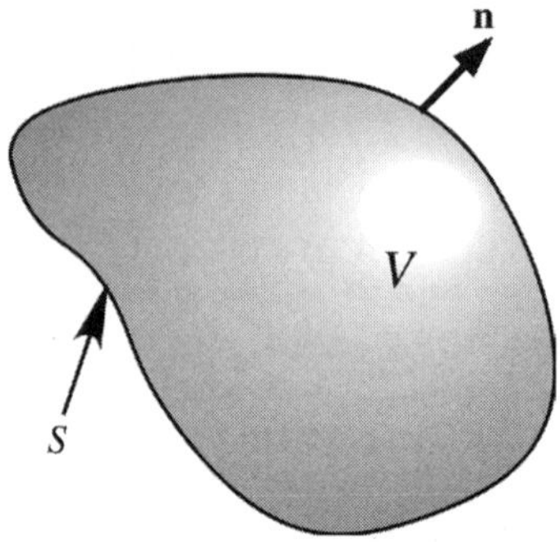

Figure 3.18. A body of volume V bounded by a surface S with an outward normal unit vector **n**.

to be computed using the unit outward normal vector of the body surface under consideration. If one is interested in evaluating the force acting at a surface element straddling an interface as shown in Figure 3.17, the force is still to be evaluated using the unit outward normal vector. In this respect, the stress tensor at the interface would arise from both sides of the interface. For the case of stress tensor discontinuity, Eq. (3.161) still holds. The implication will be illustrated through several examples.

One can expand the expression for Maxwell stress tensor, $\bar{\bar{T}}$, given by Eq. (3.156) to obtain the components of the stress tensor in Cartesian coordinates.

$$\bar{\bar{T}} = \begin{bmatrix} T_{xx} & T_{xy} & T_{xz} \\ T_{yx} & T_{yy} & T_{yz} \\ T_{zx} & T_{zy} & T_{zz} \end{bmatrix} = \epsilon \begin{bmatrix} E_x^2 - \frac{1}{2}E^2 & E_x E_y & E_x E_z \\ E_y E_x & E_y^2 - \frac{1}{2}E^2 & E_y E_z \\ E_z E_x & E_z E_y & E_z^2 - \frac{1}{2}E^2 \end{bmatrix} \tag{3.162}$$

Here, the stress component T_{ij} represents the stress on the ith plane in j-direction resulting from the electric field. For example, T_{yx} is the stress acting on the y- surface in x-direction. One should note that the stress tensor is symmetric, *i.e.*, $T_{ij} = T_{ji}$. The subscript for E signifies the component of the **E** vector and one can write

$$E^2 = E_x^2 + E_y^2 + E_z^2 \tag{3.163}$$

As the electric field strength **E** is related to the electric potential by

$$\mathbf{E} = -\nabla\psi \tag{3.164}$$

the explicit relationships in the Cartesian coordinate system become

$$E_x = -\frac{\partial\psi}{\partial x} \tag{3.165-a}$$

$$E_y = -\frac{\partial\psi}{\partial y} \tag{3.165-b}$$

$$E_z = -\frac{\partial\psi}{\partial z} \tag{3.165-c}$$

We note that all the components of the field vector **E** can be related to the scalar potential ψ.

The stress tensor can be readily written for other coordinate systems. For example, in spherical coordinates, one can write

$$\begin{bmatrix} T_{rr} & T_{r\theta} & T_{r\phi} \\ T_{\theta r} & T_{\theta\theta} & T_{\theta\phi} \\ T_{\phi r} & T_{\phi\theta} & T_{\phi\phi} \end{bmatrix} = \epsilon \begin{bmatrix} E_r^2 - \frac{1}{2}E^2 & E_r E_\theta & E_r E_\phi \\ E_\theta E_r & E_\theta^2 - \frac{1}{2}E^2 & E_\theta E_\phi \\ E_\phi E_r & E_\phi E_\theta & E_\phi^2 - \frac{1}{2}E^2 \end{bmatrix} \tag{3.166}$$

where E_r, E_θ, and E_ϕ are the components of **E** in r, θ, and ϕ directions, respectively. In spherical coordinates, Eq. (3.164) provides

$$E_r = -\frac{\partial \psi}{\partial r} \tag{3.167-a}$$

$$E_\theta = -\frac{1}{r}\frac{\partial \psi}{\partial \theta} \tag{3.167-b}$$

$$E_r = -\frac{1}{r \sin\theta}\frac{\partial \psi}{\partial \phi} \tag{3.167-c}$$

Here, E^2 is given by

$$E^2 = E_r^2 + E_\theta^2 + E_\phi^2 \tag{3.168}$$

Let us now discuss the implication of the term $(\bar{\bar{T}} \cdot \mathbf{n})$ given by Eq. (3.160). In Cartesian coordinates, the dot product of the stress tensor, $\bar{\bar{T}}$, with the unit normal vector **n** is given by

$$\begin{aligned} \bar{\bar{T}} \cdot \mathbf{n} = {} & (T_{xx}n_x + T_{xy}n_y + T_{xz}n_z)\mathbf{i}_x \\ & + (T_{yx}n_x + T_{yy}n_y + T_{yz}n_z)\mathbf{i}_y \\ & + (T_{zx}n_x + T_{zy}n_y + T_{zz}n_z)\mathbf{i}_z \end{aligned} \tag{3.169}$$

where the outward normal vector is given by

$$\mathbf{n} = n_x\mathbf{i}_x + n_y\mathbf{i}_y + n_z\mathbf{i}_z \tag{3.170}$$

If we have a planar surface whose normal is aligned along the unit vector $\mathbf{i}_x$, *i.e.*, $\mathbf{n} = n_x\mathbf{i}_x$, then Eq. (3.169) can be written as

$$\bar{\bar{T}} \cdot \mathbf{n} = T_{xx}n_x\mathbf{i}_x + T_{yx}n_x\mathbf{i}_y + T_{zx}n_x\mathbf{i}_z \tag{3.171}$$

Although for the unit normal vector **n**, the quantity n_x will be unity, we have retained it explicitly in Eq. (3.171) to ensure that the direction of the unit outward vector is clearly indicated in the analysis.

To illustrate the procedure of obtaining the net force on an interface between two dielectrics where discontinuity exists, let us consider a volume element of infinitesimal

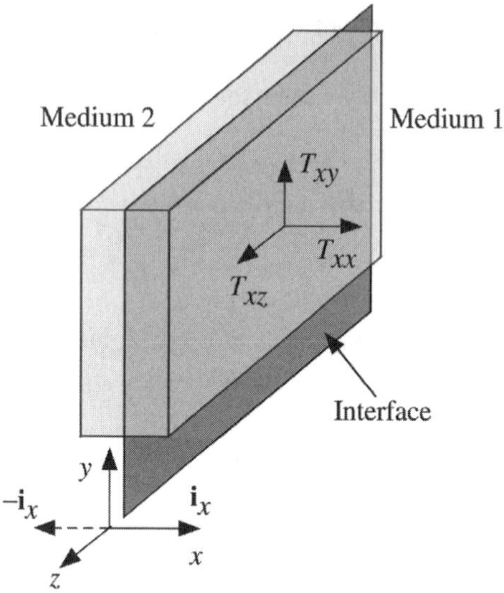

Figure 3.19. Volume element of infinitesimal thickness straddling an interface. The normal to the interface is directed along the x axis.

thickness straddling a surface normal to the x-axis as shown in Figure 3.19. The surface represents the interface between two dielectrics.

Accounting for the dielectric media on both sides of the volume element, the net stress acting on the interface is given by

$$\begin{aligned}(\bar{\bar{T}} \cdot \mathbf{n})_{\text{net}} = {} & [(T_{xx}n_x)_1 + (T_{xx}n_x)_2]\,\mathbf{i}_x \\ & + \left[(T_{yx}n_x)_1 + (T_{yx}n_x)_2\right]\mathbf{i}_y \\ & + \left[(T_{zx}n_x)_1 + (T_{zx}n_x)_2\right]\mathbf{i}_z\end{aligned} \tag{3.172}$$

Here, the subscripts 1 and 2 denote the media 1 and 2, respectively. One can think of $(\bar{\bar{T}} \cdot \mathbf{n})$ as being the local force per unit area. Consequently, one can write more specifically for the force acting on the entire surface (denoted by S) as

$$\mathbf{F} = \int_S (\bar{\bar{T}} \cdot \mathbf{n})_{\text{net}} dS \tag{3.173}$$

or,

$$\mathbf{F} = \int_S \left\{\left[\epsilon \mathbf{EE} - \frac{1}{2}\epsilon E^2 \bar{\bar{I}}\right] \cdot \mathbf{n}\right\}_{\text{net}} dS \tag{3.174}$$

It should be emphasized that the Maxwell stress tensor due to the electric field as defined above is that of the surrounding medium acting on the surface element whose normal is directed "into" the medium.

In the next chapter, we shall utilize the electrostatic equations, boundary conditions, and Maxwell stress to evaluate the electric field strength, electric potential, surface charge density, and force acting on bodies placed in an external electric field.

3.7 MAXWELL'S EQUATIONS OF ELECTROMAGNETISM

In the previous section we dealt with Maxwell's equations as related to electrostatics. For completeness, the generalized forms of Maxwell's equations for electromagnetism will be briefly presented in this section. Maxwell's equations are not limited to electrostatics, but in their general form, describe the laws of electromagnetism. These equations relate the electric field **E** and the magnetic field **H** to material properties, the charge density, and conservation principles.

The full set of Maxwell's equations in rationalized MKS (RMKS) units is

$$\nabla \cdot \mathbf{D} = \rho_f \tag{3.175}$$

$$\nabla \cdot \mathbf{B} = 0 \tag{3.176}$$

$$\nabla \times \mathbf{E} = -\frac{\partial \mathbf{B}}{\partial t} \tag{3.177}$$

and

$$\nabla \times \mathbf{H} = \frac{\partial \mathbf{D}}{\partial t} + \mathbf{J}_f \tag{3.178}$$

Here, **D** is the electric displacement vector, ρ is the charge density, **B** is the magnetic induction field (Tesla, $\mathrm{T} = \mathrm{NA}^{-1}\mathrm{m}^{-1}$), **E** is the electric field, **H** is the magnetic field (Am^{-1}), and $\mathbf{J}_f$ is the conduction (or free) current density ($\mathrm{C/m^{-2}s^{-1}}$) associated with the movement of free charge. For a linear and homogeneous medium, the following constitutive relationships can be written

$$\mathbf{D} = \epsilon \mathbf{E} \tag{3.179}$$

and

$$\mathbf{B} = \mu \mathbf{H} \tag{3.180}$$

where ϵ is the dielectric constant and μ is the magnetic permeability of the material. Utilizing Eqs. (3.179) and (3.180), the Maxwell equations can be written for a linear homogeneous medium as

$$\nabla \cdot \mathbf{E} = \frac{\rho_f}{\epsilon} \tag{3.181}$$

$$\nabla \cdot \mathbf{H} = 0 \tag{3.182}$$

$$\nabla \times \mathbf{E} = -\mu \frac{\partial \mathbf{H}}{\partial t} \tag{3.183}$$

and

$$\nabla \times \mathbf{H} = \epsilon \frac{\partial \mathbf{E}}{\partial t} + \mathbf{J}_f \tag{3.184}$$

It is clear from the above equations that when the electric field varies with time, one needs to consider the magnetic field as well. In this case, the equations governing the electric and magnetic fields are coupled. However, in the absence of time variation, the electric field equations, *i.e.*, electrostatics, can be decoupled from the magnetic equations or magnetostatics.

Another point worth noting about the above set of equations is that although these are fundamental relationships governing propagation of electromagnetic waves (*i.e.*, light), the speed of light does not appear in these equations explicitly. This important parameter is, in fact, immediately recovered by noting that

$$\epsilon_o \mu_o = \frac{1}{c^2} \tag{3.185}$$

where $\epsilon_o = 8.854 \times 10^{-12}$ $C^2/N.m^2$ and $\mu_o = 1.2566 \times 10^{-7}$ Henry/m are the permittivity and magnetic permeability of vacuum, respectively, and c is the speed of light in vacuum (2.9979×10^8 m/s). For other material media, the speed of light evolves from the dielectric constant (ϵ) and magnetic permeability (μ) in a manner analogous to Eq. (3.185).

3.8 NOMENCLATURE

A	cross sectional area of parallel plate capacitor plates, m^2
$\mathbf{B}$	magnetic induction field, Tesla
C	capacitance, Farad
c	speed of light, m/s
$\mathbf{D}$	electric displacement vector, C/m^2
$\mathbf{d}$	vector orientation between two point charges
$\mathbf{E}$	electric field strength, V/m
E	magnitude of electric field, V/m
$\mathbf{F}$	force, N
$\mathbf{F}_{12}$	force exerted by charge Q_1 on charge Q_2
$\mathbf{f}$	force per unit volume, N/m^3
$\mathbf{H}$	magnetic field, A/m
$\overline{\overline{I}}$	unit tensor
$\mathbf{i}$	unit vector
$\mathbf{i}_{12}$	unit vector in the direction of the separation vector $\mathbf{R}_{12}$
$\mathbf{J}_f$	conduction or free current density, $C/m^2.s$
K	conductance, Siemens
l	length, m
N	number of dipoles per unit volume
$\mathbf{n}$	outward unit surface normal vector

P	polarization field strength, C/m^2
Q	total electric charge or charge at a point, C
Q_f	total electric free charge on a body, C
Q_i	i^{th} point charge, C
Q_s	surface charge, C
q_s	surface electric charge density, C/m^2
q_{sf}	free surface electric charge density, C/m^2
q_{sp}	polarization surface charge density, C/m^2
R	distance between two points in space, m
	Resistance, Ohm
$\mathbf{R}_{12}$	separation or displacement vector, m
r	position vector, m
r	radial position, m
S	surface area, m^2
$\bar{\bar{T}}$	Maxwell stress tensor, N/m^2
T_{ij}	component of stress tensor (stress on i^{th} face in j-direction), N/m^2
t	unit tangent vector along a line or plane
V	volume, m^3
W_{12}	work done to move a charge along a path 1 to 2, J

Greek Symbols

ΔV	Electric potential difference, V
ϵ	dielectric permittivity of medium, C^{2}N^{-1}m^{-2} or CV^{-1}m^{-1}
ϵ_o	permittivity of vacuum (free space), C^{2}N^{-1}m^{-2} or CV^{-1}m^{-1}
ϵ_r	relative dielectric constant, dimensionless
χ	electric susceptibility of the dielectric medium, dimensionless
λ	line charge density, C/m
μ	magnetic permeability of material medium, Henry/m
μ_o	magnetic permeability of vacuum, Henry/m
ψ	electric potential, V
ψ_s	surface electric potential, V
ρ	total electric volume charge density, C/m^3
	material density, kg/m^3
ρ_f	free electric charge density, C/m^3
ρ_p	polarization charge density, C/m^3
∇	del operator, m^{-1}
∇^2	Laplacian operator, m^{-2}

3.9 REFERENCES

Archer, D. G., and Wang, P., The dielectric constant of water and Debye-Hückel limiting law slopes, *J. Phys. Chem. Ref. Data*, **19**, 371, (1990).

Eyges, L., *The Classical Electromagnetic Field*, Dover, New York, (1980).

Feynman, R. P., Leighton, R. B., and Sands, M., *The Feynman Lectures on Physics*, vol. II, Addison-Wesley, Reading, MA, (1964).

Griffiths, D. J., *Introduction to Electrodynamics*, Prentice-Hall, Upper Saddle River, NJ, (1989).

Israelachvili, J. N., *Intermolecular and Surface Forces*, Academic Press, London, (1985).

Saville, D. A., Electrohydrodynamics: The Taylor-Melcher leaky dielectric model, *Ann. Rev. Fluid Mech.*, **29**, 27–64, (1997).

Slater, J. C., and Frank, N. H., *Electromagnetism*, Dover, New York, (1969).

Whitaker, S., *Introduction to Fluid Mechanics*, Prentice-Hall, Englewood Cliffs, NJ, (1968).

CHAPTER 4

APPLICATION OF ELECTROSTATICS

In this chapter, we will present different physical problems in the area of electrostatics and seek solutions for the electric field, Maxwell electrostatic stress, and surface charge. It is hoped that through the solution of these problems we will elucidate the concept of electrostatics and provide a foundation for the study of the electrical double layer encountered in the area of electrokinetics. The concept of the electrical double layer will be discussed in Chapter 5. The problems discussed in this chapter are entirely within the scope of the principles outlined in Chapter 3, and hence, may be construed as additional examples for the previous chapter. However, these examples underscore some key electrostatic phenomena that are generally neglected in discussions of electrical double layer interactions. In particular, the scenarios discussed here are applicable to dielectric media without any free mobile charges, *i.e.*, free ions. Such systems are extremely important in a variety of situations.

4.1 TWO-DIMENSIONAL DIELECTRIC SLAB IN AN EXTERNAL ELECTRIC FIELD

Consider a flat polymer slab of permittivity ϵ_2 and thickness $2d$ located in a dielectric medium of permittivity ϵ_1. The potential is held at ψ_A at $x = -X$ and ψ_B at $x = X$. The geometry is shown in Figure 4.1. We will evaluate the electric field and potential for the system together with surface polarization charge and the Maxwell electric stresses.

Electrokinetic and Colloid Transport Phenomena, by Jacob H. Masliyah and Subir Bhattacharjee
Copyright © 2006 John Wiley & Sons, Inc.

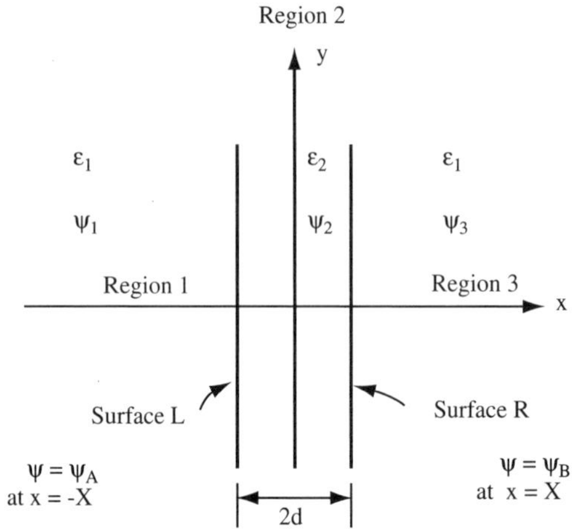

Figure 4.1. Polymer slab in a dielectric medium with an imposed electric potential.

4.1.1 Electric Potential and Field Strength

In the absence of free charge in the polymer slab and the intervening medium, the governing equation is given by the Laplace equation, Eq. (3.117). For a one-dimensional geometry, one can write Laplace's equation in one-dimension (along the x-coordinate) for the regions flanking the slab on the left and right sides and for the slab itself:

$$\frac{\partial^2 \psi_j}{\partial x^2} = 0 \tag{4.1}$$

where j signifies the region of application. The solution of Eq. (4.1) for the three regions is given by

$$\psi_1 = a_1 + b_1 x \quad \text{for} \quad -d \geq x \geq -X \tag{4.2}$$

$$\psi_2 = a_2 + b_2 x \quad \text{for} \quad d \geq x \geq -d \tag{4.3}$$

$$\psi_3 = a_3 + b_3 x \quad \text{for} \quad X \geq x \geq d \tag{4.4}$$

The applicable boundary conditions are:

$$\psi_1 = \psi_A \quad \text{at} \quad x = -X \text{ (Imposed potential)} \tag{4.5-a}$$

$$\psi_1 = \psi_2 \quad \text{at} \quad x = -d \text{ (Continuity of potential)} \tag{4.5-b}$$

$$\epsilon_1 \frac{d\psi_1}{dx} = \epsilon_2 \frac{d\psi_2}{dx} \quad \text{at} \quad x = -d \text{ (No surface free charge)} \tag{4.5-c}$$

$$\psi_2 = \psi_3 \quad \text{at} \quad x = d \text{ (Continuity of potential)} \tag{4.5-d}$$

$$\epsilon_2 \frac{d\psi_2}{dx} = \epsilon_1 \frac{d\psi_3}{dx} \quad \text{at} \quad x = d \text{ (No surface free charge)} \tag{4.5-e}$$

$$\psi_3 = \psi_B \quad \text{at} \quad x = X \text{ (Imposed potential)} \tag{4.5-f}$$

The boundary conditions provide six equations which are used to evaluate the six constants in Eqs. (4.2)–(4.4). Applying the above boundary conditions leads to

$$\psi_1 = \frac{(\psi_A + \psi_B)}{2} - Cd(\epsilon_2 - \epsilon_1)(\psi_A - \psi_B) - C\epsilon_2(\psi_A - \psi_B)x \tag{4.6}$$

$$\psi_2 = \frac{(\psi_A + \psi_B)}{2} - C\epsilon_1(\psi_A - \psi_B)x \tag{4.7}$$

$$\psi_3 = \frac{(\psi_A + \psi_B)}{2} + Cd(\epsilon_2 - \epsilon_1)(\psi_A - \psi_B) - C\epsilon_2(\psi_A - \psi_B)x \tag{4.8}$$

where

$$C = \frac{1}{2[\epsilon_2 X - d(\epsilon_2 - \epsilon_1)]} \tag{4.9}$$

For the special cases of

$$X \gg d \qquad C = \frac{1}{2\epsilon_2 X} \tag{4.10-a}$$

$$X = d \qquad C = \frac{1}{2\epsilon_1 d} \tag{4.10-b}$$

The electric potential on the left of the slab, the slab itself, and on the right of the slab are given by Eqs. (4.6)–(4.8), respectively.

In the absence of the slab, one can write $\epsilon_1 = \epsilon_2$, for which case, Eqs. (4.6)–(4.8) become identical, and are given by

$$\psi = \frac{\psi_A + \psi_B}{2} - \frac{\psi_A - \psi_B}{2X}x \tag{4.11}$$

This represents a linear variation of ψ with x. At the origin, the potential is the average of ψ_A and ψ_B and the slope of ψ *vs.* x is $-(\psi_A - \psi_B)/2X$, which is the potential gradient.

The electric field strength in the x-direction is given by Eq. (3.65) as

$$E_x = -\frac{d\psi}{dx} \tag{4.12}$$

Application of Eq. (4.12) to the three regions leads to

$$E_{1x} = C\epsilon_2(\psi_A - \psi_B) \tag{4.13}$$

$$E_{2x} = C\epsilon_1(\psi_A - \psi_B) \tag{4.14}$$

$$E_{3x} = C\epsilon_2(\psi_A - \psi_B) \tag{4.15}$$

Equations (4.13)–(4.15) show that the field strength is constant within each of the three regions. The above equations indicate that when the slab permittivity ϵ_2 is larger than the surrounding medium permittivity ϵ_1, *i.e.*, for $\epsilon_2 > \epsilon_1$, the field strength within the slab is reduced as compared to the field strength in the outside medium. For the case $\epsilon_2 < \epsilon_1$, the field strength within the slab is higher than that in the outside medium. This is simply a consequence of the polarization charge at the slab's surfaces. The evaluation of the surface charge density follows in the next section.

4.1.2 Polarization Surface Charge Density

The polarization surface charge density, q_{sp}, for a dielectric is given by Eq. (3.149). Consider the two surfaces L and R in Figure 4.2. For surface L, $\mathbf{n}_1 = -\mathbf{i}_x$ and $\mathbf{n}_2 = \mathbf{i}_x$. Application of Eq. (3.149) on surface L leads to

$$(q_{sp})_L = -(\epsilon_1 - \epsilon_o)\frac{d\psi_1}{dx} + (\epsilon_2 - \epsilon_o)\frac{d\psi_2}{dx} \tag{4.16}$$

Substituting for the derivative of the potentials, the surface charge density due to polarization for surface L becomes

$$(q_{sp})_L = -C\epsilon_o(\epsilon_2 - \epsilon_1)(\psi_A - \psi_B) \tag{4.17}$$

In a similar manner, applying Eq. (3.149) to the surface R, one obtains

$$(q_{sp})_R = C\epsilon_o(\epsilon_2 - \epsilon_1)(\psi_A - \psi_B) \tag{4.18}$$

As would be expected, the polarization surface charge density on the left and right surfaces are equal in their magnitudes but opposite in sign.

When the outer medium is free space, the permittivity ϵ_1 becomes equal to ϵ_o. Recognizing that the constant C and $(\epsilon_2 - \epsilon_o)$ are positive quantities, Eq. (4.17)

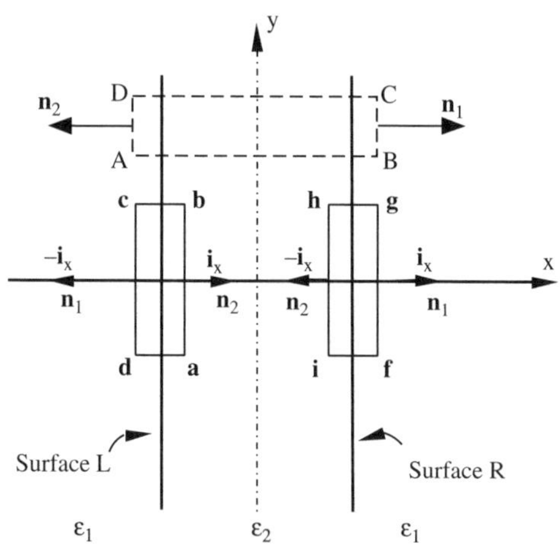

Figure 4.2. Surface elements for electrostatic force evaluation for a polymer slab in an external electric field.

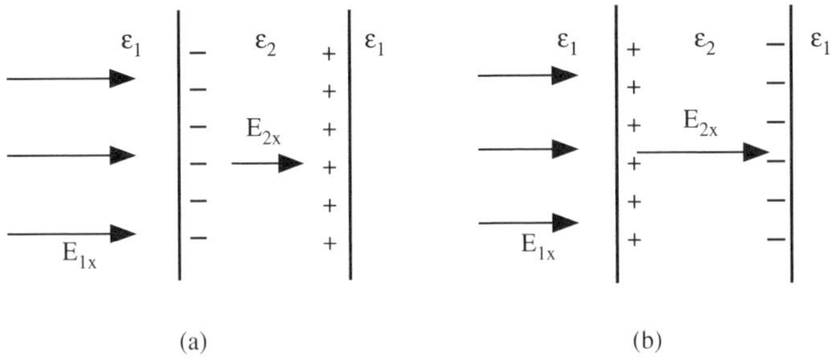

Figure 4.3. Polarization surface charge density for two dielectrics. The electric field inside the slab will be reduced for (a) $\epsilon_2 > \epsilon_1$ and enhanced for (b) $\epsilon_2 < \epsilon_1$.

would indicate that when the polymer slab is in free space, a negative charge density will always develop on the left hand side of the slab. The situation is different when the slab is in another dielectric medium. The polarization charge density on the left hand side would depend on the sign of $(\epsilon_2 - \epsilon_1)$. When $\epsilon_2 > \epsilon_1$, the left hand side surface will carry a negative charge, while for $\epsilon_2 < \epsilon_1$, it will carry a positive charge. Consequently, the electric field strength as indicated by Eqs. (4.13) and (4.14) in the slab will be

$$E_{2x} < E_{1x} \quad \text{for} \quad \epsilon_2 > \epsilon_1 \tag{4.19-a}$$

$$E_{2x} > E_{1x} \quad \text{for} \quad \epsilon_2 < \epsilon_1 \tag{4.19-b}$$

$$E_{2x} = E_{1x} \quad \text{for} \quad \epsilon_2 = \epsilon_1 \tag{4.19-c}$$

as shown in Figure 4.3.

4.1.3 Maxwell Electrostatic Stress

In this section, we will derive expressions for the Maxwell stresses for the left and right hand side surfaces. As the physical problem is one-dimensional, the analysis will be relatively straightforward since we only have to deal with the electric field component E_x and the electrostatic stress component T_{xx}. Here, the stress component T_{xx} is by definition the stress exerted by the outer environment on the x-plane in the x-direction. In all our analysis we assume that the medium permittivity is a constant.

From Eq. (3.162), the Maxwell stress, T_{xx}, is given by

$$T_{xx} = \epsilon \left(E_x^2 - \frac{E^2}{2} \right) \tag{4.20}$$

As we are dealing with a one-dimensional problem where $E_y = E_z = 0$, one can write

$$E = E_x \tag{4.21}$$

and Eq. (4.20) becomes

$$T_{xx} = \frac{\epsilon E_x^2}{2} \tag{4.22}$$

In order to evaluate the net force per unit area on a surface, it is necessary to perform a surface integration given by Eq. (3.160), which requires the Maxwell electrostatic stress components. Figure 4.2 shows the integration paths for the evaluation of the electrostatic forces.

Applying Eqs. (3.172) and (3.173) for the surface element $abcd$ located on surface L, the x-directed force per unit area, F_{xL}, is given by

$$F_{xL}\mathbf{i}_x = (\mathbf{n}_1 T_{xx})_1 + (\mathbf{n}_2 T_{xx})_2 \tag{4.23}$$

Recognizing that $\mathbf{n}_1 = -\mathbf{i}_x$ and $\mathbf{n}_2 = \mathbf{i}_x$, Eq. (4.23) becomes

$$F_{xL} = -(T_{xx})_1 + (T_{xx})_2 \tag{4.24}$$

Here, $(T_{xx})_1$ signifies the evaluation of the stress component T_{xx} in medium 1 (surface cd) and $(T_{xx})_2$ denotes evaluation of T_{xx} in medium 2 (surface ab).

Making use of Eq. (4.22), one can write for Eq. (4.24)

$$F_{xL} = -\frac{1}{2}\epsilon_1 E_{1x}^2 + \frac{1}{2}\epsilon_2 E_{2x}^2 \tag{4.25}$$

with the use of expressions for the field strength provided by Eqs. (4.13) and (4.14), Eq. (4.25) becomes

$$F_{xL} = -\frac{C^2}{2}\epsilon_1\epsilon_2^2(\psi_A - \psi_B)^2 + \frac{C^2}{2}\epsilon_2\epsilon_1^2(\psi_A - \psi_B)^2$$

Upon simplification we obtain

$$F_{xL} = -\frac{C^2}{2}\epsilon_1\epsilon_2(\epsilon_2 - \epsilon_1)(\psi_A - \psi_B)^2 \tag{4.26}$$

Clearly, as $C^2\epsilon_1\epsilon_2(\psi_A - \psi_B)^2/2$ is always positive, the direction of the force on surface L is dependent on the sign of $(\epsilon_2 - \epsilon_1)$.

Let us now evaluate the x-directed force per unit area, F_{xR}, on the surface R. Applying Eqs. (3.172) and (3.173) on surface element $fghi$, we have

$$F_{xR}\mathbf{i}_x = (\mathbf{n}T_{xx})_1 + (\mathbf{n}T_{xx})_2 \tag{4.27}$$

where the unit outward normal vectors $\mathbf{n}_1$ and $\mathbf{n}_2$ for surface R are given by

$$\mathbf{n}_1 = \mathbf{i}_x \qquad \text{and} \qquad \mathbf{n}_2 = -\mathbf{i}_x \tag{4.28}$$

The x-directed force per unit area on surface R becomes

$$F_{xR} = (T_{xx})_1 - (T_{xx})_2 \tag{4.29}$$

From Eqs. (4.22), (4.14) and (4.15), we write for Eq. (4.29)

$$F_{xR} = \frac{C^2}{2}\epsilon_1\epsilon_2(\epsilon_2 - \epsilon_1)(\psi_A - \psi_B)^2 \tag{4.30}$$

Comparing the electrostatic forces on surfaces L and R shows that $F_{xL} + F_{xR} = 0$. In other words, the net force on the polymer slab is zero. This result should not be surprising as the total charge of the slab is zero. A linear dielectric body with zero total charge does not experience a force arising from its placement in a uniform electric field strength. It should be noted that with the absence of free charge, the polarization charge represents the total charge, *cf.*, Eq. (3.101). As the net force on the slab is zero, the polymer plate would not move in any preferred direction in the external electric field and will remain stationary. A more straightforward approach to evaluate the electric force on the slab would be to use the force acting on the element ABCD as shown in Figure 4.2.

Although the net force on the polymer slab is zero, the forces acting on surfaces L and R can either be compressive or tensile. When $\epsilon_2 > \epsilon_1$ the slab is in tension and when $\epsilon_2 < \epsilon_1$, the slab is in compression. Consequently, if the polymer is made of an elastic dielectric material, then under the external electric field, the slab's thickness increases for $\epsilon_2 > \epsilon_1$ and decreases for $\epsilon_2 < \epsilon_1$.

4.2 A DIELECTRIC SPHERE IN AN EXTERNAL ELECTRIC FIELD

Consider a spherical drop of a dielectric fluid (*e.g.*, an oil) of a constant permittivity ϵ_2 being placed in another dielectric fluid of constant permittivity ϵ_1. The initial uniform electric field strength is E_o and it is parallel to the z-direction. Assume that the oil drop has no electric charge prior to its placement in the electric field. The physical configuration is shown in Figure 4.4. It is of interest to evaluate the electric field and potential, together with the polarization charge and the Maxwell electric stress for the oil drop.

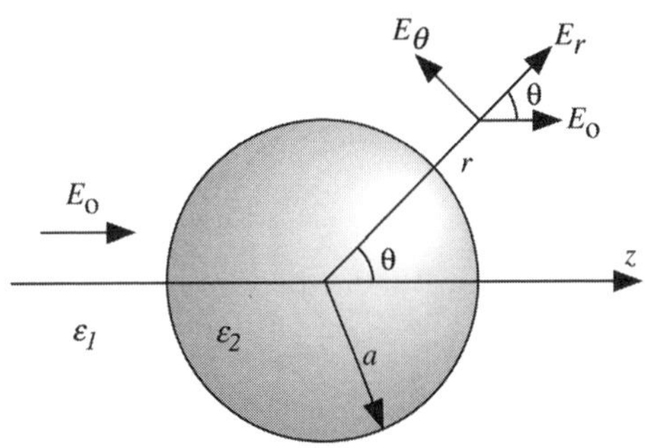

Figure 4.4. A dielectric sphere in another dielectric under an external electric field.

4.2.1 Electric Potential and Field Strength

Since there is no free charge inside and outside the sphere, Laplace's equation is applicable and it is given by

$$\nabla^2 \psi_i = 0 \tag{4.31}$$

for $i = 1, 2$ representing outside and inside the sphere, respectively. A spherical coordinate system is best suited for analyzing the problem, and we can write the Laplace equation for outside and inside the sphere as
Outside the sphere:

$$\frac{1}{r^2}\frac{\partial}{\partial r}\left(r^2 \frac{\partial \psi_1}{\partial r}\right) + \frac{1}{r^2 \sin\theta}\frac{\partial}{\partial \theta}\left(\sin\theta \frac{\partial \psi_1}{\partial \theta}\right) = 0 \tag{4.32}$$

Inside the sphere:

$$\frac{1}{r^2}\frac{\partial}{\partial r}\left(r^2 \frac{\partial \psi_2}{\partial r}\right) + \frac{1}{r^2 \sin\theta}\frac{\partial}{\partial \theta}\left(\sin\theta \frac{\partial \psi_2}{\partial \theta}\right) = 0 \tag{4.33}$$

The boundary conditions are

$$\psi_2 \quad \text{is finite at } r = 0 \tag{4.34}$$

$$\psi_1 = \psi_2 \quad \text{continuity of potential at } r = a \tag{4.35}$$

$$\epsilon_1 \frac{\partial \psi_1}{\partial r} = \epsilon_2 \frac{\partial \psi_2}{\partial r} \quad \text{no surface free charge (continuity of displacement) at } r = a \tag{4.36}$$

$$\mathbf{E} = \mathbf{i}_z E_z = \mathbf{i}_z E_o \quad \text{at} \quad r \to \infty \tag{4.37}$$

The last boundary condition states that the electric field is undisturbed at locations far from the sphere and it is in z-direction. With two second order differential equations, the four boundary conditions, Eqs. (4.34)–(4.37), define the physical problem. Before we proceed with the solution, let us examine the boundary condition far from the sphere.

The uniform electric field, E_o, at $r \to \infty$ can be decomposed into E_r and E_θ components as shown in Figure 4.4.

$$E_r = E_o \cos\theta \tag{4.38}$$

$$E_\theta = -E_o \sin\theta \tag{4.39}$$

The components of the electric field strength in spherical coordinates are given by

$$E_r = -\frac{\partial \psi_1}{\partial r} \tag{4.40-a}$$

$$E_\theta = -\frac{1}{r}\frac{\partial \psi_1}{\partial \theta} \tag{4.40-b}$$

Combining Eqs. (4.38)–(4.40-b) leads to

$$\frac{\partial \psi}{\partial r} = -E_o \cos\theta \tag{4.41-a}$$

$$\frac{\partial \psi}{\partial \theta} = E_o r \sin\theta \tag{4.41-b}$$

Solutions of Eqs. (4.40) and (4.41) are given by, respectively,

$$\psi_1 = -E_o r \cos\theta + g(\theta) \tag{4.42}$$

and

$$\psi_1 = -E_o r \cos\theta + f(r) \tag{4.43}$$

Comparison between Eqs. (4.42) and (4.43) indicates that

$$g(\theta) = f(r) = C \tag{4.44}$$

As we are normally interested in an electrical potential difference, we arbitrarily set the constant C to zero, yielding

$$\psi_1 = -E_o r \cos\theta \tag{4.45}$$

Equation (4.45) gives the boundary condition in terms of the electric potential for an undisturbed electric field. The boundary condition is applicable at $r \to \infty$ and it is more convenient to use in our analysis. This boundary condition suggests that the electric potential should vary in θ-direction as the cosine of the angular position θ. Consequently, a possible solution (see Section 4.3) is:

$$\psi_1 = \left(\frac{A_1}{r^2} + A_2 r\right)\cos\theta \tag{4.46}$$

and

$$\psi_2 = \left(\frac{B_1}{r^2} + B_2 r\right)\cos\theta \tag{4.47}$$

The constants A_1, A_2, B_1, and B_2 can now be evaluated from the boundary conditions. The solution is given as
Outside the sphere:

$$\psi_1 = E_0\left[-r + \left(\frac{G-1}{G+2}\right)\frac{a^3}{r^2}\right]\cos\theta \tag{4.48}$$

and, inside the sphere:

$$\psi_2 = -\frac{3E_o r \cos\theta}{G+2} \tag{4.49}$$

where

$$G = \frac{\epsilon_2}{\epsilon_1} \tag{4.50}$$

Recognizing that

$$\mathbf{E} = -\left(\mathbf{i}_r \frac{\partial \psi}{\partial r} + \mathbf{i}_\theta \frac{1}{r}\frac{\partial \psi}{\partial \theta}\right) \tag{4.51}$$

the electric fields are given by
Outside the sphere:

$$\mathbf{E}_1 = \mathbf{i}_r E_o \left[1 + \frac{2(G-1)}{(G+2)}\frac{a^3}{r^3}\right]\cos\theta + \mathbf{i}_\theta E_o \left[-1 + \left(\frac{G-1}{G+2}\right)\frac{a^3}{r^3}\right]\sin\theta \tag{4.52}$$

and, inside the sphere:

$$\mathbf{E}_2 = \mathbf{i}_r \frac{3E_o \cos\theta}{G+2} - \mathbf{i}_\theta \frac{3E_o \sin\theta}{G+2} \tag{4.53}$$

Close examination of Eq. (4.53) yields

$$\mathbf{E}_2 = \frac{3E_o}{(G+2)}\mathbf{i}_z \tag{4.54}$$

Equation (4.54) states that the electric field inside the sphere is uniform and it is in z-direction. Its value relative to the external electric field is dependent on G which is given by ϵ_2/ϵ_1. The inner electric field strength is smaller than E_o when the sphere's permittivity is larger than that on the outside, *i.e.*, $G > 1$. Similar conclusions were obtained for the case of the slab in the previous example.

4.2.2 Polarization Surface Charge Density

Evaluation of the polarization surface charge density is a direct application of Eq. (3.149), which is given as

$$q_{sp} = (\epsilon_1 - \epsilon_o)\frac{\partial \psi_1}{\partial n_1} + (\epsilon_2 - \epsilon_o)\frac{\partial \psi_2}{\partial n_2} \tag{4.55}$$

As shown in Figure 4.5, the unit vectors $\mathbf{n}_1$ and $\mathbf{n}_2$ are related to $\mathbf{i}_r$ leading to

$$n_1 = r \quad \text{and} \quad n_2 = -r \tag{4.56}$$

Making use of the expressions for ψ_1 and ψ_2, and Eq. (4.56), Eq. (4.55) gives the local surface polarization charge density as

$$q_{sp} = 3\epsilon_o E_o \left(\frac{G-1}{G+2}\right)\cos\theta \tag{4.57}$$

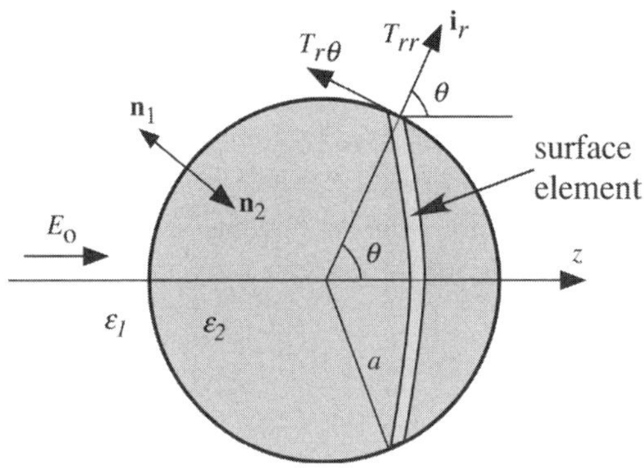

Figure 4.5. A dielectric sphere in an external electric field showing outward normal vectors, a surface element, and the Maxwell stresses.

The distribution of the polarization charge varies as the cosine of the angle θ. Furthermore, its sign is dependent on whether G is greater or smaller than unity.

The total surface polarization charge is given by

$$Q_{sp} = 3\epsilon_o E_o \left(\frac{G-1}{G+2}\right) \int_0^{\pi} 2\pi a^2 \sin\theta \cos\theta \, d\theta \tag{4.58}$$

The surface element for the integration is shown in Figure 4.5. Integration of Eq. (4.58) gives $Q_{sp} = 0$. As a whole, the oil drop remains electro-neutral after its placement in the electric field and it is in agreement with the first fundamental principle of electrostatics.

4.2.3 Maxwell Electrostatic Stress on the Dielectric Sphere

When the dielectric sphere is placed in the electric field, surface stresses develop due to the electric field. The local force can be evaluated from Eq. (3.173), where

$$\mathbf{F} = \int_{surface} (\bar{\bar{T}} \cdot \mathbf{n})_{\text{net}} dS \tag{4.59}$$

By analogy to Eq. (3.169), one can write

$$\bar{\bar{T}} \cdot \mathbf{n} = (T_{rr}n_r + T_{r\theta}n_\theta)\mathbf{i}_r + (T_{\theta r}n_r + T_{\theta\theta}n_\theta)\mathbf{i}_\theta \tag{4.60}$$

where n_r and n_θ are the components of the outer unit vector $\mathbf{n}$. Here, $\mathbf{i}_r$ and $\mathbf{i}_\theta$ are unit vectors along the r and θ coordinates, respectively.

The outer spherical surface at $r = a$ is of interest to us and it is at this surface that we need to evaluate the Maxwell stresses. For this surface, Eq. (4.60) reduces to

$$\bar{\bar{T}} \cdot \mathbf{n} = T_{rr}n_r\mathbf{i}_r + T_{r\theta}n_r\mathbf{i}_\theta \tag{4.61}$$

By analogy to Eq. (3.172) one can write the local net force per unit area acting on the outer surface ($r = a$ for all θ) as

$$(\bar{\bar{T}} \cdot \mathbf{n})_{\text{net}} = [(T_{rr})_1 - (T_{rr})_2]\,\mathbf{i}_r + [(T_{r\theta})_1 - (T_{r\theta})_2]\,\mathbf{i}_\theta \tag{4.62}$$

The term $(\bar{\bar{T}} \cdot \mathbf{n})_{\text{net}}$ represents the local net force per unit area acting on the surface.

The components of the Maxwell stress tensor are provided by Eq. (3.166)

$$T_{rr} = \epsilon\left(E_r^2 - \frac{E^2}{2}\right) \tag{4.63}$$

and

$$T_{r\theta} = \epsilon E_\theta E_r \tag{4.64}$$

In Eqs. (4.63) and (4.64), the permittivity and electric field are evaluated for the medium under consideration.

Let us now evaluate the terms in Eq. (4.62). Consider first the term

$$F_r = [(T_{rr})_1 - (T_{rr})_2] \tag{4.65}$$

where F_r represents the local net force per unit area in r-direction. From Eq. (4.63), we can substitute for the electric field components in Eq. (4.65) to give

$$F_r = \left[\epsilon\left(E_r^2 - \frac{E^2}{2}\right)\right]_1 - \left[\epsilon\left(E_r^2 - \frac{E^2}{2}\right)\right]_2 \tag{4.66}$$

Let us now consider the term

$$\left[\epsilon\left(E_r^2 - \frac{E^2}{2}\right)\right]_1$$

Setting $r = a$ and making use of Eq. (4.52), we can write

$$E_1^2 = E_o^2\left[1 + \frac{2(G-1)}{(G+2)}\right]^2 \cos^2\theta + E_o^2\left[-1 + \left(\frac{G-1}{G+2}\right)\right]^2 \sin^2\theta \tag{4.67}$$

leading to

$$E_1^2 = \left(\frac{3E_o}{G+2}\right)^2 [G^2\cos^2\theta + \sin^2\theta] \tag{4.68}$$

Now

$$E_{1r}^2 = E_o^2\left[1 + \frac{2(G-1)}{(G+2)}\right]^2 \cos^2\theta \tag{4.69}$$

leading to

$$E_{1r}^2 = \left(\frac{3GE_o}{G+2}\right)^2 \cos^2\theta \tag{4.70}$$

Consequently we can write for the first term of Eq. (4.66)

$$\left[\epsilon\left(E_r^2 - \frac{E^2}{2}\right)\right]_1 = \frac{\epsilon_1}{2}\left(\frac{3E_o}{G+2}\right)^2 \left[G^2\cos^2\theta - \sin^2\theta\right] \tag{4.71}$$

Now let us consider the term

$$\left[\epsilon\left(E_r^2 - \frac{E^2}{2}\right)\right]_2$$

From Eq. (4.53), we can write

$$E_2^2 = \left(\frac{3E_o\cos\theta}{G+2}\right)^2 + \left(\frac{3E_o\sin\theta}{G+2}\right)^2 \tag{4.72}$$

leading to

$$E_2^2 = \left(\frac{3E_o}{G+2}\right)^2 \tag{4.73}$$

Now,

$$E_{2r}^2 = \left(\frac{3E_o}{G+2}\right)\cos^2\theta \tag{4.74}$$

consequently we can write

$$\left[\epsilon\left(E_r^2 - \frac{E^2}{2}\right)\right]_2 = \frac{\epsilon_2}{2}\left(\frac{3E_o}{G+2}\right)^2 [2\cos^2\theta - 1] \tag{4.75}$$

Combining Eqs. (4.71) and (4.75) and rearranging, the local net radial force per unit area, F_r, exerted by the surrounding medium on the dielectric oil drop is given by

$$F_r = \frac{\epsilon_1}{2}\left(\frac{3E_o}{G+2}\right)^2 [(G-1) + (G-1)^2\cos^2\theta] \tag{4.76}$$

Evaluation of $\epsilon E_\theta E_r$ as given in Eq. (4.64) for the two media leads to

$$(T_{r\theta})_1 = (T_{r\theta})_2 \tag{4.77}$$

This would indicate that there is no net force in the angular direction. In other words, the oil drop does not experience a shearing force due to the electric field.

It is rather interesting to obtain a physical meaning for the net local radial force given by Eq. (4.76). The expression has two terms. The first term can be either positive or negative depending on the relative magnitude of the permittivities of the two dielectrics. Irrespective of the sign of $(G - 1)$, the first term implies a constant radial force on the spherical oil drop and it does not contribute to the drop's deformation. The second term is always positive and it implies that the radial force is dependent on angle θ. As $\cos^2\theta$ attains a maximum value at $\theta = 0$ and π, while it attains a minimum value at $\theta = \pi/2$, one can conclude from Eq. (4.76) that the oil droplet will always elongate along the electric field. Water drop deformation in a electric field is shown in Figure 4.6. Drop deformation analysis for perfect dielectrics under the influence of an electric field was conducted by O'Konski and Thatcher (1953) and Allan and Mason (1962). Their theoretical analysis showed that dielectric liquid drops always deformed to a prolate shape (elongation along the direction of the applied field) as suggested by Eq. (4.76) . Observations reported in different publications (O'Konski and Thatcher, 1953; Taylor, 1964; Garton and Krasucki, 1964; Ha and Yang, 2000; Lu, 2002) confirm such a prediction. However, for some systems, many authors noted in their experimental studies that the drops deformed to an oblate shape (Allan and Mason, 1962; Torza *et al.*, 1971; Arp *et al.*, 1980; Vizika and Saville, 1992; Ha and Yang, 1995).

In the classical papers by Taylor (1966) and Melcher and Taylor (1969), the drop deformation has been addressed by accounting for the non-zero electric conductivity of both the fluids constituting the droplet and the surrounding medium. These studies were performed by considering the redistribution of the electric field due to the migration of the charge carriers. Such an approach is referred to as the leaky dielectric model. Depending on the relationship between conducting and dielectric properties of the liquids, Taylor's theory predicts either prolate or oblate shape of the deformed droplet.

A generalized analysis of the drop deformation problem was recently proposed by Zholkovskij *et al.* (2002), who, in addition to the migration transport of the charge carriers, considered their diffusive transport. The final expression of Zholkovskij *et al.* (2002) provides a smooth bridging between the perfect dielectric (Allan and Mason, 1962) and leaky dielectric (Taylor, 1966) models.

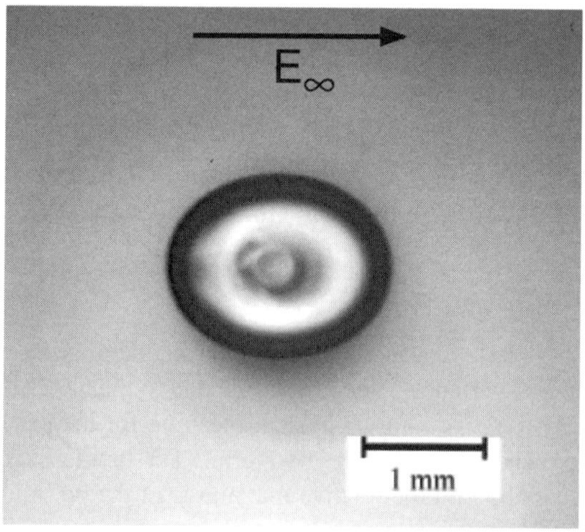

Figure 4.6. Water drop deformation in an electric field of 1.5 kV/cm. The external liquid is 0.5% by volume bitumen obtained from Alberta oil sands in toluene (Lu, 2002).

4.3 A CONDUCTING SPHERE IN AN EXTERNAL ELECTRIC FIELD

Consider a sphere made up of a conducting medium being placed in a dielectric medium in an initially uniform electric field, E_o. Assume that the sphere is grounded and prior to its placement in the electric field, it has no free charge. Figure 4.7 shows the geometry of the system. Let us evaluate the system characteristics.

4.3.1 Electric Potential and Field Strength for a Conducting Sphere

As the conducting sphere has no free charge, the Laplace equation can be used to describe the electric potential distribution. It is given by

$$\nabla^2 \psi = 0 \tag{4.78}$$

Here, the electric potential refers to the region outside the conducting sphere. As the electric field strength is zero inside the conducting sphere, it follows that in the inner region ψ is a constant and it is taken to be zero. In spherical coordinates, the Laplace equation has the form

$$\frac{1}{r^2}\frac{\partial}{\partial r}\left(r^2\frac{\partial \psi}{\partial r}\right) + \frac{1}{r^2 \sin\theta}\frac{\partial}{\partial \theta}\left(\sin\theta\frac{\partial \psi}{\partial \theta}\right) = 0 \tag{4.79}$$

The boundary conditions are

$$\psi = 0 \quad \text{at} \quad r = a \tag{4.80}$$

$$\psi = -E_o r \cos\theta \quad \text{at} \quad r \to \infty \tag{4.81}$$

As we discussed in the previous section, the boundary condition for ψ at $r \to \infty$ signified that the electric field is undisturbed. From Eq. (4.81), the form of the angular variation of ψ is suggested and one can guess the solution for ψ to be of the form

$$\psi = f(r)\ \cos\theta \tag{4.82}$$

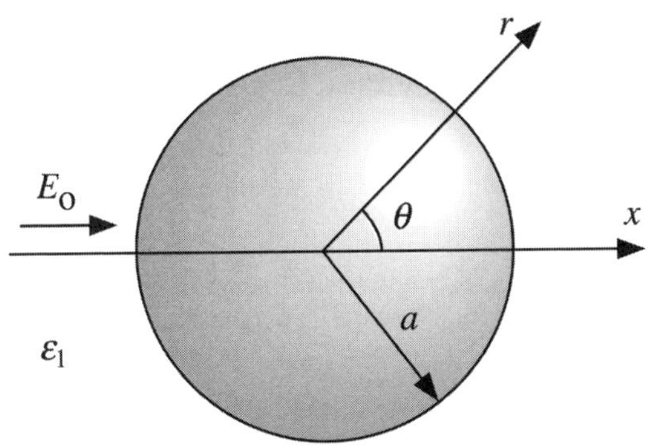

Figure 4.7. A conducting sphere in a uniform electric field E_o.

Substituting Eq. (4.82) in Eq. (4.79) leads to

$$2r\frac{df}{dr} + r^2\frac{df}{dr} - 2f = 0$$

By letting $f = r^n$, one can show that the solution of f is given by

$$f = \left(\frac{C_1}{r^2} + C_2 r\right) \tag{4.83}$$

The integration constants C_1 and C_2 can be determined by applying the boundary conditions (4.80) and (4.81), leading to

$$C_1 = a^3 E_o \quad \text{and} \quad C_2 = -E_o$$

where a is the sphere radius. The electric potential outside the sphere becomes

$$\psi = E_o\left(\frac{a^3}{r^2} - r\right)\cos\theta \tag{4.84}$$

It is interesting to note that the electric potential does not depend on the medium permittivity. The electric field strength is deduced from Eq. (4.84) where

$$\begin{aligned}\mathbf{E} &= E_r\mathbf{i}_r + E_\theta\mathbf{i}_\theta = -\nabla\psi \\ &= -\frac{\partial\psi}{\partial r}\mathbf{i}_r - \frac{1}{r}\frac{\partial\psi}{\partial\theta}\mathbf{i}_\theta\end{aligned} \tag{4.85}$$

leading to

$$\mathbf{E} = E_o\left(\frac{2a^3}{r^3} + 1\right)\cos\theta\ \mathbf{i}_r + E_o\left(\frac{a^3}{r^3} - 1\right)\sin\theta\ \mathbf{i}_\theta \tag{4.86}$$

At the sphere surface ($r = a$), the θ-component for the electric field strength is zero and we obtain from Eq. (4.86)

$$E_r|_{r=a} = 3E_o\cos\theta$$

4.3.2 Surface Charge Density for a Conducting Sphere

The polarization charge density on the conducting spherical surface is given by Eq. (3.149). With $\psi_2 = 0$ and letting $\psi_1 = \psi$, one can directly write

$$q_{sp} = (\epsilon_1 - \epsilon_o)\left.\frac{\partial\psi}{\partial r}\right|_{r=a} \tag{4.87}$$

From the solution for ψ provided by Eq. (4.84) we can deduce the polarization charge density from Eq. (4.87) to be

$$q_{sp} = -3(\epsilon_1 - \epsilon_o)E_o\cos\theta \tag{4.88}$$

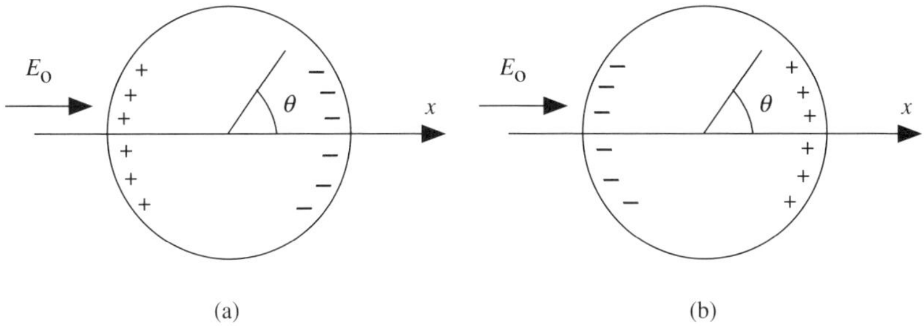

Figure 4.8. (a) Surface polarization charge on a conducting sphere. (b) Free surface charge on a conducting sphere.

where ϵ_1 is the permittivity of the external material into which the conducting sphere is placed. As $(\epsilon_1 - \epsilon_o)$ is always positive, the surface polarization charge distribution does not depend on the medium permittivity. The polarization surface charge is depicted in Figure 4.8(a).

As the sphere's material is a conductor, by placing the sphere in the electric field, unbound electrons would move to the sphere's surface. Consequently, one would expect that the total surface charge is different from the surface polarization charge. The total surface charge is given by Eq. (3.152). One can write

$$q_s = -\epsilon_o \frac{\partial \psi_1}{\partial n_1} = -\epsilon_o \frac{\partial \psi}{\partial r} \tag{4.89}$$

leading to

$$q_s = 3\epsilon_o E_o \cos\theta \tag{4.90}$$

Since the difference $(q_s - q_{sp})$ is termed free surface charge, we can write

$$q_{sf} = 3\epsilon_1 E_o \cos\theta \tag{4.91}$$

Eq. (4.91) indicates that there is a free surface charge, even though there is no free charge within the inner region of the sphere. The free surface charge is schematically depicted in Figure 4.8(b).

The total charge residing on the sphere's surface is given by

$$\begin{aligned} Q_s &= \int_S q_s dS = \int_0^{\pi} 2\pi a^2 q_s \sin\theta \, d\theta \\ &= 6\pi a^2 \epsilon_o E_o \int_0^{\pi} \sin\theta \cos\theta \, d\theta \end{aligned} \tag{4.92}$$

Upon integration, it leads to the expected result

$$Q_s = 0 \tag{4.93}$$

4.3.3 Maxwell Electrostatic Stress on the Conducting Sphere

Let us now evaluate the stress on the surface of the conducting sphere due to its placement in the external electric field. The surface stresses are shown in Figure 4.9.

The stress tensor in spherical coordinates is given by

$$\bar{\bar{T}} = \begin{pmatrix} T_{rr} & T_{r\theta} \\ T_{\theta r} & T_{\theta\theta} \end{pmatrix} = \epsilon \begin{pmatrix} E_r^2 - \frac{1}{2}E^2 & E_r E_\theta \\ E_\theta E_r & E_\theta - \frac{1}{2}E^2 \end{pmatrix} \tag{4.94}$$

At $r = a$, one can show that

$$E_\theta = 0 \tag{4.95}$$

and

$$E = E_r = 3E_o \cos\theta \tag{4.96}$$

From Eqs. (4.94) to (4.96) one can write

$$T_{r\theta} = 0 \tag{4.97}$$

and

$$T_{rr} = \frac{9\epsilon_1}{2} E_o^2 \cos^2\theta \tag{4.98}$$

For this problem, the inner electric field strength is zero, and hence, the inner T_{rr} stress component is zero.

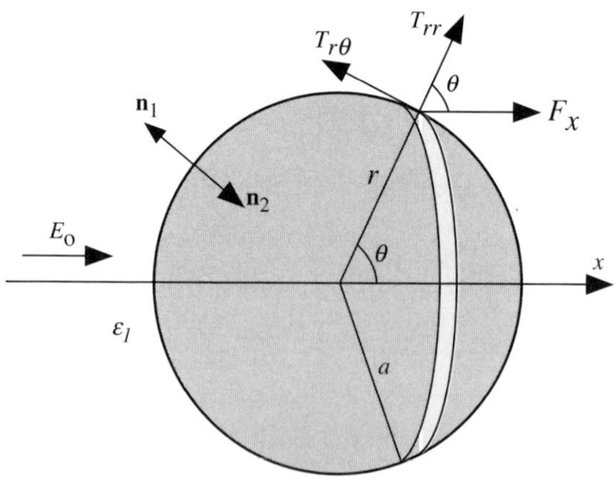

Figure 4.9. Surface stresses on a conducting sphere.

The net force in x-direction is given by

$$\begin{aligned} F_x &= \int_0^{\pi} T_{rr}(2\pi a^2 \sin\theta)\cos\theta \, d\theta \\ &= 9\pi\epsilon_1 a^2 E_o^2 \int_0^{\pi} \sin\theta \cos^3\theta \, d\theta \end{aligned} \tag{4.99}$$

Upon integration, we obtain $F_x = 0$. A physical interpretation for this result is that there is no net force on the spherical conductor when it is placed in a uniform electric field. Such a conclusion would be easily reached as the total charge, Q_s, is zero and the electric field strength is initially uniform.

4.4 CHARGED DISC AND TWO PARALLEL DISCS IN A DIELECTRIC MEDIUM

The purpose of this example is to evaluate the electric field strength due to a charged disc and to eventually evaluate the force between two infinitely large plates carrying a uniform charge density. Let us first discuss the case of the electric field strength at a point A along the z-coordinate, being at a distance z from a flat disc carrying a surface charge density q_{s1}. The geometry is shown in Figure 4.10.

Making use of Eq. (3.16) one can write for the electric field strength at point A

$$\mathbf{E} = \int_S \frac{q_{s1}\,\mathbf{i}}{4\pi\epsilon R^2} dS \tag{4.100}$$

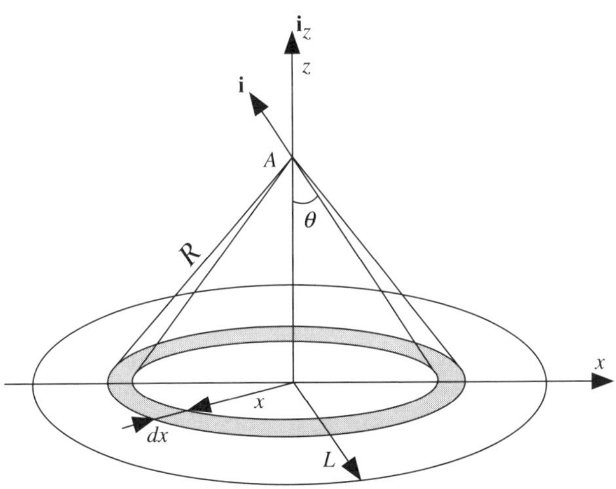

Figure 4.10. Electric field strength due to a disc carrying a surface charge density, q_{s1}.

From the geometry, the electric field strength in z-direction is given by

$$E_z = \int_0^L \frac{2\pi x q_{s1} \cos\theta}{4\pi \epsilon R^2} dx \tag{4.101}$$

where L is the disc radius, ϵ is the medium permittivity, $\mathbf{i} \cdot \mathbf{i}_z = \cos\theta$, and $dS = 2\pi x dx$. Recognizing that

$$R^2 = x^2 + z^2 \tag{4.102}$$

and

$$\cos\theta = z/R \tag{4.103}$$

Equation (4.101) becomes

$$E_z = \frac{z q_{s1}}{2\epsilon} \int_0^L \frac{x dx}{(x^2 + z^2)^{3/2}} \tag{4.104}$$

Making the substitution

$$y = x^2 + z^2$$

and performing the integration in Eq. (4.104) one obtains

$$E_z = \frac{q_{s1}[(L^2 + z^2)^{1/2} - z]}{2\epsilon (L^2 + z^2)^{1/2}} \tag{4.105}$$

This is the electric field strength at A. For the case of $L \gg z$ one can write

$$E_z = \frac{q_{s1}}{2\epsilon} \tag{4.106}$$

The above result suggests that the electric field strength is independent of the position of z along the disc axis when the radius of the circular disc is large ($L \to \infty$). At first, this appears to be a surprising result. However, this behavior can be qualitatively explained by noting that when the point A is relatively close to the disc of infinite radius, the major contribution to the field comes from the regions near the origin of the disc, while the peripheral regions of the disc have smaller contributions owing to their large separations from the point A. This is evident from the fact that the field is inversely proportional to R^2, and that R will be large even for small values of z when x is large, *cf.*, Eq. (4.102). This is analogous to stating that when the point A is close to a disc of infinitely large radius, it "sees" a smaller region of the disc. Now, when the point A moves farther away from the disc in the z-direction, the contribution to the field from the peripheral regions of the disc becomes comparable to that from the central regions – in other words, point A starts "seeing" a larger surface area of the disc. The fact that point A "sees" more of the disc surface as it moves away from the disc leads to a constant electric field strength, see Griffiths (1989) for more details.

Let us now evaluate the force between two large parallel circular planes 1 and 2, carrying surface charge densities of q_{s1} and q_{s2}, respectively. Let us locate point A on plane 2. Recognizing that a point charge of Q would experience a force of $Qq_{s1}/2\epsilon$, then by the principle of superposition, the force between the two circular planes is given by

$$F_{12} = \frac{Sq_{s1}q_{s2}}{2\epsilon} \tag{4.107}$$

or

$$\frac{F_{12}}{S} = \frac{q_{s1}q_{s2}}{2\epsilon} \tag{4.108}$$

where S is the surface area of each plate. Equation (4.108) represents the force per unit area between two large plates carrying charge densities of q_{s1} and q_{s2} in a dielectric medium with permittivity ϵ. In Chapter 5, we will compare this force per unit area with the situation of two parallel plates in an electrolyte solution where free charge is present.

4.5 POINT CHARGES IN A DIELECTRIC MEDIUM

We have discussed in Example 1 of Chapter 3 the electric field strength due to two point charges. Let us extend the analysis to include three point charges. Let us consider three point charges Q_1, Q_2, and Q_3 located along the x-axis as shown in Figure 4.11. The consecutive point charges are separated by a distance $d/2$. The medium has a constant permittivity ϵ.

In a dielectric medium, the electric potential due to a point charge at a distance r away from the point charge is given by

$$\psi = \frac{Q}{4\pi\epsilon r} \tag{4.109}$$

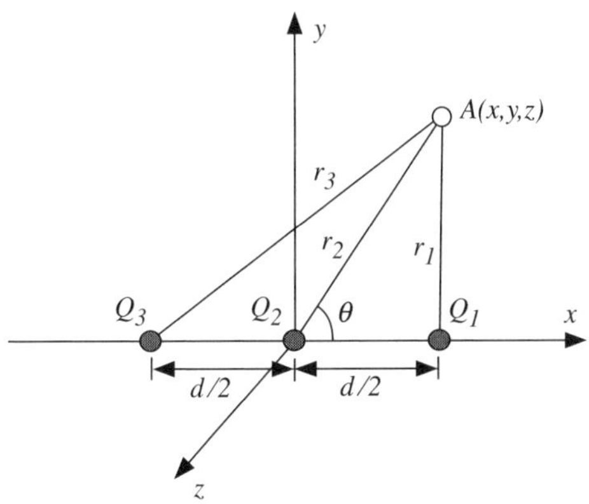

Figure 4.11. Three point charges equally spaced on x-axis.

By the superposition principle, the electric potential at A due to point charges Q_1, Q_2, and Q_3 is given by

$$\psi = \frac{1}{4\pi\epsilon}\left[\frac{Q_1}{r_1} + \frac{Q_2}{r_2} + \frac{Q_3}{r_3}\right] \tag{4.110}$$

Rearranging Eq. (4.110) gives

$$4\pi\epsilon r\psi = \left[\frac{Q_1}{r_1/r} + \frac{Q_2}{r_2/r} + \frac{Q_3}{r_3/r}\right] \tag{4.111}$$

where r denotes the distance of point A from the origin. From the geometry of Figure 4.11, making use of the cosine law and rearranging, we can write

$$r_2 = r \tag{4.112-a}$$

$$\frac{r_1}{r} = \left[1 + \left(\frac{d}{2r}\right)^2 - \frac{d}{r}\cos\theta\right]^{1/2} \tag{4.112-b}$$

$$\frac{r_3}{r} = \left[1 + \left(\frac{d}{2r}\right)^2 + \frac{d}{r}\cos\theta\right]^{1/2} \tag{4.112-c}$$

Substituting for r_1, r_2, and r_3 using the above equations, and expanding in Taylor's series, where

$$(1+x)^n = 1 + nx + \frac{n(n-1)}{2!}x^2 + \cdots$$

with $n = -1/2$ and x being the second and third terms in Eqs. (4.112-b) and (4.112-c), we can write Eq. (4.111) as

$$\begin{aligned} 4\pi\epsilon r\psi & \\ = Q_1 &\left[1 - \frac{1}{2}\left\{\left(\frac{d}{2r}\right)^2 - \frac{d}{r}\cos\theta\right\} + \frac{3}{8}\left\{\left(\frac{d}{2r}\right)^2 - \frac{d}{r}\cos\theta\right\}^2 + \cdots\right] \\ &+ Q_2 + Q_3\left[1 - \frac{1}{2}\left\{\left(\frac{d}{2r}\right)^2 + \frac{d}{r}\cos\theta\right\} + \frac{3}{8}\left\{\left(\frac{d}{2r}\right)^2 + \frac{d}{r}\cos\theta\right\}^2 + \cdots\right] \end{aligned} \tag{4.113}$$

Such an expansion is valid when $x < 1$. Upon rearrangement, Eq. (4.113) yields

$$\begin{aligned} 4\pi\epsilon\psi = &\frac{Q_1}{r}\left[1 + \frac{d}{2r}\cos\theta + \left(\frac{d}{2r}\right)^2\left(\frac{3\cos^2\theta - 1}{2}\right) + \cdots\right] \\ &+ \frac{Q_2}{r} + \frac{Q_3}{r}\left[1 - \frac{d}{2r}\cos\theta + \left(\frac{d}{2r}\right)^2\left(\frac{3\cos^2\theta - 1}{2}\right) + \cdots\right] \end{aligned} \tag{4.114}$$

Seeking a more compact form for Eq. (4.114), and letting $r \gg d$ where only two expansion terms are retained, one obtains

$$\psi(r,\theta) = \frac{1}{4\pi\epsilon}\left[\frac{1}{r}(Q_1 + Q_2 + Q_3) + (Q_1 - Q_3)\left(\frac{d}{2}\right)\frac{1}{r^2}\cos\theta \right. \\ \left. + (Q_1 + Q_3)\left(\frac{d}{2}\right)^2 \frac{1}{r^3}\left(\frac{3\cos^2\theta - 1}{2}\right)\right] \tag{4.115}$$

The electric potential at point A due to three point charges is given by Eq. (4.115). Let us discuss some of the properties of the electric potential ψ for the case of three point charges.

Clearly, the variations of $\psi(r,\theta)$ are largely dependent on the signs of the point charges. Let us consider some specific cases:

Case (i), $Q_1 = -Q_3 = Q$ and $Q_2 = 0$

$$\psi = \frac{Qd\cos\theta}{4\pi\epsilon r^2} \tag{4.116}$$

This is the electric potential due to a dipole. It is in accordance with Eq. (3.34) from Example 1 of Chapter 3, that describes the electric field strength for two point charges of $+Q$ and $-Q$. In this case the decay of the potential ψ is of the order $1/r^2$.

Case (ii), $Q_1 = Q_2 = Q_3 = Q$

$$\psi = \frac{Q}{4\pi\epsilon}\left[\frac{3}{r} + \frac{2}{r}\left(\frac{d}{2r}\right)^2\left(\frac{3\cos^2\theta - 1}{2}\right)\right]$$

or, for $d/r \ll 1$,

$$\psi = \frac{3Q}{4\pi\epsilon r} \tag{4.117}$$

In this case with $d/r \ll 1$, the potential decays quite slowly as $1/r$. One can easily generalize Case (ii) to the following scenario.

Case (iii), $\sum Q_i \neq 0$ with $d/r \ll 1$

$$\psi = \frac{\sum Q_i}{4\pi\epsilon r} \tag{4.118}$$

As long as $\sum Q_i \neq 0$, the decay of the electric potential is proportional to $1/r$.

4.6 NOMENCLATURE

d	half-thickness of a dielectric slab, m
	separation distance between point charges, m
$\mathbf{E}$	electric field strength, V/m
E	magnitude of electric field, V/m
E_o	magnitude of applied electric field, V/m
E_r, E_θ	components of electric field in spherical coordinates, V/m
E_x	electric field component along x direction, V/m
$\mathbf{F}$	force, N
F_{12}	force between two charged planes, N
F_r	radial force per unit area on a sphere, N/m^2
F_{xL}, F_{xR}	force per unit area along x on surfaces L and R, N/m^2
G	ratio of dielectric permittivities, ϵ_2/ϵ_1
$\mathbf{i}_r$	unit vector along r direction
$\mathbf{i}_x$	unit vector along x direction
L	radius of a charged disc, m
l	length, m
$\mathbf{n}$	unit outward surface normal vector
Q	total electric charge or charge at a point, C
Q_f	total electric free charge on a body, C
Q_i	i^{th} point charge, C
Q_{sp}	total surface polarization charge, C
q_s	total surface charge density, C/m^2
q_{sf}	free surface electric charge density, C/m^2
q_{sp}	polarization surface charge density, C/m^2
$\mathbf{r}$	position vector, m
r	radial position, m
S	surface area, m^2
$\overline{\overline{T}}$	Maxwell stress tensor, N/m^2
T_{rr}, $T_{r\theta}$, $T_{\theta\theta}$	components of Maxwell tensor (spherical coordinates), N/m^2
T_{xx}	component of Maxwell tensor (Cartesian coordinates), N/m^2
x, y, z	Cartesian coordinate axes

Greek Symbols

ϵ_i	dielectric permittivity of medium i, C^{2}N^{-1}m^{-2}
ϵ_o	permittivity of vacuum (free space), C^{2}N^{-1}m^{-2}
ϵ_r	relative dielectric constant, dimensionless
ψ_i	electric potential in medium i, V
ψ_A, ψ_B	surface electric potential, V
ρ	total electric volume charge density, C/m^3
θ	azimuthal coordinate in spherical coordinate system

4.7 PROBLEMS

4.1. A cylindrical polymer rod of a dielectric material is placed in another dielectric material and is subjected to an initially uniform electric field E_o as shown in Figure 4.12. The permittivity of the rod is ϵ_2 and the permittivity of the outer material is ϵ_1.

(a) Show that

$$\psi_2 = \frac{-2GE_o r \cos\theta}{(1+G)} \qquad \text{for} \quad r \le a$$

$$\psi_1 = E_o \left[\frac{a^2(1-G)}{(1+G)r} - r \right] \cos\theta \qquad \text{for} \quad r \ge a$$

where $G = \epsilon_1/\epsilon_2$.

(b) Derive an expression for the polarization charge density on the surface of the rod.

(c) Derive an expression for the radial force exerted on the rod.

4.2. Consider once again the flat dielectric slab discussed in Section 4.1. The slab has a permittivity of ϵ_2 and it is placed in a dielectric medium of permittivity ϵ_1. The slab has a "glued" surface charge density of q_s. In medium 1, where the slab is placed, the potential at $x = -X$ is ψ_A while the potential at $x = X$ is ψ_B. The slab thickness is $2d$.

(a) Discuss the reason why Laplace's equation is the governing equation for the dielectric potential inside and outside the slab.

(b) Write down the applicable boundary conditions for the Laplace equation.

(c) Derive the electric potential for the left and right hand sides of the slab, as well as for the slab itself. What are the corresponding field strengths?

(d) Derive an expression for the Maxwell electric stress on the slab surfaces.

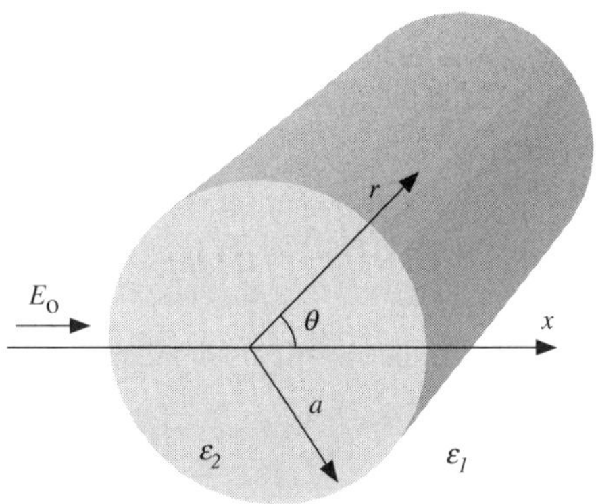

Figure 4.12. Cylindrical polymer rod.

(e) What is the net force per unit area exerted on the slab due to the electric field?

4.3. A parallel plate capacitor consists of two metal plates, separated by a distance d, with their surface charges maintained at Q and $-Q$. The gap between the two surfaces is filled with a dielectric material having a permittivity of ϵ. The capacitance, C, is defined as

$$C = \frac{Q}{V}$$

where V is the potential difference between the two plates.

(a) Show that for the case of parallel plate capacitor, the capacitance is given by

$$C = \frac{A\epsilon}{d}$$

where A is the area of a plate.

(b) For the case of two concentric metal shells, with radii a and b, show that the corresponding capacitance is given by

$$C = \frac{4\pi\epsilon ab}{(b-a)}$$

4.4. A metal rod is placed normal to an initially uniform electric field E_o. Derive expressions for

(a) Electric potential inside and outside the rod.

(b) Electric field strength inside and outside the rod.

(c) Induced surface charge.

4.5. We have discussed in Section 4.5 the electric field due to three point charges. We will now attempt to evaluate the electric field potential due to multiple point charges located at different positions.

Consider a dielectric medium of a permittivity ϵ in which N point charges Q_i are at various locations defined by (r_i, θ_i) as shown in Figure 4.13.

Here $\mathbf{R}_i = \mathbf{r} - \mathbf{r}_i$. Note that in Figure 4.13, the angle θ_i lies on the plane containing points A, O, and the location of point charge Q_i.

(a) Show that for point A

$$\frac{1}{R_i} = \frac{1}{r}\left[1 + \cos\theta_i\left(\frac{r_i}{r}\right) + \frac{1}{2}\left(3\cos^2\theta_i - 1\right)\left(\frac{r_i}{r}\right)^2 + \cdots\right]$$

Show that for $r_i/r < 1$, the electric field potential is given by

$$\psi(r) = \frac{1}{4\pi\epsilon r}\sum_{i=1}^{N} Q_i\left[1 + \frac{r_i}{r}\cos\theta_i + \frac{1}{2}(3\cos^2\theta_i - 1)\left(\frac{r_i}{r}\right)^2 + \cdots\right] \tag{4.119}$$

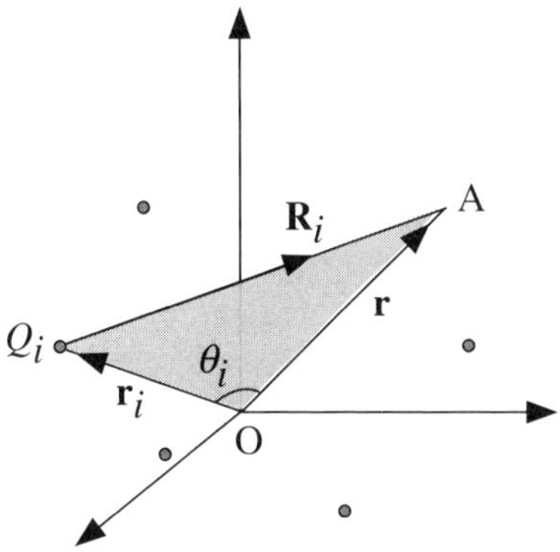

Figure 4.13. Multiple point charges.

(b) The electric potential at a point due to a spherical surface carrying a surface charge density q_s was analyzed in Example 3.3 of Chapter 3. Normalize Eq. (3.86) from that example to give

$$\frac{\psi(z)\epsilon_o}{aq_s} = \frac{a}{z} \tag{4.120}$$

Now assume that N point charges are uniformly distributed to form a charged spherical shell whose center is at the origin O. Assume that all the point charges have the same charge Q and $r_i = a$ for all i. Write a relationship between q_s and Q and show that

$$\begin{aligned}\frac{\epsilon\psi(r)}{aq_s} = \left(\frac{a}{r}\right)\Bigg[1 &+ \frac{1}{N}\sum_{i=1}^{N}\left(\frac{a}{r}\right)\cos\theta_i \\ &+ \frac{1}{N}\sum_{i=1}^{N}\frac{1}{2}(3\cos^2\theta_i - 1)\left(\frac{a}{r}\right)^2 + \cdots\Bigg]\end{aligned} \tag{4.121}$$

Compare Eqs. (4.120) and (4.121) for $N = 100$ and $r/a = 5$ and 100. What is the significance of the limit $N \to \infty$? Comment on your findings. Note that (a/r) is equivalent to (a/z).

4.8 REFERENCES

Allan, R. S., and Mason, S. G., Particle behavior in shear and electric fields: Deformation and burst of fluid drops, *Proc. Roy. Soc. Lond. A*, **267**, 45–61, (1962).

Arp, P. A., Foister, R. T., and Mason, S. G., Some electrohydrodynamic effects in fluid dispersions, *Adv. Colloid Interface Sci.*, **12**, 295–356, (1980).

Garton, C. G., and Krasucki, Z., Bubbles in insulating liquids: Stability in an electric field, *Proc. Roy. Soc. Lond. A*, **280**, 211–226, (1964).

Griffiths, D. J., *Introduction to Electrodynamics*, Prentice-Hall, Upper Saddle River, NJ, (1989).

Ha, J.-W., and Yang, S.-M., Deformation and breakup of Newtonian and non-Newtonian conducting drops in an electric field, *J. Fluid Mech.*, **405**, 131–156, (2000).

Lu, Y., *Electrohydrodynamic deformation of water drops in oil with an electric field*, M.Sc. Thesis, University of Alberta, Edmonton, Canada, (2002).

Melcher, J. R., and Taylor, G. I., Electrohydrodynamics: A review of the role of interfacial shear stresses, *Ann. Rev. Fluid Mech.*, **1**, 111–146, (1969).

O'Konski, C. T., and Thatcher Jr., H. C., The distortion of aerosol droplets by an electric field, *J. Phys. Chem.*, **57**, 955–958, (1953).

Taylor, G. I., Disintegration of water drop in an electric field, *Proc. Roy. Soc. Lond. A*, **280**, 383–390, (1964).

Taylor, G. I., Studies in electrohydrodynamics. The circulation produced by a drop in electric field, *Proc. Roy. Soc. Lond. A*, **291**, 159–166, (1966).

Torza, S., Cox, R. G., and Mason, S. G., Electrohydrodynamic deformation and burst of liquid drops, *Phil. Trans. R. Soc. Lond. A*, **269**, 295–319, (1971).

Vizika, O., and Saville, D. A., The electrohydrodynamic deformation of drops suspended in liquids in steady and oscillatory electric fields, *J. Fluid Mech.*, **239**, 1–21, (1992).

Zholkovskij, E. K., Masliyah, J. H., and Czarnecki, J., An electrokinetic model of drop deformation in an electric field, *J. Fluid Mech.*, **472**, 1–27, (2002).

CHAPTER 5

ELECTRIC DOUBLE LAYER

In Chapter 3 we have dealt with the basics of electrostatics as applied to dielectrics. By and large, we did not deal with physical problems associated with free charges, *i.e.*, ions, their spatial distribution with respect to a charged surface, and their mobility under an applied electric field. Knowledge of the spatial distribution of the free charges and their mobility forms the foundation towards the understanding of liquid flow, particle movement, and induced potential within the subject umbrella of electrokinetics. We define electrokinetics as the study of movement or flow under the influence of an electric potential and field.

5.1 ELECTRIC DOUBLE LAYERS AT CHARGED INTERFACES

So far, we have reviewed the governing equations for the electric field distribution and the corresponding electrostatic potential caused by the presence of charges in vacuum and in dielectric materials. The problems of interest in dispersions require additional considerations. In particular, in aqueous systems of interest, one needs to develop methods to deal with the redistribution of ions in the solution caused by the presence of charged surfaces such as a charged particle, surface, or microemulsion droplet. The free ions in the solution are either attracted to or repelled from a charged surface depending on the sign of the surface charges. Such a redistribution of free ions in the solution together with the surface ions give rise to what are known as *electric double layers*. The purpose of the following sections of this chapter is to develop governing equations for the electric double layers and the double

Electrokinetic and Colloid Transport Phenomena, by Jacob H. Masliyah and Subir Bhattacharjee
Copyright © 2006 John Wiley & Sons, Inc.

layer potentials and to develop the forces of interaction between two such electric double layers. Before proceeding to discuss electric double layers, we shall briefly review the origin of the charge at interfaces and present an overview of the theoretical ideas.

5.1.1 Origin of Interfacial Charge

Most substances acquire a surface electric charge when brought into contact with an aqueous medium (Everett, 1988; Probstein, 2003). The stability of colloidal dispersions is very sensitive to the addition of electrolytes and to the surface charge of the colloidal particles. Direct evidence for the existence of charge on particles comes from the phenomenon of particle movement under an applied electric field (electrophoresis) which will be dealt with at a later stage. Although, in the course of electrokinetic studies, we accept the presence of surface charges and may pay less attention to their origin, it is still important to recognize the origin of these charges. Surfaces may become electrically charged by a variety of mechanisms. Some of the important mechanisms (Hunter, 1981; Everett, 1988; Lyklema, 1995) are listed below.

1. **Ionization of Surface Groups**. If a surface contains acidic groups, their dissociation gives rise to a negatively charged surface as shown in Figure 5.1(a). Conversely, a basic surface takes on a positive charge, Figure 5.1(b). In both cases, the magnitude of the surface charge depends on the acidic or basic strengths of the surface groups and on the pH of the solution. The surface charge can be reduced to zero at the *point of zero charge*, PZC, by suppressing the surface ionization. This can be achieved by decreasing the pH for the case of a basic surface. Most metal oxides can have either a positive or a negative surface charge depending on the bulk pH.
2. **Differential Dissolution of Ions from Surfaces of Sparingly Soluble Crystals**. For example, when a silver iodide crystal (AgI) is placed in water, dissolution occurs until the product of ionic concentration equals the solubility product $[Ag^+][I^-] = 10^{-16}$ $(mol/L)^2$. If equal amounts of Ag^+ and I^- ions were to dissolve, then $[Ag^+] = [I^-] = 10^{-8}$ $(mol/L)^2$ and the surface would

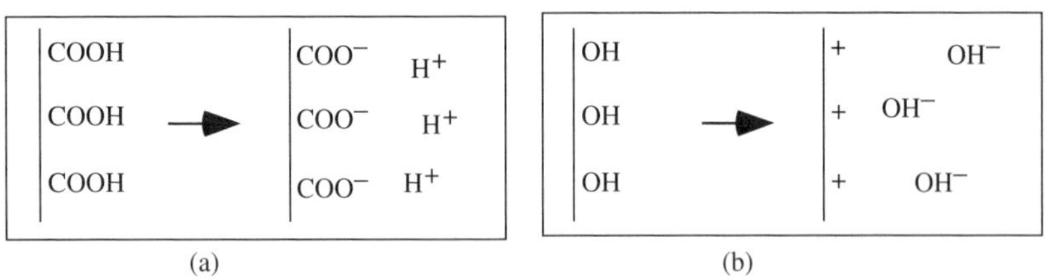

Figure 5.1. Acquisition of surface charge by ionization of (a) acidic groups and (b) basic groups at a surface.

be uncharged. However, silver ions dissolve preferentially, leaving a negatively charged surface. If Ag^+ ions now are added in the form of, say, a silver nitrate ($AgNO_3$) solution, the preferential solution of silver ions is suppressed and the charge falls to zero at a particular concentration. Further addition of $AgNO_3$ leads to a positively charged surface since it is now iodide ions that are preferentially dissolved.

3. **Isomorphic Substitution**. Clays may exchange an adsorbed intercalated or structural ion with one of lower valency, thus producing a negatively charged surface. For example, Al^{3+} may replace Si^{4+} in the surface of the clay, producing a negative surface charge. In this case, a point of zero surface charge can be reached by reducing the pH. Here, the added H^+ ions combine with the negative charges on the surface to form OH groups.
4. **Charged Crystal Surfaces**. It may happen that when a crystal is broken, surfaces with different properties are exposed. Thus, in some clays (*e.g.*, kaolinite), when a platelet is broken, the exposed edges contain $Al(OH)_3$ groups which take up H^+ ions to give a positively charged edge. This edge surface charge (positive) may coexist with negatively charged basal surfaces, leading to special properties. In this case, there will be no single PZC, but each type of surface will have its own characteristic PZC. For kaolinite, the flat surface is negatively charged and the edges are positively charged at low pH. At high pH, the positive charge on the edges decreases.
5. **Specific Ion Adsorption**. Surfactant ions may be specifically adsorbed on surfaces. Cationic surfactants can adsorb to negatively charged surfaces to yield net positive charges on the surface, while anionic surfactants can mask the positive charge of a surface by adsorbing onto them. The modification of surface charge by the adsorption of surfactants is shown in Figure 5.2. Surfactants play a major role in modifying surface charges and consequently affect the behavior of colloidal particles in terms of their stability. For instance, surfactants already present in crude oils play an important role in oil recovery. In all cases, the pH of the electrolyte solution into which a surface is immersed affects the surface electric potential.

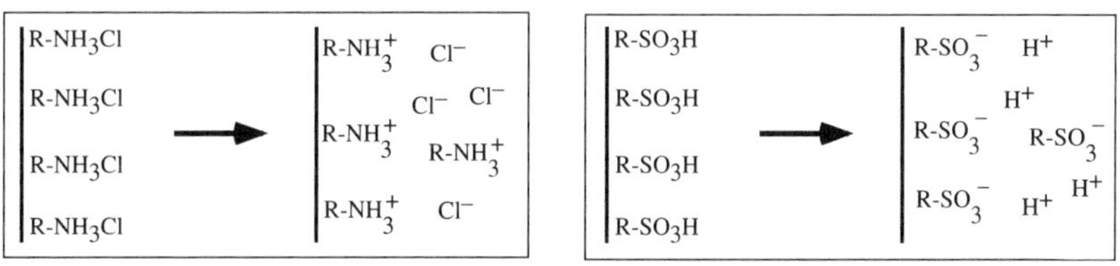

(a) (b)

Figure 5.2. Acquisition of surface charge by adsorption of (a) cationic surfactants and (b) anionic surfactants at a surface.

5.1.2 Electrical Potential Distribution Near an Interface

When electrolytes are present in water under no-flow conditions, for a sufficiently large representative volume, (*i.e.*, a volume that is several orders of magnitude larger than the volume of the ions and the water molecules), and at locations far away from the container's walls, electroneutrality condition is obeyed in an average sense. In other words, the sum of all the charges within the representative volume due to the ions is zero. The presence of electroneutrality is quite intuitive as there is no preferential spatial distribution of the ions in the bulk solution. The Brownian motion is sufficient to create a homogeneity in the spatial distribution of the ions. However, when a surface, say a glass slide is immersed in an electrolyte solution, the glass surface acquires a surface charge. Although the origin of this surface charge has been discussed earlier, in the following analysis, the reason as to why the glass surface attains a surface charge is not of immediate interest to us. For the present, we will assume that the glass surface is charged. Clearly, the charged surface will influence the distribution of the nearby ions in the electrolyte solution. Ions of opposite charge to that of the surface, called counterions, will be attracted toward the surface, while ions of like charge, called coions, will be repelled from the surface.[1] As a consequence of the attraction and repulsion between the ions in the electrolyte solution and the charged surface, there will be a non-uniform ionic distribution normal to the surface. Intuitively, one can state that there will be a higher concentration of counterions near the surface and that at a distance sufficiently far away from the charged surface, ionic electroneutrality will be reestablished. However, the manner by which the counterions and coions distribute themselves near and far away from the charged surface and the extent of the influence of the charged surface on their spatial distribution need some elaboration.

From the above discussion, the presence of a charged surface in an electrolyte solution will influence the ion distribution close to it. Such a redistribution of the free ions in solution gives rise to what we call the electric double layer. In simple terms, its name came about because of the separation of charge between the surface and the electrolyte solution. One layer is the charge on the surface, and the other, a "layer" of ions in the vicinity of the surface.

The concept of the electric double layer was introduced by Helmholtz, who envisaged an arrangement of charges in two parallel planes as shown in Figure 5.3(a), forming, in effect, a "molecular condenser". However, thermal motion causes the counterions to be spread out in space, forming a *diffuse* double layer, as shown in Figure 5.3(b). The theory for such a diffuse double layer was developed independently by Gouy and Chapman in the early 1900s (Hunter, 1981). In the *Gouy–Chapman diffuse double layer model*, the charged surface, composed of one layer of charges, has a surface potential ψ_s. The compensating ions in solution are regarded as point charges immersed in a continuous dielectric medium. The repulsion/attraction coupled with the random thermal Brownian motion of the ions within the dielectric medium gives rise to a diffuse electrical layer. Within this diffuse layer, there is no charge neutrality.

[1]The term counterions is used for the ions which have the opposite sign to the surface charge. The term coions is used for the ions which have the same sign as the surface charge.

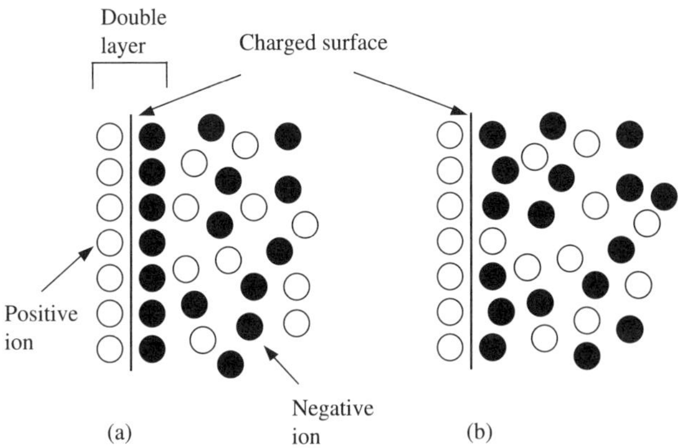

Figure 5.3. The electric double layer. (a) according to the Helmholtz model, (b) the diffuse double layer resulting from thermal motion.

The equilibrium concentration distribution of ions in the diffuse layer is established due to the forces attributed to electrostatic attraction/repulsion between the charged surface and the ions, and diffusion of ions due to concentration gradients.

The Gouy–Chapman model provides good quantitative predictions when the surface potential is low (~0.025 V) and the electrolyte concentration is not too high. A major defect of the model is that it neglects the finite size of the ions (it assumes that the ions are point charges that can approach the surface without limit) and the non-ideality of the solution. Moreover, the dielectric permittivity of the medium is taken to be constant all the way to the surface. A modification to the Gouy–Chapman model is offered by Stern (1924) and it will be elaborated upon at a later stage.

5.1.3 The Boltzmann Distribution

We have mentioned in the previous section, that due to the presence of a charged surface, there will be a spatial distribution of the ions normal to the surface. Of significance, there is equilibrium of the ionic distribution within the diffuse electric double layer. Such an equilibrium would allow one to use Boltzmann distribution to relate the ionic concentration to the electric potential. This type of a relationship is needed in order to analyze the ionic spatial distributions in electric double layers.

The Boltzmann distribution is a fundamental relation in statistical thermodynamics that describes the probability of the occurrence of microscopic states as a function of the energy of these states. Here, we shall introduce Boltzmann's distribution using simple concepts. However, at a later stage, a more formal derivation will be given. Whenever thermodynamic equilibrium exists, the probability that the system energy is confined within the range W and $W + dW$ is proportional to dW, and can be represented as $\mathcal{P}(W)dW$. The function $\mathcal{P}(W)$, which is referred to as the probability density, is given by

$$\mathcal{P} \propto \exp\left(-\frac{W}{k_B T}\right) \tag{5.1}$$

as shown by Boltzmann. In Eq. (5.1), T is the absolute temperature (K), k_B is the Boltzmann constant (J/K) given by R/N_A, where R is the universal gas constant and N_A is Avogadro's number. The above expression follows from statistical considerations (see, for example, Kittel and Kroemer 1980; McQuarrie, 1976).

In the present context, W represents the "energy" corresponding to a particular location of an ion (defined relative to a suitable reference state). The appropriate choice here is the work W required to bring one ion of valency z_i (*i.e.*, a charge $z_i e$) from infinity, where $\psi = 0$, to a given location x, having a potential ψ, given by $W = z_i e\psi$. Therefore, the probability density of finding an ion at location x can be written as[2]

$$\mathcal{P} \propto \exp\left(-\frac{z_i e\psi}{k_B T}\right) \tag{5.2}$$

Similarly, the probability density of finding the ion at the neutral state ($\psi = 0$) is

$$\mathcal{P}_0 \propto \exp(0) \tag{5.3}$$

For convenience, we take the neutral state to be at $\psi = 0$. The ratio of $\mathcal{P}$ to $\mathcal{P}_0$ is taken as being equal to the ratio of the concentrations of the species i at the respective states. Combining Eqs. (5.2) and (5.3) leads to

$$n_i = n_{i\infty} \exp\left(-\frac{z_i e\psi}{k_B T}\right) \tag{5.4}$$

where $n_{i\infty}$ is the ionic number concentration at the neutral state where $\psi = 0$ and n_i is the ionic number concentration of the ith ionic species at the state where the electric potential is ψ. The ionic valency z_i can be either positive or negative depending on whether the ion is a cation or an anion, respectively. As an example, for the case of $CaCl_2$ salt, z for the calcium ion is $+2$ and it is -1 for the chloride ion.

The Boltzmann distribution defined in Eq. (5.4) is of fundamental importance. It relates the ionic number concentration, n_i, of the ith species at a given location to the electric potential at that location. The Boltzmann distribution will be employed in evaluating the spatial variation of the electric potential and the ionic number concentration in the diffuse electric double layer.

EXAMPLE 5.1

The Barometric Equation. Use the concept of the Boltzmann distribution to develop an equation for the density of a gas in the atmosphere as a function of altitude.

---

[2]A more exact probability density function is given as

$$\mathcal{P} = \exp\left(-\frac{W_i}{k_B T}\right)$$

where $W_i = ez_i\psi + \Delta W_{ri}$, with ΔW_{ri} being the energy associated with reorganization of the medium due to the introduction of the ion. We assume here that $\Delta W_{ri}/k_B T \ll 1$.

Solution Consider the concentration of, say, helium in the atmosphere as a function of distance (height) from the earth's surface. The potential energy is given by mgx, where m is the mass of a helium atom, g is the acceleration due to gravity, and x is the distance from the earth's surface.

The probability density, $\mathcal{P}(x)$, of finding a helium atom at an altitude between x and $x + dx$ is given by

$$\mathcal{P}(x) = \alpha \exp\left(\frac{-mgx}{k_B T}\right)$$

The constant of proportionality α is independent of x. At zero altitude, the probability density is

$$\mathcal{P}(0) = \alpha \exp\left(\frac{0}{k_B T}\right) = \alpha$$

The above probability densities are proportional to the concentration $n(x)$ of He (*i.e.*, the number of He atoms per unit volume) at the appropriate altitudes. Therefore, the ratio of these probability densities is equal to the ratio of the concentrations:

$$\frac{\mathcal{P}(x)}{\mathcal{P}(0)} = \frac{n(x)}{n(0)}, \tag{5.5}$$

i.e.,

$$n(x) = n(0) \exp\left(-\frac{mgx}{k_B T}\right). \tag{5.6}$$

Clearly, the density $n(x)$ can also be expressed in other units, *e.g.*, in $\mathrm{mol/m^3}$.

Equation (5.6) is related to the *barometric equation*, which describes the pressure variation with altitude at constant temperature (see Hiemenz and Rajagopalan, 1997). ■

5.2 POTENTIAL FOR PLANAR ELECTRIC DOUBLE LAYER

In Chapter 3, we showed that the electric potential in a dielectric carrying free charge is governed by the Poisson equation, Eq. (3.116). In an electrolyte solution, the continuous phase is water which is a dielectric medium. The free charges are the ions contained in the electrolyte solution. Consequently, Poisson's equation is also the appropriate equation to be employed to analyze the electric diffuse double layers where the dielectric permittivity of water is assumed constant.

5.2.1 Gouy–Chapman Analysis

In order to facilitate the analysis, the case of a flat surface will be considered. The objective here is to obtain analytical expressions for the distribution of potential and

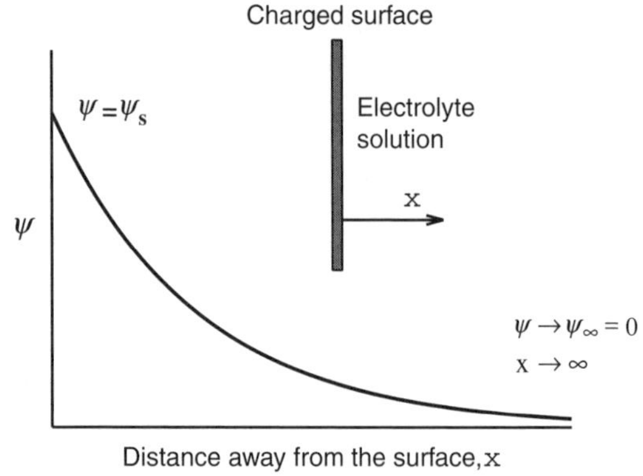

Figure 5.4. Potential distribution near a flat surface.

ion concentrations due to the presence of a charged surface in a dielectric medium having free charge. The Poisson equation is given by

$$\epsilon \nabla^2 \psi = -\rho_f \tag{5.7}$$

For a one-dimensional problem as shown in Figure 5.4, the above equation simplifies to

$$\epsilon \frac{d^2 \psi}{dx^2} = -\rho_f \tag{5.8}$$

where x is the distance normal to the charged surface. The space charge density of the mobile ("free") ions, ρ_f, can be written in terms of the number concentrations of the ions and the corresponding valencies as

$$\rho_f = \sum_{i=1}^{N} z_i e n_i \tag{5.9}$$

where n_i is the ionic number concentration of the ith species (say, in m^{-3}), z_i is the valence of the ith ionic species (with the appropriate sign), e is the magnitude of the fundamental (elementary) charge on an electron, 1.602×10^{-19} C, and N is the number of ionic species in the electrolyte solution. Equations (5.8) and (5.9) may now be combined to give

$$\epsilon \frac{d^2 \psi}{dx^2} = -\sum_{i=1}^{N} z_i e n_i \tag{5.10}$$

We can now write the right-hand side of the above equation in terms of ψ by relating the spatial distribution of the ions to ψ using the Boltzmann distribution. Substituting for n_i in the above equation through the use of the Boltzmann distribution given by

Eq. (5.4) leads to the well-known Poisson–Boltzmann equation

$$\epsilon \frac{d^2\psi}{dx^2} = -\sum_{i=1}^{N} z_i e n_{i\infty} \exp\left(-\frac{z_i e\psi}{k_B T}\right) \tag{5.11}$$

The Poisson–Boltzmann equation defines the electric potential distribution in the diffuse ionic layer adjacent to a charged surface subject to appropriate boundary conditions. In order to facilitate the solution of Eq. (5.11), the special case of a single salt dissociating into cationic and anionic species (*i.e.*, $N = 2$) will be considered with the added simplification of symmetric ($z : z$) electrolyte solution (*e.g.*, NaCl, $CuSO_4$, or AgI). In symmetric electrolytes, both the cations and anions have the same valencies.

In the case of planar electric double layers, one can obtain an analytical solution for ψ for symmetric electrolytes without any further approximations, and such a solution is known as the *Gouy–Chapman theory*. In particular, the Gouy–Chapman analysis does not require the linearization of the Boltzmann approximation and is, therefore, a nonlinear theory. We shall consider the linearization in the following subsection.

For a symmetric electrolyte, one can write

$$z_+ = -z_- = z \tag{5.12}$$

where z is the valence of the cation. Equation (5.11) can then be written as

$$\epsilon \frac{d^2\psi}{dx^2} = -zen_\infty \left[\exp\left(-\frac{ze\psi}{k_B T}\right) - \exp\left(\frac{ze\psi}{k_B T}\right)\right] \tag{5.13}$$

or

$$\epsilon \frac{d^2\psi}{dx^2} = 2zen_\infty \sinh\left(\frac{ze\psi}{k_B T}\right) \tag{5.14}$$

where $n_{+\infty} = n_{-\infty} = n_\infty$, the ionic number concentration in the bulk solution where $\psi = 0$. The appropriate boundary conditions for Eq. (5.14) are

$$x = 0 \qquad \psi = \psi_s \tag{5.15-a}$$

$$x \to \infty \qquad \psi = 0 \tag{5.15-b}$$

where ψ_s is the surface potential at $x = 0$. The solution to Eq. (5.14) under the above conditions is

$$\Psi = 2\ln\left[\frac{1 + \exp(-\kappa x)\tanh(\Psi_s/4)}{1 - \exp(-\kappa x)\tanh(\Psi_s/4)}\right] \tag{5.16}$$

where Ψ is the dimensionless potential defined as

$$\Psi = \frac{ze\psi}{k_B T} \tag{5.17}$$

from which, one has

$$\Psi_s = \frac{ze\psi_s}{k_B T} \tag{5.18}$$

In Eq. (5.16), κ^{-1} is the Debye length, which is defined as

$$\kappa^{-1} = \left(\frac{\epsilon k_B T}{2e^2 z^2 n_\infty} \right)^{1/2} \tag{5.19}$$

The Debye length, κ^{-1}, is a measure of the electric double layer thickness, and is a property of the electrolyte solution. It should be noted that this parameter contains information about the dielectric permittivity of the solvent, ϵ, as well as the valence, z, and bulk concentration, n_∞, of the ions. However, no information regarding the properties of the charged surface is present in the Debye length. Although it is normally referred to as the thickness of the electric double layer, the actual thickness of a double layer extends well beyond κ^{-1}. Typically, the Debye length represents a characteristic distance from the charged surface to a point where the electric potential decays to approximately 33% of the surface potential.

For $\Psi_s \ll 1$, *i.e.*, low surface potentials, Eq. (5.16) can be approximated to

$$\Psi = 2 \ln \left[\frac{1 + 0.25\Psi_s \exp(-\kappa x)}{1 - 0.25\Psi_s \exp(-\kappa x)} \right] \tag{5.20}$$

Figure 5.5 shows a comparison between the exact solution given by Eq. (5.16) and the approximate solution given by Eq. (5.20). For $\Psi_s = 0.5$, both the expressions yield nearly identical distributions of Ψ but, as would be expected, the two results differ for the higher surface potential of $\Psi_s = 2$. As it will be shown later, Eq. (5.20) gives a poorer approximation for the potential distribution than the Debye–Hückel approximation, Eq. (5.24), for $\Psi_s > 1$.

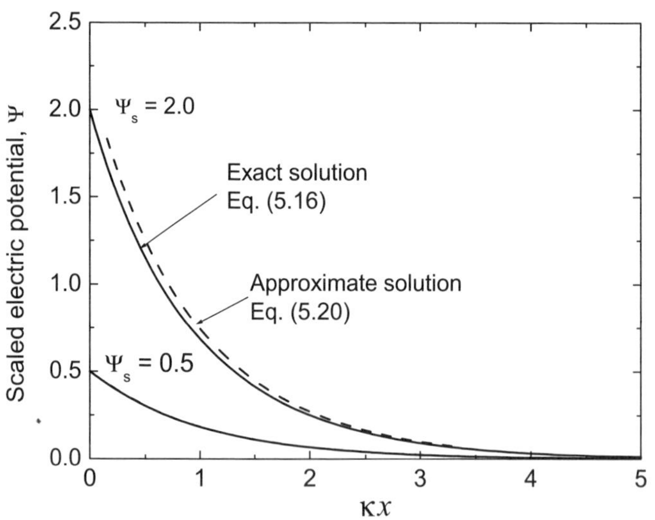

Figure 5.5. Potential distribution near a flat surface. Comparison of exact and approximate solutions.

EXAMPLE 5.2

The Debye Thickness for Symmetric Electrolytes. Calculate the Debye length for a number of values of the ionic strength for a symmetric electrolyte.

Solution The condition of electroneutrality is usually assumed to occur far away from the surface and is given as

$$\sum z_i n_i = 0$$

For a binary electrolyte, one has

$$z_- n_- + z_+ n_+ = 0$$

which, for a symmetric ($z : z$) electrolyte, leads to

$$n_- = n_+ = n_\infty$$

The Debye length κ^{-1} for a symmetric electrolyte then follows from

$$\kappa^{-1} = \left(\frac{\epsilon k_B T}{2 e^2 z^2 n_\infty} \right)^{\frac{1}{2}}$$

where $e = 1.602 \times 10^{-19}$ C and for water at $T = 298$ K, $\epsilon = 6.95 \times 10^{-10}$ $C^2N^{-1}m^{-2}$ (the value of $\epsilon_r = 78.5$).

Now, n_∞, the ionic number concentration, is given by

$$n_\infty = \left(M \frac{\text{mol}}{\text{L}} \right) \times \left(1000 \frac{\text{L}}{\text{m}^3} \right) \left(N_A \frac{1}{\text{mol}} \right)$$

or $n_\infty = 1000 N_A M$, with the Avogadro number $N_A = 6.022 \times 10^{23}$ mol^{-1} and M being the molar concentration (mol/L) of the electrolyte. The expression for the Debye length then becomes

$$\kappa^{-1} = \frac{3.04}{z\sqrt{M}} \times 10^{-10} \text{ m}$$

The ionic strength, I, of an electrolyte solution is defined as

$$I = \frac{1}{2} \sum_i z_i^2 M_i$$

where M_i is the molar concentration of the ith ionic species arising from the dissociation of the electrolyte and z_i is its valence. For a 0.1 M (1 : 1) electrolyte (such as NaCl), $M_+ = M_- = 0.1$ M. Hence, the ionic strength is given by

$$I = \frac{1}{2} \left[z_+^2 M_+ + z_-^2 M_- \right] = \frac{1}{2} \left[(1)^2 \times 0.1 + (1)^2 \times 0.1 \right] = 0.1 \text{ M}$$

Thus, for a symmetric (1 : 1) electrolyte, the ionic strength is equal to its molarity. Although the term ionic strength is frequently used in electrochemistry to denote electrolyte concentration, we will use molar and number concentrations in this book.

The values of the Debye length, κ^{-1}, for different electrolyte concentrations for the case of $z = 1$ are shown in the following table. In this case, the ionic strength is equal to the molar concentration of the electrolyte. It is clear from the tabulated results that κ^{-1} decreases as the electrolyte concentration increases. At high molarity, the electric double-layer thickness becomes very small. In a non-electrolyte system, however, the double-layer thickness can be thought of as extending to infinity (*i.e.*, a large distance from the surface).

Ionic Concentration, M	Debye Length κ^{-1}, nm
10^{-6}	304.0
10^{-4}	30.4
10^{-2}	3.04

■

5.2.2 Debye–Hückel Approximation

When the surface potential is small, say, $\psi_s \ll 0.025$ V (Hiemenz and Rajagopalan, 1997), the term $ze\psi/k_BT$ is smaller than unity and one can approximate the hyperbolic sine function as follows:

$$\sinh\left(\frac{ze\psi}{k_BT}\right) \approx \frac{ze\psi}{k_BT} \quad \text{for} \quad \frac{ze\psi}{k_BT} \ll 1 \tag{5.21}$$

Making use of the above approximation and the definition of κ given by Eq. (5.19), the Poisson–Boltzmann equation (5.14) becomes

$$\frac{d^2\psi}{dx^2} = \frac{2e^2z^2n_\infty}{\epsilon k_BT}\psi = \kappa^2\psi \tag{5.22}$$

Equation (5.22) is the linearized version of the Poisson–Boltzmann equation and is one of the most commonly used equations in the electric double layer theory (even when linearization may not be justifiable!). The linearization of the Poisson–Boltzmann equation in the above manner is generally referred to as the Debye–Hückel approximation. Its solution, with the boundary conditions $\psi = \psi_s$ at $x = 0$ and $\psi = 0$ at $x \to \infty$, follows readily and is given by

$$\psi = \psi_s \exp(-\kappa x) \tag{5.23}$$

or, in dimensionless form by

$$\Psi = \Psi_s \exp(-\kappa x) \tag{5.24}$$

The dimensionless surface potential, Ψ_s, is defined in Eq. (5.18). Equation (5.24) indicates that the potential decays exponentially as one moves away from the charged surface. It may be noted here that Eq. (5.20) can be simplified to yield

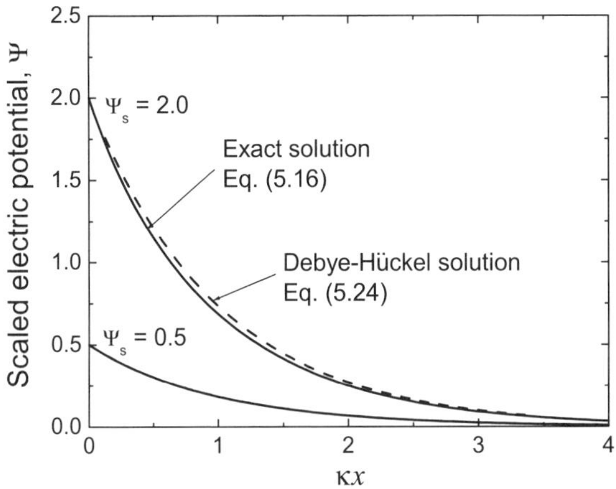

Figure 5.6. Potential distribution near a flat surface. Comparison between exact and Debye–Hückel solutions.

the Debye–Hückel potential distribution, Eq. (5.24). A comparison between the exact solution provided by Eq. (5.16) and the Debye–Hückel approximate solution, Eq. (5.24) is given in Figure 5.6 for two values of surface potentials, namely, $\Psi_s = 0.5$ and $\Psi_s = 2$. For the case of the low surface potential, $\Psi_s = 0.5$, both the exact and Debye–Hückel solutions give essentially the same dimensionless electric potential variation with κx. For the higher surface potential of $\Psi_s = 2$, there is only a slight difference between the two profiles. The Debye–Hückel solution normally gives fairly close Ψ/Ψ_s variations for Ψ_s values as high as 3. Comparing Figures 5.5 and 5.6 it becomes evident that Eq. (5.20) is less accurate than the Debye–Hückel expression, Eq. (5.24) as $\kappa x \to 0$ for large surface potentials, $\Psi_s = 2$.

Tabulation of the Debye length (κ^{-1}) for different electrolyte concentrations and valence types is given in Table 5.1. The Debye double layer thickness, κ^{-1}, varies from 9.61×10^{-9} m for a molarity of 0.001 M for a 1 : 1 electrolyte to 0.32×10^{-9} m for a 3 : 3 electrolyte at a molarity of 0.1. Increasing the molarity of the electrolyte and its valency tends to decrease the value of the Debye length κ^{-1}.

The variation of Ψ/Ψ_s with distance is shown in Figure 5.7 for a 1 : 1 electrolyte at three concentrations using the Debye–Hückel solution. Increasing the electrolyte concentration leads to a fast decay in Ψ/Ψ_s and a small electric double layer thickness. The values of κ^{-1} are indicated by solid circles. The variation of Ψ/Ψ_s with distance at 0.001 M concentration for symmetric electrolytes of three different valencies is shown in Figure 5.8 where the Debye–Hückel approximation is used. The decay in Ψ/Ψ_s is much sharper for the 3 : 3 electrolyte solution when compared to the lower valency electrolytes. As well, the electric double layer thickness (Debye length) is smaller for the 3 : 3 electrolyte. One can conclude from Figure 5.8 that for a given electrolyte concentration, the higher valency electrolytes are more effective in altering the electric potential inside the electric double layer. Consequently, higher electrolyte concentrations and electrolytes having higher valencies tend to "screen" the electric potential due to a charged surface to a larger extent.

TABLE 5.1. Values of κ and κ^{-1} for Several Different Electrolyte Concentrations and Valences with Numerical Formulas for these Quantities Given for Aqueous Solutions at 25°C.

		Symmetrical Electrolyte	
Molarity	$z_+ : z_-$	$\kappa\ (m^{-1}) = 3.29 \times 10^9 \times \lvert z\rvert M^{1/2}$	$\kappa^{-1}\ (m) = 3.04 \times 10^{-10} \times \lvert z\rvert^{-1} M^{-1/2}$
0.001	1 : 1	1.04×10^8	9.61×10^{-9}
	2 : 2	2.08×10^8	4.81×10^{-9}
	3 : 3	3.12×10^8	3.20×10^{-9}
0.01	1 : 1	3.29×10^8	3.04×10^{-9}
	2 : 2	6.58×10^8	1.52×10^{-9}
	3 : 3	9.87×10^8	1.01×10^{-9}
0.1	1 : 1	1.04×10^9	9.61×10^{-10}
	2 : 2	2.08×10^9	4.81×10^{-10}
	3 : 3	3.12×10^9	3.20×10^{-10}
		Asymmetrical Electrolyte	
Molarity	$z_+ : z_-$	$\kappa\ (m^{-1}) = 2.32 \times 10^9 \times \left(\sum_i z_i^2 M_i\right)^{1/2}$	$\kappa^{-1}\ (m) = 4.30 \times 10^{-10} \times \left(\sum_i z_i^2 M_i\right)^{-1/2}$
0.001	1 : 2, 2 : 1	1.80×10^8	5.56×10^{-9}
	1 : 3, 3 : 1	2.54×10^8	3.93×10^{-9}
	2 : 3, 3 : 2	4.02×10^8	2.49×10^{-9}
0.01	1 : 2, 2 : 1	5.68×10^8	1.76×10^{-9}
	1 : 3, 3 : 1	8.04×10^8	1.24×10^{-9}
	2 : 3, 3 : 2	1.27×10^8	7.87×10^{-10}
0.1	1 : 2, 2 : 1	1.80×10^9	5.56×10^{-10}
	1 : 3, 3 : 1	2.54×10^9	3.93×10^{-10}
	2 : 3, 3 : 2	4.02×10^9	2.49×10^{-10}

EXAMPLE 5.3

Poisson–Boltzmann Equation for Flat Surface. Derive the Poisson–Boltzmann equation for a flat charged surface immersed in an arbitrary electrolyte solution. Use the Debye–Hückel approximation.

Solution The Poisson equation is given by Eq. (5.8) and it is written as

$$\epsilon \frac{d^2\psi}{dx^2} = -\rho_f \tag{5.25}$$

where x is the Cartesian coordinate normal to the surface.

Making use of the definition of ionic free charge density, ρ_f, one can write

$$\rho_f = \sum_{i=1}^{N} n_i z_i e \tag{5.26}$$

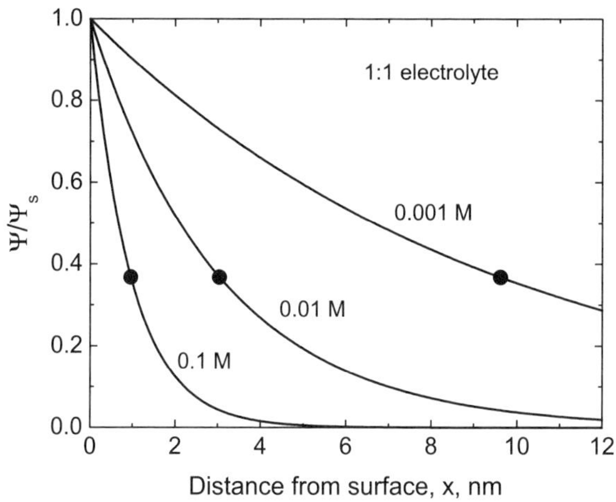

Figure 5.7. Normalized double layer potential *versus* distance from a surface according to the Debye–Hückel approximation. Curves drawn for a 1 : 1 electrolyte at three concentrations.The solid circles represent the values of κ^{-1} corresponding to each concentration.

where N is the total number of ionic species in the system. Substituting Eq. (5.26) in Eq. (5.25), one obtains

$$\epsilon \frac{d^2\psi}{dx^2} = -\sum_{i=1}^{N} n_i z_i e \tag{5.27}$$

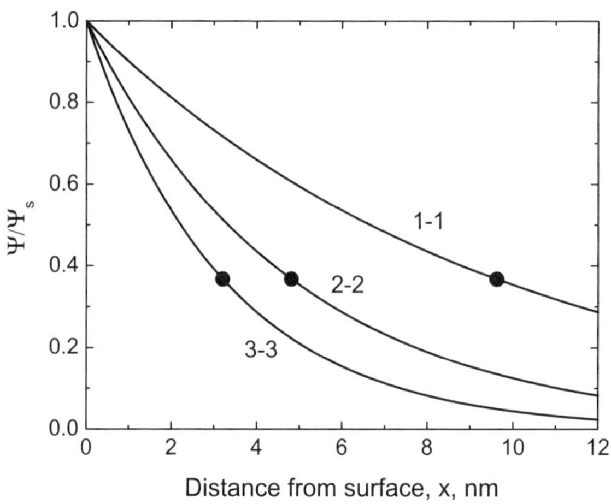

Figure 5.8. Normalized double layer potential *vs.* distance from a surface according to the Debye–Hückel approximation. Curves drawn for 0.001 M symmetrical electrolytes of three different valence types. The solid circles represent the values of κ^{-1} corresponding to each valence type.

The ionic concentration is provided by the Boltzmann equation as

$$n_i = n_{i\infty} \exp\left(\frac{-z_i e\psi}{k_B T}\right) \tag{5.28}$$

Combining Eqs. (5.27) and (5.28) leads to

$$\epsilon \frac{d^2\psi}{dx^2} = -e \sum_{i=1}^{N} n_{i\infty} z_i \exp\left(\frac{-z_i e\psi}{k_B T}\right) \tag{5.29}$$

Invoking the Debye–Hückel approximation, where

$$\left|\frac{z_i e\psi}{k_B T}\right| \ll 1 \tag{5.30}$$

and recognizing that

$$\exp(-y) \approx 1 - y \quad \text{for} \quad y \ll 1 \tag{5.31}$$

Eq. (5.29) becomes

$$\epsilon \frac{d^2\psi}{dx^2} = -e \sum_{i=1}^{N} \left[n_{i\infty} z_i \left(1 - \frac{z_i e\psi}{k_B T}\right)\right] \tag{5.32}$$

Upon rearranging, Eq. (5.32) becomes

$$\epsilon \frac{d^2\psi}{dx^2} = -e \sum_{i=1}^{N} \left[n_{i\infty} z_i - \frac{z_i^2 n_{i\infty} e\psi}{k_B T}\right] \tag{5.33}$$

From the condition of electroneutrality, we have

$$\sum_{i=1}^{N} n_{i\infty} z_i = 0 \tag{5.34}$$

which simplifies Eq. (5.33) to

$$\epsilon \frac{d^2\psi}{dx^2} = \psi \sum_{i=1}^{N} \frac{z_i^2 e^2 n_{i\infty}}{k_B T} \tag{5.35}$$

Letting

$$\kappa^2 = \sum_{i=1}^{N} \frac{z_i^2 e^2 n_{i\infty}}{\epsilon k_B T}$$

or

$$\kappa^2 = \frac{e^2}{\epsilon k_B T} \sum_{i=1}^{N} z_i^2 n_{i\infty} \tag{5.36}$$

The Poisson–Boltzmann equation for a planar surface at low potentials becomes

$$\frac{d^2\psi}{dx^2} = \kappa^2\psi \tag{5.37}$$

Equation (5.37), which describes the electrical potential in the electric double layer, is valid for an *arbitrary* electrolyte and it is not limited to the special case of a symmetric electrolyte as long as the more general definition of κ, given by Eq. (5.36), is employed. For a symmetric electrolyte ($z : z$), Eq. (5.36) reverts to Eq. (5.19). ■

EXAMPLE 5.4

Debye Length. In Table 5.1, the Debye length was given for an asymmetrical electrolyte in water at 298 K as

$$\kappa^{-1} = 4.30 \times 10^{-10}\left(\sum_i z_i^2 M_i\right)^{-1/2}$$

where z_i is the valence of the ith ionic species and M_i is its molarity. Derive the above expression.

Solution For the general case of asymmetric electrolytes, the Debye length (thickness) is given by Eq. (5.36) of **Example 5.3**, where

$$\kappa^{-1} = \left(\frac{e^2}{\epsilon k_B T}\sum_{i=1}^{N} z_i^2 n_{i\infty}\right)^{-1/2} \tag{5.38}$$

Now

$$n_{i\infty} = M_i \frac{\text{mol}}{\text{L}} \cdot \frac{1000\ \text{L}}{\text{m}^3} \cdot \frac{1}{\text{mol}} N_A$$

or

$$n_{i\infty} = 1000 M_i N_A \tag{5.39}$$

Substituting for $n_{i\infty}$ from Eq. (5.39) into Eq. (5.38) leads to

$$\kappa^{-1} = \left(\frac{1000 N_A e^2}{\epsilon k_B T}\sum_{i=1}^{N} z_i^2 M_i\right)^{-1/2} \tag{5.40}$$

Inserting known parameter values, Eq. (5.40) gives

$$\kappa^{-1} = \left(\frac{1000 \times 6.022 \times 10^{23} \times (1.602 \times 10^{-19})^2}{6.95 \times 10^{-10} \times 1.381 \times 10^{-23} \times 298}\sum_{i=1}^{N} z_i^2 M_i\right)^{-1/2} \tag{5.41}$$

Equation (5.41) leads to

$$\kappa^{-1} = 4.30 \times 10^{-10} \left(\sum_{i=1}^{N} z_i^2 M_i \right)^{-1/2} \tag{5.42}$$

The expression of the Debye length as given above is useful when one deals with the general case of a mixture of electrolytes in a solution. ■

EXAMPLE 5.5

Debye Length for a Mixture. A 400 ml of 0.01 M sodium chloride solution is mixed with 600 ml of 0.001 M sodium sulfate. What is the Debye length for this electrolyte solution?

Solution The molarity of NaCl upon mixing becomes

$$0.01 \times \frac{400}{(400 + 600)} = 0.004\,\mathrm{M}$$

The molarity of Na_2SO_4 upon mixing becomes

$$0.001 \times \frac{600}{(400 + 600)} = 0.0006\,\mathrm{M}$$

Now, we can write for NaCl and Na_2SO_4

$$\begin{aligned}\sum z_i^2 M_i &= \left(1^2 \times 0.004 + 1^2 \times 0.004\right) + \left(1^2 \times 0.0006 \times 2 + 2^2 \times 0.0006\right) \\ &= 0.008 + 0.0036 = 0.0116\end{aligned}$$

Now, substituting this value in Eq. (5.42) yields

$$\kappa^{-1} = 4.30 \times 10^{-10} \times (0.0116)^{-1/2} = 3.99 \times 10^{-9}\ \mathrm{m}$$ ■

5.2.3 Surface Charge Density

We have made use of both the Boltzmann distribution and the Poisson equation to investigate the electric potential distribution inside the electric double layer for a planar surface. We now seek to relate the electric surface potential to the surface charge density.

Let us consider a planar surface being placed in an electrolyte solution. As the surface is assumed to be electrically neutral *prior* to its placement in the solution, the acquired charge must balance the ionic charge within the electric double layer. This ionic balance is required due to the overall electroneutrality of the electrolyte solution and the surface. In other words, ionic charge conservation must be respected.

It follows then that for a planar surface

$$q_s = -\int_0^\infty \rho_f dx \tag{5.43}$$

where x is the coordinate normal to the surface, q_s is the surface charge density (C/m^2), and ρ_f is the free charge density, *i.e.*, ionic charge density, C/m^3.

Recalling the Poisson equation for one dimension,

$$\epsilon \frac{d^2\psi}{dx^2} = -\rho_f \tag{5.44}$$

one can integrate Eq. (5.44) to give

$$\epsilon \int_0^\infty \frac{d}{dx}\left(\frac{d\psi}{dx}\right) dx = -\int_0^\infty \rho_f dx \tag{5.45}$$

Combining Eqs. (5.43) and (5.45) and evaluating the integral yields

$$q_s = \epsilon \left[\frac{d\psi}{dx}\right]_0^\infty \tag{5.46}$$

Expanding Eq. (5.46) gives

$$q_s = \epsilon \left[\left.\frac{d\psi}{dx}\right|_\infty - \left.\frac{d\psi}{dx}\right|_0\right] \tag{5.47}$$

At large distances away from the surface, $x \to \infty$, the electric potential is taken as zero and Eq. (5.47) becomes

$$q_s = -\epsilon \left.\frac{d\psi}{dx}\right|_0 \tag{5.48}$$

Equation (5.48) states that the surface charge density is proportional to the gradient of the electric potential at the charged surface. The proportionality constant is simply the medium permittivity. A more general form of Eq. (5.48) is given as

$$q_s = -\epsilon \nabla\psi \cdot \mathbf{n} \tag{5.49}$$

where $\mathbf{n}$ is the outward normal to a charged surface.

Recall that the electric field strength is equal in absolute magnitude to the electric potential gradient. Therefore, for a one-dimensional case, one can write at $x = 0$

$$E_x|_0 = -\left.\frac{d\psi}{dx}\right|_0 \tag{5.50}$$

Considering Eqs. (5.48) and (5.50) and letting $E_x|_0 \equiv E_s$, one can write

$$q_s = \epsilon E_s \tag{5.51-a}$$

$$E_s = \frac{q_s}{\epsilon} \tag{5.51-b}$$

Equations (5.51-a) or (5.51-b) relate the surface charge density to the electric field strength at the surface. Equation (5.51-a) is a statement of the Gauss law.

The evaluation of the surface charge density, q_s, as provided by Eq. (5.48) would depend on the choice of the electric potential expression to be used. Let us make use of Gouy–Chapman solution, Eq. (5.16), which is valid for a symmetric electrolyte and is not limited to small surface potentials. Differentiating ψ of Eq. (5.16) with respect to x and substituting for $d\psi/dx$ in Eq. (5.48) gives

$$q_s = 2[2\epsilon k_B T n_\infty]^{1/2} \sinh(\Psi_s/2) \tag{5.52}$$

The dimensionless surface potential is defined in Eq. (5.18). For the special case of small surface potential $\Psi_s \ll 1$, one can simplify Eq. (5.52) to

$$q_s = [2\epsilon k_B T n_\infty]^{1/2} \Psi_s \tag{5.53}$$

In terms of the dimensional surface potential, Eq. (5.52) gives for $\Psi_s \ll 1$

$$q_s = \epsilon \kappa \psi_s \tag{5.54}$$

where κ is given by Eq. (5.19).

Equation (5.52) relates the surface charge density to its potential for a planar surface. Rather interesting information can be extracted from this relationship. In real systems, it is the surface charge that ultimately controls the surface potential. Consequently, in terms of the electrolyte molarity, one can rewrite Eq. (5.52) as

$$\sinh(\Psi_s/2) = \frac{(q_s/e)}{(2/e)[2000\epsilon k_B T N_A M]^{1/2}} \tag{5.55}$$

The term (q_s/e) can be thought of as the ionic surface charge density, *i.e.*, ions/m^2. Equation (5.55) indicates that the surface electric potential increases with the ionic surface charge density (q_s/e) and decreases with increasing electrolyte molarity. Figure 5.9 shows the variation of the surface electric potential, Ψ_s, with $(q_s/e)^*$ for different electrolyte molarities, where q_s^* is the surface charge in C/nm^2. A vertical line at $(q_s/e)^* = 0.2$, which represents one charge per 5 nm^2 (a reasonably ionized surface) is drawn on Figure 5.9. If the charged surface maintains its surface ionization at all electrolyte concentrations, the intersection of the the vertical dashed line with the individual solid lines would then represent the surface potential at the different electrolyte concentrations.

In real systems, however, surface ionization is a function of electrolyte concentration and pH. Therefore, to properly predict the surface potential one would need additional information on surface ionization and surface charge density characteristics. We will discuss models employed to predict surface potentials later in this chapter. In summary, for the general case, the Gouy–Chapman model relates the surface charge density to the surface potential through Eq. (5.52). For small surface potentials, say

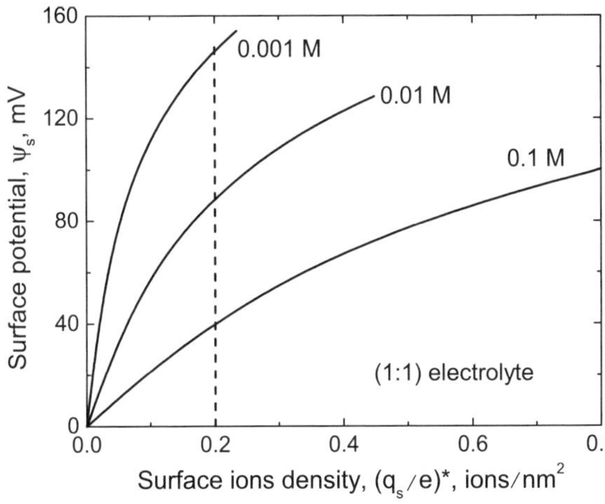

Figure 5.9. Effect of surface charge density on the surface potential of a planar surface.

$\psi_s < 0.025$ V, Eq. (5.54) provides the link between the surface charge density and the surface electric potential.

It is of interest to note that Eqs. (5.53) and (5.54) suggest the manner by which one can non-dimensionalize the surface charge density. Equation (5.53) suggests the use of $(2\epsilon k_B T n_\infty)^{1/2}$ and Eq. (5.54) suggests $\epsilon\kappa(k_B T/ze)$. Utilizing the equivalence of these two terms, we may write

$$\frac{n_\infty k_B T}{\kappa^2} = \frac{\epsilon}{2}\left(\frac{k_B T}{ze}\right)^2 \tag{5.56}$$

It is noteworthy that each term in the above expression has units of force (Newtons in SI units). These combinations of parameters can be used interchangeably to scale the interaction energy or force arising from the overlap of two electric double layers. We will revisit these terms in the later sections of this chapter.

5.2.4 Ionic Concentrations in Electric Double Layers

As we discussed earlier, the presence of a charged surface disturbs the electroneutrality of the electrolyte solution in the immediate vicinity of the charged surface. Counterions are attracted to the surface and coions are repelled from the surface. The spatial distribution of the ions normal to the surface can be provided *via* the use of the Boltzmann distribution and an expression for the electric potential. Let us derive expressions for the variation of the ion concentration ratio along the coordinate normal to the surface. The Boltzmann distribution, Eq. (5.4) gives

$$n_i = n_{i\infty} \exp\left(-\frac{z_i e \psi}{k_B T}\right) \tag{5.57}$$

Making use of the Debye–Hückel approximation, Eq. (5.24), the Boltzmann distribution, Eq. (5.57), for a symmetric ($z:z$) electrolyte provides

$$\frac{n_+}{n_\infty} = \exp[-\Psi_s \exp(-\kappa x)] \simeq 1 - \Psi_s \exp(-\kappa x) + \frac{\Psi_s^2}{2} \exp(-2\kappa x) + \mathcal{O}(\Psi_s^3) \tag{5.58-a}$$

$$\frac{n_-}{n_\infty} = \exp[\Psi_s \exp(-\kappa x)] \simeq 1 + \Psi_s \exp(-\kappa x) + \frac{\Psi_s^2}{2} \exp(-2\kappa x) + \mathcal{O}(\Psi_s^3) \tag{5.58-b}$$

where $\Psi = ze\psi/k_BT$.

Plots of Eqs. (5.58-a) and (5.58-b) are shown in Figure 5.10. The decay in the number concentration of the counterions is more steep when compared to that of the coions. In both cases, the maximum deviation from the bulk conditions occurs at the charged surface.

The difference between the counterion and coion number concentrations is referred to as the excess ionic number concentration. A plot of the variation of normalized excess number concentration, $(n_- - n_+)/n_\infty$, with the distance away from the charged surface is shown in Figure 5.11. As would be expected, the excess ion concentration has its maximum value at the surface and it decays to zero far away from the surface where the electroneutrality condition holds. One can think of the term $(n_- - n_+)/n_\infty$ as being a measure of the deviation from electroneutrality.

Noting that for a symmetric electrolyte the net excess charge will be proportional to the difference in the ionic number concentration, $(n_- - n_+)$, the normalized excess ionic concentration, $(n_- - n_+)/n_\infty$, can also be called the normalized excess charge. The variation of the normalized excess charge is shown in Figure 5.12 for different

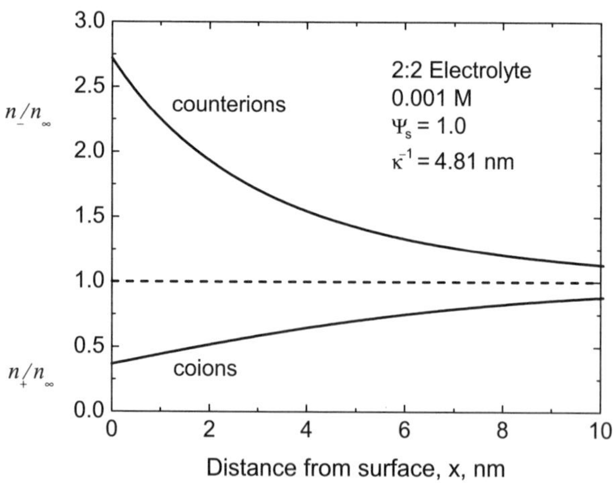

Figure 5.10. Coion and counterion distribution near a charged planar surface.

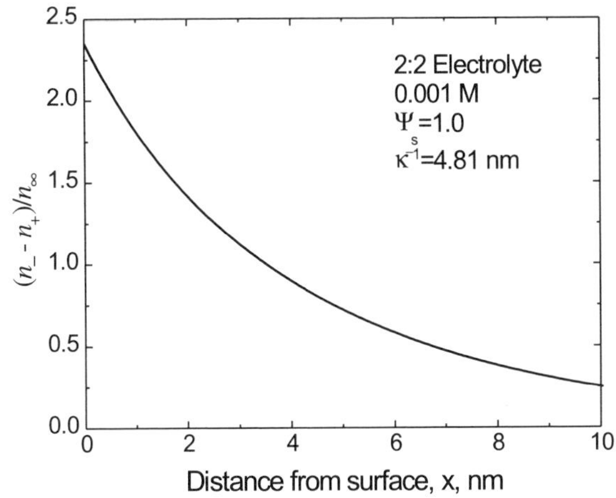

Figure 5.11. Distribution of excess ion concentration normalized with respect to the bulk ion concentration near a planar surface.

values of surface electrical potential. It is clear that as the surface potential, Ψ_s, increases, the normalized excess charge increases as well. Since both n_- and n_+ are positive quantities, such a variation in $(n_- - n_+)/n_\infty$ would imply that n_- becomes much larger than n_+ with increasing surface potential, Ψ_s. In this example, the surface potential was taken as positive and, hence, the ionic number concentration n_- is that of the counterions.

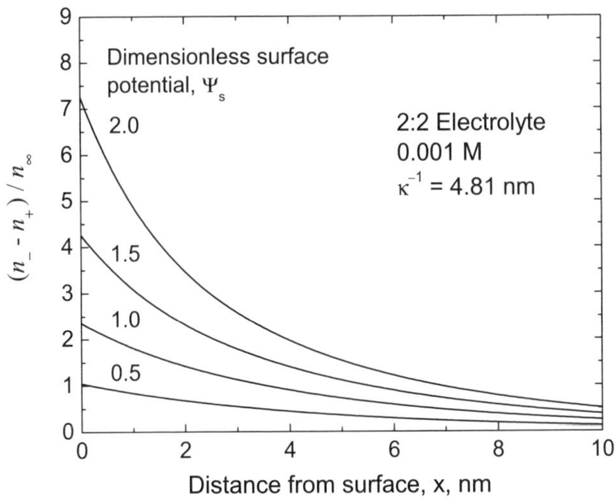

Figure 5.12. Normalized excess ion concentration near a planar surface: Effect of surface potential.

The plot of Figure 5.12 indicates that the counterions become more dominant than the coions near the surface as the surface potential becomes high. Consequently, for high surface potentials, one may only need to consider the counterion concentrations near the charged surfaces in solving the Poisson equation. A further generalization can be made in that counterions are the dominant ions in determining the surface charge behavior. Consideration of counterions alone will be considered in the next section.

5.2.5 High Surface Potentials and Counterion Analysis

We have discussed two solutions for the electric potential distribution inside the diffuse double layer. First, we presented the Gouy–Chapman analysis where no assumption for low surface potential was made but its validity was limited to symmetrical electrolytes, *i.e.*, $(z : z)$ electrolytes. Secondly, we used Debye–Hückel approximation to obtain a solution that was valid for low surface potentials and arbitrary electrolytes, *i.e.*, asymmetrical electrolytes. No explicit simple solution was presented for high surface potentials and arbitrary electrolytes.[3] For the latter case, an exact solution for ψ becomes very cumbersome. However, we can make use of the concept advanced earlier, namely, that for a high surface potential it is only the counterions that contribute in establishing the potential in the diffuse double layer, and obtain an elegant analytic solution of the Poisson–Boltzmann equation.

Let us consider the primitive form of the Poisson–Boltzmann equation, Eq. (5.11):

$$\epsilon \frac{d^2\psi}{dx^2} = -\sum_{i=1}^{N} z_i e n_{i\infty} \exp\left(-\frac{z_i e \psi}{k_B T}\right) \tag{5.59}$$

As we are only interested in the counterions, we can disregard the summation in Eq. (5.59) and write

$$\epsilon \frac{d^2\psi}{dx^2} = -zen_\infty \exp\left(-\frac{ze\psi}{k_B T}\right) \tag{5.60}$$

where z and n_∞ refer to the counterions. The term

$$-\frac{ze\psi}{k_B T} > 0 \tag{5.61}$$

[3]The electric potential can be given implicitly by

$$\kappa x = -\text{sign}(\Psi_s) \int_{\Psi_s}^{\Psi} \frac{du}{\sqrt{2\sum_k \nu_k[\exp(-uz_k) - 1]}}$$

where

$$\nu_k = \frac{n_{k\infty}}{\sum_i n_{i\infty} z_i^2}$$

Here u is an arbitrary function of integration and

$$\kappa^2 = \frac{e^2}{\epsilon k_B T} \sum_i n_{i\infty} z_i^2$$

as z and ψ have opposite signs. Consequently, for high surface potentials, one can write

$$\exp\left(-\frac{ze\psi}{k_BT}\right) \gg \exp\left(\frac{ze\psi}{k_BT}\right) \tag{5.62}$$

Utilizing the inequality presented by Eq. (5.62), one can write for Eq. (5.60)

$$\epsilon\frac{d^2\psi}{dx^2} \simeq 2zen_\infty \left[\frac{\exp\left(\frac{ze\psi}{k_BT}\right) - \exp\left(-\frac{ze\psi}{k_BT}\right)}{2}\right] \tag{5.63}$$

which can be simplified to

$$\epsilon\frac{d^2\psi}{dx^2} = 2zen_\infty \sinh\left(\frac{ze\psi}{k_BT}\right) \tag{5.64}$$

The above equation is similar to Eq. (5.14) except for the very important fact that z and n_∞ refer to the counterions. The solution of Eq. (5.64), subject to the usual boundary conditions $\psi(0) = \psi_s$ and $\psi(\infty) = 0$, is simply given by

$$\Psi = 2\ln\left[\frac{1 + \exp(-\kappa^* x)\tanh(\Psi_s/4)}{1 - \exp(-\kappa^* x)\tanh(\Psi_s/4)}\right] \tag{5.65}$$

where $\Psi_s = ze\psi_s/(k_BT)$ and

$$\kappa^* = \left(\frac{2e^2z^2n_\infty}{\epsilon k_BT}\right)^{1/2} \tag{5.66}$$

It should be noted that z is the valence of the counterions and n_∞ is the number concentration of the counterions at a location far away from the surface, *i.e.*, bulk solution. Although Eq. (5.65) is similar to that obtained for a $(z : z)$ electrolyte as given by Eq. (5.16), it is different by the manner the Debye length is calculated.

EXAMPLE 5.6

Counterion Analysis. Consider the case of a 0.01 M Na_2SO_4 solution at 298 K. Compare the electric potential in the diffuse double layer using the counterion solution, Eq. (5.65), with the Debye–Hückel solution for a (1 : 2) electrolyte.

Solution Let us first evaluate κ^* given by Eq. (5.66). We can write

$$n_\infty = 1000N_AM$$

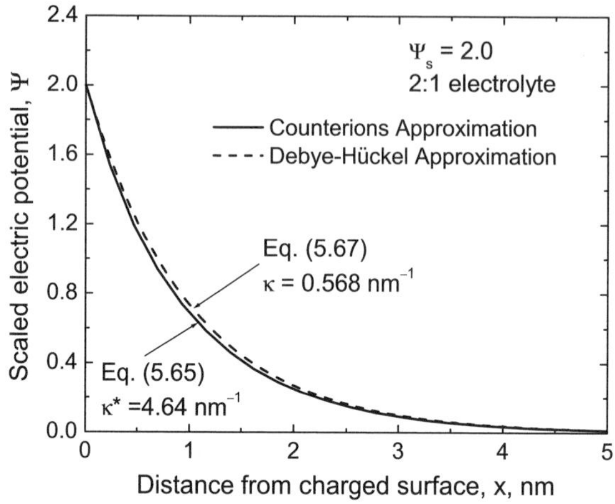

Figure 5.13. Potential distribution near a flat surface. Comparison between the Debye–Hückel and counterion approximation solutions.

and for the system at hand $T = 298$ K, $\epsilon = 6.95 \times 10^{-10}$ C^2/Nm^2, $e = 1.602 \times 10^{-19}$ C, and $N_A = 6.022 \times 10^{23}$ mol^{-1}. Using these values, we obtain

$$\kappa^* = 2.32 \times 10^9 z\sqrt{M}$$

Let the surface potential be positive, then the counterion valence is $z = 2$ and $M = 0.01$. It follows that

$$\kappa^* = 4.64 \times 10^9 \text{ m}^{-1}$$

For the case of the Debye–Hückel solution $\kappa = 0.568 \times 10^9$ m^{-1} for a (1 : 2) electrolyte. This value of κ is to be used with the Debye–Hückel approximation, which is given by

$$\Psi = \Psi_s \exp(-\kappa x) \tag{5.67}$$

Figure 5.13 shows the variation of the solutions provided by Eqs. (5.65) and (5.67) for $\Psi_s = 2$. Both solutions are fairly close to each other. This is to be expected as the Debye–Hückel approximation remains valid for Ψ_s as high as 2. The latter was shown to be true in the plot of Figure 5.6 where a $(z : z)$ electrolyte was used. For a case of high surface potential in an asymmetrical electrolyte, the solution for Ψ provided by Eqs. (5.65) and (5.66) can be readily used. ■

5.3 POTENTIAL FOR CURVED ELECTRIC DOUBLE LAYER

5.3.1 Spherical Geometry: Debye–Hückel Approximation

In contrast to the case of planar diffuse double layers, the equations for the diffuse double layers near curved surfaces are more difficult to solve. When a spherical

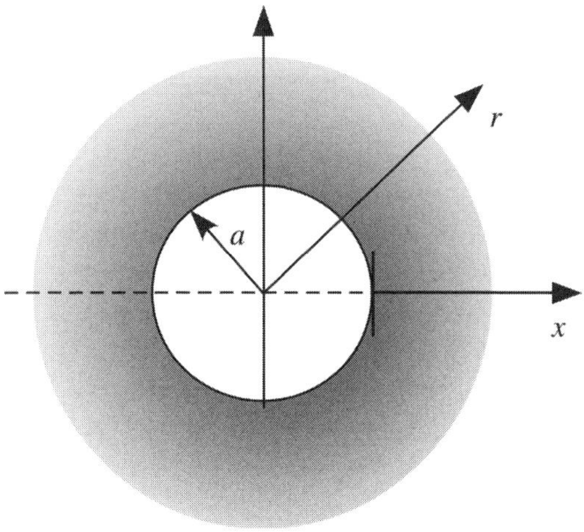

Figure 5.14. Diffuse double layer for a curved surface. Distance from the surface is given by x or $(r - a)$.

surface is considered, the corresponding Poisson–Boltzmann equation in spherical coordinates for a $(z : z)$ electrolyte solution is given as

$$\frac{1}{r^2}\frac{d}{dr}\left(r^2\frac{d\psi}{dr}\right) = \frac{2ezn_\infty}{\epsilon}\sinh\left(\frac{ze\psi}{k_BT}\right) \tag{5.68}$$

The Poisson–Boltzmann equation above assumes the variation of ψ in the radial direction only. The typical boundary conditions are

$$\psi = \psi_s \quad \text{at } r = a \tag{5.69-a}$$

$$\psi \to 0 \quad \text{as } r \to \infty \tag{5.69-b}$$

Figure 5.14 shows the geometry of the system.

In its non-linear form, the Poisson–Boltzmann Eq. (5.68) has no analytical solution subject to the boundary conditions of Eqs. (5.69-a) and (5.69-b). It is analogous to the Gouy–Chapman analysis for a planar surface. In the words of Dukhin (Dukhin and Derjaguin, 1974, p131), "No exact, or even approximate, analytical solution has been found for [Eq. (5.68)] that would be valid for all values of the parameters." Comments as to the availability of analytical solutions to Eq. (5.68) were also made nearly thirty years after Dukhin's statement. See for example Wang *et al.* (2002) and Ohshima (2002). It is more convenient to consider limiting cases for the solution of the Poisson–Boltzmann equation.

For thin double layers, *i.e.*, $\kappa a \gg 1$, one can rescale Eq. (5.68) with

$$r = a\left(1 + \frac{X}{\kappa a}\right) \tag{5.70}$$

where $X = \kappa x$ and a is a characteristic length, which is the radius in case of a spherical particle.

Letting $\Psi = ze\psi/(k_B T)$, and making use of Eq. (5.70), the Poisson–Boltzmann equation, Eq. (5.68) becomes

$$\frac{d^2\Psi}{dX^2} + \frac{2}{\kappa a[1 + X/(\kappa a)]}\frac{d\Psi}{dX} = \sinh \Psi \tag{5.71}$$

The boundary conditions of Eq. (5.69-a) and (5.69-b) transform to

$$\Psi = \Psi_s \quad \text{at } X = 0 \tag{5.72-a}$$

and

$$\Psi \longrightarrow 0 \quad \text{as } X \longrightarrow \infty \tag{5.72-b}$$

Note that for $\kappa a \to \infty$ we recover the equations corresponding to a planar diffuse double layer. As it was pointed out by Hunter (1981), it is justifiable to use the planar (flat) surface analysis for a curved surface when the local radius of curvature of a particle, a, is large compared to the double layer thickness, *i.e.*, $\kappa a \gg 1$. For example, for particles as small as 0.1 μm, the value of κa is about 10 for a 0.001 M (1 : 1) electrolyte solution. In such cases, the planar analysis of the previous sections is quite applicable to the case of a spherical geometry.

Let us now consider the Poisson–Boltzmann equation, Eq. (5.71). Although the solution of Eq. (5.71) requires a numerical route, linearization of the equation simplifies the problem enough to obtain an analytical solution. Letting $\sinh \Psi \approx \Psi$ for $\Psi \ll 1$ one can obtain a solution to Eq. (5.71) in the form

$$\Psi = \Psi_s(a/r)\exp[-\kappa(r - a)] \tag{5.73-a}$$

or

$$\Psi = \Psi_s(a/r)\exp[-\kappa a(r/a - 1)] \tag{5.73-b}$$

which shows that the decay in Ψ is of an exponential type. Although, expressions for Ψ were derived for $\Psi_s \ll 1$, in practice they are valid for Ψ_s as high as 2 to 3 for a (1 : 1) electrolyte.

The normalized electric potential variation away from the surface is plotted for different values of κa in Figure 5.15. Clearly, for $\kappa a = 0.1$ and 1, there is much difference between the solution for a sphere and a planar surface. However, for $\kappa a = 10$ the variation of the normalized potential with κx differs little from the Debye–Hückel solution for a planar surface. This is in accordance with the statements made by Hunter (1981) regarding the use of planar surface solutions for $\kappa a \geq 10$. Loeb *et al.* (1961) give a comprehensive numerical treatment of the electric potential variation for a spherical geometry for both symmetric and asymmetric electrolytes.

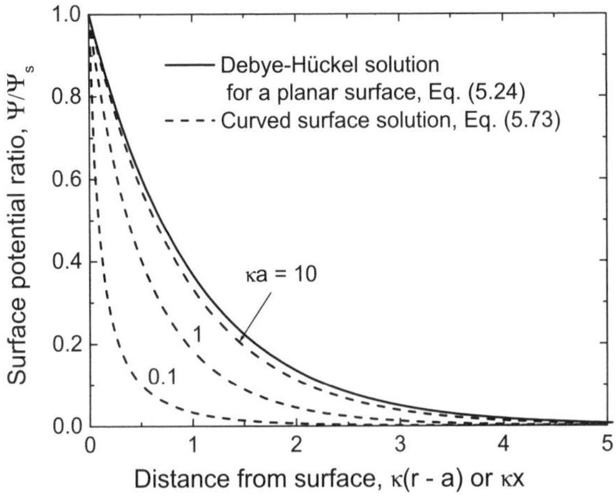

Figure 5.15. Diffuse double layer for planar and spherical surfaces.

EXAMPLE 5.7

Debye–Hückel Analysis. It is of interest to compare Debye–Hückel solutions for planar and curved surfaces. Let us consider a 0.001 M solution of a (2 : 1) electrolyte. Let the particle radius, a, vary from 2 to 100 nm. Compare the variation for ψ with distance away from a flat and a spherical surface.

Solution The Debye–Hückel approximation for a planar surface is given by

$$\Psi = \Psi_s \exp(-\kappa x) \tag{5.74}$$

Recast the Debye–Hückel solution for a spherical surface using $x = r - a$. Here, x denotes distance from the surface. Eq. (5.73-a) gives

$$\Psi = \Psi_s \left(\frac{1}{1 + \kappa x/\kappa a} \right) \exp[-\kappa x] \tag{5.75}$$

The difference between the two geometries is through the term $1/(1 + \kappa x/\kappa a)$ which is only important for small values of the radius of curvature, a, and at intermediate proximity to the surface. Plots of Eqs. (5.74) and (5.75) are shown in Figure 5.16. For the smallest particle radius, $a = 2$ nm, the curvature effect is significant as there is large departure from the planar surface. For $a = 100$ nm, there is little difference in Ψ distribution using both expressions. Here, the curvature effects are negligible. ■

The surface charge density for a charged sphere can be derived as follows. Due to electroneutrality, the total surface charge on the sphere surface is balanced by the charge in the electric double layer. Therefore,

$$Q_s = -\int_a^{\infty} 4\pi r^2 \rho_f \, dr$$

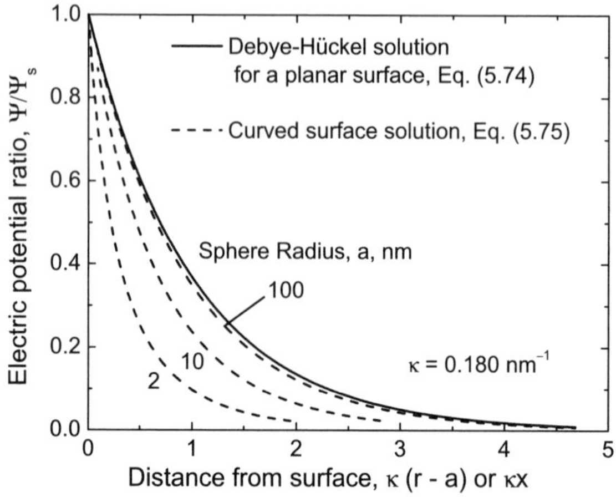

Figure 5.16. Diffuse double layer for a sphere and a planar surface.

and the surface charge density is given by

$$q_s = \frac{Q_s}{4\pi a^2} = -\frac{1}{a^2}\int_a^\infty r^2 \rho_f \, dr \tag{5.76}$$

where r is the radial component of the spherical coordinate system and a is the sphere's radius.

Poisson's equation in spherical coordinates is given by

$$\frac{1}{r^2}\frac{d}{dr}\left(r^2 \frac{d\psi}{dr}\right) = -\frac{\rho_f}{\epsilon} \tag{5.77}$$

Substituting for ρ_f using Eq. (5.77) and simplifying Eq. (5.76) gives

$$q_s = \frac{\epsilon}{a^2}\int_a^\infty d\left(r^2 \frac{d\psi}{dr}\right) \tag{5.78}$$

Upon integration, Eq. (5.78) gives

$$q_s = \frac{\epsilon}{a^2}\left[r^2 \frac{d\psi}{dr}\bigg|_\infty - r^2 \frac{d\psi}{dr}\bigg|_a \right] \tag{5.79}$$

Recognizing that

$$r^2 \frac{d\psi}{dr}\bigg|_\infty \to 0 \quad \text{as } r \to \infty$$

Eq. (5.79) gives

$$q_s = -\epsilon \frac{d\psi}{dr}\bigg|_a \tag{5.80}$$

The relationship between the surface charge density given by Eq. (5.80) is a consequence of the Gauss theorem and it is similar to that of a planar surface. Making use of the electric potential expression, Eq. (5.73-a), Eq. (5.80) gives

$$q_s = \epsilon\kappa\psi_s\left(1 + \frac{1}{\kappa a}\right) \tag{5.81}$$

Equation (5.81) relates the surface charge density to the surface potential for a charged spherical geometry. In the limit of a very small Debye length, *i.e.*, $\kappa a \to \infty$, Eq. (5.81) reduces to the case of a planar surface, where

$$q_s = \epsilon\kappa\psi_s \tag{5.82}$$

The relationship between the surface charge density, q_s, and the surface potential given by Eq. (5.81) is strictly valid for low surface potentials. A more general equation, albeit approximate, for a symmetric electrolyte is given by Loeb *et al.* (1961) as

$$q_s = \frac{\epsilon\kappa k_B T}{ez}\left[2\sinh(\Psi_s/2) + \frac{4}{\kappa a}\tanh(\Psi_s/4)\right] \tag{5.83}$$

Ohshima *et al.* (1982) provided justification for the use of Eq. (5.83). However, Lyklema (1995) pointed out that the potential distribution derived from Eq. (5.83) does not yield a very good approximation.

For $\Psi_s \ll 1$, Eq. (5.83) reduces to

$$q_s = \frac{\epsilon k_B T\kappa}{ez}\left(1 + \frac{1}{\kappa a}\right)\Psi_s \tag{5.84-a}$$

or

$$q_s = \epsilon\kappa\psi_s\left(1 + \frac{1}{\kappa a}\right) \tag{5.84-b}$$

which is identical to the previously derived expression given by Eq. (5.81).

Now let us consider Eq. (5.81) for the case of $\kappa \to 0$. Such a case is applicable for very low electrolyte concentrations and it is particularly true for non-aqueous systems where the ionic concentration is usually very low in the continuous medium. For $\kappa \to 0$, Eq. (5.81) reduces to

$$q_s = \epsilon\psi_s/a \tag{5.85-a}$$

In terms of total surface charge, the equivalent expression for the case of $\kappa \to 0$ is

$$Q_s = 4\pi a\epsilon\psi_s \tag{5.85-b}$$

The above expression is identical to the case of a charged sphere in a dielectric medium. It was derived using electrostatic procedures, see Example 3, Chapter 3.

For very large electric double layers, $\kappa^{-1} \rightarrow \infty$, expressions for surface potential and charge, derived using electrostatics arguments for a charged body in a dielectric, are useful in electrokinetics studies.

White (1977), Ohshima *et al.* (1982), and Ohshima (1995) gave improved solutions to the potential distribution for a charged spherical particle. Ohshima (2002) gave approximate expressions for the surface charge density – surface potential relationship for a spherical colloidal particle in a salt-free aqueous medium, as well as, for non-aqueous media containing only counterions. He showed the suitability of using the relationship given by Eq. (5.85-b) for the case of low potentials and large Debye lengths.

5.3.2 Cylindrical Geometry: Debye–Hückel Approximation

For a cylindrical geometry, there exists no closed form analytical solution of the Poisson–Boltzmann equation and the comments pertaining to a spherical geometry are equally valid for the case of a cylindrical geometry. The Poisson–Boltzmann equation in cylindrical coordinates (r-component only) for a $(z : z)$ electrolyte is given by

$$\frac{1}{r}\frac{d}{dr}\left(r\frac{d\psi}{dr}\right) = \frac{2ezn_\infty}{\epsilon}\sinh\left(\frac{ze\psi}{k_BT}\right) \tag{5.86}$$

For low potential, it reduces to

$$\frac{1}{r}\frac{d}{dr}\left(r\frac{d\psi}{dr}\right) = \kappa^2\psi \tag{5.87}$$

Here, r is the radial coordinate.

For the boundary conditions of $\psi(0) = \psi_s$ and $d\psi(\infty)/dr = 0$, the solution of Eq. (5.87) was given by Dube (1943) (see Lyklema, 1995) as

$$\psi = \psi_s\frac{K_0(\kappa r)}{K_0(\kappa a)} \tag{5.88}$$

where K_0 is the zeroth-order modified Bessel function. The normalized potential distribution for the case of a cylinder is shown in Figure 5.17. As in the case of a spherical geometry, large deviation from the planar type distribution is present for small values of κa. However, for $\kappa a > 10$, the potential distribution is fairly close to the Debye–Hückel solution for a planar surface.

It is of interest to compare the potential distribution for the cases of planar, cylindrical, and spherical geometries. Following Lyklema's (1995) analysis, we can write the Poisson–Boltzmann equation in the form

$$\left(\frac{d}{dr} + \frac{p}{r}\right)\frac{d\psi}{dr} = \kappa^2\psi \tag{5.89}$$

where r is the distance coordinate normal to a surface. The cases of p = 0, 1, and 2 correspond to planar, cylindrical, and spherical geometries, respectively. The values

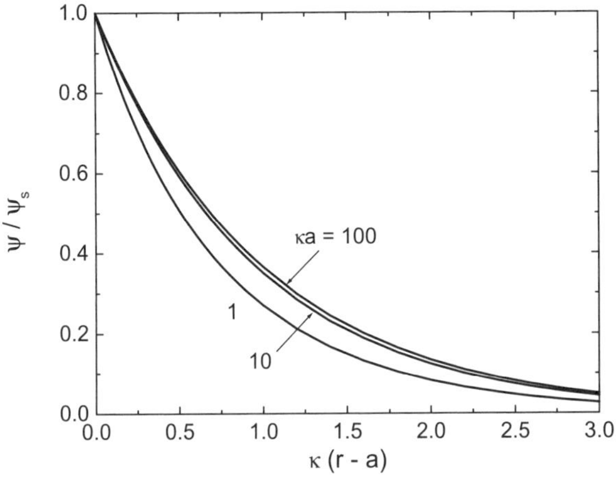

Figure 5.17. Diffuse double layer for a cylindrical surface.

of p would suggest that a spherical geometry would appear as "more curved" for a given κa value and the potential distribution for a cylindrical geometry would lie between the spherical and planar geometries.

Figure 5.18 shows the normalized potential distribution for $\kappa a = 1$ and 10 for planar, cylindrical, and spherical geometries. Clearly, the ψ/ψ_s distribution for the cylindrical geometry lies between those of the spherical and planar geometries as suggested by Lyklema (1995). All geometries considered here give the same ψ/ψ_s distribution for large values of κa, say $\kappa a > 50$.

The extension of the potential distribution outlined above for the case of a cylindrical geometry is provided by Ohshima (1998). The solutions of the Poisson–Boltzmann equation considered here were limited to simple geometries of an infinite planar

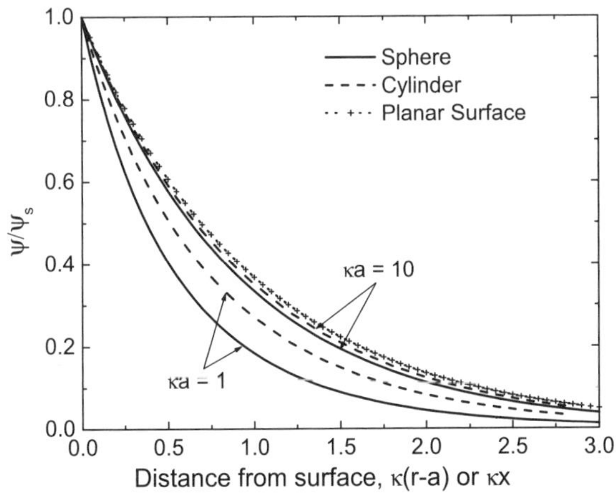

Figure 5.18. Diffuse double layer for planar, cylindrical, and spherical surfaces.

surface, long cylinders, and spheres. In essence, the solutions were provided for one-dimensional problems. For geometries requiring solution of the two-dimensional Poisson–Boltzmann equation, various analyses are available in the literature. Solutions were provided for the case of spheroids by Fair and Anderson (1989), Feng and Wu (1994), Keh and Huang (1993), and Hsu and Liu (1996). In the case of a spheroidal geometry, a rod shaped colloidal particle can be approximated by a prolate spheroid and a red blood cell shape by an oblate spheroid. Electric double layer interaction between a spherical particle and a cylinder was provided by Gu (2000).

5.4 ELECTROSTATIC INTERACTION BETWEEN TWO PLANAR SURFACES

When two charged surfaces (or two colloidal particles), each surrounded by an electrical diffuse double layer, approach one another, their respective double layers begin to overlap. As a result, the ionic and potential distributions around a given particle, when it is brought in the vicinity of a second particle, are no longer symmetrical. This causes asymmetrical stresses of electrical origin on the particle surface and, as a result, the particle experiences a force. The evaluation of the force can be made from the solution of the Poisson–Boltzmann equation for the potential coupled with the momentum conservation equation. For the sake of simplicity (and since analytical solutions of the relevant equations can be obtained readily), we shall first consider computation of the force per unit area between two charged infinite flat plates in a dielectric medium containing free charge.

5.4.1 Force between Two Charged Planar Surfaces

At the outset, it should be stated that no real system conforms to the picture of two interacting infinite parallel plates. This geometry is a highly simplified approximation of typical colloidal systems based on scaling arguments. Generally, colloidal particles are bodies with curvature (for instance spheres) suspended in a large volume of electrolyte solution. The forces that we are interested in are caused by overlap of the electric double layers when two such colloidal particles approach each other. Typically, the gaps between the particles at which overlap between their double layers becomes perceptible are of the order of nanometers. On the other hand, radii of curvature of the colloidal particles are often of the order of micrometers. Consequently, the gaps at which these forces become measurable are negligible compared to the radii of the interacting particles. In these cases, it is often possible to completely ignore the curvature of the particles, and treat them as infinite planar surfaces. As a consequence of this approximation, however, throughout the subsequent analysis, it should be borne in mind that although we are considering interaction between two infinite planar surfaces, the surfaces actually belong to finite bodies suspended in a large bath of electrolyte solution. Thus, the electrolyte solution trapped between these planar surfaces is connected to a reservoir of electroneutral electrolyte, in which, at sufficiently large distances from the charged surfaces, the ion concentrations take their bulk values, and the electric potential becomes zero.

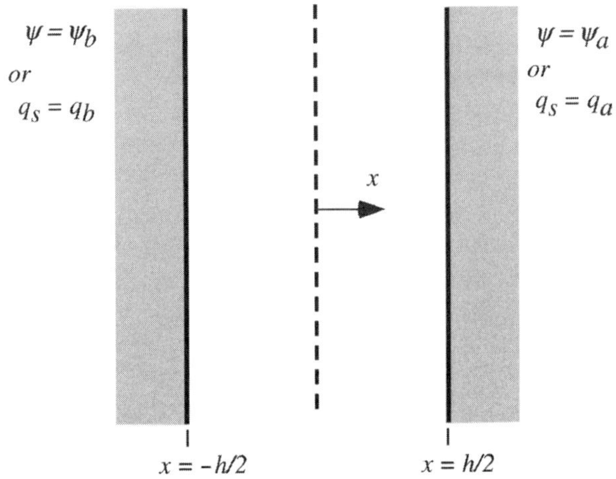

Figure 5.19. Overlap of two planar double layers.

Consider two flat plates at constant surface potentials of ψ_a and ψ_b as shown in Figure 5.19. These surface potentials will be assumed to remain constant irrespective of the separation between the plates. The Poisson equation for the case of parallel plates is given by

$$\epsilon \frac{d^2\psi}{dx^2} = -\rho_f \tag{5.90}$$

which, with the Debye–Hückel approximation, becomes

$$\frac{d^2\psi}{dx^2} = \kappa^2\psi \tag{5.91}$$

The momentum equation governing liquid flow, as given by the Navier–Stokes equation including the appropriate electric body force is given by

$$\mu\nabla^2\mathbf{u} + \rho\mathbf{g} + \rho_f\mathbf{E} = \nabla p \tag{5.92}$$

where $\mathbf{u}$ is the liquid velocity vector, μ is the liquid viscosity, p is the pressure, $\mathbf{g}$ is the gravitational acceleration, and ρ is the liquid mass density. The term $\rho_f\mathbf{E}$ represents the electric body force per unit volume, comprising of the free charge density, ρ_f, and the local electric field $\mathbf{E}$. We will discuss the Navier–Stokes equation in greater detail in Chapter 6. For a stationary system, since there is no fluid flow, we have $\mathbf{u} = \mathbf{0}$. Further, neglecting the effect of gravity, we can write Eq (5.92) as

$$\nabla p = \rho_f\mathbf{E} \tag{5.93}$$

For a one-dimensional problem (when the only space variable is x), the above equation simplifies to

$$\frac{dp}{dx} = \rho_f E_x \tag{5.94}$$

where

$$E_x = -\frac{d\psi}{dx} \tag{5.95}$$

One can combine Eqs. (5.94) and (5.95) to obtain

$$\frac{dp}{dx} + \rho_f \frac{d\psi}{dx} = 0 \tag{5.96}$$

Combining Eqs. (5.96) with the Poisson equation (5.90) leads to

$$\frac{dp}{dx} - \epsilon \frac{d^2\psi}{dx^2}\frac{d\psi}{dx} = 0 \tag{5.97}$$

The above equation relates the pressure to the electric potential ψ. The pressure arises due to variation in the ionic concentration between the plates and the surrounding electrolyte solution. Integration of Eq. (5.97) gives

$$p - \frac{\epsilon}{2}\left(\frac{d\psi}{dx}\right)^2 = C_1 \tag{5.98}$$

where C_1 is the integration constant and is the required force per unit area to keep the two plates from moving. For this particular case, it turns out that the net force per unit area F_p/A_p, which is the difference between the pressure p and the electric force per unit area, is constant everywhere within two plates.[4] Replacing C_1 by F_p/A_p, Eq. (5.98) becomes

$$\frac{F_p}{A_p} = p - \frac{\epsilon}{2}\left(\frac{d\psi}{dx}\right)^2 \tag{5.99}$$

We need to determine p and ψ in order to evaluate the force per unit area, F_p/A_p. The solution of Eq. (5.91) gives

$$\psi = A_1 \cosh \kappa x + B_1 \sinh \kappa x. \tag{5.100}$$

Making use of the boundary conditions of ψ at $x = \pm h/2$, where $\psi(h/2) = \psi_a$ and $\psi(-h/2) = \psi_b$, the constants of integration are given as

$$A_1 = \frac{\psi_a + \psi_b}{2\cosh(\kappa h/2)} \quad \text{and} \quad B_1 = \frac{\psi_a - \psi_b}{2\sinh(\kappa h/2)} \tag{5.101}$$

For the special case of $\psi_a = \psi_b,\ B_1 = 0$.

The pressure, p, in Eq. (5.99) can be derived from Eq. (5.93). Substituting the expression for the free charge density, ρ_f, in Eq. (5.93) we can write

$$\nabla p + \left[e\sum_i z_i n_{i\infty} \exp\left(-\frac{z_i e\psi}{k_B T}\right)\right]\nabla\psi = 0 \tag{5.102}$$

[4]For a derivation of the force per unit area, refer to Example 5.8.

where $n_{i\infty}$ is the bulk concentration of the ith ionic species. Equation (5.102) can be rewritten as

$$\nabla\left[p - k_B T \sum_i n_{i\infty} \exp\left(-\frac{z_i e\psi}{k_B T}\right)\right] = 0 \tag{5.103}$$

Since the term in the square brackets in Eq. (5.103) is independent of the coordinates, it is a constant. Consequently, one can equate the term at any point between the plates, where the electric potential has a finite value, with another point in the bulk electrolyte solution, where $\psi = 0$ and $p = p_\infty = 0$.[5] In other words,

$$p - k_B T \sum_i n_{i\infty} \exp\left(-\frac{z_i e\psi}{k_B T}\right) = -k_B T \sum_i n_{i\infty} \tag{5.104}$$

Rearrangement of Eq. (5.104) yields

$$p = k_B T \sum_i n_{i\infty} \left[\exp\left(-\frac{z_i e\psi}{k_B T}\right) - 1\right] \tag{5.105}$$

For low potentials, $z_i e\psi/k_B T \ll 1$, and a symmetric electrolyte, Eq. (5.105) simplifies to

$$p = n_\infty k_B T \left(\frac{ze\psi}{k_B T}\right)^2 \tag{5.106}$$

It may be noted here that in literature, the pressure term is often attributed to the osmotic pressure. This is perhaps owing to the form of the right hand side term in Eq. (5.105), which is essentially $k_B T \sum_i (n_i - n_{i\infty})$. This expression is identical to the Van't Hoff formula for osmotic pressure. From a thermodynamic viewpoint, this expression for the osmotic pressure can be derived from the Gibbs–Duhem equation as well. However, it is clear from the above derivation that the pressure is of a hydrostatic origin, and can be derived in a straightforward manner from the Navier–Stokes equation considering the electric body force. In this context, one may consider Eq. (5.93) as an equivalent form of the Gibbs–Duhem equation.

Substitution of the expression for p, Eq. (5.106), in the expression for the force per unit area, Eq. (5.99), yields

$$\frac{F_p}{A_p} = n_\infty k_B T \left(\frac{ze\psi}{k_B T}\right)^2 - \frac{\epsilon}{2}\left(\frac{d\psi}{dx}\right)^2$$

One can now use the equation for ψ, Eq. (5.100), in the above expression to obtain

$$\frac{F_p}{A_p} = \left(\frac{e^2 z^2 n_\infty}{k_B T}\right)\left[A_1^2 - B_1^2\right] \tag{5.107}$$

[5]In assuming $p_\infty = 0$ we are simply writing the hydrostatic pressure in terms of gauge pressure with the pressure in the bulk electrolyte as the reference.

Using the expressions for A_1 and B_1 from Eq. (5.101) in Eq. (5.107) yields

$$\frac{F_p}{A_p} = \frac{e^2 z^2 n_\infty}{k_B T} \left[\frac{2\psi_a \psi_b \cosh(\kappa h) - \psi_a^2 - \psi_b^2}{\sinh^2(\kappa h)} \right] \tag{5.108}$$

In terms of Debye length, the force per unit area is given as

$$\frac{F_p}{A_p} = \frac{\epsilon \kappa^2}{2} \left[\frac{2\psi_a \psi_b \cosh(\kappa h) - \psi_a^2 - \psi_b^2}{\sinh^2(\kappa h)} \right] \tag{5.109}$$

where

$$\kappa^2 = \frac{2e^2 z^2 n_\infty}{\epsilon k_B T}$$

We note here that Eqs. (5.108) and (5.109) are identical, and comparing the pre-factors (the terms on the right hand sides outside the square brackets) in these equations, we obtain

$$\frac{e^2 z^2 n_\infty}{k_B T} = \frac{\epsilon \kappa^2}{2}$$

Rearranging the above equation yields

$$\frac{n_\infty k_B T}{\kappa^2} = \frac{\epsilon}{2} \left(\frac{k_B T}{ze} \right)^2$$

which is the same relationship as Eq. (5.56). For the special case of $\psi_a = \psi_b = \psi_s$, Eqs. (5.108) and (5.109) reduce to

$$\frac{F_p}{A_p} = \left(\frac{2e^2 z^2 n_\infty \psi_s^2}{k_B T} \right) \left[\frac{\cosh(\kappa h) - 1}{\sinh^2(\kappa h)} \right] \tag{5.110}$$

Simplification of Eq. (5.110) leads to

$$\frac{F_p}{A_p} = \frac{\epsilon \kappa^2 \psi_s^2}{[\cosh(\kappa h) + 1]} \tag{5.111-a}$$

or

$$\frac{F_p}{A_p} = \frac{\epsilon \kappa^2 \psi_s^2}{2\cosh^2(\kappa h/2)} \tag{5.111-b}$$

Either Eq. (5.108) or Eq. (5.109) gives the force per unit area for the general case of two planar charged surfaces having different surface electric potentials. Equations (5.111-a) and (5.111-b) give the force per unit area for the special case of surfaces having the same surface potential.

Let us consider the limiting cases of κh for surfaces having the same potentials

(i) **Very small κh:** Making use of the expansion for $\cosh^2(\kappa h/2)$, Eq. (5.111-b) gives

$$\frac{F_p}{A_p} = \frac{\epsilon\kappa^2\psi_s^2[1 - (\kappa h)^2/4]}{2} \tag{5.112}$$

Equation (5.112) indicates that for small κh, the force per unit area is proportional to ψ_s^2 and is weakly dependent on the separation distance between the two plates.

(ii) **Very large κh:** In this limit, namely, $\kappa h \gg 1$, $h \to \infty$, Eq. (5.111-a) gives

$$\frac{F_p}{A_p} = 2\epsilon\kappa^2\psi_s^2 \exp(-\kappa h) \tag{5.113}$$

As would be expected, the interaction force is zero at very large separation distances. Additional details may be found in Russel *et al.* (1989), Probstein (2003), van de Ven (1989), Ross and Morrison (1988), and Chan *et al.* (1975, 1976).

For a given Debye length, when the two surfaces approach each other, the electric double layers associated with each surface will interact with each other. Figure 5.20 shows the normalized potential distribution for two similar surfaces for $\kappa = 1.0\ \text{nm}^{-1}$. For $h = 50$ nm, the potential at the mid-plane is zero and for $h = 10$ nm it is nearly zero. However, when $h < 10$ nm the mid-plane normalized potential, $\psi(0)/\psi_s$, is not zero and here the two electric double layers interact. Such an interaction is very important.

Surfaces with dissimilar potentials but of the same sign can give rather unexpected results. At small separations, it is possible to have an attractive force between two such

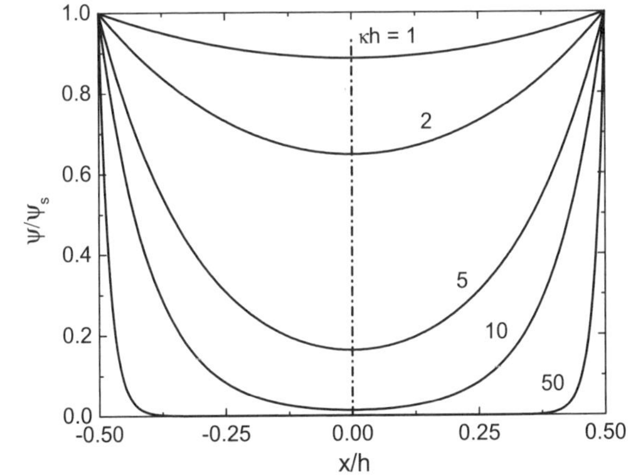

Figure 5.20. Potential distribution between two planar surfaces at various separation distances.

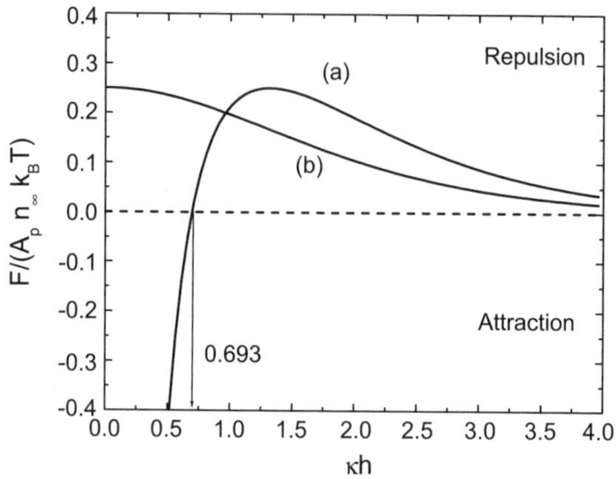

Figure 5.21. Electrostatic double layer force between two constant potential surfaces. Case (a): $\Psi_a = 0.5$ and $\Psi_b = 1.0$; Case (b): $\Psi_a = \Psi_b = 0.5$.

surfaces. This behavior at close separations is due to a change in the sign of the surface charge on one of the surfaces due to the overlapping double layers, while the surface potentials are assumed to remain constant. The very fact that surfaces with dissimilar potentials but of the same sign can encounter an attractive force on close approach has a major impact in the area of hetero-coagulation, Russel *et al.* (1989). Figure 5.21 shows the variation of the dimensionless force per unit area with separation distance, κh, for two cases. In the case of $\Psi_a = 0.5$ and $\Psi_b = 1.0$, the electrostatic force becomes attractive at a dimensionless separation distance of less than $\kappa h = 0.693$. However, for the case of $\Psi_a = \Psi_b = 0.5$, the force remains repulsive at all separation distances. In the next section we will discuss overlapping double layers in more detail.

EXAMPLE 5.8

Force between Charged Parallel Plates. Derive an expression for the force per unit area between two parallel capacitor plates having surface potentials ψ_A and ψ_B separated by a distance $2h$ in a symmetric $(z : z)$ electrolyte bath with bulk concentration n_∞.

Solution We consider two thin charged plates A and B with constant surface potentials ψ_A and ψ_B, respectively, immersed in a large electrolyte bath as shown in Figure 5.22. The electric potential and the ion concentration of the bath far away from the plates are zero and n_∞, respectively. The pressure in this bulk electrolyte region is denoted by p_∞. The origin of the Cartesian coordinate system is at the mid-plane of the two parallel plates. The plate dimensions along the y and z coordinate directions are considered to be much larger than the gap between them along the x coordinate direction.

To compute the force between the plates, it is sufficient to determine the total force acting on one of the plates. This force can be determined by integrating the total

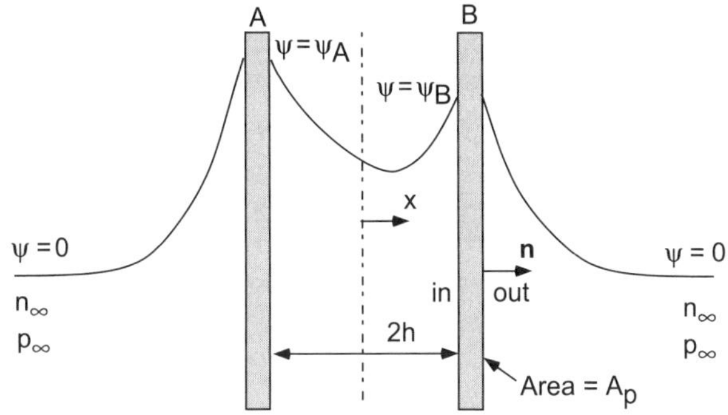

Figure 5.22. Schematic diagram for the calculation of the interaction force per unit area between two charged parallel capacitor plates.

stress tensor over the entire surface of one of the plates. Focussing on the plate B, let us consider the different stresses acting on it. The total stress, $\overline{\overline{T}}$, will have two components. First, we have the static pressure of the fluid, $-p$, acting normal to the surface of the plate at every point, the negative sign implying that the pressure is directed toward the plate (opposite to the direction of the unit surface normal, $\mathbf{n}$). This stress, $\overline{\overline{T}}_p$, is explicitly written as

$$\overline{\overline{T}}_p = -p\overline{\overline{I}} = \begin{bmatrix} -p & 0 & 0 \\ 0 & -p & 0 \\ 0 & 0 & -p \end{bmatrix}$$

where $\overline{\overline{I}}$ is the unit tensor. The second component is due to the electrical (Maxwell) stress, $\overline{\overline{T}}_e$. The electrical stress is given by the stress tensor, Eq. (3.156),

$$\overline{\overline{T}}_e = \epsilon\mathbf{E}\mathbf{E} - \frac{1}{2}\epsilon\mathbf{E}\cdot\mathbf{E}\overline{\overline{I}}$$

where $\mathbf{E}$ is the electric field vector, and ϵ is the dielectric constant of the fluid (electrolyte solution). Making use of the assumption that the plate dimensions along y and z are much larger than the gap between them, the electric field can be assumed to be one dimensional, *i.e.*, $\mathbf{E} = E_x\mathbf{i}_x$, where $\mathbf{i}_x$ is a unit vector along the positive x coordinate. Consequently, from Eq. (3.162), the electrical stress tensor becomes

$$\overline{\overline{T}}_e = \begin{bmatrix} \epsilon E_x^2/2 & 0 & 0 \\ 0 & -\epsilon E_x^2/2 & 0 \\ 0 & 0 & -\epsilon E_x^2/2 \end{bmatrix}$$

The total stress, $\overline{\overline{T}}$, is given by

$$\overline{\overline{T}} = \overline{\overline{T}}_p + \overline{\overline{T}}_e = \begin{bmatrix} \epsilon E_x^2/2 - p & 0 & 0 \\ 0 & -\epsilon E_x^2/2 - p & 0 \\ 0 & 0 & -\epsilon E_x^2/2 - p \end{bmatrix} \tag{5.114}$$

The total force on plate B is obtained by integrating the total stress over its entire surface. In other words

$$F_p = \oint_S \overline{\overline{T}} \cdot \mathbf{n}\, dS$$

where S is the total surface area of the plate. Note that this integration is similar to Eq. (3.161), except that the pressure is also included in the total stress tensor, $\overline{\overline{T}}$. Neglecting the contribution to the force arising from the extremities of the plate, one can observe that the total force can be determined by integrating the stress tensor over the plate cross sectional area A_p on the inside and outside faces of plate B (denoted by "in" and "out", respectively). Furthermore, since the stress does not depend on the y and z coordinates, we obtain

$$F_p = \int_{in} \overline{\overline{T}} \cdot \mathbf{n}\, dA_p + \int_{out} \overline{\overline{T}} \cdot \mathbf{n}\, dA_p = [\overline{\overline{T}} \cdot \mathbf{n}]_{in} A_p + [\overline{\overline{T}} \cdot \mathbf{n}]_{out} A_p \tag{5.115}$$

On the inside face of the plate B, $\mathbf{n} = -\mathbf{i}_x$, while on the outside face, $\mathbf{n} = \mathbf{i}_x$. Thus,

$$\frac{F_p}{A_p} = -[\overline{\overline{T}} \cdot \mathbf{i}_x]_{in} + [\overline{\overline{T}} \cdot \mathbf{i}_x]_{out}$$

Writing the dot product of the total stress tensor with the unit vector $\mathbf{i}_x$ explicitly yields

$$\frac{F_p}{A_p} = -\left(\frac{\epsilon E_x^2}{2} - p\right)_{in} + \left(\frac{\epsilon E_x^2}{2} - p\right)_{out}$$

The quantity $\left(\epsilon E_x^2/2 - p\right)_{out}$, evaluated on the outside surface of the plate, is a constant at every location x. Now, since $p \to p_\infty$ and $E_x \to 0$ at a large distance from the plate, $x \to \infty$ (in the bulk electrolyte solution), one can write

$$\left(\frac{\epsilon E_x^2}{2} - p\right)_{out} = -p_\infty$$

Therefore

$$\frac{F_p}{A_p} = -\left(\frac{\epsilon E_x^2}{2} - p\right)_{in} - p_\infty \tag{5.116}$$

or

$$\frac{F_p}{A_p} = p_{in} - p_\infty - \left(\epsilon \frac{E_x^2}{2}\right)_{in} \tag{5.117}$$

The above expression for the force per unit area is a slightly more generalized form of Eq. (5.99). The methodology described above gives the force on plate B. The force on plate A will be equal and opposite to the force on plate B. The force on B is repulsive if it acts along $\mathbf{i}_x$ (positive x direction), and attractive if it acts along the negative x direction. ■

5.4.2 Surface Charge Density for Planar Surfaces: Overlapping Double Layers

Similar to the analysis conducted for an isolated charged surface, we can proceed in evaluating the surface charge densities as a function of the surface potentials and separation distance. We will carry out the analysis for similar surfaces using the solution for the potential distribution obtained in the previous section under the assumption of constant surface potentials. For convenience, however, in the case of dissimilar surfaces, we will utilize the potential distribution solution under the assumption of constant surface charge densities. In the following analysis, the geometry depicted in Figure 5.19 is used.

5.4.2.1 Similar Surfaces Due to symmetry, for the case of similar surfaces only one side of the mid-plane needs to be considered. To conserve electroneutrality, the charges in the electric double layers and on the surface must be conserved. This leads to

$$q_s = -\int_0^{h/2} \rho_f dx$$

for the right hand side half-width.

Making use of the Poisson equation, we can write

$$q_s = \epsilon \int_0^{h/2} \frac{d^2\psi}{dx^2} dx = \epsilon \int_0^{h/2} d\left(\frac{d\psi}{dx}\right) \tag{5.118}$$

Upon integration, we obtain

$$q_s = \epsilon \left[\left.\frac{d\psi}{dx}\right|_{h/2} - \left.\frac{d\psi}{dx}\right|_{0} \right] \tag{5.119}$$

Due to symmetry, at the mid-point between the two surfaces, $\left.\frac{d\psi}{dx}\right|_0 = 0$ and Eq. (5.119) becomes

$$q_s = \epsilon \left.\frac{d\psi}{dx}\right|_{h/2} \tag{5.120}$$

or

$$q_s = -\epsilon E_s \tag{5.121}$$

where E_s is the electric field at the surface. To evaluate q_s we need to use an expression for the potential distribution. To that end, when the surface potentials on both the surfaces are equal, Eq. (5.100) gives

$$\psi = \frac{\psi_s \cosh(\kappa x)}{\cosh(\kappa h/2)} \tag{5.122}$$

Making use of Eq. (5.122), the surface charge distribution of Eq. (5.120) becomes

$$q_s = \epsilon\kappa\psi_s \tanh(\kappa h/2) \tag{5.123}$$

Equation (5.123) relates the surface charge density to the surface potential and the separation gap, h, between the two similar surfaces. For a constant surface potential ψ_s and given ϵ and κ, the surface charge density becomes a function of the separation gap, h. In other words, the surface charge density, q_s, is not a constant but a function of the separation distance between the two charged surfaces whose surface potentials are kept constant for all values of h. As $\tanh(\kappa h/2)$ is positive for all values of $\kappa h/2$, the surface charge density, q_s, does not change signs with changes in the gap width h or κh. As $\tanh(0) = 0$ and $\tanh(\infty) = 1$, the dimensionless surface charge density $(q_s/\epsilon\kappa\psi_s)$ varies from zero corresponding to zero gap to unity at large separations.

The force between two planar similar surfaces can be given in terms of the surface charge density q_s rather than the surface potential, ψ_s. Combining Eqs. (5.110) and (5.123) leads to

$$\frac{F_p}{A_p} = \frac{q_s^2}{\epsilon} \frac{[\cosh(\kappa h) - 1]}{\sinh^2(\kappa h)\tanh^2(\kappa h/2)}$$

Simplifying, one obtains

$$\frac{F_p}{A_p} = \frac{q_s^2}{\epsilon} \frac{1}{[\cosh(\kappa h) - 1]} \tag{5.124-a}$$

or

$$\frac{F_p}{A_p} = \frac{q_s^2}{\epsilon} \frac{1}{2\sinh^2(\kappa h/2)} \tag{5.124-b}$$

For $\kappa h \ll 1$, Eq. (5.124-b) gives

$$\frac{F_p}{A_p} = \frac{2q_s^2}{\epsilon\kappa^2 h^2} \tag{5.125}$$

and for $\kappa h \gg 1$, Eq. (5.124-a) gives

$$\frac{F_p}{A_p} = \frac{2q_s^2 \exp(-\kappa h)}{\epsilon} \tag{5.126}$$

For the case of $\kappa h \gg 1$, by holding either ψ_s or q_s constant the force per unit area approaches zero at large separation distances as indicated by Eqs. (5.113) and (5.126).

For the case of $\kappa h \ll 1$, *i.e.*, at very small separation distances, holding the surface potential constant gives a force per unit area that is weakly dependent on the separation gap as indicated by Eq. (5.112). However, holding the surface charge density constant gives a very large repulsive force at small separations as indicated by Eq. (5.125). It is accepted in literature that at small separations between two interacting surfaces, it is more likely that the surface potential holds constant rather than the surface charge remains constant. In real systems, it is likely that neither the surface charge nor the surface potential remains constant on close approach of the two surfaces. Charge regulation takes place on close approach and it plays an important role in flocculation and colloidal particle attachment to surfaces.

5.4.2.2 Dissimilar Surfaces The potential distribution for two charged planar surfaces is given by

$$\psi = A_2 \cosh(\kappa x) + B_2 \sinh(\kappa x) \tag{5.127}$$

The coefficients A_2 and B_2 can be evaluated using the boundary conditions of constant surface charge density where at $x = h/2, q_s = q_a$ and at $x = -h/2, q_s = q_b$ as shown in Figure 5.19. The constants are provided by Russel *et al.* (1989) as

$$A_2 = \frac{(q_a + q_b)}{2\epsilon\kappa \sinh(\kappa h/2)} \tag{5.128}$$

and

$$B_2 = \frac{(q_a - q_b)}{2\epsilon\kappa \cosh(\kappa h/2)} \tag{5.129}$$

Let us make use of the solution for the potential distribution given by Eqs. (5.127)–(5.129) to enable us to discuss surface charge reversal observed during interaction between dissimilar planar surfaces at close approach.

The surface potentials at $x = \pm h/2$, *i.e.*, $\psi(h/2) = \psi_a$ and $\psi(-h/2) = \psi_b$ can be given by Eqs. (5.127)–(5.129) as

$$\psi_a = \frac{(q_a + q_b)}{2\epsilon\kappa \tanh(\kappa h/2)} + \frac{(q_a - q_b)\tanh(\kappa h/2)}{2\epsilon\kappa} \tag{5.130}$$

and

$$\psi_b = \frac{(q_a + q_b)}{2\epsilon\kappa \tanh(\kappa h/2)} - \frac{(q_a - q_b)\tanh(\kappa h/2)}{2\epsilon\kappa} \tag{5.131}$$

Solving for q_a and q_b, Eqs. (5.130) and (5.131) give

$$q_a = \frac{\epsilon\kappa}{2}\left[(\psi_a + \psi_b)\tanh(\kappa h/2) + \frac{(\psi_a - \psi_b)}{\tanh(\kappa h/2)}\right] \tag{5.132}$$

and

$$q_b = \frac{\epsilon\kappa}{2}\left[(\psi_a + \psi_b)\tanh(\kappa h/2) - \frac{(\psi_a - \psi_b)}{\tanh(\kappa h/2)}\right] \tag{5.133}$$

Clearly, Eqs. (5.132) and (5.133) show that by keeping the surface potentials constant, the surface charge densities vary with the dimensionless separation gap between the plates.

Let us consider the case of dissimilar surfaces held at constant potentials where $\psi_a \neq \psi_b$ and both ψ_a and ψ_b have the same sign. Without loss of generality, let us also assume that $\psi_a < \psi_b$ and that both ψ_a and ψ_b are positive. Consideration of Eq. (5.133) would indicate that the surface charge density, q_b, will not change signs for all values of the dimensionless separation gap, κh. However, the surface charge density, q_a, can change signs. The dimensionless gap width at which q_a changes signs can be evaluated by setting $q_a = 0$ in Eq. (5.133). This leads to

$$\tanh^2(\kappa h/2) = -\left(\frac{\psi_a - \psi_b}{\psi_a + \psi_b}\right) \tag{5.134}$$

The variations of q_a and q_b for the cases of $\Psi_a = \Psi_b = 0.5$ and $\Psi_a = 0.5$ and $\Psi_b = 1.0$ are shown in Figure 5.23. For the case of similar surfaces, *i.e.*, $\Psi_a = \Psi_b = 0.5$, both q_a and q_b remain positive for all values of κh. For the case of dissimilar surfaces ($\Psi_a < \Psi_b$), q_b is positive for all κh values. However, the surface having the lower surface potential experiences a surface charge reversal that will eventually lead to an attractive force between the two dissimilar surfaces. The charge reversal for $\Psi_a = 0.5$ and $\Psi_b = 1.0$ according to Eq. (5.134) occurs at

$$\kappa h = 2\tanh^{-1}\sqrt{1/3} = 1.317$$

The location of the surface charge reversal is shown on Figure 5.23.

As the surface charge density is proportional to the gradient of the electrical potential at the surface, a surface charge reversal should occur when

$$\mathbf{n} \cdot \nabla\psi = 0 \tag{5.135-a}$$

or

$$\left.\frac{d\psi}{dn}\right|_{surface} = 0 \tag{5.135-b}$$

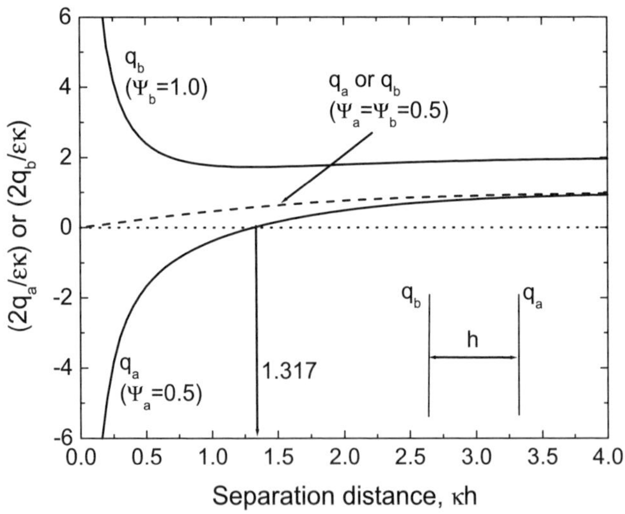

Figure 5.23. Surface charge regulation for two planar surfaces with dimensionless separation distance.

A plot of the electric potential in the gap between two parallel plates is shown in Figure 5.24 for $\Psi_a = 0.5$ and $\Psi_b = 1.0$. Clearly at $\kappa h = 1.317$, the electric potential gradient is zero at $x/h = 0.5$. The slope is positive for $\kappa h > 1.317$ and negative for $\kappa h < 1.317$. A quick review of Figures 5.21 and 5.23 would indicate that the gap location at which the charge reversal takes place does not coincide with the force reversal from repulsive to attractive.

The condition for force reversal from repulsive to attractive can be obtained by setting the force between the two dissimilar plates to zero. Setting F_p/A_p in Eq. (5.107) to zero, one obtains

$$A_1 = \mp B_1$$

For $\Psi_a < \Psi_b$ one obtains

$$\tanh(\kappa h/2) = -\left(\frac{\Psi_a - \Psi_b}{\Psi_a + \Psi_b}\right)$$

or

$$\tanh(\kappa h/2) = \frac{\Psi_b/\Psi_a - 1}{\Psi_b/\Psi_a + 1} \tag{5.136}$$

The dimensionless gap width, κh, at which the force reverses from repulsive to attractive is dependent solely on the ratio of Ψ_b/Ψ_a, Eq. (5.136), and it is not the same gap width at which the charge reversal occurs as given by Eq. (5.134). In fact, the repulsive force is maximum at the distance where the charge reverses its sign.

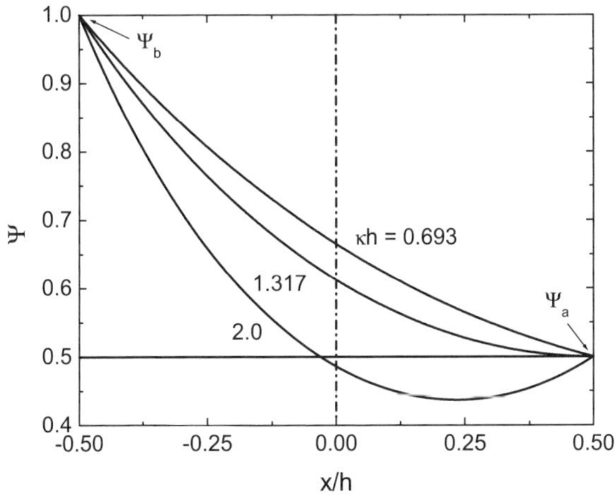

Figure 5.24. Electric potential distribution between two planar surfaces at various separation distances.

5.5 ELECTROSTATIC POTENTIAL ENERGY

The interaction between particles due to electrostatic potential forces is most conveniently characterized in terms of the electrostatic potential energy, ϕ. Recognizing that work or energy is related to force by

$$\text{Force} = -\frac{d(\text{work})}{d(\text{distance})} = -\frac{d(\text{energy})}{d(\text{distance})}$$

one can write

$$\phi = -\int_{\infty}^{h} (\text{Force})dx \tag{5.137}$$

where, by convention, the potential energy is given by bringing a particle from infinity to a separation distance h.

For the case of two parallel plates, using Eq. (5.111-b), the interaction energy per unit area, $\phi_p^*(= \phi_p/A_p)$, between the two charged parallel plates is given by

$$\phi_p^* = -\frac{\epsilon\kappa^2\psi_s^2}{2}\int_{\infty}^{h}\frac{dx}{\cosh^2(\kappa x/2)}$$

leading to

$$\phi_p^* = \epsilon\kappa\psi_s^2\,[1 - \tanh(\kappa h/2)] \tag{5.138-a}$$

or

$$\phi_p^* = 2\epsilon\kappa\psi_s^2\left[\frac{\exp(-\kappa h)}{1+\exp(-\kappa h)}\right] \tag{5.138-b}$$

For $\kappa h \ll 1$, Eq. (5.138-a) provides

$$\phi_p^* = \epsilon\kappa\psi_s^2(1 - \kappa h/2) \tag{5.139-a}$$

and for $\kappa h \gg 1$, Eq. (5.138-b) provides

$$\phi_p^* = 2\epsilon\kappa\psi_s^2\exp(-\kappa h) \tag{5.139-b}$$

EXAMPLE 5.9

Electrostatic Potential Energy. Derive the electrostatic potential energy between two identical spheres carrying the same electric surface potential. The electrostatic force between two charged spheres can be given by

$$F = 2\pi\epsilon\kappa a\psi_s^2\frac{\exp(-\kappa h)}{1+\exp(-\kappa h)} \tag{5.140}$$

where a is the sphere radius and ψ_s is their surface electric potential. Here h is the separation gap between the two spheres.

Solution The electrostatic potential energy is given by combining Eqs. (5.137) and (5.140) for the case of $\kappa a \gg 1$

$$\phi = -\int_{\infty}^{h} 2\pi\epsilon\kappa a\psi_s^2 \frac{\exp(-\kappa h)dh}{1+\exp(-\kappa h)} \tag{5.141}$$

Hence,

$$\phi = 2\pi\epsilon a\psi_s^2 \ln\left[1+\exp(-\kappa h)\right] \tag{5.142}$$

The above expression is the electrostatic interaction energy between two spheres having the same radii and surface potentials. ■

From a thermodynamic standpoint, the interaction energy between two double layers at a distance h is defined as the Gibbs free energy change[6] associated with bringing the two double layers from an infinite separation to a finite distance

$$\phi = \Delta G = G(h) - G(\infty) \tag{5.143}$$

Here, G represents the Gibbs free energy of the entire system comprised of two electric double layers.

For each double layer in isolation, the charging process yields a certain free energy, which is represented by (Overbeek, 1990)

$$G = -\int_0^{\psi_0} \sigma d\psi \tag{5.144}$$

where σ is the surface charge density and ψ is the surface potential. The integral refers to the free energy change associated with charging up a double layer from zero potential (uncharged surface) to a potential of ψ_0. Knowing a relationship between the surface charge density and the surface potential, one can evaluate the integral in Eq. (5.144) and obtain the free energy associated with an isolated electric double layer maintained at a given surface potential ψ_0. For example, considering a planar double layer with a small surface potential, such that the Debye–Hückel approximation holds, we can write, *cf.*, Eq. (5.54),

$$\sigma = \epsilon\kappa\psi \tag{5.145}$$

Substituting Eq. (5.145) in Eq. (5.144) and evaluating the integral one obtains

$$\frac{G}{A_p} = -\frac{\epsilon\kappa\psi_0^2}{2} = -\frac{\sigma_0\psi_0}{2} \tag{5.146}$$

where σ_0 and ψ_0 are the surface charge density and surface potential of the isolated planar double layer, and the term G/A_p is used to emphasize that we are dealing with the free energy per unit area for a planar geometry.

[6] Helmholtz free energy can be used instead of Gibbs free energy if the pressure volume work is zero.

EXAMPLE 5.10

Electric Potential Energy. Consider the case of two planar double layers with constant surface potentials ψ_a and ψ_b brought together to a separation distance h from infinity. Derive an expression for the interaction potential energy per unit area between the two double layers at a separation distance h. We will base the analysis on the linearized Poisson–Boltzmann equation, namely, the Debye–Hückel approximation.

Solution The geometry under consideration is given in Figure 5.25. Note that the coordinate system used in this example is slightly different from the one used in Figure 5.19. The location of the origin is shifted in this example to ensure consistency with the original derivations of Hogg *et al.* (1966), which would lead to the expressions obtained here. The solution of the linearized Poisson–Boltzmann equation for this geometry will be of the form, *cf.*, Eq. (5.100),

$$\psi = A_1 \cosh(\kappa x) + B_1 \sinh(\kappa x)$$

where the constants A_1 and B_1 can be obtained from the boundary conditions. The boundary conditions are $\psi = \psi_a$ at $x = 0$ and $\psi = \psi_b$ at $x = h$. Applying these conditions, the expression for the electrical potential distribution becomes

$$\psi = \psi_a \cosh(\kappa x) + \left(\frac{\psi_b - \psi_a \cosh(\kappa h)}{\sinh(\kappa h)} \right) \sinh(\kappa x) \tag{5.147}$$

The surface charge density at the two plates are obtained from the derivative of Eq. (5.147) evaluated at these surfaces. For the plate at $x = 0$, we have

$$\sigma_a = -\epsilon \left(\frac{d\psi}{dx} \right)_{x=0} = -\epsilon\kappa[\psi_b \text{cosech}(\kappa h) - \psi_a \coth(\kappa h)]$$

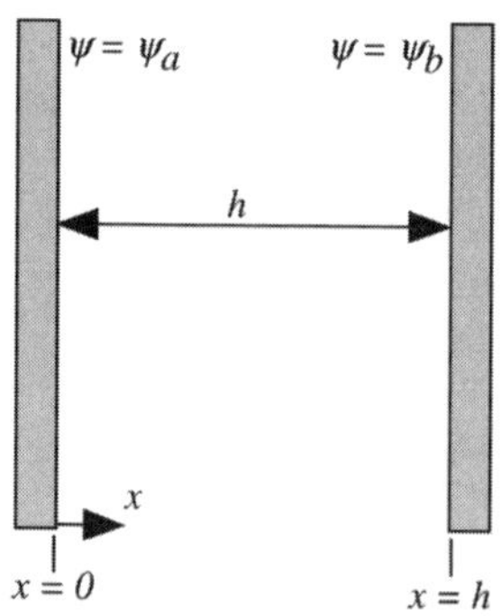

Figure 5.25. Calculation of electrostatic interaction potential between two planar surfaces separated by a distance h.

while for the plate at $x = h$ we have

$$\sigma_b = -\epsilon \left(\frac{d\psi}{dx} \right)_{x=h} = \epsilon\kappa[\psi_b \coth(\kappa h) - \psi_a \text{cosech}(\kappa h)]$$

To evaluate the interaction potential energy using Eq. (5.143), we need to determine the free energies of the system at a separation h and at infinite separation. The free energy per unit area of the system consisting of two planar double layers is given by adding the free energies per unit area of the individual double layers, Eq. (5.146). Utilizing the expressions for the electric potential and the surface charge density obtained above, one can write,

$$\frac{G(h)}{A_p} = -\frac{1}{2}(\sigma_a\psi_a + \sigma_b\psi_b) = \frac{\epsilon\kappa}{2}[2\psi_a\psi_b \text{cosech}(\kappa h) - (\psi_a^2 + \psi_b^2)\coth(\kappa h)]$$

and

$$\frac{G(\infty)}{A_p} = -\frac{1}{2}(\sigma_a\psi_a + \sigma_b\psi_b) = -\frac{\epsilon\kappa}{2}(\psi_a^2 + \psi_b^2)$$

Now, from Eq. (5.143), we obtain

$$\frac{\phi(h)}{A_p} = \frac{G(h)}{A_p} - \frac{G(\infty)}{A_p}$$
$$= \frac{\epsilon\kappa}{2}\{(\psi_a^2 + \psi_b^2)[1 - \coth(\kappa h)] + 2\psi_a\psi_b \text{cosech}(\kappa h)\} \quad (5.148)$$

■

Equation (5.148) is the well known HHF (Hogg–Healy–Fuerstenau) (Hogg *et al.*, 1966) expression for the interaction potential energy per unit area between two infinite planar surfaces. This expression applies to constant potential surfaces and is valid in the Debye–Hückel limit of low surface potentials. Integration of the interaction force, Eq. (5.109), would also yield Eq. (5.148), thus emphasizing that the approach based on the evaluation of the interaction force from integration of the stresses and the approach based on the evaluation of the free energy changes are identical. An interesting feature of this result is that the interaction energy remains finite even when one of the surfaces is assigned a zero potential. A similar derivation of the interaction energy for constant surface charge density boundary conditions on the plates can also be done, which leads to an expression for the interaction potential energy originally given by Usui (1973).

5.6 ELECTROSTATIC INTERACTIONS BETWEEN CURVED GEOMETRIES

The electrostatic force or potential energy between two charged bodies, other than parallel plates, is difficult to obtain analytically. In fact, if we note closely, the interaction force per unit area between two parallel plates obtained in the previous section was based on various assumptions about the behavior of electric double layers. Most prominently, the closed-form solutions were obtained *via* the Debye–Hückel approximation, *i.e.*, using the linearized form of the Poisson–Boltzmann equation. A general analytical solution for the electrostatic interaction force between two parallel plates based on the non-linear Poisson–Boltzmann equation, even for the case of symmetric ($z : z$) electrolytes, is non-trivial, see McCormack *et al.* (1995).

In this context, noting that analytic solutions for the electric potential fields around individual spherical and cylindrical objects were difficult to obtain except for some limiting conditions, it is clearly discernable that rigorous calculations of the electrostatic interaction forces between particles of spherical and other curved geometries based on analytic approaches are still elusive. It is, therefore, becoming more commonplace to utilize numerical procedures for solving the Poisson–Boltzmann equation to obtain the interaction forces for curved geometries. Notwithstanding this fairly recent trend, there have been numerous efforts over the past several decades to explore analytic means for estimating the electrostatic interaction force and potential energy between particles of various geometries. The most common geometries treated were those of two spherical particles, or a spherical particle interacting with a planar surface (infinite flat plate). While the sphere-sphere interactions are of interest in the context of coagulation, the sphere-plate interactions are relevant to particle deposition on surfaces. Before proceeding further with descriptions of some of the most prominent approximate analyses for such systems, let us list the approximations made in most of these approaches, so that it becomes relatively easy to establish the regimes of validity of these solutions.

The most common assumptions made in obtaining the electrostatic interaction force and potential energy for particles include:

1. Low surface potentials, and symmetric ($z : z$) electrolytes, which allow the use of the linearized form of the Poisson–Boltzmann equation, Eq. (5.91).
2. The particle surfaces are maintained either at constant surface potential or constant charge density as they approach each other.
3. Only two particles are interacting in an infinite domain. In other words, the interaction occurs in the limiting case of an infinitely dilute suspension of particles.
4. When two charged particles are at a sufficiently large separation, such that their individual double layers do not substantially overlap, the electric potential distribution around the particles may be obtained by summing the potentials due to the two individual particles. This assumption is generally known as the *linear superposition* approximation.
5. The surfaces are perfectly smooth.

The first three assumptions pertain to the governing Poisson–Boltzmann equation and the imposed boundary conditions. These limiting scenarios were explored in the context of the interaction between two overlapping planar double layers. It is not too difficult to visualize assumption 3 for planar interacting double layers, as the two charged surfaces can be considered as particles with infinitely large radii.

Assumption 4 refers to the separation distance between the two particles relative to their sizes. The last assumption pertains to the geometry of the particles. It is important to distinguish between these three classes of assumptions, namely, assumptions pertaining to the Poisson–Boltzmann equation, assumptions related to infinitely dilute systems, and assumptions related to the geometry of the interacting surfaces. In fact, ignoring these three classes of assumptions has often led to numerous erroneous analyses of colloidal behavior in the literature.

Most analytical expressions for the electrostatic double layer interaction force or potential energy for curved geometries are based on several of the aforementioned assumptions. It should be borne in mind that each of these assumptions, which is valid for certain limiting conditions, can lead to considerably erroneous results when such limiting conditions are violated. For instance, we have seen earlier that the application of the Debye–Hückel assumption at high surface potentials will provide a wrong prediction of the electrostatic potential distribution or the interaction force for planar surfaces. Similarly, if the linear superposition approximation is applied when the particles are sufficiently close such that $\kappa r \gg 1$ is violated, the resulting predictions of the potential distribution and the electrostatic force will be erroneous. Such caution should be exercised for every assumption listed above. Therefore, when considering a particular analytic expression for the electrostatic double layer interaction force between two particles, it is extremely important to consider the underlying assumptions that led to such a solution. The use of such a solution should be avoided if any of these assumptions are violated.

With this note of caution, we can now proceed with discussing a few approximate expressions for the interaction force between curved geometries. We will primarily limit our attention to two classical techniques that are most commonly used for evaluating the interaction force and potential energy for curved surfaces. These are the *Derjaguin approximation* (DA) and the *linear superposition approximation* (LSA). Following this, we will briefly discuss some of the more recent techniques for evaluating the interaction force between colloidal particles of various geometries, including rough surfaces.

5.6.1 The Derjaguin Approximation

Derjaguin (1934) recognized that in many cases two macrobodies have significant interaction between them only when the distance of closest approach between these bodies is small compared to the radii of curvature of the macrobodies. Derjaguin proposed for the case of interacting spheres, for small separations compared to the radii of the spheres, that elements on each sphere interact as parallel plane elements at the same separation, Russel *et al.* (1989). The total sphere-sphere interaction is a sum over the interaction between the planar infinitesimal elements at different distances of separation, Ross and Morrison (1988).

We will now expand on the Derjaguin approximation to arrive at a force expression between two interacting spheres. Figure 5.26 illustrates the geometrical approximation needed for developing an expression of force between two spheres separated by distance h at the closest points. Let F_{sp} be the total force on the spheres and

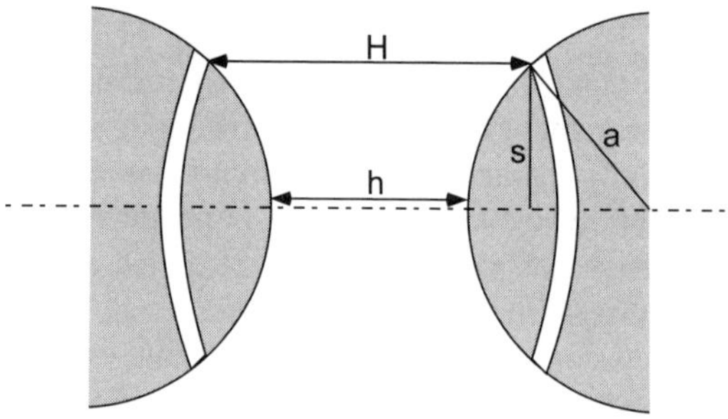

Figure 5.26. Derjaguin's geometric approximation.

$F_p^* = F_p/A_p$ be the force per unit area between two infinite parallel plates. Then, the total force between the spheres can be written as

$$F_{sp} = \int_0^a 2\pi s F_p^* ds \tag{5.149}$$

where s is the radius of a ring-shaped area on the spherical surface, see Figure 5.26.

Now, from geometrical considerations, one can write for the separation distance H between the two ring-shaped areas, one on each sphere

$$\frac{H-h}{2} = a - \sqrt{a^2 - s^2} \tag{5.150}$$

Differentiation of Eq. (5.150) leads to

$$dH = \frac{2sds}{\sqrt{a^2 - s^2}} \tag{5.151}$$

In the region of interest, *i.e.*, facing element areas, $a^2 \gg s^2$ and Eq. (5.151) reduces to

$$dH = \frac{2s}{a} ds \tag{5.152}$$

The force between the spheres then becomes[7]

$$F_{sp} = \int_h^{h+2a} 2\pi s F_p^* \left[\frac{a}{2s} dH\right] \tag{5.153}$$

or

$$F_{sp} = \pi a \int_h^{h+2a} F_p^* dH \tag{5.154}$$

It should be noted that in most texts, the upper limit for the integration is taken as infinity.

[7]The upper integration limit is traditionally taken as infinity.

A convenient expression for the electrostatic force per unit area between two interacting plates is given by Eq. (5.111-a) where

$$F_p^* = \frac{\epsilon \kappa^2 \psi_s^2}{\cosh(\kappa h) + 1} \tag{5.155}$$

Making use of Eq. (5.155), the electrostatic force between two interacting spheres is then given by

$$F_{sp} = \pi a \epsilon \kappa^2 \psi_s^2 \int_h^{h+2a} \frac{dH}{\cosh(\kappa H) + 1} \tag{5.156}$$

Upon integration, one obtains

$$F_{sp} = \pi a \epsilon \kappa \psi_s^2 \left[\tanh \frac{\kappa H}{2} \right]_h^{h+2a} \tag{5.157}$$

Recognizing that $a \gg h$, the upper limit of the integration can be set to $2a$, and setting $\tanh \kappa a$ to 1 for $\kappa a \gg 1$, one obtains

$$F_{sp} = \pi a \epsilon \kappa \psi_s^2 \left(1 - \tanh \frac{\kappa h}{2} \right) \tag{5.158}$$

or

$$F_{sp} = 2\pi a \epsilon \kappa \psi_s^2 \left(\frac{\exp(-\kappa h)}{1 + \exp(-\kappa h)} \right) \tag{5.159}$$

For $\kappa h \gg 1$, one obtains

$$F_{sp} = 2\pi a \epsilon \kappa \psi_s^2 \exp(-\kappa h) \tag{5.160}$$

The corresponding interaction potential energy for the force given in Eqs. (5.158) or (5.159) is given as

$$\phi_{sp} = 2\pi a \epsilon \psi_s^2 \ln \left[1 + \exp(-\kappa h) \right] \tag{5.161}$$

For $\kappa h \gg 1$, one obtains

$$\phi_{sp} = 2\pi a \epsilon \psi_s^2 \exp(-\kappa h) \tag{5.162}$$

The general expression for the interaction potential energy of two unequal spheres of radii a_1 and a_2 having surface potentials ψ_1 and ψ_2 was given by Hogg *et al.* (1966). They obtained this expression by applying Derjaguin's technique to the interaction energy per unit area given by Eq. (5.148). The resulting expression, commonly known as the HHF (Hogg–Healy–Fuerstenau) expression, is given by

$$\phi_{sp} = \frac{\pi \epsilon a_1 a_2}{(a_1 + a_2)} \left\{ 2\psi_1 \psi_2 \ln \left[\frac{1 + \exp(-\kappa h)}{1 - \exp(-\kappa h)} \right] + (\psi_1^2 + \psi_2^2) \ln[1 - \exp(-2\kappa h)] \right\} \tag{5.163}$$

If the two spheres are held at the same surface potential, *i.e.*, $\psi_1 = \psi_2 = \psi_s$, then the above expression simplifies to

$$\phi_{sp} = \frac{4\pi\epsilon\psi_s^2 a_1 a_2}{(a_1 + a_2)} \ln[1 + \exp(-\kappa h)] \tag{5.164}$$

Letting $a_1 = a_2 = a$, one reverts to the expression (5.161) derived earlier using the Derjaguin approximation.

Although the HHF equation is widely used, its applicability should be considered very carefully. The electric potential must be small enough, *i.e.*, $ze\psi/(k_B T) < 1$, and the radii of the spheres must be large enough as compared to the thickness of the double layer, $\kappa a > 10$. For Derjaguin's approximation to be valid, all the interaction energy should come from a small area around the point of closest approach.

White (1983) generalized Derjaguin approximation to evaluate the interaction energy between two curved bodies. Let $\phi_p^*(H)$ be the energy per unit area associated with half-space (plate) 1 interacting with half-space (plate) 2. In this approximation, the total interaction energy of bodies 1 and 2, $\phi(h)$, corresponding to a distance of closest approach h is given by the Derjaguin approximation as

$$\phi(h) = \int_{S_1} \phi_p^*(H) dS_1 \tag{5.165}$$

where S_1 is the surface area of body 1.

White very elegantly showed that the interaction energy is given by

$$\phi(h) = \frac{2\pi}{\sqrt{\lambda_1\lambda_2}} \int_h^\infty \phi_p(H) dH \tag{5.166}$$

where

$$\lambda_1\lambda_2 = \left(\frac{1}{R_1} + \frac{1}{R_1'}\right)\left(\frac{1}{R_2} + \frac{1}{R_2'}\right) + \sin^2\varphi\left(\frac{1}{R_1} - \frac{1}{R_2}\right)\left(\frac{1}{R_1'} - \frac{1}{R_2'}\right) \tag{5.167}$$

Here, R_1 and R_2 are the principal radii of curvature of body 1, R_1' and R_2' are the principal radii of curvature of body 2, and φ is the angle between the principal axes of bodies 1 and 2.

By definition, it follows that the force $F(h)$ exerted by one body on another is given by

$$F(h) = -\frac{d\phi}{dh} = \frac{2\pi}{\sqrt{\lambda_1\lambda_2}} \phi_p^*(h) \tag{5.168}$$

Equation (5.168) provides a very convenient expression for the evaluation of the interaction force between two bodies from the system geometric properties and the plate-plate interaction energy per unit area.

Equation (5.166) and consequently, Eq. (5.168) are valid approximations for *any* type of interaction energy and they are not restricted to electrostatic interaction between surfaces. One can apply Eq. (5.166) with confidence to any given interaction between two bodies provided that the conditions

$$\frac{L_0}{R_0} \ll 1 \tag{5.169-a}$$

and

$$\frac{h}{R_0} \ll 1 \tag{5.169-b}$$

are satisfied, where L_0 is the length scale on which the interaction decays to zero and R_0 is the smallest radius of curvature of the two bodies. For two interacting spheres of radii a_1 and a_2, one obtains $R_1 = R_2 = a_1$ and $R_1' = R_2' = a_2$. Using these in Eq. (5.167) gives[8]

$$\lambda_1\lambda_2 = \left(\frac{a_1 + a_2}{a_1 a_2}\right)^2 \tag{5.170}$$

The interaction energy is given by Eq. (5.166) as

$$\phi(h) = \frac{2\pi a_1 a_2}{(a_1 + a_2)} \int_h^\infty \phi_p^*(H)dH \tag{5.171}$$

and the interaction force as

$$F(h) = \frac{2\pi a_1 a_2}{(a_1 + a_2)} \phi_p^*(h) \tag{5.172}$$

The interaction energy per unit area between two plates carrying the same electric charge is given by Eq. (5.138-b) as

$$\phi_p^* = 2\epsilon\kappa\psi_s^2 \left[\frac{\exp(-\kappa h)}{1 + \exp(-\kappa h)}\right] \tag{5.173}$$

leading, for the case of two interacting spheres, to

$$\phi(h) = \frac{4\pi\epsilon\kappa a_1 a_2 \psi_s^2}{(a_1 + a_2)} \int_h^\infty \frac{\exp(-\kappa H)dH}{1 + \exp(-\kappa H)}$$

Upon integration one obtains

$$\phi(h) = \frac{4\pi\epsilon a_1 a_2 \psi_s^2}{(a_1 + a_2)} \left\{\ln\left[1 + \exp(-\kappa h)\right]\right\} \tag{5.174}$$

The expression for the force between two spheres is given by combining Eqs. (5.172) and (5.173)

$$F(h) = \frac{4\pi\epsilon\kappa a_1 a_2 \psi_s^2 \exp(-\kappa h)}{(a_1 + a_2)[1 + \exp(-\kappa h)]} \tag{5.175}$$

[8] In this case, due to symmetry, the angle $\varphi = 0$.

For the case of two equal spheres, $a_1 = a_2$, Eqs. (5.174) and (5.175) revert respectively to Eqs. (5.161) and (5.159).

For the configuration of two crossed cylinders of radii a_1 and a_2, which is of particular interest due to its use in force measurements, we have

$$R_1 = a_1, \quad R_2 = \infty \tag{5.176-a}$$

$$R_1' = a_2, \quad R_2' = \infty \tag{5.176-b}$$

and $\varphi = \pi/2$. Using these parameters, we obtain

$$\lambda_1 \lambda_2 = \frac{1}{a_1 a_2} \tag{5.177}$$

and the expressions for the interaction between two crossed cylinders are given as

$$\phi(h) = 2\pi \sqrt{a_1 a_2} \int_h^{\infty} \phi_p^*(H) dH \tag{5.178}$$

and

$$F(h) = 2\pi \sqrt{a_1 a_2} \phi_p^*(h) \tag{5.179}$$

White's approximation approach is very convenient to use to derive expressions of interaction energy and force for curved surfaces.

5.6.2 Linear Superposition Approximation

The basic principle underlying the linear superposition approximation (LSA) is that the electric potential ψ at a location far away from two charged particles can be approximated fairly well as the sum of the potentials due to the individual particles. Consider the geometry shown in Figure 5.27, which depicts two spherical particles with surface potentials ψ_1 and ψ_2 separated by a distance r. Consider a point P located on the plane S between the two spheres such that its position vectors from the centers of spheres 1 and 2 are given by $\mathbf{r}_1$ and $\mathbf{r}_2$, respectively. Provided the point P is sufficiently far away from the centers of either sphere, the linear superposition approximation allows us to write

$$\psi_P = \psi_P^{(1)} + \psi_P^{(2)} \tag{5.180}$$

where ψ_P is the total electrical potential at point P, and $\psi_P^{(1)}$ and $\psi_P^{(2)}$ are the electrical potentials at the point due to each of the spheres. In a similar manner, the electric field at location P can be expressed as

$$\mathbf{E}_P = \mathbf{E}_P^{(1)} + \mathbf{E}_P^{(2)} \tag{5.181}$$

where $\mathbf{E}_P$ is the total electric field strength at location P.

Equations (5.180) and (5.181) allow the construction of solutions for the electric potential distribution and field strength for two interacting particles from the corresponding expressions for isolated particles. Once these quantities are determined, it is straightforward to obtain the interaction force by integrating the stress tensor over the intervening plane S in Figure 5.27.

Bell *et al.* (1970) first proposed the LSA technique to obtain the interaction force and potential energy between two spherical particles. They used the solutions for the linearized Poisson–Boltzmann equation around isolated spheres, see Eqs. (5.73-a) and (5.73-b), and obtained the following expression for the magnitude of the interaction force between two spheres of radii a_1 and a_2

$$F(r) = 4\pi\epsilon\left(\frac{k_B T}{ze}\right)^2 \Psi_1\Psi_2\kappa^2 a_1 a_2 \frac{(1+\kappa r)}{(\kappa r)^2}\exp[-\kappa(r-a_1-a_2)] \qquad (5.182)$$

where Ψ_1 and Ψ_2 are the dimensionless surface potentials of the two spheres, and r is the separation distance between the centers of the two spheres. The force is directed along the line joining the centers of the two spheres. Integrating the force over r yields the interaction potential energy

$$\phi(r) = 4\pi\epsilon\left(\frac{k_B T}{ze}\right)^2 \Psi_1\Psi_2\kappa^2 \frac{a_1 a_2}{r}\exp[-\kappa(r-a_1-a_2)] \qquad (5.183)$$

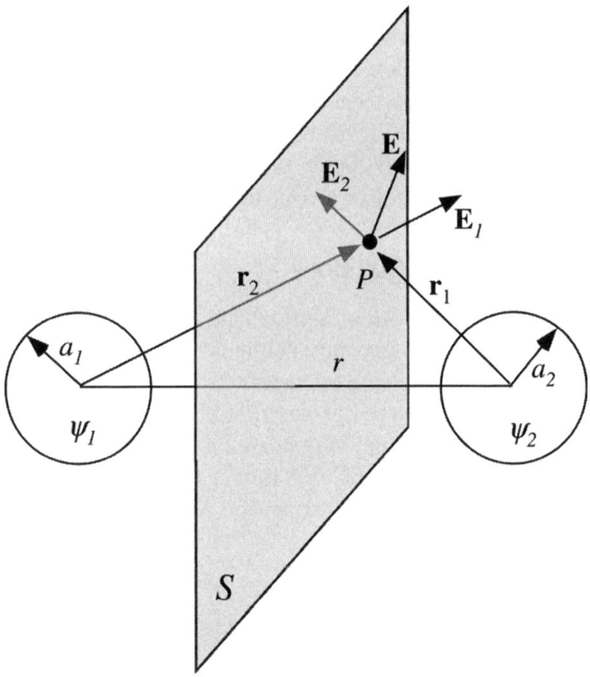

Figure 5.27. A schematic representation of the linear superposition procedure.

A more general form of the LSA expression for the interaction potential is

$$\phi(r) = 4\pi\epsilon \left(\frac{k_B T}{ze}\right)^2 Y_1 Y_2 \kappa^2 \frac{a_1 a_2}{r} \exp[-\kappa(r - a_1 - a_2)] \tag{5.184}$$

where Y_1 and Y_2 are the "effective" surface potentials on the spheres. For small potentials (Debye–Hückel limit), $Y_1 \simeq \Psi_1$ and $Y_2 \simeq \Psi_2$. However, for large surface potentials, Y_1 and Y_2 are more complex functions of the surface potentials. No exact analytical functional forms for these parameters are available, although several approximate relationships have been provided. A common approximation for the effective potential Y_i corresponding to large κa is

$$Y_i = 4\tanh(\Psi_i/4) \tag{5.185}$$

where Ψ_i represents the dimensionless potential on sphere i.

It should be noted that the linear superposition approximation can be performed for particles of different geometry, provided that analytic expressions for the potential distribution around the isolated particles are available. Furthermore, the solution around an isolated particle need not be based on the linearized Poisson–Boltzmann equation as shown above, but can involve the solution of the nonlinear version of the equation as well. One should, however, exercise caution regarding the use of the LSA expression for the interaction potential, Eq. (5.183). First, the linear superposition approximation restricts its use to the calculation of the interaction potential at large inter-particle separations. Secondly, use of the linearized version of the Poisson–Boltzmann equation restricts the LSA expression to the case of low surface potentials. In fact, it is clearly evident that Eq. (5.183) will provide incorrect results at small separation distances if one of the surface potentials is assigned a value of zero. In this case, the interaction potential predicted by Eq. (5.183) will be zero. However, we noted earlier that the interaction potential between two double layers remains finite at small separations, even when one of the surfaces has zero electrical potential.

5.6.3 Other Approximate Solutions

The Derjaguin and linear superposition approximation methods described in the previous subsections constitute two of the most commonly used leading order approximate techniques that provide simple analytic results for the interaction force and potential energy for particles with curved geometries. Numerous other results have been provided over the years that attempt to improve on the assumptions made in obtaining the interaction force and potential. A common approach was to use higher order series terms resulting from the Taylor expansion of the sinh term in the non-linear Poisson–Boltzmann equation:

$$\sinh y \simeq y + \frac{y^3}{3!} + \frac{y^5}{5!} + \cdots$$

Ohshima *et al.* (1982) used such a technique to obtain more accurate expressions for the interaction energy between two spheres. White (1977) presented a series solution of the Poisson–Boltzmann equation based on a perturbation method. Glendinning and Russel (1983) gave a series solution for the linearized Poisson–Boltzmann equation for two equal sized spheres based on a multipole expansion technique.

There has been a growing interest in obtaining estimates of the interaction energy and force between particles of arbitrary shapes, particles and surfaces with roughness, and particles with a distribution of charges. A technique for calculation of interaction energy for particles of any geometry including rough surfaces was recently developed using an extension of Derjaguin's approach (Bhattacharjee *et al.*, 1998). This technique, called surface element integration (SEI), provides a fairly good approximation for the interaction energy between a particle and a planar surface over a wide range of particle sizes and ranges of interaction. The technique, although numerical, provides a facile route for estimating the interaction energy for surfaces that have arbitrary geometric shapes.

The different analytic expressions for the interaction force and energy commonly used in literature have been extensively tabulated in texts (Russel *et al.*, 1989; Elimelech *et al.*, 1995). It should be noted that all analytic solutions will have some form of approximation embedded in it, and hence, one needs to be careful in selecting an expression that is appropriate for a given situation. As mentioned earlier, numerical approaches, based on finite difference or finite element solution of the Poisson–Boltzmann equation, are becoming more common (Carnie *et al.*, 1994; Das *et al.*, 2003) for evaluation of the electric double layer interaction forces. It seems that with enhanced computational infrastructure, such numerical solutions will be used more and more extensively over the coming years. In Chapter 14, we will discuss numerical solution of the Poisson–Boltzmann equation employing finite element analysis.

As things stand, one faces considerable challenges in numerically solving the Poisson–Boltzmann equation to a high degree of accuracy. It should be noted that finite difference or finite element techniques are based on approximate formulations of the spatial derivatives of the potential. In finite difference techniques, the derivatives are often based on Taylor expansions, while in finite element methods such formulations are based on piecewise continuous functions (orthogonal polynomials) representing the potential distribution and its spatial derivatives. In light of this, it is not too difficult to realize that the calculation of the stress tensor and the interaction force, which requires the spatial derivative of the electric field (or the second-order derivative of the potential distribution), can be quite erroneous if the numerical discretization is not of high order. The problem is generally tractable for simple geometries like spheres or cylinders. However, for arbitrary particle shapes, numerical solutions of the non-linear Poisson–Boltzmann equation in its three-dimensional form for asymmetric electrolytes is still considered to be a fairly involved task.

5.7 MODELS OF SURFACE POTENTIALS

The surface charge and the surface potential are very much dependent on the surface chemistry (and hence on the origin of the surface charges) and on the ionic concentration in the bulk solution. A number of models have been proposed for predicting the surface potentials in terms of appropriate variables (*e.g.*, Haydon, 1964; Sparnaay, 1972; Chan *et al.*, 1975, 1976; Hunter, 1981, 1991; Takamura and Wallace, 1988). Two such models will be discussed here in order to illustrate the connection between ionic bulk concentration and the surface potential. These models clearly illustrate that by changing the ionic strength, one not only changes the Debye length, but may also change the surface potential.

5.7.1 Indifferent Electrolytes

Let us consider, as an example, silver iodide particles (AgI) suspended in an aqueous solution containing AgI as the major electrolyte. Our goal here is to relate the bulk ionic concentration to the surface potential of the AgI particles by considering the chemical equilibrium between the particle and the solution.

First, the difference between the *electrochemical* potentials $\Delta\mu_e^k$ of the ions of the kth type in the solid phase and in solution is given by

$$\Delta\mu_e^k = \Delta\mu_c^k + ez_k\Delta\psi \tag{5.186}$$

where z_k is the valence of the kth ionic species, $\Delta\psi$ is the electrostatic potential difference between the surface and the bulk solution, $\Delta\mu_c^k$ is the *chemical* potential difference for the kth species between the surface and the bulk solution, and e is the elementary charge (1.602×10^{-19} C). At equilibrium $\Delta\mu_e^k = 0$, so that

$$\Delta\psi = -\frac{\Delta\mu_c^k}{ez_k} \tag{5.187}$$

The above equation relates the difference in the electrostatic potentials between the surface (of the particle) and the bulk solution to the corresponding chemical potential difference between the two. If the electrostatic potential in the bulk solution is taken as zero, then $\Delta\psi$ is simply ψ_s, the surface potential of the solid particle.

We shall now consider reasonable "models" for the chemical potentials of the liquid phase and the solid phase so that $\Delta\mu_c^k$, and hence, ψ_s can be estimated.

- Let us now assume, for convenience, that the chemical potential of species k in the bulk solution, denoted by $\mu_{c,b}^k$, can be approximated by that of an ideal solution, *i.e.*,

$$d\mu_{c,b}^k = -\frac{k_BT}{n_k}dn_k = -k_BTd\,(\ln n_k) \tag{5.188}$$

  where n_k is the ionic number concentration of the kth ionic species (say, in m^{-3}).
- For the solid phase, assume that the changes in the concentration of the solution have very little effect on the chemical potentials, $\mu_{c,p}^k$, of Ag^+ and I^- (where the subscript p stands for "particle").

Therefore, one can write

$$d\Delta\mu_c^k = d\left(\mu_{c,p}^k - \mu_{c,b}^k\right) \cong d\mu_{c,b}^k \tag{5.189}$$

so that

$$d(\Delta\psi) = d\psi_s = \frac{k_B T}{e z_k} d(\ln n_k) \tag{5.190}$$

Equation (5.190) indicates that a ten-fold increase in n_k leads to a relatively small change in the surface potential, a direct consequence of the assumption that the concentrations of the ions in the bulk have a negligible influence on the ionic composition of the particle surface.

The ions responsible for the change in the surface potential are called the *potential determining ions* (pdi). For the system chosen, namely, the AgI solution, the ions are Ag^+ and I^-. According to Eq. (5.190), the Nernst equation, the interphase potential difference (between the solid and the bulk of the liquid) changes by about 0.06 Volts when the concentration of potential determining ions changes by a factor of 10 (Russel *et al.*, 1989).

5.7.2 Ionizable Surfaces

We shall use a second example with a very different assumption to illustrate the effect of the ionic concentration in the bulk solution on surface potential (Russel *et al.*, 1989).

Suppose we have a surface that contains certain molecules that ionize as follows:

$$\mathrm{AH} \rightleftharpoons \mathrm{A}^- + \mathrm{H}^+ \tag{5.191}$$

The dissociation constant of the above ionization reaction is given by

$$K = \frac{[\mathrm{A}^-]_S [\mathrm{H}^+]_0}{[\mathrm{AH}]_S} \tag{5.192}$$

where $[\mathrm{A}^-]_S$ and $[\mathrm{AH}]_S$ denote the number concentrations (per unit area) at the surface of the solid phase and $[\mathrm{H}^+]_0$ denotes the hydronium ion number concentration in the electrolyte solution in the immediate vicinity of the surface. Note that the surface charge density q_s is simply $-e[\mathrm{A}^-]_S$. Our task here is to write $[\mathrm{A}^-]_S$, and therefore q_s, in terms of the surface potential ψ_s and other measurable quantities such as K and the bulk ionic concentration.

Let the total number concentration of the ionizable groups on the surface be

$$[n]_S = [\mathrm{A}^-]_S + [\mathrm{AH}]_S \tag{5.193}$$

By combining Eqs. (5.192) and (5.193), one can write

$$[\mathrm{A}^-]_S = \frac{K\,[n]_S}{K + [\mathrm{H}^+]_0} \tag{5.194}$$

The concentration $[\mathrm{H}^+]_0$ can be related to the concentration of H^+ in the bulk, *i.e.*, $[\mathrm{H}^+]_b$, by recognizing that the ionic concentration follows the Boltzmann distribution, which for a (1 : 1) electrolyte, is

$$[\mathrm{H}^+] = [\mathrm{H}^+]_b \exp\left(-\frac{e\psi}{k_B T}\right) \tag{5.195}$$

Since we have $\psi = 0$ in the bulk, Eq. (5.195) gives

$$[\mathrm{H}^+]_0 = [\mathrm{H}^+]_b \exp\left(-\frac{e\psi_s}{k_B T}\right) \tag{5.196}$$

The above equation relates the ionic concentration in the solution near the surface to that in the bulk solution.

Recognizing that the surface charge density is given by

$$q_s = -e[A^-]_s \tag{5.197}$$

combining Eqs. (5.194), (5.196), and (5.197) leads to

$$q_s = \frac{-eK[n]_s}{K + [H^+]_b \exp\left(-\frac{e\psi_s}{k_B T}\right)} \tag{5.198}$$

We still need to eliminate q_s from the above equation, *i.e.*, replace q_s in terms of ψ_s, so that ψ_s can be determined from $[H^+]_b$. We will do this by taking advantage of Eq. (5.52), where

$$q_s = 2(2\epsilon k_B T[H^+]_b)^{1/2} \sinh\left(-\frac{e\psi_s}{2k_B T}\right) \tag{5.199}$$

Eliminating q_s from Eqs. (5.198) and (5.199) yields

$$\sinh\left(-\frac{e\psi_s}{2k_B T}\right) = \frac{1}{2(2\epsilon k_B T[H^+]_b)^{1/2}}\left[\frac{-eK[n]_s}{K + [H^+]_b \exp\left(-\frac{e\psi_s}{k_B T}\right)}\right] \tag{5.200}$$

The surface potential, ψ_s, is related to the hydronium ion number concentration in the bulk solution, $[H^+]_b/1000N_A$. For given $[n]_s$ and K values, the solution of Eq. (5.200) provides the surface potential for different values of $[H^+]_b$. Table 5.2 gives the surface potential, ψ_s, for $[n]_s$ of 0.2×10^{18} ions/m^2, which is the equivalent of one ion per 5 nm^2. Different values for the dissociation constant, K, are used in the table. For a given K value, the absolute value of the surface potential decreases with increasing hydronium ion molarity. When the dissociation constant is very large, *e.g.*, $K = 10^2$, $[AH]_s$ becomes negligible and the total number concentration of ionizable groups, $[n]_s$, becomes equal to the surface acidic groups denoted by $[A^-]_s$. In this limiting case, the surface potential is simply given by Eq. (5.52) or Eq. (5.199) where $q_s = -e[n]_s$. The values of the surface potential for $K = 10^2$ become those denoted by the vertical dashed line of Figure 5.9.

TABLE 5.2. Surface Potential for Different Dissociation Constants and Hydronium Ion Molarities for $[n]_s = 0.2 \times 10^{18}$ ions/m^2 = 0.2 ions/nm^2.

$[H^+]_b$, M	Surface Potential, ψ_s, mV		
	$K = 10^{-4}$ M	$K = 10^{-2}$ M	$K = 10^{2}$ M
10^{-4}	−68.4	−141.5	−204.9
10^{-3}	−19.3	−83.0	−145.9
10^{-2}	−1.3	−30.6	−88.2
10^{-1}	−0.04	−3.5	−39.7

Note: The units of K in Eq. (5.198) are in ions/m^3. To convert K in units of molarity to a number concentrations multiply by $1000N_A$.

Models based on more than one dissociation reaction are given by Hunter (1981). Depending on the origin of the charge, other models can be formulated to account for the variation in the surface charge and the ionic strength of the bulk solution. For example, such models can be used to estimate bitumen and solid surface potentials at various levels of pH (Takamura and Wallace, 1988).

5.8 ZETA POTENTIAL

In our previous discussion of the Gouy–Chapman treatment of the diffuse electric double layer we have assumed that the ions are point charges and that the inner boundary of the electric double layer is located at the surface of the particle or material under consideration. The location of the outer boundary of the electric double layer is characterized by the inverse Debye length. To this end, the surface potential of a particle in a dielectric medium becomes that "seen" by the fluid surrounding the particle. For a stationary surface, the no-slip velocity occurs at the charged particle surface itself.

The Gouy–Chapman treatment of the diffuse double layer runs into some difficulties at small κx values when the surface potential, ψ_s, is large. Here x is the distance from a charged surface. The coion concentration at the surface can be evaluated from Boltzmann distribution where

$$n^- = n_\infty \exp\left(\frac{ze\psi}{k_B T}\right) \tag{5.201}$$

Setting $n = 1000N_A M$, it follows from Eq. (5.201) that

$$M_s = M_\infty \exp\left(\frac{ze\psi}{k_B T}\right) \tag{5.202}$$

where M_∞ and M_s are the solution molarities far from the surface and at the surface, respectively.

For example, at 25°C and for a 1 : 1 electrolyte at 0.001 M with a surface potential of $\psi_s = 300$ mV, Eq. (5.202) gives a surface molarity for the coions of 118 mol/L, which is not realistic. The inaccuracy of this solution stems from the assumption of point charges and the consequent neglect of the finite ionic diameters (Adamson and Gast, 1997).

In real systems, ions are of finite size and they can approach a surface to a distance not less than their radii. Stern (1924) proposed a model in which the electric double layer inner boundary is given by approximately one hydrated ion radius. This inner boundary is referred to as the Stern plane. The gap between the Stern plane and the surface is denoted as the Stern layer. The electric potential changes from the surface potential ψ_s to a Stern plane potential ψ_d within the Stern layer and it decays to zero far away from the Stern plane as shown in Figure 5.28. The centers of any ions "attached" to the surface are located within the Stern layer and they are considered to be immobile. Ions whose centers are located beyond the Stern plane form the diffuse mobile part of the electric double layer. Consequently, the mobile inner part of the electric double layer is located between one to two radii away from the surface. This boundary is referred to as the shear plane as shown in Figure 5.28(a). It is on this plane where the no-slip fluid flow boundary condition is assumed to apply. The potential at the shear plane is referred to as the electrokinetic potential, more commonly known as the *zeta potential* (ζ). The zeta potential is marginally different in magnitude from the Stern potential ψ_d as shown in Figure 5.28(b). Shaw (1980) stated that it is customary to identify ψ_d with ζ and experimental evidence suggests that the introduced errors are small. Electrophoretic potential measurements give the zeta potential of a surface. Although one at times refers to a "surface potential", strictly speaking, it is the zeta potential that needs to be specified. The Gouy–Chapman treatment of Section 5.2.1 deals with the diffuse electric layer whose inner boundary is the shear plane.

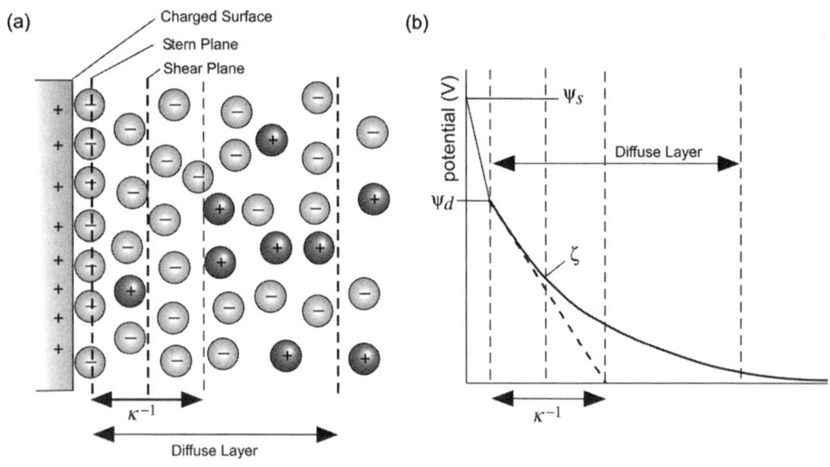

Figure 5.28. (a) Schematic representation of the electric double layer according to the Stern model. (b) Schematic representation of the electric potential profile showing the Debye length, κ^{-1}, and the overall extent of the electric double layer. The diffuse double layer starts from the Stern plane.

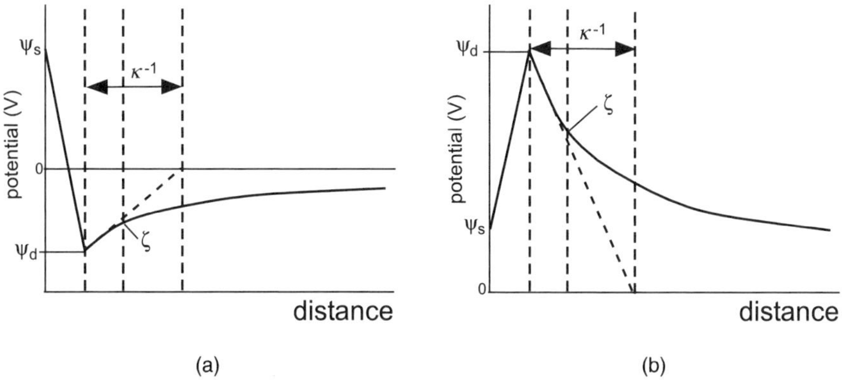

Figure 5.29. Potential variation with distance for a charged surface: (a) potential reversal due to adsorption of surface active or polyvalent counterions, (b) adsorption of surface active coions.

It should be noted that adsorbed ions can have marked effects on the zeta potential when compared with the surface potential. Adsorbed polyvalent or surface active counterions can cause reversal of charge to occur within the Stern layer as is shown in Figure 5.29(a), where ψ_s and ζ have different signs. On the other hand, adsorption of surface active coions can create a situation in which ψ_d has the same sign as ψ_s and is greater in magnitude than ψ_s. From the above discussion, it is clear that zeta potential measurements do not give direct information about the surface potential itself when adsorbed ions are present (Shaw, 1980).

The interaction of the mobile portion of the diffuse electric layer with an external or induced electric field gives rise to electrokinetic transport phenomena. For example, when an electric field is applied tangentially along a charged surface, the electric field will exert a force on the ions within the mobile diffuse electric layer close to the charged surface resulting in their motion. In turn, the moving ions will drag the surrounding liquid along, thus resulting in the liquid's flow. The various types of electrokinetic phenomena will be outlined in Chapter 7.

5.9 SUMMARY OF GOUY–CHAPMAN MODEL

5.9.1 Arbitrary Electrolyte

Charge Density:

$$\rho_f = \sum_i z_i e n_i = e \sum_i z_i n_{i\infty} \exp\left(-\frac{z_i e \psi}{k_B T}\right)$$

Electroneutrality:

$$\sum_i z_i n_i = 0$$

Poisson–Boltzmann Equation:

$$\nabla^2\psi = -\frac{e}{\epsilon}\sum_i z_i n_{i\infty} \exp\left(-\frac{z_i e\psi}{k_B T}\right)$$

Linearized Poisson–Boltzmann Equation, ($z_i e\psi/k_B T < 1$); Debye–Hückel Solution:

$$\nabla^2\psi = -\frac{e}{\epsilon}\sum_i \left[z_i n_{i\infty} - \frac{z_i^2 n_{i\infty} e\psi}{k_B T}\right] = \frac{e^2\psi}{\epsilon k_B T}\sum_i z_i^2 n_{i\infty}$$

or

$$\nabla^2\psi = \kappa^2\psi$$

where

$$\kappa^2 = \frac{2e^2}{\epsilon k_B T}\sum_i \frac{1}{2} z_i^2 n_{i\infty}$$

5.9.2 Symmetrical ($z : z$) Electrolyte

Charge Density:

$$\rho_f = ze(n_+ - n_-) = -2zen_\infty \sinh\left(\frac{ze\psi}{k_B T}\right)$$

Electroneutrality:

$$n_+ - n_- = 0$$

Poisson–Boltzmann Equation:

$$\nabla^2\psi = \frac{2zen_\infty}{\epsilon}\sinh\left(\frac{ze\psi}{k_B T}\right)$$

Linearized Poisson–Boltzmann Equation, ($z_i e\psi/k_B T < 1$); Debye–Hückel Solution:

$$\nabla^2\psi = \frac{2zen_\infty}{\epsilon}\left(\frac{ze\psi}{k_B T}\right) = \frac{2z^2e^2n_\infty}{\epsilon k_B T}\psi$$

or

$$\nabla^2\psi = \kappa^2\psi$$

where

$$\kappa^2 = \frac{2z^2e^2n_\infty}{\epsilon k_B T}$$

5.9.3 Forms of Various Notations

Various notations are used in electrostatics literature. The forms of the equations are different depending on whether number or molar concentrations are used. For example, using the molar concentrations (moles/m^3), one can write the Poisson–Boltzmann equation for a symmetric ($z:z$) electrolyte as

$$\nabla^2\psi = \frac{2z\mathcal{F}c_\infty}{\epsilon}\sinh\left(\frac{z\mathcal{F}\psi}{RT}\right)$$

with

$$\kappa^2 = \frac{2z^2\mathcal{F}^2c_\infty}{\epsilon RT}$$

where c_∞ is the bulk molar concentration of the electrolyte. This type of notation can be easily converted to the notation used in this chapter by recognizing the following relationships:

$$\mathcal{F} = eN_A \qquad \text{Faraday constant}$$

$$R = k_BN_A \qquad \text{Universal gas constant}$$

$$n = cN_A$$

and

$$\mathcal{F}c = ne$$

In this respect, we have

$$\frac{z\mathcal{F}}{RT} \equiv \frac{ze}{k_BT} \qquad \textit{i.e.,} \qquad \frac{e}{k_BT} = \frac{\mathcal{F}}{RT}$$

5.10 NOMENCLATURE

a	radius (of spherical particle), m
A_p	Area of a planar surface, m^2
c	molar concentration, mol/m^3
c_∞	bulk molar electrolyte concentration, mol/m^3
e	elementary charge, C
$\mathbf{E}$	electric field strength, V/m
E	magnitude of electric field, V/m
E_x, E_y, E_z	field components along x, y, and z directions in a Cartesian volume, respectively, V/m
$\mathcal{F}$	Faraday number, C/mol
$\mathbf{F}$	force, N
F_p	interaction force between two planar surfaces, N
F_p^*	interaction force per unit area (generally F_p/A_p) between two planar surfaces, N/m^2

F_{sp}	interaction force between spheres, N
$\mathbf{g}$	gravitational acceleration, m/s^2
h, H	separation distance, m
$\bar{\bar{I}}$	unit tensor
I	ionic strength of an electrolyte solution, mol/L
$\mathbf{i}$	unit vector
k_B	Boltzmann constant (1.38×10^{-23} J/K)
K	dissociation constant, m^{-3}
l	length, m
M	molar concentration of an electrolyte solution, mol/L
N	number of ionic species
N_A	Avogadro's number (6.022×10^{23} mol^{-1})
n_∞	ionic number concentration in the bulk solution, m^{-3}
n_+	number concentration of positive ions, m^{-3}
n_-	number concentration of negative ions, m^{-3}
$\mathbf{n}$	unit surface normal vector
p	fluid pressure, Pa
p_∞	fluid pressure in bulk electrolyte, Pa
Q	total charge, C
Q_p	total charge on a body, C
q_s	surface charge density, C/m^2
R	universal gas constant (8.314×10^3 J/(mol.K)
r	radial position, m
S	surface area, m^2
T	absolute temperature, K
$\bar{\bar{T}}$	total stress tensor, N/m^2
$\bar{\bar{T}}_e$	electric (Maxwell) stress tensor, N/m^2
$\bar{\bar{T}}_p$	fluid stress tensor (pressure in static fluid), N/m^2
$\mathbf{u}$	fluid velocity vector, m/s
V	volume, m^3
z_i	valency of ith ionic species
z	absolute value of valency of a $(z : z)$ electrolyte

Greek Symbols

ϵ	dielectric permittivity of medium, C^2/Nm^2 or C/Vm
$\phi(h)$	electrostatic potential energy at a separation of h between two bodies, J
κ	inverse Debye length, m^{-1}
κ^{-1}	Debye length, m
$\lambda_1 \lambda_2$	geometrical factor in Derjaguin approximation
μ	fluid viscosity, Pa s
$\Delta\mu_k$	electrochemical potential difference for species k, J
Ψ	dimensionless electric potential ($ze\psi/k_BT$)
ψ	electric potential, V

ψ_1, ψ_2	surface electric potentials, V
ψ_d	electric potential at Stern plane, V
ψ_s	true surface electric potential, V
ρ	fluid material density, kg/m^3
ρ_f	volumetric free charge density, C/m^3
σ	surface charge density (also denoted by q_s), C/m^2
∇	del operator, m^{-1}
∇^2	Laplacian operator, m^{-2}
ζ	zeta potential of a surface, V

5.11 PROBLEMS

5.1. A 500 ml of 0.02 M KCl is mixed with 300 ml of 0.001 M $CuSO_4$ and 100 ml of 0.001 M $AlCl_3$ at 25°C. What is the Debye length for this electrolyte mixture?

5.2. A 500 ml of water was added to 200 ml of 0.1 M KNO_3 solution at 25°C. What is the Debye length for this solution?

5.3. The exact solution for dimensionless electric potential due to a charged planar surface was provided by Eq. (5.16). Assuming the dimensionless surface potential $\Psi_s \ll 1$, show the exact solution gives Debye–Hückel expression, Eq. (5.24).

5.4. Show that for small surface potentials and for arbitrary electrolytes, the surface charge density of a planar surface is given by

$$q_s = \epsilon\kappa\psi_s$$

where ψ_s is the surface potential and κ is appropriately defined.

5.5. The solution of the Poisson–Boltzmann equation for the electric potential between two dissimilar planar surfaces held at q_a and q_b surface charge densities was given by Eq. (5.127) where the constants A_2 and B_2 are given by Eqs. (5.128) and (5.129), respectively. In order to solve the Poisson–Boltzmann equations, the boundary conditions need to be stated in terms of the surface potentials.

(a) State the boundary conditions that are equivalent to constant surface charge densities.

(b) Show that for $q_a = q_b = q_s$, the surface potential is related to the surface charge density by

$$\psi_s = \frac{q_s}{\epsilon\kappa \tanh(\kappa h/2)}$$

5.6. Evaluate the interaction force between a cylinder of radius a parallel to a plate using White's approximation.

5.7. In the text, expressions for the interactive electrostatic force between two charged plates having constant surface potentials were provided by Eqs. (5.108) and (5.109) for a $z : z$ electrolyte solution. Making use of the provided force expressions, derive the equivalent expressions for a general electrolyte solution.

5.12 REFERENCES

Adamson, A. W., and Gast, A. P., *Physical Chemistry of Surfaces*, 6th ed., Wiley-Interscience, New York, (1997).

Bell, G. M., Levine, S., and McCartney, L. N., Approximate methods of determining the double-layer free energy of interaction between two charged colloidal spheres, *J. Colloid Interface Sci.*, **33**, 335–359, (1970).

Bhattacharjee, S., Ko, C.-H., and Elimelech, M., DLVO interaction between rough surfaces, *Langmuir*, **14**, 3365–3375, (1998).

Carnie, S. L., Chan, D. Y. C., and Stankovich, J., Computation of forces between spherical colloidal particles – nonlinear Poisson–Boltzmann theory, *J. Colloid Interface Sci.*, **165**, 116–128, (1994).

Chan, D. Y. C., Perram, J. W., White, L. R., and Healy, T. W., Regulation of surface potential at amphoteric surfaces during particle-particle interaction, *J. Chem. Soc. Faraday Trans. 1*, **71**, 1046–1057, (1975).

Chan, D. Y. C., Healy, T. W., and White, L. R., Electrical double-layer interaction under regulation by surface ionization equilibria dissimilar amphoteric surfaces, *J.Chem. Soc. Faraday Trans. l*, **72**, 2844–2865, (1976).

Das, P. K., Bhattacharjee, S., and Moussa, W., Electrostatic double layer force between two spherical particles in a straight cylindrical capillary: finite element analysis, *Langmuir*, **19**, 4162–4172, (2003).

Derjaguin, B. V., Friction and adhesion. IV: The theory of adhesion of small particles, *Kolloid Z.*, **69**, 155–164, (1934).

Dube, G. P., *Ind. J. Phys.*, **17**, 189, (1943).

Dukhin, S. S., and Derjaguin, B. V., Electrokinetic Phenomena, in *Surface and Colloid Science*, vol. 7, E. Matijevic (Ed.), Wiley, (1974).

Elimelech, M., Gregory, J., Zia, X., and Williams, R. A., *Particle Deposition and Aggregation: Measurement, Modelling, and Simulation*, Butterworth, London, (1995).

Everett, D. H., *Basic Principles of Colloid Science*, Royal Society of Chemistry, London, (1988).

Fair, M. C., and Anderson, J. L., Electrophoresis of nonuniformly charged ellipsoidal particles, *J. Colloid Interface Sci.*, **127**, 388–400, (1989).

Feng, J. J., and Wu, W. Y., Electrophoretic motion of an arbitrary prolate body of revolution toward an infinite conducting wall, *J. Fluid Mech.*, **264**, 41–58, (1994).

Glendinning, A. B., and Russel, W. B., The electrostatic repulsion between charged spheres from exact solutions to the linearized Poisson–Boltzmann equation, *J. Colloid Interface Sci.*, **93**, 95–104, (1983).

Gu, Y., The electrical double-layer interaction between a spherical particle and a cylinder, *J. Colloid Interface Sci.*, **231**, 199–203, (2000).

Haydon, K. A., The electrical double-layer and electrokinetic phenomena, pp. 94–158 in *Progress in Surface Science*, vol. 1, Danelli, J. F., *et al.* (Eds.), Academic Press, London, (1964).

Hiemenz, P. C., and Rajagopalan, R., *Principles of Colloid and Surface Chemistry*, 3rd ed., Marcell Dekker, New York, (1997).

Hogg, R. I., Healy, T. W., and Fuerstenau, D. W., Mutual coagulation of colloidal dispersions, *Trans. Faraday Soc.*, **62**, 1638–1651, (1966).

Hsu, J., and Liu, B., Solution to the linearized Poisson Boltzmann equation for a spheroidal surface under a general surface condition, *J. Colloid Interface Sci.*, **183**, 214–222, (1996).

Hunter, R. J., *Zeta Potential in Colloid Science*, Academic Press, London, (1981).

Hunter, R. J., *Foundations of Colloid Science*, Vol. 1, Oxford University Press, Oxford, (1991).

Keh, H. J., and Huang, T. Y., Diffusiophoresis and electrophoresis of colloidal spheroids, *J. Colloid Interface Sci.*, **160**, 354–371, (1993).

Kittel, C., and Kroemer, H., *Thermal Physics*, 2nd ed., W. H. Freeman, San Francisco, CA, (1980).

Loeb, A. L., Overbeek, J. Th. G., and Wiersema, P. H., *The Electric Double-Layer Around a Spherical Colloid Particle*, MIT Press, Boston, (1961).

Lyklema, J., *Fundamentals of Colloid and Interface Science*, Vol. II, Academic Press, London, (1995).

McCormack, D., Carnie, S. L., and Chan, D. Y. C., Calculations of electric double-layer force and interaction free-energy between dissimilar surfaces, *J. Colloid Interface Sci.*, **169**, 177–196, (1995).

McQuarrie, D. A., *Statistical Mechanics*, Harper and Row, New York, (1976).

Ohshima, H., Healy, T. W., and White, L. R., Improvement on the Hogg-Healey-Fuerstenau formulas for the interaction of dissimilar double layers, *J. Colloid Interface Sci.*, **89**, 484–493, (1982).

Ohshima, H., Surface-charge density surface-potential relationship for a spherical colloidal particle in a solution of general electrolytes, *J. Colloid Interface Sci.*, **171**, 525–527, (1995).

Ohshima, H., Surface charge density surface potential relationship for a cylindrical particle in an electrolyte solution, *J. Colloid Interface Sci.*, **200**, 291–297, (1998).

Ohshima, H., Surface charge density/surface potential relationship for a spherical colloidal particle in a salt-free medium, *J. Colloid Interface Sci.*, **247**, 18–23, (2002).

Overbeek, J. Th. G., The role of energy and entropy in the electrical double layer, *Colloids Surf.*, **51**, 61–75, (1990).

Probstein, R. F., *Physicochemical Hydrodynamics, An Introduction*, 2nd ed., Wiley Interscience, New York, (2003).

Ross, S., and Morrison, I. D., *Colloidal Systems and Interfaces*, Wiley, New York, (1988).

Russel, W. B., Saville, D. A., and Schowalter, W. R., *Colloidal Dispersions*, Cambridge University Press, Cambridge, (1989).

Shaw, D. J., *Introduction to Colloid and Surface Chemistry*, 3rd ed., Butterworths, London, (1980).

Sparnaay, M. J., *The Electrical Double Layer*, Pergamon Press, New York, (1972).

Stern, O., Zur theorie der elektrolytischen doppelschicht, *Z. Elecktrochem.*, **30**, 508–516, (1924).

Takamura, K., and Wallace, D., The physical chemistry of the hot water process, *J. Can. Pet. Tech.*, **27**, 98–106, (1988).

Usui, S., Interaction of electrical double layers at constant surface charge, *J. Colloid Interface Sci.*, **44**, 107, (1973).

van de Ven, T. G. M., *Colloidal Hydrodynamics*, Academic Press, London, (1989).

Wang, Z. W., Li, G. Z., Guan, D. R., Yi, X. Z., and Lou, A. J., The surface potential of a spherical colloid particle: Functional theoretical approach, *J. Colloid Interface Sci.*, **246**, 302–308, (2002).

White, L. R., Approximate analytic solution of the Poisson–Boltzmann equation for a spherical colloid particle, *J. Chem. Soc. Faraday Trans. II*, **73**, 577–596, (1977).

White, L. R., On the Deryaguin approximation for the interaction of macrobodies, *J. Colloid Interface Sci.*, **95**, 286–288, (1983).

CHAPTER 6

FUNDAMENTAL TRANSPORT EQUATIONS

This chapter reviews the basic equations governing the transport phenomena of interest. First, a very brief summary of the equations relevant to single-component systems (*i.e.*, systems containing only the solvent) is presented. Following this, we consider multicomponent systems in which constituents called solutes are considered explicitly along with the solvent. Typically, the solutes considered in this context will be treated as point masses, examples of which include ions of a dissolved salt. The introduction of ions to the system requires that we consider, in addition to the equations of fluid mechanics and energy or mass transport, the conservation of ionic mass, the conservation of charge, and the current density. Moreover, the momentum equation needs to be modified in this case to account for the electrical forces arising from the presence of ions in the solution. Finally, this chapter presents a brief introduction to the hydrodynamic interactions, which are primarily an outcome of considering finite sized solute particles (colloidal entities) in a multicomponent system. In this context, the modifications of the transport equations required to assess the transport behavior of multicomponent colloidal suspensions will be presented. A number of excellent textbooks are available on the material discussed in this chapter. Bird *et al.* (1960, 2002) are classic references on transport phenomena. Newman (1991) can be consulted for information regarding transport phenomena in electrochemical systems. The basic vector calculus needed for manipulating the equations of interest is discussed in Aris (1989). A less formal and intuitively more accessible discussion of vector calculus is presented by Schey (1997). The volume by Moon and Spencer (1971)

Electrokinetic and Colloid Transport Phenomena, by Jacob H. Masliyah and Subir Bhattacharjee
Copyright © 2006 John Wiley & Sons, Inc.

is an excellent reference on coordinate systems and differential equations in thirteen orthogonal systems.

6.1 SINGLE-COMPONENT SYSTEM

For systems not undergoing nuclear reactions, one of the basic laws is that of mass conservation. It simply states that for a fixed control volume in space, the rate of mass accumulation within the control volume is the difference between the mass entering and leaving the control volume. In a differential form, the mass conservation law is given by

$$\frac{\partial \rho}{\partial t} = -\nabla \cdot (\rho \mathbf{u}) \tag{6.1}$$

The left hand side term is due to mass accumulation and the right hand side term is due to the net mass flux. Equation (6.1) is also referred to as the continuity equation for a fluid in motion. Here, ρ is the fluid density ($\mathrm{kg/m^3}$) and $\mathbf{u}$ (m/s) is the fluid velocity vector relative to a stationary observer. This velocity is measured using a pitot tube or a laser Doppler velocimeter. The symbol t represents time (s) and ∇ is the del operator ($\mathrm{m^{-1}}$).

When the fluid density is independent of time and space, *i.e.*, steady incompressible flow, the mass conservation equation becomes

$$\nabla \cdot \mathbf{u} = 0 \tag{6.2}$$

The fluid flow of a single-component system under laminar flow conditions with constant density and viscosity is governed by the momentum conservation equation and is given by the *Navier–Stokes equation* (Bird *et al.*, 2002):

$$\rho \frac{\partial \mathbf{u}}{\partial t} + \rho \mathbf{u} \cdot \nabla \mathbf{u} = -\nabla p + \mu \nabla^2 \mathbf{u} + \rho \mathbf{g} + \mathbf{f}_b \tag{6.3}$$

Equation (6.3) represents the force balance on a fluid element in space. Here p is the fluid pressure (Pa or $\mathrm{N/m^2}$), $\mathbf{g}$ is the acceleration due to gravity ($\mathrm{m/s^2}$), μ is the fluid viscosity (Pa s), and ∇^2 is the *Laplacian operator* ($\mathrm{m^{-2}}$). The term $\mathbf{f}_b$ denotes any other *body* force (force per unit volume) acting on the fluid element in addition to gravity. The gravitational force term, $\rho\mathbf{g}$, is also a body force, but is generally written explicitly in the Navier–Stokes equation to differentiate it from the other externally imposed body forces. In electrokinetic transport processes, one needs to consider the electrical body force acting on the fluid. Further discussion on the electrical body force will be presented later in this chapter.

Each term of Eq. (6.3) represents a force acting on a unit volume of the fluid. The first term on the left-hand side represents the rate of change of momentum (given by the product of the fluid density and the rate of change of the fluid velocity with time) at a given location within the flow field. For steady state conditions, this term drops out. The second term is due to the fluid inertia and is negligible for low-velocity (low Reynolds number) flows. The first term on the right-hand side represents the contribution of pressure, the second is due to viscous forces (shear stresses), the third term arises from the body force due to gravity, and the last term arises from any other body force. Equation (6.3) can be written for any orthogonal coordinate system (Bird *et al.*, 2002).

The energy equation (convection–diffusion) in the absence of a heat source and energy dissipation is given for constant physical properties (Bird *et al.*, 2002) as

$$\rho C_p \left[\frac{\partial T}{\partial t} + \mathbf{u} \cdot \nabla T \right] = k_f \nabla^2 T \tag{6.4}$$

where C_p is the fluid specific heat (J/kg °C), T is the temperature of the fluid (°C or K), and k_f is its thermal conductivity (W/m °C).

As an example, we may consider steady two-dimensional incompressible flows with negligible inertia in absence of any body force (except gravity), for which the governing equations become, for the x-directional momentum equation,

$$\frac{\partial p}{\partial x} = \mu \left[\frac{\partial^2 u_x}{\partial x^2} + \frac{\partial^2 u_x}{\partial y^2} \right] + \rho g_x \tag{6.5}$$

and, for the y-directional momentum equation,

$$\frac{\partial p}{\partial y} = \mu \left[\frac{\partial^2 u_y}{\partial x^2} + \frac{\partial^2 u_y}{\partial y^2} \right] + \rho g_y \tag{6.6}$$

for the mass conservation equation,

$$\frac{\partial u_x}{\partial x} + \frac{\partial u_y}{\partial y} = 0 \tag{6.7}$$

and, for the energy equation

$$\rho C_p \left[u_x \frac{\partial T}{\partial x} + u_y \frac{\partial T}{\partial y} \right] = k_f \left[\frac{\partial^2 T}{\partial x^2} + \frac{\partial^2 T}{\partial y^2} \right] \tag{6.8}$$

Equations (6.5) to (6.8) are a unique set of equations and are solved subject to appropriate boundary conditions as applicable to the physical problem at hand.

From the above discussion, it can be observed that for a single-component fluid one is interested in the fluid velocity, pressure, and temperature. These quantities, characterizing the flow of a single-component fluid, are consequences of different forces acting on the fluid.

6.2 MULTICOMPONENT SYSTEMS

6.2.1 Basic Definitions

While dealing with a single-component system, *e.g.*, flow of water in a pipe, there is no ambiguity as to the meaning of the water velocity at a given location and time.

However, for a fluid made up of many species, and in particular a liquid with dissolved ionic species, the concept of velocity and density needs to be clarified (Bird *et al.*, 2002; Newman, 1991).

In a multicomponent system, the local average velocity needs to be defined in terms of the individual velocity of each species that makes up the fluid itself.

The local mass average velocity of the fluid is given by

$$\mathbf{u} = \frac{\sum_{i=1}^{n} \rho_i \mathbf{v}_i}{\sum_{i=1}^{n} \rho_i} = \frac{\sum_{i=1}^{n} \rho_i \mathbf{v}_i}{\rho} \tag{6.9}$$

where $\mathbf{u}$ is the local mass average velocity vector of the bulk fluid with respect to a stationary observer as measured by a pitot tube (m/s), $\mathbf{v}_i$ is the velocity vector of the ith species with respect to a stationary observer (m/s), and ρ_i is the *mass* density of the ith species (kg/m^3) defined as

$$\rho_i = \frac{\text{mass of species } i}{\text{volume of solution}}$$

One can think of ρ_i as being a mass concentration of the ith species. The density ρ of the solution is given by

$$\rho = \sum_{i=1}^{n} \rho_i \tag{6.10}$$

The term $\rho\mathbf{u}$ represents the local rate at which mass passes through a unit area normal to the velocity vector $\mathbf{u}$. Similarly, the term $\rho_i \mathbf{v}_i$ represents the local rate at which the ith species mass passes through a unit area normal to the velocity vector $\mathbf{v}_i$.

A local *molar* average velocity $\mathbf{u}^*$ may be correspondingly defined as

$$\mathbf{u}^* = \frac{\sum_{i=1}^{n} c_i \mathbf{v}_i}{\sum_{i=1}^{n} c_i} = \frac{\sum_{i=1}^{n} c_i \mathbf{v}_i}{c} \tag{6.11}$$

where c_i is the *molar* concentration of the ith species (mol/m^3) defined as

$$c_i = \frac{\text{number of moles of species } i}{\text{volume of solution}}$$

The total molar concentration, c, is given by

$$\sum_{i=1}^{n} c_i = c \tag{6.12}$$

The term $c\mathbf{u}^*$ represents the local molar flux passing through a unit area normal to the velocity vector $\mathbf{u}^*$. Similarly, the term $c_i \mathbf{v}_i$ represents the molar flux of the ith component.

The molar concentration (mol/m^3) and mass density (kg/m^3) are related by

$$c_i = 1000\frac{\rho_i}{M_i} \tag{6.13}$$

where M_i is the molar mass (kg/kmol) of the ith species. The mass fraction of the ith species is given by

$$w_i = \frac{\rho_i}{\rho} \tag{6.14}$$

The molar (mole) fraction is given by

$$x_i \text{ (or } y_i) = \frac{c_i}{c} \tag{6.15}$$

The mean or average molar mass of a mixture is given by

$$\overline{M} = 1000\frac{\rho}{c}$$

It should be recalled that the mass of a mole of a substance is called the molar mass, $\overline{M}$, and its units are kg/kmol. It has the same magnitude as the formerly used molecular weight of g/mol (Probstein, 2003). The factor 1000 appears in order to reconcile the fact that c is given by mol/m^3 and *not* mol/L.

Finally, a number concentration n_i (m^{-3}) can also be defined as

$$n_i = \frac{\text{number of particles of species } i}{\text{volume of solution}} \tag{6.16}$$

The number concentration is related to the molar concentration as

$$n_i = c_i N_A \tag{6.17}$$

where c_i is given in mol/m^3, and N_A is the Avogadro number.

6.2.2 Mass Conservation

The mass conservation for a fluid containing several species can be written for an individual species by considering its mass density, velocity, and by taking into account chemical reactions. The mass conservation equation for the ith species in a multicomponent fluid is given by

$$\frac{\partial \rho_i}{\partial t} = -\nabla \cdot (\rho_i \mathbf{v}_i) + R_i \tag{6.18}$$

where R_i is called the source term, which gives the rate of production of species i due to chemical reaction per unit volume, kg of ith species $/m^3s$. The species mass production rate is obtained from the consideration of chemical kinetics. As the overall mass conservation must hold, one can write

$$\sum R_i = 0 \tag{6.19}$$

Summing for all the species, the mass conservation for a multicomponent system becomes

$$\frac{\partial \sum \rho_i}{\partial t} = -\nabla \cdot \left(\sum \rho_i \mathbf{v}_i\right) + \sum R_i \tag{6.20}$$

Making use of Eqs. (6.9), (6.10), and (6.19), Eq. (6.20) becomes

$$\frac{\partial \rho}{\partial t} = -\nabla \cdot (\rho \mathbf{u}) \tag{6.21}$$

Equation (6.21) indicates that the mass conservation equation of a multicomponent fluid is identical to that of a single-component fluid as long as its bulk density and its mass average velocity are utilized. In terms of a molar balance, the equivalent to Eq. (6.21) is given by

$$\frac{\partial c}{\partial t} = -\nabla \cdot \left(c\mathbf{u}^*\right) \tag{6.22}$$

Equations (6.21) or (6.22) apply to the entire multicomponent system. In practice, many multicomponent systems can be treated as "dilute", where one of the species (termed solvent) exists in large excess over other species (termed solutes). For instance, the aqueous electrolyte solutions that will be discussed in this book will have small amounts of ions obtained from dissociation of the electrolyte dissolved in a large volume of solvent. For an n-component *dilute* system, one can write the mass and molar average velocities by separating out the contributions of the solutes and the solvent as

$$\mathbf{u} = \frac{\sum_{i=1}^{n-1} \rho_i \mathbf{v}_i}{\rho} + \frac{\rho_s \mathbf{v}_s}{\rho} \simeq \mathbf{v}_s$$

and

$$\mathbf{u}^* = \frac{\sum_{i=1}^{n-1} c_i \mathbf{v}_i}{c} + \frac{c_s \mathbf{v}_s}{c} \simeq \mathbf{v}_s$$

respectively, where the subscript i refers to the ith solute species and the subscript s denotes the solvent. In other words, ρ_s, c_s, and $\mathbf{v}_s$ are the mass concentration, molar concentration, and velocity of the solvent, respectively. Thus, for dilute systems, the mass and molar average velocities can be considered identical, *i.e.*, $\mathbf{u} \simeq \mathbf{u}^*$ and both velocities are approximately equal to the solvent velocity. Under such conditions, the molar balance Eq. (6.22) can be written as

$$\frac{\partial c}{\partial t} = -\nabla \cdot (c\mathbf{u}) \tag{6.23}$$

where the mass average velocity is used.

6.2.3 Convection–Diffusion–Migration Equation

When a fluid contains an electrolyte, *i.e.*, charged ionic species, we are interested in the movement or mass transfer of the anions and the cations as well as the bulk fluid. To this end, mass transfer in an electrolytic solution requires a description of the movement of mobile ionic species, material balances, current flow, electroneutrality, and fluid mechanics.

The mass flux of the ith species in a moving fluid is given by the product of the density of the ith species ρ_i and its velocity $\mathbf{v}_i$ as measured relative to a stationary observer. The mass flux, $\rho_i \mathbf{v}_i$, has the units of kg/m^{2}s and it is a vector quantity. In a multicomponent system, as is evidenced by Eq. (6.9), the species velocity $\mathbf{v}_i$ may not be the mass average fluid velocity, $\mathbf{u}$. The deviation of the ith species velocity from the mass average fluid velocity can be due to diffusion if there is a concentration gradient. Furthermore, the ith species velocity can deviate from the mass average fluid velocity due to migration under the influence of an external force if this force acts differently on different species. For instance, gravitational forces acting on species of different densities can be different. In context of charged ions in a solvent, differently charged ions (such as positive and negative) will experience different forces under the same applied external electric field, and will have different velocities.

From the preceding discussion, it is evident that all species in a multicomponent system will be transported at a common velocity, namely, the mass average fluid velocity, $\mathbf{u}$, in absence of diffusion and other external forces. Under such conditions, the mass flux of the ith species can be given in terms of the mass average velocity as

$$\mathbf{j}_i \equiv \mathbf{j}_{i,C} = \rho_i \mathbf{u} \tag{6.24}$$

where $\mathbf{j}_{i,C}$ is termed as the convective flux. This flux is due to the purely convective transport of the species being swept along with the flowing fluid. All other types of fluxes, such as diffusion or migration, are defined in a reference frame that is relative to this convective flux. Noting that any other transport process will change the velocity of the ith species to $\mathbf{v}_i$ from the mass average fluid velocity, we can write

$$\mathbf{j}_i = \rho_i \mathbf{v}_i = \rho_i \mathbf{u} + \rho_i (\mathbf{v}_i - \mathbf{u}) = \mathbf{j}_{i,C} + \mathbf{j}_{i,Other} \tag{6.25}$$

where $\mathbf{j}_{i,Other}$ refers to the diffusive or migration flux evaluated relative to the mass average fluid velocity. Note that this flux is given as

$$\mathbf{j}_{i,Other} = \rho_i \mathbf{v}_i' = \rho_i (\mathbf{v}_i - \mathbf{u})$$

where $\mathbf{v}_i'$ can be termed as the drift velocity of the species relative to the mass average velocity.

Let us now superimpose the effects of diffusion on the flux of a species. In a multicomponent system, the diffusion flux arises from the spatial concentration gradients of different species. To simplify the discussion, we will henceforth assume a dilute solution and restrict the analysis to the solute species only. For a very dilute system, the mass flux (kg/m^{2}s) of the ith solute species due to diffusion, with respect to the mass average velocity, $\mathbf{u}$, is governed by Fick's law as

$$\mathbf{j}_{i,D} = -D_i \nabla \rho_i \tag{6.26}$$

Similarly, the molar flux (mol/m^{2}s) of the ith solute species due to diffusion, with respect to the molar average velocity, $\mathbf{u}^*$, is given by

$$\mathbf{j}^*_{i,D} = -D_i \nabla c_i \tag{6.27}$$

The flux given by Fick's law is purely due to diffusion as a result of the molecular Brownian motion. The coefficient D_i is referred to as the diffusion coefficient of the ith solute species.

One can substitute the diffusive mass flux, Eq. (6.26), for $\mathbf{j}_{i,Other}$ in Eq. (6.25), and obtain the mass flux of the ith solute species as

$$\mathbf{j}_i = \rho_i \mathbf{v}_i = \mathbf{j}_{i,C} + \mathbf{j}_{i,D} = \rho_i \mathbf{u} - D_i \nabla \rho_i \tag{6.28}$$

Equation (6.28) shows that the mass flux of the ith solute species with respect to a stationary observer is a combination of the convective flux due to bulk movement and the diffusional flux.

The molar solute flux equation for the ith species can be written as

$$\mathbf{j}^*_i = c_i \mathbf{v}_i = c_i \mathbf{u}^* - D_i \nabla c_i \tag{6.29}$$

For a very dilute system, the mass and molar average velocities are approximately equal to each other, *i.e.*, $\mathbf{u} \approx \mathbf{u}^*$, and we can write Eq. (6.29) as

$$\mathbf{j}^*_i = c_i \mathbf{v}_i = c_i \mathbf{u} - D_i \nabla c_i \tag{6.30}$$

One can define a number flux (1/m^{2}s), which is based on number concentration n_i, and relate it to the molar flux as

$$\mathbf{j}^{**}_i = N_A \mathbf{j}^*_i = N_A c_i \mathbf{v}_i = n_i \mathbf{v}_i \tag{6.31}$$

The number flux, $\mathbf{j}^{**}_i$, can be represented in terms of convective and diffusive transport as

$$\mathbf{j}^{**}_i = n_i \mathbf{v}_i = n_i \mathbf{u} - D_i \nabla n_i \tag{6.32}$$

where n_i is the number concentration of the ith solute species, expressed in units of m^{-3}.

When deriving the flux equations represented by Eqs. (6.28) to (6.32), the system was assumed to be infinitely dilute. In this context, the dissolved solutes have very small concentrations relative to the solvent. As well, the flux of a species was assumed to take place due to the bulk motion and due to diffusion only. However, when the dissolved species carry a charge, a transport flux can take place when an electric potential gradient is present. Such a mass transport (or mass transfer) is referred to as migration. This mode of transport is peculiar to a system containing charged species. We need to include a mass transport term due to migration in the flux equations for such charged species.

Consider the transport of different solute species in a multicomponent system in presence of an external force, which can act differently on different species. Let us denote the force acting on one solute particle (this can be a molecule or ion) of the ith species by $\mathbf{F}_{Ext}$. The particle will move under the influence of this external force, and one can write its velocity as

$$\mathbf{v}_i' = \omega_i \mathbf{F}_{Ext} \tag{6.33}$$

where ω_i is defined as the mobility of the particle (velocity per unit applied force). In general, the mobility is given in units of $\text{mN}^{-1}\text{s}^{-1}$. Note that the velocity of the particle is given as $\mathbf{v}_i'$ in Eq. (6.33) signifying that this is the velocity observed solely under the influence of the external force, and is determined in a reference framework of the mass average velocity of the multicomponent system [*cf.*, Eq. (6.25)]. In other words,

$$\mathbf{v}_i' = \mathbf{v}_i - \mathbf{u}$$

Multiplying the velocity of the ith species particle obtained from Eq. (6.33) by its mass concentration ρ_i, we obtain the migration flux relative to the mass average fluid velocity as

$$\mathbf{j}_{i,M} = \rho_i \mathbf{v}_i' = \rho_i \omega_i \mathbf{F}_{Ext} \tag{6.34}$$

Substituting this migration flux for $\mathbf{j}_{i,Other}$ in Eq. (6.25), we obtain the total mass flux of the ith species under the influence of convection and migration as

$$\mathbf{j}_i = \rho_i \mathbf{v}_i = \mathbf{j}_{i,C} + \mathbf{j}_{i,M} = \rho_i \mathbf{u} + \rho_i \omega_i \mathbf{F}_{Ext} \tag{6.35}$$

Assuming that the contributions to the solute transport due to diffusion and migration can be linearly superimposed, one can write the total mass flux for the ith solute species as

$$\mathbf{j}_i = \rho_i \mathbf{v}_i = \mathbf{j}_{i,C} + \mathbf{j}_{i,D} + \mathbf{j}_{i,M} = \rho_i \mathbf{u} - D_i \nabla \rho_i + \rho_i \omega_i \mathbf{F}_{Ext} \tag{6.36}$$

Equation (6.36) is widely used in assessing transport behavior of solute species in colloid and electrokinetic transport processes. Note that the unit of the flux is dictated by the unit of concentration used. In Eq. (6.36), the unit of flux is $\text{kg/m}^2\text{s}$ since the unit of concentration, ρ_i is kg/m^3. One can write the molar flux ($\text{mol/m}^2\text{s}$) as

$$\mathbf{j}_i^* = c_i \mathbf{v}_i = c_i \mathbf{u} - D_i \nabla c_i + c_i \omega_i \mathbf{F}_{Ext} \tag{6.37}$$

using the molar concentration c_i (mol/m^3) or the number flux ($1/\text{m}^2\text{s}$) as

$$\mathbf{j}_i^{**} = n_i \mathbf{v}_i = n_i \mathbf{u} - D_i \nabla n_i + n_i \omega_i \mathbf{F}_{Ext} \tag{6.38}$$

using the number concentration n_i ($1/\text{m}^3$).

At this point, we need to interpret the mobility, ω_i, which was introduced in Eq. (6.33). Normally, application of a force to a mass (here a solute particle) causes it to accelerate. In this context, assigning a constant velocity to the particle as in Eq. (6.33) seems counterintuitive. However, it should be noted that as soon as a solute particle starts to move under the influence of the external force, it encounters a counteracting drag force due to the other (predominantly solvent) particles surrounding it. Consequently, in writing Eq. (6.33), it is assumed that the solute particle moves at a terminal velocity attained under the combined influence of the external force and the frictional drag force of the surrounding medium. This is a grossly simplified view of the migration (or drift) velocity, and readers interested in further details of the physics are referred to other sources (Einstein, 1956; Dhont, 1996). In summary, using arguments of the above nature, Einstein showed that

$$\omega_i = \frac{1}{f_i} = \frac{D_i}{k_B T} \tag{6.39}$$

where f_i is the inverse of the mobility, and is known as the Stokes–Einstein friction factor, which is given as $6\pi\mu a$ for a spherical particle of radius a. In other words, the mobility of a solute species is proportional to its diffusion coefficient. Substituting Eq. (6.39) in Eq. (6.36), one obtains

$$\mathbf{j}_i = \rho_i \mathbf{v}_i = \mathbf{j}_{i,C} + \mathbf{j}_{i,D} + \mathbf{j}_{i,M} = \rho_i \mathbf{u} - D_i \nabla \rho_i + \frac{\rho_i D_i}{k_B T} \mathbf{F}_{Ext} \tag{6.40}$$

Let us now consider the migration term for ionic systems (electrolyte solutions) in presence of an electric field. For a particle of the ith species bearing a charge q_i in an electric field $\mathbf{E} = -\nabla\psi$, the electrical force is given by

$$\mathbf{F}_{El} = q_i \mathbf{E} = -q_i \nabla \psi \tag{6.41}$$

If the particle is an ion of valence z_i, then its charge is given by $q_i = z_i e$, and consequently, the electrical force on it will be

$$\mathbf{F}_{El} = -z_i e \nabla \psi \tag{6.42}$$

Substituting this electrical force given by Eq. (6.42) for $\mathbf{F}_{Ext}$ in Eq. (6.40), one obtains

$$\mathbf{j}_i = \rho_i \mathbf{v}_i = \rho_i \mathbf{u} - D_i \nabla \rho_i - \frac{z_i e \rho_i D_i}{k_B T} \nabla \psi \tag{6.43}$$

Equation (6.43) gives the mass flux of the ith ionic species. The first term on the right-hand side of Eq. (6.43) is the flux due to bulk convection, the second term on the right side is due to the concentration gradient (*i.e.*, diffusional process), and the last term is due to migration. The counterpart of Eq. (6.43) in terms of *molar flux* is given by

$$\mathbf{j}_i^* = c_i \mathbf{u} - D_i \nabla c_i - \frac{z_i e c_i D_i}{k_B T} \nabla \psi \tag{6.44}$$

It is also possible to define a flux based on number concentration in terms of the number of ions per unit time per unit area. The corresponding expression is

$$\mathbf{j}_i^{**} = n_i\mathbf{u} - D_i\nabla n_i - \frac{z_i e n_i D_i}{k_B T}\nabla\psi \tag{6.45}$$

In traditional electrochemistry, one generally uses the Faraday constant, $\mathcal{F}$, instead of the electronic charge, e, and the universal gas constant, R, instead of the Boltzmann constant, k_B. Since

$$\mathcal{F} = eN_A \tag{6.46}$$

and

$$R = k_B N_A \tag{6.47}$$

one obtains

$$\frac{e}{k_B T} = \frac{\mathcal{F}}{RT} \tag{6.48}$$

Using Eq. (6.48) in Eqs. (6.43), (6.44), and (6.45), one obtains the alternate (but completely equivalent) forms of the flux equations as

$$\mathbf{j}_i = \rho_i\mathbf{u} - D_i\nabla\rho_i - \frac{z_i\mathcal{F}\rho_i D_i}{RT}\nabla\psi \tag{6.49}$$

$$\mathbf{j}_i^{*} = c_i\mathbf{u} - D_i\nabla c_i - \frac{z_i\mathcal{F}c_i D_i}{RT}\nabla\psi \tag{6.50}$$

and

$$\mathbf{j}_i^{**} = n_i\mathbf{u} - D_i\nabla n_i - \frac{z_i\mathcal{F}n_i D_i}{RT}\nabla\psi \tag{6.51}$$

It is also customary in electrochemistry to define the mobility as the velocity of a charge carrier (ion or a charged particle) per unit electric field, or

$$\mathbf{v}_i' = -\mu_i^*\nabla\psi \tag{6.52}$$

where the ionic mobility, μ_i^*, has units of $m^2V^{-1}s^{-1}$. Considering the migration terms in Eqs. (6.43), (6.44), and (6.45), as well as in Eqs. (6.49), (6.50), and (6.51), one obtains

$$\mu_i^* = \frac{z_i e D_i}{k_B T} = \frac{z_i\mathcal{F}D_i}{RT} \tag{6.53}$$

This equation is referred to as the *Nernst–Einstein equation*. Comparing the definitions of the mobility given by Eqs. (6.39) and (6.53), we note that

$$\mu_i^* = z_i e\omega_i \tag{6.54}$$

One can rewrite the ionic flux equations in an alternate form using the above definition of the ionic mobility, μ_i^*. For example, the molar flux given by either Eq. (6.44) or (6.50) can be expressed as

$$\mathbf{j}_i^* = c_i\mathbf{u} - D_i\nabla c_i - \mu_i^* c_i \nabla\psi \tag{6.55}$$

The mass and number fluxes can also be written in a similar manner.

Clearly, one can write the ionic flux in various forms. All the forms are inter-related and usually either $\mathbf{j}_i^*$ or $\mathbf{j}_i^{**}$ is used. The flux relationships, Eqs. (6.43), (6.44), and (6.45), as well as the alternate forms given by Eqs. (6.49), (6.50), and (6.51) are called the *Nernst–Planck equations*.

As the mass, molar, or number concentration fluxes represent the ith species crossing a control volume relative to a stationary observer, the conservation laws can be written as

$$\frac{\partial \rho_i}{\partial t} = -\nabla \cdot \mathbf{j}_i + R_i \tag{6.56}$$

$$\frac{\partial c_i}{\partial t} = -\nabla \cdot \mathbf{j}_i^* + R_i^* \tag{6.57}$$

and

$$\frac{\partial n_i}{\partial t} = -\nabla \cdot \mathbf{j}_i^{**} + R_i^{**} \tag{6.58}$$

Here, the left hand side of Eqs. (6.56) to (6.58) represents the accumulation rate within a control volume element. The first term on the right hand side is the net flow into the control volume element and the last term is the production rate due to chemical reactions.

For steady state with no chemical reactions, Eqs. (6.56) to (6.58) become, respectively,

$$\nabla \cdot \mathbf{j}_i = 0 \tag{6.59}$$

$$\nabla \cdot \mathbf{j}_i^* = 0 \tag{6.60}$$

and

$$\nabla \cdot \mathbf{j}_i^{**} = 0 \tag{6.61}$$

Equations (6.59) to (6.61) constitute the continuity or material balance for the ith species under steady state with no chemical reactions.

The appropriate substitution of the flux of the ith species due to convection, diffusion, and migration into the continuity equation leads to the ith species transport equation. For example, substituting for the mass flux of the ith species given by Eq. (6.49) into Eq. (6.56) leads to

$$\frac{\partial \rho_i}{\partial t} = -\nabla \cdot \left[\rho_i\mathbf{u} - D_i\nabla\rho_i - \frac{z_i\mathcal{F}D_i\rho_i}{RT}\nabla\psi\right] + R_i \tag{6.62}$$

For steady state incompressible fluid flow one can use the fact that $\partial \rho_i / \partial t = 0$ and $\nabla \cdot \mathbf{u} = 0$ to rewrite Eq. (6.62) for constant D_i as

$$\mathbf{u} \cdot \nabla \rho_i - D_i \nabla^2 \rho_i - \frac{z_i \mathcal{F} D_i}{RT} (\nabla \cdot \rho_i \nabla \psi) = R_i \tag{6.63}$$

The first term of Eq. (6.63) represents transport due to convection, the second term represents transport due to diffusion, and the third term represents transport due to migration. Similar transport equations in terms of c_i and n_i can be written by simple substitution for ρ_i. In terms of c_i, the equivalent to Eq. (6.62) is given by

$$\frac{\partial c_i}{\partial t} = -\nabla \cdot \left[c_i \mathbf{u} - D_i \nabla c_i - \frac{z_i \mathcal{F} D_i c_i}{RT} \nabla \psi \right] + R_i^* \tag{6.64}$$

and the equivalent to Eq. (6.63), where steady state incompressible fluid flow is assumed, is given by

$$\mathbf{u} \cdot \nabla c_i - D_i \nabla^2 c_i - \frac{z_i \mathcal{F} D_i}{RT} (\nabla \cdot c_i \nabla \psi) = R_i^* \tag{6.65}$$

It should be recalled that for both Eqs. (6.63) and (6.65), it was assumed that the mobility and diffusion of the species are constant, *i.e.*, space independent. This is a reasonable assumption for very dilute systems. Throughout the analysis of this chapter, the assumption of constant ionic mobility and diffusivity will be made. Equations (6.62) to (6.65) are often referred to as the convection–diffusion–migration equations.

6.2.4 Current Density

The flow of current is a result of the individual flux of all the ionic species present in the electrolyte solution. The local current density vector (A/m^2) is given by

$$\mathbf{i} = \mathcal{F} \sum z_i \mathbf{j}_i^* \tag{6.66}$$

or

$$\mathbf{i} = e \sum z_i \mathbf{j}_i^{**} \tag{6.67}$$

In terms of ionic molar concentration, Eqs. (6.50) and (6.66) provide the current

$$\mathbf{i} = \mathcal{F} \mathbf{u} \sum z_i c_i - \mathcal{F} \sum D_i z_i \nabla c_i - \frac{\mathcal{F}^2}{RT} \nabla \psi \sum z_i^2 D_i c_i \tag{6.68}$$

Making use of Eqs. (6.45) and (6.67), one can write the current density in terms of the ionic number concentration

$$\mathbf{i} = e \mathbf{u} \sum z_i n_i - e \sum D_i z_i \nabla n_i - \frac{e^2 \nabla \psi}{k_B T} \sum z_i^2 D_i n_i \tag{6.69}$$

Equation (6.68) or Eq. (6.69) indicates that the current density is due to three contributions. The first term gives the contribution due to convection, the second term gives the contribution due to diffusion where there is an ionic concentration gradient, and the third term is the contribution due to migration where there is an electric potential gradient. The relative contribution due to a given mechanism is much dependent on the physical situation at hand.

For the special case of an electrically neutral electrolyte solution, the first term on the right-hand side of Eq. (6.69) or Eq. (6.68) drops out since

$$\sum z_i n_i = \sum z_i c_i = 0 \tag{6.70}$$

This is equivalent to stating that the bulk motion of a fluid with a zero volume charge density does not contribute to the current density.

When there is no concentration gradient and the electrolyte solution is electrically neutral, we consider only the last term on the right side of Eq. (6.68) to obtain

$$\mathbf{i} = -\frac{\mathcal{F}^2}{RT}\left[\sum z_i^2 D_i c_i\right]\nabla\psi \tag{6.71}$$

Setting σ to be the electric conductivity of the solution, one can write

$$\mathbf{i} = -\sigma\nabla\psi \tag{6.72}$$

where

$$\sigma = \frac{\mathcal{F}^2}{RT}\sum z_i^2 D_i c_i = \frac{e^2}{k_B T}\sum z_i^2 D_i n_i \tag{6.73}$$

in units of S/m, where S has the units of Siemens = Ampere/Volt. For a symmetric $(z : z)$ electrolyte, and assuming the diffusivity of both ions are identical, *i.e.*, $D_i = D$, Eq. (6.73) becomes

$$\sigma = \frac{2\mathcal{F}^2 z^2 D c_\infty}{RT} = \frac{2e^2 z^2 D n_\infty}{k_B T} \tag{6.74}$$

where c_∞ and n_∞ are the bulk molar and number concentration of the ions, respectively. In Chapters 8, 9, and 10, we will denote the bulk electrolyte solution conductivity, σ, given by Eq. (6.73), as σ^∞. This will be done to distinguish the bulk electrolyte solution conductivity from the conductivity of a concentrated suspension of charged particles. The bulk electrolyte conductivity is different from the suspension conductivity as will be seen in Chapter 9. One can consider the bulk electrolyte conductivity to be the same as the suspension conductivity in the limiting case when the particle volume fraction in the suspension approaches zero ($\phi_p \to 0$).

Equation (6.72) is an expression of Ohm's law valid for an electrically neutral electrolyte in the absence of concentration gradients (Newman, 1991).

The electric conductivity of the ith species can be written as

$$\sigma_i = \frac{\mathcal{F}^2}{RT} z_i^2 D_i c_i = \frac{e^2}{k_B T} z_i^2 D_i n_i$$

In order to compare electrolyte electric conductivities, it becomes necessary to normalize the ionic conductivity with the ith species concentration. We can define a *molar electric conductivity*, λ_i, for an ionic species (Sm^2/mol) as

$$\lambda_i = \frac{\sigma_i}{c_i} = \frac{\mathcal{F}^2}{RT} z_i^2 D_i \tag{6.75}$$

This is the conductivity of an ionic species for one mole of the ions in one cubic meter of the solution.

In some literature, a term called the *equivalent electric conductivity* is used, which is defined as λ_i / z_i. One should be careful in reading tables for the conductivity, since the symbols used can be defined differently.

The electric conductivity of a solution is given by

$$\sigma = \sum \lambda_i c_i \quad (S/m) \tag{6.76}$$

For a single salt electrolyte, we have

$$\sigma = \lambda_+ c_+ + \lambda\_c\_ \tag{6.77}$$

One can also define a *molar electric conductivity*, σ_M (Sm^2/mol), of an electrolyte in terms of the individual molar electric conductivities of the dissociated ions using $\sigma_M = \nu_+ \lambda_+ + \nu_- \lambda_-$, where ν_+ and ν_- are the respective stoichiometric numbers of cations and anions formed by dissociation of the electrolyte. For example, the molar electric conductivity of Na_2SO_4 given in Table 6.2 at 298 K and infinite dilution can be closely matched using the molar electric conductivities of the Na^+ and SO_4^{2-} ions (Table 6.4) as

$$\begin{aligned}\sigma_M = 2\lambda_+ + 1\lambda_- &= (2 \times 5.01 + 16)10^{-3} \\ &= 2.602 \times 10^{-2}\ Sm^2/mol\end{aligned}$$

Both λ_i and σ_M depend on electrolyte concentration, and one generally uses the symbols λ_i^o and σ_M^o to represent the molar ionic and molar electrolyte conductivities at infinite dilution, respectively. It should be emphasized here that σ_M refers to the

TABLE 6.1. Standard Solutions for Calibrating Conductivity Measurement Cells.

Grams of KCl per kg H_2O	Solution Conductivity, σ, (S/m)			
	0°C	18°C	25°C	25°C*
76.5829	6.514	9.781	11.131	
7.47458	0.7134	1.1163	1.2852	1.2854
0.74582	0.07733	0.12201	0.14083	0.14086
7.45510 (0.1 M)				1.28217
0.74551 (0.01 M)				0.14079

Source: Data are from Marsh (1987) except for the last column (*), which are from Wu *et al.* (1989).

TABLE 6.2. Molar Electric Conductivity, σ_M (σ_M^o at infinite dilution), of Electrolytes in Aqueous Solutions.

Compound	Concentration, mol/L							
	Infinite Dilution	0.0005	0.001	0.005	0.01	0.02	0.05	0.1
$AgNO_3$	133.29	131.29	130.45	127.14	124.70	121.35	115.18	109.09
$1/2BaCl_2$	139.91	135.89	134.27	127.96	123.88	119.03	111.42	105.14
$1/2CaCl_2$	135.77	131.86	130.30	124.19	120.30	115.59	108.42	102.41
$1/2Ca(OH)_2$	258	—	—	233	226	214	—	—
$1/2CuSO_4$	133.6	121.6	115.2	94.02	83.08	72.16	59.02	50.55
HCl	425.95	422.53	421.15	415.59	411.80	407.04	398.89	391.13
KBr	151.9	—	148.9	146.02	143.36	140.41	135.61	131.32
KCl	149.79	147.74	146.88	143.48	141.20	138.27	133.30	128.90
KI	150.31	148.2	143.32	144.30	142.11	139.38	134.90	131.50
KNO_3	144.89	142.70	141.77	138.41	132.75	132.34	126.25	120.34
KOH	271.50	—	234.0	230.0	228.0	—	219.0	213.0
LiCl	114.97	113.09	112.34	109.35	107.27	104.60	100.06	95.81
$1/2MgCl_2$	129.34	125.55	124.15	118.25	114.49	109.99	103.03	97.05
NH_4Cl	149.60	147.5	146.70	134.40	141.21	138.25	133.22	128.69
NaCl	126.39	124.44	123.68	120.59	118.45	115.70	111.01	106.69
NaI	126.88	125.30	124.19	121.19	119.18	116.64	112.73	108.73
NaOH	247.70	245.50	244.60	240.70	237.90	—	—	—
$1/2Na_2SO_4$	129.80	125.68	124.09	117.09	112.38	106.73	97.70	89.94
CH_3COONa	91.00	89.20	88.50	85.68	83.72	81.20	76.88	72.76
$1/2ZnSO_4$	132.70	121.30	114.47	95.44	84.87	74.20	61.17	52.61

The values of the molar electric conductivities are given in 10^{-4} Sm^2 mol^{-1} at a temperature of 25°C. The electrolyte solution conductivity is evaluated as the product of the *molar conductivity* of the solution and the solution molarity.

Source: Adapted from D. R. Lide (Ed.), *CRC Handbook of Chemistry and Physics*, CRC Press, 2002. Comprehensive data on electric conductivity and diffusivity are also provided in *CRC Handbook of Chemistry and Physics*.

molar electric conductivity of the electrolyte and is not to be confused with the solution electric conductivity, σ. At infinite dilution, *i.e.*, when the electrolyte concentration approaches zero, the molar conductivity σ_M approaches a maximum while the solution conductivity σ approaches zero.

The electric conductivity values are sensitive to the solution temperature, and some standard values are provided in Tables 6.1 to 6.4.

In industrial practice, for convenience, the electric conductivity is expressed in μS/cm. Figure 6.1 gives an estimate of electric conductivity of aqueous solutions. For ultrapure water, the electric conductivity is about 0.055 μS/cm and for good quality raw water it is about 50 μS/cm. In some industries, when the conductivity of water approaches that of ultrapure water, electric resistivity is used instead. It is the

TABLE 6.3. Values of Equivalent Electric Conductivities, λ_i/z_i, and Diffusion Coefficients of Selected Ions at Infinite Dilution in Water at 25°C (Newman, 1991).

Cation	z_i	$\lambda_i/z_i \times 10^4$ Sm^2/mol	$D_i \times 10^9$ m^2/s	Anion	z_i	$\lambda_i/z_i \times 10^4$ Sm^2/mol	$D_i \times 10^9$ m^2/s
H^+	1	349.8	9.312	OH^-	−1	197.6	5.260
Li^+	1	38.69	1.030	Cl^-	−1	76.34	2.032
Na^+	1	50.11	1.334	Br^-	−1	78.3	2.084
K^+	1	73.52	1.957	I^-	−1	76.8	2.044
NH_4^+	1	73.4	1.954	NO_3^-	−1	71.44	1.902
Ag^+	1	61.92	1.648	HCO_3^-	−1	41.5	1.105
Tl^+	1	74.7	1.989	HCO_2^-	−1	54.6	1.454
Mg^{2+}	2	53.06	0.7063	$CH_3CO_2^-$	−1	40.9	1.089
Ca^{2+}	2	59.50	0.7920	SO_4^{2-}	−2	80.0	1.065
Sr^{2+}	2	59.46	0.7914	$Fe(CN)_6^{3-}$	−3	101.0	0.896
Ba^{2+}	2	63.64	0.8471	$Fe(CN)_6^{4-}$	−4	111.0	0.739
Cu^{2+}	2	54.0	0.72	IO_4^-	−1	54.38	1.448
Zn^{2+}	2	53.0	0.71	ClO_4^-	−1	67.32	1.792
La^{3+}	3	69.5	0.617	BrO_3^-	−1	55.78	1.485
$Co(NH_3)_6^{3+}$	3	102.3	0.908	HSO_4^-	−1	50.0	1.330

reciprocal of the electric conductivity. For ultrapure water, the electric resistivity is about 18.2 MΩ cm.

The effect of the electrolyte concentration on the molar electric conductivity for a symmetric (1 : 1) electrolyte solution can be given by

$$\sigma_M = \sigma_M^o - (A + B\sigma_M^o)\sqrt{M} \quad \text{for } M < 10^{-3}$$

with $A = 0.00602$ and $B = 0.229$. Here σ_M^o and σ_M are the molar electric conductivities at an infinite dilution and at a molarity of M. The molar electric conductivity is given in Sm^2/mol. The relationship between the electric conductivity and molar conductivity of an electrolyte solution is given by

$$\sigma = 1000M\sigma_M$$

One can think of the difference between the molar electric conductivity at a given electrolyte concentration and the molar electric conductivity at infinite dilution as being the effect of the electrolyte concentration, *i.e.*, finite concentration.

In general, the electric conductivity of an electrolyte solution increases with temperature. For moderately high conductive electrolyte solutions, a rough estimate for the change in electric conductivity, $\Delta\sigma = \sigma_2 - \sigma_1$, over a temperature change, $\Delta T = T_2 - T_1$, is given by $K = (\Delta\sigma/\Delta T)/\sigma_1$, where T is in degrees Celsius. The

TABLE 6.4. Molar Electric Conductivity, λ_i, of Some Ions at Infinite Dilution in Water (Atkinson, 1972).

	Temperature, T		Molar conductivity $\lambda_i \times 10^3$ $Sm^2\ mol^{-1}$
	K	°C	
Cation			
H^+	288	15	30.1
	298	25	35.0
	308	35	39.7
Li^+	298	25	3.87
Na^+	288	15	3.98
	298	25	5.01
	308	35	6.15
Ca^{2+}	298	25	11.9
Cu^{2+}	298	25	10.8
La^{3+}	298	25	20.0
Anion			
OH^-	298	25	19.8
Cl^-	288	15	6.14
	298	25	7.63
	308	35	9.22
Br^-	288	15	6.33
	298	25	7.83
	308	35	9.42
NO_3^-	298	25	7.14
HCO_3^-	298	25	4.45
SO_4^{2-}	298	25	16.0

interpolation formula is given as

$$\sigma = \sigma_1[1 + K(T - T_1)]$$

where $T > T_1$ and σ_1 is the conductivity at T_1. Values of K for different electrolytes are tabulated in Table 6.5.

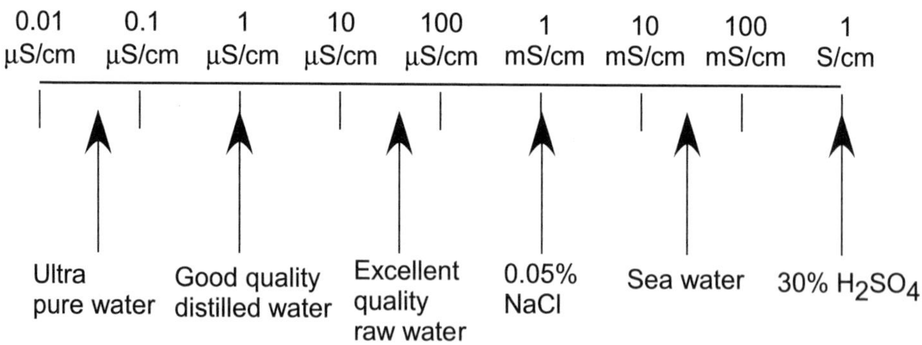

Figure 6.1. Electric conductivities of water under different conditions.

TABLE 6.5. Values of the Parameter K for Different Electrolytes.

Electrolyte	K ($\times 10^2$)	
Acids	(1.0–1.6)	1/°C
Bases	(1.8–2.2)	1/°C
Salts	(2.2–3.0)	1/°C
Fresh water	2.0	1/°C

EXAMPLE 6.1

Evaluation of Current Density. Evaluate the current density for a 0.01 molar NaCl solution at 298 K for a field strength E_x of 1000 V/m. Assume electrical neutrality and uniform ionic concentration.

Solution The current density equation is given by

$$i_x = -\sigma \frac{d\psi}{dx}$$

where x is the current flow direction. From Table 6.2, the molar conductivity is given by $118.45 \times 10^{-4}\ \mathrm{Sm^2/mol}$

$$i_x = -\left(118.45 \times 10^{-4} \frac{\mathrm{Sm^2}}{\mathrm{mol}}\right)\left(0.01 \frac{\mathrm{kmol}}{\mathrm{m^3}} \times \frac{1000\ \mathrm{mol}}{\mathrm{kmol}}\right) \times \left(-1000 \frac{\mathrm{V}}{\mathrm{m}}\right)$$

$$i_x = 118.45 \frac{\mathrm{A}}{\mathrm{m^2}}$$

■

EXAMPLE 6.2

Electric Conductivity. Evaluate the solution electric conductivity and molar conductivity of a 0.0005 M solution of Na_2SO_4 at 25°C using the ionic equivalent conductivities at infinite dilution given in Table 6.3.

Solution From Table 6.3, at infinite dilution:
For Na^+,

$$\frac{\lambda}{z} = 50.11 \times 10^{-4}\ \mathrm{Sm^2/mol}$$

For SO_4^{2-},

$$\frac{\lambda}{z} = 80.0 \times 10^{-4}\ \mathrm{Sm^2/mol}$$

The electrolyte solution conductivity is given by

$$\sigma = \sum \lambda_i c_i = (50.11 \times 10^{-4} \times 1)(1000 \times 2 \times 0.0005) + (80.0 \times 10^{-4} \times 2)(1000 \times 0.0005)$$

or

$$\sigma = 0.005011 + 0.008 = 0.0130\,\text{S/m}$$

The molar conductivity of the electrolyte is given by

$$\sigma_M = \frac{0.0130}{0.0005 \times 1000} = 0.0260\,\text{Sm}^2/\text{mol}$$

From Table 6.2, the molar conductivity at infinite dilution is given as $2 \times 129.80 \times 10^{-4} = 0.02596$ Sm^2/mol. As would be expected, this value is very close to that obtained using Table 6.3. However, the actual value using Table 6.2 for 0.0005 mol/L is

$$(125.68 \times 10^{-4} \times 2)(0.0005 \times 1000) = 0.02514\,\text{Sm}^2/\text{mol}$$

This value is slightly lower than that at infinite dilution. In our calculations, we used the individual values for λ_i/z_i at infinite dilution. ■

6.2.5 Conservation of Charge

It is a physical law of nature that electric charge is conserved. This is implicitly built into the basic transport relationships.

Multiplying Eq. (6.57), the equation for the conservation of the ith ionic species, by $z_i\mathcal{F}$ yields

$$\mathcal{F}\frac{\partial}{\partial t} z_i c_i = -\mathcal{F}\nabla \cdot (z_i \mathbf{j}_i^*) + \mathcal{F} z_i R_i^* \tag{6.78}$$

For convenience, z_i was included inside the differentiation operator. One can now sum over all species to obtain

$$\mathcal{F}\frac{\partial}{\partial t}\sum z_i c_i = -\mathcal{F}\nabla \cdot \sum z_i \mathbf{j}_i^* + \mathcal{F}\sum z_i R_i^* \tag{6.79}$$

Recognizing that the first term on the right hand side is related to the current density, [see Eq. (6.66)], one can write Eq. (6.79) as

$$\mathcal{F}\frac{\partial}{\partial t}\sum z_i c_i = -\nabla \cdot \mathbf{i} + \mathcal{F}\sum z_i R_i^* \tag{6.80}$$

Equation (6.80) represents the conservation of the current density under unsteady state conditions with chemical reactions.

The last term of Eq. (6.80) is zero as long as all homogeneous reactions that give rise to the R_i^* term are electrically neutral. With the latter in mind, Eq. (6.80) becomes

$$\mathcal{F}\frac{\partial}{\partial t}\sum z_i c_i = -\nabla \cdot \mathbf{i} \tag{6.81}$$

Noting that the term $\mathcal{F}\sum z_i c_i$ represents the volumetric free charge density (C/m^3), one can relate this term to the electric field using the Gauss law [*cf.*, Eq. (3.110)]. This gives

$$\mathcal{F}\sum z_i c_i = \nabla \cdot (\epsilon \mathbf{E}) \tag{6.82}$$

where $\mathbf{E} = -\nabla\psi$ is the electric field. Substituting Eq. (6.82) in (6.81) yields

$$\nabla \cdot \left[\frac{\partial}{\partial t}(\epsilon \mathbf{E})\right] = -\nabla \cdot \mathbf{i} \tag{6.83}$$

or

$$\nabla \cdot \left[\frac{\partial(\epsilon \mathbf{E})}{\partial t} + \mathbf{i}\right] = 0 \tag{6.84}$$

The first term in Eq. (6.84) is called the displacement current. The terminology stems from the fact that for linear dielectric materials, $\epsilon\mathbf{E} = \mathbf{D}$, where $\mathbf{D}$ is the electric displacement vector [see Eq. (3.109)]. The second term in Eq. (6.84) is referred to as the transport current. Under steady state conditions, the displacement current vanishes, leading to

$$\nabla \cdot \mathbf{i} = 0 \tag{6.85}$$

Equation (6.85) is the mathematical expression of the conservation of charge or simply the continuity of the current density under steady state conditions.

6.2.6 Binary Electrolyte Solution

The convection–diffusion–migration transport equation given by Eq. (6.64) as well as the current density given by Eq. (6.68), are valid for a very dilute system containing ionic species derived from several salts. As an example, solutions containing Na_2SO_4 and $CaCl_2$ individually or as a mixture can be represented by these equations. For the special case of electroneutrality with no chemical reaction and for a solution containing a single salt, *e.g.*, Na_2SO_4, the transport and current density equations can be greatly simplified. For clarity, let us denote the positive ions by the symbol "+" and the negative ions by the symbol "−". Next, we introduce γ_+ and γ_- as the number of positive ions and negative ions produced from the dissolution of one molecule of the single salt under consideration. The electroneutrality condition, given by

$$z_+ c_+ + z_- c_- = 0 \tag{6.86}$$

can be replaced by

$$\frac{c_+}{\gamma_+} = \frac{c_-}{\gamma_-} \tag{6.87}$$

We introduce a scaled ionic concentration C using Eq. (6.87) as

$$C = \frac{c_+}{\gamma_+} = \frac{c_-}{\gamma_-} \tag{6.88}$$

In terms of the scaled ionic concentration, the transport equation, Eq. (6.64), under the assumption of $R_i^* = 0$, electroneutrality, and incompressible flow, can be written for each ionic species as

$$\frac{\partial C}{\partial t} + \mathbf{u} \cdot \nabla C = D_+ \nabla^2 C + z_+ D_+ \frac{\mathcal{F}}{RT} \nabla \cdot (C \nabla \psi) \tag{6.89}$$

and

$$\frac{\partial C}{\partial t} + \mathbf{u} \cdot \nabla C = D_- \nabla^2 C + z_- D_- \frac{\mathcal{F}}{RT} \nabla \cdot (C \nabla \psi) \tag{6.90}$$

The divergence term containing $\nabla \cdot (C \nabla \psi)$ can be eliminated from Eqs. (6.89) and (6.90) leading to

$$\frac{\partial C}{\partial t} + \mathbf{u} \cdot \nabla C = D_{eff} \nabla^2 C \tag{6.91}$$

where

$$D_{eff} = \frac{(z_+ - z_-) D_+ D_-}{z_+ D_+ - z_- D_-} \tag{6.92}$$

Equation (6.91) governs the scaled ionic concentration of the single salt species in a solution where convective, diffusive, and migration fluxes can be present. Here, D_{eff} is the effective diffusion coefficient. Equation (6.91) is known in the mass transfer literature as the convection–diffusion equation whose solution is given for different geometries and boundary conditions. Apart from the assumption of a single salt system and no chemical reaction, the serious restriction is the electroneutrality assumption. We need to keep in mind, however, that Eq. (6.91) was derived without assuming a zero current density.

Although the current density is provided through Eqs. (6.68) or (6.85), a simplified expression can be obtained for the single salt case within the restrictions of no chemical reaction and electroneutrality. Introducing the scaled ionic concentration into Eq. (6.68) leads to

$$\mathbf{i} = -\mathcal{F}\left[D_+ z_+ \gamma_+ + D_- z_- \gamma_-\right] \nabla C - \frac{\mathcal{F}^2}{RT} C \nabla \psi \left[z_+^2 D_+ \gamma_+ + z_-^2 D_- \gamma_-\right] \tag{6.93}$$

Using Eqs. (6.86) and (6.87) to eliminate c_+ and c_- leads to

$$z_+ \gamma_+ = -z_- \gamma_- \tag{6.94}$$

Making use of the relationship (6.94), Eq. (6.93) becomes

$$\frac{\mathbf{i}}{z_+ \gamma_+ \mathcal{F}} = -(D_+ - D_-) \nabla C - (z_+ D_+ - z_- D_-) \frac{\mathcal{F}}{RT} C \nabla \psi \tag{6.95}$$

Equation (6.95) provides the current density for an electroneutral system with no chemical reactions. In the absence of a current density, *i.e.*, $\mathbf{i} = 0$, Eq. (6.95) gives

$$\frac{\mathcal{F}}{RT}\nabla\psi = -\frac{(D_+ - D_-)}{z_+D_+ - z_-D_-}\nabla \ln C \tag{6.96}$$

where

$$\frac{\nabla C}{C} = \nabla \ln C$$

Equation (6.96) shows the relationship between the electric field and concentration gradients when there is no current density at a given location within the system for an electroneutral binary electrolyte with no chemical reactions.

It should be noted that the flux equations were derived on the assumption that the solution is dilute and that the concentrations are used to establish the driving forces. However, for a semi-dilute solution (with moderate salt concentrations), the gradient of the electrochemical potential is used as the driving force for diffusion and migration. The equivalent of the Nernst–Planck equation (in molar concentration units) becomes

$$\mathbf{j}_i^* = c_i\mathbf{u} - \omega_i c_i \nabla\mu_i \tag{6.97}$$

where μ_i is the electrochemical potential of a particle of the ith species, and ω_i is the mobility defined as velocity of the ith species per unit driving force. Here, the driving force is obtained from the gradient of the electrochemical potential, comprising of the chemical potential and the electrical potential,

$$\nabla\mu_i = \nabla\left[\mu_{i,c} + \frac{z_i e\psi}{k_B T}\right]$$

The electrochemical potential μ_i depends on the composition of the electrolyte solution, represented by the chemical potential term $\mu_{i,c}$, and the local electrical potential energy. The local electrical potential energy depends on the charge of a particle of the ith species and the electric potential, ψ. The electrical potential energy is given by the term $z_i e\psi/k_B T$. For more details, the reader is referred to Newman (1991).

6.2.7 Boltzmann Distribution

The Boltzmann distribution was previously derived in Chapter 5 using heuristic arguments. One can make use of the Nernst–Planck relationship to derive the Boltzmann distribution for the ionic number concentration near a charged surface.

Consider a flat surface whose normal is in the x-direction. Equation (6.45) for the ionic flux gives

$$j_{ix}^{**} = n_i u_x - D_i\frac{dn_i}{dx} - \frac{z_i e n_i D_i}{k_B T}\frac{d\psi}{dx} \tag{6.98}$$

where the subscript x denotes the x-direction. At equilibrium, we have zero fluid velocity, $u_x = 0$, and zero flux of ions, $j_{ix}^{**} = 0$. Denoting the electric potential under these stationary (no flow) conditions by the symbol ψ^{eq}, Eq. (6.98) becomes

$$\frac{dn_i}{dx} + \left(\frac{z_i e n_i}{k_B T}\right)\frac{d\psi^{eq}}{dx} = 0 \tag{6.99}$$

or

$$\frac{d(\ln n_i)}{dx} + \frac{z_i e}{k_B T}\frac{d\psi^{eq}}{dx} = 0. \tag{6.100}$$

Let

$$n_i = n_{i\infty} \quad \text{at } \psi^{eq} = \psi^{eq}_\infty \tag{6.101}$$

Here, ψ^{eq}_∞ is the *reference* potential measured in a stationary (no flow) electroneutral electrolyte solution far away from the charged surface. The solution of Eq. (6.100) subject to boundary condition Eq. (6.101) is given by

$$n_i = n_{i\infty} \exp\left[-\frac{z_i e\left(\psi^{eq} - \psi^{eq}_\infty\right)}{k_B T}\right] \tag{6.102}$$

For convenience, one may set the reference potential $\psi^{eq}_\infty = 0$ to obtain

$$n_i = n_{i\infty} \exp\left[-\frac{z_i e \psi^{eq}}{k_B T}\right] \tag{6.103}$$

Here we recover the *Boltzmann distribution*. Note that setting $\psi^{eq}_\infty = 0$ does not necessarily mean that the electric potential is actually zero far away from the charged surface (in the stationary electroneutral electrolyte solution). One should interpret the potential ψ^{eq} in the Boltzmann distribution equation (6.103) as the difference between the actual electric potential near a charged surface and the electric potential in a stationary electrolyte reservoir located at an infinite distance from the surface. It should be noted that the Nernst–Planck relationship gives the Boltzmann distribution with the explicit assumptions of $\mathbf{j}_i^{**} = 0$ and $\mathbf{u} = 0$. In a flow system, the above assumptions are not strictly valid.

The electric double layer theory discussed in Chapter 5 was based on the Boltzmann distribution, which led to the formulation of the Poisson–Boltzmann equation. The Poisson–Boltzmann equation is a special form of the Poisson equation, and is only applicable under stationary (no flow and no ion movement) conditions. Thus, when considering a general electrokinetic transport problem, the Poisson–Boltzmann equation should be treated as the equation governing the stationary (no flow) electric potential distribution, and may be expressed as

$$\nabla^2 \psi^{eq} = -\frac{e}{\epsilon}\sum_i z_i n_{i\infty} \exp\left(-\frac{z_i e \psi^{eq}}{k_B T}\right) \tag{6.104}$$

where ψ^{eq} denotes the electric potential distribution at equilibrium near a charged surface placed in a *stationary electrolyte reservoir*. Any non-zero ionic flux ($\mathbf{j}_i^{**} \neq 0$), caused by whatsoever external means, will change the ion concentration distributions from the stationary Boltzmann distribution. In such dynamic situations, the total electric potential (generally represented as ψ) will differ from ψ^{eq}. This total electric potential, ψ, in the dynamic situation can strictly be determined from a coupled solution of the Poisson, Nernst–Planck, and Navier–Stokes equations. In many electrokinetic transport problems, the difference between the total potential under dynamic conditions, ψ, and the equilibrium potential, ψ^{eq}, is small, and can be related through a perturbation variable as

$$\psi = \psi^{eq} + \delta\psi \tag{6.105}$$

where $\delta\psi$ is a linear perturbation in the potential caused by ionic movement. We will discuss application of this perturbation approach in Chapter 9 in context of electrophoresis.

In summary, when addressing electrokinetic transport processes, one must always keep in mind that the electric potential employed in the dynamic problem is the "total" potential, ψ, and not the equilibrium or stationary potential, ψ^{eq}, obtained from a solution of the Poisson–Boltzmann equation. In this chapter, whenever we use the symbol ψ, it denotes the total electric potential in a dynamic condition.

6.2.8 Momentum Equations

As was discussed earlier, the momentum equation given by Eq. (6.3) contains a generic body force term, $\mathbf{f}_b$, which can be used to consider any type of force acting on a fluid volume element. In electrokinetic transport processes, this body force arises due to an electric field. The electric body force per unit volume on the fluid (N/m^3) is given by

$$\mathbf{f}_E = \rho_f \mathbf{E} - \frac{1}{2}\mathbf{E}\cdot\mathbf{E}\nabla\epsilon + \frac{1}{2}\nabla\left[\left(\rho\frac{\partial\epsilon}{\partial\rho}\right)_T \mathbf{E}\cdot\mathbf{E}\right] \tag{6.106}$$

where ρ is the fluid mass density and ρ_f is the free charge density (Russel *et al.*, 1989; Verbrugge and Pintauro, 1989). The force term given by Eq. (6.106) is for a variable dielectric constant.

For the special case of a constant permittivity, the electrical body force per unit fluid volume, $\mathbf{f}_E$, due to the electric field becomes

$$\mathbf{f}_E = \rho_f \mathbf{E} \tag{6.107}$$

Recognizing that the electric field, $\mathbf{E}$, is related to the electric potential, ψ, by

$$\mathbf{E} = -\nabla\psi \tag{6.108}$$

one can write for the force term

$$\mathbf{f}_E = -\rho_f \nabla\psi \tag{6.109}$$

where the units of ρ_f are $\mathrm{C/m^3}$ and the units of $\nabla\psi$ are V/m. Introducing the electric force term for the body force, $\mathbf{f}_b$, in Eq. (6.3), we obtain

$$\rho\frac{\partial \mathbf{u}}{\partial t} + \rho\mathbf{u}\cdot\nabla\mathbf{u} = -\nabla p + \mu\nabla^2\mathbf{u} + \rho\mathbf{g} - \rho_f\nabla\psi \tag{6.110}$$

Equation (6.110) is the momentum equation in the presence of an electric body force. It is the modified Navier–Stokes equation when an electric force is considered. It should be stressed here that ψ in Eq. (6.110) represents the total electric potential. The free charge density is given by

$$\rho_f = e\sum z_i n_i = \mathcal{F}\sum z_i c_i \tag{6.111}$$

where e is the fundamental charge, z_i is valence of the ith species, n_i is the ionic concentration, $\mathcal{F}$ is the Faraday constant, and c_i is the molar concentration.

For steady-state creeping flow where inertia is neglected, Eq. (6.110) becomes

$$\nabla p = \mu\nabla^2\mathbf{u} + \rho\mathbf{g} - \rho_f\nabla\psi \tag{6.112}$$

The mass conservation is still given by Eq. (6.2). For steady-state and constant density, one obtains

$$\nabla\cdot\mathbf{u} = 0 \tag{6.113}$$

Equations (6.112) and (6.113) are the equations governing the fluid mechanics for the transport of an electrolytic solution under steady-state creeping flow conditions. Additional information is needed to provide closure to these equations. To this end, the electric potential ψ is given by the Poisson equation and the species conservations are provided by the convection–diffusion–migration equation for every ionic species, Eq. (6.64).

The Navier–Stokes equation can be solved after setting appropriate boundary conditions corresponding to a given flow situation. The boundary conditions can specify the velocities or pressures at given locations of the flow boundary. Of particular interest with respect to boundary conditions for these equations is the *no-slip condition*, which is generally applied at the solid boundaries. The no-slip condition implies that the fluid in contact with a solid object moves at the same velocity as the object. If the object is stationary, such as the wall of a container, then the fluid velocity is zero at the wall. Imposition of the no-slip condition usually sets up a velocity gradient in the fluid near the solid-fluid interfaces. Such a gradient gives rise to a viscous stress. The local fluid stress tensor at a point on the surface of the solid object is given by,

$$\bar{\bar{\tau}} = -p\bar{\bar{I}} + \mu\left[\nabla\mathbf{u} + (\nabla\mathbf{u})^T\right] \tag{6.114}$$

where p is the pressure, $\bar{\bar{I}}$ is the unit tensor, and the final term on the right hand side represents the viscous stresses. This fluid stress can be integrated over the surface of the solid object to determine the total hydrodynamic force acting on the object in a manner analogous to the integration of the Maxwell stress to provide the electrical force acting on a body.

6.3 HYDRODYNAMICS OF COLLOIDAL SYSTEMS

So far in this chapter, we focussed on the transport behavior of a solute particle (ion) assuming it to be a point mass. In other words, the solute particle did not have a finite size, and in the purely hydrodynamic limit, when diffusion or migration fluxes are absent, the solute particle velocity was assumed to be equal to the mass average velocity of the fluid,

$$\mathbf{v}_i = \mathbf{u} \tag{6.115}$$

When considering colloidal particles, it is generally not prudent to consider them as point masses. For such particles, in the purely hydrodynamic limit (in the absence of diffusion and migration), the particle velocity can differ from the fluid velocity. Denoting the particle *hydrodynamic velocity* vector as $\mathbf{v}_p$ to distinguish it from the velocity vector, $\mathbf{v}_i$, which was used to define the overall particle velocity arising from the combined influence of convection, diffusion, and migration, we note that several flow situations can give rise to conditions when

$$\mathbf{v}_p \neq \mathbf{u} \tag{6.116}$$

In this section, we discuss the mechanisms causing the above mentioned deviation of the particle hydrodynamic velocity from the mass average velocity of the fluid. Following this, we will employ the formalism described here to address the modifications of the convection diffusion migration equation when applied to such finite sized colloidal particles. Before proceeding with the mathematical developments, however, let us physically inspect how a finite sized particle interacts with its surrounding fluid in the purely hydrodynamic limit. For this, one must recognize that particles are rigid bodies, which do not deform like a fluid. Under pure translation, every point on the particle will have the same velocity. When different points in the particle have different velocities, the entire particle will also undergo a rotational motion in addition to translation. The velocity of a point within the particle, $\mathbf{U}$, can be represented as a summation of the translational velocity of the particle center of mass, $\mathbf{U}_0$, and the rotational velocity, $\boldsymbol{\Omega}$, of the point with respect to the center of mass

$$\mathbf{U} = \mathbf{U}_0 + \boldsymbol{\Omega} \times (\mathbf{r} - \mathbf{r}_0) \tag{6.117}$$

where $\mathbf{r} - \mathbf{r}_0$ denotes the position vector of the point relative to the center of mass of the body.

When a rigid body (such as a colloidal particle) is placed in a viscous fluid, the no-slip boundary condition will apply at the particle-fluid interface. This means, if we consider a stationary particle in a fluid flowing with a uniform velocity, the fluid in contact with the particle will be stationary. Consequently, the fluid velocity distribution will be altered by the stationary particle. The altered fluid velocity distribution will give rise to a viscous stress, which will be felt by every point on the particle surface. Integrating the stress tensor, $\bar{\bar{\tau}}$, given by Eq. (6.114) over the surface of the particle, one can determine the total force acting on the particle as

$$\mathbf{F} = \int_S \bar{\bar{\tau}} \cdot \mathbf{n}\, dS = \int_S \left\{ -p\bar{\bar{I}} + \mu \left[\nabla \mathbf{u} + (\nabla \mathbf{u})^T \right] \right\} \cdot \mathbf{n}\, dS \tag{6.118}$$

where $\mathbf{n}$ is the unit outward surface normal vector and S is the surface area of the particle. Similarly, the torque, $\mathbf{T}$, on the particle can be determined from

$$\mathbf{T} = \int_S (\mathbf{r} - \mathbf{r}_0) \times \bar{\bar{\tau}} \cdot \mathbf{n}\, dS \tag{6.119}$$

Naturally, if the particle suspended in the fluid experiences a force $\mathbf{F}$, an equal and opposing force should be exerted by the particle on the surrounding fluid. In this context, one can determine the modification of the fluid velocity due to the particle by considering the perturbation of the fluid velocity due to a concentrated force $\mathbf{F}$ originating at the center of mass of the particle.[1] A formal approach for analyzing the fluid velocity field due to a suspended particle involves exploration of the so-called singular solutions of the Stokes creeping flow equations. The basic starting point for such an analysis is to consider the modification of the fluid velocity by a point force, $\mathbf{f}_0$, acting at the origin of a coordinate system. The body force experienced by the fluid due to this point force is represented as a delta distribution, given by

$$\mathbf{f}_b = \delta(\mathbf{r})\mathbf{f}_0 \tag{6.120}$$

where $\mathbf{f}_b$ is the body force per unit volume of the fluid, given by the last term in Eq. (6.3), $\delta(\mathbf{r})$ represents a Dirac delta function centered at the origin, and $\mathbf{r}$ is the position vector of a point in the fluid. Substituting this force in the creeping flow equations [*i.e.*, neglecting the $\mathbf{u} \cdot \nabla \mathbf{u}$ term in Eq. (6.3)], one obtains the Stokes equation,

$$\nabla \cdot \mathbf{u} = 0 \tag{6.121}$$

$$0 = -\nabla p + \mu \nabla^2 \mathbf{u} + \delta(\mathbf{r})\mathbf{f}_0 \tag{6.122}$$

where the gravity force is also neglected from the Navier–Stokes equation.

Equation (6.122) is linear, and hence, it can be shown that the solution of the creeping flow equations will provide a linear dependence of the fluid velocity vector on the point force. In other words, one may relate the fluid velocity vector to the point force through

$$\mathbf{u} = \bar{\bar{\mathbf{O}}} \cdot \mathbf{f}_0 \tag{6.123}$$

where $\bar{\bar{\mathbf{O}}}$ is the Oseen tensor. It is represented as

$$\bar{\bar{\mathbf{O}}} = \frac{1}{8\pi \mu r} \left[\bar{\bar{I}} + \frac{\mathbf{rr}}{r^2} \right] \tag{6.124}$$

[1]Strictly speaking, this is a far field approximation, valid only at sufficiently large distances from the particle.

where $r = |\mathbf{r}|$ is the distance from the point force. Equation (6.123) represents the velocity distribution of the fluid in presence of a point force at the origin. One can use this basic formalism, and apply it to any distribution of point forces by simply adding up all the contributions due to the individual point forces. Such a linear superposition is applicable in the creeping flow limit where the Stokes equations are linear.

To illustrate how the above procedure is applied to a distribution of point forces, let us consider a finite sized particle in the fluid. At each point on the surface of the particle, there is a force per unit area, $\mathbf{f}_0'$, acting due to the fluid stress, $\bar{\bar{\tau}} \cdot \mathbf{n}$. One can determine the velocity field of the fluid in presence of such a finite particle by adding up all these individual forces ($\mathbf{f}_0'$) originating from different points of the surface of the particle. In other words, the velocity of the fluid can be written as

$$\mathbf{u} = \int_S \bar{\bar{\mathbf{O}}}(\mathbf{r} - \mathbf{r}') \cdot \mathbf{f}_0'(\mathbf{r}')\, dS \tag{6.125}$$

where $\mathbf{r}'$ represents the position vectors of the distributed point forces on the surface of the particle.

Consider a sphere of radius a moving with a uniform translational velocity, $\mathbf{U}_0$, in an otherwise stationary fluid. The center of the sphere is assumed to be at the origin of the coordinate system. The fluid velocity at an infinite distance from the sphere is zero, while the velocity at the surface of the sphere is $\mathbf{U}_0$. Let us assume that the local velocity of the fluid at every point on the surface of the sphere is caused by a point force that is directly proportional to the particle velocity, *i.e.*,

$$\mathbf{f}_0'(\mathbf{r}') = c\mathbf{U}_0 \tag{6.126}$$

Here c is a constant of proportionality that needs to be determined. The criterion for determination of this proportionality constant is that a proper choice will satisfy the flow boundary conditions. Substituting this force from Eq. (6.126) and the Oseen tensor, Eq. (6.124), in Eq. (6.125), one obtains

$$\mathbf{u}(\mathbf{r}) = \frac{c}{8\pi\mu}\int_S \frac{1}{|(\mathbf{r} - \mathbf{r}')|}\left[\bar{\bar{I}} + \frac{(\mathbf{r} - \mathbf{r}')(\mathbf{r} - \mathbf{r}')}{|(\mathbf{r} - \mathbf{r}')|^2}\right] \cdot \mathbf{U}_0\, dS \tag{6.127}$$

where S is the surface of the spherical particle, and $\mathbf{r}'$ is the position vector of a point on the sphere surface. Evaluation of the above integral, after substituting a proper choice for the constant c yields

$$\mathbf{u}(\mathbf{r}) = \left\{\frac{3}{4}\frac{a}{r}\left[\bar{\bar{I}} + \frac{\mathbf{rr}}{r^2}\right] + \frac{1}{4}\left(\frac{a}{r}\right)^3\left[\bar{\bar{I}} - 3\frac{\mathbf{rr}}{r^2}\right]\right\} \cdot \mathbf{U}_0 \tag{6.128}$$

The above formulation for the velocity field satisfies the boundary conditions

$$\mathbf{u} = 0 \quad \text{as } \mathbf{r} \to \infty \tag{6.129}$$

$$\mathbf{u} = \mathbf{U}_0 \quad \text{at sphere surface, } \mathbf{r} = \mathbf{r}' \tag{6.130}$$

It is evident from the foregoing discussion that the fluid surrounding the translating spherical particle will acquire an induced velocity due to the particle. The fluid in contact with the particle will move at a velocity $\mathbf{U}_0$, and the velocity distribution will slowly decay to zero as $\mathbf{r} \to \infty$ obeying Eq. (6.128). The velocity field in the fluid surrounding the particle will decay as $\sim 1/r$ at sufficiently large distances from the particle. This implies that the presence of the particle is felt by the fluid up to a considerable distance. This is the underlying mechanism of hydrodynamic interaction between colloidal entities. The perturbations in the fluid velocity is felt up to large distances from a moving particle, and can, in turn affect the motion of another particle.

One can substitute the velocity distribution, Eq. (6.128), in Eq. (6.118) to determine the force exerted on the particle. The subsequent integration leading to the closed form expression for the force is quite laborious. However, the calculations lead to the well-known Stokes expression for the drag force exerted by the particle on the fluid as

$$\mathbf{F} = -6\pi \mu a \mathbf{U}_0 \tag{6.131}$$

EXAMPLE 6.3

Fluid Velocity Field Surrounding a Spherical Particle Translating at a Uniform Velocity. Consider a spherical particle of radius a translating at a uniform velocity, $\mathbf{U}_p$, directed along the z axis in an otherwise quiescent fluid. The particle motion is due to a force, $\mathbf{F}$, also acting along the z direction. The situation is depicted in Figure 6.2. Determine the velocity distribution in the fluid surrounding the particle.

Solution For the given geometry, the velocity vector $\mathbf{U}_p$ directed along the z coordinate is first written in the spherical coordinate system as

$$\mathbf{U}_p = U_r \mathbf{i}_r + U_\theta \mathbf{i}_\theta$$

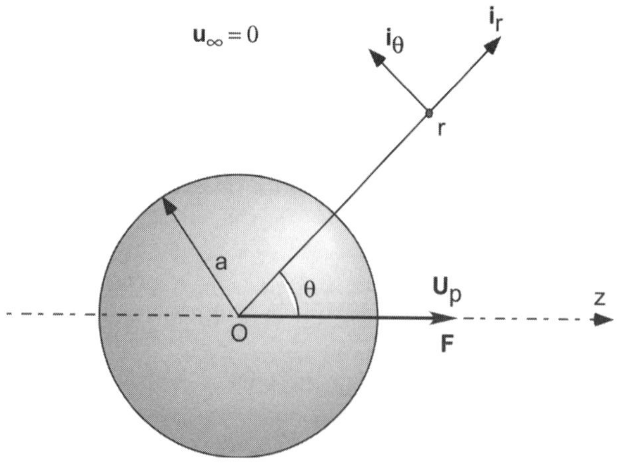

Figure 6.2. Translation of a sphere in a quiescent fluid.

with

$$U_r = U_p \cos(\theta) \quad \text{and} \quad U_\theta = -U_p \sin(\theta)$$

where $U_p = \mathbf{U}_p \cdot \mathbf{i}_z$.

The velocity distribution in the fluid surrounding the particle is given by Eq. (6.128). Using the nomenclature of Figure 6.2, we can recast the velocity distribution as

$$\mathbf{u(r)} = \left\{ \frac{3}{4}\frac{a}{r}\left[\bar{\bar{I}} + \frac{\mathbf{rr}}{r^2}\right] + \frac{1}{4}\left(\frac{a}{r}\right)^3 \left[\bar{\bar{I}} - 3\frac{\mathbf{rr}}{r^2}\right]\right\} \cdot \mathbf{U}_p \tag{6.132}$$

In spherical coordinates (r, θ, ϕ), assuming symmetry in ϕ direction, we have

$$\bar{\bar{I}} = \begin{bmatrix} 1 & 0 \\ 0 & 1 \end{bmatrix}$$

and

$$\frac{\mathbf{rr}}{r^2} = \begin{bmatrix} 1 & 0 \\ 0 & 0 \end{bmatrix}$$

Substituting these in Eq. (6.132), we obtain

$$\mathbf{u(r)} = \left\{ \frac{3}{4}\frac{a}{r}\left[\begin{pmatrix} 1 & 0 \\ 0 & 1 \end{pmatrix} + \begin{pmatrix} 1 & 0 \\ 0 & 0 \end{pmatrix}\right] + \frac{1}{4}\left(\frac{a}{r}\right)^3 \left[\begin{pmatrix} 1 & 0 \\ 0 & 1 \end{pmatrix} \right.\right.$$
$$\left.\left. -3\begin{pmatrix} 1 & 0 \\ 0 & 0 \end{pmatrix}\right]\right\} \cdot \begin{pmatrix} U_r \\ U_\theta \end{pmatrix}$$

Upon simplification, the above equation yields

$$\mathbf{u(r)} = \frac{1}{2}\frac{a}{r}\left[3 - \left(\frac{a}{r}\right)^2\right] U_r \mathbf{i}_r + \frac{1}{4}\frac{a}{r}\left[3 + \left(\frac{a}{r}\right)^2\right] U_\theta \mathbf{i}_\theta \tag{6.133}$$

Substituting for U_r and U_θ, the velocity distribution is given by

$$\mathbf{u(r)} = \frac{1}{2}\frac{a}{r}\left[3 - \left(\frac{a}{r}\right)^2\right] U_p \cos(\theta)\mathbf{i}_r - \frac{1}{4}\frac{a}{r}\left[3 + \left(\frac{a}{r}\right)^2\right] U_p \sin(\theta)\mathbf{i}_\theta \tag{6.134}$$

Equation (6.134) is the required form of the fluid velocity distribution around the translating particle.

Let us now see how the Oseen solution is recovered from the above velocity field by rendering the particle infinitesimally small. For the particle translating at a velocity U_p along the z direction under the influence of a force of magnitude F, also acting in the z direction, the velocity and and the force are related through the mobility as

$$U_p = \omega_p F = \frac{F}{6\pi\mu a} \tag{6.135}$$

where μ is the fluid viscosity and a is the particle radius. This is the well-known Stokes friction factor for a spherical particle. Substituting the above relationship in

Eq. (6.134), one obtains

$$\mathbf{u}(\mathbf{r}) = \frac{1}{2}\frac{a}{r}\left[3 - \left(\frac{a}{r}\right)^2\right]\frac{F}{6\pi\mu a}\cos(\theta)\mathbf{i}_r - \frac{1}{4}\frac{a}{r}\left[3 + \left(\frac{a}{r}\right)^2\right]\frac{F}{6\pi\mu a}\sin(\theta)\mathbf{i}_\theta \tag{6.136}$$

Taking the limit $a \to 0$, the above expression simplifies to

$$\mathbf{u}(\mathbf{r}) = \frac{1}{4\pi\mu r}F\cos(\theta)\mathbf{i}_r - \frac{1}{8\pi\mu r}F\sin(\theta)\mathbf{i}_\theta \tag{6.137}$$

This is the Oseen form of the fluid velocity field obtained under the influence of a point force F acting at the origin as given in Eqs. (6.123) and (6.124). ■

The procedure for calculating the fluid velocity field around a finite sized colloidal particle delineated above can be applied for other types of flows as well. For instance, one can determine the velocity distribution around a particle subjected to a one-dimensional shear flow. The approach is also extended to study the hydrodynamic interaction between two particles. Providing a detailed analysis of these hydrodynamic interactions is beyond the scope of this book. Several texts discuss hydrodynamic interactions in considerable detail, and the reader is referred to these books for further information (Lamb, 1932; Happel and Brenner, 1965; Russel *et al.*, 1989; Dhont, 1996). It is, however, apparent from the above discussion that a particle moving under the influence of an external force will influence the velocity of the fluid surrounding it. This, in turn, will affect the velocity of another particle when it is sufficiently close to the first particle. The modification of the particle velocity is a function of the separation distance between the particles. In colloidal phenomena of interest in this book, we will encounter such modifications of particle velocities due to hydrodynamic interactions when studying electrophoresis, sedimentation, coagulation, and particle deposition. In all these cases, we will encounter situations when one particle is sufficiently close to another particle, or to a solid surface. In analyzing the hydrodynamic velocities of the particles in close proximity, we will generally represent the particle hydrodynamic velocity as

$$\mathbf{v}_p = \overline{\overline{f}}(\mathbf{r}) \cdot \mathbf{u} \tag{6.138}$$

where the tensor $\overline{\overline{f}}(\mathbf{r})$ represents the hydrodynamic correction functions acting along different coordinate directions. These functions will all have a value of 1 as $\mathbf{r} \to \infty$, but will have different values at finite separations between the particles and surfaces. The functions will be different for different flow situations, as well as different geometries. Further details on these functions will be provided in the later chapters as needed.

Noting that the hydrodynamic interactions between finite particles can lead to a modification of the particle hydrodynamic velocity from the mass average velocity of the fluid, one can write the particle flux equation in terms of the particle hydrodynamic velocity, $\mathbf{v}_p$. Consider the number flux

$$\mathbf{j}_i^{**} = n_i\mathbf{v}_i = n_i\mathbf{v}_p + n_i(\mathbf{v}_i - \mathbf{v}_p) \tag{6.139}$$

Equation (6.139) is the analogous form of Eq. (6.25), except for the use of the particle hydrodynamic velocity instead of the mass average fluid velocity, **u**.

It should be noted that all the other fluxes of the particle, namely, diffusion and migration fluxes, are determined relative to the particle hydrodynamic velocity for finite sized particles. In other words, the drift velocity produced by a force acting on the finite sized particle is given by

$$\mathbf{v}_i' = \mathbf{v}_i - \mathbf{v}_p \tag{6.140}$$

As a consequence, the diffusion coefficients calculated for finite sized particles differ from the corresponding diffusion coefficients used for point mass solutes. In fact, the hydrodynamic correction functions are also applied to determine the diffusion coefficient of finite sized particles, and the particle diffusivity becomes a function of separation distance between the particles. In general, the particle diffusion coefficient is expressed in terms of a diffusion tensor and is anisotropic. For instance, when a particle is subjected to a shear flow parallel to a stationary planar wall, the diffusivity of the particle at small separations from the wall in the direction parallel to the wall is different from the diffusivity perpendicular to the wall. Denoting the diffusion tensor for the particle as $\bar{\bar{D}}_i$, the diffusive number flux of the particle is given by

$$\mathbf{j}_{i,D}^{**} = -\bar{\bar{D}}_i \cdot \nabla n_i \tag{6.141}$$

where n_i is the particle number concentration. Similarly, the migration flux under the influence of an external force, $\mathbf{F}_{Ext}$, is given by

$$\mathbf{j}_{i,M}^{**} = \frac{n_i}{k_B T} \bar{\bar{D}} \cdot \mathbf{F}_{Ext} \tag{6.142}$$

Using the above expressions, the total number flux of particles due to convection, diffusion, and migration in a suspension can be written as

$$\mathbf{j}_i^{**} = n_i \mathbf{v}_p - \bar{\bar{D}}_i \cdot \nabla n_i + \frac{n_i}{k_B T} \bar{\bar{D}}_i \cdot \mathbf{F}_{Ext} \tag{6.143}$$

Similar expressions for the mass and molar fluxes can also be written.

One can substitute the particle flux equations in the convection–diffusion–migration Eq. (6.58) and express the spatial particle concentration distribution as

$$\frac{\partial n_i}{\partial t} = -\nabla \cdot \mathbf{j}_i^{**} + R_i^{**} = -\nabla \cdot \left[n_i \mathbf{v}_p - \bar{\bar{D}} \cdot \nabla n_i + \frac{n_i}{k_B T} \bar{\bar{D}}_i \cdot \mathbf{F}_{Ext} \right] + R_i^{**} \tag{6.144}$$

At steady state, neglecting any source or sink term ($R_i^{**} = 0$), Eq. (6.144) simplifies to

$$\nabla \cdot \left[n_i \mathbf{v}_p - \bar{\bar{D}} \cdot \nabla n_i + \frac{n_i}{k_B T} \bar{\bar{D}}_i \cdot \mathbf{F}_{Ext} \right] = 0 \tag{6.145}$$

These transport equations for finite size colloidal particles are perfectly analogous to the transport equations for ions (point mass solutes) except for the use of the particle hydrodynamic velocities and the incorporation of the diffusion tensor containing the hydrodynamic interaction corrections.

6.4 SUMMARY OF GOVERNING EQUATIONS

In Chapters 3, 5, and 6, the equations needed to solve electrokinetic problems were derived and discussed. A summary of the governing equations is given below. In the presentation of these equations, the absence of chemical reactions, and constant physical properties (density, viscosity, and dielectric constant) are assumed.

Conservation of Mass for an Electrolyte Solution

$$\nabla \cdot \mathbf{u} = 0$$

Conservation of Ionic Species

$$\frac{\partial \rho_i}{\partial t} + \nabla \cdot \mathbf{j}_i = 0$$

or

$$\frac{\partial c_i}{\partial t} + \nabla \cdot \mathbf{j}_i^* = 0$$

or

$$\frac{\partial n_i}{\partial t} + \nabla \cdot \mathbf{j}_i^{**} = 0$$

Conservation of Current Density

$$\frac{\partial}{\partial t}\left[\mathcal{F}\sum z_i c_i\right] + \nabla \cdot \mathbf{i} = \nabla \cdot \left[\frac{\partial(\epsilon \mathbf{E})}{\partial t} + \mathbf{i}\right] = 0$$

At steady state, we have $\nabla \cdot \mathbf{i} = 0$.

Momentum Equation (Equation of Motion)

$$\rho\frac{\partial \mathbf{u}}{\partial t} + \rho \mathbf{u} \cdot \nabla \mathbf{u} = -\nabla p + \mu \nabla^2 \mathbf{u} + \rho \mathbf{g} - \rho_f \nabla \psi$$

The creeping flow (low Reynolds number) form is

$$\rho\frac{\partial \mathbf{u}}{\partial t} + \nabla p = \mu \nabla^2 \mathbf{u} + \rho \mathbf{g} - \rho_f \nabla \psi$$

where

$$\mathbf{E} = -\nabla \psi$$

and

$$\rho_f = \sum z_i e n_i = \mathcal{F}\sum z_i c_i$$

Poisson Equation

$$\epsilon \nabla^2 \psi = -\rho_f$$

Poisson–Boltzmann Equation

$$\nabla^2 \psi^{eq} = -\frac{e}{\epsilon} \sum_i z_i n_{i\infty} \exp\left(-\frac{z_i e \psi^{eq}}{k_B T} \right)$$

Ionic Flux: Nernst–Planck Equation

$$\begin{aligned} \mathbf{j}_i = \rho_i \mathbf{v}_i &= \rho_i \mathbf{u} - D_i \nabla \rho_i - \frac{z_i e \rho_i D_i}{k_B T} \nabla \psi \\ &= \rho_i \mathbf{u} - D_i \nabla \rho_i - \frac{z_i \mathcal{F} \rho_i D_i}{RT} \nabla \psi \end{aligned}$$

in kg/m^{2}s,

$$\begin{aligned} \mathbf{j}_i^* = c_i \mathbf{v}_i &= c_i \mathbf{u} - D_i \nabla c_i - \frac{z_i e c_i D_i}{k_B T} \nabla \psi \\ &= c_i \mathbf{u} - D_i \nabla c_i - \frac{z_i \mathcal{F} c_i D_i}{RT} \nabla \psi \end{aligned}$$

in mol/m^{2}s, and

$$\begin{aligned} \mathbf{j}_i^{**} = n_i \mathbf{v}_i &= n_i \mathbf{u} - D_i \nabla n_i - \frac{z_i e n_i D_i}{k_B T} \nabla \psi \\ &= n_i \mathbf{u} - D_i \nabla n_i - \frac{z_i \mathcal{F} n_i D_i}{RT} \nabla \psi \end{aligned}$$

in 1/m^{2}s.

Nernst–Einstein Relationship

$$\mu_i^* = \frac{z_i \mathcal{F} D_i}{RT} = \frac{z_i e D_i}{k_B T}$$

where μ_i^* is the mobility of an ion (defined as velocity per unit applied field).

Stokes–Einstein Relationship

$$\omega_i = \frac{1}{f_i} = \frac{D_i}{k_B T}$$

where ω_i is the mobility defined as velocity per unit applied force. f_i is the friction factor of an ion, which is given by $6\pi a \mu$ for spheres of radius a. In some literature it is denoted as λ_i. For a spherical particle of radius a,

$$D = \frac{k_B T}{6\pi a \mu}$$

The two mobilities are related by

$$\mu_i^* = z_i e \omega_i$$

Current Density

$$\mathbf{i} = \mathcal{F}\mathbf{u}\sum z_i c_i - \mathcal{F}\sum D_i z_i \nabla c_i - \frac{\mathcal{F}^2 \nabla \psi}{RT}\sum z_i^2 D_i c_i$$
$$= e\mathbf{u}\sum z_i n_i - e\sum D_i z_i \nabla n_i - \frac{e^2 \nabla \psi}{k_B T}\sum z_i^2 D_i n_i$$

Electrolyte solution electrical conductivity σ is given by

$$\sigma = \frac{\mathcal{F}^2}{RT}\sum z_i^2 D_i c_i = \frac{e^2}{k_B T}\sum z_i^2 D_i n_i$$

Some Useful Relationships

$$\mathcal{F} = eN_A$$
$$R = k_B N_A$$
$$n_i = N_A c_i$$
$$\frac{z_i e}{k_B T} = \frac{z_i \mathcal{F}}{RT}$$

Particle Hydrodynamic Velocity

$$\mathbf{v}_p = \overline{\overline{f}}(\mathbf{r}) \cdot \mathbf{u}$$

where the tensor function $\overline{\overline{f}}$ represents the hydrodynamic corrections for the particle velocity.

Particle Flux Equation

In terms of number concentration, n_i, the particle flux is represented using the particle hydrodynamic velocity as

$$\mathbf{j}_i^{**} = n_i \mathbf{v}_p - \overline{\overline{D}}_i \cdot \nabla n_i + \frac{n_i}{k_B T}\overline{\overline{D}}_i \cdot \mathbf{F}_{Ext}$$

where $\overline{\overline{D}}_i$ is the particle diffusion tensor and $\mathbf{F}_{Ext}$ is the external force.

Equations for Continuity and Momentum Conservation

Tables 6.6 to 6.9 provide the governing equations for continuity (mass conservation) and the governing equations for momentum conservation in Cartesian, cylindrical, and spherical coordinate systems for constant physical properties.

TABLE 6.6. Continuity (Conservation) Equation in Several Coordinate Systems for Constant Fluid Density: $\nabla \cdot \mathbf{u} = 0$.

Cartesian coordinates (x, y, z):

$$\frac{\partial}{\partial x}(u_x) + \frac{\partial}{\partial y}(u_y) + \frac{\partial}{\partial z}(u_z) = 0$$

Cylindrical coordinates (r, θ, z):

$$\frac{1}{r}\frac{\partial}{\partial r}(r u_r) + \frac{1}{r}\frac{\partial}{\partial \theta}(u_\theta) + \frac{\partial}{\partial z}(u_z) = 0$$

Spherical coordinates (r, θ, ϕ):

$$\frac{1}{r^2}\frac{\partial}{\partial r}(r^2 u_r) + \frac{1}{r \sin\theta}\frac{\partial}{\partial \theta}(u_\theta \sin\theta) + \frac{1}{r \sin\theta}\frac{\partial}{\partial \phi}(u_\phi) = 0$$

TABLE 6.7. Equations of Motion in Cartesian Coordinates (x, y, z).

In terms of velocity gradient for a Newtonian electrolyte solution with constant ρ, μ, and ϵ:

x-component

$$\rho\left(\frac{\partial u_x}{\partial t} + u_x\frac{\partial u_x}{\partial x} + u_y\frac{\partial u_x}{\partial y} + u_z\frac{\partial u_x}{\partial z}\right) = -\frac{\partial p}{\partial x} + \mu\left(\frac{\partial^2 u_x}{\partial x^2} + \frac{\partial^2 u_x}{\partial y^2} + \frac{\partial^2 u_x}{\partial z^2}\right) + \rho g_x - \rho_f\frac{\partial \psi}{\partial x}$$

y-component

$$\rho\left(\frac{\partial u_y}{\partial t} + u_x\frac{\partial u_y}{\partial x} + u_y\frac{\partial u_y}{\partial y} + u_z\frac{\partial u_y}{\partial z}\right) = -\frac{\partial p}{\partial y} + \mu\left(\frac{\partial^2 u_y}{\partial x^2} + \frac{\partial^2 u_y}{\partial y^2} + \frac{\partial^2 u_y}{\partial z^2}\right) + \rho g_y - \rho_f\frac{\partial \psi}{\partial y}$$

z-component

$$\rho\left(\frac{\partial u_z}{\partial t} + u_x\frac{\partial u_z}{\partial x} + u_y\frac{\partial u_z}{\partial y} + u_z\frac{\partial u_z}{\partial z}\right) = -\frac{\partial p}{\partial z} + \mu\left(\frac{\partial^2 u_z}{\partial x^2} + \frac{\partial^2 u_z}{\partial y^2} + \frac{\partial^2 u_z}{\partial z^2}\right) + \rho g_z - \rho_f\frac{\partial \psi}{\partial z}$$

TABLE 6.8. Equations of Motion in Cylindrical Coordinates (r, θ, z).

In terms of velocity gradient for a Newtonian electrolyte solution with constant ρ, μ, and ϵ:

r-component

$$\rho\left(\frac{\partial u_r}{\partial t} + u_r\frac{\partial u_r}{\partial r} + \frac{u_\theta}{r}\frac{\partial u_r}{\partial \theta} - \frac{u_\theta^2}{r} + u_z\frac{\partial u_r}{\partial z}\right) = -\frac{\partial p}{\partial r}$$

$$+ \mu\left[\frac{\partial}{\partial r}\left(\frac{1}{r}\frac{\partial}{\partial r}(ru_r)\right) + \frac{1}{r^2}\frac{\partial^2 u_r}{\partial \theta^2} - \frac{2}{r^2}\frac{\partial u_\theta}{\partial \theta} + \frac{\partial^2 u_r}{\partial z^2}\right] + \rho g_r - \rho_f\frac{\partial \psi}{\partial r}$$

θ-component

$$\rho\left(\frac{\partial u_\theta}{\partial t} + u_r\frac{\partial u_\theta}{\partial r} + \frac{u_\theta}{r}\frac{\partial u_\theta}{\partial \theta} + \frac{u_r u_\theta}{r} + u_z\frac{\partial u_\theta}{\partial z}\right) = -\frac{1}{r}\frac{\partial p}{\partial \theta}$$

$$+ \mu\left[\frac{\partial}{\partial r}\left(\frac{1}{r}\frac{\partial}{\partial r}(ru_\theta)\right) + \frac{1}{r^2}\frac{\partial^2 u_\theta}{\partial \theta^2} + \frac{2}{r^2}\frac{\partial u_r}{\partial \theta} + \frac{\partial^2 u_\theta}{\partial z^2}\right] + \rho g_\theta - \rho_f\frac{1}{r}\frac{\partial \psi}{\partial \theta}$$

z-component

$$\rho\left(\frac{\partial u_z}{\partial t} + u_r\frac{\partial u_z}{\partial r} + \frac{u_\theta}{r}\frac{\partial u_z}{\partial \theta} + u_z\frac{\partial u_z}{\partial z}\right) = -\frac{\partial p}{\partial z}$$

$$+ \mu\left[\frac{1}{r}\frac{\partial}{\partial r}\left(r\frac{\partial}{\partial r}u_z\right) + \frac{1}{r^2}\frac{\partial^2 u_z}{\partial \theta^2} + \frac{\partial^2 u_z}{\partial z^2}\right] + \rho g_z - \rho_f\frac{\partial \psi}{\partial z}$$

TABLE 6.9. Equations of Motion in Spherical Coordinates (r, θ, ϕ).

In terms of velocity gradient for a Newtonian electrolyte solution with constant ρ, μ, and ϵ:

r-component

$$\rho\left(\frac{\partial u_r}{\partial t} + u_r\frac{\partial u_r}{\partial r} + \frac{u_\theta}{r}\frac{\partial u_r}{\partial \theta} + \frac{u_\phi}{r\sin\theta}\frac{\partial u_r}{\partial \phi} - \frac{u_\theta^2 + u_\phi^2}{r}\right) = -\frac{\partial p}{\partial r}$$

$$+ \mu\left[\nabla^2 u_r - \frac{2}{r^2}u_r - \frac{2}{r^2}\frac{\partial u_\theta}{\partial \theta} - \frac{2}{r^2}u_\theta\cot\theta - \frac{2}{r^2\sin\theta}\frac{\partial u_\phi}{\partial \phi}\right] + \rho g_r - \rho_f\frac{\partial \psi}{\partial r}$$

θ-component

$$\rho\left(\frac{\partial u_\theta}{\partial t} + u_r\frac{\partial u_\theta}{\partial r} + \frac{u_\theta}{r}\frac{\partial u_\theta}{\partial \theta} + \frac{u_\phi}{r\sin\theta}\frac{\partial u_\theta}{\partial \phi} + \frac{u_r u_\theta}{r} - \frac{u_\phi^2\cot\theta}{r}\right) = -\frac{1}{r}\frac{\partial p}{\partial \theta}$$

$$+ \mu\left[\nabla^2 u_\theta + \frac{2}{r^2}\frac{\partial u_r}{\partial \theta} - \frac{u_\theta}{r^2\sin^2\theta} - \frac{2\cos\theta}{r^2\sin^2\theta}\frac{\partial u_\phi}{\partial \phi}\right] + \rho g_\theta - \rho_f\frac{1}{r}\frac{\partial \psi}{\partial \theta}$$

(continued)

TABLE 6.9. *Continued.*

ϕ-component

$$\rho\left(\frac{\partial u_\phi}{\partial t}+u_r\frac{\partial u_\phi}{\partial r}+\frac{u_\theta}{r}\frac{\partial u_\phi}{\partial \theta}+\frac{u_\phi}{r\sin\theta}\frac{\partial u_\phi}{\partial \phi}+\frac{u_\phi u_r}{r}+\frac{u_\phi u_\theta}{r}\cot\theta\right)=$$

$$-\frac{1}{r\sin\theta}\frac{\partial p}{\partial \phi}+\mu\left[\nabla^2 u_\phi-\frac{u_\phi}{r^2\sin^2\theta}+\frac{2}{r^2\sin\theta}\frac{\partial u_r}{\partial \phi}+\frac{2\cos\theta}{r^2\sin^2\theta}\frac{\partial u_\theta}{\partial \phi}\right]$$

$$+\rho g_\phi-\rho_f\frac{1}{r\sin\theta}\frac{\partial \psi}{\partial \phi}$$

In these equations,

$$\nabla^2=\frac{1}{r^2}\frac{\partial}{\partial r}\left(r^2\frac{\partial}{\partial r}\right)+\frac{1}{r^2\sin\theta}\frac{\partial}{\partial \theta}\left(\sin\theta\frac{\partial}{\partial \theta}\right)+\frac{1}{r^2\sin^2\theta}\left(\frac{\partial^2}{\partial \phi^2}\right)$$

6.5 NOMENCLATURE

a	radius of a spherical particle, m
c	total molar concentration, mol/m^3
c_i	molar concentration of the ith species, mol/m^3
c_s	molar concentration of solvent, mol/m^3
C	scaled ionic concentration ion/m^3
C_p	fluid specific heat, J/kg K
D	diffusion coefficient, m^2/s
D_i	diffusion coefficient of the ith species, m^2/s
$\overline{\overline{D}}_i$	diffusion tensor
e	elementary charge, C
$\mathbf{E}$	electric field strength vector, V/m
$\mathcal{F}$	Faraday number, C/mol
$\mathbf{F}$	force acting on a body due to fluid stress, N
$\mathbf{f}_{El}$	electrical force, N
$\mathbf{F}_{Ext}$	external force, N
$\mathbf{f}_0$	point force acting at origin
$\mathbf{f}_b$	body force per unit volume of a fluid element, N/m^3
$\mathbf{f}_E$	electric body force, N/m^3
f_i	Stokes–Einstein friction factor
$\overline{\overline{f}}(\mathbf{r})$	tensor describing the particle hydrodynamic correction functions
$\mathbf{g}$	gravitational acceleration vector, m/s^2
$\overline{\overline{I}}$	unit tensor
$\mathbf{i}$	current density vector, A/m^2
i_x	current density in the x-direction, A/m^2
$\mathbf{j}_i$	mass flux vector of ion i relative to a stationary observer, kg/m^{2}s
$\mathbf{j}_i^*$	molar flux vector of ion i relative to a stationary observer, mol/m^{2}s
$\mathbf{j}_i^{**}$	ionic flux vector of ion i relative to a stationary observer, 1/m^{2}s

k_B	Boltzmann constant, J/K
k_f	fluid thermal conductivity, W/mK
M	solution molarity, mol/dm^3
$\overline{M}$	average molar mass, kg/kmol
M_i	molar mass of ion i, kg/kmol
N_A	Avogadro number, mol^{-1}
$\mathbf{n}$	unit outward normal to a surface
n_i	number concentration of ion i, m^{-3}
$n_{i\infty}$	number concentration of ion i in the bulk solution, m^{-3}
$\overline{\overline{\mathbf{O}}}$	Oseen tensor
p	fluid pressure, Pa or N/m^2
R	gas constant, J/mol K
$\mathbf{r}$	position vector of a point, m
t	time, s
T	temperature, K
$\mathbf{T}$	torque on a rigid body, Nm
$\mathbf{u}$	mass local average velocity vector relative to a stationary observer, m/s
$\mathbf{u}_i$	velocity vector of species i relative to a stationary observer, m/s
$\mathbf{u}^*$	molar average velocity vector relative to a stationary observer, m/s
$\mathbf{v}_i$	velocity of species i, m/s
$\mathbf{v}_i'$	drift velocity ($\mathbf{v}_i - \mathbf{u}$), m/s
$\mathbf{v}_p$	particle hydrodynamic velocity, m/s
$\mathbf{v}_s$	solvent velocity, m/s
w_i	mass fraction, dimensionless
(x, y, z)	Cartesian coordinates, m
x_i, y_i	mole fraction, dimensionless
z	absolute value of the valence for a $(z : z)$ electrolyte
z_i	valence of the ith ionic species

Greek Symbols

ϵ	permittivity of a material, C/mV
λ_i	molar electric conductivity of the ith ionic species, Sm^2/mol
μ	fluid viscosity, kg/ms or Pa s
μ_i	electrochemical potential of the ith species, J/mol
μ_i^*	mobility of ith charge carrier species, m^{2}V^{-1}s^{-1}
ρ	fluid density, kg/m^3
ρ_f	free charge density, C/m^3
ρ_i	mass density of the ith species, kg/m^3
ρ_s	mass density (concentration) of the solvent, kg/m^3
σ	electrical conductivity of the electrolyte solution, S/m or A/Vm
σ_M	molar electric conductivity of an electrolyte, Sm^2/mol
σ_M°	molar electric conductivity of an electrolyte at infinite dilution, Sm^2/mol

$\bar{\bar{\tau}}$	hydrodynamic stress tensor, N/m^2
ψ	total electric potential, V
ψ^{eq}	electric potential in Poisson–Boltzmann equation for stationary system, V
ψ_∞	potential in the bulk electrolyte solution, V
ω_i	mobility of ith species, $mN^{-1}s^{-1}$
Ω	angular velocity of a point, rad/s
∇	del operator, m^{-1}
∇^2	Laplacian operator, m^{-2}

Subscripts

i	i^{th} ionic species
x	x-direction
y	y-direction
z	z-direction

6.6 PROBLEMS

6.1. Evaluate the electrolyte solution electric conductivity and its molar conductivity for a 0.5 mM $BaCl_2$ solution at 25°C using the equivalent ionic conductivities at infinite dilution provided in Table 6.3. Compare your results with the data provided in Table 6.2.

6.2. Assuming infinite dilution, evaluate the solution conductivity and its molar conductivity for a 0.3 g/L Na_2SO_4 solution at 25°C using the equivalent ionic conductivities provided in Table 6.3. Compare your results with the data in Table 6.2.

6.3. A volume of 0.5 L of 0.001 M $BaCl_2$ is added to 0.3 L of 0.0005 M KCl at 25°C. What is the final electric conductivity of the solution? Now, let us add 0.4 L of de-ionized water to the mixed salt solution. What is the electric conductivity of the diluted solution?

6.4. An electrolyte solution fills a 0.2 m long circular tube. The tube ends are placed between two electrodes having an electric potential difference of 150 V. The electrolyte solution contains 1.2 mM $CaCl_2$. What is the electric current density? Assume a temperature of 25°C and an infinitely dilute solution.

6.7 REFERENCES

Aris, R., *Vectors, Tensors, and Basic Equations of Fluid Mechanics*, Dover, New York, (1989).

Atkinson, G., Electrochemical Information, in *American Institute of Physics Handbook*, 3rd ed. Gray, D. E. (Ed.), McGraw-Hill, New York, (1972).

Bird, R. B., Stewart, W. E., and Lightfoot, E. N., *Transport Phenomena*, Wiley, New York, (1960, 2002).

Dhont, J. K. G., *An Introduction to the Dynamics of Colloids*, Elsevier, Amsterdam, (1996).

Einstein, A., *Investigations on the Theory of Brownian Motion*, Dover Publications, New York, (1956).

Happel, J., and Brenner, H., *Low Reynolds Number Hydrodynamics*, Prentice-Hall, NJ (1965).

Lamb, H., *Hydrodynamics*, Cambridge University Press, (1932).

Marsh, K. N., (Ed.), *Recommended Reference Materials for the Realization of Physicochemical Properties*, Blackwell, Oxford, (1987).

Moon, P. H., and Spencer, D. E., *Field Theory Handbook: Including Coordinate Systems, Differential Equations and Their Solutions*, 2nd ed., Springer-Verlag, New York, (1971).

Newman, J. S., *Electrochemical Systems*, 2nd ed., Prentice Hall, Englewood Cliffs, New Jersey, (1991).

Probstein, R. F., *Physicochemical Hydrodynamics, An Introduction*, 2nd ed., Wiley Interscience, New York, (2003).

Russel, W. B., Saville, D. A., and Schowalter, W. R., *Colloidal Dispersions*, Cambridge Univ. Pr., Cambridge, (1989).

Schey, H. M., *Div, Grad, Curl and All That*, 3rd ed., Norton, New York, (1997).

Verbrugge, M. W., and Pintauro, P. N., Transport Models for Ion-Exchange Membranes, *Modern Aspects of Electrochemistry*, No. 19, Conway, B. E., Bockris, J. O'M., and White, R. E., (Eds.), Plenum, New York, (1989).

Wu, Y. C., Pratt, K. W., and Koch, W. F., Determination of the absolute specific conductance of primary standard KCl solutions, *J. Solution Chem.*, **18**, 515–528, (1989).

CHAPTER 7

ELECTROKINETIC PHENOMENA

Electrokinetics is a general term associated with the relative motion between two charged phases. Electrokinetic phenomena occur when one attempts to shear off the mobile part of the electric double layer. Then, as the charged surface (plus attached material) tends to move in the appropriate direction, the ions in the mobile part of the electric double layer undergo a net migration in the opposite direction, carrying solvent along with them, thereby causing the movement of the solvent. Similarly, an electric field is created if the charged surface and the diffuse part of the double layer are made to move relative to each other (Shaw, 1980; Hiemenz and Rajagopalan, 1997).

Among the many types of phenomena that might occur as a result of relative motion between charged phases and electrolytes, four types of electrokinetic phenomena are more commonly encountered: electroosmosis, streaming potential, electrophoresis, and sedimentation potential. These four types of electrokinetic phenomena are briefly described below.

7.1 ELECTROOSMOSIS

Electroosmosis represents the movement, due to an applied electric field, of an electrolyte solution relative to a stationary charged surface (*i.e.*, a capillary tube or porous media). The pressure necessary to counterbalance electroosmotic flow is termed the *electroosmotic pressure*.

A typical electroosmotic fluid flow in a capillary tube is shown in Figure 7.1. When the capillary tube is negatively charged, the applied electric field exerts a force in the

Electrokinetic and Colloid Transport Phenomena, by Jacob H. Masliyah and Subir Bhattacharjee
Copyright © 2006 John Wiley & Sons, Inc.

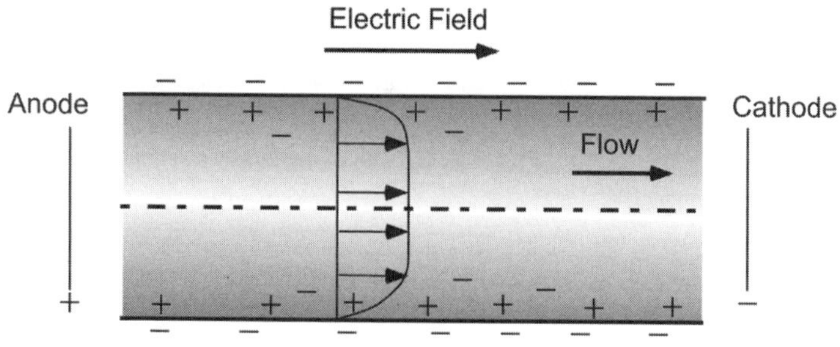

Figure 7.1. Electroosmotic flow in a capillary tube.

direction of the cathode on the excess ions of positive charge near the surface. The positively charged ions then drag the electrolyte solution along with them and cause the fluid to flow towards the cathode. Note that one should impose the experimental condition that the pressure difference between the two capillary ends be zero to study electroosmosis under the influence of an applied electric field.

Electroosmosis can be employed to drain porous media and in the evaluation of the surface charge of capillary tubes or porous media (see, *e.g.*, Hiemenz and Rajagopalan, 1997; Probstein, 2003).

The electroosmotic pressure between the two ends of the capillary can be measured when there is no flow through the capillary under the influence of the applied electric field.

7.2 STREAMING POTENTIAL

An electric field is created when an electrolyte solution is made to flow along a stationary charged surface by applying a pressure gradient. Such flows are generally encountered in narrow capillary microchannels connected to two reservoirs. When the electrolyte concentrations in the two reservoirs are identical, and when there is no net current flowing through the system, the steady-state electric field developed between the two reservoirs is called the *streaming potential*. For example, a streaming potential is set up when an electrolyte solution is pumped through a negatively charged capillary as shown in Figure 7.2. The electric field due to the flow is from right to left.

The principle of streaming potential is used in sea water desalination. The streaming potential phenomenon is considered as a reciprocal (*i.e.*, opposite) of the phenomenon of electroosmosis.

7.3 ELECTROPHORESIS

The movement of a charged surface, such as that of a colloidal particle, relative to a stationary liquid caused by an applied electric field is known as *electrophoresis*. A typical particle electrophoresis is shown in Figure 7.3. Due to the presence of the anode and cathode terminals, an electric field, E, becomes established from left to

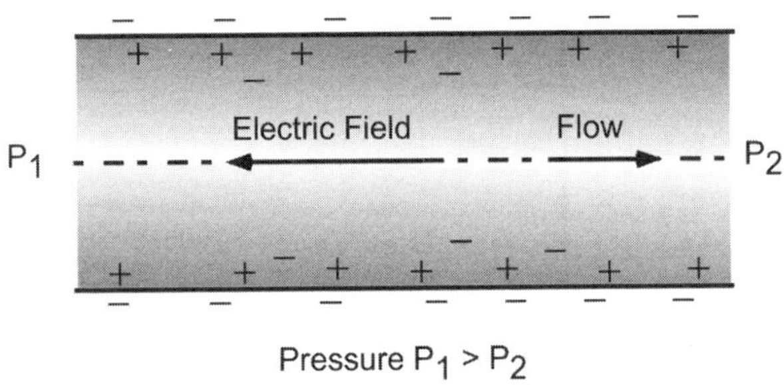

Figure 7.2. Development of streaming potential when an electrolyte is pumped through a capillary.

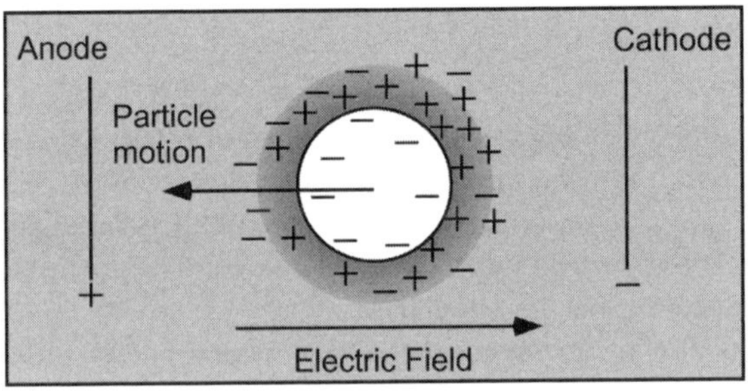

Figure 7.3. Electrophoresis of a charged particle in an external electric field.

right. Under the influence of this electric field, the negatively charged colloidal particle migrates towards the anode. Electrophoresis is usually employed in measuring the surface potential of a charged particle. Note that in electrophoresis, one does not apply any pressure gradients to cause a flow.

7.4 SEDIMENTATION POTENTIAL

An electric field is created when charged particles move relative to a liquid. The movement of the particles can be under gravitational or centrifugal fields. This phenomenon is sometimes called the *Dorn effect* or the *migration potential*. It is the least studied among the electrokinetic phenomena. The sedimentation potential of a settling suspension of charged colloidal particles under the influence of a gravitational field is illustrated in Figure 7.4. In a strict sense, the sedimentation potential is defined for the case when the current flow is zero in such processes.

7.5 NON-EQUILIBRIUM PROCESSES AND ONSAGER RELATIONSHIPS

In electrokinetic processes, one is concerned with the coupled influence of multiple types of forces or potentials (electrical, pressure, gravity, *etc.*) on the transport

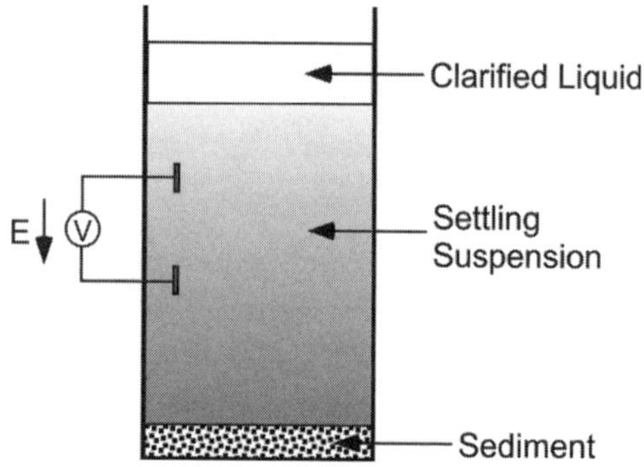

Figure 7.4. Sedimentation of charged colloidal particles under gravity setting up a sedimentation potential.

behavior of a multicomponent system. These types of processes are often described in light of non-equilibrium thermodynamic theories by means of relationships known as the *Onsager reciprocity relations* (Onsager, 1931a,b; de Groot and Mazur, 1962). A brief outline of the pertinent information on non-equilibrium thermodynamics and the Onsager relationships is presented here.

Several physical phenomena encountered in natural and engineered systems are governed by transport processes where a simple linear relationship between a flux and the corresponding driving force exists. The driving force is generally a gradient of some potential such as concentration, chemical potential, temperature, pressure, electric potential, *etc*. Some examples of transport processes governed by such a simple proportionality between a flux and its conjugate driving force are: (i) molecular diffusion, where the diffusive flux is related to the concentration gradient, (ii) thermal conduction, where the heat flux is related to the temperature gradient, (iii) fluid flow, where the flow rate is proportional to the pressure gradient, and (iv) electrical conduction, where the current is related to the applied potential gradient. The equations formulated to describe such proportionality are often called constitutive laws. For instance, the diffusive flux is related to the concentration gradient through Fick's law, Fourier's law relates the heat flux to the temperature gradient in heat conduction, Newton's law (manifested as Navier–Stokes or Darcy's equation) governs the relationship between fluid flow and pressure gradient, and finally, Ohm's law relates the current to the electric potential gradient during electrical conduction. In all the above cases, the constitutive law states that the flux is directly proportional to the conjugated driving force. In other words, one can state

$$\mathbf{J}_\alpha = L_{\alpha\alpha}\mathbf{X}_\alpha \tag{7.1}$$

where $\mathbf{J}_\alpha$ is the flux, $\mathbf{X}_\alpha$ is the conjugate driving force, and $L_{\alpha\alpha}$ is the proportionality constant or the transport coefficient. These transport processes are termed non-equilibrium thermodynamic phenomena, since the existence of the driving force in these processes signifies a deviation from equilibrium.

The electrokinetic transport processes introduced in this chapter, and further discussed in the subsequent chapters, can not be described in terms of a simple constitutive relationship between a single driving force and the corresponding flux. For instance, when an electric potential gradient is applied along the axis of a capillary containing an electrolyte solution, not only does a current flow occur, but there is also a coupled flow of the liquid along the capillary. In this case, one generally states that the electric potential gradient gives rise to a *non-conjugated* flow of the liquid, implying that flow is not related to the electric potential gradient through a direct constitutive equation. Conversely, when a pressure gradient is applied across the capillary, not only will this pressure drop cause a fluid flow, but also a coupled flow of current. A similar type of correspondence exists between electrophoresis and sedimentation. In the above instances, we observe that there exists a coupling between driving forces of one type and fluxes of another type. For sufficiently slow processes, any flow (or flux) may depend in a linear manner not only on its conjugate force, but other nonconjugate forces. The study of such coupled transport phenomena falls under the purview of non-equilibrium thermodynamics.

Mathematically, the nonconjugate or cross effects of different forces on the fluxes are described by adding new terms to the constitutive equations of the form given by Eq. (7.1). The modified constitutive equations are written in a general form relating each type of flux with all the driving forces as (Onsager, 1931a,b; de Groot and Mazur, 1962)

$$
\begin{aligned}
\mathbf{J}_1 &= L_{11}\mathbf{X}_1 + L_{12}\mathbf{X}_2 + L_{13}\mathbf{X}_3 + \cdots + L_{1n}\mathbf{X}_n \\
\mathbf{J}_2 &= L_{21}\mathbf{X}_1 + L_{22}\mathbf{X}_2 + L_{23}\mathbf{X}_3 + \cdots + L_{2n}\mathbf{X}_n \\
\cdots &= \cdots \\
\cdots &= \cdots \\
\mathbf{J}_n &= L_{n1}\mathbf{X}_1 + L_{n2}\mathbf{X}_2 + L_{n3}\mathbf{X}_3 + \cdots + L_{nn}\mathbf{X}_n
\end{aligned}
\tag{7.2}
$$

which can be written in an abbreviated form as

$$
\mathbf{J}_\alpha = \sum_{\beta=1}^{n} L_{\alpha\beta}\mathbf{X}_\beta \qquad \alpha, \beta = 1, 2, \ldots, n \tag{7.3}
$$

where $\mathbf{J}_\alpha$ stands for different fluxes, $\mathbf{X}_\beta$ represents different driving forces, and $L_{\alpha\beta}$ represents the phenomenological coefficients, which are independent of both the fluxes and the driving forces. The term phenomenological coefficients signifies that these coefficients are usually determined from experiments. Although we have used vector notations for the fluxes and driving forces, it should be noted that the fluxes can be scalars or even higher order tensors. Also, the driving forces may not strictly be regarded as forces in a Newtonian sense. These can represent any type of potential gradient causing an irreversible transport phenomenon.

The coupled irreversible transport phenomena described by Eq. (7.3) are governed by Onsager's law, which states that for a proper choice of the fluxes and driving forces, the matrix of the phenomenological coefficients $L_{\alpha\beta}$ is symmetric. In other words

$$L_{\alpha\beta} = L_{\beta\alpha} \tag{7.4}$$

These identities are referred to as *Onsager reciprocal relations* (Onsager, 1931a,b).

Let us now illustrate the above formalism for electrokinetic flow. In such processes, the current, I, and the volumetric fluid flow rate, Q, are two relevant fluxes, while the electric potential gradient ($\mathbf{E} = -\nabla\psi$), and the pressure gradient, $-\nabla p$, constitute the two corresponding associated driving forces. The transport phenomena can then be written in terms of the coupled transport equations as follows

$$\mathbf{I} = L_{11}\mathbf{E} + L_{12}(-\nabla p) \tag{7.5}$$

$$Q = L_{21}\mathbf{E} + L_{22}(-\nabla p) \tag{7.6}$$

Note that in writing the above relationships, the diagonal terms of the coefficient matrix (*i.e.*, the $L_{\alpha\alpha}$ terms) were chosen such that these phenomenological coefficients relate the fluxes to their "natural" conjugate driving forces. Clearly, for the current, the natural conjugate driving force is the potential gradient, while for the flow rate, the conjugate driving force is the pressure gradient. According to the Onsager reciprocal relations, in this case, we have $L_{12} = L_{21}$. The choice of the natural conjugate driving force for a given flux is non trivial, particularly when dealing with more than two coupled processes (de Groot and Mazur, 1962). In the next chapter, this type of relationship will be used to check the validity of the mathematical formulation for streaming potential and electroosmotic flow.

The use of reciprocity relationships allows the consideration of one type of electrokinetic transport process (for instance, electrophoresis) and the application of the results to describe the coupled transport process (*i.e.*, sedimentation) through the use of the appropriate Onsager relationships.

7.6 NOMENCLATURE

E	magnitude of electric field vector, V/m
$\mathbf{E}$	electric field, V/m
$\mathbf{I}$	current, A
$\mathbf{J}_\alpha$	flux
$L_{\alpha\beta}$	phenomenological coefficient
Q	volumetric flow rate, m^3
$\mathbf{X}_\beta$	driving force
∇p	pressure gradient, Pa/m

7.7 REFERENCES

de Groot, S. R., and Mazur, P., *Nonequilibrium Thermodynamics*, North-Holland, Amsterdam, (1962).

Hiemenz, P. C., and Rajagopalan, R., *Principles of Colloid and Surface Chemistry*, 3rd ed., Marcel Dekker, New York, (1997).

Onsager, L., Reciprocal relations in irreversible processes I, *Phys. Rev.*, **37**, 405–426, (1931a).

Onsager, L., Reciprocal relations in irreversible processes II, *Phys. Rev.*, **38**, 2265–2279, (1931b).

Probstein, R. F., *Physicochemical Hydrodynamics, An Introduction*, 2nd ed., Wiley Interscience, New York, (2003).

Shaw, D. J., *Introduction to Colloid and Surface Chemistry*, 3rd ed., Butterworths, London, (1980).

CHAPTER 8

FLOW IN MICROCHANNELS

8.1 LIQUID FLOW IN CHANNELS

Traditionally, Newtonian liquids are made to flow in channels, pipes, and porous media under the influence of a pressure gradient. When the channel dimension or the pore size is relatively large, such a method for liquid flow is quite effective. It is also possible to affect liquid flow by imposing an electric potential gradient along the flow channel or across a porous bed. Such a flow mode is possible when the channel walls carry a surface charge and the flowing liquid contains free charges. The effectiveness of a particular mode for liquid flow is dependent on the system geometry and the physical properties of the flowing liquid. In this chapter, we will refer to a liquid as being either an electrolyte solution, *e.g.*, water with dissolved ionic species, or a liquid having free charge. A pressure driven laminar flow in a straight channel is normally termed as Poiseuille flow. An electric potential driven flow is termed as electroosmotic flow.

Electroosmosis is associated with the movement of a bulk electrolyte solution or a liquid carrying a free charge, relative to a stationary charged surface, under the influence of an imposed electric field. For example, an electrolyte solution in a porous medium can be made to move when a electric field is applied.

In many devices, glass capillaries (microchannels) are used for electroosmotic flows. The charge on the glass capillary wall arises from the dissociation of surface silanol groups, $-$SiOH, or the preferential adsorption of OH^- ions onto the glass surface. In most cases, the surface charge is negative and an electric double layer

Electrokinetic and Colloid Transport Phenomena, by Jacob H. Masliyah and Subir Bhattacharjee
Copyright © 2006 John Wiley & Sons, Inc.

is present adjacent to the capillary walls. The excess charge, *i.e.*, the absence of electroneutrality within the electric double layer, is responsible for establishing the electroosmotic flow within the capillary under an external electric field.

8.2 ELECTROOSMOTIC FLOW IN A SLIT CHARGED MICROCHANNEL

8.2.1 Electric Potential

In order to understand pressure driven and electroosmotic flow in porous media, in capillary tubes of varying cross-section, or along surfaces with non-homogeneous surface charge distribution, it is first necessary to be able to analyze the simple geometry of the flow in a straight gap formed by two parallel plates. Such a flow configuration is referred to as the slit microchannel.

In this analysis, we shall present pressure driven and electroosmotic flow in the gap between two parallel surfaces having a low surface electric potential. The flow to be considered is steady-state and fully developed. In other words, there is no time dependence and there are no end effects. The analysis is that of a one-dimensional flow. By and large, the analysis is that due to Burgreen and Nakache (1964), although their analysis did not invoke the low surface potential assumption.

Consider a microchannel formed by two parallel plates, *i.e.*, a slit microchannel. The walls of the channel have a surface electric potential of ψ_s. The wall potential is assumed to be negative to facilitate plotting the results and subsequent discussions. It should be noted that assigning a negative sign for the wall potential does not impose any restriction on the analysis. Figures 8.1 and 8.2 depict the flow geometry.

The microchannel ends are subjected to an electric potential that gives rise to a uniform electric field strength of E_x where

$$E_x = -\frac{\partial \phi(x, y)}{\partial x} \tag{8.1}$$

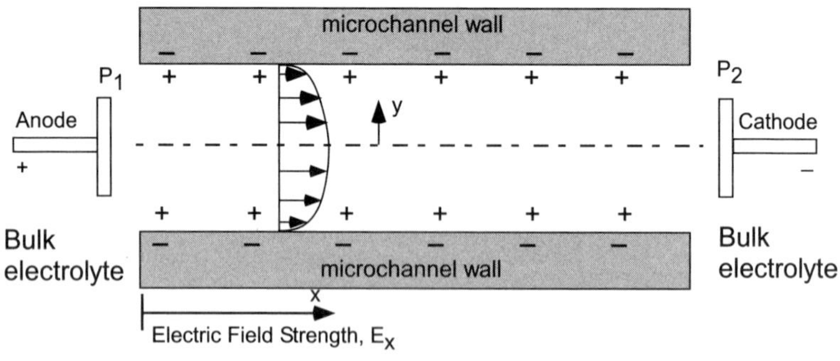

Figure 8.1. Electroosmotic and pressure driven flow in a microchannel gap formed by two parallel plates. The microchannel walls are negatively charged. The flow is purely electroosmotic when there is an electric field, E_x, acting between the electrodes along the axial direction and $P_1 = P_2$. The flow is pressure driven when $P_1 \neq P_2$ and $E_x = 0$.

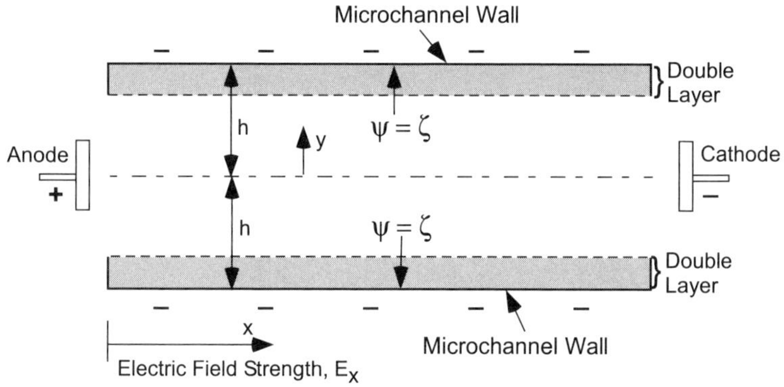

Figure 8.2. Electroosmotic flow in a microchannel gap formed by two parallel plates. Details of the two-dimensional slit microchannel geometry are shown. The walls of the microchannel have a surface potential ψ_s, which sets up an excess counterion concentration near the walls. The shaded regions near the walls represent the electric double layers where an excess counterion concentration exists.

with $\phi(x, y)$ being the local electric potential in the microchannel, x being the axial direction, and y being the coordinate normal to the axial direction, *i.e.*, the transverse direction. As discussed earlier, in the microchannel, the concentration of the counterions (in the case of a negatively charged surface, they are the positive ions or simply the cations) predominate near the charged microchannel walls, within the so called Debye sheath. This gives rise to a finite (non-zero) volumetric free charge density, ρ_f, in the fluid adjacent to the charged walls. The application of an external electric field on this charged fluid results in a net electrical body force, $\rho_f E_x$, acting on the fluid close to the microchannel surface. This electric body force results in a fluid flow through the microchannel. Such an electrically driven flow is generally referred to as electroosmotic flow.

Consider a symmetric $(z : z)$ electrolyte solution contained in two reservoirs connected by the microchannel. Let the microchannel walls have a surface potential ψ_s. In this analysis, we will not distinguish between the surface potential and the zeta potential of the microchannel surface. The electric potential at a location (x, y) in the channel, given by $\phi(x, y)$, arises due to the superposition of the applied external electric potential and the potential due to the surface charge of the microchannel walls. In other words, the electric potential at a given location is the algebraic sum of the potential due to the electric double layer and the potential due to the imposed electric field. Assuming that the potential due to the electrical double layer is independent of axial position in the microchannel (which is valid for long microchannels, neglecting any end effects), one can write

$$\phi(x, y) \equiv \phi = \psi(y) + [\phi_0 - xE_x] \tag{8.2}$$

where $\psi(y)$ is the electric potential due to the electric double layer at the equilibrium state corresponding to no fluid motion and no applied external electric field, ϕ_0 is the value of the imposed potential at $x = 0$, $\phi_0 - xE_x$ is the electric potential at a given axial location due to the electric field strength E_x in the absence of the electric double layer (*i.e.*, zero surface potential), E_x is the electric field strength being independent of position, x is the axial direction, and y is the transverse coordinate of the flow gap.

The Poisson equation defining the electric potential within the microchannel in Cartesian coordinates is given by

$$\frac{\partial^2 \phi}{\partial x^2} + \frac{\partial^2 \phi}{\partial y^2} = -\frac{\rho_f}{\epsilon} \tag{8.3}$$

Introducing the expression of the electric potential given by Eq. (8.2) into the Poisson equation, Eq. (8.3), gives

$$\frac{d^2 \psi}{dy^2} = -\frac{\rho_f}{\epsilon} \tag{8.4}$$

Equation (8.4) is the resulting Poisson equation for the slit microchannel.

The free charge density, ρ_f, is given by

$$\rho_f = \sum_k e z_k n_k \tag{8.5}$$

Now assume that the ionic number concentration, n_k, of the kth ion is given by the Boltzmann distribution. It follows that for a $(z : z)$ electrolyte

$$n_k = n_\infty \exp\left[-\frac{z_k e \psi}{k_B T}\right] \tag{8.6}$$

Here, n_∞ is the ionic number concentration in the neutral electrolyte. For a symmetric electrolyte one can write

$$z_+ = -z_- = z$$

and Eq. (8.6) provides

$$n_+ = n_\infty \exp\left[-\frac{ze\psi}{k_B T}\right] \tag{8.7}$$

and

$$n_- = n_\infty \exp\left[\frac{ze\psi}{k_B T}\right] \tag{8.8}$$

The free charge density, Eq. (8.5), can be written as

$$\rho_f = e(z_+ n_+ + z_- n_-) = ze(n_+ - n_-) \tag{8.9}$$

Making use of the Boltzmann expression, Eq. (8.9) becomes

$$\rho_f = zen_\infty \left[\exp\left(-\frac{ze\psi}{k_B T}\right) - \exp\left(\frac{ze\psi}{k_B T}\right)\right] \tag{8.10}$$

or

$$\rho_f = -2zen_\infty \sinh\left(\frac{ze\psi}{k_B T}\right) \tag{8.11}$$

In deriving Eq. (8.11), it was assumed that the local distribution of the free charges (ions) is governed solely by the electrical potential distribution due to the charged wall, namely $\psi(y)$, and not the total potential, $\phi(x, y)$. This assumption is made primarily for the sake of convenience, and should be valid if the axial variation of the electric potential is much smaller than the transverse variation of the potential. This condition will generally hold when the microchannel length is much larger than the gap, $L \gg h$, where L is the channel length.

Combining Eqs. (8.4) and (8.11) leads to

$$\frac{d^2\psi}{dy^2} = \frac{2zen_\infty}{\epsilon} \sinh\left(\frac{ze\psi}{k_B T}\right) \tag{8.12}$$

Equation (8.12) is the Poisson–Boltzmann equation that defines the electric double layer potential, $\psi(y)$, in the microchannel. Its derivation was made with the explicit assumption of the validity of the Boltzmann distribution in a flow system.

The analytical, albeit fairly complicated, solution of Eq. (8.12) was provided by Burgreen and Nakache (1964). To simplify the analysis, we will assume that $ze\psi/k_B T < 1$ so that we can write $\sinh(ze\psi/k_B T) \approx ze\psi/k_B T$. Such a linearization is known as the Debye–Hückel approximation, as discussed in Chapter 5. For $z = 1$ and $\psi = 0.025$ V, the term $(ze\psi/k_B T)$ is close to unity. Physically, such an approximation means that the electric potential is small compared to the thermal potential, *i.e.*, $|ze\psi| < k_B T$.

Therefore, for small potentials, the Poisson–Boltzmann equation (8.12), becomes

$$\frac{d^2\psi}{dy^2} = \frac{2n_\infty z^2 e^2 \psi}{\epsilon k_B T} \tag{8.13}$$

Defining the inverse double layer thickness (or inverse Debye length), κ, by

$$\kappa = \left(\frac{2n_\infty z^2 e^2}{\epsilon k_B T}\right)^{1/2} \tag{8.14}$$

Eq. (8.13) becomes

$$\frac{d^2\psi}{dy^2} = \kappa^2 \psi \tag{8.15}$$

κ is also referred to as the Debye–Hückel parameter. Equation (8.15) is a linear differential equation subject to the following boundary conditions:

$$\text{At } y = h \quad \psi = \zeta \text{ (Potential at the shear plane)} \tag{8.16-a}$$

$$\text{At } y = 0 \quad \frac{d\psi}{dy} = 0 \text{ (Axis of symmetry)} \tag{8.16-b}$$

Here, h is the half-width gap of the microchannel. The second boundary condition implicitly implies that at $y = \pm h$, $\psi = \zeta$. In other words, the confining microchannel walls carry the same surface potential and one has symmetry within the channel. It should be noted here that the potential at $y = h$ is the zeta potential (ζ) instead of the actual surface potential of the microchannel wall, ψ_s. This adjustment is necessary to account for the no-slip boundary condition for the fluid flow problem at the shear plane (*cf.*, Section 5.8), which is discussed below. At this stage, it should be borne in mind that the true surface potential of the wall does not appear in the theory for electrokinetic flows. Instead, the potential at the shear plane, the zeta potential, is used in the theory. Accordingly, the location $y = \pm h$ corresponds to the distance between the centerline of the channel and the shear plane, not the channel wall itself.

The solution to Eq. (8.15) subject to the boundary conditions of Eqs. (8.16-a) and (8.16-b) gives

$$\psi = \zeta \frac{\cosh(\kappa y)}{\cosh(\kappa h)} \tag{8.17}$$

Making use of Eqs. (8.2) and (8.17), the total electric potential is then given by

$$\phi(x, y) = \zeta \frac{\cosh(\kappa y)}{\cosh(\kappa h)} + [\phi_0 - x E_x] \tag{8.18}$$

and the free charge density for low surface potential is given by

$$\rho_f = -\epsilon \kappa^2 \psi = -\epsilon \kappa^2 \zeta \frac{\cosh(\kappa y)}{\cosh(\kappa h)} \tag{8.19}$$

At this point, it is pertinent to discuss the implications of using the Debye–Hückel approximation in the subsequent theoretical developments. The linearization of the Poisson–Boltzmann equation for a symmetric ($z : z$) electrolyte, Eq. (8.12), involves writing the sinh term on its right hand side as a Taylor series expansion in $\Psi = ze\psi/k_B T$. The expansion gives

$$\sinh(\Psi) = \Psi + \frac{\Psi^3}{3!} + \cdots$$

Note that Eq. (8.15) was obtained by retaining only the first term in the above series expansion for $\sinh(\Psi)$. This implies that the linearized form can be considered as accurate up to $\mathcal{O}(\Psi^2)$, since the next neglected term in the expansion is of $\mathcal{O}(\Psi^3)$. In the subsequent analysis, there will be at least two places where this approximation will have a bearing. In the next subsection, when considering the influence of the electrical body force, $\rho_f E_x$, on the fluid velocity distribution in the microchannel, the charge density, ρ_f, will be written in terms of the linearized form, Eq. (8.19), instead of the general form given by Eq. (8.11). Subsequently, during determination of the migration current density using the Nernst–Planck equations, we will encounter an expression of the form

$$n_+ + n_- = n_\infty[\exp(-\Psi) + \exp(\Psi)] = 2n_\infty \cosh(\Psi) \tag{8.20}$$

This expression uses the complete Boltzmann distribution. One needs to bear in mind at this step that substituting the potential distribution, Eq. (8.17), obtained from the solution of the linearized Poisson–Boltzmann equation directly in Eq. (8.20) will be mathematically inconsistent. This is because we are mixing a result that is based on at best a quadratic approximation of the Boltzmann distribution with another that is based on the complete Boltzmann distribution. Consequently, one first needs to expand the cosh term in Eq. (8.20) in a Taylor series, where one can retain up to the quadratic term

$$\cosh(\Psi) = 1 + \frac{\Psi^2}{2!} + \mathcal{O}(\Psi^4)$$

If one substitutes Eq. (8.17) in the above Taylor expansion, the resulting expression will be accurate up to $\mathcal{O}(\Psi^2)$ and will be mathematically consistent. Such a check of consistency in the entire mathematical formulation will be extremely important in correct development of the analytical models for electrokinetic flow. The above discussion also implies that the subsequent theoretical development will be somewhat approximate owing to the use of the linearized Poisson–Boltzmann equation.

8.2.2 Flow Velocity

To this point we have solved for the electric parameters by combining the Poisson and Boltzmann equations and invoking the Debye–Hückel approximation.

In order to evaluate the axial velocity, u_x, we need to consider the modified Navier–Stokes equation where the electric body force is included. For constant fluid properties under laminar flow conditions, the modified Navier–Stokes equation in the x-direction is given by [see Eq. (6.112)]:

$$\mu\frac{d^2u_x}{dy^2} = \frac{\partial p}{\partial x} + \rho_f\frac{\partial \phi}{\partial x} - \rho g_x \tag{8.21}$$

where u_x is the axial liquid velocity, μ is the liquid viscosity, ρ is the liquid mass density, g_x is the gravitational acceleration in the x-direction at location y, and p is the liquid pressure. The continuity equation for a one-dimensional flow provides

$$\frac{\partial u_x}{\partial x} = 0 \tag{8.22}$$

As the flow is fully developed, the velocity in y-direction, u_y, is taken as zero and the velocity u_x is solely a function of the transverse direction (y). The latter fact is reflected by the solution of the continuity equation, (8.22), which provides

$$u_x = f(y) \tag{8.23}$$

As the velocity in the transverse direction, u_y is zero, the modified Navier–Stokes equation for the y-direction can be written as

$$\frac{\partial p}{\partial y} + \rho_f\frac{\partial \phi}{\partial y} - \rho g_y = 0 \tag{8.24}$$

where g_y is the gravitational acceleration in the y-direction. For a horizontal microchannel, the variation in the pressure, $\partial p/\partial y$, is due to the hydrostatic head and the electrostatic body force term of Eq. (8.24). As the microchannel gap is normally very small, and $g_x = 0$ for a horizontal channel, one can assume that the liquid pressure is nearly independent of y and that the pressure is solely a function of the axial direction. Consequently, the modified Navier–Stokes equation for a horizontal channel in the x-direction becomes

$$\mu \frac{d^2 u_x}{dy^2} = \frac{dp}{dx} + \rho_f \frac{\partial \phi}{\partial x} \tag{8.25}$$

One can think of dp/dx as the axial pressure gradient in the microchannel. As we have assumed fully developed flow, dp/dx is a constant. In a normal channel flow, dp/dx is a negative quantity. For convenience, we set

$$p_x = -\frac{dp}{dx} \tag{8.26}$$

Making use of Eqs. (8.1), (8.19), and (8.26), the momentum equation as represented by the modified Navier–Stokes equation, Eq. (8.25) becomes

$$\mu \frac{d^2 u_x}{dy^2} = -p_x + \epsilon \kappa^2 E_x \zeta \frac{\cosh(\kappa y)}{\cosh(\kappa h)} \tag{8.27}$$

The momentum equation is subject to the following boundary conditions:

$$u_x = 0 \qquad \text{at } y = h \tag{8.28}$$

and

$$\frac{du_x}{dy} = 0 \qquad \text{at } y = 0 \tag{8.29}$$

The boundary condition, Eq. (8.28), is a statement that at the channel wall (or rather, at the shear plane) the electrolyte solution velocity is zero, *i.e.*, the no-slip condition is imposed. The boundary condition of Eq. (8.29) implies flow symmetry in the y-direction. This is a reasonable assumption as the electric potential was assumed to be symmetric.

Solution of Eq. (8.27) subject to the boundary conditions provided by Eqs. (8.28) and (8.29) is given by

$$u_x(y) = u_x = \frac{h^2 p_x}{2\mu}\left[1 - \left(\frac{y}{h}\right)^2\right] - \frac{\epsilon E_x \zeta}{\mu}\left[1 - \frac{\cosh(\kappa y)}{\cosh(\kappa h)}\right] \tag{8.30}$$

Equation (8.30) gives the variation of the axial velocity across the channel of an electrolyte solution or a liquid carrying free charges. The first term is the liquid velocity due to an imposed pressure gradient, p_x. It is normally referred to as Poiseuille flow. The velocity due to the imposed pressure gradient has a parabolic profile. For a given pressure gradient, p_x, and fluid viscosity, μ, the local velocity is proportional to h^2. Clearly, the flow is much affected by the channel width $2h$. As the channel width becomes very small, pressure driven flows become impractical as the pumping requirements become prohibitive.

The second term of Eq. (8.30) is due to the electroosmotic flow as a consequence of the imposed electrical potential gradient, E_x. Here, the local velocity is proportional to $(\epsilon E_x \zeta / \mu)$. For given system properties, the local velocity is directly proportional to the imposed electrical potential gradient. The channel width does not appear in the proportionality term $(\epsilon E_x \zeta / \mu)$. Consequently, electroosmotic flows become very attractive for very narrow channels. Equation (8.30) indicates that the local velocity due to electroosmotic flow is much dependent on the electric double layer thickness, κ^{-1}.

Figure 8.3 shows the variation of the dimensionless local velocity due to a pressure driven flow. The normalized local velocity due to a pressure driven flow is given as

$$\frac{u_{x,press}}{\left(h^2 p_x / 2\mu\right)} = U_{x,press} = 1 - \left(\frac{y}{h}\right)^2 \tag{8.31}$$

The normalized local velocity is parabolic when plotted against normalized transverse position, y/h, as shown in Figure 8.3.

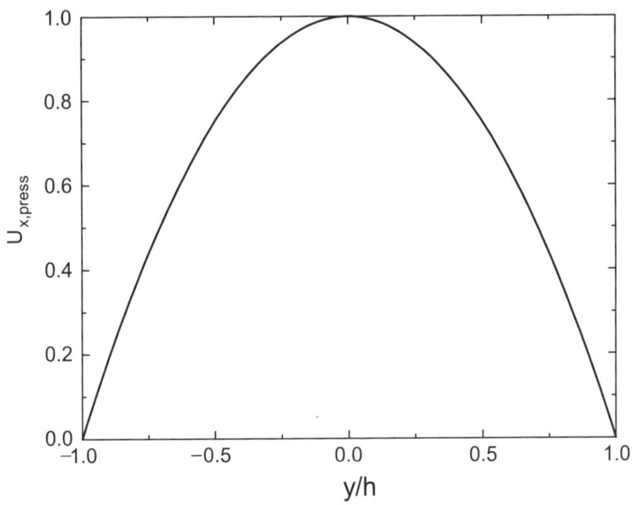

Figure 8.3. Variation of dimensionless local velocity due to a pressure driven flow with normalized transverse position in a slit microchannel.

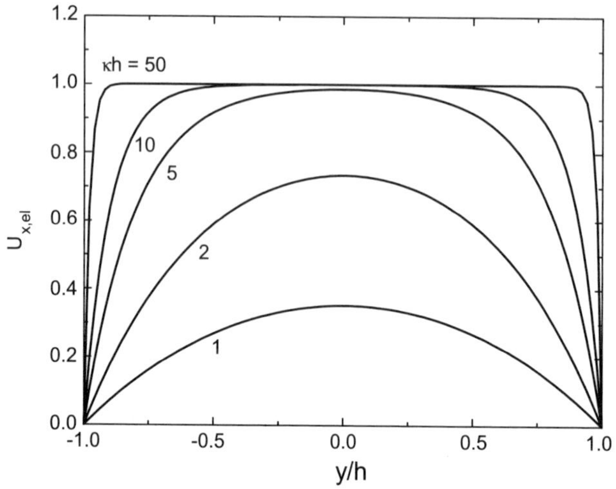

Figure 8.4. Variation of dimensionless local velocity due to electroosmotic flow with normalized transverse position in a slit microchannel.

Figure 8.4 shows the variation of the dimensionless local velocity due to electroosmotic flow, written for convenience as

$$\frac{u_{x,el}}{-(\epsilon E_x \zeta/\mu)} = \frac{u_{x,el}}{-\Omega E_x} = U_{x,el} = 1 - \frac{\cosh(\kappa h \cdot \frac{y}{h})}{\cosh(\kappa h)} \tag{8.32}$$

where

$$\Omega = \frac{\epsilon\zeta}{\mu} \tag{8.33}$$

The dimensionless electroosmotic velocity profile in the microchannel is a strong function of the dimensionless channel gap as represented by κh. For $\kappa h < 5$, the dimensionless velocity shows strong dependence on the transverse position, y/h. However, for large κh values, the velocity profile becomes fairly flat in the central region of the flow channel. In practice, for channels of $h > 10$ microns with a (1:1) electrolyte concentration as low as 10^{-6} M, the value of κh is quite large and the velocity profile is fairly flat. In the case of $h < 1$ μm coupled with free charge carrying organic liquids, one would expect to achieve low values of κh.

8.2.3 Volumetric Flow Rate

The volumetric flow per unit width of the slit microchannel is given by

$$Q = 2\int_0^h u_x(y)\,dy \tag{8.34}$$

The units of Q are $m^3/(ms)$. Substituting for $u_x(y)$ from Eq. (8.30) in Eq. (8.34) and performing the integration lead to

$$Q = \frac{2h^3 p_x}{3\mu} - \frac{2h\epsilon E_x \zeta}{\mu}\left[1 - \frac{\tanh(\kappa h)}{\kappa h}\right] \tag{8.35}$$

Denoting the channel cross-sectional area per unit width by A_c, where

$$A_c = 2h$$

Equation (8.35) becomes

$$\frac{Q}{A_c} = \frac{h^2 p_x}{3\mu} - \frac{\epsilon E_x \zeta}{\mu}\left[1 - \frac{\tanh(\kappa h)}{\kappa h}\right] \tag{8.36}$$

The term Q/A_c represents the volumetric fluid flow rate per unit width under pressure and electric potential gradients. The first term of Eq. (8.36) is the volumetric flow rate per unit width, Q_{press}/A_c, due to a pressure gradient. For a given gap and liquid viscosity, Q_{press}/A_c is directly proportional to the pressure gradient. Such a relationship is to be expected for a flow process governed by a linear differential equation. For given p_x/μ, the pressure driven volumetric flow rate per unit width is much influenced by the channel gap spacing, $2h$.

The second term of Eq. (8.36) gives the volumetric flow rate per unit width, Q_{el}/A_c, for the case of electroosmotic flow that is driven by the electric potential gradient. The term Q_{el}/A_c is directly proportional to $\epsilon\zeta E_x/\mu$, which can be considered as the effective electric driving force. Once again such a linear relationship is to be expected. The dimensionless electroosmotic flow is given by

$$\frac{Q_{el}}{(-A_c \Omega E_x)} = 1 - \frac{\tanh(\kappa h)}{\kappa h} \tag{8.37}$$

For large values of κh, $\tanh(\kappa h) \to 1$ and Eq. (8.37) indicates that the dimensionless electroosmotic volumetric flow rate approaches unity at large values of κh. For small values of $\kappa h < 1$, Eq. (8.37) gives

$$\frac{Q_{el}}{(-A_c \Omega E_x)} = \frac{(\kappa h)^2}{3} \tag{8.38}$$

indicating that there is minimal electroosmotic flow for small values of κh. Figure 8.5 depicts the scaled electroosmotic flow contributions obtained using Eqs. (8.37) and (8.38). It is clear from Figure 8.5 that the greatest advantage in having electroosmotic flow is for the case of large κh (*i.e.*, a thin electric double layer).

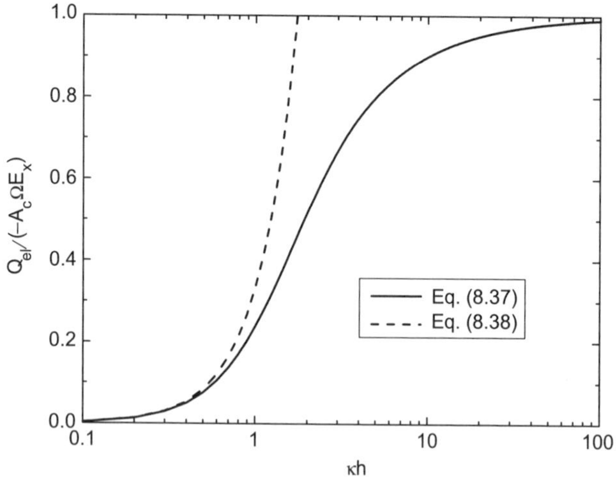

Figure 8.5. Variation of the electroosmotic volumetric flow rate per unit channel width with the electric double layer thickness κh. The dashed line represents the approximate expression valid for small values of κh.

8.3 ELECTROOSMOTIC FLOW IN A CLOSED SLIT MICROCHANNEL

Electrophoretic measurements are normally conducted in rectangular channels having a large aspect ratio. In such electrophoretic measurements, the movement of the charged particles are monitored in order to evaluate their surface charge. An electrophoretic cell is shown in Figure 8.6. The electrodes are on the extreme ends of the

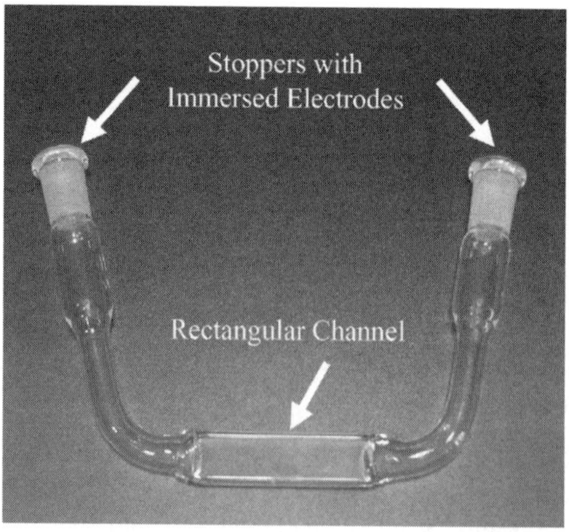

Figure 8.6. An electrophoretic cell.

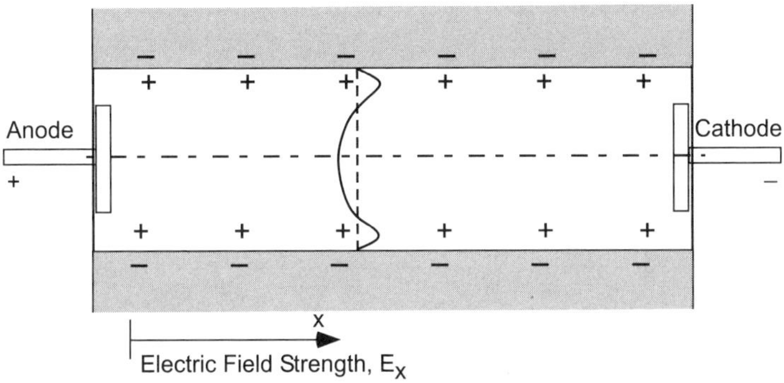

Figure 8.7. Schematic of a velocity profile in streaming potential flow in a closed slit microchannel having a negative zeta potential.

cell and the cell itself is sealed. In other words, there is no net flow within a given cross-section of the cell. As a rectangular cell with a large aspect ratio can be approximated by a one-dimensional flow problem, that of a gap formed by two parallel plates, the analysis presented here can be used to study flows in an electrophoretic cell.

For a closed narrow gap channel, as shown in Figure 8.7, the volumetric flow per unit width is zero. Setting Q to zero, Eq. (8.36) yields the pressure gradient that has to be developed by the liquid to accommodate the zero net flow,

$$p_x = \frac{3\epsilon E_x \zeta}{h^2}\left[1 - \frac{\tanh(\kappa h)}{\kappa h}\right] \tag{8.39}$$

The pressure gradient has a maximum absolute value when $\kappa h \to \infty$. Substituting for p_x in Eq. (8.30) yields the local velocity in the microchannel for the special case of $Q = 0$,

$$\frac{u_x}{(-\Omega E_x)} = -\frac{3}{2}\left[1 - \left(\frac{y}{h}\right)^2\right]\left[1 - \frac{\tanh(\kappa h)}{\kappa h}\right] + \left[1 - \frac{\cosh(\kappa y)}{\cosh(\kappa h)}\right] \tag{8.40}$$

The term $u_x/(-\Omega E_x)$ represents a convenient dimensionless local axial velocity.

Figure 8.8 shows the dimensionless local velocity in a microchannel for different values of κh. The overall velocity profile is a superposition of the electroosmotic flow and the counteracting pressure driven flow. For a negative value of the surface potential, *i.e.*, $\zeta < 0$, the flow near the wall is along the positive axial direction. Although the total flow rate is zero, the local velocities at different normal distances from the channel wall evolve from variations of the electroosmotic velocity along the positive axial direction and the counteracting pressure driven velocity along the negative axial direction. At the microchannel wall, both electroosmotic and pressure driven velocities are zero (no-slip). Within the electric double layer region adjacent to the wall, the electroosmotic velocity increases more rapidly than the pressure driven velocity as one moves away from the wall. Thus, the net velocity in the

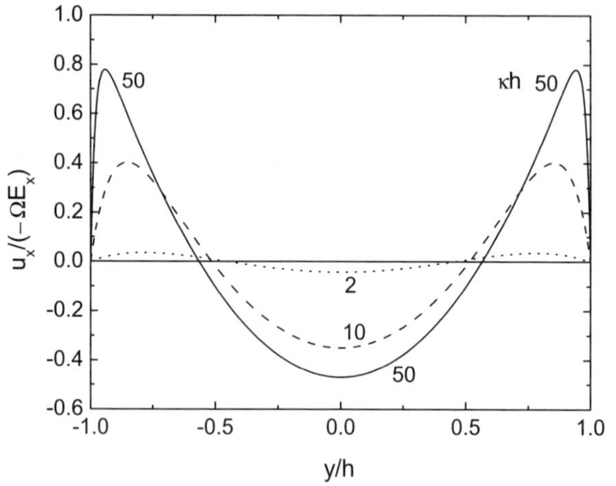

Figure 8.8. Variation of the dimensionless local velocity with transverse position in a slit microchannel during electroosmotic flow for different values of the scaled half-channel height κh.

positive axial direction increases as one moves away from the the wall. Beyond the electrical double layer, the electroosmotic velocity becomes virtually constant, although the pressure driven velocity profile continues to increase in a parabolic manner up to the axis of symmetry of the microchannel. This causes a reversal in the velocity direction, and near the central core of the microchannel, the net velocity is along the negative axial direction. Thus, near the microchannel wall, the electroosmotic flow dominates, whereas near the channel centerline, the pressure driven flow dominates.

The very fact that Q is set to zero, *i.e.*, no net flow, implies that there must exist a flow reversal in the channel gap in order to satisfy the mass conservation and, consequently, a location at which the local velocity is zero. The location of zero local velocity is referred to as the stationary surface and it is on that surface that electrophoretic measurements are made. If one is to place a charged particle at the stationary surface, then the electrophoretic velocity of the particle is not influenced by the electroosmotic velocity of the electrolyte solution. Consequently, electrophoresis devices are designed to observe particle motion along such a surface for the measurement of the electrophoretic particle velocity with the subsequent evaluation of the particle surface charge (Shaw, 1980).

The location of the stationary surface is easily obtained by setting u_x to zero in Eq. (8.40). For the special case of $\kappa h \to \infty$, one obtains

$$\frac{y}{h} = \frac{1}{\sqrt{3}} = 0.577 \tag{8.41}$$

Figure 8.9 shows the location of the stationary plane as a function of κh. It can be noted that for $\kappa h > 100$, the stationary plane is nearly $1/\sqrt{3}$.

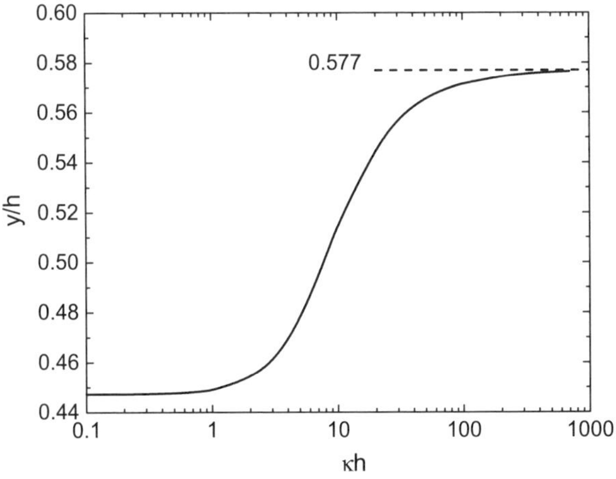

Figure 8.9. Location of the stationary plane in a slit microchannel during electroosmotic flow.

8.4 EFFECTIVENESS OF ELECTROOSMOTIC FLOW

The effectiveness of electroosmotic flow as compared to pressure driven flows can be estimated using the volumetric flow ratio, VFR, defined as

$$\text{VFR} = \frac{\text{volumetric flow rate due to electroosmosis}}{\text{volumetric flow rate due to pressure}} \tag{8.42}$$

Making use of Eq. (8.36) one can write

$$\text{VFR} = -\frac{3\epsilon E_x \zeta}{h^2 p_x}\left[1 - \frac{\tanh(\kappa h)}{\kappa h}\right] \tag{8.43}$$

For given values of ϵ, E_x, ζ, and p_x, we obtain

$$\text{VFR} \propto \frac{1}{h^2}\left[1 - \frac{\tanh(\kappa h)}{\kappa h}\right] \tag{8.44}$$

For the case of $\kappa h \gg 1$, *i.e.*, thin double layers, the term $\tanh(\kappa h)/(\kappa h)$ becomes zero and one obtains

$$\text{VFR} \propto \frac{1}{h^2}$$

The above expression clearly shows that electroosmotic flow is most effective for very narrow channels with $\kappa h \gg 1$. Consequently, in practice for the case of $\kappa h \gg 1$, one uses an applied electric potential to affect flow in a narrow channel rather than using an imposed pressure gradient.

For the case of $\kappa h \ll 1$, where the microchannel is characterized by overlapping double layers, one can expand the term $\tanh(\kappa h)$ in Eq. (8.43) in a Taylor series, yielding

$$\text{VFR} = \frac{3\epsilon E_x \zeta}{h^2 p_x}\left[1 - \frac{(\kappa h) - (\kappa h)^3/3 + \mathcal{O}[(\kappa h)^5]}{\kappa h}\right] \simeq \frac{\epsilon E_x \zeta}{p_x}\kappa^2 \tag{8.45}$$

This indicates that in the limit $\kappa h \ll 1$, the VFR becomes independent of the channel height and assumes a large constant value depending on the double layer parameter κ.

8.5 ELECTRIC CURRENT IN ELECTROOSMOTIC FLOW IN SLIT CHANNELS

In the previous sections, we have developed expressions for the electric potential and the flow velocity distributions for the case of flow in slit microchannels. The major simplification in the analysis was the assumption of low surface potentials where the Debye–Hückel approximation was employed.

During flow in a microchannel, electric current can flow within the channel itself relative to the external circuit that provided the external potential. In this case, both the potential gradient and the electric current are non-zero quantities. Such a situation occurs in purely electroosmotic flow. For purely pressure driven flows, where the two ends of a channel are not connected by an external circuit, there will be no current flow but there will be an induced electric potential gradient. Such a flow situation is referred to as streaming potential flow. In this section, we will derive an expression for the total electric current that will be used to inter-relate the various types of flows with each other.

The current density vector, in A/m^2, is obtained from Eq. (6.69) as

$$\mathbf{i} = e\mathbf{u}\sum_k z_k n_k - e\sum_k D_k z_k \nabla n_k - \frac{e^2 \nabla \phi}{k_B T}\sum_k z_k^2 D_k n_k \tag{8.46}$$

where ϕ represents the total electric potential. The flow in the slit microchannel is unidirectional where the fluid and current flows are in the x-direction only. Consequently, we need only to consider the current in the x-direction. Equation (8.46) in $x-$ direction is given by

$$i_x = eu_x \sum_k z_k n_k - e\sum_k D_k z_k \frac{\partial n_k}{\partial x} - \frac{e^2}{k_B T}\frac{\partial \phi}{\partial x}\sum_k z_k^2 D_k n_k \tag{8.47}$$

Equation (8.47) expresses the current density (A/m^2) at a given location in the channel.

As discussed earlier in Chapter 6, the first term of Eq. (8.47) is the current density due to the bulk motion of the fluid, the second term is due to ionic diffusion, and the third term is due to migration as affected by the potential gradient. In the present flow system, it is assumed that there is no variation in the ionic concentration along the length of the channel. Consequently, $\partial n_k/\partial x = 0$, and the second term of Eq. (8.47) drops out.

The total current flow per unit width of the slit channel (A/m) is given by

$$I_x = I = 2\int_0^h i_x\, dy \tag{8.48}$$

Making use of Eqs. (8.1) and (8.47), the current per unit width of the slit channel becomes

$$I = 2e\int_0^h u_x \sum z_k n_k\, dy + \frac{2e^2 E_x}{k_B T}\int_0^h \sum z_k^2 D_k n_k\, dy \tag{8.49}$$

where the local fluid velocity is given by Eq. (8.30) can be written as

$$u_x(y) = \frac{h^2 p_x}{2\mu}\left[1 - \left(\frac{y}{h}\right)^2\right] - \frac{\epsilon E_x}{\mu}[\zeta - \psi] \tag{8.50}$$

The first term in Eq. (8.49) is the contribution to the current due to convective transport and the presence of the charged channel surface where electroneutrality term $\sum z_k n_k$ is non-zero. The convection current can be thought of as being due to the flow of the excess ions in the *mobile* double layer region close to the surface caused by the pressure driven liquid flow through the channel. For the present, we will refer to it as the convection transport current, I_t. It is also referred to as the streaming current. The second term of Eq. (8.49) is due to the electric conduction within the liquid along the channel. It is referred to in the literature as the conduction current, and we will denote it by I_c. These two contributions, I_t and I_c, to the total current, I, play a major role when dealing with pressure induced flows as is the case for streaming potential flow. In this analysis we assume that there is no current due to the channel wall material itself or due to a current within the Stern layer, *i.e.*, in the layer between the channel wall and the shear plane. Note that the boundary conditions of the electrokinetic problem are applied at the shear plane in the present analysis. In other words, we are completely ignoring any electrical effects arising in the region between the channel walls and the shear plane.

Let us first consider the case of conduction current, I_c, where

$$I_c = \frac{2e^2 E_x}{k_B T}\int_0^h \sum z_k^2 D_k n_k\, dy \tag{8.51}$$

Let us consider the special case of a symmetric $(z : z)$ electrolyte where the ionic species have the same diffusion coefficients, where $D_+ = D_- = D$. For a $(z : z)$ electrolyte, one has $z_+^2 = z_-^2 = z^2$ and Eq. (8.51) becomes

$$I_c = \frac{2e^2 z^2 D E_x}{k_B T}\int_0^h [n_+ + n_-]\, dy \tag{8.52}$$

Making use of the Boltzmann distribution, Eqs. (8.7) and (8.8), Eq. (8.52) becomes

$$I_c = \frac{4e^2z^2Dn_\infty E_x}{k_BT}\int_0^h \cosh\left(\frac{ze\psi}{k_BT}\right)dy \tag{8.53}$$

One can substitute the electric potential distribution, $\psi(y)$, obtained from a solution of the Poisson–Boltzmann equation in Eq. (8.53) to obtain the conduction current. To keep the approach analytical, we will substitute for ψ using Eq. (8.17), which was obtained using the Debye–Hückel approximation, valid for low potentials. In doing so, however, we first note that the integrand on the right hand side of Eq. (8.53), namely $\cosh(ze\psi/k_BT)$, should also be expressed as a Taylor series, retaining only the leading order terms in the resulting expansion. Consequently, Eq. (8.53) simplifies to

$$I_c = \frac{4e^2z^2Dn_\infty E_x}{k_BT}\int_0^h \left[1+\frac{1}{2}\left(\frac{ze\psi}{k_BT}\right)^2 + \mathcal{O}\left\{\left(\frac{ze\psi}{k_BT}\right)^4\right\}\right]dy \tag{8.54}$$

Now, substituting Eq. (8.17) in Eq. (8.54), the expression for the conduction current becomes

$$I_c = \frac{4e^2z^2Dn_\infty E_x}{k_BT}\int_0^h \left[1+\frac{1}{2}\left(\frac{ze\zeta}{k_BT}\right)^2 \frac{\cosh^2(\kappa y)}{\cosh^2(\kappa h)}\right]dy \tag{8.55}$$

Letting $Y = y/h$, we can write for the conduction current

$$I_c = \frac{4e^2z^2Dn_\infty hE_x}{k_BT}\int_0^1 \left[1+\frac{1}{2}\left(\frac{ze\zeta}{k_BT}\right)^2 \frac{\cosh^2(\kappa h\cdot Y)}{\cosh^2(\kappa h)}\right]dY \tag{8.56}$$

We now note that for a $(z:z)$ electroneutral electrolyte solution where $n_i = n_{i\infty}$ and $D_i = D$, the electric conductivity is given by Eq. (6.74) as

$$\sigma^\infty = \frac{2e^2z^2Dn_\infty}{k_BT}$$

with units of S/m or A/Vm. Here we use the symbol σ^∞ instead of σ to emphasize that the conductivity refers to the bulk electroneutral electrolyte. Using the above equation for the electric conductivity, one can write Eq. (8.56) as

$$I_c = 2\sigma^\infty hE_xF_{cs} \tag{8.57}$$

with

$$
\begin{aligned}
F_{cs} &= \int_0^1 \left[1 + \frac{1}{2}\left(\frac{ze\zeta}{k_B T}\right)^2 \frac{\cosh^2(\kappa h \cdot Y)}{\cosh^2(\kappa h)}\right] dY \\
&= 1 + \frac{1}{4}\left(\frac{ze\zeta}{k_B T}\right)^2 \left[\frac{\tanh(\kappa h)}{\kappa h} + \frac{1}{\cosh^2(\kappa h)}\right] \qquad (8.58)
\end{aligned}
$$

where F_{cs} is a factor to account for the non-electroneutrality of the electrolyte solution due to the presence of the charged channel surfaces.

Figure 8.10 shows the values of F_{cs} for different dimensionless surface potentials, $(ze\zeta/k_B T)$ as a function of the scaled channel height, κh, where the scaling is done with respect to the Debye length, κ^{-1}. As the solution for the electric potential was obtained for $ze\zeta/k_B T < 1$ (low surface potentials), by keeping the restriction of $ze\zeta/k_B T < 1$, the value of F_{cs} can be set to unity for $\kappa h > 5$. However, if the analytical solution as obtained for $ze\zeta/k_B T < 1$ is extended for higher values of $ze\zeta/k_B T$, then one cannot set F_{cs} to unity except for large values of κh, say $\kappa h > 50$.

Now we can turn our attention to the first term of Eq. (8.49), and deal with the convection transport current. The convection transport current, I_t, is given by

$$
I_t = 2e \int_0^h u_x \sum z_k n_k \, dy \qquad (8.59)
$$

Recognizing that

$$
e \sum z_k n_k = \rho_f \qquad (8.60)
$$

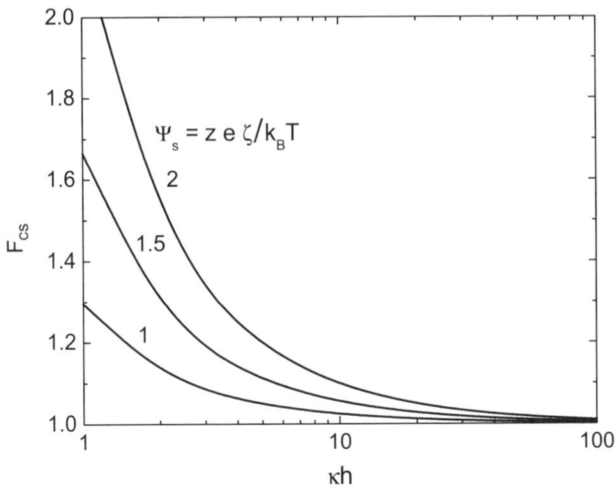

Figure 8.10. Variation of the function F_{cs} given in Eq. (8.58) with κh for different scaled surface potentials.

and using the Poisson equation, Eq. (8.4), the convection transport current becomes

$$I_t = -2\epsilon \int_0^h u_x \frac{d^2\psi}{dy^2}\, dy \tag{8.61}$$

A straightforward approach would be to substitute the expressions for u_x and ψ in the integral of Eq. (8.61) and to perform the integration. A different approach, based on integration by parts, may be followed to simplify the problem considerably (Hunter, 1981; Erickson *et al.*, 2000).

Integration by parts of Eq. (8.61) leads to

$$I_t = -2\epsilon \left[\left\{ u_x \frac{d\psi}{dy} \right\}_{y=0}^{y=h} - \int_0^h \frac{d\psi}{dy}\, du_x \right] \tag{8.62}$$

The first term of Eq. (8.62) is zero since

$$\frac{d\psi}{dy} = 0 \quad \text{at } y = 0 \qquad \text{and} \qquad u_x = 0 \quad \text{at } y = h$$

Equation (8.62) can be written as

$$I_t = 2\epsilon \int_0^h \frac{du_x}{dy}\, d\psi \tag{8.63}$$

Differentiation of the velocity field given by Eq. (8.50) and its substitution in Eq. (8.63) yield

$$I_t = \frac{2\epsilon}{\mu} \int_0^h \left[-p_x y + \epsilon E_x \frac{d\psi}{dy} \right] \cdot \frac{d\psi}{dy}\, dy$$

leading to

$$I_t = \frac{2\epsilon}{\mu} \int_0^h \left[-p_x y \frac{d\psi}{dy} + \epsilon E_x \left(\frac{d\psi}{dy} \right)^2 \right] dy \tag{8.64}$$

The integrals of Eq. (8.64) are given by

$$\int_0^h y \frac{d\psi}{dy} dy = h\zeta \left[1 - \frac{\tanh(\kappa h)}{\kappa h} \right] \tag{8.65}$$

and

$$\int_0^h \left(\frac{d\psi}{dy} \right)^2 dy = \frac{\kappa \zeta^2}{2} \left[\tanh(\kappa h) - \frac{\kappa h}{\cosh^2(\kappa h)} \right] \tag{8.66}$$

The convection transport current is then given by

$$I_t = -\frac{2\epsilon h p_x \zeta}{\mu}\left(1 - \frac{\tanh(\kappa h)}{\kappa h}\right) + \mu\kappa^2 h\Omega^2 E_x\left(\frac{\tanh(\kappa h)}{\kappa h} - \frac{1}{\cosh^2(\kappa h)}\right) \tag{8.67}$$

The total current can now be expressed as

$$I = -\frac{2\epsilon h p_x \zeta}{\mu}\left(1 - \frac{\tanh(\kappa h)}{\kappa h}\right) + \mu\kappa^2 h\Omega^2 E_x\left(\frac{\tanh(\kappa h)}{\kappa h} - \frac{1}{\cosh^2(\kappa h)}\right) + 2\sigma^\infty h E_x F_{cs} \tag{8.68}$$

For convenience, let

$$\alpha_1 = 1 - \frac{\tanh(\kappa h)}{\kappa h} \tag{8.69-a}$$

and

$$\alpha_2 = \frac{\tanh(\kappa h)}{\kappa h} - \frac{1}{\cosh^2(\kappa h)} \tag{8.69-b}$$

Using the above, the total current expression becomes

$$I = -\frac{2\epsilon h p_x \zeta}{\mu}\alpha_1 + \mu\kappa^2 h\Omega^2 E_x\alpha_2 + 2h\sigma^\infty E_x F_{cs} \tag{8.70}$$

In the limit of $\kappa h \gg 1$, F_{cs} and α_1 approach unity and α_2 approaches zero.

Noting that the current is caused by two driving forces, namely, the electric field, E_x, and the pressure gradient, p_x, Eq. (8.70) can be written in the terminology of non-equilibrium thermodynamics as

$$I = L_{11}E_x + L_{12}p_x \tag{8.71}$$

where

$$L_{11} = 2h\sigma^\infty\left[F_{cs} + \frac{\mu\kappa^2\Omega^2\alpha_2}{2\sigma^\infty}\right] \tag{8.72}$$

and

$$L_{12} = -\frac{2\epsilon h\zeta}{\mu}\alpha_1 = -\frac{2\epsilon h\zeta}{\mu}\left[1 - \frac{\tanh(\kappa h)}{\kappa h}\right] \tag{8.73}$$

Let us now recast the expression for Q, Eq. (8.35), as

$$Q = \frac{2h^3 p_x}{3\mu} - \frac{2\epsilon h E_x \zeta}{\mu}\alpha_1 \tag{8.74}$$

Once again, noting that the volumetric flow is caused by the two driving forces E_x and p_x, we can write Eq. (8.74) in the form of a non equilibrium thermodynamic relationship as

$$Q = L_{21}E_x + L_{22}p_x \tag{8.75}$$

where

$$L_{21} = -\frac{2\epsilon h\zeta}{\mu}\alpha_1 = -\frac{2\epsilon h\zeta}{\mu}\left[1 - \frac{\tanh(\kappa h)}{\kappa h}\right] \tag{8.76}$$

and

$$L_{22} = \frac{2h^3}{3\mu} \tag{8.77}$$

Comparing Eqs. (8.73) and (8.76) we immediately note that the coefficients L_{12} and L_{21} are identical. This validates the Onsager reciprocal relationship between the current and the volumetric flow rate. We will revisit these Onsager relationships in the next section.

For the case of a flow purely driven by an electric potential gradient, E_x, in the absence of a pressure gradient, *i.e.*, $p_x = 0$, Eqs. (8.71) and (8.75) provide

$$\left(\frac{Q}{I}\right)_{p_x=0} = \frac{L_{21}}{L_{11}} = \frac{L_{12}}{L_{11}} = -\frac{\epsilon\zeta}{\mu\sigma^\infty}\frac{[1 - \tanh(\kappa h)/(\kappa h)]}{\left[F_{cs} + \mu\kappa^2\Omega^2\alpha_2/(2\sigma^\infty)\right]} \tag{8.78}$$

where the final expression was obtained by using Eqs. (8.73) and (8.72) for L_{12} and L_{11}, respectively. The term $(Q/I)_{p_x=0}$ is often referred to as the electroosmotic coefficient. After substituting the expressions for F_{cs} and α_2 in Eq. (8.78), and expanding the resulting expression as a Taylor series in $ze\zeta/k_BT$, one can obtain a simplified relationship for the electroosmotic coefficient. The resulting expression, neglecting terms of the order of $\mathcal{O}[(ze\zeta/k_BT)^3]$ is

$$\left(\frac{Q}{I}\right)_{p_x=0} \simeq -\frac{\epsilon\zeta}{\mu\sigma^\infty}\left[1 - \frac{\tanh(\kappa h)}{(\kappa h)}\right] + \mathcal{O}\left[\left(\frac{ze\zeta}{k_BT}\right)^3\right] \tag{8.79}$$

Equation (8.79) relates the liquid volumetric flow to the current flow for a purely electroosmotic flow, a situation depicted in Figure 8.11.

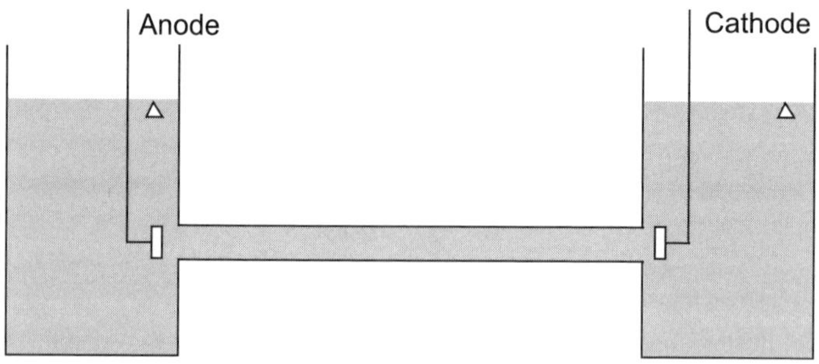

Figure 8.11. Purely electroosmotic flow in a slit microchannel. There is no pressure gradient across the channel. The flow is caused solely by the electric field gradient set up between the electrodes at the channel entrance and exit.

8.6 STREAMING POTENTIAL IN SLIT CHANNELS

When a liquid is forced through a channel, the ionic charges in the mobile part of the electric double layer near the surface of the channel walls are convected in the flow direction toward the channel exit. Such a current constitutes the streaming current. However, as there is no external electric connection between the channel inlet and the exit, the accumulation of charge sets up an electric field such that a current flows in the opposite direction through the bulk of the liquid. This current is called the conduction current. At steady state, the net current is zero whereby the streaming and conduction currents sum up to zero. The electric potential that is induced by the flow is called the streaming potential and the pressure induced flow is referred to as streaming potential flow. To analyze the streaming potential flow, we simply set the current to zero. Setting $I = 0$ in Eq. (8.71) provides,

$$\left(\frac{E_x}{p_x}\right)_{I=0} = -\frac{L_{12}}{L_{11}} \tag{8.80}$$

Comparing Eqs. (8.78) and (8.80) shows that

$$\left(\frac{E_x}{p_x}\right)_{I=0} = -\left(\frac{Q}{I}\right)_{p_x=0} \tag{8.81}$$

This is a fundamental relationship originally shown by Mazur and Overbeek (1951) to be a consequence of the Onsager principle of reciprocity for irreversible phenomena. Substituting the expressions for L_{11} and L_{12} in Eq. (8.81) provides

$$\left(\frac{E_x}{p_x}\right)_{I=0} = -\left(\frac{Q}{I}\right)_{p_x=0} = \frac{\epsilon\zeta}{\mu\sigma^\infty}\frac{[1-\tanh(\kappa h)/(\kappa h)]}{[F_{cs}+\mu\kappa^2\Omega^2\alpha_2/(2\sigma^\infty)]} \tag{8.82}$$

Once again, to the leading order in $ze\zeta/k_BT$, the above expression simplifies to

$$\left(\frac{E_x}{p_x}\right)_{I=0} = -\left(\frac{Q}{I}\right)_{p_x=0} \simeq \frac{\epsilon\zeta}{\mu\sigma^\infty}\left[1-\frac{\tanh(\kappa h)}{(\kappa h)}\right] + \mathcal{O}\left[\left(\frac{ze\zeta}{k_BT}\right)^3\right] \tag{8.83}$$

In the limit of $\kappa h \gg 1$, we can write

$$\left(\frac{E_x}{p_x}\right)_{I=0} = -\left(\frac{Q}{I}\right)_{p_x=0} = \frac{\epsilon\zeta}{\mu\sigma^\infty} \tag{8.84}$$

Due to the nature of the functions F_{cs} and α_2, although Eq. (8.83) is strictly valid for small values of $ze\zeta/k_BT < 1$, Eq. (8.84) is valid for all values of $ze\zeta/k_BT$ as long as $\kappa h \gg 1$.

Note that the streaming potential is manifested as an electric potential difference, $\Delta V = V_{inlet} - V_{outlet}$, (Volts) over a certain length, L, of the microchannel. Using

$$E_x = \frac{\Delta V}{L}$$

one can obtain from Eq. (8.83)

$$\Delta V = \frac{\epsilon \zeta L}{\mu \sigma^{\infty}} \left[1 - \frac{\tanh(\kappa h)}{(\kappa h)} \right] p_x \tag{8.85}$$

Note that the potential difference is measured under conditions that the total current, I, is zero.

8.7 ELECTROVISCOUS FLOW IN SLIT MICROCHANNELS

When a liquid is forced through a microchannel under an applied pressure gradient (in the absence of an externally applied electric field), an induced streaming electric field, E_x, is established and the volumetric flow rate is given by Eq. (8.74). The current flow is zero, as there is no external electric connections between the inlet and outlet of the slit microchannel. The streaming electric field is now given by Eq. (8.82) as E_x is no longer an independent variable imposed by the experimentalist. The field is related to the pressure gradient, p_x, as

$$E_x = \frac{\epsilon \zeta}{\mu \sigma^{\infty}} \frac{[1 - \tanh(\kappa h)/(\kappa h)]}{[F_{cs} + \mu \kappa^2 \Omega^2 \alpha_2/(2\sigma^{\infty})]} p_x \quad \text{for } I = 0 \tag{8.86}$$

Substituting the above expression for E_x in Eq. (8.74), the volumetric flow rate per unit width is given by

$$Q = \frac{2h^3 p_x}{3\mu} \left[1 - \frac{3\epsilon^2 \kappa^2 \zeta^2}{\mu \sigma^{\infty} (\kappa h)^2} \frac{\alpha_1^2}{[F_{cs} + \mu \kappa^2 \Omega^2 \alpha_2/(2\sigma^{\infty})]} \right] \tag{8.87}$$

Once again, to a leading order term in $ze\zeta/k_B T$, one can simplify Eq. (8.87) to obtain

$$Q = \frac{2h^3 p_x}{3\mu} \left[1 - \frac{3\epsilon^2 \zeta^2 \kappa^2}{\mu \sigma^{\infty} (\kappa h)^2} \alpha_1^2 \right] \tag{8.88}$$

The first term of Eq. (8.88), including the multiplier $(2h^3 p_x/3\mu)$ is due to the imposed pressure gradient. It is a positive quantity. The second term on the right hand side of the above expression is due to the induced potential (streaming potential). This term is positive irrespective of the sign of the surface potential ζ. Consequently, the induced potential gives rise to a reduced volumetric flow rate for a given applied pressure, irrespective of its sign. This reduction in volumetric flow rate gives the appearance of an increased liquid viscosity, hence, the term "electroviscous effects".

The reduced rate of flow results in an apparent viscosity μ_a defined by

$$Q = \frac{2h^3 p_x}{3\mu_a} \tag{8.89}$$

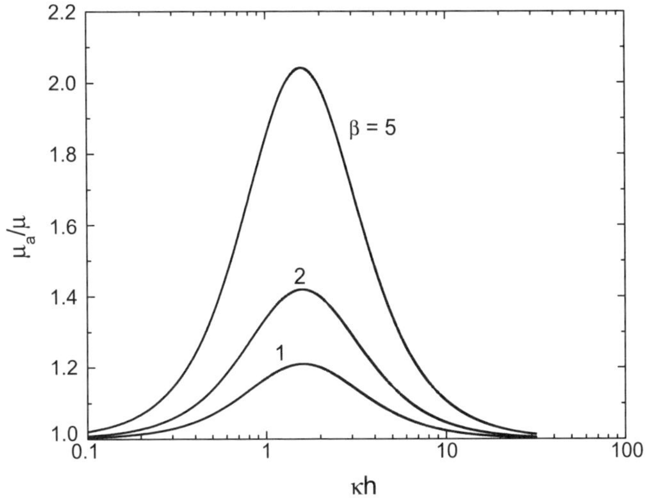

Figure 8.12. Variation of the apparent viscosity, μ_a/μ, with κh for different values of the parameter β. The results are depicted for a scaled surface potential of $\Psi_s = ze\zeta/k_BT = 1.0$ on the walls of the slit microchannel.

Equating for Q in Eqs. (8.88) and (8.89) leads to

$$\frac{\mu_a}{\mu} = \left[1 - \frac{3\epsilon^2\zeta^2\kappa^2}{\mu\sigma^\infty(\kappa h)^2}\alpha_1^2\right]^{-1} \simeq \left[1 + \frac{3\epsilon^2\zeta^2\kappa^2}{\mu\sigma^\infty(\kappa h)^2}\alpha_1^2\right] \tag{8.90}$$

where the final term was obtained using a Taylor expansion.

Figure 8.12 shows the variation of μ_a/μ with κh for different values of the parameter β, where

$$\beta = \frac{\epsilon^2\zeta^2\kappa^2}{\mu\sigma^\infty} = \frac{\Omega^2\mu\kappa^2}{\sigma^\infty} \tag{8.91}$$

The increase in μ_a/μ is prominent at small and intermediate values of κh.

8.8 ELECTROOSMOTIC FLOW IN A CIRCULAR CHARGED CAPILLARY

The analysis for slit microchannels described so far in this chapter delineates the fundamental principles of electroosmotic flow in narrow microchannels. However, for realistic scenarios involving electrokinetic transport in microchannels, a more relevant modelling approach would be to employ circular cylindrical geometries to model capillary flows. For example, an electrolyte solution in a porous medium can be made to move when an electric field is applied. In order to understand electroosmotic flow in such a porous medium or in capillary tubes of varying cross-sections, it is

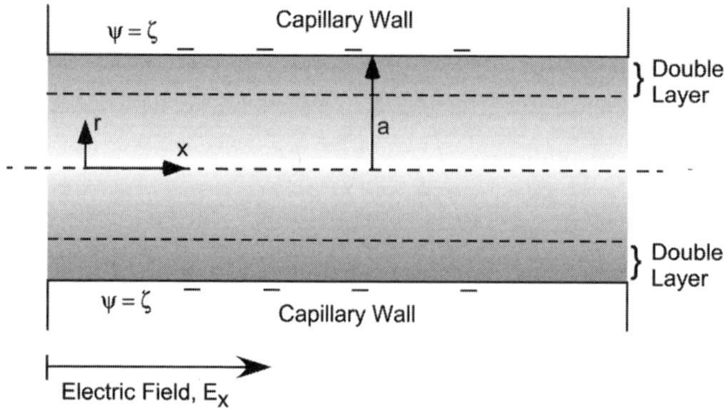

Figure 8.13. Geometry of a circular cylindrical microchannel used for modelling electroosmotic flow.

first necessary to be able to analyze the simple situation of electroosmotic flow in a straight circular capillary tube or a circular cylindrical microchannel.

In this analysis, we shall present electroosmotic flow in a straight tube having a low surface potential. The flow to be considered is fully developed; *i.e.*, no end effects are present. By and large, the analysis is that given by Rice and Whitehead (1965). Improvements to include a high surface potential are given by Levine *et al.* (1975). Furthermore, since the details of the theoretical principles are quite similar to the slit microchannel analysis, except that the present analysis is based on a cylindrical coordinate system, we will present a condensed version of the pertinent derivations.

Consider a circular cylindrical microchannel of radius a with a negatively charged surface bearing a surface potential of ζ,[1] as illustrated in Figure 8.13. The coordinate system used is cylindrical with r representing the radial direction and x representing the axial direction. Consider a symmetric $(z : z)$ electrolyte flowing in the capillary having a surface potential ζ. Let the total potential at a point (r, x) be

$$\phi = \phi(r, x) = \psi(r) + (\phi_o - xE_x) \tag{8.92}$$

where $\psi(r)$ is the potential due to the double layer at the equilibrium state corresponding to no fluid motion and no applied external field, ϕ_o is the value of the imposed potential at $x = 0$, and $\phi_o - xE_x$ is the potential at any axial location in the capillary due to the applied electric field E_x in the absence of the double layer. Equation (8.92) is identical to Eq. (8.2) except that it is written for a cylindrical coordinate system. The Poisson equation defining the potential distribution in cylindrical coordinates is given by

$$\nabla^2\phi = -\frac{\rho_f}{\epsilon}$$

[1]The surface potential ζ represents the potential at the shear plane, as in the case of a slit microchannel.

or

$$\frac{1}{r}\frac{\partial}{\partial r}\left(r\frac{\partial \phi}{\partial r}\right)+\frac{\partial^2 \phi}{\partial x^2}=-\frac{\rho_f}{\epsilon} \tag{8.93}$$

When Eq. (8.92) is substituted, one obtains

$$\frac{1}{r}\frac{d}{dr}\left(r\frac{d\psi}{dr}\right)=-\frac{\rho_f}{\epsilon} \tag{8.94}$$

where the free charge density, ρ_f, is a function of r, and it is given by

$$\rho_f=\sum_k e z_k n_k \tag{8.95}$$

For a symmetric $(z:z)$ electrolyte, invoking the Boltzmann distribution, Eqs. (8.7) and (8.8), for the ion concentrations, one obtains the Poisson–Boltzmann equation, written as

$$\frac{1}{r}\frac{d}{dr}\left(r\frac{d\psi}{dr}\right)=\kappa^2\sinh\left(\frac{ze\psi}{k_BT}\right) \tag{8.96}$$

where κ is the inverse Debye screening thickness given by Eq. (8.14).

If $ze\psi/k_BT$ is small (*i.e.*, $\psi \leq 25$mV), we can write $\sinh(ze\psi/k_BT)\approx ze\psi/k_BT$, whereby the Debye–Hückel approximation is invoked and one obtains the linearized Poisson–Boltzmann equation:

$$\frac{1}{r}\frac{d}{dr}\left(r\frac{d\psi}{dr}\right)=\kappa^2\psi \tag{8.97}$$

The above Poisson–Boltzmann equation can be solved subject to the boundary conditions

$$\text{at } r=0,\ \frac{d\psi}{dr}=0 \quad \text{(axisymmetry)} \tag{8.98}$$

and

$$\text{at } r=a,\ \psi=\zeta \tag{8.99}$$

where a is the radius of the capillary tube.

The solution to Eq. (8.97) satisfying condition (8.98) is

$$\psi=BI_o(\kappa r) \tag{8.100}$$

where I_o is the zeroth-order modified Bessel function of the first kind and B is a constant of integration. The second boundary condition, namely, $\psi=\zeta$ at $r=a$, leads to

$$\psi=\zeta\frac{I_o(\kappa r)}{I_o(\kappa a)} \tag{8.101}$$

With the above solution, one has for the free charge density distribution

$$\rho_f(r) = -\epsilon\kappa^2\psi = -\epsilon\kappa^2\zeta\frac{I_o(\kappa r)}{I_o(\kappa a)} \tag{8.102}$$

Up to this point we have solved for the electric parameters using the Boltzmann distribution for the ionic concentration along the radial direction.

Having obtained the expressions for the electrostatic potential in the radial direction and for the charge distribution, we are prepared to solve the flow equation for the velocity profile. The modified Navier–Stokes equation in the axial flow direction is given by

$$\mu\frac{1}{r}\frac{d}{dr}\left(r\frac{du_x}{dr}\right) = \frac{dp}{dx} + \rho_f\frac{\partial\phi}{\partial x} \tag{8.103}$$

with

$$E_x = -\frac{\partial\phi}{\partial x} \tag{8.104}$$

Using Eq. (8.102), the momentum equation becomes (setting $p_x = -dp/dx$)

$$\mu\frac{1}{r}\frac{d}{dr}\left(r\frac{du_x}{dr}\right) = -p_x + \epsilon\kappa^2\zeta E_x\frac{I_o(\kappa r)}{I_o(\kappa a)} \tag{8.105}$$

with the boundary conditions

$$u_x = 0 \quad \text{at } r = a \tag{8.106}$$

and

$$\frac{du_x}{dr} = 0 \quad \text{at } r = 0 \tag{8.107}$$

The solution of Eq. (8.105), subject to Eqs. (8.106) and (8.107), is given by

$$u_x(r) \equiv u_x = \frac{a^2p_x}{4\mu}\left[1 - \left(\frac{r}{a}\right)^2\right] - \frac{\epsilon\zeta}{\mu}\left[1 - \frac{I_o(\kappa r)}{I_o(\kappa a)}\right]E_x \tag{8.108}$$

The above result shows that the velocity of the electrolyte solution is the sum of the Poiseuille flow term and the electrokinetic term. Equations (8.101) and (8.108) define the solution to the electroosmotic flow in the circular capillary.

We can now make use of the general solution given by Eq. (8.108) and derive interesting limiting cases. Consider the case of $p_x = 0$, *i.e.*, there is no imposed pressure gradient as is the case for a capillary connecting two large reservoirs as shown in Figure 8.14. Equation (8.108) then gives for $p_x = 0$

$$u_{x,el} = -\frac{\epsilon\zeta}{\mu}\left[1 - \frac{I_o(\kappa r)}{I_o(\kappa a)}\right]E_x \tag{8.109}$$

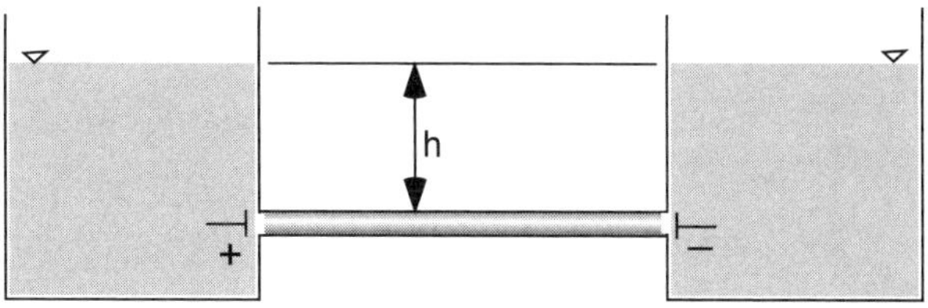

Figure 8.14. A circular cylindrical microchannel connecting two reservoirs having the same hydrostatic head (same liquid depth). In this case, one obtains a purely electroosmotic flow in the capillary microchannel when an electric field is applied across the electrodes.

For given κa and ϵ, $u_{x,el}$ increases linearly with the surface potential ζ and the electric field strength E_x. Also, $u_{x,el}$ is inversely proportional to the fluid viscosity (other parameters being constant).

Let

$$\frac{I_o(\kappa r)}{I_o(\kappa a)} = A \tag{8.110}$$

The term in the square brackets of Eq. (8.109) is always positive. For a negative surface potential the axial velocity $u_{x,el}$ becomes positive.

8.8.1 Thin Double Layers: Helmholtz–Smoluchowski Equation, $\kappa a \gg 1$

In the absence of a pressure gradient, *i.e.*, $p_x = 0$, and for thin double layers, $\kappa a \gg 1$, the term $I_o(\kappa r)/I_o(\kappa a) = 0$, in which case Eq. (8.109) becomes

$$u_{x,el} = -\frac{\epsilon \zeta E_x}{\mu} = \text{constant} \neq f(r) \tag{8.111}$$

a result known as the *Helmholtz–Smoluchowski equation* – a classical equation for the flow of an electrolyte past a charged surface under the influence of an electric field along the surface. Here the fluid moves as a plug, as if the fluid slips at the wall. A plot of Eq. (8.109) is shown in Figure 8.15 for various values of κa. To keep the nomenclature consistent with the slit microchannel analysis presented earlier, we will define

$$\Omega = \frac{\epsilon \zeta}{\mu}$$

8.8.2 Thick Double Layers, $\kappa a \ll 1$

For the special case of $\kappa a \ll 1$, *i.e.*, overlapping electric double layers with $p_x = 0$, one can write a series expansion for I_o in terms of κa and κr to obtain

$$u_{x,el} = -\frac{\Omega}{4} E_x (\kappa a)^2 \left[1 - \left(\frac{r}{a}\right)^2\right] \tag{8.112}$$

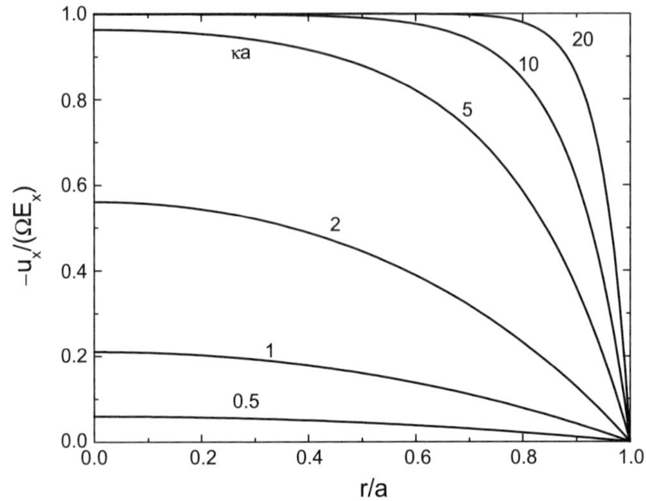

Figure 8.15. Dimensionless electroosmotic velocity profiles for $p_x = 0$ for a circular capillary, where $\Omega = \epsilon\zeta/\mu$.

Note that if $\kappa a \rightarrow 0$ then $u_{x,el} \rightarrow 0$ and we have no electroosmotic flow for substantially overlapping double layers. Equation (8.112) gives a parabolic velocity profile. The term $-\Omega E_x(\kappa a)^2/4$ is equivalent to $(a^2 p_x/4\mu)$, which characterizes the pressure driven Poiseuille flow. The Poiseuille type flow characteristics at low values of κa are due to the large overlap of the double layer with a virtually constant net free charge density across the capillary. This situation gives rise to a liquid body force similar to that of a pressure gradient.

The volumetric flow rate is given by

$$Q = 2\pi \int_0^a r u_x \, dr \tag{8.113}$$

For the special case of $p_x = 0$, we can write

$$Q_{el} = -\Omega A_c E_x \left[1 - \frac{2I_1(\kappa a)}{(\kappa a) I_0(\kappa a)} \right] \tag{8.114}$$

where I_1 is first-order modified Bessel function of the first kind, A_c is the cross-sectional area of the capillary tube equal to πa^2, and Q_{el} is the volumetric flow rate due to the electroosmotic flow.

A dimensionless form of Q_{el} is shown in Figure 8.16. As can be observed from Figure 8.16, the greatest advantage in having electroosmotic flow is for the case of large κa (*i.e.*, a thin double layer).

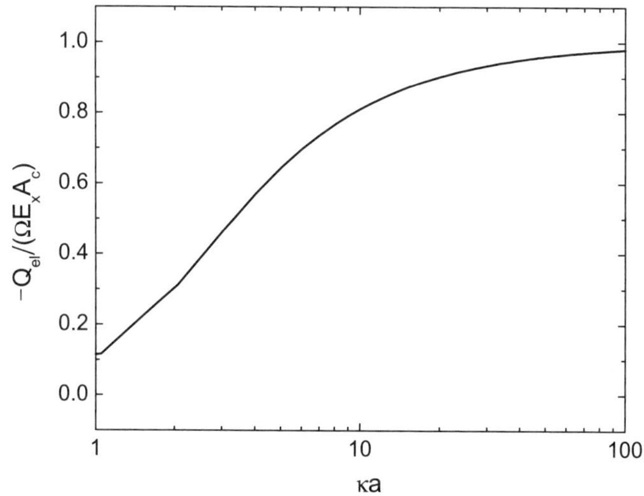

Figure 8.16. Variation of the scaled electroosmotic volumetric flow rate for a circular capillary with κa for $p_x = 0$.

For the general case ($p_x \neq 0$), the total volumetric flow rate Q is given by

$$Q = \frac{p_x a^2}{8\mu} A_c - \Omega A_c E_x \left[1 - \frac{2I_1(\kappa a)}{(\kappa a) I_0(\kappa a)} \right] \tag{8.115}$$

When $Q = 0$, *i.e.*, there is no net flow, the ratio of the developed pressure gradient to the electric field is given by

$$\frac{p_x}{E_x} = \left(\frac{8\epsilon\zeta}{a^2} \right) \left[1 - \frac{2I_1(\kappa a)}{(\kappa a) I_0(\kappa a)} \right] \tag{8.116}$$

The pressure gradient p_x becomes the electroosmotic pressure. For $\kappa a \gg 1$ (*i.e.*, a thin double layer), Eq. (8.116) becomes

$$\frac{p_x}{E_x} = \frac{8\epsilon\zeta}{a^2} \tag{8.117}$$

For the special case of $Q = 0$, one can make use of Eq. (8.116) to eliminate p_x and evaluate the local axial velocity profile as given by Eq. (8.108) to obtain

$$\frac{u_x}{E_x} = \Omega \left\{ 2 \left[1 - \left(\frac{r}{a} \right)^2 \right] \left[1 - \frac{2I_1(\kappa a)}{(\kappa a) I_0(\kappa a)} \right] - \left[1 - \frac{I_0(\kappa r)}{I_0(\kappa a)} \right] \right\} \tag{8.118}$$

For the special case of a thin double layer, *i.e.*, $\kappa a \gg 1$, Eq. (8.118) gives

$$\frac{u_x}{E_x} = \Omega \left[1 - 2 \left(\frac{r}{a} \right)^2 \right] \tag{8.119}$$

Equation (8.119) is valid for $(r/a) < 1$.

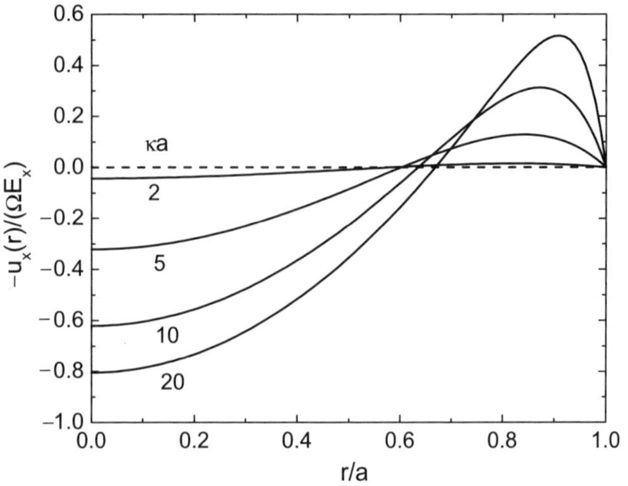

Figure 8.17. Variation of the dimensionless axial velocity with radial position for the special case of zero volumetric flow rate at various values of κa for a circular capillary.

The dimensionless velocity profiles,

$$-\frac{\mu u_x}{\epsilon \zeta E_x} = -\frac{u_x}{\Omega E_x}$$

given by Eq. (8.118) are shown in Figure 8.17. As the volumetric flow rate Q is zero, the axial velocity u_x has to change direction in order to conserve mass balance.

As described in the case of a slit microchannel, the overall velocity profile evolves from the competition between the electroosmotic flow and the counteracting pressure driven flow. The electroosmotic flow has a dominant effect near the microcapillary wall, inside the electric double layer, while the pressure driven flow dominates in the core region.

An important feature of Figure 8.17 is the presence of a cylindrical shell within the capillary tube where the electrolyte solution velocity is zero. The radial position of this shell is usually referred to as the stationary surface, which is similar to the stationary plane obtained for the case of a slit microchannel. If one places a charged particle at the stationary surface, then the electrophoretic velocity of the particle is not influenced by the velocity of the electrolyte solution. Consequently, electrophoresis devices are designed to observe particle movements along such a surface for the measurement of electrophoretic particle velocity and the subsequent evaluation of the particle surface charge (Shaw, 1980).

Table 8.1 gives the values of the stationary radial position, R_{st} (made dimensionless with a). In the limit of $\kappa a \to \infty$, the dimensionless radial position becomes $R_{st} = \sqrt{1/2}$. For a capillary radius of about 1 mm, κa can be considered to be very large for most electrolyte solutions and $R_{st} = 0.7071$.

TABLE 8.1. Dimensionless Radial Position, R_{st}, of the Stationary Surface for a Low Surface Potential Capillary.

κa	R_{st}
1	0.57867
2	0.58241
5	0.60216
10	0.63500
20	0.66851
50	0.69238
100	0.69989
200	0.70354
500	0.70569
∞	0.70711

For the general case of electroosmotic flow, the effectiveness of the electric field can be estimated using the *volumetric flow ratio*, VFR, defined as

$$\text{VFR} = \frac{\text{flow rate due to electroosmosis}}{\text{flow rate due to pressure}} \tag{8.120}$$

From Eq. (8.115) one can write

$$\text{VFR} = -\frac{8\epsilon\zeta E_x}{a^2 p_x}\left[1 - \frac{2I_1(\kappa a)}{\kappa a I_0(\kappa a)}\right] \tag{8.121}$$

For given values of ζ, ϵ, E_x and p_x, Eq. (8.121) shows that

$$\text{VFR} \propto \frac{1}{a^2}\left[1 - \frac{2I_1(\kappa a)}{\kappa a I_0(\kappa a)}\right] \tag{8.122}$$

For the case of $\kappa a \gg 1$, *i.e.*, thin double layers, the numerator on the right side of Eq. (8.122) becomes unity and

$$\text{VFR} \propto \frac{1}{a^2} \tag{8.123}$$

In this case of thin double layers, the electric field becomes very effective in driving the flow.

For the case of $\kappa a \ll 1$ where the capillary tube is characterized by overlapping double layers, the VFR attains a large value independent of the capillary radius. This can be demonstrated by expanding the Bessel functions of Eq. (8.121) in a Taylor series of κa, and simplifying the resulting expression. The procedure yields

$$\text{VFR} \simeq -\frac{\epsilon E_x \zeta \kappa^2}{p_x} \tag{8.124}$$

The same limiting cases were encountered for the slit microchannel.

8.8.3 Current Flow in Electroosmosis

The total current due to the electroosmotic flow is given by

$$I = 2\pi \int_0^a i_x r\, dr \tag{8.125}$$

where i_x is the local current density (A/m^2). Making use of Eq. (6.69) one can write for the total current in x-direction

$$\begin{aligned} I = {} & 2\pi \int_0^a u_x \sum_k e z_k n_k r\, dr - 2\pi e \int_0^a \sum_k \left(D_k z_k \frac{\partial n_k}{\partial x} \right) r\, dr \\ & - \frac{e^2}{k_B T} 2\pi \int_0^a \frac{\partial \phi}{\partial x} \sum_k \left(z_k^2 D_k n_k \right) r\, dr \end{aligned} \tag{8.126}$$

For a fully developed flow, there is no axial variation in the ionic concentration, and hence, the second term of Eq. (8.126) becomes zero. Recognizing that the charge density is given by

$$\rho_f = e \sum_k z_k n_k$$

and setting

$$\frac{\partial \phi}{\partial x} = -E_x$$

Equation (8.126) becomes

$$I = 2\pi \int_0^a u_x \rho_f r\, dr + \frac{2\pi e^2}{k_B T} E_x \int_0^a \left(\sum_k z_k^2 D_k n_k \right) r\, dr \tag{8.127}$$

Let $D_k = D$, *i.e.*, $D_+ = D_- = D$. For a $z : z$ electrolyte, one has $z_+^2 = z_-^2 = z^2$. Using these simplifications in Eq. (8.127) yields

$$I = 2\pi \int_0^a u_x \rho_f r\, dr + \frac{2\pi e^2 z^2 D E_x}{k_B T} \int_0^a \sum_k n_k r\, dr \tag{8.128}$$

Note that

$$\sum_k n_k = 2n_\infty \cosh\left(\frac{ze\psi}{k_B T} \right) \simeq 2n_\infty \left[1 + \frac{1}{2} \left(\frac{ze\psi}{k_B T} \right)^2 \right] \tag{8.129}$$

where the final expression is obtained by expanding the cosh term in a Taylor series for low potentials, $ze\psi/k_BT \ll 1$. Using the above simplification, Eq. (8.128) becomes

$$I = 2\pi \int_0^a u_x \rho_f r\, dr + \frac{4\pi e^2 z^2 D n_\infty E_x}{k_B T} \int_0^a \left[1 + \frac{1}{2}\left(\frac{ze\psi}{k_B T}\right)^2 \right] r\, dr \tag{8.130}$$

The first term on the left hand side of Eq. (8.130) represents the current due to convection and the second term represents the current due to migration. Recognizing that for a $z:z$ electroneutral electrolyte solution, the electric conductivity is given by Eq. (6.74) as

$$\sigma^\infty = \frac{2e^2 z^2 D n_\infty}{k_B T} \tag{8.131}$$

and that the potential, ψ, due to the electric double layer is given by Eq. (8.101), one can rewrite the current, Eq. (8.130) as

$$I = 2\pi \int_0^a u_x \rho_f r\, dr + A_c \sigma^\infty E_x F_{cc} \tag{8.132}$$

The term F_{cc} in Eq. (8.132) is given by

$$F_{cc} = 1 + \left(\frac{ze\zeta}{k_B T}\right)^2 \frac{1}{I_0^2(\kappa a)} \int_0^1 I_0^2(\kappa a \cdot R)\, R\, dR \tag{8.133}$$

where $R = r/a$ and $A_c = \pi a^2$. The term F_{cc} is a factor that accounts for the non-electroneutrality of the electrolyte solution due to the presence of the charged channel surface. It is equivalent to the factor derived for the slit microchannel, F_{cs} [see Eq. (8.58)]. Figure 8.18 depicts the variation of the function F_{cc} with κa for different values of dimensionless capillary surface potentials, $(ze\zeta/k_BT)$. It is observed that F_{cc} deviates considerably from unity at low κa values, particularly for high surface potentials. Many earlier works (such as Rice and Whitehead, 1965) use $F_{cc} = 1$. One would expect that F_{cc} approaches unity only for $\kappa a > 5$.

The contribution to the total current given by the first term on the right side of Eq. (8.132) is due to the liquid bulk flow that is influenced by the pressure gradient. This is the current due to convection. The contribution of the second term on the right side of Eq. (8.132) to the total current is due to the electric conductivity of the electrolyte solution within the capillary tube. This is the current due to migration, and is often referred to as the conduction current.

The total current given by Eq. (8.132) can be evaluated using expressions for $\rho_f(r)$ and $u_x(r)$ as provided by Eqs. (8.102) and (8.108), respectively. Upon integration,

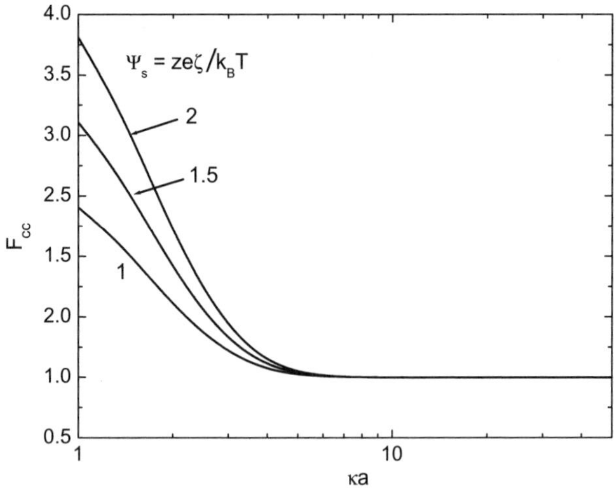

Figure 8.18. Variation of the function F_{cc} for a circular capillary given by Eq. (8.133) with κa for different scaled surface potentials.

Eq. (8.132) provides

$$
\begin{aligned}
I &= -\Omega p_x A_c\left(1-\frac{2A_1}{\kappa a}\right) - E_x A_c \Omega^2 \mu\kappa^2\left(1-\frac{2A_1}{\kappa a}-A_1^2\right) + E_x A_c \sigma^\infty F_{cc} \\
&= E_x A_c \sigma^\infty F_{cc}\left[1-\frac{\Omega^2\mu\kappa^2}{\sigma^\infty F_{cc}}\left(1-\frac{2A_1}{\kappa a}-A_1^2\right)\right] - \Omega p_x A_c\left(1-\frac{2A_1}{\kappa a}\right) \qquad (8.134)
\end{aligned}
$$

where $\Omega = \epsilon\zeta/\mu$ and $A_1 = I_1(\kappa a)/I_o(\kappa a)$. The first term in Eq. (8.134) is due to the applied electric field. The second term is due to the applied pressure. As discussed in Section 8.5, one can express the current given by Eq. (8.134) in terms of a non-equilibrium thermodynamic relationship as

$$I = L_{11}E_x + L_{12}p_x \qquad (8.135)$$

In a similar manner, the volumetric flow rate, Q, given by Eq. (8.115) can be expressed as

$$Q = L_{21}E_x + L_{22}p_x \qquad (8.136)$$

Comparing the coefficients of the above non-equilibrium thermodynamic forms of the current and volumetric flow rate equations, one can satisfy the Onsager reciprocal relationship, $L_{12} = L_{21}$, where

$$L_{12} = L_{21} = -\Omega A_c\left(1-\frac{2A_1}{\kappa a}\right) = -\frac{\epsilon\zeta A_c}{\mu}\left(1-\frac{2A_1}{\kappa a}\right) \qquad (8.137)$$

For the case of $p_x = 0$, *i.e.*, when the current and the volumetric flow rate are due to the applied electric field only, Eqs. (8.114) and (8.134) can be combined to give

$$\left(\frac{Q}{I}\right)_{p_x=0} = \frac{L_{21}}{L_{11}} = \frac{L_{12}}{L_{11}} = -\frac{\Omega}{\sigma^\infty}\left(1 - \frac{2A_1}{\kappa a}\right) f\left(\kappa a, \beta, F_{cc}\right) \tag{8.138}$$

where

$$f(\kappa a, \beta, F_{cc}) = \frac{1}{F_{cc} - \beta\left[1 - 2A_1/(\kappa a) - A_1^2\right]} \tag{8.139}$$

and

$$\beta = \frac{\epsilon^2 \zeta^2 \kappa^2}{\mu \sigma^\infty} = \frac{\Omega^2 \mu \kappa^2}{\sigma^\infty} \tag{8.140}$$

Equation (8.138) represents the ratio of the volumetric flow rate to the applied current at zero pressure gradient. This ratio is directly related to the function $f(\kappa a, \beta, F_{cc})$. It turns out that the function only contributes to the $\mathcal{O}[(ze\zeta/k_B T)^3]$ terms in Eq. (8.138). Consequently, neglecting this term, one obtains to a leading order in $ze\zeta/k_B T$,

$$\left(\frac{Q}{I}\right)_{p_x=0} \simeq -\frac{\Omega}{\sigma^\infty}\left(1 - \frac{2A_1}{\kappa a}\right) = -\frac{\epsilon\zeta}{\mu\sigma^\infty}\left(1 - \frac{2A_1}{\kappa a}\right) \tag{8.141}$$

For thin double layers, $\kappa a \gg 1$, Eq. (8.141) becomes

$$\left(\frac{Q}{I}\right)_{p_x=0} = -\frac{\Omega}{\sigma^\infty} = -\frac{\epsilon\zeta}{\mu\sigma^\infty} \tag{8.142}$$

The ratio of (Q/I) at zero pressure gradient is inversely proportional to the conductivity of the electrolyte solution.

8.8.4 Streaming Potential Analysis

The streaming potential is the steady potential which builds up across a capillary in the presence of an applied pressure gradient and is just sufficient to prevent any net current flow. Here, of course, we do not have an *applied* electric field, but we have an *induced* electric field. The streaming potential occurs when flow takes place in a capillary under an applied pressure gradient (Rice and Whitehead, 1965).

Setting $I = 0$ in Eq. (8.134) and rearranging, one obtains

$$\left(\frac{E_x}{p_x}\right)_{I=0} = \frac{\Omega}{\sigma^\infty}\left(1 - \frac{2A_1}{\kappa a}\right) f(\kappa a, \beta, F_{cc}) \tag{8.143}$$

Equation (8.143) gives the ratio of (E_x/p_x) at zero electric current. Comparison between Eqs. (8.138) and (8.143) leads to

$$-\left(\frac{Q}{I}\right)_{p_x=0} = \left(\frac{E_x}{p_x}\right)_{I=0} = \frac{\Omega}{\sigma^\infty}\left(1 - \frac{2A_1}{\kappa a}\right) f(\kappa a, \beta, F_{cc}) \tag{8.144}$$

which can be further simplified by retaining only the leading order term in $ze\zeta/k_BT$, yielding

$$-\left(\frac{Q}{I}\right)_{p_x=0} = \left(\frac{E_x}{p_x}\right)_{I=0} \simeq \frac{\Omega}{\sigma^\infty}\left(1 - \frac{2A_1}{\kappa a}\right) \tag{8.145}$$

For the case of $\kappa a \gg 1$,

$$-\left(\frac{Q}{I}\right)_{p_x=0} = \left(\frac{E_x}{p_x}\right)_{I=0} = \frac{\Omega}{\sigma^\infty} = \left(\frac{\epsilon\zeta}{\mu\sigma^\infty}\right) \tag{8.146}$$

Equation (8.144) relates electroosmosis to streaming potential flows and is in accordance with the Onsager principle of reciprocity for irreversible phenomena. Equation (8.84) for a slit microchannel is identical to Eq. (8.146) derived for a microcapillary tube.

8.8.5 Electroviscous Effect

When a liquid is forced through a narrow capillary under an applied pressure gradient (in the absence of an externally applied electric field), an induced streaming potential gradient E_x is established and the volumetric flow rate is given by Eq. (8.115). Note that the term E_x is now given by Eq. (8.144) as E_x is no longer an independent variable imposed by the experimentalist.

The volumetric flow rate under a zero current condition is given by combining Eq. (8.115) with Eq. (8.143):

$$\begin{aligned} Q &= \frac{p_x a^2 A_c}{8\mu} - \frac{p_x A_c \Omega^2}{\sigma^\infty}\left(1 - \frac{2A_1}{\kappa a}\right)^2 f(\kappa a, \beta, F_{cc}) \\ &= \frac{p_x a^2 A_c}{8\mu}\left[1 - \frac{8\beta}{(\kappa a)^2}\left(1 - \frac{2A_1}{\kappa a}\right)^2 f(\kappa a, \beta, F_{cc})\right] \end{aligned} \tag{8.147}$$

The first term on the right side of the above expression is due to the imposed pressure gradient. It is positive in value. The second term on the right side of the above expression is due to the induced potential (streaming potential). This term is positive irrespective of the sign of the surface potential ζ. Consequently, the induced potential gives rise to a reduced volumetric flow rate for a given applied pressure. The reduction in the volumetric flow rate gives the appearance of an increased viscosity, hence the term "electroviscous effect".

The reduced rate of flow results in an apparent viscosity μ_a defined by

$$Q = \frac{p_x a^2 A_c}{8\mu_a} \tag{8.148}$$

Equating the volumetric flow rates provided by Eqs. (8.147) and (8.148) gives

$$\frac{\mu_a}{\mu} = \left[1 - \frac{8\beta}{(\kappa a)^2}\left(1 - \frac{2A_1}{\kappa a}\right)^2 f(\kappa a, \beta, F_{cc})\right]^{-1}$$
$$\simeq 1 + \frac{8\beta}{(\kappa a)^2}\left(1 - \frac{2A_1}{\kappa a}\right)^2 f(\kappa a, \beta, F_{cc}) \tag{8.149}$$

Neglecting the higher order contributions due to $f(\kappa a, \beta, F_{cc})$, Eq. (8.149) can be simplified, yielding

$$\frac{\mu_a}{\mu} \simeq 1 + \frac{8\beta}{(\kappa a)^2}\left(1 - \frac{2A_1}{\kappa a}\right)^2 \tag{8.150}$$

or

$$\mu_a = \mu + \frac{8\epsilon^2\zeta^2}{\sigma^\infty a^2}\left(1 - \frac{2A_1}{\kappa a}\right)^2 \tag{8.151}$$

A plot of Eq. (8.150) is given in Figure 8.19. It clearly shows that for large values of β and small values of κa, μ_a/μ can be substantially larger than unity. Figure 8.19 also shows that packed beds having large β with κa close to 2 would not be very permeable to the electrolyte flow.

For the case of $\kappa a \gg 1$, we have

$$1 - \frac{2A_1}{\kappa a} \longrightarrow 1$$

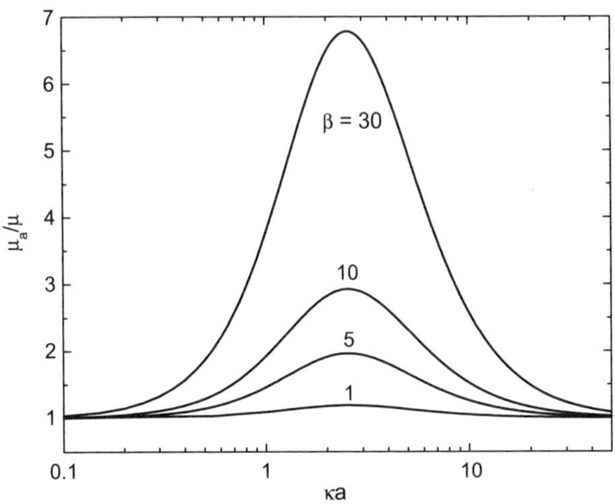

Figure 8.19. Variation of normalized apparent viscosity with κa for different values of the parameter β.

Consequently, Eq. (8.151) provides

$$\mu_a \simeq \mu + \frac{8\epsilon^2\zeta^2}{\sigma^\infty a^2} \tag{8.152}$$

This is the result of Elton (1948). The fluid acts as if it has a higher viscosity because of the additional force opposing the flow.

8.9 HIGH SURFACE POTENTIAL

The study of Levine *et al.* (1975) extended the analysis presented here which was given by Rice and Whitehead (1965) for $ze\psi/k_BT = \Psi \ll 1$. This was possible through the use of the approximation given by Philip and Wooding (1970), where

$$\sinh\Psi \approx \Psi \qquad \text{for } \Psi < 1$$

and

$$\sinh\Psi \approx 1/2\exp(\Psi) \qquad \text{for } \Psi > 1$$

The equivalent equation for the axial velocity as given by Eq. (8.108) becomes

$$u_x(r) = \frac{a^2 p_x}{4\mu}\left[1 - \left(\frac{r}{a}\right)^2\right] - \frac{\epsilon}{\mu}\left[\zeta - \psi(r)\right]E_x \tag{8.153}$$

The total volumetric flow rate as given by Levine *et al.* (1975) is

$$Q = \frac{p_x a^2 A_c}{8\mu} - \Omega A_c(1 - G)E_x \tag{8.154}$$

where

$$G = \frac{2}{a^2\zeta}\int_0^a r\psi(r)\,dr \tag{8.155}$$

The function $(1 - G)$ is shown in Figure 8.20. The curve shown in Figure 8.16 is identical to that for $\Psi_s = ze\zeta/K_BT \to 0$ of Figure 8.20. At $\kappa a = 1$ with $\Psi_s \to 0$, Figure 8.20 indicates that $(1 - G) \cong 0.11$. This is in agreement with Figure 8.16 which shows that the term in the square brackets of Eq. (8.115) is also ~ 0.11. However, for $\Psi_s = 10$, $(1 - G)$ is $= 0.55$. Consequently, a large deviation occurs when a high value of Ψ_s is used in Rice and Whitehead's (1965) analysis which was derived for $\Psi_s < 1$. It should be noted, however, that the effect of high surface potential becomes less important for large κa, say $\kappa a > 100$. Levine *et al.* (1975) also showed that large differences in (μ_a/μ) are involved for $\Psi_s > 1$ when their results were compared with Rice and Whitehead's (1965) study.

The electrokinetic parameters describing electroosmosis and streaming potential for long capillaries having different cross-sectional areas, *e.g.*, circle, ellipse, and

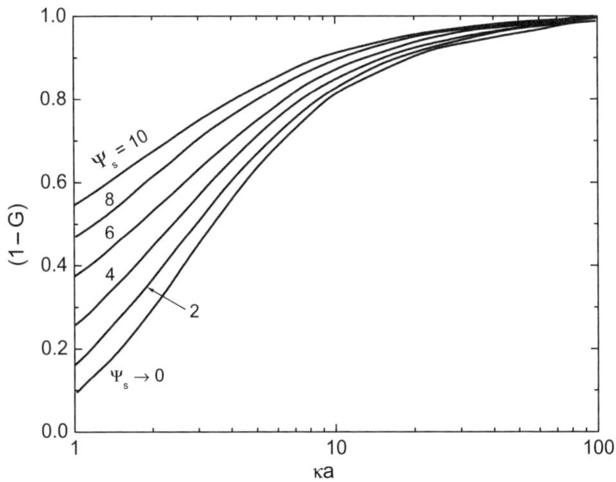

Figure 8.20. Plot of $(1 - G)$ versus κa for various values of the dimensionless surface potential (Levine *et al.*, 1975).

infinite slit were treated by Anderson and Koh (1977). Based on principles similar to those outlined for the electroosmosis phenomenon, Sasidhar and Ruckenstein (1981, 1982) studied electrolyte osmosis through capillaries, and Jacazio *et al.* (1972) analyzed the process of electrokinetic salt rejection in hyperfiltration through porous materials. Using a similar type of analysis, salt rejection in sinusoidal capillary tubes was studied by Masliyah (1994).

By the turn of the century, there has been a large increase in the number of publications dealing with flow in microchannels. This increase was prompted by the fast development of microfluidic micro-electromechanical systems (MEMS) where transport processes on the micro-scale take place in microchannel flows (NSTI, 2003; Erickson and Li, 2004; Yang, 2004; Li, 2004). Lab-on-chip devices that utilize electroosmotic flows have become a promising microfluidic technology where various molecular or colloidal entities can be fractionated, *e.g.*, proteins, bacteria, and polymers. Some of these electroosmotic flow driven microfluidic operations in lab-on-chip devices was pioneered by Harrison and co-workers (Harrison *et al.*, 1993; Hu *et al.*, 1999; Cheng *et al.*, 2001; Tang *et al.*, 2002; Li *et al.*, 2002; Jemere *et al.*, 2002; Ocvirk *et al.*, 2003).

Analysis of electroosmotic flow went well beyond the basic analysis presented here. Studies dealing with electroosmotic entry flow in microchannels were conducted by Yang *et al.* (2001). Transient electroosmotic flows was studied by Keh and Tseng (2001). Oscillatory microchannel flows were analyzed by Yang *et al.* (2003) and Bhattacharyya *et al.* (2003). The effect of slip at the microchannel surface, where the surface is hydrophobic was investigated, among others, by Yang and Kwok (2003), Nagayama and Cheng (2004) and Yang (2004).

As particulate matter move under the electroosmotic flow, they experience dispersion, *i.e.*, spreading due to diffusion and electroosmotic flow. The effect of dispersion plays an important role in particulate matter fractionation. The role of hydrodynamic dispersion on microchannel flow was studied by Martin and Guiochon (1984), Datta (1990), Griffiths and Nilson (1999), Zholkovskij *et al.* (2003), and Zholkovskij and Masliyah (2004). We will discuss hydrodynamic dispersion in microchannel electroosmotic flows later in this chapter.

The studies mentioned above make use of the Poisson–Boltzmann equation in describing the electrokinetic transport phenomena. However, the Poisson–Boltzmann equation assumes ideal solution behavior of both the solute and the solvent. Electrokinetic models have been developed to account for finite ion size, ion hydration effects, and electric field dependent solvent dielectric constant (Gur *et al.*, 1978a,b; Babchin and Masliyah, 1993; Lyklema, 1991; Attard, 2001). The resulting modified Poisson–Boltzmann (MPB) equation was used to show that ions of identical charge but different ionic size can yield different concentrations, dielectric constant profiles, and electric potentials within a capillary tube. The MPB equation was also used by Gur *et al.* (1978a) and Ravina and Gur (1985) to study electroviscous effects during capillary transport in ion-exchange membranes. The difference in the results obtained by the use of the Poisson–Boltzmann equation and its modified form is significant when the surface potential is high and the electrolyte concentration is large (Lyklema, 1991).

8.10 SURFACE CONDUCTANCE

In the analysis conducted for both the slit and the circular microchannels, attention was given to the flow region extending from the shear plane (see Figure 5.28) to the inner regions of the channel. In this context, no fluid and ionic flow were assumed to take place between the channel wall and the shear plane, namely, the Stern layer. The no-slip boundary condition and the electric surface potential in terms of zeta potential, ζ, were imposed at the shear plane. Upon imposition of a potential difference across the length of the channel, an electric current will flow. According to the analysis presented in the previous sections, the electric current is simply due to the convection and conduction in the channel within a region bounded by the shear planes. In the context of the electric double layer model used in this book, the analysis is correct.

In the literature, the concept of "surface conductance" appears quite frequently. We shall attempt to clarify this concept. To begin with, we are not considering the electric conductance of the material that makes up the channel walls. In our discussion, we assume that the channel material is a perfect insulator, and therefore, no current flows within the material that forms the channel walls. Here, the conductance of the channel material is zero. Consequently, "surface conductance" has no relation to the channel material in terms of electrical conductivity. Let us now discuss surface conductance within the framework of the electric double layer.

Historically, the discussion of surface conductance correction arises in context of the Smoluchowski result for the streaming potential

$$\left(\frac{E_x}{p_x}\right)_{I=0} = \frac{\epsilon\zeta}{\mu\sigma^{\infty}} = \frac{\Omega}{\sigma^{\infty}} \tag{8.156}$$

where the effect of the electric double layer is ignored. Here, σ^∞ is the *bulk* conductivity of the electrolyte solution in the absence of a surface bearing a charge. As stated by Lyklema (2003), a striking feature of Eq. (8.156) is that it does not contain the radius of the microchannel. Smoluchowski, among other authors, devoted much attention to it. The fact that Eq. (8.156) does not contain the radius of the microchannel is simply because the equation is approximate, and is valid only in the limit of large $\kappa a \to \infty$. A more general form of the streaming potential, albeit valid only at low zeta potentials, for the case of circular microchannels is (*cf.*, Eq. 8.145)

$$\left(\frac{E_x}{p_x}\right)_{I=0} = \frac{\epsilon\zeta}{\mu\sigma^\infty}\left(1 - \frac{2A_1}{\kappa a}\right) \tag{8.157}$$

The additional term implicitly contains the effect of the electric double layer and hence the microchannel radius. For large κa, $1 - 2A_1/(\kappa a)$ becomes unity, and Eq. (8.157) becomes identical to Eq. (8.156). In terms of modern electrokinetics, there is no issue regarding incorporation of surface conductance correction as long as Eq. (8.157) is used for low zeta potentials where the effect of the electric double layer is accounted for as was done in Section 8.8.3.

Before further discussion, let us consider the physical picture of current transport in the capillary corresponding to large and small values of κa. Figure 8.21 schematically shows the radial distribution of co- and counter-ion in the capillary under these conditions. For large values of κa, the electric double layer is very thin as shown in Figure 8.21(a), and only a small region near the capillary wall has any significant difference in the ion concentrations. The core of the capillary microchannel has essentially an electroneutral ($n_+ = n_-$) electrolyte solution flowing through it. Consequently, the streaming current is zero in the capillary core. The streaming current is perceptible only when the co- and counter-ion concentrations differ, and hence, is measurable only within the electric double layer. Smoluchowski's approximate expression for the

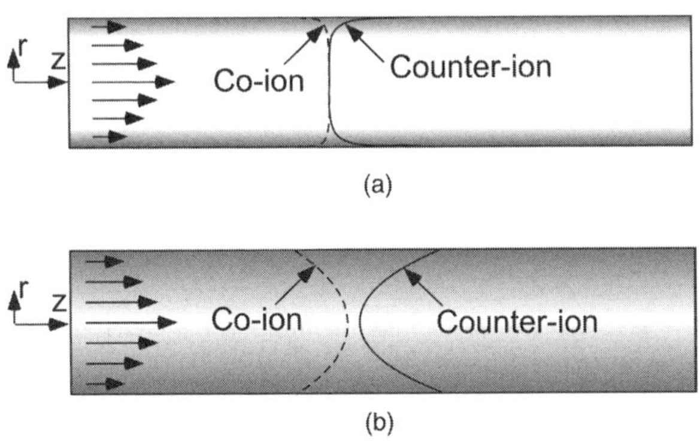

Figure 8.21. Schematic representation of the radial co- and counter-ion concentration distributions in a capillary for (a) large κa and (b) small κa.

streaming current is

$$I_t = -\frac{\epsilon \zeta p_x A_c}{\mu} \tag{8.158}$$

Referring back to Eq. (8.134), one immediately notes that this is an approximation of the streaming current for $\kappa a \to \infty$ and considers only the effect of the pressure gradient on the current. In contrast, the conduction or migration current is dominant in the core region of the capillary, and was approximated by Smoluchowski as

$$I_c = \sigma^{\infty} E_x A_c \tag{8.159}$$

where A_c is the capillary cross-sectional area. Once again, comparing this with Eq. (8.134), one notes that the Smoluchowski expression omits the function F_{cc}, which accounts for the conduction current in the double layer. The rationale for this approximation is that for large κa, the double layer is vanishingly thin, and hence, its contribution to the migration current will be insignificant. Setting $I_t = I_c$, one recovers Eq. (8.156) from the above expressions.

When κa is small, the electrical double layer extends well within the capillary core, and the co-ion and counter-ion concentrations differ at every radial position in the capillary (Figure 8.21b). Such a difference in the ion concentrations affects both the convective (streaming) and the migration (conduction) current through the capillary. This is clearly evident from the expression for the total current given by Eq. (8.134).

The term "surface conductance" was originally introduced to incorporate the effect of a finite conductance that arises from a vanishingly thin electric double layer considered in the Smoluchowski relationship, Eq. (8.156) (Bickerman, 1940). The concept is that the total electric current due to conduction arises from two contributions. The first contribution is due to the bulk electrolyte and the second contribution is due to the surface conductance that accounts for the presence of the electric double layer. As a consequence, $\sigma^{\infty} A_c$ in Eq. (8.159) is replaced by $\sigma^{\infty} A_c + \sigma_{S,dl} S$, yielding

$$I_c = E_x(\sigma^{\infty} A_c + \sigma_{S,dl} S) = \sigma^{\infty} A_c E_x \left(1 + \frac{\sigma_{S,dl} S}{\sigma^{\infty} A_c}\right) \tag{8.160}$$

where S is the microchannel wetted perimeter and $\sigma_{S,dl}$ is the surface conductance due to the presence of the electric double layers. For a circular channel, $A_c = \pi a^2$ and $S = 2\pi a$. Introducing the correction term for $\sigma^{\infty} A_c$, Eq. (8.156) becomes (Hunter, 2001)

$$\frac{E_x}{p_x} = \frac{\epsilon \zeta}{\mu \sigma^{\infty}} \left[1 + \frac{2\sigma_{S,dl}}{a\sigma^{\infty}}\right]^{-1} \tag{8.161}$$

where $\sigma_{S,dl}/a\sigma^{\infty}$ is referred to as the Dukhin number, Du (see Lyklema, 1995, 2003). For $Du \ll 1$, one reverts to Eq. (8.156) where there is no dependence on the channel radius. Several studies attempted to evaluate the $\sigma_{S,dl}$ term from electroosmotic measurements. The incorporation of surface conductance in the manner described above is extremely helpful when addressing electrokinetic flows in the limit of large $\kappa a \to \infty$.

Comparing the conduction current term, $A_c E_x \sigma^\infty F_{cc}$, in Eq. (8.134) with the form given by Eq. (8.160) it is evident that for a circular capillary,

$$F_{cc} = 1 + \frac{2\sigma_{S,dl}}{a\sigma^\infty} \tag{8.162}$$

Noting that $F_{cc} \to 1$ as $\kappa a > 5$ from Figure 8.18, it is clearly evident that the surface conductance correction introduced in Eq. (8.161) is only manifested for small values of κa. This brings us to an important issue regarding surface conductance correction as presented by Eq. (8.161). Comparing Eqs. (8.157) and (8.161), one might erroneously infer that the term $1 - 2A_1/(\kappa a)$ in Eq. (8.157) accounts for the surface conductance correction. However, this term originated from the streaming current in the detailed analysis, *cf.*, Eq. (8.134). Smoluchowski's expression for the streaming current, Eq. (8.158), does not contain this term. An analysis of the model presented in Section 8.8 indicates that for small κa, the dependence of the streaming current on κa is more prominent than the corresponding dependence of the conduction current on κa. The latter contribution actually does not even show up in the leading order solution for the streaming potential using the linearized Poisson–Boltzmann equation! Therefore, attempting to incorporate a surface conductance correction in the general theory of streaming potential at low κa without modifying the streaming current from the Smoluchowski form is conceptually incorrect.

The surface conduction term becomes pertinent in the electrokinetic flow models only when the complete Poisson–Boltzmann equation is used in the development of the theory. Such an attempt was made by Stein *et al.* (2004) for a slit microchannel geometry. They used an exact analytical solution of the Poisson–Boltzmann equation (Ninham and Parsegian, 1971; Behrens and Borkovec, 1999) and developed the subsequent theoretical model for the electrokinetic flow. Their results indicate that for small κh, where h is the channel gap, the overall conductance becomes significantly different from the bulk electrolyte conductance. Furthermore, the conductance becomes independent of the channel depth and electrolyte concentration at very small κh. These observations suggest that the conductance through the electric double layer becomes the controlling parameter in the electrokinetic transport process under such conditions. The striking feature of this study is that the mathematical model developed by Stein *et al.* (2004) was entirely within the context of the electrical double layer theory as presented in this chapter. The only difference between their model and the one delineated in this Chapter is in the use of different forms of the Poisson–Boltzmann equation. In this context, the remarkable agreement of the model developed by Stein *et al.* (2004) with experiments indicates that the general theoretical construct is capable of accounting for all the pertinent physical processes associated with electrokinetic transport in microchannels, without any need to introduce the concept of surface conductance.

The above discussion of the conduction current was made within the framework of the electric double layers. Consequently, "surface conductance" is confined to the ions close to the shear plane. However, there is electric current transport within the Stern layer that is not accommodated for in our analysis using the framework of the electric double layer theory. Electric current transport within the Stern layer is referred to as "anomalous surface conduction", (Hunter, 2001). Dukhin and Derjaguin (1974) introduced the term "anomalous surface conduction" to designate the tangential transfer of charge within the Stern layer. Ion movement within the layer by electromigration and diffusion is governed by the Nernst–Planck equation where the ion mobility may be different from that in the bulk solution, Saville (1998).

Subsequently, Zukoski and Saville (1986) introduced the concept of a dynamic Stern layer (DSL) to model the "anomalous surface transfer", *i.e.*, the tangential electric current transport in the Stern layer. Such an analysis was an attempt to reconcile the differences observed between zeta potentials obtained from mobility and conductivity measurements, Mangelsdorf and White (1990).

The tangential transport of ions in the hydrodynamically immobilized liquid within the Stern layer seems at first sight contradictory. As pointed out by Lyklema and Minor (1998), the situation can be compared to that of gels, *e.g.*, gelatine in water, where a little gelling agent can immobilize water without substantially reducing the electric conductivities and diffusivities. It is quite feasible that ions actually move tangentially within the Stern layer or they simply "hop" to the diffuse layer and then return to the Stern layer, and in so doing, move along the surface of the capillary. Details of the ion transport within the Stern layer still have to be resolved.

If one is to include the Stern layer surface conductance, then Eq. (8.161) takes the form

$$\frac{E_x}{p_x} = \frac{\epsilon\zeta}{\mu\sigma^{\infty}}\left[1 + \frac{2(\sigma_{S,dl} + \sigma_{ss})}{a\sigma^{\infty}}\right]^{-1} \tag{8.163}$$

where σ_{ss} is surface conductance due to the mobile ions in the Stern layer. When anomalous surface conduction occurs, *i.e.*, $\sigma_{ss} \neq 0$, it leads to an under-estimation of the zeta potential, ζ, using electrophoresis measurements, as is evident from Eq. (8.163).

8.11 SOLUTE DISPERSION IN MICROCHANNELS

In our analysis of electroosmotic flow in narrow channels thus far, we have neglected solute dispersion effects. In this section, we will briefly describe the phenomena of hydrodynamic and diffusional dispersion (Taylor, 1953; Aris, 1956), and assess their importance in electroosmotic flow.

When a fluid flows in a narrow channel, for instance, in the case of a one-dimensional parabolic pressure driven flow in a cylindrical capillary, the axial velocity of the fluid varies with radial position. The velocity is zero at the channel wall (owing to the no-slip condition) and attains a maximum value at the center of the channel. If we consider a solute being convected by this fluid, naturally, the solute molecules at different radial positions in the channel will travel with different convective velocities. We also know that molecules undergo random thermal agitation, or diffusion, which leads to a diffusive flux of the solute from high to low concentration regions.

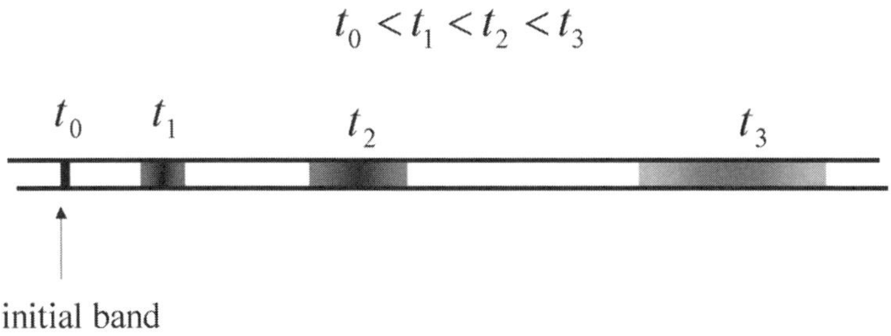

Figure 8.22. The concept of dispersion in a microchannel flow. An initially uniform solute concentration band translates with the flow and broadens due to diffusion and hydrodynamic effects.

Due to this diffusive motion, the solute molecules can diffuse from one flow streamline to another neighboring flow streamline having a different velocity. The net effect of these two phenomena is a broadening of an initially uniform band of solute. This process is generally termed dispersion. In other words, diffusion, coupled with a spatially non-uniform velocity field, tends to enhance the transport of solutes, causing an enhanced mixing of the solutes. Figure 8.22 schematically depicts the process of dispersion.

The dispersion of solutes plays an important role in many technological processes. For example, dispersion is always detrimental in separation and fractionation processes, since it leads to the mixing of those substances that are being separated from each other. Conversely, some chemical technologies require substances to be homogenized within a certain volume and dispersion can provide the homogenization.

In the case of transport through a straight channel, dispersion occurs due to (i) longitudinal diffusion of the solute (diffusional dispersion) and (ii) non-uniformity in the liquid velocity within a channel cross-section (hydrodynamic dispersion). Being a consequence of the thermal motion of molecules, the diffusional dispersion is unavoidable. As for the hydrodynamic dispersion, it can be substantially reduced by employing a hydrodynamic flow with a nearly uniform velocity distribution within the channel cross-section. For example, a substantial reduction of dispersion is observed when, for transportation of a solute, electroosmosis is used instead of pressure driven flow. As it was shown earlier in this chapter, when $\kappa a \gg 1$, the electroosmotic velocity is nearly uniform over the cross-section except at the thin interfacial electric double layer region. In contrast, the parabolic type pressure driven (Poiseuille) flow is strongly non-uniform. Consequently, electroosmosis produces much weaker dispersion than the Poiseuille flow. Due to such a property, electroosmotic flow of a liquid through a capillary is widely employed in microfluidic devices as a means for solute transport.

8.11.1 Diffusional and Hydrodynamic Dispersion

To understand the mechanism of purely diffusional dispersion, we will consider the time evolution of a solute concentration band in a uniform flow as shown in

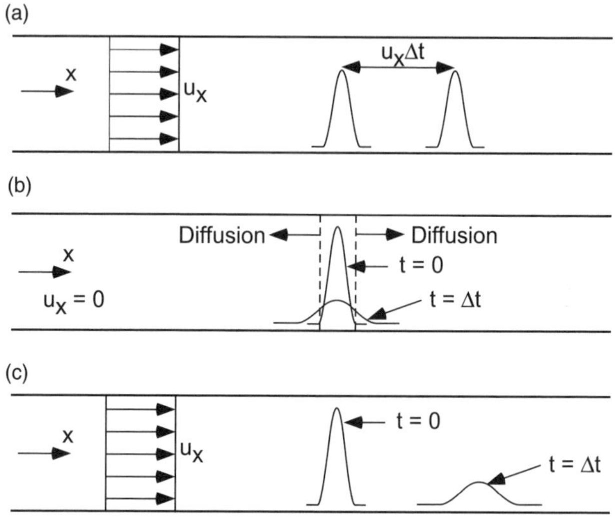

Figure 8.23. Dispersion of a concentration band for (a) a uniform velocity field but zero diffusion, (b) zero velocity but non-zero diffusion, and (c) uniform velocity and non-zero diffusion.

Figure 8.23. Each of the solute molecules participates in a translational motion with the liquid flow and in a random thermal motion that results in diffusion. It is first noted that when the thermal motion is vanishingly weak (negligible diffusion), the solute molecules have a common and time independent velocity in the uniform flow field. Consequently, the solute band is translated as a solid body *i.e.*, without any deformation [Figure 8.23(a)].

To analyze the role of thermal motion, consider a reference system linked to the concentration band in the liquid which moves with velocity u_x (Figure 8.23b). In this reference system, the molecules participate solely in thermal motion, giving rise to the solute diffusion fluxes that are directed toward the lower solute concentrations (outward from the concentration band). Thus, the longitudinal diffusion leads to a permanent transport of the molecules out from the concentration band. Combining the translational (convective) and the diffusive effects, one can observe a superposition of the behavior shown in Figure 8.23(a) and (b): the band is translated with the liquid velocity, u_x, and is broadened [Figure 8.23(c)].

To understand the mechanism of hydrodynamic dispersion, consider a solute band in a flow with a non-uniform velocity distribution over a channel cross-section (Figure 8.24). For the limiting case of vanishingly slow diffusion, the solute molecules participate in a purely convective motion with different velocities. Owing to the difference in velocities, different parts of an initial solute band are transported to different distances. As a result, the band is deformed [Figure 8.24(a)], and its longitudinal dimension, W_h, becomes longer than that of the initial band, W_0. At a given moment, and for a given channel geometry, such a longitudinal hydrodynamic spreading becomes stronger with increasing mean cross-sectional velocity.

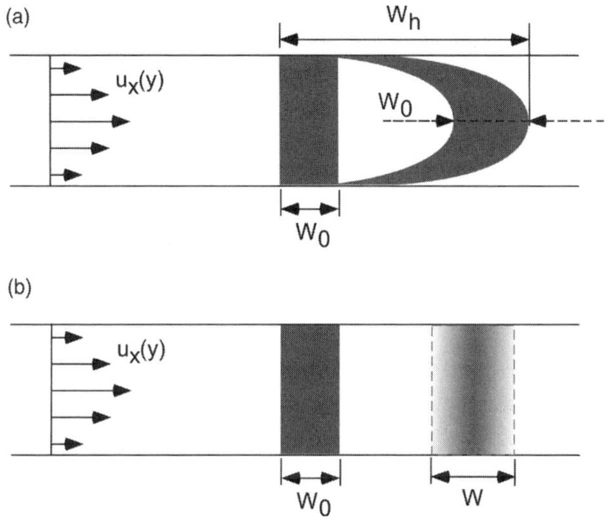

Figure 8.24. Dispersion of concentration band (a) for non-uniform velocity in a channel (purely hydrodynamic dispersion), and (b) for the general case of convection in a non-uniform velocity field and diffusion.

Longitudinal and transverse diffusion, when superimposed on the hydrodynamic dispersion, lead to qualitatively different results. Longitudinal diffusion manifests itself as diffusional dispersion. Basically, its role remains the same as for the case of a uniform flow, *i.e.*, it amounts to additional broadening of the band. As for transverse diffusion, it results in decreasing the hydrodynamic dispersion. The non-uniformity of a velocity distribution leads to the non-uniformity of the concentration distribution within channel cross-sections [Figure 8.24(a)] and, consequently, to transverse diffusion fluxes. Due to the transverse diffusion fluxes, the solute concentration is levelled within any cross-section, and, for sufficiently narrow channels, becomes nearly (but not perfectly) constant. As a result, the solute concentration only varies in the longitudinal direction, thereby forming a longitudinal concentration profile [Figure 8.24(b)]. Such a profile is wider than the initial concentration band ($W > W_0$). However, it is always narrower than the longitudinal distribution which corresponds to the hydrodynamically deformed band shown in Figure 8.24(a) ($W < W_h$).

To explain this fact, we will consider three cross-sections A, B, and C, shown in Figure 8.25. Within cross-section A, which is chosen in the rear part of the deformed band, the solute molecules diffuse from a stagnant zone of the band into the region where the liquid has a high velocity (Figure 8.25). Consequently, these molecules move with the liquid toward the center of the band (cross-section C). Simultaneously, within cross-section B, which is chosen at the leading part of the band, the molecules diffuse from the fast moving part of the band into the slower moving zone of the liquid. These molecules also approach cross-section C. Thus, due to transverse diffusion, the mean cross-sectional concentration decreases at the ends (cross-sections A and B) and increases in the middle of the band (cross-section C). This causes the

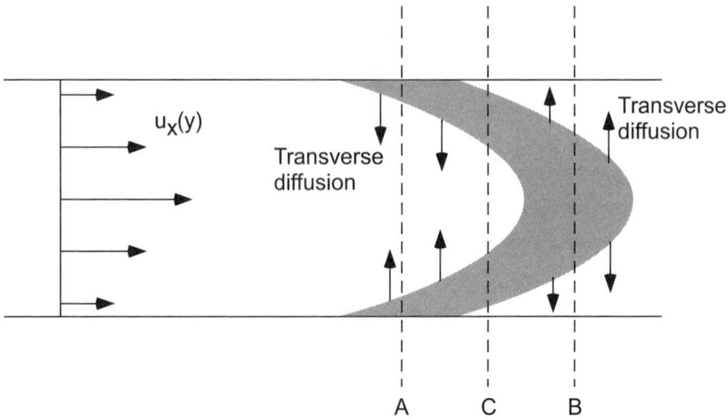

Figure 8.25. Narrowing of longitudinal concentration band due to transverse diffusion.

longitudinal distribution to become narrower (Figure 8.25). Consequently, the band width W satisfies the inequality $W_0 < W < W_h$ (Figure 8.24).

The above discussion demonstrates that transverse diffusion strongly affects hydrodynamic dispersion – a stronger transverse diffusion leading to a weaker hydrodynamic dispersion. It is remarkable that, for the limiting case of infinitely fast diffusion, when the transverse concentration distribution has time to become perfectly uniform within a cross-section, the hydrodynamic dispersion does not manifest itself.

The main properties of diffusional and hydrodynamic dispersion are summarized below:

Diffusional Dispersion

- is a consequence of the thermal motion of molecules;
- exists in the case of a uniform flow;
- becomes stronger for solutes having larger diffusion coefficients.

Hydrodynamic Dispersion

- is a consequence of the non-uniformity in liquid velocity within a cross-section;
- does not exist in the case of a uniform flow;
- becomes stronger with increasing mean cross-sectional velocities;
- becomes weaker for solutes having higher diffusion coefficients.

In the general case, one deals with a combination of both dispersion types.

8.11.2 Convective-Diffusional Transport Through Channels

To analyze the dispersion behavior of solutes, one needs to track the spatio-temporal evolution of the solute concentration band flowing through a channel. The time

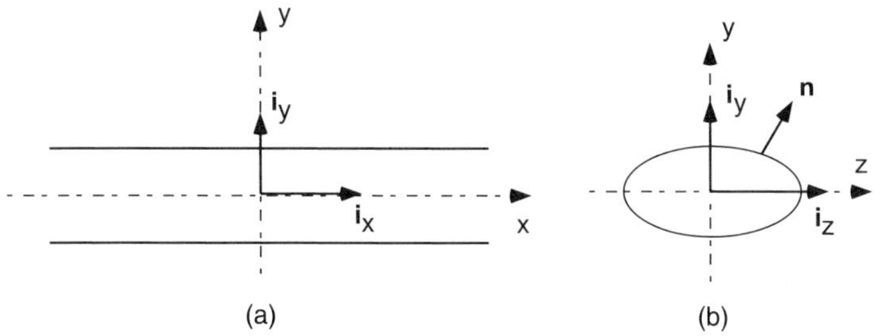

Figure 8.26. Microchannel geometry and coordinate framework employed to analyze hydrodynamic and diffusional dispersion in a one-dimensional flow. (a) Left: side view, (b) right: cross-sectional view.

dependent spatial distribution of the local solute concentration, $c = c(x, y, z, t)$, can be expressed in terms of the convective diffusion equation as

$$\frac{\partial c}{\partial t} = D\nabla^2 c - \mathbf{u} \cdot \nabla c \tag{8.164}$$

where D is the diffusion coefficient of the solute and the velocity field $\mathbf{u}$ is provided from an appropriate solution of the Navier–Stokes and continuity equations.

The general geometry under consideration is depicted in Figure 8.26. Considering a steady flow in a long straight channel and neglecting end effects, we make the following simplifying assumptions regarding the flow structure: (i) the velocity is directed along the x-axis, and (ii) at all cross-sections, the velocity distributions are equal, *i.e.*, the velocity distribution does not depend on the longitudinal (axial) coordinate, x. Consequently, one can represent the velocity as

$$\mathbf{u} = \mathbf{i}_x u_x(y, z) \tag{8.165}$$

where $\mathbf{i}_x$ is the unit vector attributed to the x-axis. Combining Eqs. (8.164) and (8.165) one obtains

$$\frac{\partial c}{\partial t} = D\frac{\partial^2 c}{\partial x^2} + D\nabla_{yz}^2 c - u_x(y, z)\frac{\partial c}{\partial x} \tag{8.166}$$

where the 2D Laplace operator is given by

$$\nabla_{yz}^2 = \frac{\partial^2}{\partial y^2} + \frac{\partial^2}{\partial z^2}$$

One can now define a cross-sectional average concentration as

$$\bar{c}(x, t) = \frac{1}{A}\int_A c(x, y, z, t)\, dA \tag{8.167}$$

where A is the cross-sectional area and $dA = dy\,dz$. In a similar manner, one can also define a cross-sectional average velocity $\bar{u}_x$. Furthermore, the local variations in concentration and velocity from their cross-sectional average values can be represented as

$$c = \bar{c} + \delta c \quad \text{and} \quad u_x = \bar{u}_x + \delta u_x \tag{8.168}$$

Employing Eqs. (8.167) and (8.168), and applying a set of averaging rules, the convective–diffusion equation (8.166) can be written as

$$\frac{\partial \bar{c}}{\partial t} = D\frac{\partial^2 \bar{c}}{\partial x^2} - \bar{u}_x \frac{\partial \bar{c}}{\partial x} - \frac{\partial \overline{\delta u_x \delta c}}{\partial x} \tag{8.169}$$

The last term in Eq. (8.169) becomes zero when the velocity and/or the concentration are uniformly distributed over the cross-section ($\delta u_x = 0$ and/or $\delta c = 0$, respectively). According to the earlier discussion, for these cases, one can expect purely diffusional dispersion of the solute band. Consequently, omitting the final term Eq. (8.169) describes the regime of purely diffusional dispersion.

8.11.2.1 Dispersion in Uniform Flow Assuming that the flow is uniform ($\delta u_x = 0$), Eq. (8.169) simplifies to

$$\frac{\partial \bar{c}}{\partial t} = D\frac{\partial^2 \bar{c}}{\partial x^2} - \bar{u}_x \frac{\partial \bar{c}}{\partial x} \tag{8.170}$$

The cross-sectional mean concentration should satisfy the boundary conditions at infinity given by

$$\bar{c}(x = -\infty) = \bar{c}(x = \infty) = 0 \tag{8.171}$$

and the initial condition (at $t = 0$) given by

$$\bar{c}(x, t = 0) = \bar{c}_0(x) \tag{8.172}$$

On the right hand side of Eq. (8.172), the function $\bar{c}_0(x)$ is the longitudinal distribution of the cross-sectional mean concentration at $t = 0$.

Due to the incompressibility of liquid, for a straight channel, the mean cross-sectional velocity $\bar{u}_x$ does not depend on the longitudinal coordinate, x. Thus, the homogeneous partial differential equation (8.170) depends on constant coefficients D and $\bar{u}_x$. Applying the initial and boundary conditions, Eq. (8.170) can be solved, yielding,

$$\bar{c}(x, t) = \frac{1}{2(\pi D t)^{1/2}} \int_{-\infty}^{\infty} \bar{c}_0(x') \exp\left[-\frac{(x - x' - \bar{u}_x t)^2}{4Dt}\right] dx' \tag{8.173}$$

Equation (8.173) can be used to predict the longitudinal concentration profile at any time, provided the concentration profile $\bar{c}_0(x)$ at $t = 0$ is a known function.

To assess the dispersion behavior in a quantitative manner, several parameters can be defined. The most common geometrical and kinematic parameters quantifying the dispersion behavior are, the mass center coordinate, the variance and band width, the velocity of mass center, and the dispersion coefficient. These terms are defined below:

Mass Center Coordinate: This quantity provides the instantaneous location of the center of mass of the solute band. When the solute molecules have a common molecular weight, the mass center coordinate can be represented as

$$x_c(t) = \frac{\int_{-\infty}^{\infty} x\bar{c}(x,t)dx}{\int_{-\infty}^{\infty} \bar{c}(x,0)dx} \tag{8.174}$$

The above function bears information about the translational movement of the band. Employing the concentration distribution given by Eq. (8.173) in Eq. (8.174), one observes

$$x_c = x_{c0} + \bar{u}_x t \tag{8.175}$$

where x_{c0} denotes the mass center coordinate of the initial solute band at $t = 0$. This implies that the center of mass of the concentration band translates as a linear function of time in the case of purely diffusional dispersion in a uniform velocity field.

Variance and Band Width: Dispersion is often described using the parameter known as the band width, W. It should be noted that, usually, a band does not have sharp external borders. A universal approach to describing the band width of dispersion is to use the variance, σ, which is the root mean square deviation of the molecule coordinate, x, from the mass center coordinate, x_c.

$$\sigma^2(t) = \frac{\int_{-\infty}^{\infty} [x - x_c(t)]^2 \bar{c}(x,t)dx}{\int_{-\infty}^{\infty} \bar{c}(x,0)dx} \tag{8.176}$$

The band width is expressed in terms of the variance as

$$W = \beta\sigma \tag{8.177}$$

where the dimensionless factor β depends on the shape of the concentration profile. For a Gaussian distribution $\beta = 4$, whereas for a rectangular distribution $\beta = 6$. For the present case, using the concentration profile from Eq. (8.173) in Eq. (8.176), we obtain

$$\sigma^2(t) = \sigma_0^2 + 2Dt \tag{8.178}$$

where the variance of the initial concentration distribution is denoted by σ_0. The mass center coordinate, x_c, and the variance, σ (or the band width, W), characterize the geometry of the solute concentration band at a given instant. These geometrical parameters change with time. To address the rate of these changes, a set of relevant kinematical parameters is usually employed.

Velocity of Mass Center: The mass center velocity, u_c, is defined as

$$u_c = \frac{dx_c}{dt} \tag{8.179}$$

For the uniform velocity field, the mass center velocity is simply given by

$$u_c = \bar{u}_x \tag{8.180}$$

Dispersion Coefficient: The dispersion coefficient, K, describes the rate of the change in the squared variance with time. It is defined as

$$K = \frac{1}{2}\frac{d\sigma^2(t)}{dt} \tag{8.181}$$

Employing Eq. (8.178) in Eq. (8.181) yields

$$K = D \tag{8.182}$$

Thus, for purely diffusional dispersion in a uniform velocity field, the dispersion coefficient is equal to the diffusion coefficient of the solute.

8.11.2.2 Dispersion in Non-Uniform Flow: Taylor–Aris Theory

We now consider the transport of the solute concentration band in a flow of the general type. Accordingly, we will deal with a non-uniform velocity distribution, for which, the last term of Eq. (8.169), $\partial\overline{\delta u_x \delta c}/\partial x$, takes a non-zero value (except for the case of a perfectly uniform solute concentration, $\delta c = 0$).

In contrast to the case for uniform velocity field, Eq. (8.170), where the only unknown was the cross-sectional average solute concentration, $\bar{c}(x, t)$, solution of the general transport equation, Eq. (8.169), involves determination of two unknowns, namely, $\bar{c}(x, t)$ and $\delta c(x, y, z, t)$. Therefore, it is necessary to obtain an additional equation, which, together with Eq. (8.169), would form an equation set for obtaining the two unknown functions. This additional equation can be derived by substituting Eq. (8.168) in Eq. (8.166). This second partial differential equation pertains to the concentration fluctuation, $\delta c(x, y, z, t)$. After assigning appropriate boundary conditions for the second equation, one can solve the two equations employing a perturbation approach. For details of the approach, one is referred to the original works of Taylor (Taylor, 1953) and Aris (Aris, 1956; Aris, 1959). Their approach leads to an equation for the spatio-temporal evolution of the cross-sectional average concentration of the form

$$\frac{\partial \bar{c}}{\partial t} = \left(D + \frac{\bar{u}_x^2 H_*^2}{16D}\right)\frac{\partial^2 \bar{c}}{\partial x^2} - \bar{u}_x \frac{\partial \bar{c}}{\partial x} \tag{8.183}$$

where H_* is a length-scale parameter, often referred to as the minimum plate height, that depends on the cross-sectional geometry of the channel. Thus, for an arbitrary distribution of the flow velocity within a channel cross-section, the behavior of the mean cross-sectional concentration $\bar{c}(x, t)$ is described by Eq.(8.183) subject to the boundary conditions, Eq. (8.171), and the initial condition, Eq. (8.172).

Let us now compare the equation describing the behavior of the concentration band derived for a uniform flow, namely Eq. (8.170), with Eq. (8.183), which is derived for an arbitrary velocity distribution. Remarkably, the equations have a common structure and differ from each other solely due to the difference in the coefficients of the second derivative term. In Eq. (8.170) it is the diffusion coefficient D, whereas, in Eq. (8.183), it is a quantity, $(D + \bar{u}_x^2 H_*^2/16D)$, which depends on the mean cross-sectional velocity, $\bar{u}_x$. According to Eq. (8.182), for uniform flow, when purely diffusional dispersion is observed, the dispersion coefficient K is equal to the diffusion coefficient D. In contrast, for a flow of general type, the dispersion coefficient is given by (*cf.*, Eq. 8.183),

$$K = D\left[1 + \left(\frac{\bar{u}_x H_*}{4D}\right)^2\right] \tag{8.184}$$

Equation (8.184) yields the dispersion coefficient for a flow of the general type when both diffusional and hydrodynamic dispersions manifest themselves. Such an expression for the dispersion coefficient, in a slightly different form, was obtained by Taylor (1953) who considered Poiseuille flow through a channel with a circular cross-section. Accordingly, Taylor used $H_* = a/\sqrt{3}$, where a is the radius of the channel in Eq. (8.183). Later, the Taylor theory was generalized by Aris (1956) to address dispersion for an arbitrary channel cross-section. The approach of Aris leads to Eq. (8.184).

8.11.3 Dispersion in a Slit Microchannel

We now assess the dispersion of solutes in a slit microchannel. Two types of flow will be considered, namely pressure driven Poiseuille type flow and electroosmotic flow. It should be noted at the outset, that for the given geometry, the variables (concentration and velocity) will be independent of the z-coordinate.

8.11.3.1 Pressure Driven Flow For the case of purely pressure driven flow, the velocity distribution, $u_x(y)$, is given by the expression

$$u_x(y) = \frac{3}{2}\bar{u}_x\left[1 - \left(\frac{y}{h}\right)^2\right] \tag{8.185}$$

where h is the channel height. It can be shown that the minimum plate height, H_*, is related to the channel height by

$$H_*^2 = \frac{32}{105}h^2 \tag{8.186}$$

Using the above relationship, one can now easily obtain the dispersion coefficient for this type of flow

$$K = D\left[1 + \frac{2}{105}\left(\frac{\bar{u}_x h}{D}\right)^2\right] \tag{8.187}$$

This expression for the dispersion coefficient was obtained by Aris (1959).

8.11.3.2 Electroosmotic Flow For purely electroosmotic flow, which is observed in absence of the longitudinal pressure gradients, the velocity distribution, $u_x(y)$, is given by the second term of Eq. (8.30)

$$u_x(y) = -\frac{\epsilon E_x \zeta}{\mu}\left[1 - \frac{\cosh(\kappa y)}{\cosh(\kappa h)}\right] \tag{8.188}$$

The cross-sectional mean velocity, obtained from Eq. (8.188), is

$$\bar{u}_x = -\frac{\epsilon E_x \zeta}{\mu}\left[1 - \frac{\tanh(\kappa h)}{(\kappa h)}\right] \tag{8.189}$$

Griffiths and Nilson (1999) used these expressions to obtain for the minimum plate height, H_*, the following expression

$$H_*^2 = \frac{4}{3\kappa^2} f(\kappa h) \tag{8.190}$$

where

$$f(\kappa h) = \frac{4[6 + (\kappa h)^2]\sinh^2(\kappa h) - 9(\kappa h)\sinh(2\kappa h) - 6(\kappa h)^2}{[(\kappa h)\cosh(\kappa h) - \sinh(\kappa h)]^2} \tag{8.191}$$

The variation of the normalized minimum plate height, κH_*, with the scaled channel width, κh, is depicted in Figure 8.27. Now, substituting the average velocity and the minimum plate height in Eq. (8.184), the dispersion coefficient for purely electroosmotic flow in a slit microchannel can be determined as

$$K = D\left\{1 + \frac{1}{12}\left(\frac{\epsilon E_x \zeta}{\mu \kappa D}\right)^2 \left[1 - \frac{\tanh(\kappa h)}{\kappa h}\right]^2 f(\kappa h)\right\} \tag{8.192}$$

For $\kappa h \ll 1$, the function $f(\kappa h)$ in Eq. (8.191) can be approximated as

$$f(\kappa h) = \frac{8}{35}(\kappa h)^2 + \mathcal{O}[(\kappa h)^4] \tag{8.193}$$

Substituting, Eq. (8.193) in Eq. (8.190), and retaining the leading term, we arrive at the expression given by Eq. (8.186). Thus, for electroosmotic flow through a channel with small κh, the asymptotic expression for the minimum plate height is the same as that obtained for pressure driven flow. This happens because for $\kappa a \ll 1$, the electroosmotic velocity profile becomes similar to the pressure driven flow velocity profile.

For large values of the parameter $\kappa h \gg 1$, the electroosmotic velocity profile becomes uniform (independent of y), given by the Helmholtz–Smoluchowski expression

$$u_x = -\frac{\epsilon E_x \zeta}{\mu} \tag{8.194}$$

In this case of $\kappa h \gg 1$, the expression for the function $f(\kappa h)$ becomes

$$\lim_{\kappa h \to \infty} f(\kappa h) = 4 \tag{8.195}$$

and the minimum plate height, H_*, is given by

$$\lim_{\kappa h \to \infty} H_*^2 = \frac{16}{3\kappa^2} \tag{8.196}$$

This asymptotic value is depicted in Figure 8.27. Hence, for $\kappa h \to \infty$, the normalized minimum plate height, κH_*, approaches the constant value of $4/\sqrt{3}$. The dispersion coefficient for this type of flow simply becomes

$$K = D\left[1 + \left(\frac{\epsilon E_x \zeta}{\sqrt{3}\mu\kappa D}\right)^2\right] \tag{8.197}$$

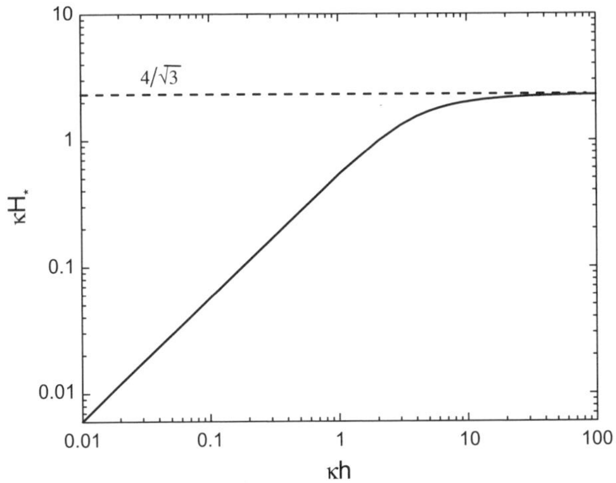

Figure 8.27. Variation of the scaled minimum plate height, κH_*, with the scaled width of the slit microchannel, κh.

In the limit $\kappa \to \infty$, one recovers from Eq. (8.197) the purely diffusional dispersion behavior for the uniform velocity profile

$$K = D \tag{8.198}$$

To summarize, the influence of hydrodynamic and diffusional dispersion is perceptible in electroosmotic flows in channels where $\kappa h \ll 1$. In this case, however, the dispersion behavior is similar to that observed for a purely pressure driven parabolic velocity profile. For large values of κh, *i.e.*, $\kappa h \gg 1$, the hydrodynamic dispersion has no influence on the solute transport, and the dispersion process is solely dictated by diffusion.

8.12 NOMENCLATURE

a	capillary tube radius, m
A_c	cross-sectional area of slit microchannel per unit width, m^2/m
	capillary tube cross-sectional area, m^2
c	solute concentration
$\bar{c}$	cross-sectional average solute concentration
D	solute diffusion coefficient, m^2/s
D_k	diffusivity of kth ionic species, m^2/s
Du	Dukhin number
e	elementary charge, C
E_x	electric field in axial direction, V/m
h	half width of a slit microchannel, m
H_*	minimum plate height, m
$\mathbf{i}$	current density vector, A/m^2
i_x	component of the current density vector along axial flow direction, A/m^2
I_x	total current per unit width for slit microchannel, A/m
	total current for capillary microchannel, A
I_c	conduction current per unit width for slit microchannel, A/m
	conduction current for capillary microchannel, A
I_t	transport or streaming current per unit width for slit microchannel, A/m
	transport current for capillary microchannel, A
I_o	zero-order modified Bessel function
I_1	first-order modified Bessel function
I	total current
k_B	Boltzmann constant, J/K
K	dispersion coefficient, m^2/s
n_+	ionic number concentration of the cations, m^{-3}
n_-	ionic number concentration of the anions, m^{-3}
n_k	ionic number concentration of kth species, m^{-3}
n_∞	ionic number concentration in the bulk solution, m^{-3}
p	fluid pressure, Pa

p_x	negative pressure gradient, $-dp/dx$, Pa/m
Q	volumetric flow rate per unit width in a slit microchannel, m^3/ms
	total volumetric flow rate in capillary microchannel, m^3/s
Q_{el}	volumetric flow rate per unit width due to electrical field in a slit microchannel, m^3/ms
	volumetric flow rate due to electrical field in capillary microchannel, m^3/s
r	radial coordinates, m
T	absolute temperature, K
$\mathbf{u}$	fluid velocity vector, m/s
u_c	velocity of mass center, m/s
u_x	local axial fluid velocity, m/s
$\bar{u}$	cross-sectional average fluid velocity, m/s
VFR	volumetric flow ratio
x	axial coordinate, m
x_c	center of mass coordinate of a solute concentration band
z	absolute value of a $(z:z)$ electrolyte solution valency
z_k	valency of the kth species

Greek Symbols

β	$\Omega^2\mu\kappa^2/\sigma^\infty$
ϵ	dielectric permittivity of solvent, C/Vm
κ	inverse Debye length, m^{-1}
μ	electrolyte solution viscosity, Pa s
μ_a	apparent electrolyte solution viscosity, Pa s
ρ_f	free charge density, C/m^3
σ^∞	electrolyte solution conductivity, S/m or A/Vm
$\sigma_{S,dl}$	surface conductance, S or A/V
σ	variance of the solute concentration band, m^2
ϕ	total potential, V
ϕ_o	total potential at $x=0$, V
ψ	potential due to electric double layer, V
Ψ_s	scaled surface potential (at the shear plane) $ze\zeta/k_BT$
Ψ	dimensionless potential, $ze\psi/k_BT$
ζ	zeta potential of surface, V
Ω	$\epsilon\zeta/\mu$

8.13 PROBLEMS

8.1. For the problem of electroosmotic flow in a slit microchannel discussed in Section 8.2, the boundary condition given at the channel mid-plane, Eq. (8.16-b), is to be replaced by

$$\psi = 0 \quad \text{at } y = 0$$

This boundary condition has been used quite frequently in the literature. Clearly, it is only valid for large κh. Such a boundary condition leads to

$$\psi = \zeta \frac{\sinh(\kappa y)}{\sinh(\kappa h)}$$

Derive an expression for $-u_{x,el}/(\Omega E_x)$ and compare your solution with the corresponding Eq. (8.32) and Figure 8.4.

8.2. In Section 8.2.2, we assumed that for a slit microchannel, the pressure gradient is fairly independent of the transverse y-direction. Let us examine this approximation.

(a) To avoid the assumption of a narrow channel where the hydrostatic head is not important, let us deal with the dynamic pressure, defined as $p + \rho g y$, where g is the acceleration due to gravity. Show that the dynamic pressure gradient for a $(z : z)$ electrolyte using the Debye–Hückel approximation is given by

$$\frac{1}{\epsilon \kappa^2 \zeta^2} \frac{\partial (p + \rho g y)}{\partial y} = \frac{\cosh(\kappa y) \sinh(\kappa y)}{\cosh^2(\kappa h)}$$

(b) Making use of the identities

$$\sinh(2x) = 2 \sinh(x) \cosh(x)$$

and

$$\cosh(2x) = 1 + 2 \sinh^2(x)$$

show that the dynamic pressure is given by

$$p + \rho g y = \left(\frac{\epsilon \kappa^2 \zeta^2}{4} \right) \left[\frac{1 + 2 \sinh^2(\kappa y)}{\cosh^2(\kappa h)} \right] + G(x)$$

Show that for $\kappa h \to 0$, the term in the square brackets become unity and the dynamic pressure is constant across the channel width for a given axial location.

Show that for $\kappa h \to \infty$, the dynamic pressure is given by

$$p + \rho g y = \frac{\epsilon \kappa^2 \zeta^2}{2} \left[\frac{\sinh(\kappa y)}{\cosh(\kappa h)} \right]^2 + G(x)$$

The term in the square brackets in the above equation is zero except very close to the channel wall, indicating the dynamic pressure is fairly constant across the channel width.

8.3. The factor F_{cc} is given as

$$F_{cc} = 2 \int_0^1 \cosh \left[\Psi_s \cdot \frac{I_0(\kappa a R)}{I_0(\kappa a)} \right] R dR$$

where $R = r/a$ and $\Psi_s = ze\zeta/k_B T$.

The modified Bessel function of the first kind is given by an integral expression

$$I_n(x) = \frac{2}{\sqrt{\pi}\Gamma\left(n + \frac{1}{2}\right)} \left(\frac{x}{2}\right)^n \int_0^{\pi/2} \cosh(x \sin\theta)(\cos\theta)^{2n} d\theta$$

where Γ is the Gamma function and n is the order of the modified Bessel function. Evaluate F_{cc} for $\Psi_s = 1$, 1.5, and 2. Compare with the integral F_{cs} for the case of a slit microchannel.

8.4. Solve the electroosmotic flow problem for the slit microchannel where symmetry at $y = 0$ is not assumed. Here the surface potentials are given as

$$\psi = \zeta_1 \quad \text{at } y = h$$

and

$$\psi = \zeta_2 \quad \text{at } y = -h$$

(a) Obtain an expression for $\psi(y)$ that is equivalent to Eq. (8.17).

(b) Obtain an expression for $u_x(y)$ that is equivalent to Eq. (8.30).

(c) Compare the volumetric flow rates, Q_{el}, for the cases (i) $\zeta_1 = \zeta_2 = 1.5$ and (ii) $\zeta_1 = 2$ and $\zeta_2 = 1.0$ for $\kappa h = 1$, 5, and 50.

Assume Debye–Hückel assumption to hold.

8.5. A circular capillary has an electric surface potential of −0.05 V, a diameter of 10 μm and a length of 3 cm. The capillary tube is connected to two reservoirs containing 0.01 M $CaSO_4$. Electrodes are immersed in the two reservoirs and the applied potential difference is 100 V. Assume that the solution viscosity is 10^{-3} Pa.s at 20°C.

(a) What is the volumetric flow rate of the electrolyte solution?

(b) What would be the required pressure drop to affect the volumetric flow rate in part (a) without an electric field?

8.6. Consider water at 20°C containing 0.001 M NaCl flowing in a slit microchannel under the influence of an external electric field of 1000 V/m. The channel wall zeta potential is 50 mV. The material of the channel wall is not conductive.

(a) Plot the volumetric flow rate per unit area (m^3/m^2s) versus channel half height, h (nm).

(b) Evaluate the pressure gradient required to achieve the electroosmotic flow rate for a given channel half height.

(c) Evaluate the current per unit area as a function of channel half height.

Comment on the plots obtained in parts (a) to (c).

8.7. It is of interest to compare dispersion between electroosmotic and pressure driven flows for the same average velocity and dimensionless slit channel half height, κh.

(a) Show that

$$\frac{(K/D-1)_{pressure}}{(K/D-1)_{electroosmotic}} = \frac{105 f(\kappa h)}{24(\kappa h)^2}$$

The function $f(\kappa h)$ is that given by Eq. (8.191).

(b) Plot the ratio $(K/D-1)_{pressure}/(K/D-1)_{electroosmotic}$ against κh.

(c) From the plot under (b) one observes that the dispersion ratio is nearly unity when $\kappa h < 1$. Would there be any advantage in using a channel with $\kappa h < 1$ to minimize hydrodynamic dispersion under electroosmotic flow?

(d) What is the value of the dispersion ratio for $\kappa h = 100$? Can one minimize hydrodynamic dispersion for $\kappa h = 100$ under electroosmotic flow?

(e) Why is the dispersion ratio so different at small and large values of κh?

(f) Plot the variation of the dispersion ratio with h for water flowing in slit channels with different heights at 20°C. The electrolyte is $CaCl_2$ at 0.001 M.

8.14 REFERENCES

Anderson, J. L., and Koh, W. H., Electrokinetic parameters for capillaries of different geometries, *J. Colloid Interface Sci.*, **59**, 149–158, (1977).

Aris, G., On the dispersion of a solute in fluid flowing through a tube, *Proc. Roy. Soc. Lond.*, **235A**, 67–77, (1956).

Aris, G., On the dispersion of a solute by diffusion, convection and exchange between phases, *Proc. Roy. Soc. Lond.*, **252A**, 538–550, (1959).

Attard, P., Recent advances in the electric double layer in colloid science, *Curr. Opin. Colloid Interface Sci.*, **6**, 366–371, (2001).

Babchin, A. J., and Masliyah, J.H., Modified Nernst–Planck equation for hydration effects, *J. Colloid Interface Sci.*, **160**, 258–259, (1993).

Behrens, S. H., and Borkovec, M., Exact Poisson–Boltzmann solution for the interaction of dissimilar charge-regulating surfaces, *Phys. Rev. E.*, **60**, 7040–7048, (1999).

Bhattacharyya, A., Masliyah, J. H., and Yang, J., Oscillating laminar electrokinetic flow in infinitely extended circular microchannels, *J. Colloid Interface Sci.*, **261**, 12–20, (2003).

Bickerman, J. J., Electrokinetic equations and surface conductance. A survey of the diffuse double layer theory of colloidal solutions, *Trans. Faraday Soc.*, **36**, 154–160, (1940).

Burgreen, D., and Nakache, F. R., Electrokinetic flow in ultrafine capillary slits, *J. Phys. Chem.*, **68**, 1084, (1964).

Cheng, S. B., Skinner, C., Taylor, J., Attiya, S., Lee, W. E., Picelli, G., and Harrison, D. J., Development of a multi-channel microfluidic analysis system employing affinity capillary electrophoresis for immunoassay, *Anal. Chem.*, **73**, 1472–1479, (2001).

Datta, R., and Kotamarthi, V. R., Electrokinetic dispersion in capillary electrophoresis, *AIChE J.*, **36**, 916–926, (1990).

Dukhin, S. S., and Derjaguin, B. V., Electrokinetic phenomena, in *Surface and Colloid Science*, vol. 7, E. Matijevic (Ed.), Wiley, (1974).

Elton, G. A. H., Electroviscosity I. The flow of liquids between surfaces in close proximity, *Proc. Roy. Soc. Lond.*, **194A**, 259–274, (1948).

Erickson, D., Li, D. Q., and Werner, C., An improved method of determining the zeta-potential and surface conductance, *J. Colloid Interface Sci.*, **232**, 186-197, (2000).

Erickson, D., and Li, D. Q., Integrated microfluidic devices, *Analytica Chim. Acta*, **507**, 11–26, (2004).

Griffiths, S. K., and Nilson, R. H., Hydrodynamic dispersion of a neutral non-reacting solute in electroosmotic flow, *Anal. Chem.*, **71**, 5522–5529, (1999).

Gur, Y., Ravina, I., and Babchin, A. J., On the electrical double layer theory. I. Numerical method for solving a generalized Poisson–Boltzmann equation, *J. Colloid Interface Sci.*, **64**, 326–332, (1978a).

Gur, Y., Ravina, I., and Babchin, A. J., On the electrical double layer theory. II. The Poisson–Boltzmann equation including hydration forces, *J. Colloid Interface Sci.*, **64**, 333–341, (1978b).

Guzman-Garcia, A. G., Pintauro, P. N., Verbrugge, M. W., and Hill, R. F., Development of a space-charge transport model for ion-exchange membranes, *AIChE J.*, **36**, 1061–1074, (1990).

Harrison, D. J., Fluri, K., Seiler, K., Fan, Z. H., Effenhauser, C. S., and Manz, A., Micromachining a miniaturized capillary electrophoresis-based chemical-analysis system on a chip, *Science*, **261**, 895–897, (1993).

Hiemenz, P. C., and Rajagopalan, R., *Principles of Colloid and Surface Chemistry*, 3rd ed., Marcel Dekker, New York, (1997).

Hu, L. G., Harrison, D. J., and Masliyah, J. H., Numerical model of electrokinetic flow for capillary electrophoresis, *J. Colloid Interface Sci.*, **215**, 300–312, (1999).

Hunter, R. J., *Zeta Potential in Colloid Science*, Academic Press, London, (1981).

Hunter, R. J., *Foundations of Colloid Science*, 2nd ed., Oxford University Press, Oxford, (2001).

Jacazio, A., Probstein, R. F., Sonin, A. A., and Ying, E., Electrokinectic salt rejection in hyperfiltraion through porous material. Theory and experiment, *J. Phys. Chem.*, **76**, 4015–4023, (1972).

Jemere, A. B., Oleschuk, R., Ouchen, F., Fajuyigbe, F., and Harrison, D. J., An integrated solid phase extraction system for sub-pico molar detection, *Electrophoresis*, **23**, 3537–3544, (2002).

Keh, H. J., and Tseng, H. C., Transient electrokinetic flow in fine capillaries, *J. Colloid Interface Sci.*, **242**, 450–459, (2001).

Levine, S., Marriott, J. R., Neale, G., and Epstein, N., Theory of electrokinetic flow in fine cylindrical capillaries at high zeta-potentials, *J. Colloid Interface Sci.*, **52**, 136–149, (1975).

Li, D. Q., *Electrokinetics in Microfluidics*, Elsevier, Amsterdam, (2004).

Li, J., LeRiche, T., Tremblay, T. L., Wang, C., Bonneil, E., Harrison, D. J., and Thibault, P., Application of microfluidic devices to proteomics research, *Mol. and Cell. Proteomics*, **1.2**, 157–168, (2002).

Lyklema, J., *Fundamentals of Interface and Colloid Science*, vol. 1, Academic Press, London, (1991).

Lyklema, J., *Fundamentals of Colloid and Interface Science*, vol. II, Academic Press, London, (1995).

Lyklema, J., and Minor, M., On surface conduction and its role in electrokinetics, *Colloids Surf. A*, **140**, 33–41, (1998).

Lyklema, J., Electrokinetics after Smoluchowski, *Colloids Surf. A*, **222**, 5–14, (2003).

Mangelsdorf, C. S., and White, L. R., Effects of Stern-layer conductance on electrokinetic transport properties of colloidal particles, *J. Chem. Soc. Faraday Trans.*, **86**, 2859–2870, (1990).

Martin, M., and Guiochon, G., The axial-dispersion in open-tubular capillary liquid-chromatography with electroosmotic flow, *Anal. Chem.*, **56**, 614–620, (1984).

Masliyah, J. H., Salt rejection in a sinusoidal capillary tube, *J. Colloid Interface Sci.*, **166**, 383–393, (1994).

Mazur, P., and Overbeek, J. Th. G., On electro-osmosis and streaming-potentials in diaphragms .2. general quantitative relationship between electro-kinetic effects, *J. Roy. Netherlands Chem. Soc.*, **70**, 83–91, (1951).

Nagayama, G., and Cheng, P., Effects of interface wettability on microscale flow by molecular dynamics simulation, *Int. J. Heat Mass Transfer*, **47**, 501–513, (2004).

Nano Science and Technology Institute (NSTI), Technical Proceedings of the 2003 Nanotechnology Conference and Trade Show, NSTI, Cambridge, MA, (2003).

Ninham, B. W., and Parsegian, V. A., Electrostatic potential between surfaces bearing ionizable surface groups in ionic equilibrium with physiologic saline solution, *J. Theor. Biol.*, **31**, 405–428, (1971).

Ocvirk, G., Salimi-Moosavi, H., Szarka, R. J., Arriaga, E. A., Andersson, P. E., Smith, R., Dovichi, N. J., and Harrison, D. J., β-galactosidase assays of single cell lysates on a microchip: A complementary method for enzymatic analysis of single cells, *IEEE Spl. Iss. on Biomedical Applications for MEMS and Microfluidics*, (2003).

Philip, J. R., and Wooding, R. A., Solution of the Poisson–Boltzmann equation about a cylindrical particle, *J. Chem. Phys.*, **52**, 953–959, (1970).

Pintauro, P. N., and Verbrugge, M. W., The electric-potential profile in ion-exchange membranes, *J. Membrane Sci.*, **44**, 197–212, (1989).

Ravina, I., and Gur, Y., Rheological model of pore water, *J. Rheology*, **29**, 131–145, (1985).

Rice, C. L., and Whitehead, R., Electrokinetic flow in a narrow cylindrical capillary, *J. Phys. Chem.*, **69**, 4017–4024, (1965).

Sasidhar, V., and Ruckenstein, E., Electrolyte osmosis through capillaries, *J. Colloid Interface Sci.*, **82**, 439–457, (1981).

Sasidhar, V., and Ruckenstein, E., Anomalous effects during electrolyte osmosis across charged porous membranes, *J. Colloid Interface Sci.*, **85**, 332–362, (1982).

Saville, D. A., Electrokinetic phenomena and anomalous conduction, *Croat. Chem. Acta*, **74**, 1039–1047, (1998).

Shaw, D. J., *Introduction to Colloid and Surface Chemistry*, 3rd ed., Butterworths. London, (1980).

Stein, D., Kruithof, M., and Dekker, C., Surface-charge governed ion transport in nanofluidic channels, *Phys. Rev. Lett.*, **93**, 035901, (2004).

Tang, T., Badal, M. Y., Ocvirk, G., Lee, W. E., Bader, D. E., Bekkaoui, F., and Harrison, D. J., Integrated microfluidic electrophoresis system for analysis of genetic materials using signal amplification methods, *Anal. Chem.*, **74**, 725–733, (2002).

Taylor, G., Dispersion of soluble matter in solvent flowing slowly through a tube, *Proc. Roy. Soc. Lond.*, **219A**, 186–203, (1953).

Yang, J., Bhattacharyya, A., Masliyah, J. H., and Kwok, D. Y., Oscillating laminar electrokinetic flow in infinitely extended rectangular microchannels, *J. Colloid Interface Sci.*, **261**, 21–31, (2003).

Yang, J., and Kwok, D. Y., Effect of liquid slip in electrokinetic parallel plate microchannel flow, *J. Colloid Interface Sci.*, **261**, 225–233, (2003).

Yang, J., *Surface effects of microchannel wall on microfluidics*, Ph.D. Thesis, Department of Mechanical Engineering, University of Alberta, (2004).

Yang, R. J., Fu, L. M., and Hwang, C. C., Electroosmotic entry flow in a microchannel, *J. Colloid Interface Sci.*, **244**, 173–179, (2001).

Zholkovskij, E. K., Masliyah, J. H., and Czarnecki, J., Electroosmotic dispersion in microchannels with a thin double layer, *Anal. Chem.*, **75**, 901–909, (2003).

Zholkovskij, E. K., and Masliyah, J. H., Hydrodynamic dispersion due to combined pressure-driven and electroosmotic flow through microchannels with a thin double layer, *Anal. Chem.*, **76**, 2708–2718, (2004).

Zukoski, C. F., and Saville, D. A., The interpretation of electrokinetic measurements using a dynamic-model of the stern layer. I. the dynamic-model, *J. Colloid Interface Sci.*, **114**, 32–44, (1986).

CHAPTER 9

ELECTROPHORESIS

9.1 INTRODUCTION

In the previous chapter, we discussed the flow behavior of an electrolyte past a stationary charged surface in presence of an external electric field. A similar, albeit, somewhat more complicated scenario arises when a colloidal particle bearing a surface charge immersed in an electrolyte solution is subjected to an external electric field. In this case, the particle is also susceptible to move in response to the imposed electric field, and a relative motion between the particle and the electrolyte solution is developed. Broadly, when an electric field is applied to a suspension of charged particles (or a single particle), the particles migrate along the field owing to the presence of surface charge on the particles. This phenomenon is referred to as electrophoresis. Electrophoresis is one of the most avidly studied electrokinetic transport phenomena, and provides a highly practical basis for separation of a mixture of charged colloidal particles and macromolecules, such as proteins, employing an electrical field.

The literature on electrophoresis is rich in seminal contributions from numerous researchers dealing with the field of electrokinetic phenomena. These studies range from the earliest works of Smoluchowski (1918) and Henry (1931), where the attention was devoted to the motion of a single spherical colloidal particle under certain limiting conditions, to modern theories detailing the motion of a swarm of particles in a concentrated dispersion employing cell models (Levine and Neale, 1974a,b; Saville, 1977; Shilov *et al.*, 1981; Kozak and Davis, 1986; Kozak and Davis, 1989a,b; Ohshima, 1997a; Dukhin *et al.*, 1999; Carrique *et al.*, 2001). In this chapter, we will

Electrokinetic and Colloid Transport Phenomena, by Jacob H. Masliyah and Subir Bhattacharjee
Copyright © 2006 John Wiley & Sons, Inc.

briefly track the development of the theory of electrophoresis over the past century, primarily emphasizing the approaches that yield tractable analytical results. A brief overview of the numerical procedures will be provided in Chapter 14.

The analysis to be presented is aimed at evaluation the electrophoretic velocity of a charged particle when it is placed in an electric field. First, we will consider the motion of a single charged spherical particle under the influence of an external electric field. After presenting the general mathematical construct describing such a situation, two limiting cases of the problem will be presented, namely, when the particle is very small or in an almost perfect dielectric with few mobile charges, *i.e.*, in the limit $\kappa a \ll 1$, and when the particle is large or in a highly concentrated electrolyte, *i.e.*, $\kappa a \gg 1$. We will then bridge these two limiting cases with a detailed analysis for intermediate values of κa (Henry, 1931). All the above mentioned analyses will be confined to particles having low surface potentials. A brief overview of perturbation approaches employed to predict the electrophoresis of highly charged particles will be presented. Following the analysis for single particles, we will present methodologies for addressing higher particle concentrations, representing dilute or highly concentrated colloidal dispersions. In the entire analysis, the particle will be assumed to be rigid and electrically non-conducting.[1]

9.2 ELECTROPHORESIS OF A SINGLE CHARGED SPHERE

In this section, we will first provide a general description of the different types of transport phenomena encountered during electrophoresis. Following this, the electrophoretic mobility of a charged particle will be derived for two limiting cases. First, we will consider the electrophoresis of a single charged particle in the limit of $\kappa a \ll 1$, when the particle is small or when the electrolyte solution is infinitely dilute, containing vanishingly few mobile ions (free charges). Following this, we will consider the limit of $\kappa a \gg 1$, when the particle is large or the electrolyte solution is highly concentrated.

9.2.1 Transport Mechanisms in Electrophoresis

A qualitative understanding of the general transport mechanisms encountered during electrophoresis is an important prerequisite for developing a mathematical framework to correctly predict the electrophoretic mobility. To develop this understanding, let us first consider a spherical particle of radius a bearing a charge Q_s suspended in a pure dielectric fluid (containing no free charge or ions). When subjected to a uniform external electric field, E_∞, the particle will translate under the influence of the electric force acting on it. Since there is no free charge in the dielectric fluid, there will be no flow of this fluid under the influence of the uniform external electric field as long

[1]Note that even a conducting particle will be rapidly polarized in the external electric field and will have no inner field. Consequently, such particles will also prevent passage of current through them, and appear as non-conducting (Probstein, 2003).

as there is no pressure gradient. The net electrical force on the charged particle will simply be

$$F_E = Q_s E_\infty \tag{9.1}$$

As soon as the particle starts to move under the influence of this electrical force, it encounters an oppositely directed fluid drag force given by

$$F_H = 6\pi \mu a U \tag{9.2}$$

where μ is the fluid viscosity, and U is the particle velocity. Equating the electrical and drag forces for steady-state translation, one obtains

$$U = \frac{Q_s E_\infty}{6\pi \mu a} \tag{9.3}$$

The direction of the particle velocity will be governed by the sign of the particle charge. The electrophoretic mobility of the particle can be defined as velocity per unit applied electric field, and is given by

$$\eta = \frac{U}{E_\infty} = \frac{Q_s}{6\pi \mu a} \tag{9.4}$$

Consider now an electroneutral electrolyte solution instead of a pure dielectric fluid subjected to a uniform external electric field, E_∞. First, consider the case when the charged particle is not immersed in this fluid. In this case, just as noted for the pure dielectric fluid, since the volumetric charge density, ρ_f, is zero everywhere in the electrolyte solution, there will be no electrical body force acting on the fluid. Consequently, there will be no flow of the electrolyte solution. However, there will be a migration or conduction current through the fluid that will obey Ohm's law for ionic conductors.

Let us next immerse the spherical particle of radius a carrying a total charge Q_s in this electrolyte solution. The charged particle will polarize the electrolyte solution surrounding it, resulting in the formation of an electric double layer. The electric double layer will have a spatially varying volumetric charge density. As soon as a charge density is developed in the electrolyte solution, it will experience an electric body force under the influence of the overall electric field near the particle. The overall electric field is a superimposition of two fields: (i) the (generally unidirectional) external field and (ii) a spherically symmetric field in the double layer due to the charged particle in absence of the external field. The electric body force will cause a motion in the fluid immediately surrounding the particle. Noting that the net accumulation of the charge in the electric double layer will be opposite in sign to that of the particle charge, it is evident that the fluid flow will be opposite to the direction of the particle movement. In addition to the above mode of fluid flow, the concentration gradients of the ions, as well as the electric potential gradients will give rise to ionic fluxes, which are usually described in terms of the Nernst–Planck equations.

It is therefore evident that a complete analysis of electrophoresis of a charged particle in an electrolyte solution involves consideration of three coupled physical processes. These are:

(i) Interaction of the charged particle with the external electric field giving rise to an electrical force on the particle.
(ii) Formation of the electric double layer surrounding the particle giving rise to a spatially inhomogeneous charge distribution, which results in an electrical body force driven fluid flow, and,
(iii) Transport of ions relative to the charged particle under the combined influence of convection, diffusion, and migration.

9.2.2 General Governing Equations

The governing equations describing the three coupled processes mentioned above are presented here. The steady state equations presented here apply to the steady state velocity of a single charged particle under an imposed electric field, $\mathbf{E}_\infty$. The geometry is depicted in Figure 9.1. Note that the particle velocity, $\mathbf{U}$, and the electric field are represented as vector quantities. To facilitate the problem solution, we translate the problem to a particle fixed reference frame, and consider the flow of an

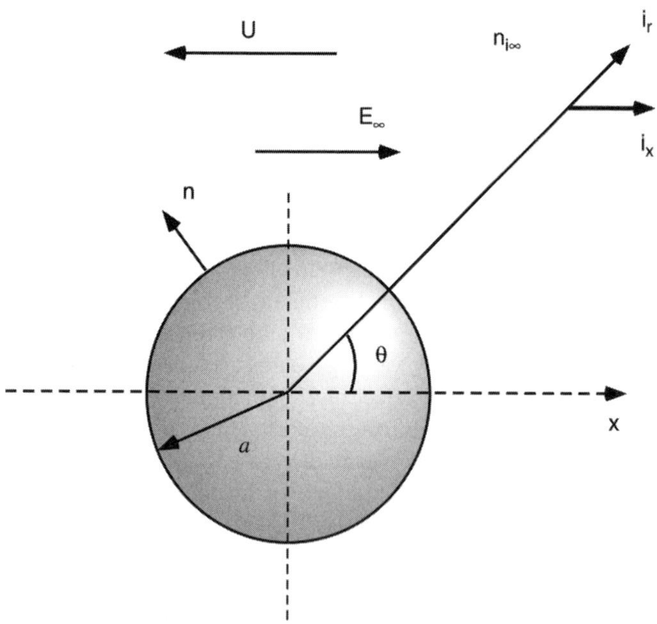

Figure 9.1. Geometrical details required for writing the generalized governing equations for the electrophoretic mobility of a spherical particle of radius a held stationary in a uniform electric field $\mathbf{E}_\infty$.

electrolyte solution around a stationary charged particle of radius a. In this case, far from the particle, the velocity of the electrolyte solution is given by $\mathbf{u}_\infty = -U\,\mathbf{i}_x$, where U is the particle velocity along the x direction and $\mathbf{i}_x$ is a unit vector along the x coordinate. The velocity U represents the electrophoretic particle velocity when the particle is allowed to move in a quiescent fluid. Consequently, $u_\infty = -U$ represents the fluid velocity necessary to keep the particle stationary. The electrolyte solution is a Newtonian fluid having a viscosity μ. The electrolyte solution contains N ionic species of valency z_i with a bulk concentration $n_{i\infty}$ and diffusion coefficient D_i.

The electric potential, ψ, in the electrolyte solution surrounding the charged particle and the space ionic charge density, ρ_f, in the electrolyte solution are related by Poisson's equation, (5.7), as

$$\epsilon \nabla^2 \psi = -\rho_f \tag{9.5}$$

where the space charge density [Eq. (5.9)] is given by

$$\rho_f = \sum_{i=1}^{N} z_i e n_i \tag{9.6}$$

For small Reynolds number and neglecting the effect of gravity, the electrolyte solution velocity field, $\mathbf{u}$, is given by the modified momentum equation, as was presented by Eq. (6.112)

$$\mu \nabla^2 \mathbf{u} - \nabla p - \rho_f \nabla \psi = 0 \tag{9.7}$$

where ∇p is the pressure gradient and $\rho_f \nabla \psi$ is the electrical body force on the electrolyte solution caused by the overall potential gradient, $\nabla \psi$. The steady-state continuity equation for the electrolyte solution under dilute conditions, *i.e.*, constant density, Eq. (6.113), provides

$$\nabla \cdot \mathbf{u} = 0 \tag{9.8}$$

The ionic fluxes are given by the Nernst–Planck equations, which can be written in terms of the number concentration, n_i, of the ith ionic species as [*cf.*, Eq. (6.45)]

$$\mathbf{j}_i^{**} = n_i \mathbf{v}_i = n_i \mathbf{u} - D_i \nabla n_i - \frac{z_i e n_i D_i}{k_B T} \nabla \psi \tag{9.9}$$

where $\mathbf{v}_i$ and D_i are the velocity and diffusivity of the ith ionic species, respectively. For a steady state transport process, the ion conservation equation then yields

$$\nabla \cdot \mathbf{j}_i^{**} = \nabla \cdot \left[n_i \mathbf{u} - D_i \nabla n_i - \frac{z_i e n_i D_i}{k_B T} \nabla \psi \right] = 0 \tag{9.10}$$

9.2.3 Boundary Conditions

The shear plane of the electric double layer is assumed to coincide with the surface of the sphere. At the sphere surface, the electrical boundary condition for the Poisson equation can be defined as

$$-\epsilon\, \mathbf{n} \cdot \nabla\psi = q_s \tag{9.11}$$

where q_s is the surface charge density of the particle. Typically, for an isolated spherical particle, one can relate the surface charge density to the surface potential using independent expressions such as Eq. (5.83). Assuming the spherical particle to be stationary, the no slip condition for the Navier–Stokes equations becomes

$$\mathbf{u} = 0 \tag{9.12}$$

Finally, for the Nernst–Planck equations, we impose the condition that no electrolyte ions penetrate the sphere surface

$$\mathbf{v}_i \cdot \mathbf{n} = \left[D_i \nabla n_i + \frac{z_i e n_i D_i}{k_B T} \nabla\psi \right] \cdot \mathbf{n} = 0 \quad \text{and} \quad \mathbf{u} \cdot \mathbf{n} = 0 \tag{9.13}$$

In the above equations, $\mathbf{n}$ is a unit outward normal vector on the sphere's surface.

Far away from the sphere, the following boundary conditions hold. The electrical boundary condition is

$$\nabla\psi = -\mathbf{E}_\infty \tag{9.14}$$

For the Navier–Stokes equations, the appropriate far field boundary condition is that the overall fluid stress (both pressure and viscous stresses) vanishes. Accordingly, the total fluid stress at every point of the far field boundary is expressed using

$$\bar{\bar{\sigma}} \cdot \mathbf{n} = \left\{ -p\bar{\bar{I}} + \mu \left[\nabla\mathbf{u} + (\nabla\mathbf{u})^T \right] \right\} \cdot \mathbf{n} = 0 \tag{9.15}$$

where p is the hydrostatic pressure. Finally, for the Nernst–Planck equations, the ion concentrations in the far field are given by their bulk values

$$n_i = n_{i\infty} \tag{9.16}$$

which should satisfy the bulk electroneutrality condition, $\sum z_i e n_{i\infty} = 0$.

Note that the formulation of the problem does not specify the far field electrolyte solution velocity. The fluid velocity far away from the particle, $\mathbf{u}_\infty$, is determined from the solution of the governing equations in the particle fixed coordinate system. From this far field fluid velocity, the particle velocity, U, can be deduced noting that

$$\mathbf{u}_\infty = -U\, \mathbf{i}_x \tag{9.17}$$

For an N component electrolyte system, a self-contained solution of the general governing equations will consist of the fluid velocity vector $\mathbf{u}$, the pressure, the electric potential ψ, and the ionic concentration, n_i, for all the N species. There are thus $N + 3$ unknowns, assuming the velocity vector to be a single quantity, and that the surface charge density of the particle, q_s, is an independently known constant. Equations (9.7) and (9.8) provide the equations corresponding to the velocity vector and pressure, Eq. (9.10) provides an equation for each of the ion concentrations (there will be N such equations) while Eqs. (9.5) and (9.6) provide the potential distribution, ψ. The overall solution of these equations (for steady-state flow) will self-consistently satisfy the force balance on the particle, which involves equalizing the electrical (Maxwell) and hydrodynamic forces acting on the particle. Unfortunately, a complete solution of the above general equations is only possible through numerical techniques. We will discuss such a numerical solution methodology in Chapter 14. Approximate analytical solutions of the above transport model are generally based on perturbation approaches and involve several approximations.

When considering the general problem formulation described above, it should be noted that the formulation is only based on the "outer problem" pertaining to the electrolyte solution surrounding the particle. This forced us to assume the surface charge density (or the surface potential) to remain unaltered in the presence of the external electric field. Such a simplification does not lead to difficulties as long as the externally imposed electric field is small compared to the electric field surrounding the particle due to the electric double layer. However, the formulation as stated above becomes inadequate when considering large external fields. For a rigorous modelling of the problem, one additionally needs to obtain the electrical potential distribution and fields inside the particle. This is usually derived from the Laplace equation written within the particle when there is no free charge inside the particle. The electric boundary conditions at the particle-fluid interface then evolve from the conditions of continuity of the potential and the discontinuity of electric displacement as discussed in Chapters 3 and 4.

The simplification of the governing equations to render them amenable to analytical solution techniques requires a qualitative understanding of some key features of the electrophoretic transport. The simplifications of the model equations should account for these transport mechanisms, otherwise the electrophoretic mobility predictions will be of limited accuracy. For a finite-thickness electric double layer around a charged particle, three effects need to be considered (Shaw, 1980). These are:

Electrophoretic Retardation: A charged particle will *polarize* the electrolyte solution in its immediate vicinity giving rise to a countercharged ion cloud. When this charge cloud with an effective volumetric charge density, ρ_f, is subjected to an externally imposed electric field, E_∞, it will create a body force in the fluid of magnitude $\rho_f E_\infty$. The fluid surrounding the particle will move under the influence of this body force. Since the ion cloud will be oppositely charged compared to the particle, the overall fluid motion will be opposite to the direction of the particle motion. The oppositely directed motion of the fluid tends to retard the movement of the particle, hence the term electrophoretic retardation. Solving the Navier–Stokes equation together with the electric force term would account for this effect.

Relaxation: The charge cloud surrounding the particle in presence of the external electric field is not symmetric as in the case of an equilibrium electric double layer around a stationary particle. Since, in general, the particle has a different dielectric permittivity and conductivity than the surrounding fluid, the external electric field will give rise to a polarization charge (induced charge) and an associated field at the particle surface. Such an induced charge relaxes through diffusion and migration mechanisms. Furthermore, due to the movement of the particle under consideration relative to the mobile ions in the double layer, the distribution of the ions around the particle is not symmetric. A finite time (relaxation time) is required for the original symmetry to be restored by convection, diffusion, and migration of the ions. This asymmetry gives rise to a retardation force called the relaxation effect. If one takes into account the Nernst–Planck equation in the analysis, *i.e.*, non-zero Peclet numbers, the relaxation effects would be included. When $\kappa a < 0.1$, the relaxation effects are not important. For large $ze\zeta / k_B T$, the relaxation effect is important. The shape of the electric double layer with the relaxation effect is schematically shown in Figure 9.2.

Surface Conductance: Due to the presence of a charged surface, the distribution of the ions in the mobile region of the electric double layer gives rise to a higher conductivity close to the surface than that in the bulk electrolyte. When ζ and κa are small, the electric conductivity in the double layer is close to that of the bulk electrolyte. When κa or ζ is not small, the electrophoretic mobility of a particle is affected by the different electric conductivity in the double layer. This effect is called the surface conductance. If the surface conductance is important (large κa and/or ζ), the calculated zeta potentials may be quite low (Shaw, 1980).

The general governing equations account for all the transport effects mentioned above. However, approximate solutions often neglect one or more of the above mentioned transport mechanisms. Typically, all approximate solutions start from the assumption that the external field is much smaller in magnitude compared to the field produced by the charged particle. Thus, all the approximate analytical solutions

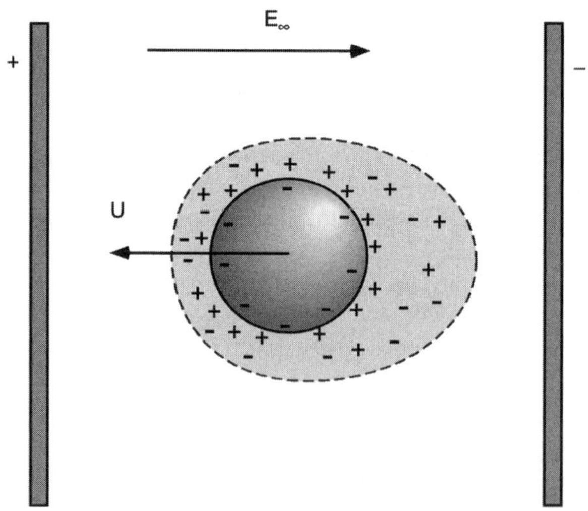

Figure 9.2. Double layer distortion due to relaxation effect. Translation of a negatively charged particle.

consider small perturbations in the equilibrium double layer surrounding the charged particle due to the external field. The manner in which the external field is considered to perturb the equilibrium electric double layer results in different analytic results of different range of applicability. In the following, we will first consider two of the simplest approaches that provide some limiting perspective of the electrophoretic particle mobility for the limiting cases of small and large κa.

9.2.4 Electrophoresis in the Limit $\kappa a \ll 1$

Consider a spherical non-conducting colloidal sphere of radius a placed in an electrolyte solution subject to an external electrical field E_∞ as shown in Figure 9.3. Although we consider presence of free charge in the form of mobile ions, the concentration of these ions are sufficiently low to make the Debye screening length κ^{-1} very large compared to the particle radius. In this limit, $\kappa a \ll 1$, the expression for the electrophoretic mobility of the particle will be derived.

Figure 9.3 shows the geometry under consideration. Using a spherical coordinate system, and utilizing symmetry in the angular directions, the potential distribution around the sphere can be represented by the Poisson–Boltzmann equation. Assuming a symmetric ($z : z$) electrolyte, the governing Poisson–Boltzmann equation, (5.68), in the Debye–Hückel limit becomes

$$\frac{1}{r^2}\frac{d}{dr}\left(r^2\frac{d\psi}{dr}\right) = \kappa^2\psi \tag{9.18}$$

where κ is the inverse Debye length, given by

$$\kappa = \left(\frac{2e^2z^2n_\infty}{\epsilon k_B T}\right)^{1/2}$$

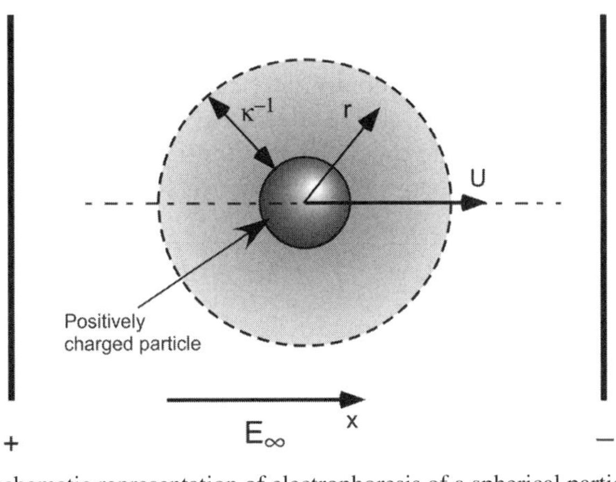

Figure 9.3. A schematic representation of electrophoresis of a spherical particle for $\kappa a \ll 1$.

Let

$$\xi = r\psi \tag{9.19}$$

which renders Eq. (9.18) as

$$\frac{d^2\xi}{dr^2} = \kappa^2\xi \tag{9.20}$$

The general solution of the above equation is

$$\xi = Ae^{-\kappa r} + Be^{\kappa r} \tag{9.21}$$

where A and B are constants, which can be evaluated from the boundary conditions.

The boundary conditions are

$$\psi \rightarrow \psi_\infty = 0 \quad \text{as } r \rightarrow \infty \tag{9.22}$$

and

$$\psi \rightarrow \psi_s = \zeta \quad \text{at } r = a \tag{9.23}$$

Note the use of the zeta potential, ζ, at the particle surface ($r = a$) instead of the actual surface potential ψ_s of the particle in Eq. (9.23).

Substituting the boundary conditions, Eqs. (9.22) and (9.23) in Eq. (9.21) yields the potential distribution as

$$\psi = \zeta\frac{a}{r}\exp[-\kappa(r-a)], \tag{9.24}$$

where a is the particle radius. Use of the Poisson–Boltzmann equation to obtain the potential distribution implies that the electrolyte solution is stationary. It should be recalled that the assumption of a zero Peclet number was used to obtain the Boltzmann distribution and hence Eq. (9.18). This assumption implies that the electrolyte velocity does not affect the ionic equilibrium.

We are now in a position to determine the charge on the particle surface, Q_s. In Chapter 5, we noted that the total charge of a system comprised of a charged particle and the surrounding electrolyte should be zero (owing to electroneutrality). Using this information, the surface charge, Q_s, of a spherical particle of radius a is given by

$$Q_s = -\int_a^\infty (4\pi r^2)\rho_f dr \tag{9.25}$$

where the right hand side of Eq. (9.25) denotes the total amount of free charge in the electrolyte solution. Using Poisson's equation,

$$\frac{1}{r^2}\frac{d}{dr}\left(r^2\frac{d\psi}{dr}\right) = -\frac{\rho_f}{\epsilon} \tag{9.26}$$

Eq. (9.25) becomes

$$Q_s = 4\pi\epsilon \int_a^{\infty} r^2 \left[\frac{1}{r^2} \frac{d}{dr} \left(r^2 \frac{d\psi}{dr} \right) \right] dr = 4\pi\epsilon \left[r^2 \frac{d\psi}{dr} \right]_a^{\infty} \tag{9.27}$$

Noting that

$$\text{at } r \to \infty, \quad \frac{d\psi}{dr} \to 0$$

we obtain

$$Q_s = -4\pi a^2 \epsilon \left. \frac{d\psi}{dr} \right|_{r=a} \tag{9.28}$$

The term $d\psi/dr$ can be readily obtained by taking the radial derivative of ψ with respect to r, Eq. (9.24). Using this derivative in Eq. (9.28) gives

$$Q_s = 4\pi a^2 \epsilon \zeta \frac{(1 + \kappa a)}{a} \tag{9.29}$$

The electrical force acting on the particle in an external field E_∞ is given by

$$F_E = Q_s E_\infty = 4\pi a^2 \epsilon \zeta E_\infty \frac{(1 + \kappa a)}{a} = 4\pi a \epsilon \zeta E_\infty (1 + \kappa a) \tag{9.30}$$

As soon as the sphere starts to translate due to the electrical force, an electrolyte solution induced drag force counteracts this motion. However, to consider the drag force acting on the particle, one needs to first note that the charged fluid in the electric double layer surrounding the particle will acquire an oppositely directed velocity under the influence of the external electric field. The velocity of the center of the spherically symmetric charge cloud can be computed using the singular solution of Stokes equation discussed earlier in Section 6.3. Recall that the Oseen formulation is used to calculate the velocity field due to a point force acting at a given location in the fluid, and that the overall velocity field due to distributed point forces is obtained by integrating the singular solutions of the Stokes equation. Here, the point force is given as the product of the external field, E_∞, and the differential charge, $dQ = \rho_f \, dV$, of a small volume element dV in the fluid. From this analysis, it can be shown that the velocity component acting along the field direction is given by

$$U_R = \int_V \frac{E_\infty \rho_f}{8\pi\mu r} \left[1 + \cos^2\theta \right] dV \tag{9.31}$$

where U_R is the velocity of the center of the spherically symmetric charge cloud, V is the volume of the charge cloud, r is the radial position, and θ is the polar angle. Performing the integration over the entire volume of the charge cloud, one obtains the velocity of the center of the charge cloud as

$$U_R = -\frac{2}{3} \frac{\epsilon \zeta}{\mu} (\kappa a) E_\infty \tag{9.32}$$

This is the retardation velocity of the fluid surrounding the particle, and it approaches zero as $\kappa a \to 0$.

In an infinitely dilute colloidal dispersion (where the particle volume fraction, $\phi_p \to 0$), the Stokes hydrodynamic drag force on the spherical particle of radius a is given by

$$F_H = 6\pi \mu a (U - U_R) \tag{9.33}$$

where $U - U_R$ is the velocity of the particle relative to the electrolyte solution and μ is the viscosity of the electrolyte solution. For a non-accelerating particle, balancing the electrical and drag forces, *i.e.*, $F_E = F_H$, results in

$$U = \frac{2}{3} \frac{\epsilon \zeta E_\infty}{\mu} \tag{9.34}$$

This is the Hückel (1924) solution for the electrophoretic velocity valid for small κa. Equation (9.34) is usually written as

$$\eta = \frac{U}{E_\infty} = \frac{2}{3} \frac{\epsilon \zeta}{\mu} \tag{9.35}$$

where $\eta = U/E_\infty$ is the velocity per unit field strength, and is referred to as the *electrophoretic mobility*. Its unit is ($m^2\ V^{-1}\ s^{-1}$) or ($C\ s\ kg^{-1}$).

As would be expected, increasing the surface potential ζ or decreasing the electrolyte solution viscosity μ would increase the electrophoretic mobility. The Hückel equation is valid for large Debye length and, in particular, for non-electrolyte systems, *e.g.*, organic liquids, where κ^{-1} is generally large relative to the particle radius, a. The Hückel solution accounts for the retardation effect, but assumes a spherically symmetric double layer, and hence, does not consider any relaxation effect.

9.2.5 Electrophoresis in the Limit $\kappa a \gg 1$

When the particle radius is large, or more specifically, when the Debye screening length κ^{-1} is small relative to the particle radius, *i.e.*, $\kappa a \gg 1$, the electric double layer becomes extremely thin compared to the particle radius, and we can neglect the curvature effects of the particle. In this limiting case, one can consider the relative movement of the ions with respect to a planar surface. In other words, the electrophoretic problem reverts to the case of an electrolyte flowing past a planar surface with the externally imposed field aligned parallel to the surface as shown in Figure 9.4. The analysis based on this assumption is generally referred to as the Helmholtz–Smoluchowski analysis.

Consider the governing Navier–Stokes equation for the fluid flow past a horizontal flat surface, with the surface moving at a velocity U. Referring to Figure 9.4(b), we consider a coordinate system fixed on the particle surface. In this particle fixed coordinate frame, the fluid velocity at the particle surface is assumed to be zero, and the fluid velocity far away from the particle surface is given by $-U$. Based on the

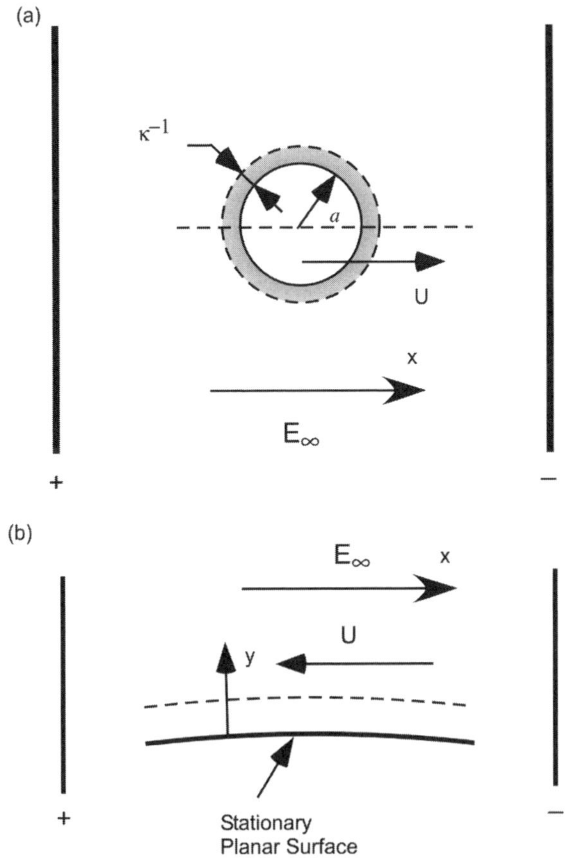

Figure 9.4. A schematic representation of electrophoresis of a positively charged spherical particle for $\kappa a \gg 1$. In this case, the relative motion between the electrolyte and the particle shown in (a) can be recast as the electroosmotic transport problem of the flow of the electrolyte past a stationary planar surface with the electric field aligned parallel to the surface (b). The direction of the fluid flow in (b) is opposite to the direction of the particle motion in (a).

geometry of Figure 9.4 and Table 6.7, we obtain

$$\mu \frac{d^2 u_x}{dy^2} = -\rho_f E_\infty \tag{9.36}$$

Here u_x is the velocity of the fluid tangential to the plate, ρ_f is the charge density in the fluid, and E_∞ is the externally imposed electric field.

Combining with the Poisson equation,

$$\epsilon \frac{d^2 \psi}{dy^2} = -\rho_f \tag{9.37}$$

Eq. (9.36) becomes

$$\mu \frac{d^2 u_x}{dy^2} = \epsilon E_\infty \frac{d^2 \psi}{dy^2} \tag{9.38}$$

The boundary conditions for the problem are as follows:

$$\psi \to 0 \quad \text{and} \quad u_x \to -U \qquad \text{as } y \to \infty \tag{9.39-a}$$

$$\frac{d\psi}{dy} \to 0 \quad \text{and} \quad \frac{du_x}{dy} \to 0 \qquad \text{as } y \to \infty \tag{9.39-b}$$

$$\psi = \zeta \quad \text{and} \quad u_x = 0 \qquad \text{at } y = 0 \tag{9.39-c}$$

Integrating Eq. (9.38) from infinity to an arbitrary distance y from the particle surface, and employing the boundary conditions (9.39-a) and (9.39-b), yields

$$u_x(y) = \frac{\epsilon E_\infty \psi(y)}{\mu} - U \tag{9.40}$$

Substituting the boundary condition at the particle surface, Eq. (9.39-c), in Eq. (9.40) results in

$$U = \frac{\epsilon \zeta E_\infty}{\mu}$$

or, expressed in terms of the electrophoretic mobility,

$$\eta = \frac{U}{E_\infty} = \frac{\epsilon \zeta}{\mu} \tag{9.41}$$

Equation (9.41) is referred to as the Helmholtz–Smoluchowski electrophoretic mobility equation, and is valid in the limit $\kappa a \gg 1$ and $ze\zeta / k_B T < 1$. This equation is also referred to as Smoluchowski's electrophoretic mobility equation in the literature.

It should be noted that the Hückel and the Helmholtz–Smoluchowski results, given by Eqs. (9.34) and (9.41), respectively, provide the two limiting values of the electrophoretic velocity of a single colloidal particle as $\kappa a \to 0$ and $\kappa a \to \infty$, respectively. These two expressions differ by a factor of $2/3$. Note that both the Hückel and Helmholtz–Smoluchowski (H–S) equations are independent of particle size. Moreover, since the H–S equation (for large κa) was derived by assuming a flat surface, *i.e.*, no curvature effect, we can conclude that it should also be valid for any shape provided that the electric double layer is very thin everywhere and that the particle is non-conducting.

9.3 IMPROVED SOLUTIONS: ARBITRARY DEBYE LENGTH

The electrophoretic velocities derived so far are for either $\kappa a \ll 1$ or $\kappa a \gg 1$. In particular, the H–S equation was derived with the particle surface being treated as locally flat. On the other hand, in the derivation of the Hückel equation, any perturbation of the electric double layer due to the influence of the external electrical field (the relaxation effect) was neglected. While neglecting the curvature of the particle and the ionic movement effects in the two limiting solutions is quite reasonable, these effects cannot be ignored when considering the electrophoretic velocities of a charged particle at intermediate values of the parameter κa. Numerous solutions of the electrophoretic velocity of a charged particle for intermediate values of κa have been developed.

In the following, we will discuss the solution methodologies accounting for the electrophoretic retardation and relaxation effects employing the linear perturbation approach. First, the general governing equations pertaining to the perturbation technique are presented. We then briefly describe the approach due to Henry (1931), which allows calculation of the electrophoretic mobility for the entire range of κa ($0 < \kappa a < \infty$). The approach of Henry (1931) is however limited to low surface potentials, and ignores the relaxation effect arising from ionic convection. A brief overview of the perturbation approaches accounting for the relaxation effects and higher surface potentials on the particle is also presented in this section (Overbeek, 1943; Booth, 1950; Wiersema *et al.*, 1966; O'Brien and White, 1978).

9.3.1 Perturbation Approach

The coupled non-linear governing equations outlined in Section 9.2.1 pose a formidable challenge in terms of obtaining a direct analytic solution. For the cases where the applied electrical field $\mathbf{E}_\infty$ is small compared to the electrical fields inside the double layer, one can assume that the electrical double layer is only slightly distorted from the equilibrium configuration due to the applied field and the ensuing particle motion. In this case, the governing equations, Eqs. (9.5)–(9.10), as well as their associated boundary conditions, can be replaced by their approximate linearized forms using a perturbation approach. To derive these linearized equations, the unknown variables in the governing equations are represented as a sum of the equilibrium value in absence of the external field (see Section 6.2.7) and a perturbation due to the external field. The perturbation term is considered to be directly proportional to the external field. Thus, one can write

$$\psi = \psi^{eq} + \delta\psi \tag{9.42-a}$$

$$\mathbf{u} = \mathbf{0} + \delta\mathbf{u} \tag{9.42-b}$$

$$p = p^{eq} + \delta p \tag{9.42-c}$$

$$n_i = n_i^{eq} + \delta n_i \tag{9.42-d}$$

Substituting the above variables in the general governing equations, (9.5)–(9.10), and retaining up to the linear terms, one obtains two sets of equations. First, retaining only the equilibrium terms, one obtains

$$\epsilon\nabla^2\psi^{eq} = -\rho_f = -\sum z_i e n_i^{eq} \tag{9.43}$$

for the Poisson equation,

$$0 = -\nabla p^{eq} + \epsilon \nabla^2 \psi^{eq} \nabla \psi^{eq} \tag{9.44}$$

for the modified Navier–Stokes equation, and

$$\nabla \cdot \left[\nabla n_i^{eq} + \frac{z_i e n_i^{eq}}{k_B T} \nabla \psi^{eq} \right] = 0 \tag{9.45}$$

for the Nernst–Planck equation.

The boundary conditions for the equilibrium problem are set as follows: At the particle surface, Eq. (9.11) gives

$$-\epsilon \mathbf{n} \cdot \nabla \psi^{eq} = q_s \tag{9.46}$$

while Eq. (9.13) yields

$$\left[\nabla n_i^{eq} + \frac{z_i e}{k_B T} n_i^{eq} \nabla \psi^{eq} \right] \cdot \mathbf{n} = 0 \tag{9.47}$$

Far away from the particle, $\mathbf{r} \to \infty$, we have

$$n_i^{eq} = n_{i\infty} \quad \text{and} \quad \psi^{eq} = 0 \tag{9.48}$$

Solution of the equilibrium equations (9.43) to (9.45) subject to the above boundary conditions provides the Poisson–Boltzmann description of the electrostatic problem around the stationary particle in absence of any external electric field, and hence, any ionic or fluid movement. Solution of the above stationary problem was discussed in detail in Chapter 5.

The equations containing the linear perturbation terms are written next. In writing these equations, only the linear perturbation terms are retained, and the products of the perturbation terms are neglected. The Poisson equation in terms of the perturbed variables becomes

$$\epsilon \nabla^2 \delta\psi = -\sum z_i e \delta n_i \tag{9.49}$$

The Navier–Stokes equation becomes

$$\mu \nabla^2 \delta\mathbf{u} - \nabla \delta p - e \sum z_i n_i^{eq} \nabla \delta\psi - e \sum z_i \delta n_i \nabla \psi^{eq} = 0 \tag{9.50}$$

with

$$\nabla \cdot \delta\mathbf{u} = 0 \tag{9.51}$$

for the continuity equation. Finally, the perturbed form of the Nernst–Planck equation becomes

$$\nabla \cdot \left[n_i^{eq} \delta\mathbf{u} - D_i \nabla \delta n_i - \frac{z_i e D_i}{k_B T} (\delta n_i \nabla \psi^{eq} + n_i^{eq} \nabla \delta\psi) \right] = 0 \tag{9.52}$$

For this problem, the boundary conditions at the particle surface are

$$\delta \mathbf{u} = 0 \tag{9.53-a}$$

$$-\mathbf{n} \cdot \nabla \delta \psi = 0 \tag{9.53-b}$$

$$\left[\nabla \delta n_i + \frac{z_i e}{k_B T} \left(n_i^{eq} \nabla \delta \psi + \delta n_i \nabla \psi^{eq} \right) \right] \cdot \mathbf{n} = 0 \tag{9.53-c}$$

Far from the particle, the boundary conditions at $\mathbf{r} \to \infty$ are

$$\nabla \delta \psi = -\mathbf{E}_{\infty} \tag{9.54-a}$$

$$\delta n_i = 0 \tag{9.54-b}$$

$$\left\{ -\delta p \bar{\bar{I}} + \mu \left[\nabla \delta \mathbf{u} + (\nabla \delta \mathbf{u})^T \right] \right\} \cdot \mathbf{n} = 0 \tag{9.54-c}$$

Equations (9.49)–(9.52) are the linearized forms of the governing transport equations, and are traditionally the starting points for theoretical investigations employing perturbation methods in electrophoresis. Simplifications are made along the way, which usually involve expanding the perturbation variables as a Taylor series in terms of the scaled particle surface potential, $\Psi_s = ze\zeta / k_B T$.

Henry (1931) derived an analytical expression for the electrophoresis of a single spherical particle that is valid for small zeta potentials (the Debye–Hückel approximation) and for an arbitrary double-layer thickness which considers the electrophoretic retardation effect. However, the relaxation effects arising from ionic convection was neglected.

Overbeek (1943) and Booth (1950) were the first to correctly account for the ionic convection effect about a spherical particle but their results are only valid for small zeta potentials $(ze\zeta / k_B T)$.

Wiersema *et al.* (1966) relaxed the constraint of a small zeta potential and numerically solved the governing equations for the electrophoresis of a non-conducting sphere. Ionic convection was considered. Numerical difficulties at high values of ζ were encountered. Subsequently, O'Brien and White (1978) numerically solved the governing equations for arbitrary ζ and κa. Their work represents a proper solution of the electrophoresis problem for a sphere (Kozak and Davis, 1989a).

9.3.2 Henry's Solution

As stated earlier, the limiting results for $\kappa a \gg 1$ and $\kappa a \ll 1$ given by the Helmholtz–Smoluchowski and the Debye–Hückel expressions, respectively, were bridged by Henry (1931) for intermediate values of the parameter κa. One can derive Henry's result starting from the linear perturbation equations, (9.49)–(9.54-c). Such an approach requires expanding the perturbation variables discussed in Section 9.3.1 in terms of the scaled particle surface potential, and retaining the leading order terms. However, we will not discuss the perturbation approach here. Instead, we will simply reiterate the approach originally presented by Henry (1931), reserving some comments on its validity for subsequent discussion. Henry developed the theory from a general set of fluid mechanical and electrostatic equations describing the electrophoretic transport problem. Two main assumptions were made in Henry's approach:

1. The electric double layers are not distorted by the flow. The total potential within the double layer is a linear combination of the electric double layer potential and the potential due to the externally imposed electric field.
2. The Debye–Hückel approximation for low surface potential is assumed.

Henry's approach for solving the governing transport equations of Section 9.3.1 was to consider the equilibrium shape of the electrical double layer (no distortion of the ionic cloud due to the movement of the particle). Due to Henry's assumptions, effects arising from convective relaxation and surface conductance are not accounted for in the analysis. In this case, the governing equations for ion transport, Eq. (9.10) simplify to provide the equilibrium Boltzmann distribution of ions.

Consider a sphere of radius a, electrical conductivity σ' and surface potential ζ immersed in an infinite volume of a Newtonian electrolyte solution with an electrical conductivity σ^{∞}, dielectric permeability ϵ, and viscosity μ. If an external field E_{∞} is applied to the system in a direction parallel to the x-axis, the particle will be subject to an electrophoretic velocity U along the positive x-direction. Henry considered the situation when the particle is immobile (*i.e.*, stationary) and the electrolyte solution flows past it with a velocity U along the negative x-direction. The geometry of the problem is depicted in Figure 9.5. It is assumed that the problem can be decomposed into two parts: One part is to solve for a spherically-symmetric potential distribution around the charged sphere in an electrolyte solution containing free charges. The

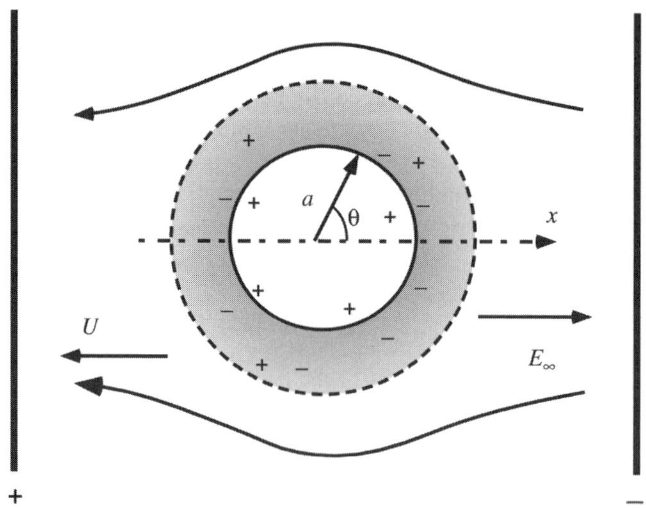

Figure 9.5. Electrophoresis with no relaxation effects. A stationary particle with the solvent (electrolyte) flowing along the negative x direction past it with a velocity U.

second part is to solve for a charged particle in a dielectric having no free charge but with an external electric field being applied.

At the outset, we note that Henry developed his mathematical formulation assuming finite conductivities for the particle and the suspending medium. However, to facilitate comparison with the Smoluchowski and Hückel results, his final results were provided for a non-conducting sphere, $\sigma' = 0$. Accordingly, we will derive his results here assuming a non-conducting spherical particle.

The potential due to the charged spherical particle is given by[2]

$$\epsilon \nabla^2 \psi = -\rho_f \tag{9.55}$$

Neglecting θ-variation, making use of the fact that $ze\zeta/k_B T \ll 1$ and assuming that the Boltzmann distribution is valid, Eq. (9.55) becomes

$$\frac{1}{r^2}\frac{d}{dr}\left(r^2\frac{d\psi}{dr}\right) = \kappa^2\psi \tag{9.56}$$

where

$$\kappa = \left(\frac{2e^2z^2n_\infty}{\epsilon k_B T}\right)^{1/2}$$

The boundary conditions are

$$r \to \infty \quad \psi \to 0$$

$$r = a \quad \psi = \zeta$$

The solution is given by

$$\psi = \zeta\left(\frac{a}{r}\right)e^{-\kappa(r-a)} \tag{9.57}$$

The potential given by Eq. (9.57) is solely due to the electric double layer. No applied electric field is being considered.

Consider now a particle in a charge-free dielectric. The potential due to the external field is ϕ. The governing equation becomes

$$\nabla^2\phi = 0 \tag{9.58}$$

with

$$\mathbf{E} = -\nabla\phi \tag{9.59}$$

Equation (9.58) can be written as (see Table 6.9)

$$\frac{1}{r^2}\frac{\partial}{\partial r}\left(r^2\frac{\partial\phi}{\partial r}\right) + \frac{1}{r^2\sin\theta}\frac{\partial}{\partial\theta}\left(\sin\theta\frac{\partial\phi}{\partial\theta}\right) = 0 \tag{9.60}$$

[2]In this section, we denote the potential due to the stationary electric double layer as ψ instead of ψ^{eq}. We do this to simplify the notation used for the mathematical development.

The boundary conditions are

$$\text{BC-1: } \phi = -E_\infty r \cos\theta \quad \text{for } r \to \infty \tag{9.61}$$

$$\text{BC-2: } \frac{\partial \phi}{\partial r} = 0 \quad \text{at } r = a \tag{9.62}$$

BC-2 is obtained by recognizing that the dielectric permittivity of the particle is insignificant compared to that of the surrounding fluid (which is generally the case). Also, at the particle surface, $j_{ir}^{**} = 0$, $u_r = 0$ and $n_i = 0$. E_∞ is the undisturbed electric field strength. Solution of this electrostatic problem was provided in great detail in Section 4.2. The reader is advised to refer to that solution at this point.

The solution of Eq. (9.60) subject to the boundary conditions of Eqs. (9.61) and (9.62) is given by

$$\phi = -E_\infty \left[r + \frac{a^3}{2r^2} \right] \cos\theta \tag{9.63}$$

It is now necessary to evaluate the hydrodynamic and electric forces exerted on the charged particle. The hydrodynamic forces can be evaluated from the solution of the Navier–Stokes equation modified for the electric force.

The momentum equation for creeping flow is given by

$$\mu \nabla^2 \mathbf{u} - \nabla p = \rho_f \nabla(\phi + \psi) \tag{9.64}$$

which needs to be solved together with the continuity equation

$$\nabla \cdot \mathbf{u} = 0 \tag{9.65}$$

The following boundary conditions apply for the fluid flow problem

$$u_r = -U \cos\theta \quad \text{as } r \to \infty \tag{9.66}$$

$$u_\theta = U \sin\theta \quad \text{as } r \to \infty \tag{9.67}$$

and

$$u_r = u_\theta = 0 \quad \text{at } r = a \tag{9.68}$$

As mentioned earlier, the spherical particle is held stationary with the fluid flowing at a velocity of U along the negative x direction. The potentials ψ and ϕ are given by, respectively, Eqs. (9.57) and (9.63). The solution to Eqs. (9.64) and (9.65) subject to the boundary conditions of Eqs. (9.66), (9.67) and (9.68) was obtained by Henry (1931). The mathematical manipulations are quite involved. The solution provides expressions for p, u_r, and u_θ as functions of r and θ.

The hydrodynamic force F_H exerted by the fluid on the particle can be written as

$$F_H = 2\pi a^2 \int_0^\pi [-\tau_{rr} \cos\theta + \tau_{r\theta} \sin\theta]_{r=a} \sin\theta \, d\theta \tag{9.69}$$

where

$$\tau_{rr} = -p + 2\mu \frac{\partial u_r}{\partial r} \tag{9.70}$$

and

$$\tau_{r\theta} = \mu \left[\frac{\partial u_\theta}{\partial r} - \frac{u_\theta}{r} + \frac{1}{r} \frac{\partial u_r}{\partial \theta} \right] \tag{9.71}$$

In Eq. (9.69) τ_{rr} and $\tau_{r\theta}$ represent the normal and tangential stresses exerted by the fluid on the particle. Figure 9.6 shows the hydrodynamic surface forces. Note that F_H is taken as positive when measured along the velocity U (in opposite direction to x).

The electric force due to the electric field is obtained by integrating the Maxwell stress tensor over the particle surface. This force can also be represented as the product of the total particle charge and the electric field strength E_∞. Let us derive the electrical force term. Consider a charged spherical particle of radius a located in a dielectric containing free charge; the electric force in the x-direction, F_E, exerted on the particle is given by

$$F_E = \int_S [q_s E_x]_a \, dS \tag{9.72}$$

where q_s is the surface charge density (charge per unit surface area, $\mathrm{C/m^2}$). The subscript x denotes x-direction and subscript a is for $r = a$. Let us first determine q_s and E_x on the particle surface.

For the special case of no angular variation in ψ, the charge density q_s follows from Eq. (9.28), and is written as

$$q_s = \frac{Q_s}{4\pi a^2} = -\epsilon \left. \frac{d\psi}{dr} \right|_{r=a} \tag{9.73}$$

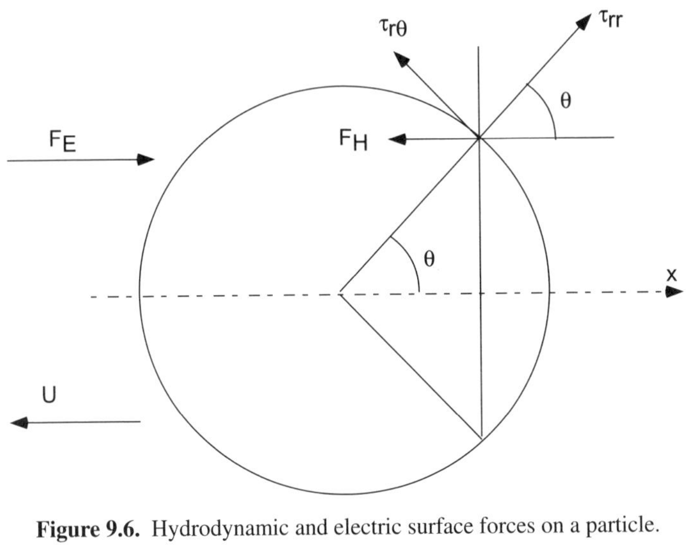

Figure 9.6. Hydrodynamic and electric surface forces on a particle.

The electric field strength E_r is given by

$$E_r = -\frac{\partial}{\partial r}(\psi(r) + \phi(r, \theta)) \tag{9.74}$$

Since at $r = a$, $\partial\phi/\partial r = 0$ [Eq. (9.62)], we have at the surface of the sphere

$$E_r = -\frac{d\psi}{dr} \quad \text{(independent of } \theta\text{)} \tag{9.75}$$

Now

$$E_\theta = -\frac{1}{r}\frac{\partial}{\partial\theta}(\psi + \phi) = -\frac{1}{r}\frac{\partial\phi}{\partial\theta} \tag{9.76}$$

The term $\partial\psi/\partial\theta = 0$ because of the assumption that there is no θ-dependence for ψ. The electric field strength in x-direction is given by

$$E_x = -E_\theta \sin\theta + E_r \cos\theta \tag{9.77}$$

Making use of Eqs. (9.75) and (9.76), we obtain

$$E_x = \frac{1}{r}\frac{\partial\phi}{\partial\theta}\sin\theta - \frac{d\psi}{dr}\cos\theta \tag{9.78}$$

Making use of Eqs. (9.73) and (9.78), the electric force in x-direction as provided by Eq. (9.72) is given by

$$F_E = \int_0^\pi -\left[\epsilon\frac{d\psi}{dr}\right]\left[\frac{1}{a}\frac{\partial\phi}{\partial\theta}\sin\theta - \frac{d\psi}{dr}\cos\theta\right][2\pi a \sin\theta]\, a\, d\theta \tag{9.79}$$

As the integral

$$\int_0^\pi \cos\theta \sin\theta \, d\theta = 0$$

The term $(d\psi/dr)\cos\theta$ of Eq. (9.79) does not contribute to the force term, and consequently, Eq. (9.79) reduces to

$$F_E = -2\pi a\epsilon \left.\frac{d\psi}{dr}\right|_a \int_0^\pi \left.\frac{\partial\phi}{\partial\theta}\right|_a \sin^2\theta \, d\theta \tag{9.80}$$

From Eq. (9.63), we obtain

$$\left.\frac{\partial \phi}{\partial \theta}\right|_a = \frac{3a}{2} E_\infty \sin\theta$$

and Eq. (9.80) becomes

$$F_E = -3\pi a^2 \epsilon E_\infty \left.\frac{d\psi}{dr}\right|_a \int_0^\pi \sin^3\theta \, d\theta \tag{9.81}$$

The electric force on the particle in x-direction is then given by

$$F_E = \left[-4\pi a^2 \epsilon \left.\frac{d\psi}{dr}\right|_{r=a} \right] E_\infty \tag{9.82}$$

Equation (9.82) gives the electric force in Henry's solution. Here the term in brackets is the total surface charge. Note that ϕ does not contribute to the force F_E, as it acts in a charge free dielectric.

At steady state, the sum of the hydrodynamic and electric forces is zero as the particle does not have any acceleration. This leads to

$$F_H = F_E \tag{9.83}$$

and

$$U = \left[\frac{2\epsilon\zeta E_\infty}{3\mu} \right] f(\kappa a) \tag{9.84}$$

where

$$\begin{aligned} f(\kappa a) = 1 &+ \frac{1}{16}(\kappa a)^2 - \frac{5}{48}(\kappa a)^3 - \frac{1}{96}(\kappa a)^4 + \frac{1}{96}(\kappa a)^5 \\ &+ \frac{1}{8}(\kappa a)^4 e^{\kappa a}\left(1 - \frac{(\kappa a)^2}{12}\right) \int_{\kappa a}^{\infty} \left(\frac{e^{-t}}{t}\right) dt \end{aligned} \tag{9.85}$$

Equation (9.84) can be rearranged to express the electrophoretic mobility as

$$\eta = \frac{U}{E_\infty} = \left[\frac{2\epsilon\zeta}{3\mu} \right] f(\kappa a) \tag{9.86}$$

Note that the integral in the last term of Eq. (9.85) is an exponential integral, and a simple numerical quadrature formula (like trapezoidal or Simpson's rules) is generally not sufficient to evaluate it. One should either employ appropriate numerical integration schemes specially designed for evaluating exponential integrals, or should use asymptotic series expansions (which are available for large values of κa). A plot of the function f, usually called Henry's function, is given in Figure 9.7.

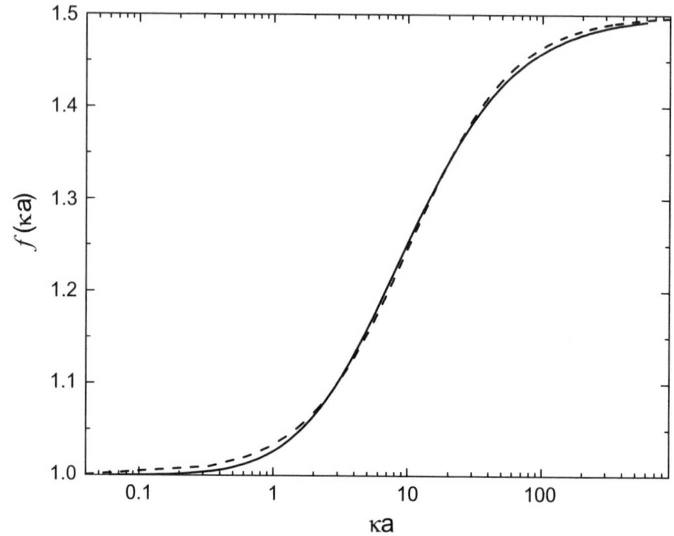

Figure 9.7. Variation of Henry's function with κa. Solid line, Eq. (9.85); dashed line, Eq. (9.87).

From Figure 9.7 it can be seen that for $\kappa a \ll 1$, $f(\kappa a) \to 1$, and the electrophoretic velocity given by Eq. (9.84) reduces to Hückel's expression, Eq. (9.35). For the case of $\kappa a \gg 1$, *i.e.*, a thin electric double layer $f(\kappa a) \to 3/2$ and Eq. (9.84) gives the electrophoretic velocity as given by Eq. (9.41) for the Helmholtz–Smoluchowski expression.

Strictly speaking, Henry's equation is valid for small values of surface potential: *i.e.*, $ze\zeta/k_BT \ll 1$ for all κa values. It should also be noted that in Henry's derivation, it was assumed that the fluid moves with a velocity U while the particle is held stationary in the electric field. Table 9.1 gives values for Henry's function as calculated using Eq. (9.85).

A curve fit equation for $f(\kappa a)$ is given by

$$f(\kappa a) = \frac{3}{2} - \frac{1}{2[1 + a_1(\kappa a)^{a_2}]} \tag{9.87}$$

where $a_1 = 0.072$ and $a_2 = 1.13$.

It is perhaps pertinent at this point to identify certain conceptual inconsistencies with Henry's approach described above. It is conceptually incorrect to assume that the potential distribution engendered by the external field obeys the Laplace equation as discussed in this section leading to the potential distribution given by Eq. (9.63). It is also incorrect to assume that the external potential, ϕ, can be linearly superimposed on the potential due to the charged particle, ψ. In fact, the perturbation equations described in Section 9.3.1 do not lead to such a linear superposition of the electrical potentials. Instead, after representing the ionic concentrations in terms of

TABLE 9.1. Values for Henry's Function Evaluated Numerically Using Eq. (9.85).

κa	$f(\kappa a)$
0	1.000
0.5	1.009
1.0	1.027
2.0	1.065
3.0	1.101
4.0	1.132
5.0	1.160
10.0	1.253
25.0	1.365
50.0	1.423
100.0	1.458
∞	1.500

the electrochemical potentials through

$$\mu_i = \mu_{i0} + k_B T \ln n_i + z_i e \psi \tag{9.88}$$

where μ_i is the ith ion electrochemical potential, one can derive a perturbation variable of the form

$$\delta\phi_i = \frac{\delta\mu_i}{z_i e} = \frac{k_B T}{z_i e}\frac{\delta n_i}{n_i^{eq}} + \delta\psi \tag{9.89}$$

It is this perturbation variable, $\delta\phi_i$, which eventually leads to a Laplace equation

$$\nabla^2 \delta\phi_i = 0 \tag{9.90}$$

when substituted in the perturbed form of the Nernst–Planck equation. Such a formulation leads to a solution for $\delta\phi_i$ that is identical in form to Eq. (9.63). The subsequent solution of the perturbation equations is quite similar to the approach originally developed by Henry, leading to the same result for the electrophoretic mobility as given by Eq. (9.86). Thus, in retrospect, although Henry's formulation might be construed as conceptually flawed, the analysis leads to a correct final result for the electrophoretic mobility. Indeed, if one performs the linear perturbation analysis, it becomes evident that Henry's solution accounts for the diffusion-migration based relaxation effects (it only excludes the convective relaxation effect).

9.3.3 Effect of Particle Conductivity and Shape

The results of Henry (1931) discussed above dealt with a non-conducting sphere. The conductivity of a particle affects (distorts) the applied electric field. In this context, one should first note that the subsequent discussion refers to particles in which the conductivity arises due to the presence of free charge carriers (ions). An example of such a system is a porous particle, with ions penetrating its bulk. The discussion below does not pertain to metallic (electronic) conductors. As stated earlier, most metallic conductors will behave as non-conducting particles in an electrolyte solution. Dukhin and Derjaguin (1974) give a full discussion on the effect of the particle conductivity. The expression for the case of a spherical particle with a finite conductivity is given by

$$U = \frac{2\epsilon\zeta E_\infty}{3\mu} F(\kappa a, K') \tag{9.91}$$

where

$$K' = \frac{\sigma'}{\sigma^\infty} \tag{9.92}$$

and

$$F(\kappa a, K') = 1 + 2\frac{(1 - K')}{(2 + K')}[f(\kappa a) - 1] \tag{9.93}$$

Here σ' and σ^∞ are the electric conductivities of the particle and bulk electrolyte medium, respectively, and $f(\kappa a)$ is Henry's function given by Eq. (9.85).

For a non-conducting spherical particle $\sigma' = 0$ and $K' = 0$. This leads to

$$\frac{(1 - K')}{(2 + K')} = \frac{1}{2}$$

and

$$F(\kappa a, K') \equiv f(\kappa a)$$

Equation (9.91) then becomes

$$U = \frac{2}{3}\frac{\epsilon\zeta E_\infty}{\mu} f(\kappa a)$$

which is Eq. (9.84).

For a perfectly conducting sphere, $\sigma' \to \infty$ leads to

$$K' \to \infty \quad \text{and} \quad \frac{(1 - K')}{(2 + K')} \to -1$$

Hence

$$U = \frac{2\epsilon\zeta E_\infty}{3\mu}[3 - 2f(\kappa a)]$$

As $f(\kappa a) = 1$ for $\kappa a \to 0$ (large Debye length) the above equation indicates that $[3 - 2f(\kappa a)] \to 1$. Consequently, for $\kappa a \to 0$, both the conducting and non-conducting spheres behave in a similar manner. This is quite reasonable as, for a large Debye length, κ^{-1}, the electric field is undisturbed by the presence of a particle. However, for

$\kappa a \to \infty$ the situation is different for the conducting sphere. Here $f(\kappa a) \to 3/2$ as $\kappa a \to \infty$ and $[3 - 2f(\kappa a)] \to 0$. Hence $U \to 0$ for $\kappa a \to \infty$ (small Debye length).

For the case of an arbitrary particle shape having $K' = 1$ (*i.e.*, $\sigma' = \sigma^{\infty}$)

$$\frac{(1 - K')}{(2 + K')} = 0$$

leading to $F(\kappa a, 1) = 1$. This is Hückel's model where the electric field is distorted by the double layer but not by the presence of the particle.

For the case of non-conducting cylinders aligned parallel to the electric field, the electric field E_{∞} is always parallel to the particle surface for all κa values. Here Smoluchowski's model is valid: *i.e.*, $F = 3/2$ for all κa values.

Gorin (1939) extended Henry's solution for the case of non-conducting cylinders oriented normal to the electric field. The F values are given in Figure 9.8. The above analysis includes all Henry's assumptions cited earlier. It should be remembered, however, that due to polarization, all particles act as non-conductors and hence having $K' \neq 0$ is an academic curiosity (Hunter, 1981).

For randomly oriented cylinders, Keizer *et al.* (1975) showed that

$$U_{\text{effective}} = \frac{1}{3}\left(U_{\parallel} + 2U_{\perp}\right) \tag{9.94}$$

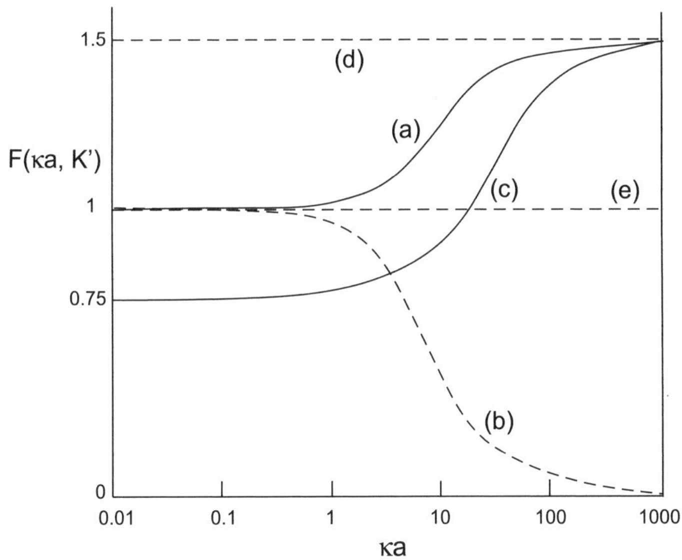

Figure 9.8. Values of Henry's function for particles of various shapes and conductivities. (a) Non-conducting sphere ($K' = 0$); (b) conducting sphere ($K' = \infty$); (c) and (d) non-conducting cylinder ($K' = 0$) with axis parallel and perpendicular to the applied field, respectively; (e) particle of any shape with $K' = 1$. Here $K' = \sigma'/\sigma^{\infty}$, *i.e.*, the ratio of the particle conductivity to the electrolyte solution conductivity.

where $U_{\parallel}$ and $U_{\perp}$ are the electrophoretic mobilities of cylinders parallel and normal to the electric field, respectively.

For more details on electrophoresis of cylinders and arbitrary shape particles, the reader can refer to Hunter (1981), Morrison (1970), and Van der Drift *et al.* (1979).

9.3.4 Alternate Forms of the Electrophoretic Velocities

The solutions given so far are expressed in terms of a constant potential at the particle surface. We shall now discuss the case of electrophoretic velocity of spherical particles expressed in terms of a constant surface charge.

For a particle carrying a surface charge Q_s in an electric field E_∞, the electrical force is given by

$$F_E = Q_s E_\infty \tag{9.95}$$

Equating this force with the drag force on the particle, Eq. (9.33) and using Eq. (9.29), we obtain

$$U = \frac{Q_s E_\infty}{6\pi a \mu}(1 - \kappa a) \quad \text{for } \kappa a \ll 1 \tag{9.96}$$

This is the electrophoretic velocity of a particle with constant surface charge in the Hückel limit ($\kappa a \ll 1$).

For the case of a thin electric double layer, $\kappa a \gg 1$, the electrophoretic velocity of the particle is given by the Helmholtz–Smoluchowski equation, (9.41). Now, using Eq. (9.29), the surface charge on the particle can be related to its zeta potential at large κa as

$$Q_s = 4\pi a^2 \epsilon \zeta \kappa \quad \text{for } \kappa a \gg 1 \tag{9.97}$$

Combining Eq. (9.41) with Eq. (9.97) to eliminate ζ, one obtains

$$U = \frac{Q_s E_\infty}{4\pi a^2 \mu \kappa} \quad \text{for } \kappa a \gg 1 \tag{9.98}$$

Equations (9.96) and (9.98) are the Hückel and Helmholtz–Smoluchowski expressions for constant surface charge, respectively.

Following the non-dimensionalization given by Russel *et al.* (1989), we can write

$$\beta^* = \frac{aeE_\infty}{k_B T} \quad \text{electric field strength} \tag{9.99}$$

$$q_s^* = \frac{eQ_s}{4\pi a \epsilon k_B T} \quad \text{surface charge density} \tag{9.100}$$

and

$$U^* = \frac{\mu a}{\epsilon}\left(\frac{e}{k_B T}\right)^2 U \quad \text{electrophoretic velocity} \tag{9.101}$$

Making use of Eqs. (9.99) to (9.101), the dimensionless electrophoretic velocity U^* can be given for the constant surface charge boundary condition as

Constant surface charge:

$$U^* = \frac{2}{3}\beta^* q_s^* \quad \text{for } \kappa a \ll 1 \tag{9.102}$$

and

$$U^* = \beta^* \frac{q_s^*}{\kappa a} \quad \text{for } \kappa a \gg 1 \tag{9.103}$$

Putting $\zeta^* = e\zeta/k_B T$, the electrophoretic velocities corresponding to Eqs. (9.34) and (9.41) are given for the constant surface potential condition in dimensionless form as

Constant surface potential:

$$U^* = \frac{2}{3}\beta^* \zeta^* \quad \text{for } \kappa a \ll 1 \tag{9.104}$$

and

$$U^* = \beta^* \zeta^* \quad \text{for } \kappa a \gg 1 \tag{9.105}$$

It is possible to generalize the limiting cases given above in terms of Henry's solution for both constant surface potential and surface charge.

The dimensionless form of Eq. (9.29) is

$$q_s^* = \zeta^*(1 + \kappa a) \tag{9.106}$$

Making use of the above expression, Henry's solution becomes (Russel *et al.*, 1989),

Constant surface charge:

$$U^* = \left(\frac{2}{3}\right)\left(\frac{\beta^* q_s^*}{1 + \kappa a}\right) f(\kappa a) \tag{9.107}$$

Constant surface potential:

$$U^* = \frac{2}{3}\beta^* \zeta^* f(\kappa a) \tag{9.108}$$

Figure 9.9 shows the scaled electrophoretic mobility $(3U^*/2\beta^* q_s^*)$ for the case of a constant surface charge and $(3U^*/2\beta^*\zeta^*)$ for the case of a constant surface potential. It is evident from the figure that changes in κa affect the scaled electrophoretic mobility when either the charge or the potential is held constant. While the Hückel limit ($\kappa a \ll 1$) is unaffected by the electrical boundary condition on the particle surface, the scaled electrophoretic mobilities in the Helmholtz–Smoluchowski limit ($\kappa a \gg 1$) are vastly different for the two types of boundary conditions. In the case of constant surface potential, the scaled electrophoretic mobility stays non-zero throughout the

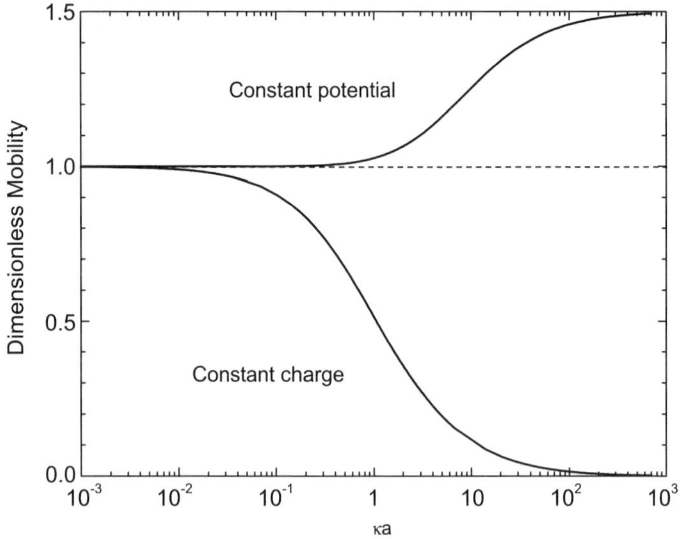

Figure 9.9. Variation of the scaled electrophoretic mobility with κa for constant potential and constant surface charge particles based on Henry's solution.

range of κa because as the double-layer thickness diminishes, the particle charge increases. In case of constant charge particles, however, the scaled electrophoretic mobility becomes vanishingly small at large κa values.

It should be remembered that although we treated the case of holding the surface charge or the potential constant, the particle charge or its potential and the diffuse-layer thickness are closely related to each other in an experimental situation. For a given particle, changes in ionic strength alter the particle charge as well as the diffuse-layer thickness (Russel *et al.*, 1989).

9.3.5 Solutions Accounting for Relaxation Effects

Henry's solution did not consider the effects arising from the distortion of the electrical double layer around a charged particle as it migrates along the imposed electrical field (*cf.*, Figure 9.2). Strictly speaking, the convective relaxation effect was not accounted for in Henry's solution. Since the charged ions are also dynamic in presence of the electrical field, the double layer becomes distorted. The ions move under the coupled influence of the external electrical field as well as the field due to the charged particle to restore the shape of the electric double layer. The ionic movement occurs due to convection, diffusion, and migration. Of these three modes, only the diffusion-migration based ion movements are considered in Henry's result for low particle surface potentials. The asymmetry in the electrical double layer also gives rise to a convective migration of the ions in the vicinity of the particles. This process of relaxation becomes very important at higher particle surface potentials.

The early works of Overbeek (1943) and Booth (1950) based on perturbation techniques provided some insight into the relaxation problem, although these solutions were limited to low surface potentials owing to the complexity of the problem. Wiersema *et al.* (1966) provided a numerical technique to solve the governing mobility equations for a binary electrolyte. Their study provided tabulated information on the mobility of a particle with different zeta potentials. However, their iterative numerical scheme failed to converge for zeta potentials greater than 150 mV for a 1 : 1 electrolyte. A more robust numerical scheme to solve the governing mobility equations was provided by O'Brien and White (1978). All of the above studies start from a set of perturbation equations pertaining to the mobility of the particle and the ions in an applied electrical field as outlined in Section 9.3.1.

For the general case of arbitrary Debye length and surface potential, O'Brien and White (1978) developed a computational scheme to solve the governing perturbation equations (Section 9.3.1) for the electrophoretic problem corresponding to a constant surface potential on the particle. Figures 9.10 and 9.11 summarize the solutions given by O'Brien and White. Figure 9.10 shows the variation of dimensionless electrophoretic mobility with the dimensionless zeta potential for small and large κa. Employing Eqs. (9.99) and (9.101), the dimensionless electrophoretic mobility is written as

$$\eta^* = \frac{3\mu e}{2\epsilon k_B T}\frac{U}{E_\infty} = \frac{3\mu e}{2\epsilon k_B T}\eta = \frac{3U^*}{2\beta^*}$$

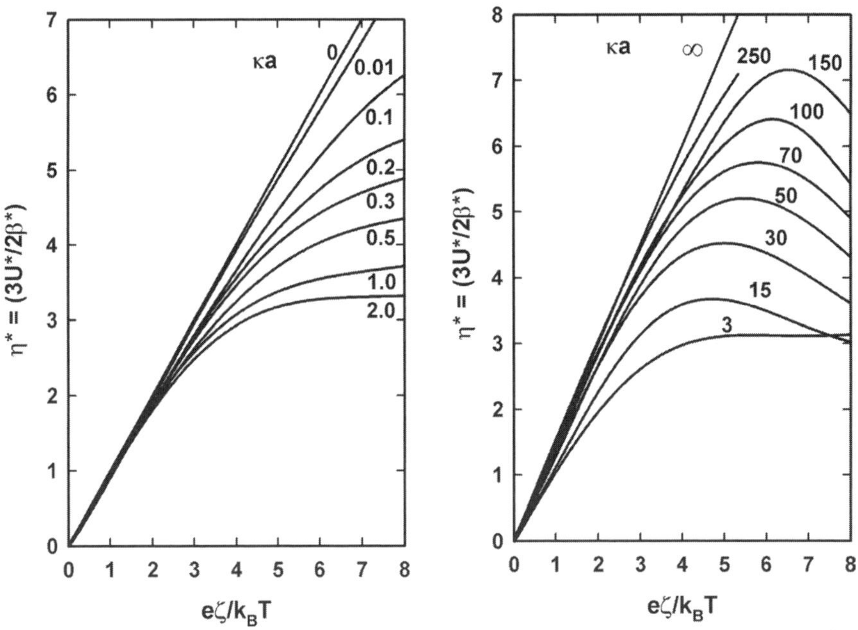

Figure 9.10. Dimensionless electrophoretic mobility of a spherical particle in KCl. Left: $\kappa a < 2$; Right: $\kappa a > 3$ (O'Brien and White, 1978).

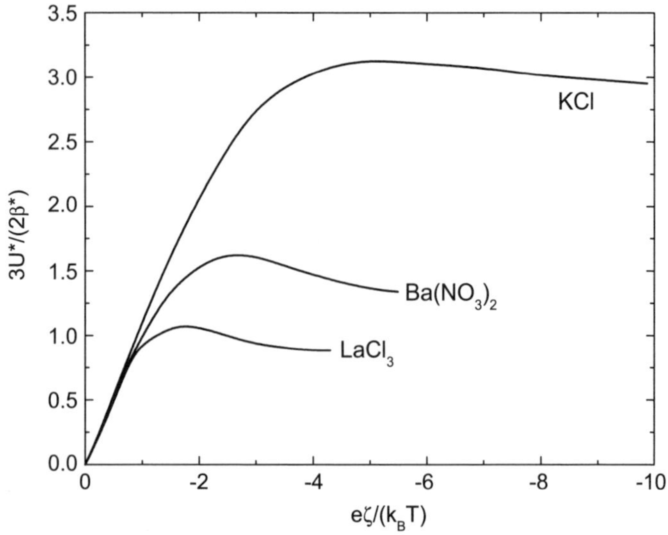

Figure 9.11. Effect of counterion valence on dimensionless electrophoretic mobility of a spherical particle at $\kappa a = 5$ (O'Brien and White, 1978).

It is observed from Figure 9.10 that for $e\zeta/k_B T = \zeta^* \leq 2$, for $\kappa a \rightarrow 0$, the slope of the dimensionless electrophoretic mobility, $3U^*/2\beta^*$, versus ζ^* is unity as would be expected for Eq. (9.104). For $\kappa a \rightarrow \infty$, the slope given by Figure 9.10 is $3/2$ in accordance with Eq. (9.105). Deviation from Henry's solution occurs for $\zeta^* \geq 2$ where the dimensionless electrophoretic mobility, $3U^*/2\beta^*$, versus ζ^* curves are no longer linear. For $0.01 < \kappa a < 2.75$, the increase in $3U^*/2\beta^*$ with ζ^* becomes more gradual and for $\kappa a \approx 2$, $3U^*/2\beta^*$ becomes nearly independent of ζ^*. In other words, the particle mobility becomes unaffected by its surface potential. For $\kappa a > 3$, the dimensionless electrophoretic mobility $3U^*/2\beta^*$ has a maximum value which becomes particularly pronounced at high κa values. This maximum value occurs at $\zeta^* \cong 5 - 7$, which corresponds to $\zeta \cong 0.125 - 0.175\ V$ for a (1 : 1) electrolyte solution. Dukhin and Semenikhin (1970), using a perturbation analysis, also showed the presence of maxima in the electrophoretic mobility arising from the convective relaxation effects.

Figure 9.11 shows effects due to the valence of the electrolyte. The electrolyte solutions correspond to (1 : 1): KCl, (2 : 1): $Ba(NO_3)_2$, and (3 : 1): $LaCl_3$. The maximum in the dimensionless mobility moves from $\zeta^* \cong -5$ for the (1 : 1) electrolyte to $\zeta^* \cong -2.5$ for the (2 : 1) electrolyte and to $\zeta^* \cong -1.7$ for the (3 : 1) electrolyte. These values are in the ratios of $1 : 1/2 : 1/3$, which are inversely proportional to the valencies.

From Figure 9.10, it is apparent that the maxima exhibited by the dimensionless electrophoretic mobility $3U^*/2\beta^*$ above $(e\zeta/k_B T) \sim 3$ introduces difficulty in the estimation of U^*. This is owing to the fact that same value of the mobility is obtained for two different zeta potentials. This is shown schematically in Figure 9.12 where two possible values for ζ^* exist.

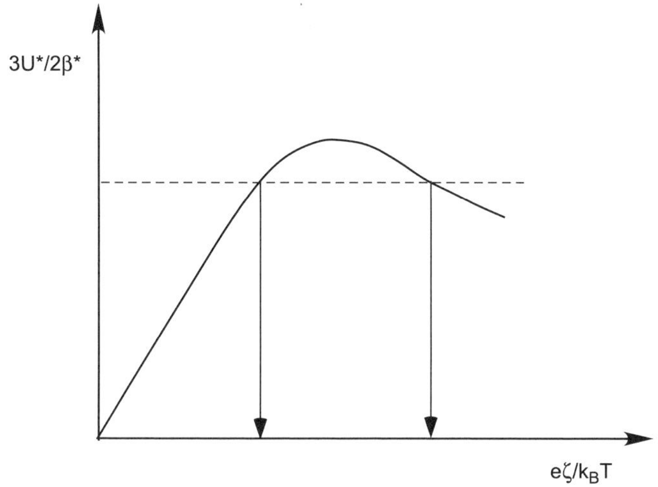

Figure 9.12. Evaluation of dimensionless electrophoretic mobility at large values of ζ potential.

The solution given by O'Brien and White (1978) takes into consideration the relaxation, the electrophoretic retardation, and high surface potential effects. However, this solution still applies for a single colloidal particle in an infinitely large electrolyte reservoir, in other words, in an infinitely dilute colloidal suspension. The results of O'Brien and White cannot be applied for concentrated colloidal suspensions, since the solution does not account for the presence of the neighboring particles. In the next section, we will discuss approaches for modelling electrophoretic mobility of a swarm of particles representing a colloidal suspension with a finite particle concentration *i.e.*, particle volume fraction.

9.4 ELECTROPHORETIC MOBILITY IN CONCENTRATED SUSPENSIONS

Under most practical circumstances, measurements of electrophoretic mobility, or conversely, electroosmotic flow of an electrolyte through a bed of stationary colloidal particles must consider the effects of numerous charged colloidal particles. A colloidal suspension with a finite concentration of particles will clearly modify the electrophoretic mobility of a particle from the mobilities computed in the limit of infinite dilution, as discussed in the previous section. This difference in mobilities is affected by two factors. First, the presence of neighboring particles affects the fluid velocity field in the immediate vicinity of each particle. Secondly, the proximity of the charged particles will modify the ion distribution, and hence, the structure of the electrical double layer surrounding each colloidal particle. It is therefore important to know how the particle concentrations affect the electrophoretic mobilities of the particles. Incidentally, the same mathematical construct used to obtain the electrophoretic mobility of a particle in a suspension can be employed to determine the electroosmotic flow of an electrolyte through the interstices of a porous medium consisting of stationary charged colloidal particles, since electrophoresis and electroosmosis are simply two different manifestations of the relative velocity between the particle and the electrolyte solution.

9.4.1 Cell Models for the Hydrodynamic Problem

It is clearly discernable that a colloidal suspension representing a swarm of particles is generally unstructured, implying that the distance between any two neighboring particles will be randomly distributed. Rigorous analysis of such unstructured suspensions are often only possible through statistical mechanics. However, in colloid science literature, a class of simplified models called the *cell models* has been frequently employed with considerable success. The cell model makes a simplified assumption regarding the suspension structure. Instead of considering the overall structure, or arrangement of the particles in the suspension, these models focus on a single particle and a representative volume of the fluid phase enclosing the particle. The representative fluid volume surrounding each particle is determined from the overall volume fraction (or conversely, the porosity) of the swarm of particles in the suspension. Assuming spherical colloidal particles of radius a, the volume fraction of particles in the suspension, ϕ_p, is related to the particle number concentration, N_0 (m^{-3}), as

$$\phi_p = \frac{4}{3}\pi a^3 N_0$$

The porosity of the suspension, *i.e.*, the volume fraction of the electrolyte solution in the suspension is given by

$$f = 1 - \phi_p$$

The cell model for a suspension of spherical colloidal particles of radius a assigns a spherical fluid shell surrounding the sphere, where the radius of the shell is related to the particle volume fraction (or porosity) as

$$\frac{a}{b} = \phi_p^{1/3} = (1 - f)^{1/3} \tag{9.109}$$

Here, b is referred to as the radius of the cell. Equation (9.109) simply assigns a radius b to the spherical cell such that the particle to cell volume ratio equals the particle volume fraction throughout the suspension. A typical cell model representation of a suspension of spherical colloidal particles is depicted schematically in Figure 9.13.

Once a colloidal suspension is modelled in terms of a cell, one needs to describe the governing equations for fluid flow, electric potential distribution, and ionic fluxes in the cell to solve an electrokinetic problem. While the basic governing equations remain mostly unchanged for a cell compared to their infinite dilution counterparts (see Section 9.3), the boundary conditions at the outer cell surface (at radius b) become

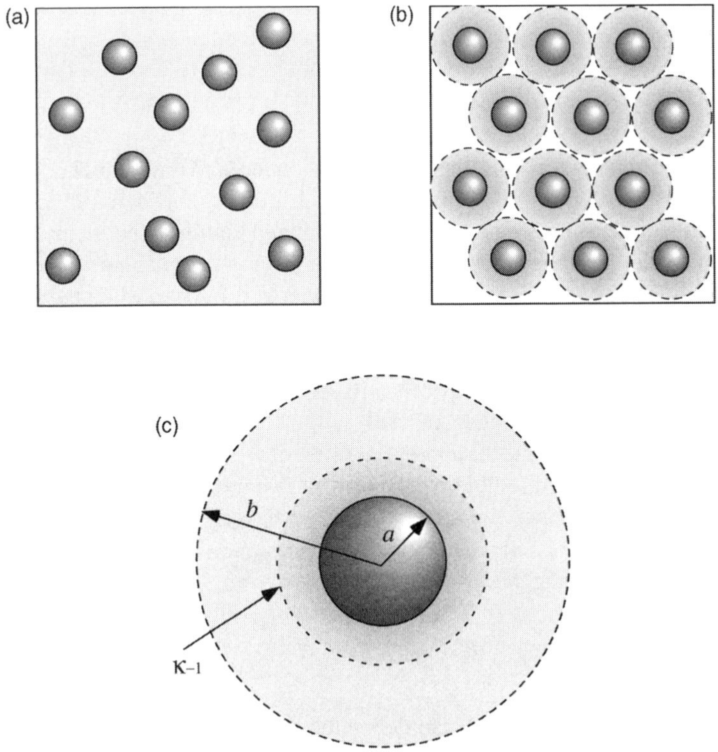

Figure 9.13. Schematic representation of cell model of a colloidal suspension. The colloidal dispersion (a) is represented as a cluster of cells (b). Each cell contains one particle surrounded by a fluid envelope (c). The radius of the envelope is determined from the particle volume fraction in the suspension.

the most important factors determining the suitability and accuracy of the model. Specification of this boundary condition in different manners can lead to different results. For the following discussion, we will consider a spherical coordinate system (r, θ, ϕ).

Let us first consider the fluid mechanics part of the problem. The Stokes equation, applicable for low Reynolds number flow, for the relative motion of the fluid with respect to the particle (assuming a particle fixed coordinate system) is

$$\mu \nabla^2 \mathbf{u} = \nabla p + \mathbf{f}_b \tag{9.110}$$

where $\mathbf{u}$ is the fluid velocity vector, $\mathbf{f}_b$ is the body force per unit volume, which may contain the gravitational force ($\rho \mathbf{g}$) and the electrical body force, $\rho_f \nabla \psi$. The Stokes equation needs to be solved along with the continuity equation $\nabla \cdot \mathbf{u} = 0$ to obtain the velocity distribution in the cell. Employing the vector identity

$$\nabla^2 \mathbf{u} = \nabla(\nabla \cdot \mathbf{u}) - \nabla \times \nabla \times \mathbf{u}$$

we can rewrite the Stokes equation as

$$\mu \nabla \times \nabla \times \mathbf{u} + \nabla p = -\mathbf{f}_b \tag{9.111}$$

Use of different body forces in the above equation leads to different classes of problems. For instance, when only the gravitational force is considered in Eq. (9.111), the cell model pertains to the problem of hindered settling (sedimentation) of an uncharged colloidal suspension. When the body force is solely comprised of the electrical force, the cell model pertains to electrophoresis. Finally, when the body force contains both gravity and electrical forces, the problem is that of sedimentation of charged particles in a gravitational field. Of course, the problems of electrophoresis and sedimentation of charged particles will require solution of additional equations for the electric potential and the ionic fluxes.

One now needs to provide the appropriate boundary conditions for the Stokes equation, (9.111), governing the fluid mechanical problem. Once again, we will assume that the particle is stationary and the fluid flow in the cell is $\mathbf{U} = -U\mathbf{i}_x$. Figure 9.14 depicts the relevant geometry. The fluid boundary conditions at the particle surface will be that of zero slip, given by

$$u_r(r, \theta) = u_\theta(r, \theta) = 0 \quad \text{at } r = a \tag{9.112}$$

At the cell boundary, $r = b$, one can define the radial velocity in terms of the mean relative velocity of the bulk fluid with respect to the particle, U, as

$$u_r(r, \theta) = \mathbf{n} \cdot \mathbf{u} = -U \cos\theta \quad \text{at } r = b \tag{9.113}$$

where $\mathbf{n}$ is the unit outward normal to the spherical cell surface ($\mathbf{n} = \mathbf{i}_r$).

The final boundary condition at the outer boundary is stated differently in different cell models. Happel (Happel, 1958; Happel and Brenner, 1965, 1983) suggested that a plausible condition at the outer boundary is the absence of any tangential shear stress, which can be written as

$$\tau_{r\theta} = \mu \left(r \frac{\partial (u_\theta / r)}{\partial r} + \frac{1}{r} \frac{\partial u_r}{\partial \theta} \right) = 0 \quad \text{at } r = b \tag{9.114}$$

When the governing fluid dynamic problem is solved with this set of boundary conditions, we refer to it as the Happel cell model.

An alternative approach of defining the boundary condition at $r = b$ was provided by Kuwabara (1959), who stated that the azimuthal vorticity be zero at the outer cell envelope:

$$\omega_\theta = \frac{1}{r} \frac{\partial (r u_\theta)}{\partial r} - \frac{1}{r} \frac{\partial u_r}{\partial \theta} = 0 \quad \text{at } r = b \tag{9.115}$$

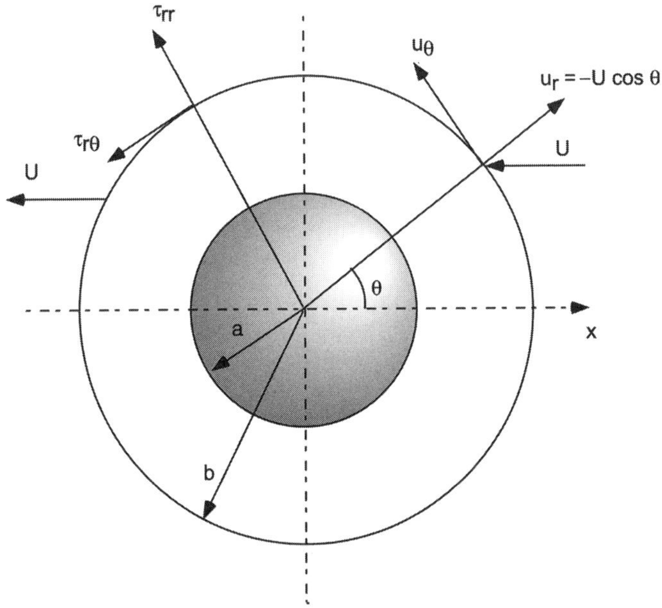

Figure 9.14. Geometrical considerations used for assigning boundary conditions for the cell model. In a particle fixed reference frame, the fluid velocity is U along the negative x direction at the outer cell boundary $r = b$. Happel cell model imposes an outer cell surface boundary condition employing the tangential shear stress, while Kuwabara cell model imposes a boundary condition based on the vorticity.

A general vectorial notation of Kuwabara's boundary condition is

$$\omega = \nabla \times \mathbf{u} = 0 \quad \text{at } r = b \tag{9.116}$$

When the fluid mechanics problem is solved with the Kuwabara boundary condition, the model is referred to as the Kuwabara cell model.

Using the solution of a given cell model, the drag on a particle within the cell is calculated, and its hindered settling velocity is then compared with Stokes free settling velocity. The term hindered settling velocity is used for the settling velocity of a suspension. The word "hindered" refers to the lowering of the settling velocity of a particle in a suspension due to the presence of neighboring particles.

When one solves the fluid mechanics problem for the hindered settling of a monodisperse uncharged colloidal suspension (that is, in absence of any electrical body force), Happel and Kuwabara cell models yield slightly different expressions for the velocity field and the hindered settling velocities. For instance, the ratio of the hindered settling velocity to the free settling velocity (Stokes settling velocity) obtained by solving the above cell model using the Happel boundary condition (stress free cell surface) yields

$$\frac{U_{HS}^{Happel}}{U_{St}} = \frac{1 - \frac{3}{2}\phi_p^{1/3} + \frac{3}{2}\phi_p^{5/3} - \phi_p^2}{1 + \frac{2}{3}\phi_p^{5/3}}$$

where U_{HS} and U_{St} refer to the hindered and Stokes free settling velocities, respectively. The corresponding expression for the settling velocity ratio obtained using the Kuwabara boundary condition (zero vorticity at the cell boundary) is

$$\frac{U_{HS}^{Kuwabara}}{U_{St}} = 1 - \frac{9}{5}\phi_p^{1/3} + \phi_p - \frac{1}{5}\phi_p^2$$

The results from these two cell models are depicted in Figure 9.15. One can note from this figure that the two results are usually within a few percent of each other over a wide range of volume fractions ($0 < \phi_p < 0.74$), and are in excellent agreement at very low ($\phi_p \to 0$) as well as very high ($\phi_p > 0.5$) volume fractions. This comparison provides a benchmark regarding the extent by which the predictions of the Kuwabara and Happel cell models can differ in terms of the fluid mechanical problem in absence of any electrical effects.

In the following, we discuss cell model based theories of electrophoretic mobility in concentrated particle suspensions.

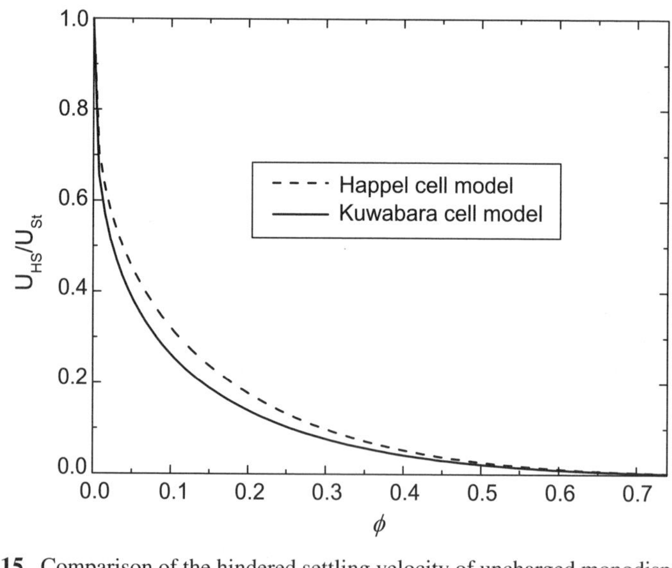

Figure 9.15. Comparison of the hindered settling velocity of uncharged monodisperse spherical particles at different volume fractions obtained employing the Happel and Kuwabara cell models.

9.4.2 The Levine–Neale Cell Model

It is reasonable to state that in a regime where the double-layer thickness is very small relative to the particle size, *i.e.*, $\kappa a \gg 1$, one would expect that the electrophoretic velocity of a given particle is not affected by the presence of other particles, especially at low particle concentrations. Here, the particle electrophoretic mobility U/E_∞ would not be concentration dependent. For the case of $\kappa a \gg 1$, one would expect Smoluchowski's equation to hold (Reed and Morrison 1976; Hunter 1981).

For the case of either very high particle concentrations or small κa, the double layers of adjacent particles overlap and one would not expect Henry's equation for a single particle to hold. Moreover, for $\kappa a \ll 1$, one would also expect Hückel's equation to be inappropriate even for fairly dilute suspensions.

In order to study the effect of particle concentration on the electrophoretic mobility of a particle, it is necessary to model the particle suspension accounting for the presence of neighboring particles. Among the early models is that of Möller *et al.* (1961). They developed a model which led to an expression for the variation of electrophoretic mobility of monodisperse spheres of uniform and constant surface charge as the double layers overlap. In this model, each particle is enclosed by a spherical shell of the electrolyte medium which contains just the right amount of excess ions to neutralize the total charge on the surface of the particle.

Möller *et al.* (1961) gave an expression for the electrophoretic mobility of a particle in a suspension of the form

$$\eta = \frac{U}{E_\infty} = \frac{\epsilon\zeta}{\mu} \cdot g_1(\kappa a, \phi_p)$$

where $g_1(\kappa a, \phi_p)$ is a function of κa and ϕ_p, μ is the fluid viscosity, and η is the electrophoretic mobility of a particle. The limits of the applicability of the above expression is given by Levine and Neale (1974b), who provide the maximum values of κa corresponding to a given particle volume fraction for which the mobility expression of Möller *et al.* (1961) is valid. This is shown in Table 9.2

Levine and Neale (1974a) extended Henry's solution and the model given by Möller *et al.* (1961) for the case of suspension of spherical particles employing both the Happel and Kuwabara cell models. Henry's theory (1931) for the electrophoresis of a single isolated sphere was used. The sole modification in this regard was that the hydrodynamic and electrical boundary conditions at $r \to \infty$ in Henry's approach was replaced by appropriate conditions at the outer shell boundary $r = b$ in the Levine and Neale approach. A brief outline of their cell model is given below.

TABLE 9.2. Range of Validity of the Electrophoretic Mobility Expression of Möller *et al.* (1961) According to Levine and Neale (1974b).

ϕ_p	0.0001	0.001	0.01	0.05	0.2	0.4	0.6
κa	0.5	0.7	1.2	2.3	5	8	15

As in Henry's approach, the total potential is assumed to be $(\phi + \psi)$, where ψ arises from the charged particle[3] and ϕ from the applied field. The following equations are solved:

$$\epsilon \nabla^2 \psi = -\rho_f \tag{9.117}$$

where $\psi = \psi(r)$, *i.e.*, ψ depends on radial distance only, and ρ_f is the free charge density. The boundary conditions are

$$\frac{d\psi}{dr} = -\frac{q_s}{\epsilon} \quad \text{at } r = a \tag{9.118}$$

and

$$\frac{d\psi}{dr} = 0 \quad \text{at } r = b \tag{9.119}$$

where q_s is the surface charge density. Boundary condition (9.118) relates the ψ gradient to the surface charge density q_s. As an equivalent to this boundary condition one can use $\psi = \zeta$ at $r = a$, see Eqs. (9.28) and (9.73) relating the zeta potential to the surface charge. The boundary condition (9.119) indicates that there is no current entering the particle; *i.e.*, the surface is non-conducting or that the cell is isolated from other cells.

The potential due to the external field $\phi(r, \theta)$ is governed by[4]

$$\nabla^2 \phi = 0 \tag{9.120}$$

with the boundary conditions

$$\frac{\partial \phi}{\partial r} = 0 \quad \text{at } r = a \tag{9.121}$$

and

$$\frac{\partial \phi}{\partial r} = -E_\infty \cos\theta \quad \text{at } r = b \tag{9.122}$$

Making use of the Debye–Hückel approximation, the solutions of the potentials are given by

$$\psi(r) = \left(\frac{q_s a}{\epsilon}\right)\left(\frac{a}{r}\right)\left[\frac{\sinh(\kappa b - \kappa r) - \kappa b \cosh(\kappa b - \kappa r)}{(1 - \kappa^2 ab)\sinh(\kappa b - \kappa a) - (\kappa b - \kappa a)\cosh(\kappa b - \kappa a)}\right] \tag{9.123}$$

[3]Once again, in this section, we denote the stationary electric double layer potential as ψ.

[4]Recall the conceptual inconsistency with this approach pointed out while presenting Henry's calculations. Once again, we simply reproduce the approach originally presented by Levine and Neale (1974a), noting that the application of these equations do not lead to a correct description of the electrical potential distribution around the particle. However, the final results for the electrophoretic mobility turn out to be accurate.

and

$$\phi(r,\theta) = -\frac{E_\infty}{(1-\alpha_p)}\left[r + \frac{a^3}{2r^2}\right]\cos\theta \tag{9.124}$$

As stated earlier, the fluid flow around the spherical particle is governed by the Stokes equation for creeping flows. In the present electrophoresis problem, the body force in Eq. (9.110) is given by $\rho_f \nabla(\psi + \phi)$, which yields

$$\mu\nabla^2\mathbf{u} = \nabla p + \rho_f \nabla(\phi + \psi) \tag{9.125}$$

with the continuity equation

$$\nabla \cdot \mathbf{u} = 0 \tag{9.126}$$

The boundary conditions are given by Eqs. (9.112), (9.113), and, depending on whether Happel or Kuwabara cell models are employed, either Eq. (9.114) or (9.115), respectively. The solution of Eqs. (9.125) and (9.126) subject to the boundary conditions Eqs. (9.112), (9.113), and (9.115) was given by Levine and Neale (1974a).

The electrophoretic mobility, η, was defined as

$$\eta = \frac{U}{E_\infty} = \left(\frac{\epsilon\zeta}{\mu}\right) g_2(\kappa a, \phi_p)$$

Making use of a force balance similar to Henry's approach (Section 9.3.2), the function $g_2(\kappa a, \phi_p)$, which accounts for the presence of neighboring particles, was evaluated employing the Kuwabara cell model (Levine and Neale, 1974a). The function g_2 was derived assuming $ze\zeta/k_B T < 1$ (Debye–Hückel limit). It was of the form

$$g_2(\kappa a, \phi_p) = \frac{2}{3(1-y^3)}[-I_1 + I_2] \tag{9.127}$$

where

$$I_1 = -\left[1 + R\kappa a + \frac{(\kappa a)^2}{16} - \left(\frac{5Q}{48} - \frac{Ry}{4} + \frac{Ry^3}{12}\right)(\kappa a)^3 - \frac{(\kappa a)^4}{96} - \left(\frac{Q}{96} - \frac{Ry}{48}\right)(\kappa a)^5 + \left(\frac{(\kappa a)^4}{8} - \frac{(\kappa a)^6}{96}\right)\right] I$$

and

$$I_2 = -\left[1 + \frac{3}{(\kappa a)^2} + \frac{3Q}{(\kappa a)} + \left(\frac{R}{y^3} - \frac{R}{10} + \frac{Q}{10}\right)(\kappa a) - \frac{(\kappa a)^2}{40} + \left(\frac{Q}{120} - \frac{Ry^3}{30}\right)(\kappa a)^3 - \frac{(\kappa a)^4}{240} + \left(\frac{Q}{240} - \frac{Ry}{120}\right)(\kappa a)^5 - \frac{(\kappa a)^6}{240} I\right]$$

In the above expressions, $y = \phi_p^{1/3} = a/b$,

$$\frac{1}{R} = \sinh(\kappa b - \kappa a) - \kappa b \cosh(\kappa b - \kappa a)$$

$$Q = \frac{1 - \kappa b \tanh(\kappa b - \kappa a)}{\tanh(\kappa b - \kappa a) - \kappa b}$$

and

$$I = \int_{\kappa a}^{\kappa b} \frac{A \cosh t - B \sinh t}{t} dt$$

The parameters A and B are expressed as

$$A = R[\sinh(\kappa b) - \kappa b \cosh(\kappa b)]$$
$$B = R[\cosh(\kappa b) - \kappa b \sinh(\kappa b)]$$

One should note that the evaluation of the integral I is non-trivial for large values of $\kappa a > 10$. In that case, one should express the integral I in terms of the exponential integrals. It is sometimes more convenient to use an alternative form of the Levine–Neale expression for the electrophoretic mobility, given by (Ohshima, 1997a)

$$\eta = -\frac{2}{3}\frac{\epsilon\zeta}{\mu(1-\phi_p)} \times \int_a^b \left\{1 - \left(\frac{a}{r}\right)^3 + \frac{3}{2}\left(\frac{a}{r}\right)^5 + \frac{\phi_p}{10}\left[1 - 10\left(\frac{r}{a}\right)^3 - 6\left(\frac{a}{r}\right)^5\right]\right\} \frac{1}{\zeta}\frac{d\psi^{(0)}}{dr} dr \tag{9.128}$$

where the function $\psi^{(0)}$ is given by

$$\psi^{(0)} = \zeta \frac{a}{r} \left(\frac{\sinh[\kappa(b - r)] - \kappa b \cosh[\kappa(b - r)]}{\sinh[\kappa(b - a)] - \kappa b \cosh[\kappa(b - a)]} \right) \tag{9.129}$$

The function $g_2(\kappa a, \phi_p)$ obtained using the Kuwabara cell model is plotted in Figure 9.16. For $\phi_p = 0$, g_2 becomes $(2/3)f(\kappa a)$, where $f(\kappa a)$ is Henry's function. The plot for $g_2(\kappa a, \phi_p)$ against κa for various particle volume fractions ϕ_p clearly shows that at $\kappa a > 10^2$, the volume fraction has little influence on the electrophoretic mobility, *i.e.*, $g_2(\kappa a, \phi_p) \cong 1.0$, and the Smoluchowski equation can be used. For $\kappa a \ll 1$, $g_2(\kappa a, \phi_p)$ becomes 2/3 only when $\phi_p = 0$. In this limit of small κa, the particle concentration has a marked effect on $g_2(\kappa a, \phi_p)$.

The Kuwabara cell model has been preferred over the Happel cell model following the work of Levine and Neale (1974a), who noted that the Happel cell model predicts that the electrophoretic velocity U slightly increases as the volume fraction

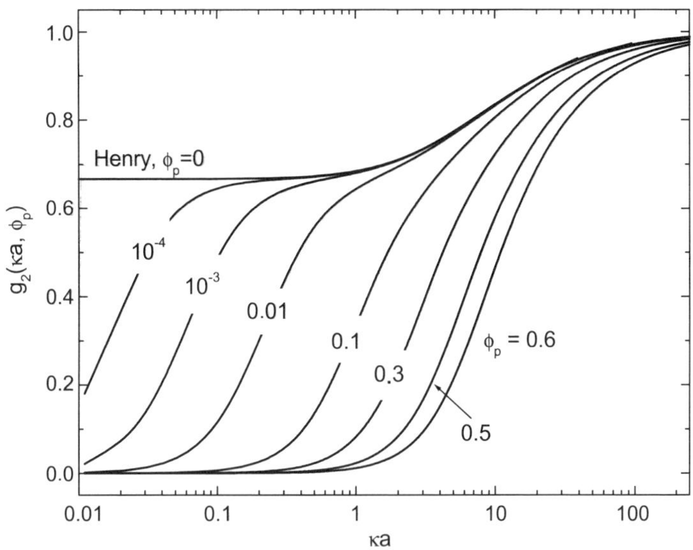

Figure 9.16. Variation of the function $g_2(\kappa a, \phi_p)$ with κa for different particle volume fractions.

of the particles increases. This behavior is considered counter intuitive and it is not observed experimentally. On the other hand, as noted from Figure 9.16, Kuwabara cell model predicts that the electrophoretic velocity corresponding to large κa values is independent of volume fraction. Consequently, the Kuwabara cell model is generally employed in the analysis of electrophoresis in concentrated suspensions.

For the case of high surface potentials, Kozak and Davis (1989a) used Kuwabara's model to include the relaxation effect as well. They used the perturbation approach of O'Brien and White (1978) to obtain the potential distribution around the spherical particle, albeit with the far field ($r \to \infty$) condition replaced by the appropriate boundary conditions at the cell surface $r = b$. Some of their results are shown in Figures 9.17 to 9.19 for large values of κa. Deviation from Smoluchowski's equation is shown on these plots. Their dimensionless mobility is defined such that the dimensionless mobility for the Smoluchowski equation is equal to $2(ze\zeta/k_BT)$ where $(ze\zeta/k_BT)$ is the dimensionless surface potential. In other words, the dimensionless mobility is defined by Kozak and Davis (1989a) as

$$M^* = \frac{2\mu U ze}{\epsilon E_\infty k_B T} = \frac{2U^*}{\beta^*}$$

The Smoluchowski equation in dimensionless form is then given by

$$M^* = 2\zeta^*$$

Figure 9.17 shows Kozak and Davis work for $\phi_p = 0$ (or porosity $f = 1$). For a single sphere, it compares well with O'Brien and White (1978) where high potential

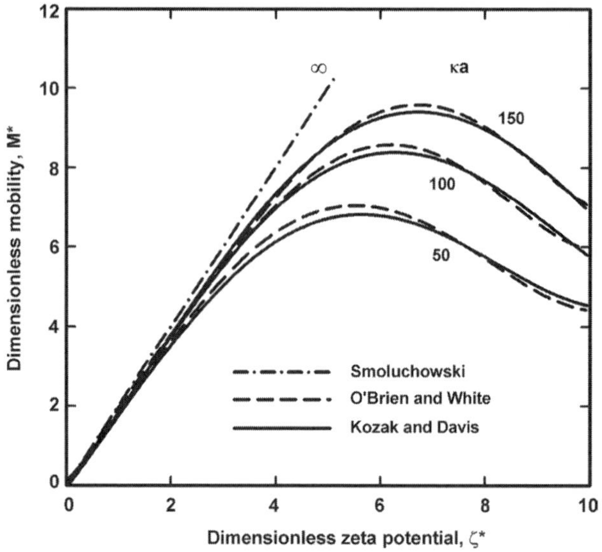

Figure 9.17. A comparison among the analyses of O'Brien and White (1978), Smoluchowski (1918), and Kozak and Davis (1989a,b) for single spheres in a $(z : z)$ electrolyte.

and relaxation effects are taken into account in the analysis. Figures 9.18 and 9.19 are given for $f = (1 - \phi_p)$ in the range of 0.5 to 1.0. The deviation from the Smoluchowski analysis becomes even greater at a higher surface potential. The results of Kozak and Davis (1989a,b) when adjusted by a factor of 3/4, reduce to the results provided by Carrique *et al.* (2001) for the electrophoretic mobility. Various asymptotic expansion expressions are given by Kozak and Davis (1989b).

It should be noted, however, that although the analysis was made using a higher potential, polarization usually occurs at a high potential leading to a variable dielectric constant ϵ. Moreover, the above models neglect the effect of a finite ion size. The above simplifications would lead to a greater deviation form Henry's curve. For models dealing with a variable dielectric constant and finite size of ions, the reader is referred to Gur *et al.* (1978a,b), Pintauro and Verbrugge (1989), and Datta and Kotamarthi (1990).

An experimental study on electrophoresis of porous aggregates was conducted by Miller and Berg (1993). They obtained reasonable agreement between the experimentally measured electrophoretic mobility of porous aggregates and the aforementioned theoretical analysis of Levine and Neale (1974a,b) and with the theoretical study of Miller *et al.* (1992). The theoretical analysis tends to underestimate the effect of the porosity of the aggregates.

Although the Levine and Neale cell model provides accurate estimates of the electrophoretic mobility of a concentrated suspension for low surface potentials, there seems to be an apparent controversy regarding one of the boundary conditions used in their approach. Referring back to the boundary condition Eq. (9.122), we note that the

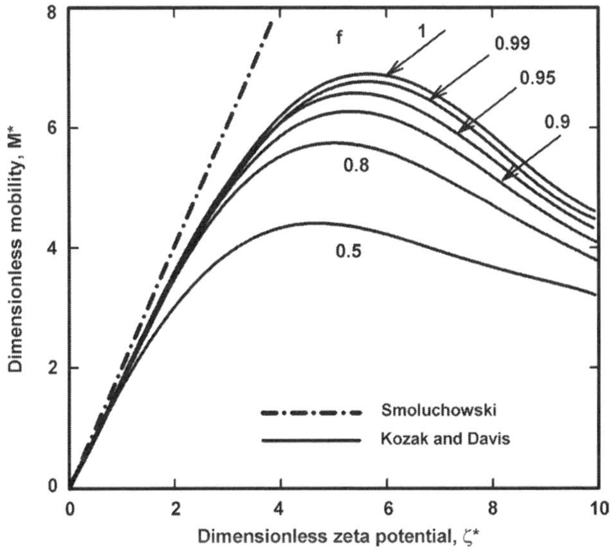

Figure 9.18. The effects of porosity, $f = 1 - \phi_p$, and zeta potential on the dimensionless mobility for $\kappa a = 50$ (Kozak and Davis, 1989b).

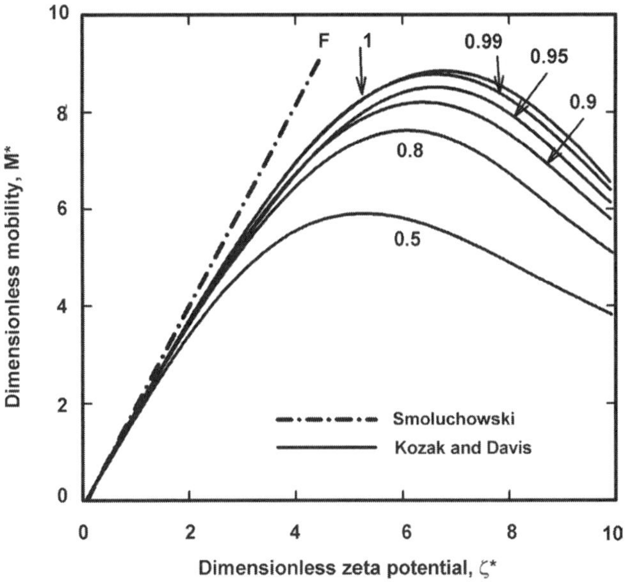

Figure 9.19. The effects of porosity and zeta potential on the dimensionless mobility for $\kappa a = 150$ (Kozak and Davis, 1989b).

electric field at the outer cell boundary ($r = b$) is given in terms of E_∞, which is the undisturbed external electric field. One may ask the question as to whether the external electric field exists at the cell boundary. In other words, does the electric field applied between two external electrodes bounding a colloidal dispersion remain unchanged at the boundary of a cell, which exists inside the swarm of particles representing the dispersion? We will revisit this problem regarding the cell surface boundary condition later in this chapter.

9.4.3 The Ohshima Cell Model

Significant contribution has been made in the field of electrophoresis by Ohshima (Ohshima, 1997a,b; Ohshima, 1999; Ohshima, 2000). He employed the Kuwabara cell model to develop the theory of electrophoresis in concentrated suspensions based on a perturbation approach. Here, we first summarize the general construct of Ohshima's model.

Considering a spherical coordinate system (r, θ, ϕ) with one of the particle centers as the origin, and the imposed external electric field aligned parallel to the polar axis ($\theta = 0$), the governing electrokinetic equations are written for the fluid flow, ionic fluxes, and the electrochemical potential. The system comprises of N mobile ion species with valence z_i and drag coefficient $\lambda_i = k_B T / D_i$. The bulk concentration of the ions in the electroneutral solution is denoted by $n_{i\infty}$. The governing electrokinetic equations underlying Ohshima's model are:

$$\mu \nabla \times \nabla \times \mathbf{u} + \nabla p = -\rho_f \nabla \psi \quad \text{Stokes equation} \tag{9.130}$$

$$\nabla \cdot \mathbf{u} = 0 \quad \text{fluid continuity} \tag{9.131}$$

$$\mathbf{v}_i = \mathbf{u} - \frac{1}{\lambda_i} \nabla \mu_i \quad \text{ionic velocity} \tag{9.132}$$

$$\nabla \cdot (n_i \mathbf{v}_i) = 0 \quad \text{conservation of ions} \tag{9.133}$$

$$\rho_f = \sum_{i=1}^{N} z_i e n_i \quad \text{charge density} \tag{9.134}$$

$$\mu_i = \mu_i^0 + z_i e \psi + k_B T \ln n_i \quad \text{electrochemical potential of ion} \tag{9.135}$$

and

$$\nabla^2 \psi = -\frac{\rho_f}{\epsilon} \quad \text{Poisson equation} \tag{9.136}$$

In the above equations, n_i represents the local number concentration of the ith ionic species, μ_i is the chemical potential of the ith ionic species, and μ_i^0 is the constant term in μ_i (the so-called reference chemical potential).

Assuming that the electrical double layer around the reference colloidal particle in the spherical cell is only slightly distorted due to the applied electric field, the ionic concentration, n_i, the electric potential, ψ, and the electrochemical potential, μ_i, can be represented as small perturbations from their equilibrium values:

$$n_i = n_i^{(0)}(r) + \delta n_i(r, \theta) \tag{9.137}$$

$$\psi = \psi^{(0)}(r) + \delta\psi(r, \theta) \tag{9.138}$$

$$\mu_i = \mu_i^{(0)}(r) + \delta\mu_i(r, \theta) \tag{9.139}$$

where the superscript (0) refers to the equilibrium state and δ denotes a small perturbation from equilibrium. Employing Eq. (9.135), the perturbation variables δn_i, $\delta\psi$, and $\delta\mu_i$ can be related as

$$\delta\mu_i = z_i e\delta\psi + k_B T \ln\left[1 + \frac{\delta n_i}{n_i^{(0)}}\right] \simeq z_i e\delta\psi + k_B T \frac{\delta n_i}{n_i^{(0)}} \tag{9.140}$$

where the final expression in Eq. (9.140) is obtained using a Taylor expansion of the logarithmic term, $\ln[1 + \delta n_i/n_i^{(0)}]$. Note that the perturbation approach is similar to the one described in Section 9.3.1. In this case, however, the total electrochemical potential is employed as an additional perturbation parameter. The substitution of the perturbation expressions, Eqs. (9.137)–(9.139) in the governing equations leads to the following equations:

$$\mu\nabla \times \nabla \times \nabla \times \mathbf{u} = \sum_{i=1}^{N} \nabla\delta\mu_i \times \nabla n_i^{(0)} \tag{9.141}$$

and

$$\nabla \cdot \left[n_i^{(0)}\mathbf{u} - n_i^{(0)}\frac{\nabla\delta\mu_i}{\lambda_i}\right] = 0 \tag{9.142}$$

Furthermore, assuming that the concentration of the ions at equilibrium obeys Boltzmann distribution, the distribution of $\psi^{(0)}$ is obtained from the Poisson–Boltzmann equation

$$\frac{1}{r^2}\frac{d}{dr}\left(r^2\frac{d\psi^{(0)}}{dr}\right) = -\frac{1}{\epsilon}\sum_{i=1}^{N} z_i e n_{i\infty} \exp\left(-\frac{z_i e\psi^{(0)}}{k_B T}\right) \tag{9.143}$$

Let us now define the boundary conditions for Eqs. (9.141) to (9.143). The boundary conditions at the particle surface are same as described in Section 9.3.1. These are zero slip, no ion penetration into the particle, and a known electric potential at the shear plane (or particle surface). The no-slip and no ion penetration boundary conditions are given by Eqs. (9.12) and (9.13) respectively. At the outer cell boundary, $r = b$, the boundary conditions for the Stokes equation are given by Eqs. (9.113) and (9.116) following the Kuwabara cell model. The boundary condition for the electrostatic problem, Eq. (9.143) at the outer cell surface $r = b$ was given by Ohshima (1997a) as

$$\left.\frac{d\psi^{(0)}}{dr}\right| = 0 \tag{9.144}$$

This condition is identical to the cell surface boundary condition employed by Levine and Neale (1974a), and simply states that the cell is electroneutral for the unperturbed (no external field) case. The stationary problem, Eq. (9.143), can be solved subject to the boundary conditions of a constant surface potential at $r = a$ and Eq. (9.144) at $r = b$.

It should be noted that although the electrophoretic velocity U appears in the boundary conditions, it is actually unknown. Hence, one needs to specify another boundary condition for the perturbation problem. This equation follows from the assumption that the net force on a cell must be zero. In other words, for the electrophoresis problem, the total electrical force applied to the dispersed system is zero, and accordingly, the hydrodynamic force acting on the cell must be zero. This condition is given by

$$\mathbf{F}_H = \int_S \bar{\bar{\tau}} \cdot \mathbf{n}\, dS = 0 \tag{9.145}$$

where the integral is evaluated over the surface S representing the outer cell boundary $r = b$, and $\bar{\bar{\tau}}$ is the hydrodynamic stress tensor with the following components:

$$\tau_{rr} = -p + 2\mu \frac{\partial u_r}{\partial r}$$

$$\tau_{r\theta} = \mu \left[\frac{1}{r}\frac{\partial u_r}{\partial \theta} + \frac{\partial u_\theta}{\partial r} - \frac{u_\theta}{r} \right]$$

Note that pressure is specified in this formulation through the boundary condition, Eq. (9.145). The electrophoretic velocity U is implicitly embedded in the hydrodynamic stress.

The final boundary condition at $r = b$ involves specifying the electrochemical potential perturbation. Using the specification of the boundary condition originally given by Levine and Neale (1974a) for the potential gradient,

$$\nabla\delta\psi \cdot \mathbf{n} = -\mathbf{E}_\infty \cdot \mathbf{n} = -E_\infty \cos\theta \quad \text{at } r = b \tag{9.146}$$

as well as, using

$$\nabla\delta n_i \cdot \mathbf{n} = 0 \quad \text{at } r = b$$

the electrochemical potential perturbation at the cell boundary can be written as

$$\nabla\delta\mu_i \cdot \mathbf{n} = -z_i e \mathbf{E}_\infty \cdot \mathbf{n} = -z_i e E_\infty \cos\theta \quad \text{at } r = b \tag{9.147}$$

Note that Eqs. (9.146) and (9.147) are alternate but equivalent statements of the outer cell surface electrochemical boundary condition, with the only difference that the latter equation uses the electrochemical potential instead of the electric potential. In subsequent discussions, we will refer to either of these boundary conditions as the Levine–Neale boundary condition.

The perturbation analysis of Ohshima (1997a) involves solving the governing perturbation equations, Eqs. (9.141) and (9.142), using the above boundary conditions. The solution of the perturbation problem yields the electrophoretic mobility (Ohshima, 1997a)

$$\eta(\kappa a, \phi_p) = \eta = \frac{U}{E_\infty} = \frac{2\epsilon\zeta}{3\mu}\left[\int_a^b P_1(r)\left(1 + \frac{1}{2}\frac{a^3}{r^3}\right)dr + Q_1(b)\right] \tag{9.148}$$

where

$$P_1(r) = -\frac{(\kappa a)^2}{6(1-\phi_p)}\left[1 - \frac{3r^2}{a^2} + \frac{2r^3}{a^3} - \frac{a^3}{b^3}\left(\frac{2}{5} - \frac{r^3}{a^3} + \frac{3r^5}{5a^5}\right)\right]\frac{1}{\zeta}\frac{d\psi^{(0)}}{dr}$$

and

$$Q_1(b) = \frac{(\kappa a)^2}{3\zeta(1-\phi_p)}\psi^{(0)}(b)\left(1 + \frac{a^3}{2b^3}\right)\left(1 + \frac{b^3}{a^3} - \frac{9b^2}{5a^2} - \frac{a^3}{5b^3}\right)$$

Here the unperturbed electrical potential based on the solution of the Poisson–Boltzmann equation (9.143) is

$$\psi^{(0)}(r) = \zeta\frac{a}{r}\frac{\kappa b\cosh[\kappa(b-r)] - \sinh[\kappa(b-r)]}{\kappa b\cosh[\kappa(b-a)] - \sinh[\kappa(b-a)]}$$

Equation (9.148) requires numerical evaluation of the mobility. It is similar to the Levine and Neale (1974a) electrophoretic mobility expression. Equation (9.148) is valid for low ζ-potentials, which allows the linearization of the Poisson–Boltzmann equation required to arrive at the given expression. Although both solutions are for constant surface potential on the particles, the Levine–Neale solution uses the surface charge density boundary condition on the sphere, while the Ohshima approach directly uses the surface potential condition on the particle. This leads to slightly different forms for the electrical potential expression in the two solutions.

For low surface potentials and weak overlap of the electrical double layers, Ohshima (1997a) derived an approximate analytical expression for the electrophoretic mobility from Eq. (9.148), which is in excellent agreement with the Levine and Neale (1974a) result, (Ohshima, 1997a). The expression for the mobility is (Ohshima, 1997a)

$$\eta = \frac{\epsilon\zeta}{\mu}G(\kappa a, \phi_p) \tag{9.149}$$

where

$$G(\kappa a, \phi_p) = \left[\frac{2}{3}\left\{1 + \frac{1}{2[1+\delta/(\kappa a)]^3}\right\} M_1(\kappa a, \phi_p) + M_2(\kappa a, \phi_p)\right] \tag{9.150}$$

In the above equation,

$$M_1 = 1 - \frac{3\phi_p(1+\kappa a Q)}{(\kappa a)^2(1-\phi_p)} + \frac{(\kappa a)^2}{5P\phi_p^{2/3}}\left(1 + \frac{1-\phi_p}{3} - \frac{3}{1-\phi_p} + \frac{3\phi_p^{1/3}}{1-\phi_p}\right)$$

$$M_2 = \frac{2(\kappa a)^2(1+\phi_p/2)}{9P(1-\phi_p)}\left(\phi_p^{1/3} + \frac{1}{\phi_p^{2/3}} - \frac{9}{5\phi_p^{1/3}} - \frac{\phi_p^{4/3}}{5}\right)$$

$$\begin{aligned} P &= \cosh[\kappa(b-a)] - (\kappa b)^{-1}\sinh[\kappa(b-a)] \\ &= \cosh\left[\kappa a(\phi_p^{-1/3}-1)\right] - \frac{\phi_p^{1/3}}{(\kappa a)}\sinh\left[\kappa a(\phi_p^{-1/3}-1)\right] \end{aligned}$$

$$Q = \frac{1-\kappa b\tanh[\kappa(b-a)]}{\tanh[\kappa(b-a)]-\kappa b} = \frac{1-\kappa a\phi_p^{-1/3}\tanh\left[\kappa a(\phi_p^{-1/3}-1)\right]}{\tanh\left[\kappa a(\phi_p^{-1/3}-1)\right]-\kappa a\phi_p^{-1/3}}$$

and

$$\delta = \frac{2.5}{1+2\exp(-\kappa a)}$$

The interesting feature of this analytical expression is that although it was derived assuming weak overlap of double layers, it is in excellent agreement with the Levine and Neale (1974a) solution even for $\kappa a < 1$, or substantial double layer overlap. Figure 9.20 shows a comparison of the dimensionless electrophoretic mobility, $\eta\mu/\epsilon\zeta$, where $\epsilon\zeta/\mu$ is the Helmholtz–Smoluchowski electrophoretic mobility, obtained using the expression of Levine and Neale (1974a) and Eq. (9.150) for selected values of the particle volume fraction, ϕ_p.

9.4.4 Suspension Electric Conductivity

Ohshima also provided an expression for the suspension electrical conductivity in terms of the particle volume fraction (Ohshima, 1999). The suspension conductivity, σ^{susp}, is a function of the particle volume fraction, and differs from the bulk electrolyte conductivity. The bulk electrolyte conductivity is given by

$$\sigma^\infty = \sum_{i=1}^{N} z_i^2 e^2 n_{i\infty}/\lambda_i$$

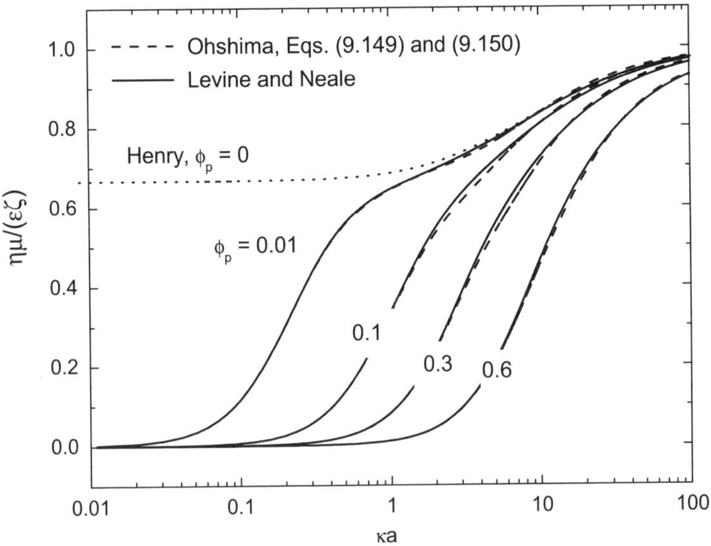

Figure 9.20. Variation of the dimensionless electrophoretic mobility with κa obtained using Ohshima's approximate expression, Eq. (9.150) compared with the results of Levine and Neale (1974a).

where the Stokes–Einstein friction factor, λ_i, is defined as

$$\lambda_i = \frac{k_B T}{D_i}$$

Note that we denoted the Stokes–Einstein friction factor as f_i in Chapter 6, *cf.*, Eq. (6.39).

Ohshima's expression for the suspension conductivity applies for low surface potentials and non-overlapping double layers. It is given by

$$\sigma^{susp} = \sigma^{\infty} \frac{1 - \phi_p}{1 + \phi_p/2} \left(1 - 3\phi_p \left(\frac{e\zeta}{k_B T} \right) L(\kappa a, \phi_p) \frac{\sum_{i=1}^{N} z_i^3 n_{i\infty}/\lambda_i}{\sum_{i=1}^{N} z_i^2 n_{i\infty}/\lambda_i} \right) \tag{9.151}$$

where

$$L(\kappa a, \phi_p) = -\frac{1}{3a^3\zeta(1 - \phi_p)(1 + \phi_p/2)} \int_a^b \left(\frac{a^3}{2} + r^3 \right) \left(1 - \frac{a^3}{r^3} \right) \frac{d\psi^{(0)}}{dr} dr$$

An approximate expression for $L(\kappa a, \phi_p)$ was also provided by Ohshima (1999), and it is given as

$$L(\kappa a, \phi_p) = \frac{1 + \kappa a Q}{(\kappa a)^2 (1 - \phi_p)(1 + \phi_p/2)} \left[1 + \frac{1}{2[1 + \delta/(\kappa a)]^3} \right]$$

Here Q and δ are the same functions used in Eq. (9.150). The limiting form of Eq. (9.151) for uncharged particles ($\zeta = 0$), as well as, for infinitesimally thin double layers, $\kappa a \to \infty$, is

$$\frac{\sigma^{susp}}{\sigma^{\infty}} = \frac{1-\phi_p}{1+\phi_p/2} \tag{9.152}$$

Further simplification of Eq. (9.152) yields for very dilute suspensions, $\phi_p \to 0$ (Russel *et al.*, 1989),

$$\frac{\sigma^{susp}}{\sigma^{\infty}} = \left[1 - \frac{3}{2}\phi_p\right] \tag{9.153}$$

9.4.5 The Shilov–Zharkikh Cell Model

A large body of studies on electrophoresis is based on the Shilov–Zharkikh cell model (Shilov *et al.*, 1981; Carrique *et al.*, 2001). The basic construct of the model is identical to the perturbation approach delineated so far in this section. A major point of departure in the Shilov–Zharkikh model is the application of the boundary condition at the outer cell surface ($r = b$), which is given by:

$$\delta\psi = -b\mathbf{E}_{susp}\cdot\mathbf{n} = -bE_{susp}\cos\theta \quad \text{at } r = b \tag{9.154}$$

The boundary condition of Eq. (9.154) is different from the outer cell surface boundary conditions used in the Levine–Neale cell model and the Ohshima cell model [given by Eqs. (9.122) and (9.146), respectively]. The Levine–Neale and Ohshima boundary conditions are identical, and in the following discussion, we will refer to both these conditions as the Levine–Neale boundary condition. There are two differences between the Shilov–Zharkikh boundary condition, Eq. (9.154), and the Levine–Neale boundary condition. First, the notation $\mathbf{E}_{susp}$ in Eq. (9.154) refers to a different electric field compared to the Levine–Neale boundary condition, which uses the external field, $\mathbf{E}_{\infty}$ as in Eq. (9.146). Here, we will refer to $\mathbf{E}_{susp}$ as the internal electric field or the electric field within the suspension. Secondly, Eq. (9.154) specifies the perturbed potential, $\delta\psi$ at the outer cell boundary, whereas the Levine–Neale boundary condition specifies the gradient of the perturbed potential, $\nabla\delta\psi$, as in Eq. (9.146).

Let us first address the difference between the internal and external electric fields. Figure 9.21 depicts a suspension between two external reservoirs of electrolyte solution. When a potential difference between the two external electrodes is set up, a current will flow between these electrodes. For a steady state process, the total current at every cross section in the volume between the two electrodes will be the same. The current densities (A/m^2) at every cross section will also be the same, since the cross sectional area is the same everywhere. Denoting the spatial average current density vector inside the suspension as $\langle\mathbf{i}\rangle$ and the current density in the external electrolyte solution as $\mathbf{i}_{\infty}$, we can therefore write

$$\langle\mathbf{i}\rangle = \mathbf{i}_{\infty} \tag{9.155}$$

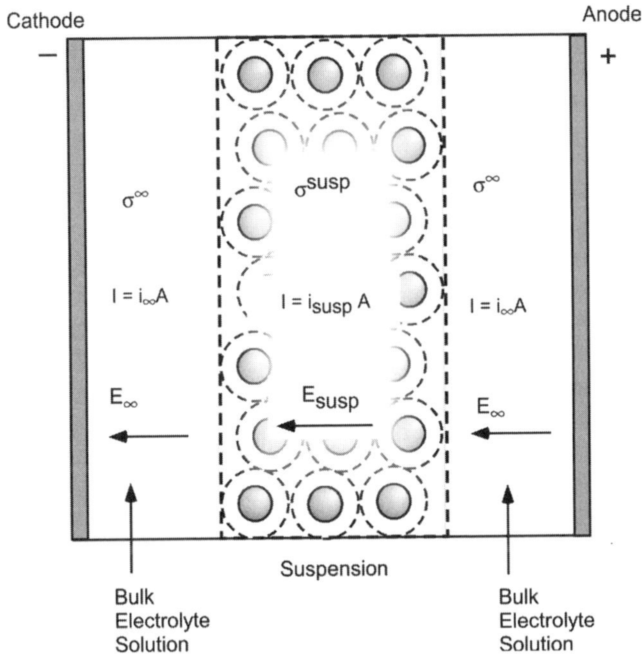

Figure 9.21. Schematic representation of electrophoresis in a concentrated suspension showing the internal and external electric field and conductivities. The total current I across any cross section, A, is constant, which implies that the current densities in the external electrolyte solutions and the suspension are the same.

Considering Figure 9.21, there is a slight difference in how the current flows through the suspension and through the external electrolyte. The current in the suspension is only due to the flowing electrolyte (the particles are non-conducting) and hence, in the suspension, the current passes only through the interstitial volume between the solid particles. This implies that the specific conductivity in the suspension, σ^{susp}, is different from the specific conductivity of the external electrolyte solution, σ^{∞}. One can apply Ohm's law [see Eq. (6.72)] locally in the suspension and the external electrolyte reservoirs to relate the current density and the electric field. For the suspension, the relationship is

$$\langle \mathbf{i} \rangle = \sigma^{susp} \mathbf{E}_{susp} \tag{9.156}$$

and for the external electrolyte, the corresponding relationship is

$$\mathbf{i}_{\infty} = \sigma^{\infty} \mathbf{E}_{\infty} \tag{9.157}$$

In Eqs. (9.156) and (9.157), $\mathbf{E}_{susp}$ and $\mathbf{E}_{\infty}$ refer to the local electric field in the suspension and the external electric field, respectively. The local electric field in the suspension can be formally defined as the volume average of the potential gradient within the suspension

$$\mathbf{E}_{susp} = -\frac{1}{V_{susp}} \int_{V_{susp}} \nabla\psi \, dV_{susp} = -\langle \nabla\psi \rangle \tag{9.158}$$

where V_{susp} is the volume of the suspension (which takes into consideration both the volume occupied by the particles as well as the volume of the electrolyte solution in the interstices). It is of interest to note that Eq. (9.158) is also used to define the sedimentation potential.

The relationships between the current and the field given above are conceptually identical to what would be expected in an electrical circuit with resistances in series: The total current through each resistor is the same, although the potential difference across each resistor is different.

Combining Eqs. (9.155), (9.156), and (9.157), one obtains

$$\sigma^{susp}\mathbf{E}_{susp} = \sigma^{\infty}\,\mathbf{E}_{\infty} \tag{9.159}$$

Denoting the unit vector along the direction normal to the electrode planes in Figure 9.21 as $\mathbf{i}_x$, one can obtain the magnitudes of the electric fields acting along the direction of $\mathbf{i}_x$ as

$$E_{susp} = \mathbf{E}_{susp} \cdot \mathbf{i}_x \quad \text{and} \quad E_{\infty} = \mathbf{E}_{\infty} \cdot \mathbf{i}_x$$

Using the above components of the field acting along the direction $\mathbf{i}_x$ in Eq. (9.159), one can write

$$\sigma^{susp}\,E_{susp} = \sigma^{\infty}\,E_{\infty}$$

or

$$\frac{E_{susp}}{E_{\infty}} = \frac{\sigma^{\infty}}{\sigma^{susp}} \tag{9.160}$$

Referring to Eq. (9.151), which provides the relationship between the suspension conductivity and the external conductivity, one immediately observes that since the two conductivities are different, Eq. (9.160) would lead to a difference between the suspension and the external electric fields. Consequently, it is more appropriate to use the internal or suspension electric field, E_{susp}, to define the cell surface boundary condition. One should note, however, that as the suspension becomes dilute, $\phi_p \to 0$, the internal and external electric fields tend to become identical, $E_{susp} \to E_{\infty}$.

We now turn our attention to the second difference between the Shilov–Zharkikh and Levine–Neale outer cell boundary conditions, namely, the use of the perturbed potential instead of the gradient of the perturbed potential at the outer cell boundary. Equation (9.154) was originally proposed by Shilov *et al.* (1981) based on an extensive analysis of the electrical properties of the cell. A simpler heuristic approach for rationalizing the use of the potential instead of the potential gradient in Eq. (9.154) is suggested here based on the approach of Kozak and Davis (1986). Consider the potential distribution obtained by Henry (1931), Eq. (9.63),

$$\phi = -E_\infty \left[r + \frac{a^3}{2r^2} \right] \cos\theta$$

Henry's solution was obtained in the infinite dilution limit, $\phi_p \to 0$, where $E_{susp} = E_\infty$. Using the above equation to determine the potential at the outer cell boundary, $r = b$, we obtain,

$$\phi = -E_{susp} \left[b + \frac{a^3}{2b^2} \right] \cos\theta$$

or

$$\phi = -b\, E_{susp} \left[1 + \frac{1}{2} \left(\frac{a}{b} \right)^3 \right] \cos\theta \tag{9.161}$$

Note that in applying Eq. (9.63) at the outer cell boundary, we have used the internal electric field in the suspension, E_{susp}. Equation (9.161), in the limiting case of $a/b \ll 1$, reduces to

$$\phi = -bE_{susp} \cos\theta \tag{9.162}$$

The similarity between Eqs. (9.154) and (9.162) suggests that the use of the perturbed potential instead of the gradient of the perturbed potential leads to an appropriate representation of the boundary condition at $r = b$.

Use of the Shilov–Zharkikh boundary condition in the cell model instead of the Levine–Neale boundary condition leads to an apparently different result for the electrophoretic mobility of a particle in a concentrated dispersion. Carrique *et al.* (2001) solved the cell model for electrophoresis employing both Levine–Neale and Shilov–Zharkikh boundary conditions. The expression for the electrophoretic mobility obtained by Carrique *et al.* (2001) employing the Shilov–Zharkikh boundary condition was

$$\eta_{susp}(\kappa a, \phi_p) \equiv \eta_{susp} = \frac{U}{E_{susp}} = \frac{2\epsilon\zeta}{3\mu} \left[\int_a^b H(r) \left(1 + \frac{1}{2} \frac{a^3}{r^3} \right) dr + F(r) \right] \tag{9.163}$$

where

$$H(r) = -\frac{(\kappa a)^2}{6(1 + \phi_p/2)} \left[1 - \frac{3r^2}{a^2} + \frac{2r^3}{a^3} - \frac{a^3}{b^3} \left(\frac{2}{5} - \frac{r^3}{a^3} + \frac{3r^5}{5a^5} \right) \right] \frac{1}{\zeta} \frac{d\psi^{(0)}}{dr}$$

and

$$F(r) = \frac{(\kappa a)^2}{3\zeta} \psi^{(0)}(b) \left(1 + \frac{b^3}{a^3} - \frac{9b^2}{5a^2} - \frac{a^3}{5b^3} \right)$$

Note that in Eq. (9.163), the electrophoretic mobility of a particle in the suspension, η_{susp}, is defined as U/E_{susp} based on the internal electric field, E_{susp}. The definition used in Eq. (9.163) is different from the definition of the mobility employed earlier in this chapter, for instance in Eq. (9.148), which is based on the external electric field, $\eta = U/E_\infty$. The expression for the mobility, η_{susp}, obtained by Carrique *et al.* (2001), Eq. (9.163), differs from the mobility, η, obtained by Ohshima (1997a), Eq. (9.148), owing to a different dependence on the solute volume fraction. To emphasize this difference, we denote the mobility in Eq. (9.163) as η_{susp}.

There is considerable confusion in the electrophoresis literature, particularly among the studies pertaining to the calculation of electrophoretic mobility of concentrated suspensions, owing to somewhat misleading use of the two alternate definitions of the electric field, and hence, due to the use of the Shilov–Zharkikh and Levine–Neale boundary conditions. We note that the electric field used in the Levine–Neale boundary condition (also used in Ohshima's model) was the external field, E_∞. However, it should have been the internal electric field $\mathbf{E}_{susp}$. The implication of these two electric fields on the evaluated electrophoretic mobility will become clear once we consider the two commonly adopted definitions for the mobility. Smoluchowski's original definition of electrophoretic mobility is given by

$$\eta = \frac{U}{E_\infty} = \frac{\mathbf{U} \cdot \mathbf{i}_x}{\mathbf{E}_\infty \cdot \mathbf{i}_x} \tag{9.164}$$

which is based on the external electric field, $\mathbf{E}_\infty$. Another definition of the electrophoretic mobility is possible, on the basis of the internal electric field, $\mathbf{E}_{susp}$ (Dukhin *et al.*, 1999)

$$\eta_{susp} = \frac{U}{E_{susp}} = \frac{\mathbf{U} \cdot \mathbf{i}_x}{\mathbf{E}_{susp} \cdot \mathbf{i}_x} \tag{9.165}$$

This mobility can be termed as the internal or suspension mobility. The two definitions of the electrophoretic mobility, given by Eqs. (9.164) and (9.165) are perfectly legitimate and are equivalent, as demonstrated by Dukhin and Shilov (1974), owing to Eq. (9.160). Combining Eqs. (9.160), (9.164), and (9.165), one obtains

$$\frac{\eta_{susp}}{\eta} = \frac{E_\infty}{E_{susp}} = \frac{\sigma^{susp}}{\sigma^\infty} \tag{9.166}$$

In the limit of an infinitely dilute suspension, $\phi_p \to 0$, the two conductivities become identical, and accordingly, the external and internal electric fields and mobilities will also become identical.

Carrique *et al.* (2001) noted that the external electrophoretic mobility, η, obtained by solving the cell model with the Levine–Neale boundary condition approaches the Smoluchowski result, $\eta = \epsilon\zeta/\mu$, as $\kappa a \to \infty$ for all particle volume fractions, ϕ_p. In contrast, the internal (or suspension) mobility, η_{susp}, based on the Shilov–Zharkikh cell model, Eq. (9.163), has significantly different values for different volume fractions as $\kappa a \to \infty$. However, if one converts the internal mobility to the external mobility using the relation

$$\eta = \eta_{susp} \frac{\sigma^\infty}{\sigma^{susp}} \tag{9.167}$$

which follows from Eq. (9.166), the Shilov–Zharkikh cell model provides values of the external mobility, which are identical to the Levine–Neale mobility. The conductivity ratio in Eq. (9.167) can be obtained from Eq. (9.151).

Figure 9.22 depicts the difference between the internal electrophoretic mobility, U/E_{susp}, obtained using Carrique's expression, Eq. (9.163), and the external mobility, U/E_{∞}, obtained using the Levine–Neale boundary condition, Eq. (9.148). Notably, converting the suspension mobility to the external mobility using Eq. (9.167) leads to identical results as given by Eq. (9.148). It is therefore evident that employing the Shilov–Zharkikh boundary condition we obtain the suspension mobility, which strongly varies with particle volume fraction for large κa. However, when we determine the corresponding external mobility, the results based on the Shilov–Zharkikh cell model become identical to those obtained using the Levine–Neale cell model. Both cell models predict that the external mobility approaches the Smoluchowski limit as $\kappa a \to \infty$. Similar consequences arising from the use of the Shilov–Zharkikh and the Levine–Neale electrophoretic mobilities have also been discussed by Ding and Keh (2001).

Disregarding the implications of the two definitions of electrophoretic mobility has sometimes provided misleading conclusions regarding the validity of the Shilov–Zharkikh cell model. For instance, Kozak and Davis (1986) criticized the

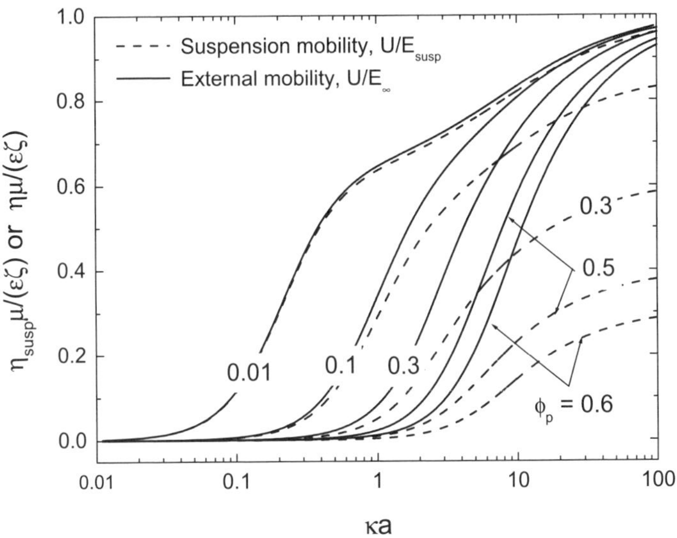

Figure 9.22. Dimensionless suspension electrophoretic mobility plots obtained by Carrique *et al.* (2001) based on Shilov–Zharkikh cell model (dashed lines) at different particle volume fractions. When one converts these internal mobilities to external mobilities using Eq. (9.167), the solid lines are obtained. These solid lines are identical to the results obtained using the Levine–Neale cell model, see Figure 9.20. All the solid lines, obtained for different particle volume fractions, ϕ_p, approach the Helmholtz–Smoluchowski electrophoretic mobility as $\kappa a \to \infty$.

Shilov–Zharkikh model after noting that the calculated mobility based on the Shilov–Zharkikh boundary condition did not approach the Smoluchowski limit at large κa. They arrive at this conclusion due to the fact that the correct definition of the mobility was not employed while assessing the validity of the Shilov–Zharkikh cell model. However, from the foregoing results, it is clear that the Shilov–Zharkikh and Levine–Neale cell models provide identical results for the external electrophoretic mobility, η, and both models appear to be correct.

9.4.6 Accuracy of the Cell Model Predictions

In the foregoing discussion on two types of cell models (Levine–Neale and Shilov–Zharkikh), we resolved one of the two discrepancies arising from the use of different boundary conditions. The first discrepancy stems from the use of different definitions of the electric field (external and internal) in the two cell model approaches. At the end, we observed that the two types of cell models provide identical results for electrophoretic mobility once a consistent definition of electric field is used in either approach. There is a second discrepancy in the formulation of the outer cell surface boundary condition between the two approaches. This discrepancy arises from use of the gradient of the perturbed potential, $\nabla\delta\psi$, in the Levine–Neale boundary condition, as opposed to use of the perturbed potential, $\delta\psi$, in the Shilov–Zharkikh boundary condition. This implies that one of the cell models still incorrectly specifies the outer cell surface boundary condition. Yet, once the first discrepancy between the two cell models is resolved, the predictions of the electrophoretic mobility obtained using both the cell models become identical. This leads to the question as to why, in spite of the second discrepancy between the boundary conditions, the two cell models yield identical predictions of the electrophoretic mobilities? Furthermore, one is faced with the question as to which of the two cell models use the correct boundary condition?

The questions raised above have tremendous ramifications in the literature concerned with electrophoresis in concentrated dispersions, as well as sedimentation of concentrated suspensions. Broadly, two criteria are employed to test the validity of a model for electrophoretic mobility. A successful model should abide by both these criteria. These are:

(i) The *Smoluchowski criterion*, which states that the electrophoretic mobility (more precisely, the external mobility, η) should approach the Smoluchowski limit result, $\eta = \epsilon\zeta/\mu$, as $\kappa a \to \infty$ for every particle volume fraction, and,

(ii) The *Onsager reciprocity criterion*, which implies that the electrophoresis and sedimentation being reciprocal processes, should lead to a specified linear relationship between the sedimentation potential and electrophoretic mobility.

Violation of the Smoluchowski criterion implies that the predicted electrophoretic mobility will not become independent of ϕ_p as $\kappa a \to \infty$. Violation of the Onsager reciprocity criterion implies that when one uses the same cell model and perturbation technique to obtain the electrophoretic mobility and the sedimentation potential independently, the two quantities will not be related by the Onsager reciprocal relationship.

The Onsager reciprocal relationship between the sedimentation potential, $\mathbf{E}_{sed}$, and the electrophoretic mobility, η, is (see Chapter 10 for details)

$$\mathbf{E}_{sed} = -\frac{\phi_p(\rho_p - \rho)\eta}{\sigma^{\infty}}\mathbf{g} \tag{9.168}$$

Here, ϕ_p is the particle volume fraction, ρ_p and ρ are the particle and fluid material densities, respectively, σ^{∞} is the bulk electrolyte solution conductivity, and $\mathbf{g}$ is the gravitational acceleration.

The electrophoretic mobility computed by both the Levine–Neale and the Shilov–Zharkikh cell models satisfy the Smoluchowski criterion. However, the Levine–Neale cell model does not provide the correct Onsager relationship between the sedimentation potential and the electrophoretic mobility (Shilov *et al.*, 1981), thereby violating the Onsager reciprocity criterion. In other words, the sedimentation potential obtained by substituting η calculated using the Levine–Neale cell model for electrophoresis in Eq. (9.168) will not be the same as $\mathbf{E}_{sed}$ obtained by directly solving the governing electrochemical transport equations for sedimentation using the Levine–Neale cell model. To satisfy the Onsager reciprocity criterion, however, both procedures should provide the same result. The Shilov–Zharkikh cell model, on the other hand, satisfies the Onsager reciprocity criteria.

The Levine–Neale cell model violates the Onsager relationship due to the use of the perturbed potential gradient, $\nabla\delta\psi$, in defining the outer cell surface boundary condition. In other words, Eqs. (9.122) and (9.146) are inappropriate for defining the electrical conditions at the outer cell boundary. Fortunately, despite the use of this boundary condition, the Levine–Neale perturbation solution for the electrophoretic mobility turns out to be accurate. The Levine–Neale perturbation approach represents a first order perturbation expansion in zeta potential. Up to that order, the model provides correct results for the electrophoretic mobility. If higher order perturbations are included, the error in the Levine–Neale cell model would manifest itself in the calculated electrophoretic mobility. The inconsistency with the Levine–Neale boundary condition becomes more prominent when applying the cell model based perturbation approach for obtaining the sedimentation potential. As will be discussed in Chapter 10, direct application of the perturbation technique using the Levine–Neale boundary condition for the sedimentation problem will provide an incorrect dependence of the sedimentation potential on volume fraction, ϕ_p.

To summarize, the Levine–Neale cell model results, as well as the model of Ohshima presented in this section fortuitously provide the correct dependence of the electrophoretic mobility on volume fraction and κa for concentrated suspensions. However, it is not advisable to apply this model directly to determine the sedimentation potential. Consequently, to obtain a correct prediction of the sedimentation potential of concentrated suspensions employing the Levine–Neale or Ohshima cell models for low zeta potential, one can first calculate the electrophoretic mobility and then use the Onsager reciprocal relationship, Eq. (9.168), to determine the sedimentation potential from the electrophoretic mobility.

9.5 CIRCULAR CYLINDERS NORMAL TO THE ELECTRIC FIELD

Electrophoresis of a swarm of fibres (circular cylinders) is important for the study of the electrokinetic transfer phenomenon of colloidal particles in fibrous media. For the case of the electric field normal to a circular cylinder, Henry (1931) analyzed the electrophoretic mobility of an isolated cylinder. Abramson *et al.* (1942) further developed Henry's analysis of an isolated cylinder to obtain results for intermediate values of κa where "a" represents the circular cylinder radius. Van der Drift *et al.* (1979) applied Henry's approach, relaxing the constraint of linearization of the Boltzmann equation that describes the charge density.

The extension of the analysis for the electrophoretic mobility for the case of a mat of cylinder fibers was conducted by Guzy *et al.* (1983), and Kozak and Davis (1986). The latter combined Henry's approach with Kuwabara's cell model to analyze the electrophoretic mobility of a mat of cylinders normal to the electric field strength. The governing equations are similar to those of Levine and Neale (1974a), except that they are cast in cylindrical coordinates. The Debye Hückel approximations are employed.

For the case of the cell model, the fibrous mat concentration is given by

$$\phi_p = \left(\frac{a}{b}\right)^2 = 1 - f \tag{9.169}$$

where a is the cylinder radius, b is the outer shell radius, ϕ_p and f are the mat volume fraction and porosity, respectively.

Strictly speaking, the cell model is applicable to an ordered array of fibres as shown in Figure 9.23. The effect of the mat porosity is given in Figure 9.24. As is for the case of a suspension of spheres, the effect of particle concentrations is important when the electric double layers overlap, $\kappa a \ll 1$ (Kozak and Davis, 1986). The correction factor is shown in Figure 9.24 and is given by

$$U = \frac{\epsilon \zeta E_\infty}{\mu} g_3(\kappa a, f) \tag{9.170}$$

where U is the electrophoretic velocity of a cylinder within the mat.

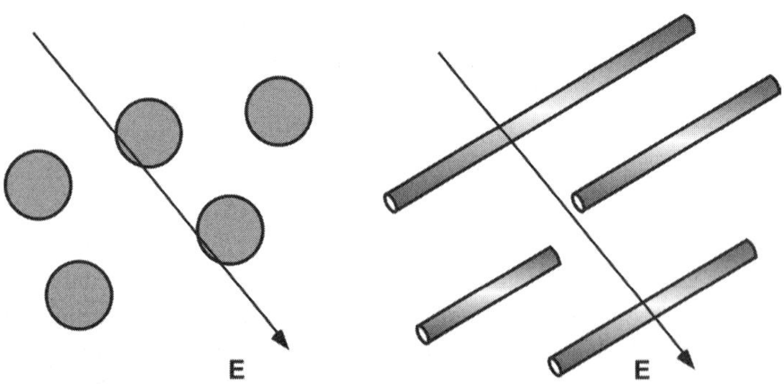

Figure 9.23. Electrophoretic mobility of a mat of cylinders. Cell model schematic.

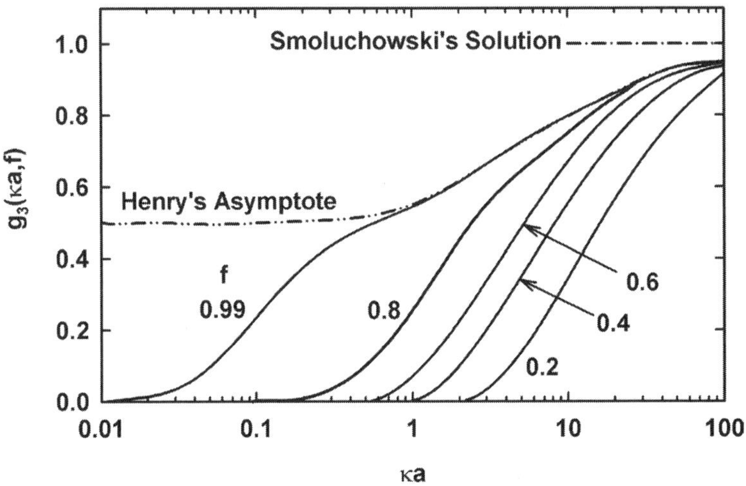

Figure 9.24. Electrophoretic mobility of a mat of cylinders moving normal to their axis. Data are due to Kozak and Davis (1986), redrawn. Henry's asymptote refers to porosity of unity (single cylinder); Smoluchowski asymptote is for large κa and a single cylinder.

For large values of κa [$\kappa a > 20$ and for $(\kappa b - \kappa a) > 4$] the solution for the electrophoretic mobility is given by

$$\eta = \frac{\epsilon\zeta}{\mu}\left[1 - \frac{(2+f)}{\kappa a f} + \frac{6.9984 + 11.4984 f}{(\kappa a)^2 f} - \frac{55.7865 + 71.845 f}{(\kappa a)^3 f} + O(\kappa a)^{-4}\right] \tag{9.171}$$

For an isolated cylinder $f = 1$ and Eq. (9.171), which is valid for $\kappa a > 20$ becomes

$$\eta = \frac{\epsilon\zeta}{\mu}\left[1 - \frac{3}{\kappa a} + \frac{18.5}{(\kappa a)^2} - \frac{127.6}{(\kappa a)^3} + \cdots\right] \tag{9.172}$$

The above expression for an isolated cylinder is very similar to the corresponding expression due to Henry (1931) for the case of an isolated sphere, which is given as (Hunter, 1981):

$$\eta = \frac{\epsilon\zeta}{\mu}\left[1 - \frac{3}{\kappa a} + \frac{25}{(\kappa a)^2} - \frac{220}{(\kappa a)^3} + \cdots\right] \quad \text{for } \kappa a \geq 20 \tag{9.173}$$

For the limit of small $\kappa a (\kappa a < 0.01)$, the asymptotic solution for all f for the case of cylinder is given by

$$\eta = \frac{\epsilon\zeta}{2\mu}\left[1 - \frac{2(1-f)}{2(1-f) - \kappa a \ln \kappa a}\right] \tag{9.174}$$

In the limit of $f \to 1$ and $\kappa a \ll 1$, Eq. (9.174) reduces to Hückel's expression for the case of an isolated cylinder with the electric field normal to the axis of the cylinder.

Equation (9.174) together with Figure 9.24 clearly show that when the double layer is thick, the presence of the surrounding fibers (cylinders) has a significant effect on the electrophoretic mobility of the cylinders. For $\kappa a < 0.1$, the overlapped electric layers would reduce the electroosmotic velocity even in a very dilute mat of cylinders.

An important observation is that the expressions derived for the electrophoretic velocity for a swarm of spheres or a mat of cylindrical fibers are equally valid for electroosmotic flow within a rigid bed of spheres or a mat of fibres. This is because both processes involve the relative notion of solid particles and electrolyte under the influence of an externally applied field, in the absence of an applied pressure gradient (Levine and Neale, 1974a).

9.6 NOMENCLATURE

a	radius of a sphere or a cylinder, m
A, A_c	cross-sectional flow area, m^2
b	outer radius of the cell model
D	particle diffusion coefficient, m^2/s
E_r	electric field in the radial direction, V/m
E_x	electric field in x-direction, V/m
E_y	electric field in y-direction, V/m
E_θ	electric field component in the angular direction, V/m
E_∞	electric field magnitude far from a particle, V/m
$\mathbf{E}_{sed}$	sedimentation potential, V/m
$\mathbf{E}_\infty$	electric field vector far from the particle, V/m
$\mathbf{E}_{susp}$	suspension (or internal) electric field, V/m
e	magnitude of electronic charge, C
$f(\kappa a)$	Henry's function, dimensionless
f	porosity, $1 - \phi_p$
$\mathbf{f}_b$	body force (force per unit volume) acting on fluid, N/m^3
F_E	electric force on a particle, N
F_H	force on a particle due to hydrodynamic drag, N
$F(\kappa a, K')$	generalized Henry's function defined by Eq. (9.93)
$\mathbf{g}$	gravitational acceleration, m/s^2
$g_1(\kappa a, \phi_p)$	Möller *et al.* (1961) function to account for κa and spherical particle volume fraction, dimensionless
$g_2(\kappa a, \phi_p)$	Levine and Neale (1974a,b) function to account for κa and spherical particle volume fraction, dimensionless
$g_3(\kappa a, f)$	Kozak and Davis (1986) function to account for κa and cylindrical fibre mat porosity, dimensionless
$G(\kappa a, \phi_p)$	Ohshima's function to account for κa and particle volume fraction
$\langle \mathbf{i} \rangle$	suspension current density, A/m^2
$\mathbf{i}_\infty$	external current density in electrolyte solution, A/m^2
$\mathbf{j}_i^{**}$	flux of ith ionic species, $m^{-2}s^{-1}$
k_B	Boltzmann constant, J/K

K'	ratio of a particle to medium electric conductivity defined by Eq. (9.92)
M^*	dimensionless mobility $= 2U^*/\beta^*$
n_i	number concentration of ith ionic species, m^{-3}
n_∞	ionic number concentration in the bulk solution, m^{-3}
$\mathbf{n}$	unit surface normal vector
N_0	number concentration of particles in suspension, m^{-3}
p	pressure, Pa
P_e	particle Peclet number, Ua/D
q_s	surface charge density, C/m^2
q_s^*	dimensionless surface charge density defined by Eq. (9.100)
Q_f	total free charge, C
Q_s	total surface charge, C
r	radial coordinate, m
t	dummy variable
T	absolute temperature, K
u_r	local radial fluid velocity, m/s
u_x	local fluid velocity in x-direction, m/s
u_θ	local angular fluid velocity, m/s
$\mathbf{u}$	local fluid velocity vector, m/s
$\mathbf{u}_\infty$	velocity of fluid far from the particle, m/s
U	electrophoretic particle velocity, m/s; fluid velocity far from the particle; fluid velocity at the outer surface of a cell
U^*	dimensionless electrophoretic velocity defined by Eq. (9.101)
U_{St}	Stokes free settling velocity, m/s
U_{HS}	hindered settling velocity, m/s
$\mathbf{v}_i$	ionic species velocity, m/s
(x, y)	Cartesian coordinates, m
z_i	valence of ith ionic species
z	absolute value of the valency for a $(z : z)$ electrolyte

Greek Symbols

β^*	dimensionless electric field strength defined by Eq. (9.99)
$\delta\mu_i$	perturbation in i^{th} species chemical potential
$\delta\psi$	perturbation in electric potential, V
ϵ	dielectric permittivity of a material, C/mV
ϵ_p	porosity of a suspension $= 1 - \alpha_p$
η	electrophoretic mobility based on external electric field, E_∞, $m^2V^{-1}s^{-1}$
η_{susp}	electrophoretic mobility based on suspension electric field, E_{susp}, $m^2V^{-1}s^{-1}$
κ	inverse Debye length, m^{-1}
λ_i	Stokes Einstein friction factor, k_BT/D_i
ζ	zeta potential, V
ζ^*	dimensionless zeta potential $= ze\zeta/k_BT$

θ	angular coordinate
μ	fluid viscosity, Pa s
μ_i	electrochemical potential of ith ionic species
ρ_f	free charge density, C/m^3
$\bar{\bar{\sigma}}$	total hydrodynamic stress tensor, N/m^2
σ'	particle electric conductivity, S/m
σ^{∞}	bulk electrolyte solution conductivity, S/m
σ^{susp}	suspension conductivity, S/m
ϕ	potential due to external electric field, V
ϕ_p	particle volume fraction in a suspension
ψ	total electric potential, V
$\psi^{eq}, \psi^{(0)}$	potential due to stationary double layer, V
∇	del operator, m^{-1}
∇^2	Laplace operator, m^{-2}
ω	vorticity, s^{-1}

9.7 PROBLEMS

9.1. Calculate the electrophoretic velocity U of a charged spherical particle of radius 0.1 μm and zeta potential 35 mV suspended in an aqueous NaCl solution at 300 K when an electric field of 10^5 V/m is applied. Plot the electrophoretic mobility of the particle as a function of κa for $0.1 < \kappa a < 100$.

9.2. We intend to separate the components of a very dilute aqueous binary colloidal suspension using electrophoresis. The suspension contains two types of spherical colloidal particles, one having a radius of 50 nm and a zeta potential of −30 mV, while another of radius 20 nm and unknown zeta potential. Figure 9.25 shows the separation scheme. Both particles start at the same location and traverse the length L at different times when an electric field is applied between the electrodes as shown.

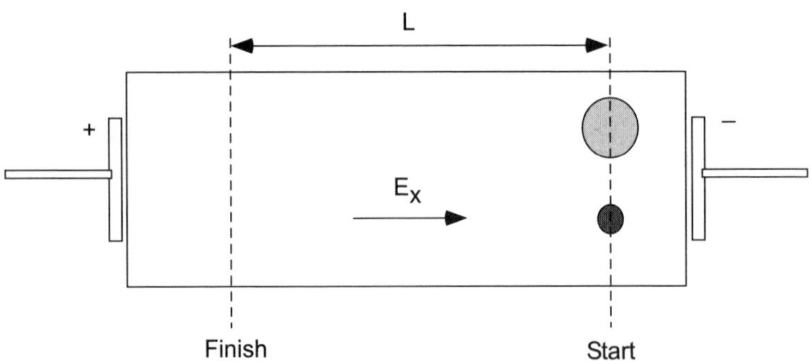

Figure 9.25. Electrophoresis of two spherical colloidal particles of different size and surface potential.

The length L is 200 μm. Under the external field, the larger particle travels the distance from start to finish in 12 seconds. What is the zeta potential of the smaller particle if it reaches the finish line in 26.5 seconds? The electrolyte (1 : 1) concentration is 0.01 M and the solvent viscosity is 0.001 Pa s. Since the suspension is very dilute, you can assume that the particle mobility can be adequately represented by the single particle mobility expressions. Use the approximate curve fit formula, Eq. (9.87) to calculate Henry's function.

9.3. Consider the suspension conductivity expression given by Eq. (9.151). Assume that the aqueous solution is made up of a $(z : z)$ electrolyte with equal diffusion coefficients. Plot the variation of $\sigma^{susp}/\sigma^{\infty}$ with particle volume fraction for $e\zeta/k_BT = 1.5, -1.5$ and $\kappa a = 10, 20,$ and 100. As well, plot the corresponding variation for $e\zeta/k_BT = 0$ and very large κa.

9.4. Evaluate the variation of the normalized conductivity of a suspension, given by $\sigma^{susp}/\sigma^{\infty}$, with particle volume fraction for different aqueous systems. Assume that the charged particles have a zeta potential of 35 mV. The normalized conductivity ratio, $\sigma^{susp}/\sigma^{\infty}$, gives a measure of the change in the electric conductivity due to the presence of the charged particles relative to the solution conductivity in the absence of particles. (a) Sodium chloride solution at 0.001 M at 25°C. (b) Hydrogen chloride solution at 0.001 M at 25°C. (c) Sodium hydroxide solution at 0.001 M at 25°C. (d) Calcium chloride solution at 0.001 M at 25°C.

9.5. Evaluate the variation of the relative conductivity ratio of a suspension, $\sigma^{susp}/\sigma^{susp}_{\zeta=0}$, with volume fraction for different aqueous systems. Assume that the charged particles have a zeta potential of 35 mV. Here $\sigma^{susp}_{\zeta=0}$ is defined as the conductivity of a suspension made up of particles having zero electric surface charge. Consequently, $\sigma^{susp}/\sigma^{susp}_{\zeta=0}$ gives a measure of the suspension electric conductivity due particles surface charge. (a) Sodium chloride solution at 0.001 M at 25°C. (b) Hydrogen chloride solution at 0.001 M at 25°C. (c) Sodium hydroxide solution at 0.001 M at 25°C. (d) Calcium chloride solution at 0.001 M at 25°C.

9.8 REFERENCES

Abramson, H. A., Moyer, L. S., and Gorin, M. H., *The Electrophoresis of Proteins*, Reinhold, New York, (1942).

Booth, F., The cataphoresis of spherical, solid non-conducting particles in a symmetrical electrolyte, *Proc. Roy. Soc. Lond.*, **203A**, 514, (1950).

Carrique, F., Arroyo, F. J., and Delgado, A. V., Electrokinetics of concentrated suspensions of spherical colloidal particles: Effect of a dynamic Stern layer on electrophoresis and DC conductivity, *J. Colloid Interface Sci.*, **243**, 351–361, (2001).

Datta, R., and Kotamarthi, V. R., Electrokinetic dispersion in capillary electrophoresis, *AIChE J*, **36**, 916–925, (1990).

Ding, J. M., and Keh, H. J., The electrophoretic mobility and electric conductivity of a concentrated suspension of colloidal spheres with arbitrary double-layer thickness, *J. Colloid Interface Sci.*, **236**, 180–193, (2001).

Dukhin, S. S., and Derjaguin, B. V., Electrokinetic phenomena, in *Surface and Colloid Science*, vol. 7, Matijevic, E. (Ed.), Wiley, New York, (1974).

Dukhin, S. S., and Semenikhin, N. M., Theory of double layer polarization and its effect on the electrokinetic and electroptical phenomena and te dielectric constants of dispersed systems, *Kolloid Z.*, **32**, 360–368, (1970).

Dukhin, A. S., Shilov, V., Borkovskaya, Y., Dynamic electrophoretic mobility in concentrated dispersed systems. Cell model, *Langmuir*, **15**, 3452–3457, (1999).

Dukhin, S. S., and Shilov, V. N., Dielectric phenomena and the double layer in dispersed systems and polyelectrolytes, John Wiley and Sons, New York, (1974).

Gorin, M. H., *in* Abramson, H. A., Gorin, M. H., and Moyer, L. S., *Chem. Rev.*, **24**, 345–366, (1939).

Gur, Y., Ravina, I., and Babchin, A. J., On the electrical double layer theory. i. a numerical methods for solving a generally set Poisson–Boltzmann equation, *J. Colloid Interface Sci.*, **64**, 326–332, (1978a).

Gur, Y., Ravina, I., and Babchin, A. J., On the electrical double layer theory. II. The Poisson–Boltzmann equation including hydration forces, *J. Colloid Interface Sci.*, **64**, 333–341, (1978b).

Guzy, C. J., Bonano, E. J., and Davis, D. J., The analysis of flow and colloidal particle retention in fibrous porous media, *J. Colloid Interface Sci.*, **95**, 523–543, (1983).

Happel, J., Viscous flow in multiparticle systems, *AIChE J.*, **4**, 197–201, (1958).

Happel, J., and Brenner, H., *Low Reynolds Number Hydrodynamics*, Prentice Hall, Englewood Cliffs, New Jersey, (1965).

Henry, D. C., The cataphoresis of suspended particles, Part 1. The equation of cataphoresis, *Proc. Roy. Soc. Lond.*, **133A**, 106–129, (1931).

Hückel, E., Die kataphorese der kugel, *Phys. Z.*, **25**, 204–210, (1924).

Hunter, R. J., *Zeta potential in Colloid Science*, Academic Press, London, (1981).

Keizer, A. D. E., van der Drift, W. P. J. T., and Overbeek, J. Th. G., Electrophoresis of randomly oriented cylindrical particles, *Biophys. Chem.*, **3**, 107–108, (1975).

Kozak, M. W., and Davis, E. J., Electrokinetic phenomena in fibrous porous media, *J. Colloid Interface Sci.*, **112**, 403–411, (1986).

Kozak, M. W., and Davis, E. J., Electrokinetics of concentrated suspensions and porous media. I. Thin electrical double layers, *J. Colloid Interface Sci.*, **127**, 497–510, (1989a).

Kozak, M. W., and Davis, E. J., Electrokinetics of concentrated suspensions and porous media. 2. Moderately thick electrical double layers, *J. Colloid Interface Sci.*, **129**, 166–174, (1989b).

Kuwabara, S., The forces experienced by randomly distributed parallel circular cylinders or spheres in a viscous flow at small Reynolds numbers, *J. Phys. Soc. Japan*, **14**, 527–532, (1959).

Levine, S., and Neale, G. H., The prediction of electrokinetic phenomena within multiparticle systems. I. Electrophoresis and electroosmosis, *J. Colloid Interface Sci.*, **47**, 520–529, (1974a).

Levine, S., and Neale, G. H., Electrophoretic mobility of multiparticle systems, *J. Colloid Interface Sci.*, **49**, 330–332, (1974b).

Miller, N. P., and Berg, J. C., Experiments on the electrophoresis of porous aggregates, *J. Colloid interface Sci.*, **159**, 253–254, (1993).

Miller, N. P., Berg, J. C., and O'Brien, R. W., The electrophoretic mobility of a porous aggregate, *J. Colloid Interface Sci.*, **153**, 234–243, (1992).

Möller, J. H. N., Van Os, G. A. J., and Overbeek, J. Th. G., Interpretations of the conductance and transference of bovine serum albumin solutions, *Trans. Faraday Soc.*, **57**, 325–337, (1961).

Morrison, F. A., Electrophoresis of a particle of arbitrary shape, *J. Colloid Interface Sci.*, **34**, 210–214, (1970).

O'Brien, R. W., and White, L. R., Electrophoretic mobility of a spherical colloidal particle, *J. Chem. Soc. Faraday Trans.*, II, **74**, 1607–1626, (1978).

Ohshima, H., Electrophoretic mobility of spherical colloidal particles in concentrated suspensions, *J. Colloid Interface Sci.*, **188**, 481–485, (1997a).

Ohshima, H., Dynamic electrophoretic mobility of spherical colloidal particles in concentrated suspensions, *J. Colloid Interface Sci.*, **195**, 137–148, (1997b).

Ohshima, H., Electrical conductivity of a concentrated suspension of spherical colloidal particles, *J. Colloid Interface Sci.*, **212**, 443–448, (1999).

Ohshima, H., Cell model calculation for electrokinetic phenomena in concentrated suspensions: an Onsager relation between sedimentation potential and electrophoretic mobility, *Adv. Colloid Interface Sci.*, **88**, 1–18, (2000).

Overbeek, J. Th. G., Theory of the relaxation effect in electrophoresis, *Kolloide Beihefte*, **54**, 287–364, (1943).

Probstein, R. F., *Physicochemical Hydrodynamics, An Introduction*, 2nd ed., Wiley Interscience, New York, (2003).

Pintauro, P. N., and Verbrugge, M. W., The electric-potential profile in ion-exchange membrane pores, *J. Membrane Sci.*, **44**, 197–212, (1989).

Reed, L. D., and Morrison, F. A., Hydrodynamic interaction in electrophoresis, *J. Colloid Interface Sci.*, **54**, 117–33, (1976).

Russel, W. B., Saville, D. A., and Schowalter, W. R., *Colloidal Dispersions*, Cambridge University Press, Cambridge, (1989).

Saville, D. A., Electrokinetic effects with small particles, *Ann. Rev. Fluid Mech.*, **9**, 321–337, (1977).

Shaw, D. J., *Introduction to Colloid and Surface Chemistry*, 3rd ed., Butterworths, London, (1980).

Shilov, V. N., Zharkikh, N. I., and Borkovskaya, Y. B., Theory of non-equilibrium electrosurface phenomena in concentrated disperse system. 1. Application of non-equilibrium thermodynamics to cell model, *Colloid J.*, **43**, 434, (1981).

Smoluchowski, M. von, Versuch einer mathematischen theorie der koagulation kinetic kolloider lösungen, *Z. Phys. Chem.*, **93**, 129, (1918).

Van der Drift, W. P. J. T., Keizer, A. D. E., and Overbeek, J. Th. G., Electrophoretic mobility of a cylinder with high surface charge density, *J. Colloid Interface Sci.*, **71**, 67–78, (1979).

Verbrugge, M. W., and Pintauro, P. N., Transport models for ion-exchange membranes, *Modern Aspects of Electrochemistry*, No. 19, Conway, B. E., Bockris, J. O'M., and White, R. E. (Eds.), Plenum, New York, (1989).

Wiersema, P. H., Loeb, A. L., and Overbeek, J. Th. G., Calculation of the electrophoretic mobility of a spherical colloid particle, *J. Colloid Interface Sci.*, **22**, 78–99, (1966).

CHAPTER 10

SEDIMENTATION POTENTIAL

10.1 SEDIMENTATION OF UNCHARGED SPHERICAL PARTICLES

Sedimentation of colloidal particles in liquids is of great industrial importance. In the absence of colloidal forces, for instance, when electric double layer interactions are absent between uncharged particles, their sedimentation rate is uniquely defined by the volumetric concentration of the particles, particle and liquid densities, liquid viscosity, and particle shape and size. For the case of a single sedimenting uncharged spherical particle in an infinite medium, the sedimenting velocity is given by

$$U_{St} = 2a^2 \frac{(\rho_p - \rho)g}{9\mu} \tag{10.1}$$

where U_{St} is the Stokes sedimentation velocity, a is the radius of the spherical particle, ρ_p and ρ are the particle and liquid densities, respectively, μ is the liquid density, and g is the gravitational acceleration. Equation (10.1) is known as the Stokes settling velocity and it is strictly valid for a Newtonian fluid with the particle Reynolds number being much less than unity, *i.e.*,

$$Re_p = \frac{2aU_{St}\rho}{\mu} \ll 1 \tag{10.2}$$

Equation (10.1) was derived by Stokes (1851) by solving the flow momentum equations in the creeping flow limit. In Eq. (10.1), it is assumed that gravity acts downward

Electrokinetic and Colloid Transport Phenomena, by Jacob H. Masliyah and Subir Bhattacharjee
Copyright © 2006 John Wiley & Sons, Inc.

and that the density of the particle is higher than that of the liquid, thereby rendering the direction of the sedimentation or settling velocity downward.

Settling of hard spheres in a suspension, *i.e.*, non-interacting uncharged spherical particles, has been the subject of a great many studies, to name a few, Brinkman (1947), Richardson and Zaki (1954), Happel (1958), Kuwabara (1959), Happel and Brenner (1983), Masliyah (1979), Batchelor (1982), Batchelor and Wen (1982), and Wang and Wen (1998).

The settling of a sphere within a suspension is normally referred to as hindered settling. The settling rate of a particle within a suspension, or simply the suspension settling rate, is a strong function of the particle concentration. Experimental data have shown that for hard spheres, the hindered settling velocity, U_{HS}, can be approximately given by

$$\frac{U_{HS}}{U_{St}} = (1 - \phi_p)^{4.6} \tag{10.3}$$

The velocity ratio (U_{HS}/U_{St}) is a measure of the effect of the particle volume fraction, ϕ_p, on the settling rate of a sphere in a monodisperse suspension. Equation (10.3) is normally referred to as hindered settling in a closed container (batch settling) and it is attributed to Richardson and Zaki (1954). The suspension hindered settling velocity, U_{HS}, is the velocity of the suspension interface as observed by a stationary observer as shown in Figure 10.1.

Due to the difficulty in tackling the problem of sedimentation of monodisperse suspensions, several methods were developed to evaluate U_{HS}. The most used models are those of Kuwabara's and Happel's cell models. The cell model involves the concept that a suspension can be divided into a number of identical cells where one sphere occupies each cell. The fluid mechanics of sedimentation of a suspension of spherical particles is thus reduced to the consideration of a single sphere with its cell bounding envelope. For example, making use of the cell model approach, Kuwabara's cell model

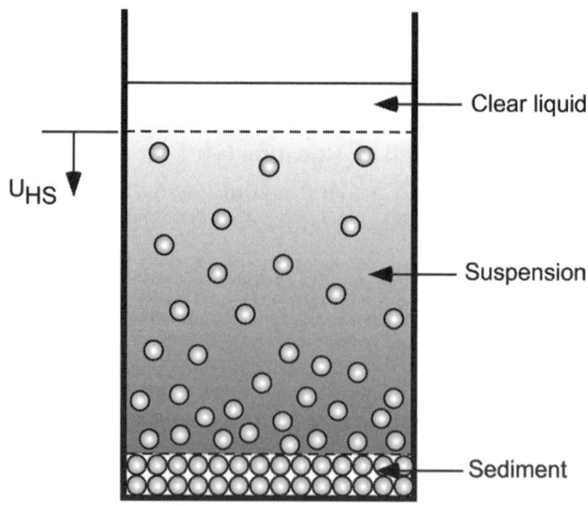

Figure 10.1. Settling of particles in a suspension in a closed container.

provides

$$\frac{U_{HS}^{Kuwabara}}{U_{St}} = 1 - \frac{9}{5}\phi_p^{1/3} + \phi_p - \frac{1}{5}\phi_p^2 \tag{10.4}$$

where ϕ_p is the volume fraction of the spheres. It is the ratio of the volume of the spherical particle to the total volume of the suspension.

Determination of the settling velocity of the spheres, U_{St}, and the suspension settling velocity, U_{HS} can be made by using the fluid continuity and momentum conservation equations subject to appropriate boundary conditions, as was discussed in Section 9.4.1 (Happel and Brenner, 1983).

10.2 CONCEPT OF SEDIMENTATION POTENTIAL AND VELOCITY

Sedimentation of charged spherical colloidal particles in electrolyte solutions has received much more attention in the last decade. This type of electrokinetic transport is more complex than its counterpart of uncharged spherical colloidal particles. In a suspension of charged spherical particles suspended in an electrolyte solution sedimenting under gravity, the electrical double layer surrounding each particle is distorted by the liquid flow around the particle. As discussed earlier, the deformation of the electric double layer as a result of the liquid motion is referred to as the relaxation effect. The liquid (electrolyte solution) behind each particle therefore carries an excess of counterions compared to the liquid ahead of the particle. As the total current must be zero, an induced electric field is set up such that the net current becomes zero. Consequently, when electrodes are placed near the top and bottom of the settling tube containing a suspension of sedimenting charged particles, an induced electrical potential gradient can be measured. This electric potential is referred to as the migration potential or the Dorn effect (Dorn, 1878). It is the least investigated electrokinetic phenomenon discussed so far. Figure 10.2 depicts the sedimentation of a charged spherical particle in an electrolyte solution.

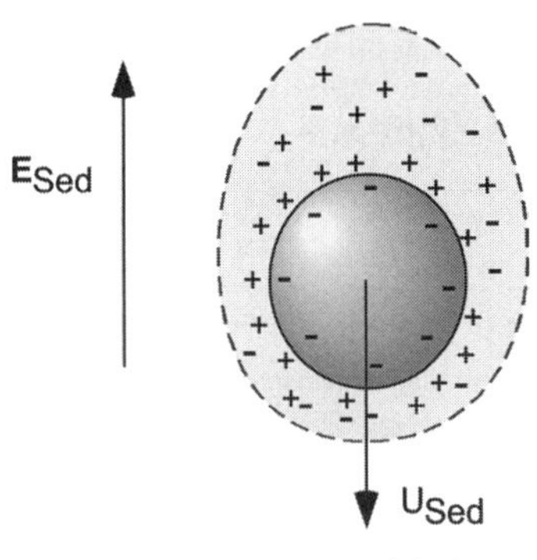

Figure 10.2. Sedimentation of a charged particle in an electrolyte solution.

A simple derivation of the induced sedimentation potential can be made as follows. One can think of the sedimentation potential as the converse of the streaming potential that we discussed in Chapter 8. For the case of very thin electric double layers, Eq. (8.146) provides

$$\left(\frac{E_x}{p_x}\right)_{I=0} = \frac{\epsilon\zeta}{\mu\sigma^\infty} \tag{10.5}$$

where ζ is the zeta potential of the particles and σ^∞ is the bulk electrolyte solution electrical conductivity, Eq. (6.73), given as

$$\sigma^\infty = \frac{e^2}{k_B T}\sum z_i^2 D_i n_{i\infty} \tag{10.6}$$

Equation (10.5) can be re-written as

$$E_{Sed} = \frac{\epsilon\zeta}{\mu\sigma^\infty} p_{Sed} \tag{10.7}$$

The term E_{Sed} is the induced sedimentation electric field due to particle sedimentation and p_{Sed} denotes the "driving pressure" gradient causing sedimentation.[1] Although Eq. (10.5) was derived for electroosmotic flow in a straight channel, it is assumed to be valid for an arbitrary cross section, especially for the case of non-overlapping electrical double layers and low zeta (ζ) potentials.

Let us consider a box containing N_p charged spheres of radius a as depicted in Figure 10.3. The pressure gradient for the case of sedimentation within the box can

[1]The driving pressure in case of streaming potential is purely hydrostatic pressure, p, while in case of sedimentation potential, it is a combination of hydrostatic and gravitational pressures. This can be seen from the Stokes equation (for creeping flow), written for streaming potential and sedimentation potential. For streaming potential, we have

$$-\nabla p + \mu\nabla^2\mathbf{u} + \rho_f\mathbf{E} = 0$$

while for sedimentation potential, we have

$$-\nabla p + \rho\mathbf{g} + \mu\nabla^2\mathbf{u} + \rho_f\mathbf{E} = 0$$

where $\mathbf{E}$ is the electric field. One can write

$$\Pi_g = p - \rho\mathbf{g}\cdot\mathbf{r}$$

where $\mathbf{r}$ is the position vector. Substituting the above expression in the Stokes equation for sedimentation gives

$$-\nabla\Pi_g + \mu\nabla^2\mathbf{u} + \rho_f\mathbf{E} = 0$$

Thus, although the Stokes equation for sedimentation and streaming potential appear to be identical, the interpretation of the driving pressure gradient is different in the two cases.

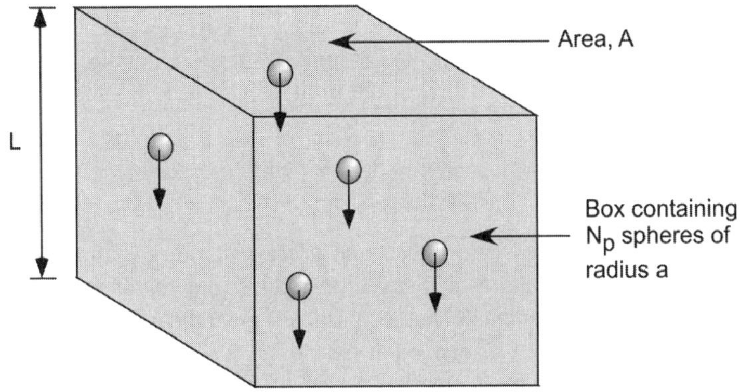

Figure 10.3. A volume element containing sedimenting spheres.

be evaluated as follows:

$$\text{Total driving force:} \quad \frac{4}{3}\pi a^3 N_p(\rho_p - \rho)g \tag{10.8}$$

$$\text{Driving pressure:} \quad \frac{4}{3}\pi a^3 N_p(\rho_p - \rho)g/A \tag{10.9}$$

$$\text{Driving pressure gradient:} \quad \frac{4}{3}\pi a^3 N_p(\rho_p - \rho)g/(AL) \tag{10.10}$$

where A is the cross-sectional area of the box containing a suspension of height L.

Now, the number of particles per unit volume is given by $N_0 = N_p/(AL)$. Hence, the particle volume fraction, ϕ_p, is given by $\phi_p = (4/3)\pi a^3 N_0$. Here, a is the radius of the spherical particle, g is the acceleration due to gravity, and ρ_p and ρ are the particle and the electrolyte solution (liquid) densities, respectively.

From the definitions of N_0 and ϕ_p, the expression (10.10) provides the driving pressure gradient, p_{Sed}, as

$$p_{Sed} = (\rho_p - \rho)\phi_p g \tag{10.11}$$

Combining Eqs. (10.7) and (10.11) leads to

$$E_{Sed} = \frac{\epsilon\zeta(\rho_p - \rho)\phi_p g}{\mu\sigma^\infty} \tag{10.12}$$

Equation (10.12) provides a measure of the sedimentation potential due to the settling of the charged spheres. Strictly speaking, one would expect that Eq. (10.12) be valid for non-overlapping electric double layers, $\kappa a \gg 1$, and for dilute suspensions, $\phi_p \ll 1$. This expression is that stated by Smoluchowski (1903, 1921). Experimental results show that Eq. (10.12) is obeyed for Pyrex glass powder ($a \simeq 50$ μm) settling in a KCl solution (Booth, 1954). An important statement provided by Eq. (10.12) is that the sedimentation potential, E_{Sed}, is independent of the particle radius and that $E_{Sed} \to 0$ as $\phi_p \to 0$ (a single particle).

It is of interest at this stage to compare Smoluchowski's electrophoretic mobility expression, Eq. (9.41), which is valid in the limit of $\kappa a \to \infty$, and Smoluchowski's sedimentation potential, Eq. (10.12), in the limit of $\kappa a \to \infty$. Comparing the two expressions leads to

$$E_{Sed} = \frac{\phi_p(\rho_p - \rho)\eta}{\sigma^\infty} g \tag{10.13}$$

where the directions of the vectors $\mathbf{E}_{Sed}$ and $\mathbf{g}$ are ignored. Equation (10.13) is a statement of the Onsager relationship between an induced potential due to sedimenting charged particles and movement of charged particles under the influence of an electric field. Although Eq. (10.13) was arrived at for $\kappa a \to \infty$, it is found to be valid for intermediate κa values and for $\phi_p \gg 0$.

The microscopic electric field due to the distortion of the electric double layer, shown in Figure 10.2, reduces the sedimentation velocity of the particles. The superposition of the individual electric fields from the particles in the suspension gives rise to the macroscopic electric sedimentation potential, which is uniform for a homogeneously dispersed suspension, Ohshima *et al.* (1984).

Smoluchowski (1921) recognized the reduction in the sedimentation rate of particles due to the presence of the induced electric field. Smoluchowski's expression for the reduction in the Stokes settling velocity for a single sphere due to the distorted electric double layer is given by

$$\frac{U_{Sed}}{U_{St}} = \left[1 + \frac{\epsilon^2\zeta^2}{a^2\sigma^\infty\mu}\right]^{-1} \tag{10.14}$$

for $\kappa a \gg 1$ and equal ionic mobilities (Booth, 1954; Sengupta, 1968).

For small values of the term $\epsilon^2\zeta^2/(a^2\sigma^\infty\mu)$ in Eq. (10.14), one can write

$$\frac{U_{Sed}}{U_{St}} = 1 - \frac{\epsilon^2\zeta^2}{a^2\sigma^\infty\mu} \tag{10.15}$$

Let us make some estimate as to the magnitude and effect of the induced sedimentation electric potential using two examples.

EXAMPLE 10.1

Sedimentation Potential. Consider a sedimenting suspension of a volume fraction of 0.1, of spherical particles with a radius of 10 μm. The zeta potential is 50 mV. The liquid medium is an electrolyte solution of 0.01 M KCl at 20°C. The particle density is 2600 kg/m^3. Neglect the effect of KCl on the water density and viscosity. Evaluate the sedimentation electric potential.
Data: $\epsilon = 80.2 \times 8.854 \times 10^{-12}$ C/Vm; $g = 9.81$ m/s^2; $\mu = 1.00 \times 10^{-3}$ Pa s; $\rho = 998.2$ kg/m^3; and $z = 1$

Solution From Table 6.1 $\sigma = 0.122$ S/m at 18°C. $\sigma = 0.1408$ S/m at 25°C.

The coefficient K is given by

$$K = \frac{1}{0.122} \frac{0.1408 - 0.122}{25 - 18} = 0.022$$

Making use of the interpolation formula provided in Section 6.2.4:

$$\sigma = \sigma_1[1 + K(T - T_1)]$$

The electric conductivity at 20 °C is given by

$$\sigma = 0.122[1 + 0.022(20 - 18)]$$

or

$$\sigma = \sigma^{\infty} = 0.127\,\mathrm{S/m} \quad \text{at } 20°\mathrm{C}$$

From definition, for a 0.01 M (1 : 1) electrolyte solution, $\kappa = 3.3 \times 10^8$ leading to

$$\kappa a = (3.3 \times 10^8)(10 \times 10^{-6}) = 3300$$

As $\kappa a \gg 1$, Eq. (10.12) can be used. Recall, Eq. (10.12)

$$E_{Sed} = \frac{\epsilon\zeta(\rho_p - \rho)\phi_p g}{\mu\sigma^{\infty}}$$

Using the given data in the above expression, we obtain

$$E_{Sed} = \frac{(80.2 \times 8.854 \times 10^{-12}) \times 0.05(2600 - 998.2) \times 0.1 \times 9.81}{0.001 \times 0.127}$$

or

$$E_{Sed} = 0.00044\,\mathrm{V/m} \quad \text{at } 20°\mathrm{C}.$$

This is a weak electrical field strength; however, with a high impedance millivolt meter, one can measure this potential (Hunter, 1981). ■

EXAMPLE 10.2

Settling Velocity. Evaluate the decrease in the settling velocity of a single sphere in an unbounded medium using the data of Example 10.1.

Solution From Eq. 10.14 we have

$$\frac{U_{Sed}}{U_{St}} = \left[1 + \frac{\epsilon^2\zeta^2}{a^2\sigma^{\infty}\mu}\right]^{-1}$$

Hence,

$$\frac{U_{Sed}}{U_{St}} = \left[1 + \frac{(80.2 \times 8.854 \times 10^{-12})^2(0.05)^2}{(10 \times 10^{-6})^2 \times 0.127 \times 0.001}\right]^{-1}$$

or

$$\frac{U_{Sed}}{U_{St}} = (1 + 9.9 \times 10^{-8})^{-1}$$

or

$$\frac{U_{Sed}}{U_{St}} \simeq 1 - 9.9 \times 10^{-8}$$

This decrease in the Stokes settling velocity is very small under the present conditions of the example where κa is very large. ■

It is rather curious that the electric field strength, E_{Sed}, that is induced by the settling charged particles is referred to as a "potential". In our earlier chapters we reserved the term "potential" to indicate an electric potential measured in Volts. However, for the sedimentation of charged particles, we use the term "potential" to mean the "electric field strength" with units of Vm^{-1}. In doing so, we are adhering to the traditional nomenclature employed in the literature pertaining to studies on sedimentation of charged particles.

10.3 DILUTE SUSPENSIONS: OHSHIMA'S MODEL

Following Smoluchowski's studies in early 1900s on electrophoresis and sedimentation potential, Booth (1954) provided a general theory of the sedimentation phenomenon and it was closely related to the electrophoresis theory advanced by Overbeek (1943). Without considering particle-particle interaction, Booth solved a set of electrokinetic equations using regular perturbation methods to derive formulae for the sedimentation velocity and potential as a Taylor series in the zeta potential of the charged spheres. Stigter (1980) extended Booth's analysis to allow for higher surface potentials.

In this presentation, we shall follow the approach advanced by the classical treatise of Ohshima *et al.* (1984), which is based on O'Brien and White (1978), to analyze the sedimentation velocity of a single charged sphere and the sedimentation potential of a dilute suspension.

The momentum equation governing the flow past the sphere and the equations governing the ions surrounding the charged sphere will be utilized to develop the sedimentation velocity. The sedimentation potential for a dilute suspension will be extracted using the analysis of a single sedimenting charged sphere, Ohshima *et al.* (1984).

10.3.1 Fundamental Governing Equations

Consider a charged rigid spherical particle having a radius a, sedimenting steadily in an electrolyte solution with a downward velocity, $\mathbf{U}_{sed}$, relative to a stationary

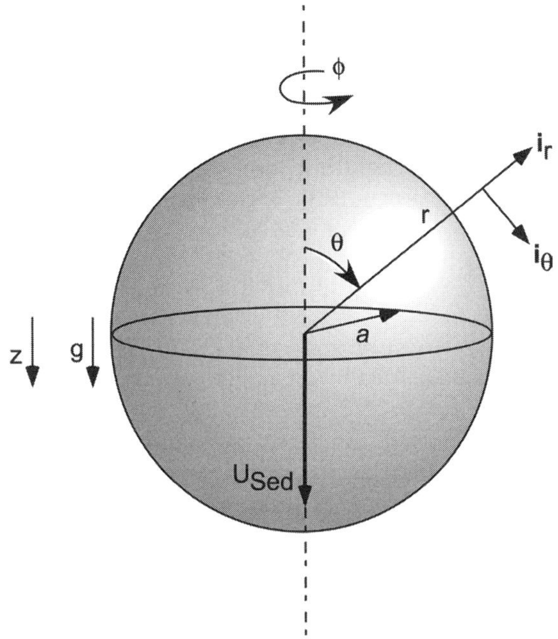

Figure 10.4. Geometry of a sedimenting spherical particle.

observer. The velocity is in the same direction as the gravitational field, **g**. The electrolyte solution is a Newtonian fluid having viscosity μ and density ρ. The electrolyte solution contains N ionic species of valency z_i with a bulk concentration $n_{i\infty}$ and a diffusion coefficient D_i. A spherical coordinate system (r, θ, ϕ) is used for the analysis. The origin of the coordinate system is fixed at the center of the spherical particle. The reference frame is taken to travel with the particle. The geometry of the sedimenting spherical particle is shown in Figure 10.4.

The governing steady-state equation for low Reynolds number, where the inertial term is neglected, is given by Eq. (6.112)

$$-\nabla p + \mu \nabla^2 \mathbf{u} + \rho \mathbf{g} - \rho_f \nabla \psi = 0 \tag{10.16}$$

The steady-state continuity equation for the electrolyte solution having a constant density under dilute conditions, Eq. (6.113), provides

$$\nabla \cdot \mathbf{u} = 0 \tag{10.17}$$

The ionic flux for a dilute electrolyte solution is given by the Nernst–Planck equation, which is a statement of force balance on an ionic species, Eq. (6.45)

$$\mathbf{j}_i^{**} = n_i \mathbf{v}_i = n_i \mathbf{u} - D_i \nabla n_i - \frac{z_i e n_i D_i}{k_B T} \nabla \psi \tag{10.18}$$

To conform with the analysis of Ohshima *et al.* (1984), we write the ionic velocity, $\mathbf{v}_i$ in Eq. (10.18) as

$$\mathbf{v}_i = \mathbf{u} - \frac{D_i}{k_B T} \nabla \mu_i = \mathbf{u} - \frac{1}{\lambda_i} \nabla \mu_i \tag{10.19}$$

where λ_i is the drag coefficient of the ith ionic species, and is identical to the term f_i used in Eq. (6.39). The term μ_i is the electrochemical potential of the ith ionic species. The electrochemical potential is given by

$$\mu_i = \mu_i^{(o)} + k_B T \ln n_i + z_i e \psi \tag{10.20}$$

Here, $\mu_i^{(o)}$ is a constant term in μ_i. Equation (10.19) governs the velocity of the ith ionic species, $\mathbf{v}_i$ as influenced by the liquid velocity $\mathbf{u}$ and the gradient of the electrochemical potential μ_i.

The local electric potential ψ and the space ionic charge density, ρ_f, are related by Poisson's equation, Eq. (5.7), as

$$\epsilon \nabla^2 \psi = -\rho_f \tag{10.21}$$

The space free charge density is given by Eq. (5.9) as

$$\rho_f = \sum_{i=1}^{N} z_i e n_i \tag{10.22}$$

The steady state continuity (mass conservation) of the ith ionic species in the absence of a chemical reaction is given by Eqs. (6.31) and (6.61)

$$\nabla \cdot (n_i \mathbf{v}_i) = 0 \tag{10.23}$$

Due to axial symmetry, we need only to consider the variation in the r and θ directions. The electrolyte solution velocity $\mathbf{u}$ is then provided by its components, u_r and u_θ.

The unknowns in this system of equations are u_r, u_θ, p, v_{ir}, $v_{i\theta}$, n_i, ψ, and U_{Sed}. Consequently, the number of unknowns are $5 + 3N$. The fundamental governing equations provide

Eq. (10.16):	2 equations
Eq. (10.17):	1 equation
Eq. (10.19):	2N equations
Eq. (10.21):	1 equation
Eq. (10.23):	N equations

This gives a total of $4 + 3N$ equations.

The above accounting indicates that the number of unknowns exceeds the number of governing equations by one. Consequently, we will need to develop an additional governing equation. This will be done after our discussion of the boundary conditions pertaining to this problem.

10.3.2 Boundary Conditions

Let us assume that the shear plane of the electric double layer is assumed to coincide with the surface of the sphere, $r = a$. As the coordinate system is fixed on the particle center, and travels downward at a velocity U_{Sed}, the velocity at the particle surface becomes zero relative to the moving coordinate frame of reference.
At the sphere surface, $r = a$,
No slip condition

$$\mathbf{u} = 0 \quad (u_r = u_\theta = 0) \tag{10.24}$$

No electrolyte ions penetrate the sphere surface

$$\mathbf{v}_i \cdot \mathbf{i}_r = 0 \tag{10.25}$$

where $\mathbf{i}_r$ is a unit normal outward vector on the sphere's surface, *i.e.*, $\mathbf{r} = r\mathbf{i}_r$.

At the slip or shear plane, one may impose the condition of constant electric potential

$$\psi_a = \zeta \tag{10.26}$$

or a constant surface charge density

$$-\epsilon \mathbf{n} \cdot \nabla \psi = q_f$$

One should, however, note that such specifications of the particle surface boundary conditions are an artifact of an incomplete problem formulation based on neglecting the electrical fields within the particle. For a complete solution of the problem, one requires the Poisson or Laplace equation to be written within the particles, with appropriate conditions of continuity of electric potential and discontinuity of electric displacement imposed at the particle surface.

For radial positions far away from the sphere, $r \to \infty$, $\mathbf{u} \to -\mathbf{U}_{Sed} = -U_{Sed}\mathbf{i}_z$, the following boundary conditions hold:

$$\mathbf{u} \cdot \mathbf{i}_r = -U_{Sed} \cos\theta \tag{10.27-a}$$

$$\mathbf{u} \cdot \mathbf{i}_\theta = U_{Sed} \sin\theta \tag{10.27-b}$$

In other words, the radial velocity component is given as

$$u_r = -U_{Sed} \cos\theta \tag{10.28}$$

and the angular velocity component is given by

$$u_\theta = U_{Sed} \sin\theta \tag{10.29}$$

where u_r and u_θ are the radial and angular components of the fluid velocity vector, respectively.

At large distances from the sphere, the influence of the sphere is not felt, leading to

$$\psi = 0 \tag{10.30}$$

and

$$n_i = n_{i\infty} \tag{10.31}$$

The boundary conditions for $\mathbf{v}_i$ follows from Eqs. (10.18), (10.30), and (10.31), leading to $\mathbf{v}_i = \mathbf{u}$ at $r \to \infty$.

Although the sedimentation velocity U_{Sed} appears in the boundary conditions, it is an unknown quantity. We need now to provide an additional governing equation to account for U_{Sed}. The sedimentation velocity can be evaluated by recognizing that the net force acting on the spherical particle or an arbitrary volume enclosing the particle is zero. To this end, we consider a large spherical shell, S, of radius r enclosing the particle. The radius r of the surface S is taken to be large enough so that the net electric charge within S is zero (Ohshima *et al.*, 1984). Consequently, one needs to consider the gravitational force, F_g, and hydrodynamic force, F_H, where

$$F_g + F_H = 0 \tag{10.32}$$

The gravity force acting on the material enclosed by the spherical shell, S, is given by

$$F_g = \frac{4}{3}\pi a^3 \rho_p g + \frac{4}{3}\pi (r^3 - a^3)\rho g \tag{10.33}$$

The hydrodynamic force, F_H, is given by

$$F_H = \int_S \bar{\bar{\tau}} \cdot \mathbf{n}\, dS \tag{10.34}$$

The hydrodynamic stress tensor, $\bar{\bar{\tau}}$ has the following components:

$$\tau_{rr} = -p + 2\mu \frac{\partial u_r}{\partial r} \tag{10.35}$$

and

$$\tau_{r\theta} = \mu \left[\frac{1}{r}\frac{\partial u_r}{\partial \theta} + \frac{\partial u_\theta}{\partial r} - \frac{u_\theta}{r} \right] \tag{10.36}$$

where τ_{rr} and $\tau_{r\theta}$ are the normal and tangential components of the hydrodynamic stress, respectively. The sedimentation velocity, U_{Sed}, is implicitly contained in the hydrodynamic stress components, τ_{rr} and $\tau_{r\theta}$.

10.3.3 Perturbation Approach

The governing equations for the charged sedimenting sphere are non-linear and difficult to solve. To obtain a solution, Ohshima *et al.* (1984) assumed that the electric potential, ion concentrations, and the electrochemical potentials in the double layer around the spherical particle are only slightly perturbed from their symmetric equilibrium distribution due to the superimposition of the gravitational field. One can then write

$$n_i = n_i^{(0)}(r) + \delta n_i(r, \theta) \tag{10.37}$$

$$\psi = \psi^{(0)}(r) + \delta\psi(r, \theta) \tag{10.38}$$

$$\mu_i = \mu_i^{(0)} + \delta\mu_i(r, \theta) \tag{10.39}$$

where the superscript (0) refers to the equilibrium state and δ signifies the perturbation from equilibrium. Through the definition of μ_i, given by Eq. (10.20), the perturbed variables, δn_i, $\delta\psi$, and $\delta\mu_i$ are inter-related by

$$\delta\mu_i = z_i e \delta\psi + k_B T \delta n_i / n_i^0 \tag{10.40}$$

In the absence of a gravitational field, $\mathbf{g}$, there is no flow, *i.e.*, $\mathbf{u} = \mathbf{v}_i = 0$, and one can obtain the classical solution for the electric field problem of a charged sphere in an electrolyte solution. Consequently, $n_i^{(0)}$ obeys the Boltzmann equation where

$$n_i^{(0)} = n_{i\infty} \exp(-z_i \Psi^{(0)}) \tag{10.41}$$

and we obtain the conventional Poisson–Boltzmann equation

$$\frac{1}{r^2}\frac{d}{dr}\left(r^2 \frac{d\Psi^{(0)}}{dr}\right) = -\kappa^2 \frac{\sum_{i=1}^{N} n_{i\infty} z_i \exp(-z_i \Psi^{(0)})}{\sum_{i=1}^{N} n_{i\infty} z_i^2} \tag{10.42}$$

where the Debye length κ^{-1} is given by

$$\kappa^{-1} = \sqrt{\frac{\epsilon k_B T}{\sum_{i=1}^{N} n_{i\infty} z_i^2 e^2}} \tag{10.43}$$

and

$$\Psi^{(0)} = \frac{e\psi^{(0)}(r)}{k_B T} \tag{10.44}$$

Solution of Eq. (10.42) subject to the boundary conditions of

$$\Psi^{(0)} = \frac{e\zeta}{k_B T} \quad \text{at } r = a \tag{10.45}$$

and

$$\Psi^{(0)} \to 0 \quad \text{at } r \to \infty \tag{10.46}$$

provides the distribution of $\psi^{(0)}(r)$ and $n_i^{(0)}(r)$ as a function of the radial position (noting that it is independent of angular position).

Recognizing that the curl ($\nabla\times$) of a scalar is zero and using the identity

$$\nabla^2\mathbf{u} = \nabla(\nabla\cdot\mathbf{u}) - \nabla\times(\nabla\times\mathbf{u}) \tag{10.47}$$

the substitution of Eqs. (10.37) to (10.39) into the fundamental governing equations leads to

$$\mu\nabla\times\nabla\times\nabla\times\mathbf{u} = \sum_{i=1}^{N}\nabla\delta\mu_i\times\nabla n_i^{(0)} \tag{10.48}$$

and

$$\nabla\cdot\left[n_i^{(0)}\mathbf{u} - n_i^{(0)}\nabla\delta\mu_i/\lambda_i\right] = 0 \tag{10.49}$$

In the above derivation, the products of $\mathbf{u}$, δn_i, $\delta\psi$, and $\delta\mu_i$ are neglected.

The sedimentation problem reduces to solving Eqs. (10.48) and (10.49). The boundary conditions for $\mathbf{u}$ are those given by Eqs. (10.24), (10.28), and (10.29). Making use of the boundary condition of Eq. (10.25) and the no-slip condition at $r = a$, one can write

$$(\nabla\mu_i)\cdot\mathbf{i}_r = \nabla(\mu_i^0 + \delta\mu_i)\cdot\mathbf{i}_r = 0 \tag{10.50}$$

leading to

$$\nabla(\delta\mu_i)\cdot\mathbf{i}_r = 0 \quad \text{at } r = a \tag{10.51}$$

At a location far from the sphere, Ohshima *et al.* (1984) set $\delta\psi\to 0$, and $\delta n_i\to 0$. With the aforementioned conditions, Eq. (10.40) leads to

$$\delta\mu_i \to 0 \quad \text{at } r\to\infty \tag{10.52}$$

Equations (10.51) and (10.52) are the boundary conditions for $\delta\mu_i$.

Ohshima *et al.* (1984) obtained solutions for $\psi^{(0)}$, $n^{(0)}$, $\mathbf{u}$, and $\delta\mu_i$, which were subsequently used to evaluate the sedimentation velocity and potential. They provided analytical expressions for both the sedimentation velocity and potential for low zeta potentials for highly dilute suspensions of charged spheres.

10.3.4 Sedimentation Velocity: Single Charged Sphere

Ohshima *et al.* (1984), in their classical paper, provided numerical calculations for the sedimentation velocity of a single sphere that is valid for low zeta potentials. As well, for ease of computations, they also provided an approximate expression for the sedimentation velocity that is given by

$$\begin{aligned}\frac{U_{Sed}}{U_{St}} = 1 - A_1\left(\frac{e\zeta}{k_BT}\right)^2 &\{\exp(2\kappa a)[3E_4(\kappa a) - 5E_6(\kappa a)]^2 \\ &+ 8\exp(\kappa a)\,[E_3(\kappa a) - E_5(\kappa a)] \\ &- \exp(2\kappa a)\,[4E_3(2\kappa a) + 3E_4(2\kappa a) - 7E_8(2\kappa a)]\} + \mathcal{O}(\zeta^3)\end{aligned} \tag{10.53}$$

where

$$E_n(x) = x^{n-1} \int_x^\infty \frac{\exp(-t)}{t^n} dt \tag{10.54}$$

is an exponential integral of order n, Abramowitz and Stegun (1965), and

$$A_1 = \frac{1}{8} \left(\frac{\sum_{i=1}^{N} z_i^4 n_{i\infty} m_i}{\sum_{i=1}^{N} z_i^2 n_{i\infty}} \right) \tag{10.55}$$

where

$$m_i = \frac{2\epsilon k_B T}{3\mu z_i^2 e^2} \lambda_i \tag{10.56}$$

Here, m_i is the scaled drag coefficient of the ith ionic species and λ_i is the drag coefficient given as

$$\lambda_i = \frac{k_B T}{D_i} \tag{10.57}$$

Equation (10.53) gives the ratio of the sedimentation velocity of a charged sphere to that of an uncharged sphere, *i.e.*, Stokes sedimentation velocity given by Eq. (10.1). Figure 10.5 shows the variation of the velocity ratio, U_{Sed}/U_{St}, for a single charged sedimenting sphere with the dimensionless surface zeta potential of $e\zeta/k_B T$. Both the complete numerical solution and the approximate analytical expression, Eq. (10.53) are plotted in Figure 10.5. For a very thin electric double layer, $\kappa a \to \infty$, as well

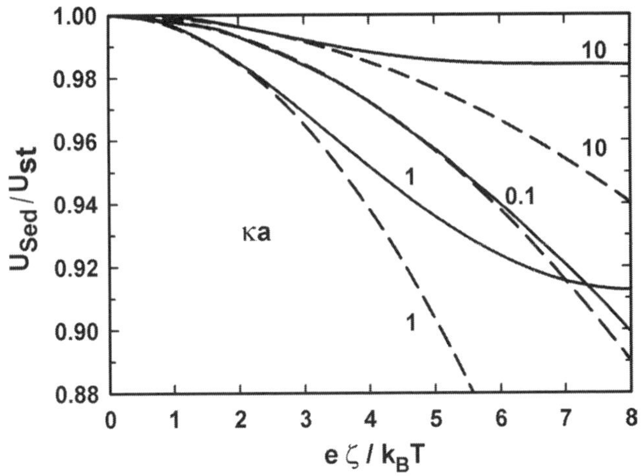

Figure 10.5. Ratio of sedimentation velocity to Stokes velocity, U_{Sed}/U_{St}, as a function of dimensionless zeta potential, $e\zeta/k_B T$. Solid lines represent the complete solution of Ohshima *et al.* (1984), and the dashed lines show the approximate solution, Eq. (10.53). Physical conditions are for KCl electrolyte solution at 25°C.

as for a very thick double layer, $\kappa a \to 0$, the velocity ratio is unity and the surface charge has a negligible influence on the Stokes sedimentation velocity. The effect of the surface charge on the sedimenting velocity is evident only for the intermediate κa values and at large zeta potentials. From Figure 10.5, it can be observed that the approximate analytical expression is valid for $e\zeta/k_BT \leq 3$ for the case of (1 : 1) KCl electrolyte solution.

10.3.5 Sedimentation Potential: Dilute Suspensions

Ohshima *et al.* (1984) considered a dilute suspension made up of identical spherical particles in an electrolyte solution. The electric field arising from each charged particle is superimposed to give the macroscopic electric sedimentation potential, E_{Sed}. One can think of E_{Sed} as being a uniform electric field strength and being the average of the gradient of the potential ψ within the suspension over the suspension volume, V_{susp}. Consequently, one can write

$$\mathbf{E}_{Sed} = -\frac{1}{V_{susp}}\int_{V_{susp}} \nabla\psi(r)dV_{susp} = -\langle\nabla\psi\rangle \tag{10.58}$$

The above definition of the sedimentation potential is equivalent to that of $\mathbf{E}_{susp}$ used in Eq. (9.158). Making use of Eq. (10.38) and the fact that the volume average of $\nabla\psi^{(0)}(r)$ is zero, the sedimentation potential is given by

$$\mathbf{E}_{Sed} = -\frac{1}{V_{susp}}\int_{V_{susp}} \nabla\delta\psi(r,\theta)dV_{susp} \tag{10.59}$$

Equation (10.59) clearly illustrates that the sedimentation potential arises from the distortion of the electric double layer and it is macroscopic.

Making use of the solution for $\delta\psi(r,\theta)$ and the fact that the average current density, $\langle\mathbf{i}\rangle$, is zero where

$$\langle\mathbf{i}\rangle = \frac{1}{V_{susp}}\int_{V_{susp}} \mathbf{i}(r,\theta)dV_{susp} \tag{10.60}$$

Ohshima *et al.* (1984) obtained a general solution for $\mathbf{E}_{Sed}$ that would require a numerical procedure for its evaluation.

For the special case of dilute suspensions and low zeta potentials, Ohshima *et al.* (1984) provided a useful approximate analytical expression for the sedimentation potential that is valid for low zeta potentials, *i.e.*, $e\zeta/k_BT \leq 2$.

$$\mathbf{E}_{Sed}(\kappa a) = \mathbf{E}_{Sed} = -\frac{\epsilon\zeta(\rho_P - \rho)\phi_p}{\mu\sigma^{\infty}}\mathbf{g}H(\kappa a) + \mathcal{O}(\zeta^2) \tag{10.61}$$

where

$$H(\kappa a) = 1 + \exp(\kappa a)\,[2E_5(\kappa a) - 5E_7(\kappa a)] \tag{10.62}$$

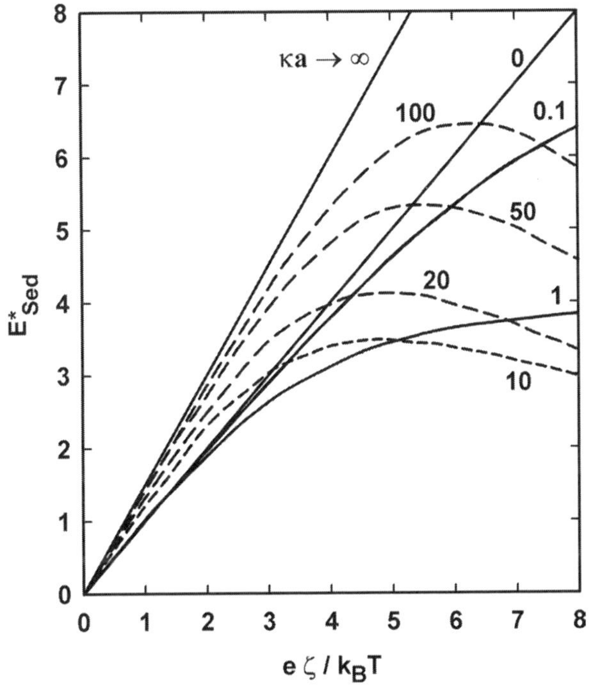

Figure 10.6. Variation of dimensionless sedimentation potential, E^*_{Sed}, with dimensionless zeta potential for a dilute sedimenting suspension of spheres in a KCl solution at 25°C. As a result of the Onsager relationship, the curves are similar to those obtained for electrophoresis, see Figure 9.10.

The exponential integral $E_n(\kappa a)$ is given by Eq. (10.54). The expression for the sedimentation potential, Eq. (10.61), agrees with the study of Saville (1982). As stated by Ohshima *et al.* (1984), the negative sign on the right hand side of Eq. (10.61) implies that for positively charged spheres, the electric field acts upward in the opposite direction of gravity along which sedimentation takes place. Such a result is to be expected as the counterion charge cloud for positively charged particles is swept to the trailing edge of the particle.

Figure 10.6 shows the variation of the dimensionless sedimentation potential, E^*_{Sed}, with the dimensionless zeta potential, $e\zeta / k_B T$, obtained using numerical calculations, where

$$E^*_{Sed} = \frac{3\mu e \sigma^\infty}{2\epsilon k_B T(\rho_p - \rho)\phi_p} \frac{|\mathbf{E}_{Sed}|}{|\mathbf{g}|} \tag{10.63}$$

For both the limiting cases of $\kappa a \to \infty$ and $\kappa a \to 0$, E^*_{Sed} increases with the surface zeta potential. As with the case of electrophoretic mobility, O'Brien and White (1978), the E^*_{Sed} curve exhibits a maximum at $e\zeta / k_B T \approx 5$ for $\kappa a \geq 3$. See Figure 9.10 for the corresponding behavior of the electrophoretic mobility. For low zeta potentials, we can make use of the analytical expression for $\mathbf{E}_{Sed}$, Eq. (10.61), to give

$$E^*_{Sed} = \frac{3}{2}\left(\frac{e\,\zeta}{k_B T}\right) H(\kappa a) \tag{10.64}$$

Equation (10.64) is strictly valid for low zeta potentials except for the special limiting cases of $\kappa a \to \infty$ and $\kappa a \to 0$, where it is valid for higher zeta potentials.

It is interesting to note that in the limit of $\kappa a \to \infty$, $H(\kappa a) \to 1$, and Eq. (10.61) degenerates to the expression given by Eq. (10.12), which was obtained using heuristic arguments.

The function $H(\kappa a)$ is related to Henry's function, $f(\kappa a)$ used in electrophoretic analysis. The relationship is given as

$$H(\kappa a) = \frac{2}{3} f(\kappa a) \tag{10.65}$$

In the limits of $\kappa a \to 0$, $H(\kappa a) = 2/3$ and $f(\kappa a) \to 1$, while for $\kappa a \to \infty$, $H(\kappa a) \to 1$ and $f(\kappa a) \to 3/2$.

Recalling Henry's analysis for the electrophoretic mobility at low zeta potentials, Eq. (9.86),

$$\eta = \frac{2}{3}\frac{\epsilon\zeta}{\mu} f(\kappa a) \tag{10.66}$$

we can write the electrophoretic mobility of a single spherical particle as

$$\eta(\kappa a) = \frac{\epsilon\zeta}{\mu} H(\kappa a) \tag{10.67}$$

where the electrophoretic mobility is defined without any ambiguity as

$$\eta(\kappa a) = \eta = \frac{U}{E_\infty}$$

Combining Eqs. (10.61) and (10.67) leads to

$$\mathbf{E}_{Sed}(\kappa a) = \mathbf{E}_{Sed} = -\frac{\phi_p(\rho_p - \rho)\eta}{\sigma^\infty}\mathbf{g} \tag{10.68}$$

Equation (10.68) is a statement of the Onsager relationship between the sedimentation potential and the electrophoretic mobility. For the present, let us state that Eq. (10.68) is correct only for a dilute system, $\phi_p \to 0$. Equation (10.68) was also derived by de Groot *et al.* (1952) using irreversible thermodynamics. Although the derivation presented here was for the case of small values of zeta potential, where both the sedimentation potential and the electrophoretic mobility were linked together through the function, $H(\kappa a)$, Ohshima *et al.* (1984) showed that Eq. (10.68) is also valid for high values of zeta potential.

In conformity with the manner used to non-dimensionalize electrophoretic mobility, we can write

$$\eta^*(\kappa a) = \frac{3\mu e}{2\epsilon k_B T}\eta = \frac{3U^*}{2\beta^*} \tag{10.69}$$

where η^* is the dimensionless electrophoretic mobility that was previously defined as $3U^*/2\beta^*$, Section 9.3.5. From Eqs. (10.63), (10.68), and (10.69), we can write

$$E^*_{Sed}(\kappa a) = \eta^*(\kappa a) \tag{10.70}$$

The equality of Eq. (10.70) is another statement of the Onsager relationship, which is also valid for higher zeta potentials. In Figure 9.10, the dimensionless electrophoretic mobility, $3U^*/2\beta^*$, was plotted against the scaled zeta potential, $e\zeta/k_BT$, whereas in Figure 10.6, the dimensionless sedimentation potential, E^*_{Sed}, is plotted against $e\zeta/k_BT$. Comparison between Figures 9.10 and 10.6 would indicate the reciprocity between E^*_{Sed} and η^*, where E^*_{Sed} is equivalent to η^* as indicated by Eq. (10.70).

10.4 SEDIMENTATION POTENTIAL OF CONCENTRATED SUSPENSIONS

It was mentioned in Section 10.1 that sedimentation of uncharged monodisperse concentrated suspensions still poses theoretical challenges toward determination of the sedimentation rate. The physical problem becomes more complex when the particles are charged. In order to facilitate the analysis, many workers have resorted to the use of cell models. In Chapter 9, we showed the use of Kuwabara's cell model by Levine and Neale (1974) and Ohshima (1997) for the study of electrophoresis of spheres in concentrated suspensions. In a similar manner, Levine *et al.* (1976) combined Henry's approach with Kuwabara's cell model to evaluate the sedimentation velocity and potential for non-overlapping double layers and low zeta potentials.

To facilitate the analysis of the problem of sedimentation potential in concentrated suspensions of charged spherical particles, the Kuwabara cell model was first employed by Levine *et al.* (1976). In a similar manner, Ohshima (1998) carried out sedimentation potential analysis using the approach of Levine *et al.* (1976), where he derived an analytical expression for the sedimentation potential that is applicable for low zeta potentials and non-overlapping electric double layers of adjacent particles. The primary modification of Ohshima (1998) to the original cell model approach of Levine *et al.* (1976) pertains to the introduction of the total electrochemical potential of the ions, μ_i, and its perturbation, $\delta\mu_i$, in presence of external forces (in this case gravity). Carrique *et al.* (2001) extended the study of Ohshima (1998) for higher zeta potentials. Their study utilized the Shilov–Zharkikh cell model (Shilov *et al.*, 1981), which, according to the discussion in Chapter 9, provides a self-consistent specification of the outer cell surface boundary condition that obeys Onsager reciprocal relationship between electrophoretic mobility and sedimentation potential.

The solution to the sedimentation potential problem for concentrated systems, *i.e.*, $\phi_p \gg 0$, can be analyzed in a similar manner as to the problem of electrophoresis using a perturbation analysis. The governing continuity, momentum and electrokinetic equations are similar to those presented in Section 10.3. With the use of Kuwabara's cell model, the outer cell boundary at $r = b$ determines the system volume fraction, ϕ_p where $\phi_p = (a/b)^3$. Consequently, the outer boundary conditions are set at $r = b$ rather than at $r \to \infty$ as was the case for dilute suspensions. To that end, the boundary conditions at the outer cell envelope $r = b$ can be written as

$$\text{Parallel flow:} \quad u_r = -U_{Sed}\cos\theta \tag{10.71}$$

$$\text{Zero vorticity:} \quad \nabla \times \mathbf{u} = 0 \tag{10.72}$$

$$\text{Electrically neutral cell:} \quad \frac{d\psi^{(0)}}{dr} = 0 \tag{10.73}$$

$$\text{Electric field at cell boundary:} \quad \delta\mu_i = -z_i e b\, \mathbf{n}\cdot\mathbf{E} \quad \text{and} \quad \delta n_i \simeq 0 \tag{10.74}$$

The last boundary condition specified above, Eq. (10.74), is a source of controversy in literature. As stated here, the equation corresponds to the Shilov–Zharkikh specification, with the electric field **E** referring to the suspension (or internal) electric field, $\mathbf{E}_{susp}$ [*cf.*, Eq. (9.154)]. Ohshima (1998) specified this condition as $\delta\mu_i = 0$ for the sedimentation problem, which is analogous to setting the suspension electric field to zero. The interpretation of this outer cell surface boundary condition is subject to considerable debate (Dukhin *et al.*, 1999). A detailed discussion on the implications of this electrochemical boundary condition in the results of Ohshima (1998) and Carrique *et al.* (2001) is provided in Zholkovskij *et al.* (2006). A direct solution of the perturbation problem for sedimentation potential was obtained by Zholkovskij *et al.* (2006). To provide a correct description of the perturbation problem for sedimentation potential, they used the boundary condition Eq. (10.74) based on the Shilov–Zharkikh cell model in accordance with the discussion presented by Dukhin *et al.* (1999).

To avoid ambiguity and any confusion arising from the use of different electrochemical boundary conditions at the outer cell surface in the perturbation analysis of sedimentation potential, we will present here a different approach for calculation of the sedimentation potential. Following the discussion on the Onsager reciprocity criterion in Chapter 9 (see Section 9.4.6), it may be noted that predictions of the electrophoretic mobility can be used in the Onsager relationship between sedimentation potential and electrophoretic mobility to determine $\mathbf{E}_{Sed}$. Thus, if one can determine the electrophoretic mobility of a concentrated suspension accurately, the sedimentation potential for the suspension at the same particle volume fraction can be evaluated by

$$\mathbf{E}_{Sed}(\kappa a, \phi_p) = -\frac{\phi_p(\rho_p - \rho)\,\eta(\kappa a, \phi_p)}{\sigma^{\infty}}\mathbf{g} \tag{10.75}$$

The term $\eta(\kappa a, \phi_p)$ in Eq. (10.75) refers to the external electrophoretic mobility. In Chapter 9, it was observed that all the perturbation theories (based either on the Levine–Neale or the Shilov–Zharkikh cell models) eventually led to the correct expressions for the external electrophoretic mobility over the entire range of particle volume fractions and κa. Consequently, substituting the external electrophoretic mobility obtained from the Levine–Neale or Shilov–Zharkikh cell model in Eq. (10.75) will lead to the correct prediction of the sedimentation potential. This was indeed found to be the case (Carrique *et al.*, 2001; Zholkovskij *et al.*, 2006). The expression for the external electrophoretic mobility, $\eta(\kappa a, \phi_p)$, based on the Levine–Neale cell model is given by Eq. (9.128) or (9.148).

There are thus two approaches for evaluating the sedimentation potential in concentrated suspensions. The first approach involves direct solution of the perturbation equations for sedimentation potential based on the Shilov–Zharkikh cell model (Zholkovskij *et al.*, 2006). The second approach involves using the external electrophoretic mobility calculated employing either the Levine–Neale or the Shilov–Zharkikh cell models in the Onsager relationship, Eq. (10.75), to indirectly evaluate the sedimentation potential. In the following, we will compare the sedimentation potentials obtained using the direct and indirect approaches, and discuss the implications of these comparisons. However, before this comparison, we present some pertinent non-dimensionalizations required for a facile representation of the results.

In conformity with our previous analysis, we can non-dimensionalize the electrophoretic mobility as

$$\eta^*(\kappa a, \phi_p) = \frac{3\mu e}{2\epsilon k_B T}\eta(\kappa a, \phi_p) \tag{10.76}$$

and the sedimentation potential as

$$E^*_{Sed}(\kappa a, \phi_p) = \frac{3\mu e \sigma^\infty}{2\epsilon k_B T(\rho_p - \rho)\phi_p}\frac{|\mathbf{E}_{Sed}|}{|\mathbf{g}|} \tag{10.77}$$

Combining Eqs. (10.76) and (10.77), and making use of the Onsager relationship, Eq. (10.75), leads to

$$E^*_{Sed}(\kappa a, \phi_p) = \eta^*(\kappa a, \phi_p) \tag{10.78}$$

In terms of the approximate analytical solution that is valid for low zeta potential, given by Eq. (9.149) and the function $G(\kappa a, \phi_p)$ of Eq. (9.150), one can express the dimensionless electrophoretic mobility as

$$\eta^*(\kappa a, \phi_p) = \frac{3}{2}\left(\frac{e\zeta}{k_B T}\right)G(\kappa a, \phi_p) \tag{10.79}$$

Making use of Eqs. (10.78) and (10.79), we obtain an approximate analytical expression for evaluating E^*_{Sed},

$$E^*_{Sed}(\kappa a, \phi_p) = E^*_{Sed} = \frac{3}{2}\left(\frac{e\zeta}{k_B T}\right)G(\kappa a, \phi_p) \tag{10.80}$$

Of interest, for a dilute suspension, in the limit of $\phi_p \rightarrow 0$, comparing Eqs. (10.64), (10.65), and (10.79), one obtains

$$G(\kappa a, \phi_p) = H(\kappa a) = \frac{2}{3}f(\kappa a) \tag{10.81}$$

where $H(\kappa a)$ is given by Eq. (10.62) and $f(\kappa a)$ is Henry's function given by Eq. (10.65) or (9.85). In the limit of $\phi_p \to 0$ and $\kappa a \to \infty$, the various functions yield:

$$H(\kappa a) \to 1, \quad f(\kappa a) \to \frac{3}{2}, \quad \text{and} \quad G(\kappa a, \phi_p) \to 1$$

and in the limit of $\phi_p \to 0$ and $\kappa a \to 0$, one obtains

$$H(\kappa a) \to \frac{2}{3}, \quad f(\kappa a) \to 1, \quad \text{and} \quad G(\kappa a, \phi_p) \to \frac{2}{3}$$

The latter limit is reached only when ϕ_p is very close to zero. This asymptotic behavior is evident from Figures 9.16 and 9.20.

Figure 10.7 shows the variation of E_{Sed}^{**} with volume fraction, ϕ_p, where E_{Sed}^{**} is defined as

$$E_{Sed}^{**} = \left[\frac{\mu \sigma^{\infty}}{g(\rho_p - \rho)\epsilon\zeta} \right] E_{Sed} \tag{10.82}$$

Using Eq. (10.75) for E_{Sed}, we can obtain

$$E_{Sed}^{**} = \frac{\mu \phi_p}{\epsilon \zeta} \eta(\kappa a, \phi_p) \tag{10.83}$$

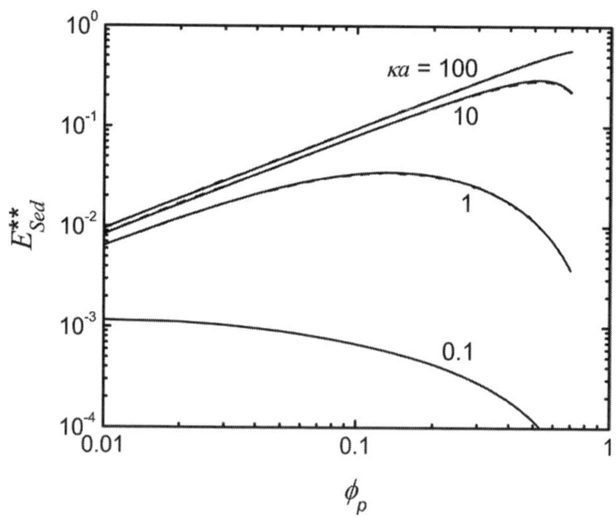

Figure 10.7. Variation of scaled sedimentation potential, E_{Sed}^{**}, with particle volume fraction, ϕ_p, for different values of scaled Debye length (κa). Solid lines: Direct solution of the perturbation problem for sedimentation potential using the Shilov–Zharkikh cell model (Zholkovskij *et al.*, 2006). Dashed lines: Predictions obtained by substituting the electrophoretic mobility in the Onsager reciprocal relationship. Equations (10.84) and (9.150) (Ohshima, 1997) were employed for obtaining the sedimentation potential using this indirect approach.

where $\eta(\kappa a, \phi_p)$ is the external electrophoretic mobility. Here, we use Eq. (9.149) to obtain the electrophoretic mobility. Substitution of Eq. (9.149) in (10.83) yields

$$E_{Sed}^{**} = \phi_p G(\kappa a, \phi_p) \tag{10.84}$$

where the function $G(\kappa a, \phi_p)$ is given by Eq. (9.150).

The solid lines in Figure 10.7 are obtained using perturbation analysis for the sedimentation potential given by Zholkovskij *et al.* (2006). The dashed lines are obtained by substituting Ohshima's approximate expression for electrophoretic mobility in terms of $G(\kappa a, \phi_p)$, given by Eq. (9.150), in Eq. (10.84). The two results are virtually indistinguishable, which is a clear demonstration of two facts, namely (i) the Shilov–Zharkikh cell model conforms to the Onsager reciprocity criterion, and (ii) the electrophoretic mobility determined using the Levine–Neale cell model can be used in the Onsager relationship, Eq. (10.75), to provide an accurate estimate of the sedimentation potential. Figure 10.7 also shows that E_{Sed}^{**} exhibits a maximum when plotted against ϕ_p. The maximum in the scaled sedimentation potential, E_{Sed}^{**}, shifts to higher values of ϕ_p with increasing κa. For a given particle volume fraction, ϕ_p, higher scaled sedimentation potentials are observed at higher values of κa.

Figure 10.8 shows the variation of the scaled sedimentation potential, E_{Sed}^{**} with κa for different values of particle volume fraction, ϕ_p. The solid lines are from the

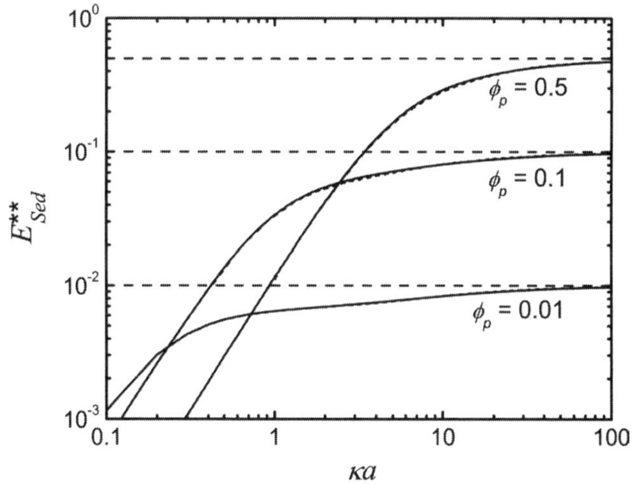

Figure 10.8. Variation of scaled sedimentation potential, E_{Sed}^{**}, with scaled Debye length (κa) for different particle volume fractions. Solid lines: Direct solution of the perturbation problem for sedimentation potential using the Shilov–Zharkikh cell model (Zholkovskij *et al.*, 2006). Dashed lines: Predictions obtained by substituting the electrophoretic mobility in the Onsager reciprocal relationship. Equations (10.84) and (9.150) (Ohshima, 1997) were used for obtaining this indirect estimate of the sedimentation potential. The horizontal dashed lines represent the limiting values of the sedimentation potential as $\kappa a \to \infty$. It is evident that $E_{Sed}^{**} \to \phi_p$ as $\kappa a \to \infty$.

direct solution of the perturbation problem for sedimentation potential according to Zholkovskij *et al.* (2006). The dashed lines are obtained using the indirect approach based on the Onsager reciprocal relationship [Eqs. (10.84) and (9.150)]. Once again, the agreement between the two procedures of evaluating sedimentation potential is excellent. For large κa, the asymptotic value reached by E^{**}_{Sed} is ϕ_p. This can be clearly observed from Eq. (10.84). Here, $G(\kappa a, \phi_p)$ approaches unity as $\kappa a \to \infty$ and consequently, $E^{**}_{Sed} \to \phi_p$ as $\kappa a \to \infty$. For the case of a dilute suspension, $\phi_p \to 0$, the conclusion of $E^{**}_{Sed} \to \phi_p$ as $\kappa a \to \infty$ could have been easily reached from the dilute suspension analysis. One can easily show that

$$\frac{E^{*}_{Sed}}{(e\zeta/k_B T)} = \frac{3}{2\phi_p} E^{**}_{Sed} \tag{10.85}$$

For the case of a dilute suspension for $\kappa a \to \infty$ one can observe that $E^{*}_{Sed}/(e\zeta/k_B T) \to 3/2$ as $\kappa a \to \infty$. The plot of Figure 10.8 would indicate that the asymptotic behavior of $E^{**}_{Sed} \to \phi_p$ as $\kappa a \to \infty$ is equally valid for non-dilute suspensions.

One should note that Eq. (9.150) represents an approximate analytic solution for the electrophoretic mobility function, $G(\kappa a, \phi_p)$ (Ohshima, 1997). Despite the "approximate" nature of this expression, the virtual overlap between the solid and dashed lines in Figures 10.7 and 10.8 is remarkable. In fact, if one uses Eq. (9.128) or Eq. (9.148) for the electrophoretic mobility in Eq. (10.83) to obtain the dashed lines, then the solid and dashed lines in these figures will be identical.

From the foregoing discussion, it is evident that analysis of sedimentation potential in concentrated suspensions is non-trivial and still poses significant difficulties. Such difficulties are owing to non-availability of direct means of measuring the sedimentation potential, as well as controversy related to application of proper boundary conditions in the perturbation analysis. One can, however, evaluate the sedimentation potential in concentrated suspensions quite accurately by employing an accurate expression for the electrophoretic mobility in the Onsager reciprocal relationship. The electrokinetic phenomena dealing with sedimentation velocity and potential still pose considerable challenges in analysis of multi-species systems having particles of different surface potentials, densities, and sizes.

10.5 NOMENCLATURE

a	radius of a spherical particle, m
A	cross section area, m^2
D_i	diffusion coefficient of ith ionic species, m^2/s
e	elementary charge, C
E_{susp}	electric field strength measured within the suspension, V/m
E_{Sed}	induced electric field referred to as sedimentation potential, V/m
E_θ	electric field strength in the angular direction, V/m
F_g	force on a particle due to gravity, N
F_h	force on a particle due to hydrodynamic drag, N
$\mathbf{g}$	acceleration due to gravity, m/s^2

g	magnitude of acceleration due to gravity, $\|\mathbf{g}\|$, m/s^2
$\mathbf{i}_r$	unit normal outward vector on the sphere surface
I	total current, A
k_B	Boltzmann constant, J/K
n_i	ionic number concentration of the ith species, m^{-3}
$n_{i\infty}$	bulk ionic number concentration of the ith species, m^{-3}
N	total number of ionic species
N_p	total number of charged particles
N_0	number of charged particles per unit volume, m^{-3}
p	pressure, Pa
p_{Sed}	induced sedimentation pressure gradient, Pa/m
r	radial coordinate (in spherical coordinate system), m
$\mathbf{r}$	position vector, m
Re_p	particle Reynolds number
T	absolute temperature, K
u_r	local radial fluid velocity, m/s
u_θ	local angular fluid velocity, m/s
$\mathbf{u}$	local fluid velocity, m/s
U	electrophoretic particle velocity, m/s; fluid velocity far from the particle or fluid velocity at the outer shell of a cell
U_{Sed}	sedimentation velocity of a single charged particle or a suspension of charged particles, m/s
U_{St}	Stokes settling velocity, m/s
U_{HS}	Hindered settling velocity, m/s
$\mathbf{v}_i$	velocity of the ith ionic species, m/s
v_{ir}	radial velocity component of the ith ionic species, m/s
$v_{i\theta}$	angular velocity component of the ith ionic species, m/s
z_i	valency of the ith ionic species

Greek Symbols

δ	symbol for a perturbation variable
ϵ	electric permittivity of continuous phase electrolyte solution, C V^{-1} m^{-1}
λ_i	drag coefficient of ith ionic species, Js/m^2
ζ	zeta potential, V
ζ^*	dimensionless zeta potential $= ze\zeta/k_BT$
η_{el}	electrophoretic mobility of a charged particle defined as U/E_∞, m^2/Vs
σ^∞	electric conductivity of continuous phase electrolyte solution, S/m
θ	angular coordinate
ϕ_p	particle volume fraction
κ	inverse Debye length, m^{-1}
μ	fluid viscosity, Pa s
μ_i	electrochemical potential of ith ionic species, J
μ_i^o	constant term in μ_i
ρ_f	ionic free charge density, C/m^3

ρ	continuous medium density, kg/m^3
ρ_p	particle density, kg/m^3
τ_{rr}	normal stress acting on the r-plane in r-direction, N/m^2
$\tau_{r\theta}$	shear stress acting on the r-plane in θ-direction, N/m^2
θ, ϕ	angular and azimuthal coordinate directions in spherical coordinates
$\psi^{(0)}$	equilibrium (unperturbed) electric potential in absence of flow, V
$\Psi^{(0)}$	scaled equilibrium electric potential, $e\psi^{(0)}/k_B T$
ω_i	mobility of ith ionic species, m^2/J s

10.6 PROBLEMS

10.1. For the case of concentrated suspensions with particles having low zeta potentials, show that the electric potential in the absence of flow, for the sedimentation potential problem, using Ohshima perturbation approach, becomes

$$\frac{1}{r^2}\frac{d}{dr}(r^2\psi^{(0)}) = \kappa^2\psi^{(0)}$$

subject to the boundary conditions

$$\psi^{(0)} = \zeta \quad \text{sphere surface}$$

and

$$\frac{d\psi^{(0)}}{dr} = 0 \quad \text{at } r = b \text{ (cell outer boundary).}$$

(a) Show that the solution of the governing differential equation subject to the two boundary conditions is given by

$$\psi^{(0)}(r) = \zeta\left(\frac{a}{r}\right)\left[\frac{\kappa b\cosh[\kappa(b-r)] - \sinh[\kappa(b-r)]}{\kappa b\cosh[\kappa(b-a)] - \sinh[\kappa(b-a)]}\right]$$

(b) Show that

$$\int_V \nabla\psi^{(0)}(r)dV = 0$$

where dV is an element in the spherical coordinate system.

(c) What conclusion can you draw from the result of part (b)?

10.2. Evaluate the sedimentation velocity ratio U_{Sed}/U_{St} for sedimenting spheres of 1.2 μm radius with a density of 2600 kg/m^3 in a very dilute aqueous suspension containing 10^{-5} M HCl at 25°C. The zeta potential of the spheres is 80 mV.

10.3. A suspension of charged spherical particles is allowed to settle in a 0.01 M KCl solution at 20°C in a graduated cylinder. The particles have a radius of 10 μm, a zeta potential of 35 mV and a density of 2600 kg/m^3. Assume a solution density of 998 kg/m^3 and a viscosity of 0.001 Pa s.

(a) Evaluate the sedimentation potential for such a settling suspension as a function of volume fraction.

(b) How would the result change for particles having a radius of 2 μm in 10^{-3} M KCL solution?

10.7 REFERENCES

Abramowitz, M., and Stegun, I. A., *Handbook of Mathematical Functions*, US Dept. of Commerce, (1970).

Batchelor, G. K., Sedimentation in a dilute polydisperse system of interacting spheres, Part I: General theory, *J. Fluid Mech.*, **124**, 379, (1982).

Batchelor, G. K., and Wen, C. S., Sedimentation in a dilute polydisperse system of interacting spheres, Part II: Numerical results, *J. Fluid Mech.*, **124**, 495, (1982).

Booth, F., Sedimentation potential and velocity of solid spherical particles, *J. Chem. Phys.*, **22**, 1956–1968, (1954).

Brinkman, H. C., A calculation of the viscous force exerted by a flowing fluid on a dense swarm of particles, *Appl. Sci. Res.*, **A1**, 27–34, (1947).

Carrique, F., Arroyo, F. J., and Delgado, A. V., Sedimentation velocity and potential in a concentrated colloidal suspension: Effect of a dynamic Stern layer, *Colloids Surf. A*, **195**, 157–169, (2001).

de Groot, R., Mazur, P., and Overbeek, J. Th. G., Nonequilibrium thermodynamics of the sedimentation potential and electrophoresis, *J. Chem. Phys.*, **20**, 1825–1829, (1952).

Dorn, E., *Ann. Physik*, **3**, 20, (1878).

Dukhin, A. S., Shilov, V., and Borkovskaya, Y., Dynamic electrophoretic mobility in concentrated dispersed systems. Cell model *Langmuir*, **15**, 3452–3457, (1999).

Happel, J., and Brenner, H., *Low Reynolds Number Hydrodynamics*, Martinus Nijhoff, The Hague, (1983).

Happel, J., Viscous flow in multiparticle systems: Slow motion of fluids relative to beds of spherical particles, *AIChE J.*, **4**, 197–201, (1958).

Hunter, R. J., *Zeta Potential in Colloid Science*, Academic Press, London, (1981).

Kuwabara, S., The forces experienced by randomly distributed parallel circular cylinders or spheres in a viscous flow at small Reynolds numbers, *J. Phys. Soc. Japan*, **14**, 527–532, (1959).

Levine, S., and Neale, G. H., The prediction of electrokinetic phenomena within multiparticle systems. I. Electrophoresis and electroosmosis, *J. Colloid Interface Sci.*, **47**, 520–529, (1974).

Levine, S., Neale, G. H., and Epstein, N., The prediction of electrokinetic phenomena within multiparticle systems. II. Sedimentation potential, *J. Colloid Interface Sci.*, **57**, 424–437, (1976).

Masliyah, J. H., Settling of multi-species particle system, *Chem. Eng. Sci.*, **34**, 1166–1168, (1979).

O'Brien, R. W., and White, L. R., Electrophoretic mobility of a spherical colloidal particle, *J. Chem. Soc. Faraday Trans. II*, **74**, 1607–1626, (1978).

Ohshima, H., Healy, T. W., and White, L. R., Sedimentation velocity and potential in a dilute suspension of charged spherical particles, *J. Chem. Soc. Faraday Trans. II*, **80**, 1299–1317, (1984).

Ohshima, H., Electrophoretic mobility of spherical colloidal particles in concentrated suspensions, *J. Colloid Interface Sci.*, **188**, 481–485, (1997).

Ohshima, H., Sedimentation potential in a concentrated suspension of spherical colloidal particles, *J. Colloid Interface Sci.*, **208**, 295–301, (1998).

Overbeek, J. Th. G., Theorie der electrophorese, *Kolloidchem. Beih.*, **54**, 287–364, (1943).

Richardson, J. F., and Zaki, W. N., Sedimentation and fluidization. Part I, *Trans. Inst. Chem. Eng.*, **32**, 35–53, (1954).

Saville, D. A., The sedimentation potential in a dilute suspension, *Adv. Colloid Interf. Sci.*, **16**, 267–279, (1982).

Sengupta, M., The sedimentation of non-conducting solid spherical particles, *J. Colloid Interface Sci.*, **26**, 240–243, (1968).

Shilov, V. N., Zharkikh, N. I., and Borkovskaya, Y. B., Theory of non-equilibrium electrosurface phenomena in concentrated disperse system. 1. Application of non-equilibrium thermodynamics to cell model, *Colloid J.*, **43**, 434, (1981).

Smoluchowski, M. von, Contribution a la theorie de l'endosmose electrique et de quelques phenomenes correlatifs, *Bull. International de l'Academie des Sciences de Cracovie*, **8**, 182–200, (1903).

Smoluchowski, M. von, in *Handbuch der Electrizitat und des Magnetismus*, (Graetz), **11**, 336, Barth, Leipzig, (1921).

Stigter, D., Sedimentation of highly charged colloidal spheres, *J. Phys. Chem.*, **84**, 2758–2762, (1980).

Stokes, G. G., On the effect of the internal friction of fluids on the motion of pendulums, *Trans. Cambridge Phil. Soc.*, **9**, 8–106, (1851).

Wang, H., and Wen, C. S., Interparticle potential and sedimentation of monodisperse colloid system, *AIChE J.*, **44**, 2520, (1998).

Zholkovskij, E. K., Masliyah, J. H., Shilov, V. N., and Bhattacharjee, S., Notes on electrokinetic cell model approach (submitted), (2006).

CHAPTER 11

LONDON–VAN DER WAALS FORCES AND THE DLVO THEORY

11.1 DISPERSION FORCES BETWEEN BODIES IN VACUUM

So far, we have solely considered the presence of electrostatic interactions between colloidal entities. For charged entities in vacuum, the interaction was expressed in terms of the Coulomb potential, while in a dielectric medium containing free charges, the interaction between charged particles was described in terms of a screened electric double layer potential. All the electrokinetic phenomena described so far considered only one type of interparticle force, which was electrostatic in origin. Forces of a more fundamental and more ubiquitous nature exist between every colloidal entity. Such forces between colloidal entities are always attractive in vacuum. Microscopic observations of colloidal particles reveal their tendency to form persistent aggregates induced by Brownian motion, clearly indicating the presence of an attractive force. Consequently, to understand the behavior of charged colloidal particles in a suspension, and colloidal transport phenomena, one needs to consider these forces in addition to the electrostatic forces.

The mutual attraction of particles in vacuum is a consequence of dispersion forces, often called the London–van der Waals forces. The dispersion attraction forces are due to the spontaneous fluctuation of the electronic cloud in one material causing a corresponding fluctuation in neighboring material, leading, on the average, to an attractive force (Ross and Morrison, 1988; Morrison and Ross, 2002). These forces are strictly quantum mechanical in nature and they are expressed in terms of the

---

Electrokinetic and Colloid Transport Phenomena, by Jacob H. Masliyah and Subir Bhattacharjee
Copyright © 2006 John Wiley & Sons, Inc.

same oscillator strengths as appear in the equations for the dispersion of light or electromagnetic waves.

When we consider monatomic molecules (such as noble gases, *e.g.*, Helium and Argon), a fair idea about the origin of the dispersion forces can be obtained by considering the atoms to have a positively charged nucleus and a negatively charged electron cloud, which is in a constant state of oscillation around the nucleus. Such a system can behave as an electromagnetic dipole. When another atom of the noble gas experiences the influence of such a dipole, the oscillations of its electron cloud are also affected, and so is its dipolar nature. Fluctuating dipoles can also be induced in a non-polar atom or molecule by absorption of photons from the background electromagnetic radiation field. Interaction between such a pair of atomic dipoles gives rise to the attractive dispersion forces. When the atoms are brought close together, such that their electron clouds overlap, they experience a strong repulsive force. The classical Lennard–Jones potential summarizes the interaction between such a pair of atoms through an empirical equation of the form

$$u_{LJ} = \epsilon_D \left[\left(\frac{\sigma}{r} \right)^{12} - \left(\frac{\sigma}{r} \right)^{6} \right] \tag{11.1}$$

where ϵ_D is a characteristic energy of the dipolar interaction, and σ is the distance of neutral approach, both being empirical constants. The second term in Eq. (11.1) represents the dependence of the attractive dispersion force on the separation distance r between two atoms. The attractive force with the inverse r^6 dependence arises due to several types of dipolar interactions, namely, London interactions, dipole-dipole (Keesom) interactions, dipole-induced dipole (Debye) interactions, *etc.* (Israelachvili, 1991). The Lennard–Jones potential forms the basis for the calculation of non-bonded interactions in most atomic calculations like molecular mechanics simulations. These interactions are considered to be pairwise additive, and can be employed to determine interaction between polyatomic molecules.

The situation becomes somewhat more complex when we consider colloidal particles of much larger dimensions comprising millions of atoms. Calculation of the dispersion interactions between such large particles is primarily achieved through two techniques: One is based on a microscopic molecular model, which is attributed to Hamaker (1937) and the other is based on a macroscopic continuum model of condensed media, attributed to Lifshitz (1956).

In the microscopic model, the attraction between particles is calculated by summing the attractive energies between all pairs of atoms in the separate particles, ignoring multibody perturbations. This approximation is equivalent to predicting the spectra of condensed media as the sum of the molecular spectra. Corrections have been introduced to account for such factors as third body perturbations, the effect of intervening material, and retardation of the dispersion forces due to the finite speed of light.

In the macroscopic model of particle-particle dispersion attractions, the lowering of the zero-point energy of a particle, due to the coordinator of its instantaneous electric moments with those of a nearby particle, is calculated by quantum electrodynamics

(Ross and Morrison, 1988). The expressions derived require as data the dielectric susceptibilities of the particles as a function of frequency and are more complex than those from the Hamaker theory. The complexity of Lifshitz formulae and the difficulty of obtaining the necessary material constants have hampered its use. Ninham and Parsegian (1970) have, however, developed a numerical method to approximate the necessary material functions from a few, readily obtained values, and have made the use of this theory applicable for some common materials.

In the Hamaker model, the free energy of interaction separates into a material-dependent constant (called the *Hamaker constant*) and a geometry-dependent integral. The Lifshitz theory, on the other hand, gives such a separation of terms only for the special case of interactions between parallel plates. In this chapter we will primarily restrict our discussion of the van der Waals interactions within the context of Hamaker's microscopic approach. Readers interested in the electrodynamic approach are referred to the original work of Lifshitz (1956), the subsequent developments made by Mahanty and Ninham (1976), and a more recent summary of the theory given by Russel *et al.* (1989).

11.2 HAMAKER'S APPROACH

A brief description of Hamaker's approach is given here. The reader is referred to Russel *et al.* (1989), Morrison and Ross (2002), and Hunter (1991) for details and for further references. The approach of Hamaker (1937) is based on the assumption that the dispersion potential between two colloidal particles can be represented as summation of the dispersion interaction between pairs of atoms located within the two particles. The procedure starts by considering the attractive Lennard–Jones potential [*cf.*, Eq. (11.1)]

$$u_{LJ,att} = -\frac{c}{r^6} \tag{11.2}$$

where c is a constant characterizing the strength of the dispersion attraction between the atoms ($c = \epsilon_D \sigma^6$). Denoting the two interacting colloidal particles by 1 and 2, and assuming the density of atoms (atoms per unit volume) within these particles to be ρ_1 and ρ_2, respectively, the total interaction energy, dU_{12}, between all the atoms in a volume element dV_1 in particle 1 and a volume element dV_2 in particle 2 is

$$dU_{12} = -\frac{c\rho_1\rho_2}{r^6}\, dV_1\, dV_2 \tag{11.3}$$

where r is the separation between the volume elements dV_1 and dV_2. The situation is depicted in Figure 11.1.

Integrating Eq. (11.3) over all the volume elements in the two particles provides the dispersion energy of attraction between them. The work done by the attractive forces in bringing the two particles from infinity to a given separation distance is the

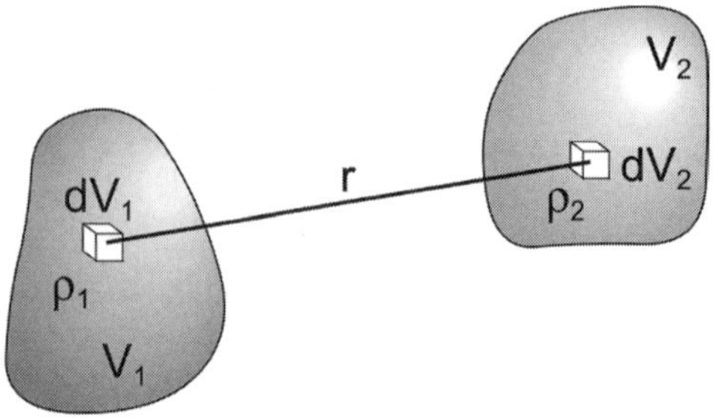

Figure 11.1. Schematic representation of Hamaker's approach applied to two colloidal particles.

potential due to the dispersion interaction. At a constant temperature, we obtain

$$U = -c \int_{V_1} \int_{V_2} \frac{\rho_1 \rho_2}{r^6} \, dV_1 \, dV_2 \tag{11.4}$$

where U is the total interaction energy between the volumes V_1 and V_2 of the particles 1 and 2, respectively.

Hamaker's approach is based on the following assumptions (Ross and Morrison, 1988):

1. The interactions can be considered pairwise; *i.e.*, many body interactions are ignored.
2. The bodies are assumed to have uniform density up to the interfaces.
3. The interactions are instantaneous.
4. The intervening medium is a vacuum.
5. All the dispersion force attractions are due to one dominant frequency.
6. Effects of free charge and permanent dipoles are negligible.
7. The bodies (particles) are not distorted by the attractive forces.

As a consequence of these assumptions, the term $c\rho_1\rho_2$ is a constant.

We now illustrate the application of this approach to obtain the dispersion interaction energy per unit area between two infinite planar slabs (half-spaces) separated by a distance h. Consider the geometry shown in Figure 11.2. The slabs are represented in cylindrical coordinates, with the r coordinate extending to infinity. The z-coordinate starting from the origin of slab 1 (denoted by O) extends to $+\infty$ [Figure 11.2(a)]. The z'-coordinate of the infinite slab 2 with origin at O' extends to $-\infty$ [Figure 11.2(b)]. For this geometry, the interaction energy between slab 1 and an arbitrarily chosen point P in slab 2 is first evaluated. In cylindrical coordinates, the volume element dV_1 in slab 1 is represented by

$$dV_1 = 2\pi r dr dz \tag{11.5}$$

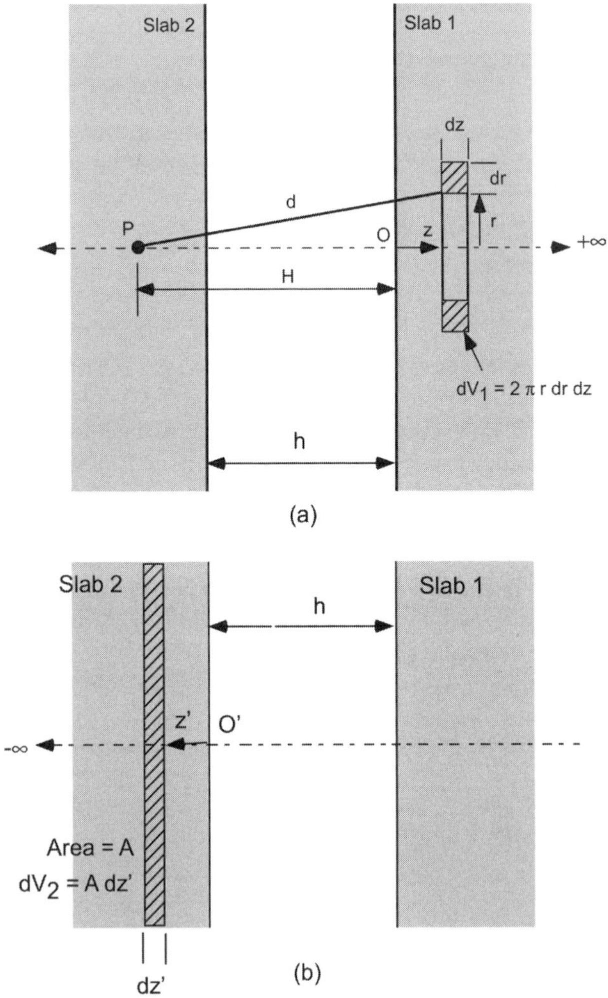

Figure 11.2. Evaluation of van der Waals interaction energy per unit area between two infinite flat slabs.

Denoting the distance of this volume element from point P on slab 2 by d [Figure 11.2(a)], we can write

$$d = \sqrt{r^2 + (z + H)^2} \tag{11.6}$$

where H is the distance of the point P from the surface of slab 1. The total interaction between point P and volume element dV_1 is then given by

$$du_{1P} = -\frac{c\rho_1}{d^6}dV_1 = -\frac{2\pi c\rho_1}{[r^2 + (z + H)^2]^3}rdrdz \tag{11.7}$$

Integrating over the volume of the slab 1, this yields for the total interaction energy between the slab 1 and point P as

$$u_{1P} = -2\pi c\rho_1 \int_{z=0}^{\infty} \int_{r=0}^{\infty} \frac{r dr dz}{[r^2 + (z + H)^2]^3} \tag{11.8}$$

Evaluation of the two nested integrals provides

$$u_{1P} = -\frac{\pi c\rho_1}{6H^3} \tag{11.9}$$

The term u_{1P} denotes the interaction energy between slab 1 and the point P located in slab 2.

To evaluate the total interaction energy between the two infinite slabs, we now need to integrate the interaction energy between slab 1 and point P given by Eq. (11.9) over every point in slab 2. Toward this, let us first consider the point P to be in a differential volume element dV_2 in slab 2 as shown in Figure 11.2(b). This volume element is a solid cylindrical disc of cross sectional area A and thickness dz'. Note that since slab 1 is infinite, the interaction energy between any point located inside the volume element dV_2 at a fixed z' will by given by Eq. (11.9). In other words, the differential energy between slab 1 and the volume element dV_2 is

$$dU_{12} = \rho_2 u_{1P} dV_2 = -\frac{\pi c\rho_1\rho_2}{6H^3} A dz' \tag{11.10}$$

The total interaction energy per unit area of slab 2 is now given by integrating Eq. (11.10) over the entire thickness of slab 2, yielding,

$$\frac{U_{12}}{A} = -\frac{\pi}{6} c\rho_1\rho_2 \int_{z'=0}^{\infty} \frac{1}{(h + z')^3} dz' \tag{11.11}$$

where U_{12}/A is the interaction energy per unit area between the two slabs, and the distance H in Eq. (11.10) is represented by

$$H = h + z' \tag{11.12}$$

Evaluation of the integral provides

$$\frac{U_{12}}{A} = -\frac{A_{12}}{12\pi h^2} \tag{11.13}$$

where

$$A_{12} = \pi^2 c\rho_1\rho_2 \tag{11.14}$$

is defined as the Hamaker constant.

The above illustration demonstrates how evaluation of the dispersion interactions is separated into two components. The Hamaker constant accounts for all the material properties providing the magnitude of the dispersion interactions, while the geometrical dependence is embedded in the various nested integrals that were explicitly evaluated. A similar approach can be adopted to evaluate the interaction energy for colloidal entities of various shapes. For example, for the case of planar parallel slabs of thickness δ_1 and δ_2 separated by a distance h in vacuum, one has

$$\frac{U_{12}}{A} = -\left(\frac{A_{12}}{12\pi}\right)\left[\frac{1}{h^2} + \frac{1}{(h+\delta_1+\delta_2)^2} - \frac{1}{(h+\delta_1)^2} - \frac{1}{(h+\delta_2)^2}\right] \tag{11.15}$$

where A_{12} is the Hamaker constant for bodies 1 and 2. For $h \ll \delta_1, \delta_2$, Eq. (11.15) becomes identical to Eq. (11.13). The interaction energy per unit area in Eqs. (11.13) and (11.15) has units of J/m^2. The negative sign in these expressions signifies an attractive force. The Hamaker constant generally has values in the range of 10^{-21}–10^{-19} J.

Hamaker's approach can be employed to obtain the van der Waals interaction energy between particles of various geometries. Some pertinent expressions of van der Waals interaction energy between particles of common geometrical shapes is given in Table 11.1. Note that we use the term A_H to denote the effective Hamaker constant in the expressions provided in this and subsequent tables. The effective Hamaker constant can assume different values depending on the combination of interacting materials and the intervening medium. The methodologies for calculating the effective Hamaker constant will be described later. Expressions for various geometries can be found in Mahanty and Ninham (1976) and Russel *et al.* (1989). One should note that the Hamaker expressions are based on exact integration of the attractive Lennard–Jones potential over two bodies, and are deemed more accurate than the corresponding estimates based on different types of approximations like Derjaguin approximation, which will be discussed later.

A typical problem associated with the Hamaker expressions is that the interaction energy diverges (goes to $-\infty$) as the surfaces of the bodies come in contact ($h \to 0$). This unphysical divergence of the van der Waals energy stems from the fact that the repulsive (r^{12}) part of the Lennard–Jones potential, Eq. (11.1), is not considered during the integration of the atom-atom interaction energy. In reality, the strong r^{12} repulsion, termed as the Born repulsion, which arises due to overlap of the electron clouds of atoms, renders the interaction between two particles strongly repulsive at finite separations ranging from 0.1 to 0.3 nm. Considering the difficulty in integrating the repulsive Lennard–Jones potential over two macrobodies, one simply accounts for this repulsion in the Hamaker expressions by employing a cut-off distance of about 0.16 nm as the contact separation between two surfaces. This minimum cut-off distance prevents the divergence of van der Waals interaction between objects of colloidal dimensions.

A notable feature of the van der Waals interactions is that integrating the extremely short range attractive Lennard–Jones interactions with a $1/r^6$ decay behavior over two particle volumes changes the distance dependence of these interactions dramatically.

TABLE 11.1. Expressions for Unretarded van der Waals Interaction Energy (or Energy Per Unit Area) Between Bodies of Common Geometrical Shapes Obtained from Hamaker's Approach. A_H is the Hamaker Constant of the System and h is the Distance of Closest Approach Between the Bodies.

Two spheres of radius a_1 and a_2, J	$-\frac{A_H}{6}\left[\frac{2a_1a_2}{h^2+2a_1h+2a_2h} + \frac{2a_1a_2}{h^2+2a_1h+2a_2h+4a_1a_2} + \ln\left(\frac{h^2+2a_1h+2a_2h}{h^2+2a_1h+2a_2h+4a_1a_2}\right)\right]$
Two equal spheres of radius a, J	$-\frac{A_H}{6}\left[\frac{2a^2}{h^2+4ah} + \frac{2a^2}{h^2+4ah+4a^2} + \ln\left(\frac{h^2+4ah}{h^2+4ah+4a^2}\right)\right]$
Sphere of radius a and an infinite flat plate, J	$-\frac{A_H}{6}\left[\frac{a}{h} + \frac{a}{h+2a} + \ln\left(\frac{h}{h+2a}\right)\right]$
Sphere of radius a on the axis of a straight cylindrical capillary of radius b, J	$-\frac{4A_H}{3}\frac{a^3}{(b^2-a^2)^{3/2}}$
Two infinite plates of equal thickness δ, J/m^2	$-\frac{A_H}{12\pi}\left[\frac{1}{h^2} + \frac{1}{(h+2\delta)^2} - \frac{2}{(h+\delta)^2}\right]$
Two semi-infinite parallel plates (half-spaces), J/m^2	$-\frac{A_H}{12\pi h^2}$

For instance, the interaction energy per unit area between two infinite planar slabs (or half-spaces) decays as the inverse square of the separation between the slabs. The macroscopic van der Waals interactions are thus considerably long-range, and are felt between particles that are separated by even several hundreds of nanometers.

11.2.1 Approximate Expressions for van der Waals Interaction

Evaluating two volume integrals is relatively straightforward for simple geometries like two planar surfaces or two spheres, where utilization of symmetry allows considerable reduction of the computational burden involved in the integrations, even enabling analytic evaluation of the integrals. For complicated geometries, application of Hamaker's approach is computationally burdensome, and approximate techniques are often employed to evaluate the interaction energies. Here we will discuss a classical technique, namely the Derjaguin approximation, and also briefly present a modified Derjaguin type technique called surface element integration (SEI) for approximately evaluating the van der Waals interaction energy with very little computational effort.

11.2.1.1 Derjaguin Approximation The essential principle of Derjaguin's technique as generalized by White (1983), which was applied for evaluation of electric double layer interactions between curved geometries in Chapter 5, remains the same when one applies it for calculation of van der Waals energy. Under this approximation, one integrates the interaction energy per unit area between two infinite planar surfaces (half-spaces) over the surfaces of the interacting particles through an expression of the form (*cf.*, Chapter 5)

$$U(h) = \frac{2\pi}{\sqrt{\lambda_1\lambda_2}} \int_h^\infty \frac{U_p(H)}{A}\, dH \tag{11.16}$$

where U is the interaction energy between two curved surfaces separated by h (distance of closest approach between their surfaces), U_p/A is the interaction energy per unit area between two infinite planar surfaces at a separation H, and

$$\lambda_1\lambda_2 = \left(\frac{1}{R_1} + \frac{1}{R_1'}\right)\left(\frac{1}{R_2} + \frac{1}{R_2'}\right) + \sin^2\varphi\left(\frac{1}{R_1} - \frac{1}{R_2}\right)\left(\frac{1}{R_1'} - \frac{1}{R_2'}\right) \tag{11.17}$$

Here, R_1 and R_2 are principal radii of curvature of body 1, R_1' and R_2' are the principal radii of curvature of body 2, and φ is the angle between the principal axes of bodies 1 and 2. For van der Waals interaction, the interaction energy per unit area based on Hamaker's approach is given by Eq. (11.13). Using this expression in Eq. (11.16), one can obtain the interaction energy between two particles as

$$U(h) = -\frac{A_H}{6\sqrt{\lambda_1\lambda_2}} \int_h^\infty \frac{1}{H^2}\, dH = -\frac{A_H}{6\sqrt{\lambda_1\lambda_2}}\frac{1}{h} \tag{11.18}$$

It should be noted that we substituted A_{12} in Eq. (11.13) by the effective Hamaker constant, A_H, to write Eq. (11.18). Introducing the principal radii of curvature for the geometry under consideration in Eq. (11.17) yields an expression for the geometrical factor, which, when substituted in Eq. (11.18) provides a simple approximate form for the van der Waals interaction.

For two interacting spheres of radius a, all the principal radii of curvatures equal a, and the geometrical factor is

$$\lambda_1\lambda_2 = \frac{4}{a^2} \tag{11.19}$$

Hence, Eq. (11.18) gives

$$U(h) = -\frac{A_H a}{12h} \tag{11.20}$$

For a sphere and an infinite flat plate, $\lambda_1\lambda_2 = 1/a^2$, which yields

$$U(h) = -\frac{A_H a}{6h} \tag{11.21}$$

TABLE 11.2. Expressions for Unretarded van der Waals Interaction Energy Between Bodies of Common Geometrical Shapes Obtained using Derjaguin's Approximation. A_H is the Hamaker Constant of the System and h is the Distance of Closest Approach Between the Bodies.

Two spheres of radius a_1 and a_2	$-\frac{A_H}{6}\left(\frac{a_1 a_2}{a_1 + a_2}\right)\frac{1}{h}$
Two equal spheres of radius a	$-\frac{A_H a}{12h}$
Sphere of radius a and an infinite flat plate	$-\frac{A_H a}{6h}$
Two crossed cylinders of radius a_1 and a_2	$-\frac{A_H\sqrt{a_1 a_2}}{6h}$
Two parallel cylinders of length L and radii a_1 and a_2	$-\frac{A_H L}{12\sqrt{2}h^{3/2}}\left(\frac{a_1 a_2}{a_1 + a_2}\right)^{1/2}$

A list of expressions for the approximate van der Waals interaction energy obtained using Derjaguin's technique for different geometries is provided in Table 11.2.

Recalling the assumptions inherent in Derjaguin's approximation, primarily the fact that the separation between the interacting bodies must be much smaller than the principal radii of curvature of the bodies, it becomes apparent that Derjaguin's technique is only applicable for prediction of the van der Waals interaction between quite large objects at very small separations. In fact, the expressions based on Derjaguin's approximation significantly overestimate the van der Waals interaction energy at separation distances greater than about 20% of the particle radius.

11.2.1.2 Surface Element Integration Starting from the interaction energy per unit area, one can obtain a much better approximation for the van der Waals interaction energy compared to Derjaguin's technique by using the surface element integration (SEI) (Bhattacharjee and Elimelech, 1997; Bhattacharjee *et al.*, 1998). In this technique, the interaction energy between two bodies is obtained by integrating the interaction energy per unit area over the exact surfaces of the bodies. While computationally somewhat expensive compared to Derjaguin approximation, the procedure can provide the interaction energy more accurately for two bodies of arbitrary geometrical shapes, and even provides the exact Hamaker interaction energy between a body and an infinite flat plate. SEI is thus quite useful in evaluating the van der Waals interactions between a particle and a planar surface, a situation commonly encountered in particle deposition processes.

In surface element integration, the interaction energy between two curved surfaces is represented by the integral

$$U(h) = \int\int_A \frac{U_p(H)}{A}\frac{\mathbf{n}\cdot\mathbf{k}}{|\mathbf{n}\cdot\mathbf{k}|}\,dA \qquad (11.22)$$

Here, U is the interaction energy between the two bodies, $U_p(H)/A$ is the interaction energy per unit area between the materials of the two bodies, A is the projected area of the body normal to the line of closest distance between the bodies. The term $\mathbf{n} \cdot \mathbf{k}/|\mathbf{n} \cdot \mathbf{k}|$ provides a measure of the angle made by the curved surface of the body with the projected normal area A, and assumes a value of $+1$ or -1 depending on whether the surfaces of the bodies face each other or they face away from each other.

When applied for the case of a sphere interacting with an infinite flat surface, SEI provides an analytic expression for the unretarded van der Waals interaction given by

$$U^{SEI} = -\frac{A_H}{6}\left[\frac{a}{h} + \frac{a}{h+2a} + \ln\left(\frac{h}{h+2a}\right)\right] \tag{11.23}$$

This expression is identical to that for the interaction energy between a sphere of radius a and an infinite planar surface shown in Table 11.1. In other words, SEI provides the exact unretarded van der Waals interaction between a colloid and an infinite planar surface. The results are not exact for two curved surfaces. However, even for such cases, SEI provides remarkably accurate predictions of the interaction energy.

Clearly, SEI is much superior to Derjaguin's approximation when it comes to evaluation of the van der Waals interaction energy between small colloidal particles. The technique can be readily generalized for surfaces of arbitrary geometries (Bhattacharjee *et al.*, 1998; Hoek *et al.*, 2003).

11.2.2 Cohesive Work and Hamaker's Constant

The concept of cohesion work can be utilized to estimate the Hamaker constant, A_{12}, between bodies 1 and 2. The work of cohesion consists of producing two new interfaces from a given material. It measures the attraction between the molecules of the two portions being produced. The cohesion work due to the dispersion force for material 1 is given by

$$W_{11}^d = 2\gamma^d \tag{11.24}$$

where γ^d is the theoretical surface tension due to dispersion energies. For a given material the attraction potential is

$$U_A = -\frac{A_{11}}{12\pi r_1^2} \tag{11.25}$$

where A_{11} is the Hamaker constant for material 1 that is being separated to produce two parts and r_1 is the intermolecular distance. However, the cohesion work is related to the attraction potential by

$$W_{11}^d = -U_A \tag{11.26}$$

and we obtain

$$A_{11} = 24\pi\gamma^d r_1^2 \tag{11.27}$$

Equation (11.27) was originally derived by Fowkes (1964) and was used to calculate the Hamaker constant using surface tension data (Hiemenz and Rajagopalan, 1997; Morrison and Ross, 2002). A system for which the Fowkes equation is apt to work best is for nonpolar materials such as alkanes, where the attractive forces are due to the dispersion forces. For alkanes, with $\gamma^d \cong 25$ mJ/m^2 and $r_1 = 0.2$ nm, one obtains $A_{11} = 7.5 \times 10^{-20}$ J.

For the attractive force between materials 1 and 2, the cohesion work is approximated by

$$W_{12}^d = -U_A = -\frac{A_{12}}{12\pi r_{12}^2} = 2\left(\gamma_1^d \gamma_2^d\right)^{1/2} \tag{11.28}$$

where

$$r_{12} = (r_1 r_2)^{1/2} \tag{11.29}$$

Here, r_{12} is the intermolecular distance approximated by the root mean square of the individual intermolecular distances. Equation (11.28) leads to

$$A_{12} = 24\pi r_{12}^2 \left(\gamma_1^d \gamma_2^d\right)^{1/2} \tag{11.30}$$

The above equation can be used jointly with Eqs. (11.27) to (11.29) to develop a useful relation for A_{12} in terms of A_{11} and A_{22}:

$$A_{12} = (A_{11} A_{22})^{1/2} \tag{11.31}$$

Equations (11.27) and (11.31) give the means to evaluate the Hamaker constant for two non-polar materials in vacuum using surface tension data. Values for Hamaker constant of several common materials are given in Table 11.3.

11.2.3 Electromagnetic Retardation

The expressions for the attractive forces were derived with the assumption that the speed of light is infinite. The energy of the interaction is usually decreased at large distances because the time of electric field propagation from one body to another body and back is such that the fluctuating electric moments become slightly out of phase. The correction due to the finite speed of light is called the retardation correction. This correction is unity for $h = 0$ and it decreases monotonically with increasing h.

For the case of two parallel plates, Gregory (1981) gives the retarded potential energy per unit area as

$$\frac{U_p}{A} = -\frac{A_{12}}{12\pi h^2}\left[\frac{1}{1 + 5.32(h/\lambda)}\right] \tag{11.32}$$

where $\lambda = 10^{-7}$ m (London wavelength). Note that the bracketed term of Eq. (11.32) is unity for $h = 0$ and tends to 0 as $h \to \infty$. For the case of two equal-sized spheres,

TABLE 11.3. Commonly Reported Hamaker Constants, A_{11}, for Unretarded Interaction for Some Common Materials in Vacuum.

Substance	Hamaker Constant A_{11} J ($\times 10^{20}$)	Substance	Hamaker Constant A_{11} J ($\times 10^{20}$)
Acetone	4.10, 4.20	Octane	4.5, 5.02
Aluminum	14.0, 15.4, 15.5	Oxides (most)	10.0–20.0
Benzene	5.0	Pentane	3.8, 3.94
Carbon tetrachloride	4.78, 5.5	Polybutadiene	8.20
Cyclohexane	4.82, 5.20	Polydimethylsiloxane	6.27
Decane	5.45	Polyethene oxide	7.51
Dodecane	5.0, 5.84	Polyisobutylene	10.10
Ethanol	4.2	Polypropylene oxide	3.8, 3.95
Ethyl acetate	4.17	Polystyrene	6.4, 9.80
Gold	37.6, 45.3, 45.5	Polyvinyl acetate	8.91
Graphite	47.0	Polyvinyl chloride	7.5
n-Hexadecane	5.1	Quartz	7.93
Hexane	4.32	Rutile (TiO_2)	43
Magnesia	10.5	Silicon	25.5
Methyl ethyl ketone	4.53	Silver	39.8, 40.0
Mica	10.0, 10.8	Toluene	5.40
		Water	3.7, 4.35, 4.38

Source: Adapted from Ross and Morrison (1988); Morrison and Ross (2002).

the effect of retardation on the interparticle dispersion potential is given by Schenkel and Kitchener (1960) as

$$U = -\frac{aA_{12}}{12h}\left[\frac{1}{1+1.77p}\right] \quad p \leq 0.57, \quad h/a \ll 1 \tag{11.33}$$

$$U = -\frac{aA_{12}}{h}\left[\frac{2.45}{60p} - \frac{2.17}{180p^2} + \frac{0.59}{420p^3}\right] \quad p \geq 0.57, \quad h/a \ll 1 \tag{11.34}$$

where $p = 2\pi h/\lambda$.

Other expressions are given by Schenkel and Kitchener (1960), Gregory (1981), and Hunter (1991), and several of these are summarized in Elimelech *et al.* (1995).

11.3 EFFECTS OF INTERVENING MEDIUM

Hamaker's approach for the evaluation of the dispersion force is used when the medium between the two materials is vacuum. A correction to the Hamaker constant is needed when the intervening medium is a dielectric.

A common approximation is to estimate the effect of an intervening medium 1 between two bodies of composition 2 and 3 by considering a pseudo-chemical reaction as shown:

$$\{2\}\,(1) + (1)\,[3] \rightarrow \{2\}\,[3] + (1)\,(1) \tag{11.35}$$

where the numbers enclosed within different types of brackets represent different materials. The change in potential energy $\Delta\phi$ for the case of the above reaction is

$$\Delta\phi = \phi_{23} + \phi_{11} - \phi_{21} - \phi_{13} \tag{11.36}$$

Each term on the right side of Eq. (11.36) depends in the same way on the size and distance parameters and differs only in molecular parameters which are fully contained in the Hamaker constant (Hiemenz and Rajagopalan, 1997). One can, therefore, write

$$A_{213} = A_{23} + A_{11} - A_{21} - A_{13}, \tag{11.37}$$

where A_{213} is the overall Hamaker interaction parameter for bodies 2 and 3 being intervened by medium 1. As was pointed out earlier, the interaction between dissimilar bodies is given by the geometrical mean of the homogeneous interactions for the bodies; *i.e.*,

$$A_{12} = (A_{11}A_{22})^{1/2}, \tag{11.38}$$

where A_{11} is the overall Hamaker constant for two bodies of the same material in vacuum. Making use of Eq. (11.38), A_{213} of Eq. (11.37) becomes

$$A_{213} = (A_{22}A_{33})^{1/2} + (A_{11}A_{11})^{1/2} - (A_{11}A_{22})^{1/2} - (A_{11}A_{33})^{1/2} \tag{11.39}$$

or

$$A_{213} = \left(A_{33}^{1/2} - A_{11}^{1/2}\right)\left(A_{22}^{1/2} - A_{11}^{1/2}\right).$$

For the special case of $A_{22} \equiv A_{33}$, Eq. (11.39) gives

$$A_{212} = \left(A_{11}^{1/2} - A_{22}^{1/2}\right)^2 \tag{11.40}$$

Equation (11.40) defining A_{212} indicates that for all values of A_{11} and A_{22}, the parameter A_{212} is always positive. This means that the potential energy U is always negative (*i.e.*, attractive) for two similar bodies with an intervening medium. However, the general expression (11.39) shows that A_{213} can have negative values. Here, a repulsive potential can develop between two dissimilar bodies with an intervening medium. The dispersion force between a particle and an air bubble (air–water–solid) can therefore be repulsive depending on the particle and the intervening medium. Table 11.4 shows values of the Hamaker constant for interaction between common materials when the intervening medium is water.

In the literature, it is sometimes customary to lump the Hamaker parameter with the term due to the retardation effects as one variable denoted as $A_{ret}(h)$, with $A_{ret}(0)$ being the Hamaker constant with no retardation effects. Thus, $A_{ret}(0)$ is identical to the term A_H used earlier for the effective unretarded Hamaker constant. Table 11.5 gives values for $A_H = A_{ret}(0)$ for various materials. Figure 11.3 depicts variations of A_{ret} with distance h for polystyrene parallel plates in pure water and salt water. $A_{ret}(h)$ accounts for the retardation effects, where the speed of light is not taken to be infinity.

TABLE 11.4. Effective Hamaker Constants A_{212} for Unretarded Interaction Between Common Materials (denoted by 2) Immersed in Water (denoted by 1).

Substance	Hamaker Constant A_{212} J ($\times 10^{20}$)	Substance	Hamaker Constant A_{212} J ($\times 10^{20}$)
Alkanes:		Fused quartz	0.63,0.833
Pentane	0.336	Fused silica	0.849
Hexane	0.360	Germanium	16.0, 17.7
Heptane	0.386	Gold	27, 33.4, 33.5
Octane	0.410	Magnesia	1.60, 1.76
Decane	0.462	Metals	3–33.4
Dodecane	0.50	Mica	2.0, 2.33
Tetradecane	0.514	Oxides	1.76–4.17
Hexadecane	0.50		
		Polymethyl methacrylate	1.05
Air	3.7	Polystyrene	0.27–1.01
Alumina	4.12, 4.17	Polyvinyl chloride	1.30
Calcite	2.23	Quartz	1.08
Copper	17.5	Rutile (TiO_2)	26
Crystalline quartz	1.70	Sapphire	5.32
		Silicon	11.9, 13.4
		Silver	26.6, 28.2

Source: Adapted from Ross and Morrison (1988); Morrison and Ross (2002).

TABLE 11.5. Effective Hamaker Constants A_H for Unretarded Interaction Between Common Materials in Vacuum (A_{22}) and in Water (A_{212}).

	A_H ($\times 10^{20}$), J		
	Vacuum	Water	
Calcite	10.1	2.23	
Copper	40	30	
Decane	4.8	0.46	
Gold	40	30	
Hexadecane	5.2	0.54	
Hexadecane	5.4	—	(Parsegian and Weiss, 1981)
Pentane	3.8	0.34	
Polyisoprene	6.0	0.74	
Polymethyl methacrylate	7.1	1.05	
Polystyrene	7.9	1.3	
Poly tetrafluoroethylene	3.8	0.33	
Polyvinyl chloride	7.8	1.30	
Quartz (crystalline)	8.8	1.70	
Quartz (fused)	6.5	0.83	
Sapphire	15.6	5.32	
Silica (fused)	6.6	0.85	
Silver	50	40	
Water	3.7	—	(Hough and White, 1980)

Source: Adapted from Russel *et al.* (1989).

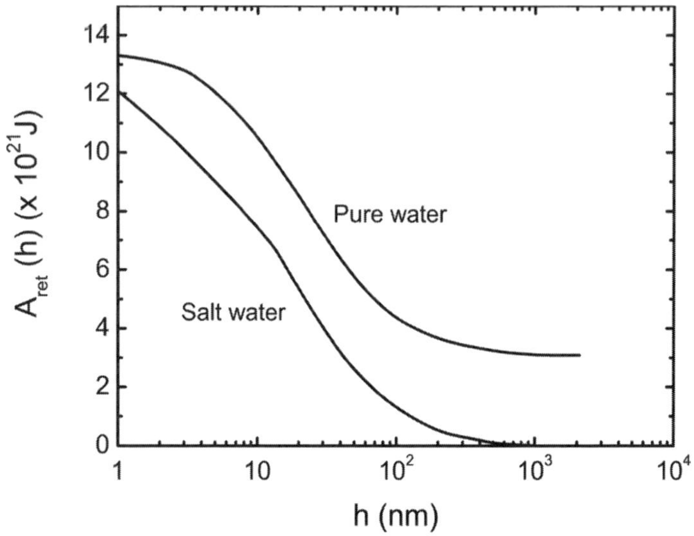

Figure 11.3. Calculated values for the effective Hamaker constant for interaction between polystyrene blocks in pure and salt water (Parsegian, 1975).

Noting that the Hamaker constant is calculated differently for different combinations of interacting bodies and intervening media, leading to descriptive nomenclature such as A_{212}, A_{132}, A_{12}, *etc.*, it should be borne in mind that the symbol A_H used for the effective Hamaker constant in several expressions in this and subsequent chapters should be calculated appropriately for each combination of interacting materials.

11.4 DLVO THEORY OF COLLOIDAL INTERACTIONS

In a colloidal system, whether colloidal particles flocculate or coalesce depends on the net interaction resulting from the combined attractive and repulsive forces to which the colloidal particles are subjected. For a system where electrostatic and van der Waals forces are dominant, the stability of the system, *i.e.*, whether coagulation occurs or not, is dependent on the electrostatic repulsion due to the presence of the electric double layer and the dispersion attractive forces (London–van der Waals). This concept was developed independently by Derjaguin and Landau (1941) in the USSR, and Verwey and Overbeek (1948) in the Netherlands. It is known as the DLVO theory.

In order to appreciate the contribution of potential energy due to the electrostatic and London–van der Waals forces, it is best to consider the sum of the interactive potential energies due to these two types of forces for a case of two flat parallel plates.

The repulsive electric double layer force between two parallel plates having similar surface potentials was derived previously in Chapter 5 [see, for instance, Eq. (5.113)] and it is given, with usual Debye–Hückel approximation of low potential, by

$$F_R/A = 2\epsilon\kappa^2\psi_s^2 \exp(-\kappa h) \tag{11.41}$$

where F_R/A is the repulsive force ($\mathrm{N/m^2}$) due to the surface charge and, ψ_s is the surface potential.

The potential energy per unit area due to the repulsive force is defined by the energy required to bring two plates from infinity to a finite separation distance, h. The potential energy is related to the force by

$$\frac{dU_R/A}{dh} = -\frac{F_R}{A} \tag{11.42}$$

Integration of Eq. (11.42) leads to

$$\int_{\infty}^{U_R/A} \frac{dU_R}{A} = \int_{\infty}^{h} -\frac{F_R}{A}\, dh \tag{11.43}$$

Setting $(U_R/A)_\infty$ as zero, Eqs. (11.41) and (11.43) give

$$\frac{U_R}{A} = -\int_{\infty}^{h} 2\epsilon\kappa^2\psi_s^2 \exp(-\kappa h)\, dh \tag{11.44}$$

The potential for the case of two parallel plates is then given by (in $\mathrm{J/m^2}$)

$$\frac{U_R}{A} = 2\epsilon\kappa\psi_s^2 \exp(-\kappa h) \tag{11.45}$$

The potential (energy) per unit area due to the dispersion force between two similar plates (2) in a medium (1) is given by

$$\frac{U_A}{A} = -\frac{A_{212}}{12\pi h^2} \tag{11.46}$$

The total free energy per unit area of the interaction between the two flat parallel plates as a function of separation distance h is given by combining Eqs. (11.45) and (11.46):

$$\frac{U}{A} = \frac{U_R}{A} + \frac{U_A}{A} = 2\epsilon\kappa\psi_s^2 \exp(-\kappa h) - \frac{A_{212}}{12\pi h^2} \tag{11.47}$$

The above expression for the total energy of interaction U/A determines the stability of the two parallel plates. This stability analysis for a colloidal dispersion is in essence the DLVO theory for colloidal stability. The total energy U/A given by Eq. (11.47) shows that U/A is dependent on the system properties such as ϵ and A_{212} and on the "manipulated" physical characteristics of the system as manifested by the electrolyte solution properties represented in the inverse Debye length parameter, κ, as well as the surface electric potential, ψ_s.

In a similar fashion, for $\kappa a \gg 1$, we can write the total interaction energy for two equal spheres as

$$U = 2\pi a\epsilon\psi_s^2 \exp(-\kappa h) - \frac{aA_{212}}{12h} \tag{11.48}$$

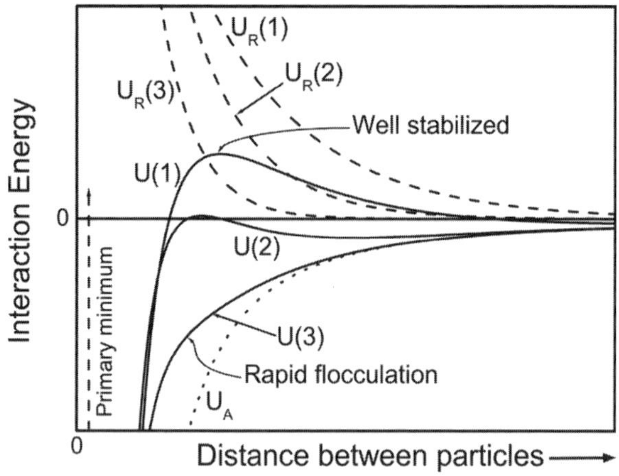

Figure 11.4. Total interaction energy curves (solid lines) obtained by the summation of an attractive potential curve (dotted line) with different repulsive potential curves (dashed lines).

where $(h + 2a)$ is the distance between the sphere centers, a is the sphere radius, and κ is the inverse Debye length. Here, the van der Waals interaction is assumed to be unretarded.

Figure 11.4 shows three different total interaction energy curves for the case of two parallel plates, each obtained by combining the repulsive electrostatic double layer interaction with an attractive van der Waals interaction. The repulsive energy is an exponential function of the interparticle distance h with a range of the order of κ^{-1}. On the other hand, the van der Waals interaction energy decreases as the inverse power of the interparticle distance. Consequently, the van der Waals attraction energy dominates at very small (near contact) and very large interparticle distances, whereas the electric double layer repulsion dominates at intermediate distances (Probstein, 2003).

In Figure 11.4, curve $U(1)$ represents a well-stabilized colloidal system with a repulsive energy maximum. This represents the energy barrier preventing or hindering the approach of two colloidal particles to contact. If this repulsive maximum is large compared to the thermal energy k_BT of the colloidal particles, the system should be stable at least when no other forces (like shear forces) exist.

Curve $U(3)$ in Figure 11.4, on the other hand, represents a case where the repulsive barrier is absent, implying that the dispersion is unstable. In this scenario, the colloidal particles will coagulate rapidly, as they will be attracted to a deep attractive energy minimum at contact. Curve $U(2)$ represents the transition between stability and coagulation at the primary minimum.

An interesting additional feature of the potential energy versus distance plot is the presence of a secondary minimum at relatively large interparticle distance as given by curve $U(2)$. If this minimum is relatively deep (several k_BT values), it should give rise to loose flocs. However, this type of a coagulated dispersion can be easily redispersed by agitation.

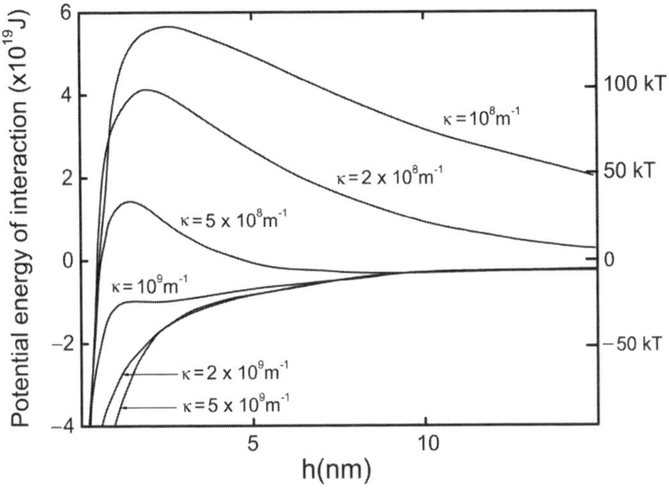

Figure 11.5. Influence of electrolyte concentration on the total potential energy of interaction of two spherical particles.

It should be noted that at very small distances, repulsion due to the overlapping of electron clouds (Born repulsion) predominates. Consequently, at such small distances there is a deep minimum (primary minimum) in the potential energy curves. This region of the potential energy curves is shown only for curve $U(1)$ of Figure 11.4.

Figure 11.5 shows the influence of the Debye length κ^{-1} on the total energy interaction of two similar spheres. The variation in κ can be thought of in terms of electrolyte concentration. Large values of κ indicates higher electrolyte concentrations. Figure 11.5 clearly shows that for $\kappa = 10^8\ \text{m}^{-1}$ a stable colloidal system is obtained. Increasing κ to $10^9\ \text{m}^{-1}$ gives an unstable system where coagulation will occur. The electrolyte causes the diffuse part of the double layer to compress. When the double layer is reduced in thickness, the colloidal particles coagulate because the particles can get close enough to each other for the London–van der Waals forces to dominate.

11.5 SCHULZE–HARDY RULE

One of the earlier generalizations regarding the effect of added electrolyte is a result known as the Schulze–Hardy Rule (1900) (Hiemenz and Rajagopalan, 1997). This rule states that it is the valence of the ions of opposite charge to the colloidal particle (counterions) that have the main effect in the stability of the colloidal particle. The critical flocculation concentration (CFC) value for a particular electrolyte is essentially determined by the valence of the counterion regardless of the nature of the ions having the same charge as the surface (*i.e.*, coions). The CFC is the critical concentration of electrolyte required to flocculate the colloidal particles. In the above rule it is understood that the electrolyte is a non-adsorbing indifferent electrolyte where its addition has no effect on the surface potential ψ_s.

The essentials of the Schulze–Hardy rule can be derived from the DLVO theory. The primary maximum of curve $U(2)$ is the demarcation point between stability and instability. At this point

$$U_R + U_A = 0 \tag{11.49}$$

and

$$\frac{d}{dh}(U_R + U_A) = 0 \tag{11.50}$$

Making use of the potential energy between two identical spheres valid for higher surface potential and for a $(z : z)$ electrolyte (Overbeek, 1972), one can write

$$U = \frac{32\pi a\epsilon (k_B T)^2}{z^2 e^2} \tanh^2\left[\frac{ze\psi_s}{4k_B T}\right]\exp(-\kappa h) - \frac{aA_{212}}{12h} \tag{11.51}$$

with

$$\kappa = \left(\frac{2e^2 z^2 n_\infty}{\epsilon k_B T}\right)^{1/2}$$

where the first term of Eq. (11.51) is the double-layer repulsive potential energy, U_R, and the second term is the attractive dispersion potential energy, U_A.

Application of Eqs. (11.49) and (11.50) to (11.51) leads to having the maximum in U occur at a value equal to κ^{-1} and

$$n_{crit} = \frac{49.63}{z^6 I_b^3}\left[\frac{k_B T}{A_{212}}\right]^2 \tanh^4\left(\frac{ze\psi_s}{4k_B T}\right) \tag{11.52}$$

where n_{crit} is the number concentration of ions and CFC is given in mol/liter (molarity); *i.e.*,

$$\frac{1}{1000N_a} n_{crit} = CFC \tag{11.53}$$

and

$$I_b = \frac{e^2}{4\pi\epsilon k_B T} \tag{11.54}$$

For large potential ψ_s, $\tanh(ze\psi_s/4k_B T)$ becomes $\approx$1 and

$$CFC \propto \frac{1}{z^6} \tag{11.55}$$

which explains the Schulze–Hardy rule (Russel *et al.*, 1989). Here, the coagulating effectiveness of the dominating ion increases with its valency in the proportion of 1 : 64 : 729 for uni-, di-, and trivalent ions, respectively. Thus, if the univalent ion requires 729 units of concentration for coagulation, the divalent ion would require 64 units, and the trivalent ion would require only 1 unit.

For small potentials,

$$\tanh \frac{ze\psi_s}{4k_BT} \approx \frac{ze\psi_s}{4k_BT} \tag{11.56}$$

and

$$\text{CFC} \propto \frac{\psi_s^4}{z^2} \tag{11.57}$$

which also partially explains the Schulze–Hardy rule. Here the surface potential becomes important as well.

The effect of the surface potential on the critical flocculation concentration (CFC) for $z = 1$, 2, and 3 are given in Figure 11.6 (Shaw, 1980).

The reason that the valency of the counterions is important and not the valency of the coion can be explained by making use of the Boltzmann distribution. For a positive surface potential, ψ_s is positive and the counterion distribution is given by

$$n_- = n_\infty \exp\left[\frac{e\psi_s|z_-|}{k_BT}\right] \tag{11.58}$$

where z_- is the valency of the counterions. For the case of coions:

$$n_- = n_\infty \exp\left[-\frac{|z_-|e\psi_s}{k_BT}\right] \tag{11.59}$$

For the case of large ψ_s, Eqs. (11.58) and (11.59) indicate that $n_- \gg n_+$ within the double layer close to the surface. Hence the counterions become the controlling ions.

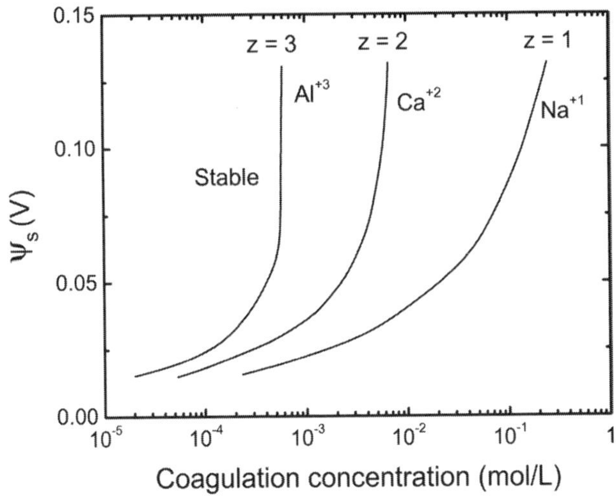

Figure 11.6. Coagulation concentrations calculated by taking $A_{212} = 10^{-19}$ J, for counterion charge numbers 1, 2, and 3. The colloidal system is predicted to be stable above and to the left of each curve and coagulated below and to the right of each curve (Shaw, 1980).

The DLVO theory predicts that ions having the same valency have the same coagulation capability. However, experimented results indicate that the ion effectiveness is as follows (Hiemenz and Rajagopalan, 1997):

- For monovalent cations

$$Cs^+ > Rb^+ > NH_4^+ > K^+ > Na^+ > Li^+$$

- For monovalent anions

$$F^- > Cl^- > Br^- > NO_3^- > I^-$$

11.6 VERIFICATION OF THE DLVO THEORY

Many attempts have been made to verify the DLVO theory taking into account both the dispersion and electrostatic forces. The major difficulty was the limitation due to the roughness of the surfaces used. It was Tabor and Winterton (1969) who first recognized that sheared muscovite mica provides a molecularly smooth surface ideal for such measurements. Figure 11.7 shows the comparison between the experimental observations and theoretical predictions of the retarded Hamaker constants of molecularly smooth mica plates. The experiments were conducted in a vacuum. Allowing for experimental errors, the agreement between the theory and experiment can be brought to within 10%.

For the case of the force between mica plates in aqueous electrolyte solutions, Figure 11.8 shows the force between two mica plates divided by the radius of curvature R of the plates (Israelachvili, 1985). Here, the agreement between the experimental

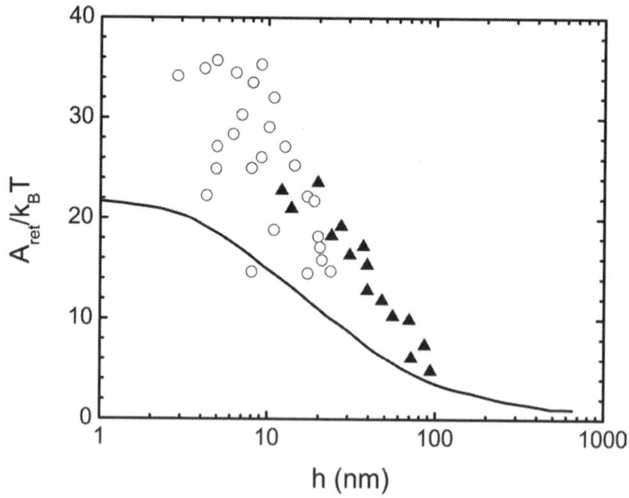

Figure 11.7. Retarded Hamaker constant versus separation for mica plates interacting across a vacuum (Chan and Richmond, 1977): ○, data from Tabor and Winterton (1969); ▲, data from Israelachvili and Tabor (1973); ——, calculation (Russel *et al.*, 1989).

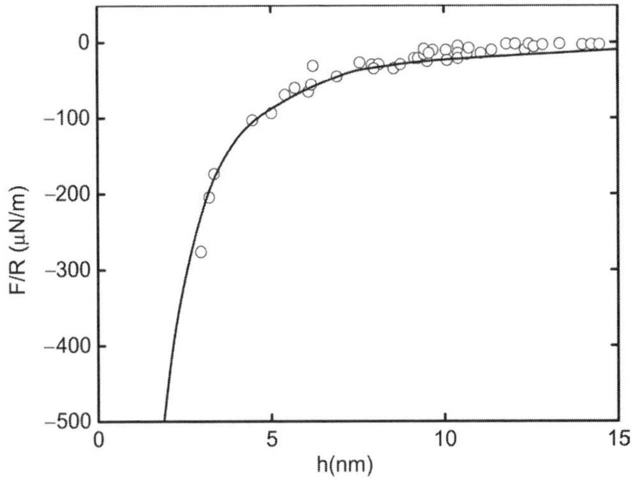

Figure 11.8. Force measured between mica plates (crossed cylinders) in aqueous electrolyte solutions compared with theoretical predictions; R denotes the radius of curvature of the mica plates (Israelachvili, 1985).

results and the theoretical predictions is excellent. The experimental device used in the measurements is described by Israelachvili and Adams (1978). Such an apparatus is shown in Figure 11.9. The force was measured between the macroscopic surfaces, in a configuration of crossed cylinders at nanometer separations. The surfaces consisted of 1 μm thick mica sheet silvered on the back and glued to quartz pieces with a radius of curvature of 10 mm. The separation was measured optically using a spectrometer. The results shown in Figure 11.8 are restricted to situations where there are no hydration effects arising from the presence of hydrated metal ions on the mica surface.

The study of Israelachvili and Adams (1978) and the later studies of Pashley (1981a,b) and of Pashley and Israelachvili (1984) indicated that counterion hydration on the mica surface can give rise to an additional short-range repulsive force between mica surfaces in aqueous solutions of K^+ and Na^+ of given molarity. This hydration force is not affected by an increase in temperature (up to 65°C) and it is completely absent in pure water where H_3O^+ is the counterion.

Figure 11.10 shows a case where force measurements were made in a 4×10^{-5} M KCl electrolyte solution. The agreement with the DLVO theory is excellent and the primary maximum occurs at a separation distance of 2 mm. However, at higher KCl concentrations, say 3×10^{-4} and 10^{-3} M, no primary maximum was observed and there appears to be an additional repulsive force which cannot be accounted for by the DLVO theory. If one is to account for the solvent structure, then the total interaction potential energy can be given by

$$U = U_A + U_R + U_S \tag{11.60}$$

where U_S is the potential due to the solvent structure. This term is necessitated in order to reconcile experimental measurements with the classical DLVO theory. At present,

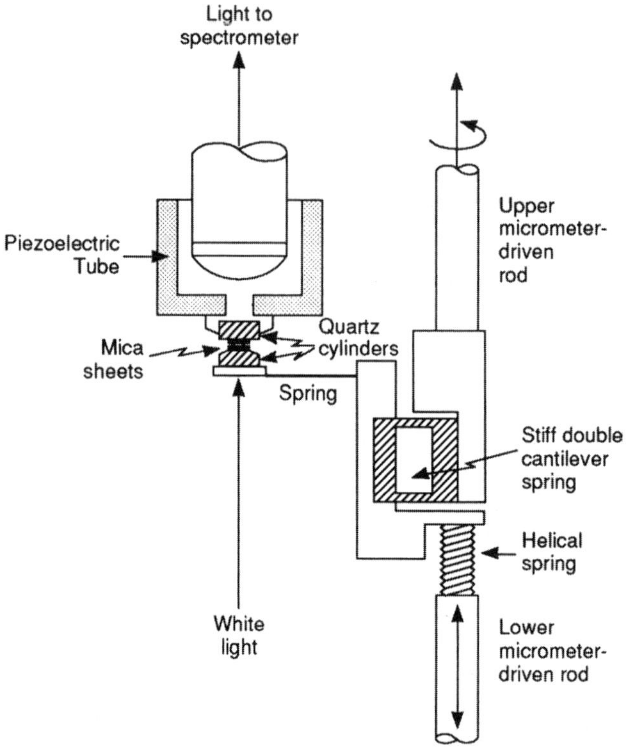

Figure 11.9. Schematic drawing of apparatus to measure long-range forces between two crossed cylindrical sheets of mica (of thickness 1 μm and radius of curvature 12 mm) immersed in an electrolyte. By use of white light and multiple beam interferometry, the shapes of and separation between the two mica surfaces may be independently measured. The separation between the two mica surfaces can be controlled by use of two micrometer-driven rods and a piezoelectric crystal tube to better than 0.1 nm (Israelachvili and Adams, 1978).

the lack of an adequate theory for the structure of water prevents theoretical evaluation of the potential U_S. Its magnitude for mica can be estimated using Israelachvili and Adams type measurements. The difference between the measured total potential and that given by the DLVO theory would then be the U_S term (Hunter, 1991). In general, discrepancy in the region of $0 < h < 5 \times 10^{-9}$ m occurs when the term U_S is non-zero.

In aqueous solutions, both positive and negative values of U_S can arise due to the presence of "hydration forces", where positive interactions pertain to hydrophilic repulsion and negative interactions are believed to be due to hydrophobic attraction. Both types of interaction decay exponentially with a decay length typically on the order of several nanometers (Hunter, 1991; Xu and Yoon, 1989, 1990).

For electrokinetic modeling, where the hydration of the ions in the electric double layer occurs, it may become necessary to modify the Boltzmann distribution and to

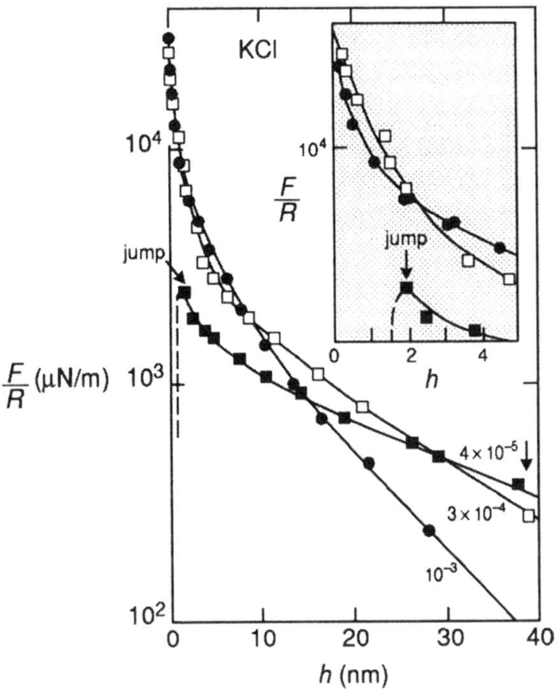

Figure 11.10. Forces measured between mica surfaces in KCl solutions at pH 5.7. 10^{-3} M; 3×10^{-4} M; 4×10^{-5} M. In 4×10^{-5} M no hydration forces were present, hence the surfaces jumped to a primary minimum from $h \simeq 2$ nm. At the higher salt concentrations, strong hydration repulsion forces prevented the primary minimum adhesion. The inset shows the short-range forces in more detail (Pashley, 1981b).

assume a variable dielectric constant within the electric double layer. The reader can refer to the studies by Gur *et al.* (1978a,b) and Guzman–Garcia *et al.* (1990) for details.

In the above discussions, we only dealt with aqueous systems. A non-aqueous medium is usually characterized with a low dielectric constant and poor dissociation of the electrolytes. By and large the ionic strengths are much lower than 10^{-6} M and the inverse Debye length is fairly large. Electrostatic stabilization in non-aqueous media is reviewed by van der Hoeven and Lyklema (1992).

11.7 LIMITATIONS OF DLVO THEORY

Several approximations were made in arriving at the DLVO theory of colloidal interactions. When one systematically revisits these assumptions and approximations, one might tend to wonder as to why the theory provides such a good qualitative and often quantitative description of stability of such a wide array of charged colloidal dispersions. In this section, we will first outline some common approximations made in arriving at the DLVO theory, which might result in deviations of the predictions based on the theory from reality. Following this, we will briefly discuss some of the common approaches that have been adopted to overcome these limitations of the DLVO theory.

11.7.1 Major Assumptions in DLVO Model of Colloidal Interactions

The fundamental assumption in the DLVO theory is that the total interaction energy between two colloidal entities is comprised solely of the electrostatic double layer interaction and the van der Waals interaction. Furthermore, it is assumed that these two interactions can be computed separately and added up as in Eq. (11.47). Both these assumptions are questionable. First, other types of interactions, for instance, the hydration interactions (hydrophobic attraction and hydrophilic repulsion) mentioned in the previous section may be present in colloidal systems. Secondly, when we compute the electric double layer interactions based on the Poisson–Boltzmann equation, the Boltzmann distribution of the ions is obtained solely on the basis of the electrical potential energy. Strictly speaking, the ionic distribution should be based on the total energy or potential of mean force (Russel *et al.*, 1989). This potential of mean force should consist of both the electrical potential as well as the potential due to the dispersion interactions.

There are other types of fundamental assumptions inherent in the DLVO theory, which stem from the manner in which we compute the individual components of the potential. In this theory, the solvent is treated as a continuum, and its presence is accounted for through an effective Hamaker constant and an effective dielectric constant. In particular, the granular nature of the solvent is totally ignored in the theory. Consideration of the solvent explicitly should naturally modify the predictions of the interaction energy (Israelachvili, 1985; Hirata, 2003), particularly when the interacting surfaces are close enough such that the intervening distance is of the order of the solvent molecular dimensions (less than 1 nm). In using the Poisson–Boltzmann model for the electrostatic double layer interaction, we also ignore the finite dimensions of the ions, treating them as point charges. Once again, such an assumption will render the electrostatic double layer interactions based on the Poisson–Boltzmann model inaccurate. Particularly, when the bulk ion concentrations are greater than 0.5 M, the ion distributions around charged interfaces are no more governed by the Boltzmann distribution. Instead, at high salt concentrations, one observes ion density fluctuations with multiple peaks (Kjellander *et al.*, 1992; Kjellander and Greberg, 1998; Hirata, 2003). There is another implication of this assumption. For highly charged confinements of nanometer scale dimensions (like a cylindrical capillary pore), neglecting the finite size of the ions can result in unrealistically high counterion concentrations in the confined domains. For instance, if the surface potential of a wall is -125 mV, the concentration of a univalent counterion (*e.g.*, Na^+) at the wall will be approximately 150 times greater than the bulk concentration. Now, assuming that the bulk concentration of the electrolyte is 0.2 M, we will have a local ion concentration of about 30 M at the wall (Das and Bhattacharjee, 2004). This is unrealistically high, particularly considering the wall to be that of a narrow capillary. Given that a hydrated sodium ion has a diameter of about 0.4–0.5 nm, one simply cannot pack 30 M sodium ions inside a few tens of nanometer diameter capillary.

Barring the above fundamental type of assumptions, we make several geometrical simplifications when calculating the DLVO potential. For instance, we assume all surfaces to be geometrically smooth and uniformly charged. Naturally occurring surfaces are inherently rough and charges appear on most surfaces due to functional groups that are discrete. Influence of roughness and chemical heterogeneity of the surfaces on the DLVO interactions is a widely studied area, and over the years numerous approaches for studying the influence of these heterogeneities on the interaction energy have been proposed (Czarnecki, 1986; Elimelech and O'Melia, 1990; Herman and Papadopoulos, 1991; Lenhoff, 1994; Suresh and Walz, 1996; Bhattacharjee *et al.*, 1998; Das and Bhattacharjee, 2004). A common consensus from these studies is that the magnitude and range of the interaction energy become considerably suppressed in presence of roughness.

When we compute the two components of the DLVO interaction, we often employ different approximations for calculating the van der Waals and electrostatic double layer components. A common error stems for the use of Derjaguin approximation. Notably, Derjaguin approximation introduces a large error in the computation of the van der Waals interaction (Russel *et al.*, 1989; Bhattacharjee and Elimelech, 1997). The error becomes significantly large as the interparticle separation increases. When we incorporate the retardation effects, quite often the particle-particle retarded van der Waals interaction is obtained employing Derjaguin approximation (Schenkel and Kitchener, 1960). Thus, even though our intention is to correct Hamaker's results by incorporating retardation for large separations, we actually grossly overestimate the interaction energy by using Derjaguin approximation. The error in the DLVO potential becomes quite large when we combine a reasonably accurate expression for the electrostatic double layer interaction with a highly incorrect estimate of the van der Waals attraction. Finally, in calculating the electrostatic interaction, we generally employ constant potential or constant charge density boundary conditions, which are rarely found in real surfaces. A more realistic boundary condition should be the charge regulation type boundary condition, which arises from the interactions between the ions and the surface functional groups (Behrens and Borkovec, 1999).

To summarize, when one applies the DLVO theory, one should note that the theory simply states that the attractive and repulsive interaction potentials exert a combined influence on the stability of a colloidal system. There is no specific recipe for obtaining these components of the interactions. In this context, one should be careful about which expressions for the van der Waals and electrostatic interactions are employed in the DLVO potential, and ensure that these are valid for the problem at hand. For instance, employing an electrostatic double layer potential based on the linear superposition approximation will not be accurate when the interparticle separation is small (implying significant double layer overlap). In this case, the repulsive energy barrier height calculated using the linear superposition approximation will be incorrect.

Finally, one should note that the DLVO potential is an approximate potential of mean force between two charged particles suspended in an infinitely large medium. Application of this potential to concentrated colloidal systems without considering the many body effects on phenomena such as coagulation or disorder-order transition (Russel *et al.*, 1989) is untenable. There have been several approaches where different aspects of the many body interactions between colloidal particles in concentrated suspensions are considered within the framework of the DLVO type potential of mean force. A typical modification of the theory in concentrated dispersions involves consideration of a modified Debye length to account for the large particle volume fractions (Chaikin *et al.*, 1982). There are also approaches that explicitly consider many body interactions through pairwise summation of the DLVO interaction between two particles in integral equation theories (Vlachy and Prausnitz, 1992; Bhattacharjee *et al.*, 1999). Such approaches often predict properties of concentrated colloidal dispersions quite accurately.

There are several occasions when the DLVO theory has been criticized as inaccurate. However, the inability of the DLVO theory in predicting the behavior of colloidal systems might be an artifact of either the oversimplified nature of the DLVO potential, which fails to account for the complexities inherent in the system studied, or due to inaccurate usage of the DLVO theory. The inaccurate usage of the theory can result from employing expressions of the electrostatic interaction energies that are valid for high electrolyte concentrations (large κa) to predict the behavior of colloidal systems at low electrolyte concentrations (small κa). Similarly, expressions based on the linearized Poisson–Boltzmann equation, if employed to predict coagulation and deposition behavior of highly charged particles ($ze\psi/k_BT > 3$), might result in erroneous predictions. Finally, as mentioned earlier, roughness or charge heterogeneity of real interfaces can significantly modify the interactions from those predicted by the DLVO theory assuming smooth geometrical shapes. Consequently, when the DLVO theory fails to predict colloidal phenomena accurately, one should systematically explore the causes of such failure before considering the model to be inaccurate. The DLVO theory was a significant breakthrough in the field of colloidal stability prediction, and has survived six decades of intense scrutiny. When it is applied to a system that respects all the assumptions inherent in the model, it provides a remarkably accurate prediction of colloidal stability. To quote Ninham (1999), "The genius of DLVO lies not in complicated models that add more and more parameters, but in its extraction of the essential physics of the problem of lyophobic colloid stability."

11.7.2 When DLVO Theory Falls Short

Numerous other interactions are invoked when DLVO theory is unable to resolve the behavior of a colloidal system. The most common interactions that are referred to in this context are hydrophobic or hydrophilic interactions (Israelachvili, 1985; van Oss and Giese, 2004). In addition, depletion interactions, structural forces, steric interactions, *etc.*, have been employed to describe colloid stability for many systems (Israelachvili, 1985).

At the outset, it should be mentioned that most of the other forces mentioned earlier are ramifications of neglecting the granularity of the solvent and the ions in the DLVO theory. To assess the nature of these other forces in aqueous media, one should first take a closer look at water, the most common solvent studied in colloid science. Water is a complex solvent. It exhibits hydrogen bonding, and has unusual properties arising from its strong dipolar nature. A very simple depiction of water's anomalous behavior is in its large surface tension of 72.8 mJ/m^2. In Section 11.2.2, we noted that the surface tension of apolar liquids can be related to their Hamaker constants. Employing this approach, the surface tension of water can be computed from its Hamaker constant as 21.8 mJ/m^2. The remaining 51 mJ/m^2 (*i.e.*, 72.8–21.8 mJ/m^2) is unaccounted for. This contribution to the surface tension of water is ascribed to the hydrogen bonding effects (van Oss *et al.*, 1988). Nowhere in the calculation of the van der Waals and electrostatic double layer interactions have these hydrogen bonding interactions been taken into account.

When two colloidal particles approach separations less than a few nanometers, the water molecules in the intervening space have to be removed to facilitate closer approach between the particles. If the particles have an affinity toward the water molecules, a large amount of energy must be spent to remove the water. In other words, hydrophilic particles will experience a strong hydrophilic repulsion. On the other hand if the particles do not have an affinity for water, it is energetically favorable to remove the intervening water molecules. This is often referred to as hydrophobic attraction between the particles. Typically, hydrophilic or hydrophobic forces are assigned an exponential decay behavior of the form (van Oss, 1993)

$$\Delta G(d) = \Delta G(d_0) \exp\left(\frac{d_0 - d}{\lambda}\right) \tag{11.61}$$

Here, ΔG is the hydration free energy per unit area between the two interacting surfaces separated by a distance d, d_0 is the minimum equilibrium distance corresponding to contact between the surfaces, and λ is the decay length of the hydration interactions. $\Delta G(d_0)$ is the hydration interaction energy at contact, which is usually obtained from the interfacial tension between the interacting surfaces and the solvent. It should be noted that the exponential decay behavior, including the decay length λ, are completely empirical.

Depletion forces can arise due to the finite size of ions. Recall that for a particle with constant surface potential approaching another charged surface, the osmotic pressure contribution to the electrostatic double layer force will vanish upon integration over the particle surface. This happens because the ion concentration at the constant potential surface of the particle will be uniform everywhere. When we consider ions of a finite size, this condition of uniform surface concentration cannot be satisfied if the particle approaches too close to another charged surface. In this case, the ions in the intervening space between the charged surfaces will have to be removed to bring the particle close to the surface. This will set up an imbalance in osmotic stress around the particle surface. The net effect of this imbalance is manifested as an attractive force, often referred to as depletion interaction. Depletion forces are generally prominent in polydisperse colloidal suspensions, where the smaller colloidal particles (or polymeric entities) contribute to the osmotic stresses around larger particles (Russel *et al.*, 1989). Imbalance in the concentration of the smaller particles around the larger particles gives rise to the depletion forces in these systems (Weronski and Walz, 2003; Weronski *et al.*, 2003).

Structural forces and steric interactions are other manifestations of granularity of the solvent and microions in a colloidal system. These types of forces also appear due to adsorbed polyelectrolytes or tethered chains on particle surfaces (Wang and Denton, 2004). When dealing with bio-colloids, such as bacteria, steric forces must be accounted for to accommodate the effects of an exo-polymer (or lipo-polysaccharide, LPS) layer on the bacterial surfaces.

Quite often, various other interactions (like hydration, depletion, steric, *etc.*) are simply added to the DLVO interaction potential to yield the extended DLVO (XDLVO) potential (van Oss, 1993). Equation (11.60) is an example of such an approach. While it is extremely useful in many applications involving polymeric and biological systems, such an approach may be questioned from a fundamental point of view, since all the additional interactions are ramifications of the granularity of the solvent and the ions, or other factors neglected in the formulation of the DLVO theory. In this context, there is a gradual, but systematic shift toward development of better models of colloidal interactions based on explicit consideration of the solvent and ion structures. Computer simulations and statistical mechanics based theories are now common for prediction of the colloidal interactions and forces (Levin, 2002; Hirata, 2003).

11.8 NOMENCLATURE

a, a_1, a_2	radius of a spherical body, m
A_H	effective Hamaker constant, J
$A_{ret}(h)$	Hamaker constant accounting for retardation effects, J
$A_{ret}(0)$	Hamaker constant not accounting for retardation effects, $(=A_H)$ J
A_{11}	Hamaker constant for similar bodies (or material) in vacuum, J
A_{12}	Hamaker constant for bodies 1 and 2 in vacuum, J
A_{312}	Hamaker constant for materials 3 and 2 with an intervening medium 1, J
c	material constant in Lennard–Jones equation $(\epsilon_D \sigma^6)$
CFC	critical flocculation concentration of electrolyte solution required to flocculate a colloidal system, mol/L
d	separation distance between interacting surfaces, m
d_0	minimum equilibrium cut-off separation between interacting surfaces, m
F_R	electrostatic repulsive force, N (N/m^2 for the case of parallel plates)
ΔG	Gibbs free energy, J
h	gap between two bodies (surface to surface distance), m
k_B	Boltzmann constant, J/K
n_{crit}	critical ion number concentration, m^{-3}
n_+	coions number concentration, m^{-3} (assuming $\psi_s > 0$)
n_-	counterions number concentration, m^{-3} (assuming $\psi_s > 0$)
n_∞	ion number concentration in the bulk solution, m^{-3}
r	distance separating centers of two bodies, m

r_1	intermolecular distance, m
T	absolute temperature, K
u_{LJ}	Lennard Jones interaction potential between two atoms, J
U	total interaction potential, J (J/m^2 for the case of parallel plates)
U_A	attractive potential energy due to London–van der Waals forces, J (or J/m^2 for the case of parallel plates)
U_R	electrostatic repulsive potential, J (J/m^2 for the case of parallel plates)
U_S	potential due to ionic structure, J (J/m^2 for the case of parallel plates)
V_1, V_2	volumes of bodies 1 and 2, respectively, m^3
W_{11}^d	cohesion work for material 1 due to dispersion forces, J
z	absolute value of the valency for a $(z : z)$ electrolyte
z_-	valency of counterions (assuming $\psi_s > 0$)
z_+	valency of coions (assuming $\psi_s > 0$)

Greek Symbols

γ^d, γ_1^d, γ_2^d	surface tension due to dispersion energies, N/m
ϵ	dielectric constant of a material, C/mV
ϵ_D	characteristic energy of dipolar interactions in the Lennard–Jones equation
ζ	zeta potential, V
κ	inverse Debye length, m^{-1}
ρ_1, ρ_2	molecular density of bodies 1 and 2, respectively, mol/m^3
σ	diameter of an atom in the Lennard–Jones equation, m
ψ_s	surface electric potential, V

11.9 PROBLEMS

11.1. Using the values of the unretarded Hamaker constants in vacuum from Tables 11.3 and 11.5, determine the effective Hamaker constants, A_{121} or A_{132}, in Joules for the following materials: (a) Two polystyrene particles in water. (b) Polystyrene and poly-tetraflouroethylene in hexane. (c) two poly-tetrafluoroethylene particles in hexane. What is the implication of a negative value of the effective Hamaker constant?

11.2. Staring from the attractive Lennard–Jones potential and using Hamaker's approach, derive an expression for the van der Waals interaction energy between two spherical particles of same radius a, with their surfaces separated by a distance h. Clearly show the geometrical details of your calculations.

11.3. Plot the variation of the scaled van der Waals interaction energy, U_A/A_H, where A_H is the effective Hamaker constant, with scaled separation distance between two particles of equal radius, a_p, calculated using Hamaker's approach and Derjaguin's approximation. How do the two results compare for three different particle radii (a_p = 10 nm, 100 nm, and 1 μm)?

11.4. Consider the van der Waals and electrostatic double layer interaction energies per unit area between two infinite planar surfaces to be

$$U_A = -\frac{A_H}{12\pi D^2}$$

and

$$U_R = \frac{64 n_\infty k_B T}{\kappa} Z^2 \exp(-\kappa D)$$

respectively, where D is the separation distance between the surfaces and $Z = \tanh(ze\psi_s/k_B T)$, with ψ_s being the surface potential. For the above interactions, the total interaction potential energy and the force can become zero at some separation distance. The conditions for the energy and the force to become zero are given by

$$U_{DLVO} = U_A + U_R = 0$$

and

$$\frac{dU_{DLVO}}{dD} = 0$$

Show that this occurs when $\kappa D = 2$.

11.10 REFERENCES

Behrens, S. H., and Borkovec, M., Exact Poisson–Boltzmann solution for the interaction of dissimilar charge-regulating surfaces, *Phys. Rev. E.*, **60**, 7040–7048, (1999).

Bhattacharjee, S., and Elimelech, M., Surface element integration: A novel technique for evaluation of DLVO interaction between a particle and a flat plate, *J. Colloid Interface Sci.*, **193**, 273–285, (1997).

Bhattacharjee, S., Kim, A. S., and Elimelech, M., Concentration polarization of interacting solute particles in cross flow membrane filtration, *J. Colloid Interface Sci.*, **212**, 81–99, (1999).

Bhattacharjee, S., Ko, C. H., and Elimelech, M., DLVO Interaction between rough surfaces, *Langmuir*, **14**, 3365–3375, (1998).

Chaikin, P. M., Pincus, P., Alexander, S., and Hone, D., BCC-FCC, melting, and reentrant transitions in colloidal crystals, *J. Colloid Interface Sci.*, **89**, 555–562, (1982).

Chan, D., and Richmond, P., van der Waals forces for mica and quartz: Calculations from complete dielectric data, *Proc. Roy. Soc. Lond.*, **353A**, 163–176, (1977).

Czarnecki, J., The effects of surface inhomogeneities on the interactions in colloidal systems and colloid stability, *Adv. Colloid Interface Sci.*, **24**, 283–319, (1986).

Das, P. K., and Bhattacharjee, S., Electrostatic double layer interaction between spherical particles inside a rough capillary, *J. Colloid Interface Sci.*, **273**, 278–290, (2004).

Elimelech, M., and O'Melia, C. R., Effect of particle size on collision efficiency in the deposition of Brownian particles with electrostatic energy barriers, *Langmuir*, **6**, 1153–1163, (1990).

Fowkes, F. M., Attractive forces at interfaces, *Ind. Eng. Chem.*, **56**, 40–52, (1964).

Gregory, J., Approximate expressions for retarded van der Waals interaction, *J. Colloid Interface Sci.*, **84**, 138–145, (1981).

Gur, Y., Ravina, I., and Babchin, A. J., On the electrical double layer theory. 1. A numerical method for solving a generalized Poisson–Boltzmann equation, *J. Colloid Interface Sci.*, **64**, 326–332, (1978a).

Gur, Y., Ravina, I., and Babchin, A. J., On the electrical double layer theory. 2. The Poisson–Boltzmann equation including hydration forces, *J. Colloid Interface Sci.*, **64**, 333–341, (1978b).

Guzman–Garcia, A. G., Verbrugge, M. W., and Hill, R. F., Development of a space-charge transport model for ion-exchange membranes, *AIChE J.*, **36**, 1061–1074, (1990).

Hamaker, H. C., London–van der Waals attraction between spherical particles, *Physica*, **4**, 1058–1072, (1937).

Herman, M. C., and Papadopoulos, K. D., A method for modeling the interactions of parallel flat-plate systems with surface-features, *J. Colloid Interface Sci.*, **142**, 331–342, (1991).

Hiemenz, P. C., and Rajagopalan, R., *Principles of Colloid and Surface Chemistry*, 3rd ed., Marcel Dekker, New York, (1997).

Hirata, F., Theory of molecular liquids, in *Molecular Theory of Solvation*, Hirata F. (Ed.), Kluwer, Dordrecht, (2003).

Hoek, E. M. V., Bhattacharjee, S., and Elimelech, M., Effect of membrane surface roughness on colloid-membrane DLVO interactions, *Langmuir*, **19**, 4836–4847, (2003).

Hough, D. B., and White, L. R., The calculation of Hamaker constants from Lifshitz theory with applications to wetting phenomena, *Adv. Colloid Interface Sci.*, **14**, 3–41, (1980).

Hunter, R. J., *Foundations of Colloid Science*, vol. 1., Oxford University Press, Oxford, (1991).

Israelachvili, J. N., *Intermolecular and Surface Forces*, Academic Press, London, (1985).

Israelechvili, J. N., and Adams, G. E., Measurement of forces between two mica surfaces in aqueous electrolyte solutions in the range 1–100 nm, *J. Chem. Soc., Faraday Trans. 1*, **74**, 975–1001, (1978).

Israelachvili, J. N., and Tabor, D., van der Waals forces theory and experiment, *Prog. Surface Membrane Sci.*, **7**, 1–55, (1973).

Kjellander, R., and Greberg, H., Mechanisms behind concentration profiles illustrated by charge and concentration distributions around ions in double layers, *J. Electroanal. Chem.*, **450**, 233–251, (1998).

Kjellander, R., Akesson, T., Jonsson, B., and Marcelja, S., Double layer interactions in monovalent and divalent electrolytes – a comparison of the anisotropic hypernetted chain theory and Monte-Carlo simulations, *J. Chem. Phys.*, **97**, 1424–1431, (1992).

Lenhoff, A. M., Contributions of surface features to the electrostatic properties of rough colloidal particles, *Colloids Surf. A*, **87**, 49–59, (1994).

Levin, Y., Electrostatic correlations: From plasma to biology, *Rep. Progr. Phys.*, **65**, 1577–1632, (2002).

Lifshitz, E. M., The theory of molecular attractive forces between solids, *Soviet Physics JETP*, **3**, 73–83, (1956).

Mahanty, J., and Ninham, B. W., *Dispersion Forces*, Academic Press, London, (1976).

Morrison, I. D., and Ross, S., *Colloidal Dispersions, Suspensions, Emulsions and Foams*, Wiley Interscience, New York, (2002).

Ninham, B. W., and Parsegian, V. A., van der Waals forces. Special characteristics in lipid–water systems and a general method of calculation based on the Lifshitz theory, *Biophys. J.*, **10**, 646–663, (1970).

Ninham, B. W., On progress in forces since the DLVO theory, *Adv. Colloid Interface Sci.*, **83**, 1–17, (1999).

Overbeek, J. Th. G., *Colloid and Surface Chemistry. A Self-Study Course. Part 2, Lyophobic Colloids*, Cambridge, Mass., MIT Center For Advanced Eng. Study, (1972).

Parsegian, V. A., Long range van der Waals forces, in *Physical Chemistry: Enriching Topics in Colloid and Surface Science*, Van Olphen, H. and Mysels, K. J. (Eds.) (1975).

Parsegian, V. A., and Weiss, G. H., Spectroscopic parameters for computation of van der Waals forces, *J. Colloid Interface Sci.*, **81**, 285–289, (1981).

Pashley, R. M., Hydration forces between mica surfaces in aqueous electrolyte solutions, *J. Colloid Interface Sci.*, **80**, 153–162, (1981a).

Pashley, R. M., DLVO and hydration forces between mica surfces in Li^+, Na^+ and Cs^+ electrolyte solutions, *J. Colloid Interface Sci.*, **83**, 531–546, (1981b).

Pashley, R. M., and Israelachvili, J. N., DLVO and hydration forces between mica surfaces in Mg^{2+}, Ca^{2+}, Sr^{2+} and Ba^{2+} chloride solutions, *J. Colloid Interfce Sci.*, **97**, 446–455, (1984).

Probstein, R. F., *Physicochemical Hydrodynamics, An Introduction*, 2nd ed., Wiley Interscience, New York, (2003).

Ross, S., and Morrison, I. D., *Colloidal Systems and Interfaces*, John Wiley, New York, (1988).

Russel, W. B., Saville, D. A., and Schowalter, W. R., *Colloidal Dispersions*, Cambridge University Press, Cambridge, (1989).

Schenkel, J. M., and Kitchener, J. A., A test of the Derjaguin-Landau-Verwey-Overbeek theory with a colloidal suspension, *Trans. Faraday Soc.*, **56**, 161–173, (1960).

Shaw, D. J., *Introduction to Colloid and Surface Chemistry*, 3rd ed., Butterworths, London, (1980).

Suresh, L., and Walz, J., Effect of surface roughness on the interaction energy between a colloidal sphere and a flat plate, *J. Colloid Interface Sci.*, **183**, 199–213, (1996).

Tabor, D., and Winterton, R. M., Direct measurements of normal and retarded van der Waals forces, *Proc. Roy. Soc. Lond.*, **312A**, 435–450, (1969).

van der Hoeven, P. H. C., and Lyklema, J., Electrostatic stabilization in non-aqueous media, *Adv. Colloid Interface Sci.*, **42**, 205–277, (1992).

van Oss, C. J., Chaudhuri, M. K., and Good, R. J., Interfacial Lifshitz–van der Waals and polar interactions in macroscopic systems, *Chem. Rev.*, **88**, 927–941, (1988).

van Oss, C. J., Acid–base interfacial interactions in aqueous media, *Colloids Surf. A*, **78**, 1–49, (1993).

van Oss, C. J., and Giese, R. F., Role of the properties and structure of liquid water in colloidal and interfacial systems, *J. Dispersion Sci. Technol.*, **25**, 631–655, (2004).

Vlachy, V., and Prausnitz, J. M., Donnan equilibrium: Hypernetted-chain study of one-component and multicomponent models for aqueous polyelectrolyte solutions, *J. Phys. Chem.*, **96**, 6465–6469, (1992).

Wang, H., and Denton, A. R., Effective electrostatic interactions in suspensions of polyelectrolyte brush coated colloids, *Phys. Rev. E*, **70**, Art. No. 041404, (2004).

Weronski, P., and Walz, J. Y., An approximate method for calculating depletion and structural interactions between colloidal particles, *J. Colloid Interface Sci.*, **263**, 327–332, (2003).

Weronski, P., Walz, J. Y., and Elimelech, M., Effect of depletion interactions on transport of colloidal particles in porous media, *J. Colloid Interface Sci.*, **262**, 372–383, (2003).

White, L. R., On the Deryaguin approximation for the interaction of macrobodies, *J. Colloid Interface Sci.*, **95**, 286–288, (1983).

Xu, Z., and Yoon, R. H., The role of hydrophobic interactions in coagulation, *J. Colloid Interface Sci.*, **132**, 532–541, (1989).

Xu, Z., and Yoon, R. H., A study of hydrophobic coagulation, *J. Colloid Interface Sci.*, **134**, 427–434, (1990).

CHAPTER 12

COAGULATION OF PARTICLES

12.1 INTRODUCTION

In the previous chapter, we dealt with van der Waals forces and discussed the DLVO theory, which represents the summation of the van der Waals and the electric double layer interactions between colloidal particles. In this context, we were able to discern whether two particles, when brought together, would coagulate. Our arguments were based solely on the variations of the DLVO interaction potential with separation distance between the colloidal particles. Consequently, we were only able to predict that coagulation might occur if the attachment of two particles is energetically favorable. In addressing coagulation in terms of the energetically favorable state, we ignored the dynamic processes involved in bringing the two particles close together. When suspended colloidal particles in a dispersion are not stable, the rate at which they coagulate depends on the frequency with which they encounter, or collide with, each other. This frequency of collision is a function of the velocities of the fluid and the particles, Brownian motion, and colloidal forces, *e.g.*, electrostatic and van der Waals forces.

When two particles of comparable size collide with each other, the collision may lead to the formation of a doublet. This process is termed coagulation. Here, it is usually necessary to know the velocity field of the interacting particles and the prevailing interaction forces. On the other hand, when a particle collides with a much larger particle or a surface, usually called a collector, such a process is often termed

Electrokinetic and Colloid Transport Phenomena, by Jacob H. Masliyah and Subir Bhattacharjee
Copyright © 2006 John Wiley & Sons, Inc.

deposition. The study of deposition usually involves a knowledge of the prevailing forces and flow field around the collector.

The physics of particle-particle interaction in air as well as in a liquid are similar. Dissimilarities between interactions in aerosols (gas) and hydrosols (liquid) arise from the magnitudes of the prevailing forces and the ratio of the diameter of an aerosol particle to the mean-free path of the gas molecules. For example, charged aerosols are strongly influenced by electrostatic forces at comparatively large distances from a surface. One the other hand, electrostatic forces in an aqueous medium are restricted by the presence of the electrical double layers (Spielman, 1977).

For submicron particles, the primary collision mechanism is due to the Brownian motion. When coagulation is caused solely by the Brownian motion, it is referred to as *perikinetic* coagulation. On the other hand, when coagulation is caused solely by deterministic forces, *e.g.*, hydrodynamic forces, it is called *orthokinetic* coagulation.

For two colloidal particles that are similarly charged, the interparticle interaction potential typically includes an attractive part due to the London–van der Waals forces and a repulsive part due to the electrostatic forces. When the electrostatic repulsive forces are absent, particles undergo coagulation and the coagulation process is termed *rapid* coagulation. When the electrostatic repulsive forces or any other repulsive forces, *e.g.*, steric forces are present, the process is termed *slow* coagulation. The rate of coagulation is a measure of the colloidal system stability.

12.2 DYNAMICS OF COAGULATION

The dynamics of coagulation address how rapidly or slowly a suspension of colloidal particles will coagulate. The rate of coagulation has immense practical implications. Addition of alum causes suspended colloidal matter in water to coagulate and settle during water treatment in clarifiers. The rate of coagulation plays an important role in the design of clarifiers in water purification plants. A colloidal dispersion is considered stable if the coagulation rate is extremely slow. On the other hand, a dispersion is considered unstable if the particles in it coagulate rapidly.

In a colloidal suspension, the suspended particles are in a state of continuous random motion. It is conceivable that such a random motion of particles will result in collision between the particles. The probability of a collision can be increased by stirring the suspension vigorously. If there is an electrostatic repulsion between the particles, then the repulsive interaction will hinder the collision between the particles. On the other hand, if the electrostatic repulsion does not exist, the particles become "sticky" due to the van der Waals attraction. In such a scenario, all the collisions between two particles will be "successful", resulting in the formation of doublets, triplets, quadruplets, and successively larger aggregates.

Brownian motion, hydrodynamic interactions, and other interparticle interactions determine the rate of collision between particles. Smoluchowski (1917) was the first to study the rate of coagulation of particles due solely to Brownian motion with the diffusion coefficient of the particles taken as constant. His model was essentially that of diffusion of non-interacting spheres relative to a reference sphere. The diffusion dynamics were described by the radial component of the diffusion equation in spherical coordinates with a uniform diffusivity (Feke and Schowalter, 1983). Smoluchowski evaluated the rate of diffusion of the particles to a reference sphere and derived an expression for the collision frequency.

Smoluchowski also evaluated the collision frequency of spherical particles in a shear flow in the absence of the Brownian motion. Here, it was assumed that the particles follow undisturbed streamlines and that no interparticle forces are present. Comparison of the collision frequency of the two asymptotic cases clearly demonstrated the enhancement of coagulation due to shear flow.

Fuchs (1934) generalized Smoluchowski's analysis to include an arbitrary interaction potential, assuming again that the diffusion coefficient of the particles is constant. Derjaguin and Landau (1941) and Verwey and Overbeek (1948) incorporated electrostatic and van der Waals forces into the arbitrary interaction potential derived by Fuchs to show the effect of these forces on the collision frequency between colloidal particles. Further improvements of the Fuchs derivation were made by Spielman (1970) and Honig *et al.* (1971), who included the variation of the particle diffusion coefficient due to the presence of another particle. Subsequently, van de Ven and Mason (1976), Zeichner and Schowalter (1977), and van de Ven (1989) evaluated collision frequency, in the absence of the Brownian motion, by incorporating hydrodynamic interactions and interaction potentials into the flow equations.

In the following sections, we shall explore, in some detail, the historical developments together with the more recent studies.

12.3 BROWNIAN MOTION

The coefficient of diffusion, D, is an important transport property of a colloidal particle and it represents the role of Brownian motion on the movement of the particle. One normally associates the diffusion coefficient with Fick's law,

$$\mathbf{j}_D^{**} = -D\nabla n \tag{12.1}$$

where $\mathbf{j}_D^{**}$ is the particle diffusive flux, and ∇n is the gradient of the number concentration of particles. In other words, one typically associates diffusion with the transport of a particle from a region of high concentration to a region of low concentration due to Brownian motion. This, however, does not imply that diffusion and Brownian motion cease to exist when there is no concentration gradient. Fick's law simply states that a directional flux of particles is developed due to diffusion when there is a concentration gradient. When the concentration gradient vanishes, the flux also vanishes. However, Brownian motion still exists in a suspension of uniform particle concentration. The random Brownian motion still causes a particle to "wander" around in the suspension volume, but in a macroscopic sense, this does not cause a change in the concentration of the particles.

When considering diffusion of a colloidal particle, it should be noted that the random Brownian motion is imparted to the colloidal particle by the thermal motion of the solvent molecules.

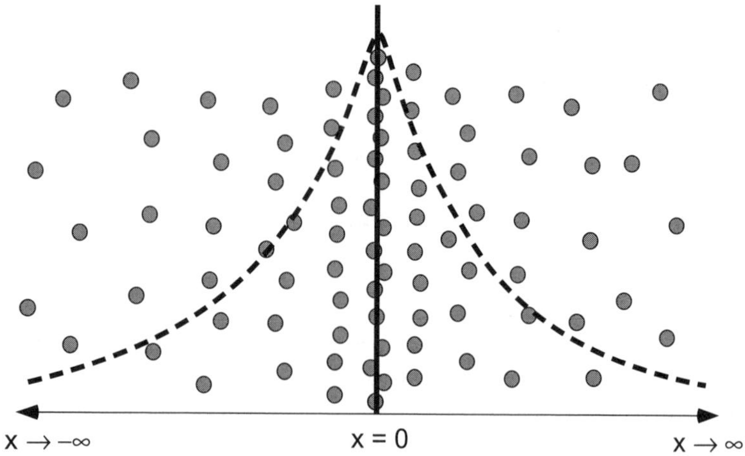

Figure 12.1. Diffusion in a one-dimensional plane.

The coefficient of diffusion D can be expressed in terms of the particle size. Consider one-dimensional diffusion of a swarm of colloidal particles. Assume that such a swarm is released over a narrow slit at $x = 0$ and $t = 0$ such that the number of released particles is N_i per unit area. The number of released colloidal particles per unit area is N_i. At time $t = 0$, the concentration of the particles for $x \neq 0$ is zero. With increasing time, the colloidal particles will diffuse due to the Brownian motion within the medium. Figure 12.1 shows the spread of the particles at time t in one dimension. The governing equation for the spread of the particles is given by the convection-diffusion equation, while dropping out the convection term,

$$\frac{\partial n}{\partial t} = D\frac{\partial^2 n}{\partial x^2} \tag{12.2}$$

where $n(x, t)$ is the number of particles per unit volume at distance x and at time t. The boundary conditions are

$$n(\infty, t) = 0 \quad \text{for all } t$$

and

$$n(-\infty, t) = 0 \quad \text{for all } t$$

Initially all the particles are located at $x = 0$ and $n(x, 0) = 0$ for $|x| > 0$. The solution of Eq. (12.2) is given by the Gaussian distribution, *viz.*,

$$n(x, t) = \frac{N_i}{2(\pi Dt)^{1/2}} \exp\left(-\frac{x^2}{4Dt}\right) \tag{12.3}$$

where N_i is the number of particles released at $x = 0$ per unit cross-sectional area. It is given by

$$N_i = \int_{-\infty}^{\infty} n(x,t)dx$$

The mean square displacement at time t of the particles from $x = 0$ is given by

$$\overline{x^2} = \frac{1}{N_i}\int_{-\infty}^{\infty} x^2 n(x,t)dx \tag{12.4}$$

If one makes use of the expression for $n(x,t)$ given by Eq. (12.3), the mean square displacement becomes

$$\overline{x^2} = 2Dt \tag{12.5}$$

Equation (12.5) states that the mean square displacement of the diffusing particles is proportional to the lapsed time t.

For a particle of mass m with velocity v, one can use Newton's laws of motion to write a force balance as

$$m\frac{dv}{dt} = -fv + F(t) \tag{12.6}$$

an equation known as the *Langevin equation*. Equation (12.6) balances the inertial force arising from the acceleration (dv/dt) against the dissipative force represented by $-fv$ (where f is the friction coefficient) and a stochastic thermal noise $F(t)$ of the fluid around the particle. At infinite dilution and in the absence of charge, the friction coefficient for spherical particles of radius a in a fluid of viscosity μ is given by the *Stokes equation*

$$f = 6\pi\mu a \tag{12.7}$$

where, as one can readily see, the units of f are force per unit velocity, *i.e.*, Ns/m. The second term on the right hand side of Eq. (12.6) represents the fluctuating force resulting from the thermal motion of the molecules of the fluid. The random trajectories of the particles are the result of the random fluctuation in the forces that the colliding fluid molecules collectively exert on the particles. The fluctuating force $F(t)$ is assumed to be independent of the particle velocity v and its mean value over a long time t is zero.

Equation (12.6) can be written as

$$\frac{d(vx)}{dt} + \beta vx = v^2 + xA(t) \tag{12.8}$$

where $\beta = f/m$, $A(t) = F(t)/m$, and $v = dx/dt$. Integrating Eq. (12.8) between $t = (0, t)$ and recognizing that at $t = 0$, $vx = 0$, one obtains

$$vx = \exp(-\beta t)\int_0^t v^2 \exp(\beta\tau)\,d\tau + \exp(-\beta t)\int_0^t A(\tau)x\exp(\beta\tau)\,d\tau \tag{12.9}$$

where τ is a variable of integration. Making use of averaging arguments, Friedlander (2000) gave at large times

$$\frac{\overline{x^2}}{2} = \frac{\overline{v^2}t}{\beta} \tag{12.10}$$

We can now make use of a physical argument made by Einstein that relates the Brownian motion of a colloidal particle to the molecular motion of the medium. As the particles share the molecular-thermal motion of the fluid, the principle of equipartition of energy is assumed to apply to the translational energy of the colloidal particles and leads to

$$\frac{m\overline{v^2}}{2} = \frac{k_B T}{2} \tag{12.11}$$

Combining Eq. (12.10) with (12.11) and making use of the definition of β allows one to relate the mean-square deviation in position to the viscous dissipation

$$\overline{x^2} = 2\frac{k_B T}{f}t \tag{12.12}$$

A comparison of Eq. (12.5) with (12.12) reveals that

$$D = \frac{k_B T}{f} \tag{12.13}$$

the well-known *Stokes–Einstein equation* for the coefficient of diffusion of an inert spherical particle. A discussion on the various forms of the diffusion coefficient as influenced by the proximity of other particles, the particle surface potential, the Debye length, and the size of the ions is given in van de Ven (1989). At infinite dilution, the diffusion coefficient of an inert particle in a fluid is given by

$$D = D_\infty = \frac{k_B T}{6\pi\mu a} \tag{12.14}$$

where k_B is Boltzmann constant and T is the absolute temperature. In the above analysis, the diffusion coefficient D was assumed to be constant and not affected by the presence of surrounding particles.

12.4 COLLISION FREQUENCY

In this section, we shall derive the kinetic equation for coagulation. Here particle collisions lead to coagulation and hence a reduction in the total number of particles and an increase in the particle average size. We shall assume that every collision leads to coagulation.

Consider a colloidal dispersion containing an infinite number of species ($i = 1, 2, 3, \ldots, \infty$) characterized solely by its volume v_i. Let J_{ij} be the collision frequency per unit volume occurring between two classes of spherical particles characterized by volume v_i and v_j, respectively. When particle i collides with particle j a third particle is formed having a volume of $(v_i + v_j)$. The collision frequency per unit volume J_{ij} can be given as

$$J_{ij} = \beta(v_i, v_j)n_i n_j \tag{12.15}$$

where n_i is the number concentration of species i. Here $\beta(v_i, v_j)$ is the collision frequency function. This function is dependent on the system properties and on the prevailing interparticle forces.

The rate of formation per unit volume of particles of size k by collision of particles of size i and j is given by

$$\frac{1}{2}\sum_{i+j=k} J_{ij} \tag{12.16}$$

The notation $i + j = k$ means that the summation is taken over those collisions for which a volume v_k can be produced; *i.e.*,

$$v_i + v_j = v_k \tag{12.17}$$

The factor 1/2 in Eq. (12.16) is present because each collision is counted twice in the summation. Equation (12.16) is due to the generation of the kth species. In turn, particle k collides with other particles leading to particle volumes larger than v_k. The disappearance term is given by

$$\sum_{i=1}^{\infty} J_{ik} \tag{12.18}$$

The net balance equation for the kth species is given by

$$\frac{dn_k}{dt} = \frac{1}{2}\sum_{i=1;i+j=k}^{i=k-1} J_{ij} - \sum_{i=1}^{\infty} J_{ik} \quad \text{for } k = 1, 2, 3, \ldots, \infty \tag{12.19}$$

With the use of Eq. (12.15), the above equation becomes

$$\frac{dn_k}{dt} = \frac{1}{2}\sum_{i=1;i+j=k}^{i=k-1} \beta_{ij} n_i n_j - n_k \sum_{i=1}^{\infty} \beta_{ik} n_i \quad \text{for } k = 1, 2, 3, \ldots, \infty \tag{12.20}$$

where $\beta_{ij} = \beta(v_i, v_j)$.

Equation (12.20) is the dynamic equation for the evolution of the population of particle k with time. This equation, called the population balance equation, is a fairly general equation that accounts for the generation and loss of a population. The solution of Eq. (12.20) depends on the collision frequency function β_{ij}, which is determined solely by the mechanism of the particle collisions which is dependent on the prevailing interparticle forces.

12.5 BROWNIAN COAGULATION

12.5.1 The Smoluchowski Solution without a Field Force

First, we shall deal with the diffusion of colloidal particles due to their Brownian motion. To begin with, the particles are assumed to be non-interacting and the only prevailing forces are due to the Brownian motion. In other words, forces due to electrostatic and dispersion are not accounted for. This analysis was put forward by M. von Smoluchowski in the early 1900s.

Consider a diffusional flux of spherical particles of volume v_j diffusing towards a particle of volume v_i. Particle v_i is the reference or test particle fixed at the coordinate system. In the absence of a convective flux, the diffusion equation is given by

$$\frac{\partial n_j}{\partial t} = D\frac{1}{r^2}\frac{\partial}{\partial r}\left(r^2\frac{\partial n_j}{\partial r}\right) \tag{12.21}$$

where n_j is the number concentration of particle v_j and r is the radial coordinate of the spherical coordinate system. Angular symmetry is assumed. The initial condition is given as

$$\text{At } t = 0 \quad n_j = n_{j\infty} \quad \text{for } r > a_i + a_j$$

The boundary conditions are

$$\text{at } r = a_i + a_j \quad n_j = 0 \quad \text{for } t > 0$$

and

$$\text{at } r \to \infty \quad n_j = n_{j\infty} \quad \text{for } t > 0$$

Here a_i is the radius of the particles having volume v_i and $n_{j\infty}$ represents the uniform number concentration of the jth particle. One can solve Eq. (12.21) by making use of the transformation (Friedlander, 2000)

$$w = \left(\frac{n_{j\infty} - n_j}{n_{j\infty}}\right)\left(\frac{r}{a_i + a_j}\right) \tag{12.22}$$

and

$$y = \frac{r - (a_i + a_j)}{a_i + a_j} \tag{12.23}$$

The transformed form of Eq. (12.21) is

$$\frac{\partial w}{\partial t} = D'\frac{\partial^2 w}{\partial y^2} \tag{12.24}$$

where $D' = D/(a_i + a_j)^2$. The solution of Eq. (12.24) is given by

$$w = 1 - \text{erf}\left(\frac{y}{2(D't)^{1/2}}\right) \tag{12.25}$$

where erf is the function which has the property of $\mathrm{erf}(0) = 0$. At $t \to \infty$, Eq. (12.25) gives

$$n_j = n_{j\infty}\left[1 - \frac{a_i + a_j}{r}\right] \tag{12.26}$$

This is the steady-state solution which can be obtained by setting $\partial n_j/\partial t = 0$ in Eq. (12.21). It is interesting to note that steady state is reached in relatively short times (Probstein, 2003). For particles of radius a, one may define the characteristic time for Brownian diffusion as

$$\frac{a^2}{D} = \frac{a^2}{[k_B T/(6\pi\mu a)]} = \frac{6\pi\mu a^3}{k_B T} \tag{12.27}$$

For a colloidal particle with $a = 10^{-7}$ m in water at room temperature, the characteristic time is about 5×10^{-3} s indicating that steady state is reached relatively quickly.

The total rate (particles per unit time) at which particles j arrive at the surface $r = a_i + a_j$ is given by

$$D\left[4\pi r^2 \frac{\partial n_j}{\partial r}\right]_{r=a_i+a_j} = 4\pi(a_i + a_j)Dn_{j\infty}\left[1 + \frac{a_i + a_j}{(\pi Dt)^{1/2}}\right] \tag{12.28}$$

The steady-state rate at which particles j arrive at the collision radius of the test particle i is given by

$$4\pi(a_i + a_j)Dn_{j\infty}$$

The collision rate frequency per unit volume between particles i and j is then given by

$$J_{ij} = 4\pi(a_i + a_j)Dn_{i\infty}n_{j\infty} \tag{12.29}$$

Making use of Eq. (12.15), β_{ij} for the case of the Brownian coagulation is given by

$$\beta_{ij} = 4\pi(a_i + a_j)D \tag{12.30}$$

It is necessary now to quantify the diffusion coefficient D. When the test particle i is also in Brownian motion, the diffusion coefficient D should be relative to the motion of the two particles given by $(x_i - x_j)$ where x_i and x_j are the displacement of the particles in x-direction. From Eq. (12.5), one can write

$$\overline{(x_i - x_j)^2} = 2Dt \tag{12.31}$$

Equation (12.31) can be expanded to give

$$\overline{x_i^2} - 2\overline{x_i x_j} + \overline{x_j^2} = 2Dt \tag{12.32}$$

As the motions of the two particles are independent of each other, $\overline{x_i x_j} = 0$. Equations (12.15) and (12.32) give

$$D_i + D_j = D \tag{12.33}$$

Making use of Eqs. (12.29) and (12.33), we can write for the collision rate per unit volume

$$J_{ij} = 4\pi(a_i + a_j)(D_i + D_j)\, n_{i\infty} n_{j\infty} \tag{12.34}$$

The Stokes–Einstein relation, Eq. (12.14), when combined with Eq. (12.34) gives

$$J_{ij} = \left[\frac{2k_B T}{3\mu}(a_i + a_j)\left(\frac{1}{a_i} + \frac{1}{a_j}\right)\right] n_{i\infty} n_{j\infty} \tag{12.35}$$

and

$$\beta_{ij} = \frac{2k_B T}{3\mu}(a_i + a_j)\left(\frac{1}{a_i} + \frac{1}{a_j}\right) \tag{12.36}$$

For convenience, the ∞ subscript will be dropped and n_i and n_j become the prevailing concentrations. Substituting for β_{ij} from Eq. (12.36) in the population balance, Eq. (12.20) yields the kinetic equation describing the change in the population of the kth particle. Equation (12.35) represents the collision rate caused solely by the Brownian motion.

In order to obtain an analytical solution to Eq. (12.20), a simplifying assumption is necessary. Smoluchowski assumed that $a_i = a_j$, which leads to

$$(a_i + a_j)\left(\frac{1}{a_i} + \frac{1}{a_j}\right) = 4 \tag{12.37}$$

For an initially monodispersed colloidal system, the assumption of $a_i = a_j$ is valid. However, at long times it is not true. To a large extent, its justification rests on experimental evidence. Making use of Eq. (12.37), the collision frequency function of Eq. (12.36) becomes

$$\beta(v_i = v_j) = \frac{8k_B T}{3\mu} = K \tag{12.38}$$

where K is a constant independent of particle size. The population balance equation, (12.20), becomes

$$\frac{dn_k}{dt} = \frac{K}{2}\sum_{i=1;i+j=k}^{i=k-1} n_i n_j - K n_k \sum_{i=1}^{\infty} n_i \quad k = 1, 2, \ldots, \infty \tag{12.39}$$

Let the total number of particles per unit volume in a closed system at time t be $N_{tot}(t)$; then

$$N_{tot}(t) = \sum_{i=1}^{\infty} n_i$$

Summing Eq. (12.39) over all particles $k = 1, 2, \ldots, \infty$, one obtains

$$\frac{dN_{tot}}{dt} = \frac{K}{2}\sum_{k=1}^{\infty}\sum_{i=1;i+j=k}^{i=k-1} n_i n_j - KN_{tot}^2$$

The first term on the right side of the above equation is given by $(K/2)N_{tot}^2$, leading to

$$\frac{dN_{tot}}{dt} = -\frac{K}{2}N_{tot}^2 \tag{12.40}$$

The above equation for the Brownian coagulation is analogous to second-order reaction kinetics. The condition is given by $N_{tot} = N_0$ at $t = 0$. Integration of Eq. (12.40) leads to

$$N_{tot}(t) = \frac{N_0}{1 + KN_0 t/2} \tag{12.41}$$

The above equation shows the decay in the total number of particles with time. Making use of Eq. (12.39), it can be shown (*e.g.*, Friedlander, 2000) that

$$n_k(t) = N_0 \frac{(t/t_{Br})^{k-1}}{(1 + t/t_{Br})^{k+1}} \tag{12.42}$$

where

$$t_{Br} = \frac{3\mu}{4N_0 k_B T} = \frac{2}{KN_0} \tag{12.43}$$

The characteristic time t_{Br} is known as the coagulation time and it is the time for the concentration to halve itself (Zeichner and Schowalter, 1979; Probstein, 2003). Figure 12.2 shows the variation of n_k/N_0 for $k = 1, 2, \ldots, 4$ with dimensionless time, t/t_{Br}. It is interesting to note that (N_{tot}/N_0) and (n_1/N_0) decay monotonically; however, (n_k/N_0) with $k \geq 2$ exhibits a maximum before its final decay.

One can rewrite the expression for the characteristic time t_{Br} as

$$t_{Br} = \frac{\pi \mu a^3}{\alpha_p k_B T} \tag{12.44}$$

where α_p is the colloidal particles volume fraction. Table 12.1 shows the time scale for the formation of a doublet under Brownian motion. It is clear from Table 12.1 that Brownian coagulation is not important for particles much larger than 10 μm where the characteristic time becomes fairly large.

12.5.2 Effect of a Field Force

Fuchs (1934) was the first to introduce an interparticle field force to the Smoluchowski Brownian coagulation analysis. Once again, we consider a test particle of radius a_i to which particles a_j are diffusing. Here particles i and j exert a force on each other.

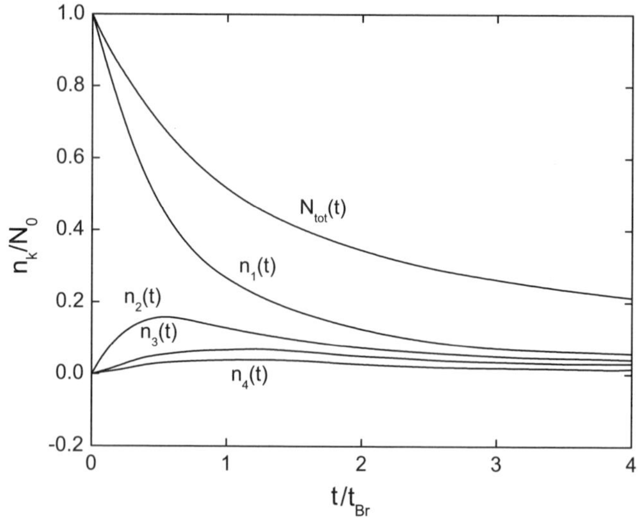

Figure 12.2. The variations in N_{tot}, n_1, n_2, ... with time for an initially monodisperse colloidal system. The total number concentration, N_{tot}, and the concentration of the species n_1 both decrease monotonically with increasing time. The concentrations of $n_2(t)$, $n_3(t)$, $n_4(t)$, *etc.* pass through a maximum (Smoluchowski, 1917).

Let this force be $F(r)$ and it is a function of position. The steady-state collision rate per unit area between particles j and the surface of the test particle i is given by

$$-\left[-D\frac{dn_j}{dr} + \frac{F(r)}{f}n_j\right]$$

The diffusion coefficient D is relative to the motion of the two particles and for an infinite medium it is given by Eqs. (12.14) and (12.33).

The total collision rate (particle per unit time) between particles j and the test particle i is given by

$$-\left[-D\frac{dn_j}{dr} + \frac{F(r)}{f}n_j\right]4\pi r^2 = \text{constant}$$

TABLE 12.1. Characteristic Time, t_{Br} for Brownian Coagulation of Spherical Particles in Water at 293 K for $\alpha_p = 0.1$.

a (m)	t_{Br} (s)
10^{-8}	7.77×10^{-6}
10^{-7}	7.77×10^{-3}
10^{-6}	7.77
10^{-5}	7.77×10^{3}

Defining the force $F(r)$ by its potential $U(r)$ leads to

$$\left[D\frac{dn_j}{dr} + \frac{1}{f}\frac{dU}{dr}n_j\right]4\pi r^2 = \text{constant}$$

Making use of Eq. (12.13) one obtains for the total collision rate between particles j and the test particle i,

$$4\pi r^2 D\left[\frac{dn_j}{dr} + \frac{n_j}{k_B T}\frac{dU}{dr}\right] = \text{constant} \tag{12.45}$$

Differentiation of Eq. (12.45) with respect to r gives

$$\frac{d}{dr}\left[r^2 D\left(\frac{dn_j}{dr} + \frac{n_j}{k_B T}\frac{dU}{dr}\right)\right] = 0 \tag{12.46}$$

where $n_j = 0$ at $r = a_i + a_j$ and $n_j = n_{j\infty}$ at $r =\rightarrow \infty$

Solution of Eq. (12.46) subject to the above boundary conditions gives

$$n_j = n_{j\infty}\left[\exp\left(-\frac{U(r)}{k_B T}\right)\int_{a_1+a_2}^{r}\exp\left(\frac{U(r)}{k_B T}\right)\frac{dr}{r^2}\right]\Big/\left[\int_{a_i+a_j}^{\infty}\exp\left(\frac{U(r)}{k_B T}\right)\frac{dr}{r^2}\right] \tag{12.47}$$

The total rate at which particles j arrive at surface $r = a_i + a_j$ is given by

$$D\left[4\pi r^2\frac{dn_i}{dr}\right]_{r=a_i+a_j}$$

Making use of Leibnitz's rule for the differentiation of integrals, the collision rate per unit volume between particles i and j is then given by

$$J_{ij} = \left[\frac{4\pi D(a_i + a_j)}{(a_i + a_j)\int_{a_i+a_j}^{\infty}\frac{1}{r^2}\exp\left(\frac{U(r)}{k_B T}\right)dr}\right]n_{i\infty}n_{j\infty} \tag{12.48}$$

The term in the square brackets in β_{ij}. The denominator of the term within the square brackets is the stability ratio W. Making use of Eq. (12.33) and the Stokes–Einstein equation for D_i, setting $n_{i\infty} = n_{j\infty} = n$, $a_i = a_j = a$ for the case of a monodisperse system and putting $J = J_{ij}$, one obtains

$$J = \frac{8k_B T}{3\mu}\cdot\frac{1}{2a\int_{2a}^{\infty}\frac{1}{r^2}\exp\left(\frac{U(r)}{k_B T}\right)dr} \tag{12.49}$$

For coagulation solely due to Brownian motion, *i.e.*, in the absence of any external force, Eq. (12.35) can be written for $a_i = a_j = a$ and $n_{i\infty} = n_{j\infty} = n$ as

$$J_{0,Br} = \frac{8k_B T}{3\mu}n^2 \tag{12.50}$$

where $J_{0,Br}$ denotes the Smoluchowski collision rate per unit volume due solely to Brownian motion, Combining Eqs. (12.49) and (12.50) leads to

$$W = \frac{J_{0,Br}}{J} \tag{12.51}$$

where W is the stability ratio. For the case of constant diffusion coefficient, it is given as

$$W = 2a \int_{2a}^{\infty} \frac{1}{r^2} \exp\left(\frac{U(r)}{k_B T}\right) dr \tag{12.52}$$

When $U(r)$ is a positive function, it signifies repulsion between the particles and it leads to $J < J_{0,Br}$ where $W > 1$. A large value for W means a stable system. In general, the potential $U(r)$ can have the form of the London–van der Waals (attractive) potential or an electrostatic (repulsive) potential. W is equal to unity for $U = 0$.

Figure 12.3 shows the effect of attractive London–van her Waals forces on the rapid Brownian coagulation J, normalized with the Smoluchowski Brownian coagulation $J_{0,Br}$. When $A_H/k_BT = 0$, the stability ratio is unity and it decreases with A_H/k_BT indicating rapid coagulation. However, the results as presented in Figure 12.3, which are based on the analysis presented here, are not in agreement with the experimental data. The major simplifying assumption lies in the fact that the diffusion coefficient is taken to be constant and independent of the gap width between the two particles. Recall the discussion in Section 6.3 (*e.g.*, Eq. 6.138) regarding the modification of the particle

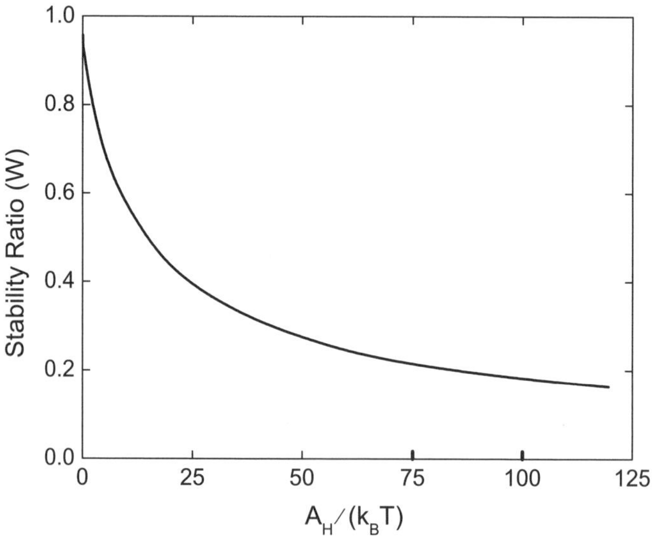

Figure 12.3. Variation of stability ratio W with A_H/k_BT. The van der Waals interaction potential is given by $U_A = -A_H a/[12(r - 2a)] = -A_H/(12h)$. Here, A_H is the Hamaker constant. The diffusion coefficient is a constant.

hydrodynamic velocities and diffusion coefficients owing to hydrodynamic interactions. For finite size colloidal particles, incorporation of the hydrodynamic interaction effects becomes important. Incorporation of these hydrodynamic interactions in the model for Brownian coagulation is briefly elaborated next.

Independently, Derjaguin and Muller (1967), Spielman (1970), and Honig *et al.* (1971) modified the analysis of the Brownian motion coagulation with interparticle potential to account for the decrease in the diffusion coefficient which occurs when two particles are very close.

The diffusion coefficient tensor $\bar{\bar{D}}$ for the case of two spheres is given by Brenner (1961) and Van de Ven (1989) as

$$\bar{\bar{D}} = (D_i + D_j)_\infty \begin{bmatrix} d_\parallel(h) & 0 \\ 0 & d_\perp(h) \end{bmatrix} \tag{12.53}$$

where $d_\parallel(h)$ and $d_\perp(h)$ are correction factors for the diffusion coefficient of sphere i relative to sphere j. Symbols $\parallel$ and $\perp$ signify the relative diffusion parallel and normal to the two surfaces, respectively.

The correction factors, $d_\parallel(h)$ and $d_\perp(h)$, are due to hydrodynamic interactions between the two spheres (van de Ven, 1989). Here $d_\parallel(h)$ signifies the effect of hydrodynamic interactions on the diffusion coefficient due to the motion of the spheres parallel to each other. This component acts perpendicular to the line joining the centers of the two spheres. The term $d_\perp(h)$ signifies the effect of hydrodynamic interactions on the diffusion coefficient due to the motion of the spheres along the line joining their centers. Figure 12.4 schematically depicts the directions of the two components of the hydrodynamic correction to the diffusion coefficient.

The terms $D_{i\infty}$ and $D_{j\infty}$ are the particle diffusion coefficients at infinite dilution and they are given by Eq. (12.14). For the case of spheres of equal radii, we set $d_\perp(h) = 1/G(h)$. Here h is the dimensionless gap width between the sphere, where

$$\frac{r}{a} = 2 + h \quad \text{with } a_i = a_j = a$$

and r is the dimensional distance between the sphere centers. $G(h)$ is a complex function of h. A simplified form for $G(h)$ is given by Honig *et al.* (1971) for the case of equal-sized spheres:

$$G(h) = \frac{6h^2 + 13h + 2}{6h^2 + 4h} \tag{12.54}$$

The above expression indicates that

$$G(h) \to 1 \quad \text{as } h \to \infty$$

and

$$G(h) \to 1/(2h) \quad \text{as } h \to 0$$

For $h \to 0$, the diffusion coefficient D becomes much smaller than D_∞ indicating a retardation in the Brownian coagulation process.

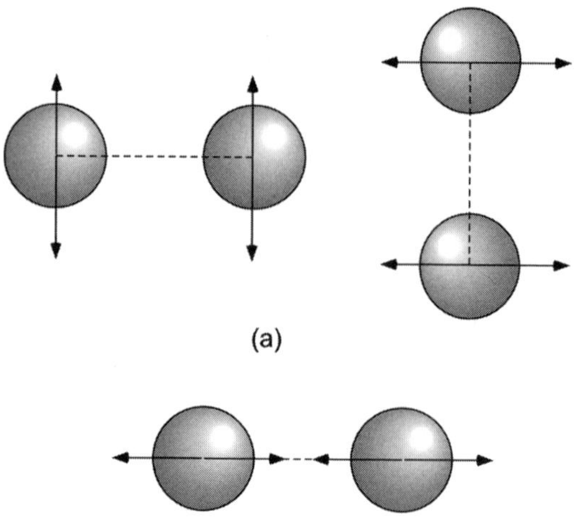

Figure 12.4. Schematic depiction of the directions of the hydrodynamic correction components of the diffusion tensor (a): $d_{\|}(h)$ and (b): $d_{\perp}(h)$. The arrows indicate the direction of motion in each case. The dashed line represents the line joining the sphere centers.

Following an approach similar to that adopted previously, the coagulation rate per unit volume, for equal-sized spherical particles having a variable diffusion coefficient, becomes

$$J = \frac{8k_BT}{3\mu}\left[2\int_0^{\infty}\frac{G(h)\exp\left(\frac{U(h)}{k_BT}\right)}{(2+h)^2}dh\right]^{-1} n^2 \tag{12.55}$$

with

$$W = \frac{J_{0,Br}}{J} \tag{12.56}$$

where W is modified to

$$W = 2\int_0^{\infty}\frac{G(h)\exp\left(\frac{U(h)}{k_BT}\right)}{(2+h)^2}dh \tag{12.57}$$

The characteristic time becomes

$$t_{Br} = \frac{\pi\mu a^3 W}{\alpha_p k_B T} = \frac{3\mu W}{4N_0 k_B T} \tag{12.58}$$

where α_p is the volume fraction of the particles and N_0 is their initial number concentration.

If one is to assume that $U(h) = 0$, *i.e.*, a complete absence of an external potential and that the diffusion coefficient is affected by the gap width, then $W \to \infty$ indicating that Brownian motion alone cannot lead to coagulation. In other words, when modification of the diffusion coefficient engendered by hydrodynamic interactions is taken into account, the above analysis predicts that no coagulation should occur at all because viscous friction forces become infinitely large as the particles approach each other. Thus, unless an attractive interaction is present between the particles, a colloidal suspension is inherently stable. To this end, one should recognize the importance of the London–van der Waals attraction forces in the coagulation process.

Figure 12.5 shows the variation of the stability ratios in the absence of electrostatic repulsion for the case of a constant and a variable diffusion coefficient. With D being a constant, the stability ratio decreases with (A_H/k_BT) and it has a value of unity at $A_H/k_BT \to 0$ as was shown earlier in Figure 12.3. However, when correction for D is considered, the value of the stability ratio is higher than unity for $A_H/k_BT \leq 75$ indicating the effects of hydrodynamic interactions in retarding the coagulation process. This is strictly due to the decrease of the particle mobility close to a rigid surface. The expression for $U(h)$ used in Figure 12.5 is given by

$$U(h) = -\frac{A_H}{6}\left[\frac{2}{h^2+4h} + \frac{2}{(h+2)^2} + \ln\frac{h^2+4h}{(h+2)^2}\right] \tag{12.59}$$

where h is the dimensionless gap width normalized using the particle radius, a. Experimental results for rapid coagulation confirm the analysis summarized by Eq. (12.57). Figure 12.6 and Table 12.2 show the stability ratio for rapid coagulation of polystyrene

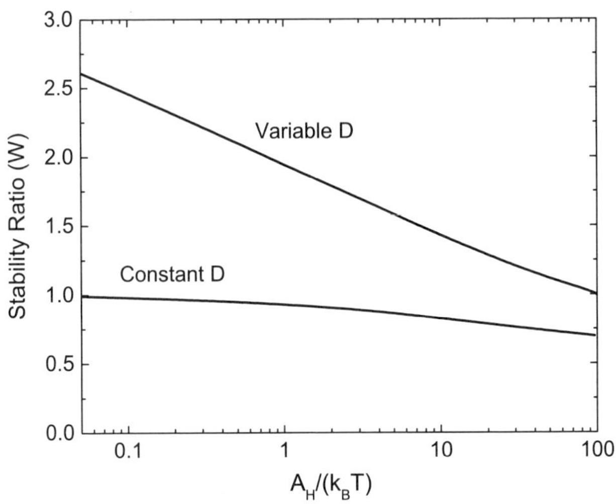

Figure 12.5. Variation of stability ratio W with A_H/k_BT for constant and variable D. The London–van der Waals interaction potential is given by Eq. (12.59).

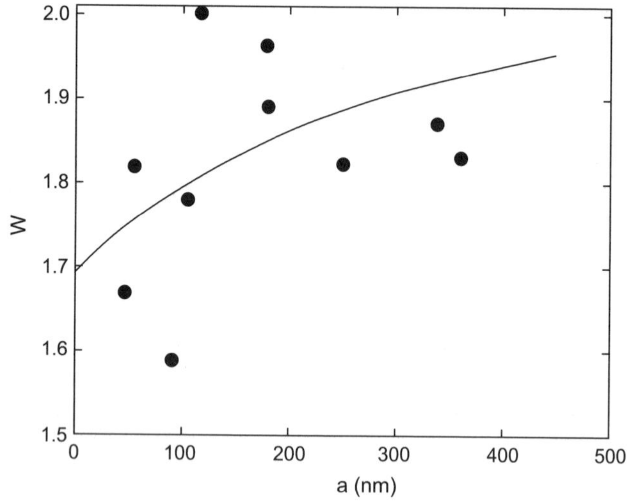

Figure 12.6. Stability ratio for rapid coagulation of polystyrene in water as a function of the sphere radius. Solid line: prediction from retarded van der Waals potential. Symbols: data from Table 12.2. (Russel *et al.*, 1989).

particles in water as a function of initial radius. The theoretical curve uses a retarded dispersion potential (Russel *et al.*, 1989).

Higashitani and Matsuno (1979) modified the population balance equation given by Eq. (12.20) to account for the interparticle potential and a variable diffusion coefficient. For the initial stage of the coagulation of a monodisperse system where a_i and a_j

TABLE 12.2. Data for Rapid Coagulation Experiments with Polystyrene Particles (Russel *et al.*, 1989).

	$2a$ (μm)	W	A_H/k_BT
Mathews and Rhodes (1970)	0.714	1.83	1.5
Lips and Willis (1973)	0.207	1.78	1.9
	0.357	1.89	1.2
	0.500	1.82	1.7
Lichtenbelt *et al.* (1974)	0.091	1.67	2.7
	0.109	1.82	1.5
	0.176	1.59	4.1
	0.234	2.00	0.7
	0.357	1.96	1.7
Feke and Schowalter (1983)	0.675	1.87	1.1

are not too widely different, the modified population balance equation takes the form

$$\frac{dn_k}{dt} = \frac{1}{2W} \sum_{i+j=k;i=1}^{i=k-1} \beta_{ij} n_i n_j - \frac{1}{W} n_k \sum_{i=1}^{\infty} \beta_{ik} n_i \quad k = 1, 2, 3, \ldots, 10 \tag{12.60}$$

where β_{ij} is defined by Eq. (12.36), W is defined by Eq. (12.57), and $G(h)$ is given by Eq. (12.54).

The experimental results of Higashitani and Matsuno (1979) for (n_k/N_0) variation with time are given in Figure 12.7 for polystyrene particles. The experimental conditions were such that the electrostatic repulsion forces were absent. Excellent agreement with theory is shown. This plot is similar to the case of the Smoluchowski Brownian motion coagulation shown in Figure 12.2 where $W = 1$ and the diffusion coefficient is constant. The dispersion potential was given by Eq. (12.59) with $A_H = 9 \times 10^{-21}$J. The excellent agreement with the solution of Eq. (12.60) is a further verification of the theoretical analysis presented for rapid coagulation.

The dimensionless groups that control coagulation due to Brownian motion with added interparticle potential can be arrived at by making use of the dimensionless

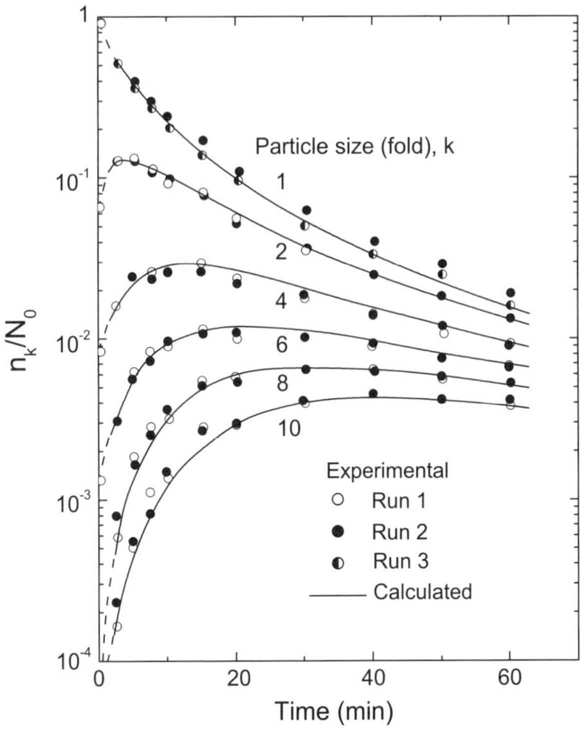

Figure 12.7. Rapid Brownian flocculation for polystyrene spheres with $a = 0.487$ μm and $W = 1.74$. The calculated curves are from Eq. (12.60) (Higashitani and Matsuno, 1979).

form of the flux or rate equation. The collision rate between the j particles and the i test particle, including transfer by convection (dropping the index j), is given for steady state in Cartesian coordinates by

$$\frac{d}{dx}\left[D\left(-\frac{dn}{dx}-\frac{n}{k_BT}\frac{dU}{dx}+u_x n\right)\right]=0 \tag{12.61}$$

where u_x is the fluid velocity in x-direction. Strictly speaking, u_x is equal to the particle hydrodynamic velocity v_x.

The total interaction potential, U, appearing in Eq. (12.61) can be represented as a sum of the attractive van der Waals interaction, U_A, and an electric double layer interaction, U_R. The non-retarded van der Waals interaction potential is usually given by

$$U_A = A_H\, f(X) \tag{12.62}$$

where A_H is the Hamaker constant.

The electric double layer interaction potential for identical charged spheres is given by

$$U_R = 2\pi\epsilon\, \psi_s^2\, a\, g(X, \kappa a) \tag{12.63}$$

The functions f and g are dimensionless functions and

$$X = \frac{x}{a} \tag{12.64}$$

Let

$$U_x = \frac{u_x}{a\dot{\gamma}} \tag{12.65}$$

where $a\dot{\gamma}$ is the characteristic velocity and $\dot{\gamma}$ is the shear rate. Note that U_x is the dimensionless velocity and is different from the interaction potential, U. Making use of Eqs. (12.62) to (12.65), Eq. (12.61) becomes

$$\frac{d}{dX}\left[\frac{D}{a}\left(-\frac{dn}{dX}-\frac{n}{k_BT}A_H\frac{df(X)}{dX}-\frac{n}{k_BT}2\pi\epsilon\psi_s^2 a\frac{dg(X,\kappa a)}{dX}\right)+a\dot{\gamma}\,U_x\,n\right]=0 \tag{12.66}$$

Assuming D to be a constant that is equal to $2D_\infty$ and rearranging, Eq. (12.66) becomes

$$\frac{d}{dX}\left[-\frac{dn}{dX}-\left(\frac{A_H}{k_BT}\right)n\frac{df(X)}{dX}-\frac{n}{2}\left(\frac{4\pi\epsilon\psi_s^2 a}{k_BT}\right)\frac{dg(X,\kappa a)}{dX}+\frac{1}{2}\left(\frac{a^2\dot{\gamma}}{D_\infty}\right)U_x\,n\right]=0 \tag{12.67}$$

Let

$$N_A = \frac{A_H}{k_BT} \tag{12.68}$$

$$N_R = \frac{4\pi\epsilon\psi_s^2 a}{k_BT} \tag{12.69}$$

and

$$Pe = \frac{a^2 \dot{\gamma}}{D_\infty} \tag{12.70}$$

where *Pe* is the Peclet number which is a measure of convection to diffusion transport, N_A is a dimensionless parameter that describes the relative importance of the dispersion potential to Brownian motion, N_R is a dimensionless parameter that describes the relative importance of the electrostatic potential to Brownian motion.

Particle coagulation is governed by the dimensionless groups given by Eq. (12.71) in addition to the scaled inverse Debye length κa. The governing equation is

$$\frac{d}{dX}\left[-\frac{dn}{dX} - nN_A\frac{df}{dX} - \frac{n}{2}N_R\frac{dg}{dX} + \frac{Pe}{2}U_x n\right] = 0 \tag{12.71}$$

It is possible to define other dimensionless groups. They are given by

$$\begin{aligned} Pe &= \frac{6\pi a^3 \dot{\gamma} \mu}{k_B T} \\ &= \text{hydrodynamic potential/Brownian motion} \end{aligned} \tag{12.72}$$

$$\begin{aligned} N_r &= \frac{N_R}{N_A} = \frac{4\pi \epsilon \psi_s^2 a}{A_H} \\ &= \text{electrostatic potential/dispersion potential} \end{aligned} \tag{12.73}$$

and

$$\begin{aligned} N_f &= \frac{P_e}{N_A} = \frac{6\pi a^3 \dot{\gamma} \mu}{A_H} \\ &= \text{hydrodynamic potential/dispersion potential} \end{aligned} \tag{12.74}$$

The dimensionless parameters of Eqs. (12.72), (12.73) and (12.74) represent one possible form of non-dimensionalization of the governing equations. Different research groups use different combinations of variables to obtain the dimensionless parameters. For instance, dimensionless parameters of the form

$$\begin{aligned} C_R &= \left(\frac{N_r}{N_f}\right) = \frac{2\epsilon \psi_s^2}{3a^2 \mu \dot{\gamma}} \\ &= \text{electrostatic force/hydrodynamic force} \end{aligned} \tag{12.75}$$

and

$$\begin{aligned} C_A &= \frac{1}{6N_f} = \frac{A_H}{36\pi \mu a^3 \dot{\gamma}} \\ &= \text{dispersion potential/hydrodynamic potential} \end{aligned} \tag{12.76}$$

are also used in the literature. It should be recognized that additional dimensionless groups would enter into the analysis when retarded dispersion expressions, unequal sphere sizes, ion size, gravitational and inertial forces, surface tension, variable diffusion coefficient, and dielectric constant are considered. It is clear from Eq. (12.71) that by setting $N_A = N_R = Pe = 0$, the coagulation is due solely to Brownian motion. We turn our attention now to coagulation due to a non-zero Peclet number.

12.6 COAGULATION DUE TO SHEAR

12.6.1 The Smoluchowski Solution in the Absence of Brownian Motion

Smoluchowski (1917) was the first to study particle coagulation due to shear. This represents the case of the Peclet number being very large ($Pe \to \infty$). The simplest case is that of a laminar shear flow having a constant shear rate $\dot{\gamma}$. It is assumed that the particles follow straight streamlines. This situation is equivalent to the case of negligible inertia. The geometry of the flow is shown in Figure 12.8.

For the purpose of the analysis, we consider a stationary test spherical particle of radius a_i located at the origin with particles having a radius of a_j moving towards it along the flow streamlines. When the centers of the two spheres are less or equal to $(a_i + a_j)$, coagulation occurs. Equivalently, contact between the two spheres occurs when $y \leq (a_i + a_j) \sin \theta$ as shown in Figure 12.9. The velocity in x-direction u_x is given by

$$u_x = \dot{\gamma} y \tag{12.77}$$

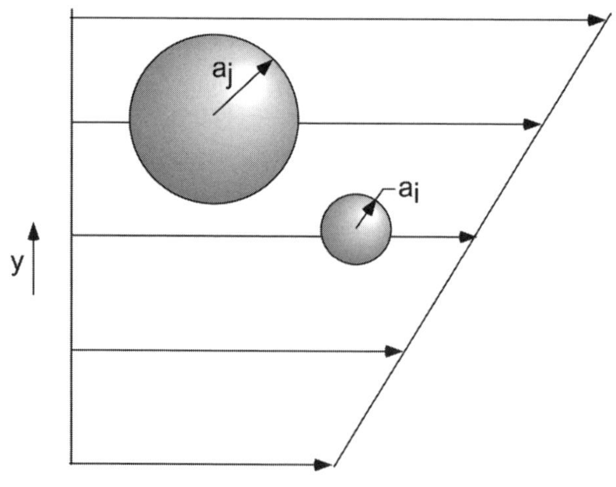

Figure 12.8. Coagulation due to shear for particles of radii a_i and a_j.

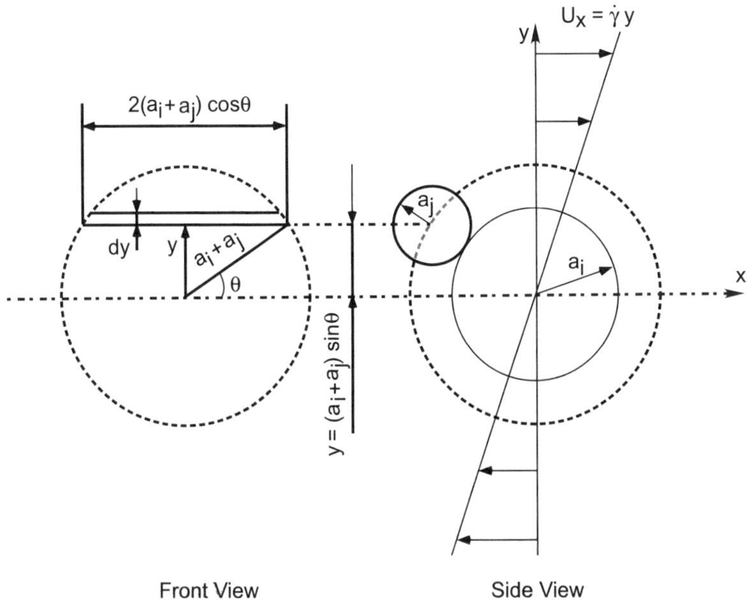

Figure 12.9. Flow geometry for Smoluchowski coagulation under shear flow (adapted from Probstein, 2003).

where $\dot{\gamma}$ is the constant shear rate. The number of a_j particles per unit time entering element dy is given by

$$2[(a_i + a_j)\cos\theta](\dot{\gamma}y)\, n_j\, dy$$

The total number of collisions per unit time between the j particles and the test particle is given by

$$2(2)\int_0^{a_i+a_j} [(a_i + a_j)\cos\theta](\dot{\gamma}y)n_j\, dy \tag{12.78}$$

where the factor 2(2) in Eq. (12.78) accounts for the four quadrants of the test sphere. Substituting for y and including the number concentration of all the test particles, Eq. (12.78) yields

$$J_{ij} = 4n_i n_j (a_i + a_j)^3 \dot{\gamma} \int_0^{\pi/2} \cos^2\theta \sin\theta d\theta$$

Hence

$$J_{ij} = \frac{4}{3}(a_i + a_j)^3 \dot{\gamma} n_i n_j \tag{12.79}$$

with

$$\beta_{ij} = \frac{4}{3}(a_i + a_j)^3 \dot{\gamma} \tag{12.80}$$

where J_{ij} is the collision frequency per unit volume.

For a monodisperse system, at the initial stage of coagulation, one can assume that $a_i = a_j = a$, $n_i = n_j = n$, and $J_{ij} = J_{0,sh}$ leading to

$$J_{0,sh} = \left(\frac{32}{3}\right) a^3 \dot{\gamma} n^2 \quad \text{and} \quad \beta = \left(\frac{32}{3}\right) a^3 \dot{\gamma}$$

where $J_{0,sh}$ is Smoluchowski collision frequency per unit volume due solely to shear.

The population balance Eq. (12.20) becomes

$$\frac{dn_k}{dt} = \frac{1}{2} \sum_{i=1}^{i=k-1} \frac{32}{3} a^3 \dot{\gamma} n_i n_j - n_k \sum_{i=1}^{\infty} \frac{32}{3} a^3 \dot{\gamma} n \tag{12.81}$$

Summing over all k, the above equation gives

$$\frac{dN_{tot}}{dt} = -\frac{16}{3} a^3 \dot{\gamma} N_{tot}^2 \tag{12.82}$$

where

$$N_{tot} = \sum_{i=1}^{\infty} n_i$$

The volume fraction of the particles α_p at the initial stage of coagulation is given by

$$\alpha_p = \frac{4}{3} \pi a^3 N_{tot} = \text{constant}$$

Making use of α_p definition, Eq. (12.82) becomes

$$\frac{dN_{tot}}{dt} = -\frac{4}{\pi} \dot{\gamma} \alpha_p N_{tot} \tag{12.83}$$

The initial condition is given by

$$N_{tot}(0) = N_0 \quad \text{at } t = 0$$

The solution of Eq. (12.83) is

$$N_{tot}(t) = N_0 \exp\left(-\frac{t}{t_{sh}}\right) \tag{12.84}$$

where

$$t_{sh} = \frac{\pi}{4 \dot{\gamma} \alpha_p} \tag{12.85}$$

t_{sh} is a characteristic coagulation time for coagulation due solely to shear. Comparing Eqs. (12.40) and (12.82) leads to

$$\frac{\text{rate of coagulation due to shear}}{\text{rate of coagulation due to Brownian motion}} = \frac{4a^3\dot{\gamma}\mu}{k_B T}$$

One can observe that the above ratio increases with increasing particle radius, shear rate, and fluid viscosity. For $a = 10^{-5}$ m (10 μm), $\dot{\gamma} = 1\mathrm{s}^{-1}$, $\mu = 10^{-3}$ Pa s, and $T = 300$ K, the ratio of shear coagulation to diffusion becomes about 1000. Reducing the particle size to one micron, the ratio becomes about unity.

Coagulation in the absence of the Brownian motion represents the case of a very high Peclet number ($Pe \rightarrow \infty$).

12.6.2 Coagulation due to Shear in the Absence of Brownian Motion: With Hydrodynamic and Field Forces

In the previous section, we dealt with particle coagulation where the flow field is unaffected by the presence of the particles. Here we shall deal with the case of coagulation due to shear with the full description of the flow field. Simple shear will be assumed where $u_x = \dot{\gamma}y$.

The simplest case of laminar shear flow past a sphere is given by Cox *et al.* (1968). The streamlines of the flow are shown in Figure 12.10. The streamlines can be thought of as the tracing of the trajectory of a neutral particle with a zero radius. Figure 12.10(b) clearly shows that there exist open and closed streamlines separated by a limiting streamline or surface.

The case of two spheres in a simple shear flow, the creeping flow problem was solved by Lin *et al.* (1970), Batchelor and Green (1972), Kao *et al.* (1977), and Kim and Mifflin (1985). With the origin of the coordinates placed at the center of sphere 1,

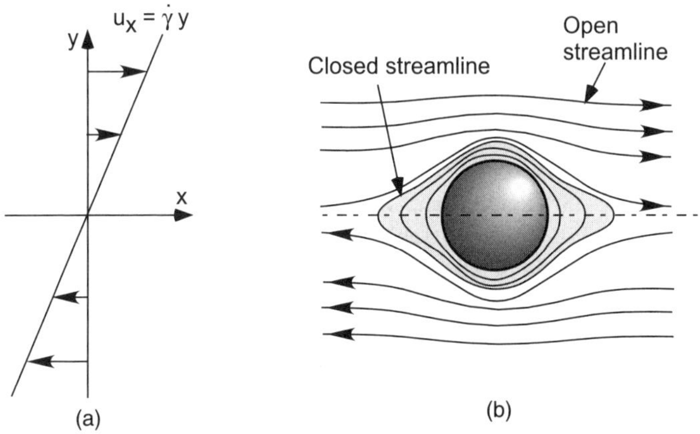

Figure 12.10. Laminar shear flow about a sphere: (a) flow field far from sphere; (b) equatorial streamlines around a sphere (x-y) plane (Cox *et al.*, 1968).

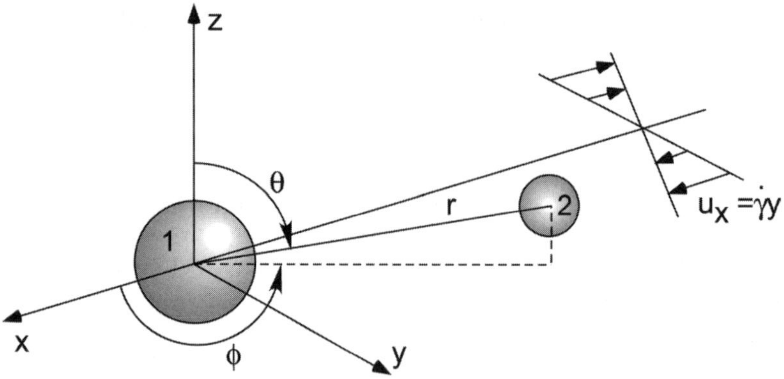

Figure 12.11. Coordinates for trajectory equations for two spheres in a simple shear flow.

the relative position of sphere 2 is given by the spherical coordinates (r, θ, ϕ) as shown in Figure 12.11.

For a simple shear far from the spheres, the flow velocities are given by

$$u_x = \dot{\gamma} y \tag{12.86}$$

with

$$u_y = u_z = 0 \tag{12.87}$$

The relative trajectories of sphere 2 with respect to sphere 1 are given by

$$\frac{dr}{dt} = \dot{\gamma} r (1 - \mathcal{A}) \sin^2 \theta \sin \phi \cos \phi \tag{12.88}$$

$$\frac{d\theta}{dt} = \dot{\gamma} (1 - \mathcal{B}) \sin \theta \cos \theta \sin \phi \cos \phi \tag{12.89}$$

and

$$\frac{d\phi}{dt} = -\dot{\gamma} \left[\sin^2 \phi + \frac{1}{2} \mathcal{B} \left(\cos^2 \phi - \sin^2 \phi \right) \right] \tag{12.90}$$

where t is time. Coefficients $\mathcal{A}$ and $\mathcal{B}$ are functions of radial distance r and size ratio a_1/a_2. They are tabulated in Table 12.3 for equal-sized spheres $(a_1 = a_2)$ with $R\ (= r/a)$, where r is the separation distance between the spheres centers and a is the radius of the equal-sized spheres 1 and 2. $\mathcal{A}$ and $\mathcal{B}$ assume different values for unequal spheres.

Figure 12.12 shows the equatorial trajectories for two spheres of radii a_1 and a_2. It is clear that the trajectories are similar to the case around a sphere in simple shear as is shown in Figure 12.10. The flow is characterized by open and closed trajectories. In the absence of the Brownian motion and external forces, two spheres approaching each other from infinity are unable to penetrate the region of closed trajectories and

TABLE 12.3. Values of $\mathcal{A}$, $\mathcal{B}$, and $\mathcal{C}$ for Equal Sized Spheres. (Taken from Zeichner and Schowalter, 1977.)

R	$\mathcal{A}$	$\mathcal{B}$	$\mathcal{C}$
2.0000	1.0000	0.4060	0.0000
2.0001	0.9996	0.3213	0.0004
2.0025	0.9900	0.2762	0.0098
2.0100	0.9619	0.2461	0.0374
2.0401	0.8679	0.1996	0.1310
2.0907	0.7505	0.1608	0.2510
2.1621	0.6313	0.1275	0.3777
2.2553	0.5214	0.0988	0.5017
2.3709	0.4248	0.0748	0.6190
2.5103	0.3424	0.0553	0.7287
2.6750	0.2735	0.0399	0.8306
2.8662	0.2167	0.0281	0.9254
3.0862	0.1704	0.0193	1.0133
3.3370	0.1331	0.0130	1.0950
3.6213	0.1033	0.0086	1.1709
4.7048	0.0468	0.0023	1.3666
6.2149	0.0204	0.0006	1.5207
11.1139	0.0036	0.0000	1.7310
20.1353	0.0006	0.0000	1.8512

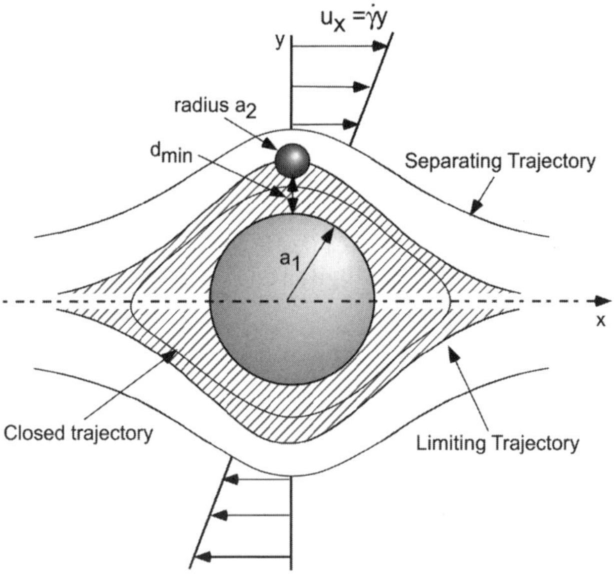

Figure 12.12. Equatorial trajectories of two spheres in a simple shear (schematic). The solid lines are the relative trajectories of a sphere of radius a_2 with respect to a reference sphere of radius a_1. Two kinds of trajectories exist: separating (or open) and closed ones, separated by a limiting trajectory. The shaded area is the region of closed trajectories (van de Ven, 1989).

hence they are unable to approach each other closer than distance $d_{\min}$ where $d_{\min}$ is the minimum distance of approach of two spheres on the limiting trajectory (van de Ven, 1989). For equal-sized spheres, $d_{\min}/a_1 = 4.2 \times 10^{-5}$ (Arp and Mason, 1977) and for $a_2/a_1 = 0$, *i.e.*, a single sphere, $d_{\min}/a_1 = 0.16$ (Cox *et al.*, 1968). It becomes clear, then, that when the hydrodynamic interactions are taken into consideration, no coagulation takes place between two particles under a simple shear flow in the absence of the Brownian motion and van der Waals dispersion (attractive) forces. This is in contrast to Smoluchowski's analysis. In order for coagulation to occur under a simple shear flow, an attractive force needs to be present between the particles. In the above discussion, inertial effects are assumed to be negligible.

We consider now the case of coagulation of two equal-sized spheres in laminar shear flow under the influence of interaction forces. Equation (12.88) can be regarded as a force balance along the line between sphere centers where the hydrodynamic force acts on the spheres; *i.e.*,

$$\text{Force} = 6\pi a\mu \left(\frac{dr}{dt}\right)$$

For the case of a non-zero interaction force, Eq. (12.88) becomes

$$\begin{aligned} 6\pi a\mu \frac{dr}{dt} &= 6\pi a\mu \left[\dot{\gamma} r(1 - \acute{A}) \sin^2\theta \sin\phi \cos\phi\right] \\ &\quad + C(r/a) \sum \text{interaction forces in } r\text{-direction} \end{aligned} \tag{12.91}$$

The implicit assumption made in writing Eq. (12.91) is that the interaction forces act only along the line of centers and are balanced by the component of the hydrodynamic force in that direction, so that the net force on each sphere is zero and there is no additional torque (van de Ven and Mason, 1976), Function $C(r/a)$ is a correction parameter to Stokes law due to the presence of a second sphere.

When the interaction forces are due to the London–van der Waals (attractive) and to electrostatic (repulsive) forces, one can write

The London–van der Waals force is given by

$$F_A(r) = -\frac{dU_A}{dr} \tag{12.92}$$

where for the case of unretarded dispersion potential U_A between two equal-sized spheres, it can be given by

$$U_A = -\frac{A_H}{6}\left[\frac{2}{R^2 - 4} + \frac{2}{R^2} + \ln\left(\frac{R^2 - 4}{R^2}\right)\right]$$

with $R = r/a$. In dimensionless form, Eq. (12.92) becomes

$$f_A(R) = \frac{a}{A_H} F_A(r) = -\frac{1}{A_H}\frac{dU_A}{dR} \tag{12.93}$$

The electrostatic repulsive force between the two spheres is given by

$$F_R(r) = -\frac{dU_R}{dr} \tag{12.94}$$

where for a constant low surface potential, the electrostatic potential U_R for two equal-sized spheres is given by Eq. (5.142) as

$$U_R = 2\pi\epsilon a\psi_s^2 \ln[1 + \exp(-\kappa a(R-2))]$$

In dimensionless form, Eq. (12.94) becomes

$$f_R = -\frac{1}{4\pi\epsilon a\psi_s^2}\frac{dU_R}{dR} \tag{12.95}$$

leading to

$$f_R = \frac{1}{4\pi\epsilon\psi_s^2}F_R(R)$$

Making use of Eqs. (12.93) and (12.95), Eq. (12.91) becomes

$$\frac{dr}{dt} = [\dot{\gamma}r(1-\mathcal{A})\sin^2\theta\sin\phi\cos\phi] + \frac{\mathcal{C}(R)}{6\pi a\mu}\left[4\pi\epsilon\psi_s^2 f_R(R) + \frac{A_H}{a}f_A(R)\right] \tag{12.96}$$

Non-dimensionalizing the time t using $\dot{\gamma}$ as

$$\tau = t\dot{\gamma}$$

and making use of Eqs. (12.73) and (12.74), Eq. (12.96) becomes in dimensionless form

$$\frac{dR}{d\tau} = R(1-\mathcal{A})\sin^2\theta\sin\phi\cos\phi + \frac{\mathcal{C}(R)}{N_f}(N_r f_R + f_A) \tag{12.97}$$

The corresponding trajectory equation represented by Eqs. (12.89) and (12.90) becomes

$$\frac{d\theta}{d\tau} = (1-\mathcal{B})\sin\theta\cos\theta\sin\phi\cos\phi \tag{12.98}$$

and

$$\frac{d\phi}{d\tau} = -\left[\sin^2\phi + \frac{1}{2}\mathcal{B}(\cos^2\phi - \sin^2\phi)\right] \tag{12.99}$$

The function $\mathcal{C}(R)$ is related to the parameter $G(h)$ given by Eq. (12.54) for the case of diffusion coefficient, where

$$G(R) \cong \frac{2}{\mathcal{C}(R)} \tag{12.100}$$

and

$$\mathcal{C}(R) \cong \frac{4(3R^2 - 10R + 8)}{6R^2 - 11R} \tag{12.101}$$

where $R = r/a = 2 + h$.

Tabulation of $\mathcal{C}(R)$ is given in Table 12.3. Various expressions for $\mathcal{A}$, $\mathcal{B}$, and $\mathcal{C}$ are given by Higashitani *et al.* (1982) and van de Ven and Mason (1976) for $R \to 2$ and $R \to \infty$.

Solution of the trajectory equation provides the trajectories of the colliding spheres. For the case of $N_r = 0$ (*i.e.*, absence of a repulsive force) and $N_f = 1$, Feke and Schowalter (1983) provide the trajectories at $z = 0$ plane (equatorial plane) as shown in Figure 12.13. There are three types of trajectories. The first type is an open trajectory initiated from downstream and extending to upstream infinity. Here, the two spheres do not coagulate. The second type of trajectory leads to coagulation. It crosses the x-axis and reverses direction prior to collision. This is shown by the shaded area of Figure 12.13. The third type of trajectory leads to coagulation before crossing the x-axis which is the plane of shear.

Takamura *et al.* (1981) visualized the trajectories of shear-induced collisions between two equal-sized latex spheres for three cases: (a) repulsion is dominant, (b) weak attraction, and (c) strong attraction. Figure 12.14 shows the three cases. The two curves in each figure represent the projection on the xy-plane of the paths of the centers of the two spheres. At the origin is the exclusion sphere which cannot be penetrated by either colliding sphere. Figure 12.14(a) demonstrates the effect of a net repulsive force on the spheres trajectories. The spheres initially approach each other

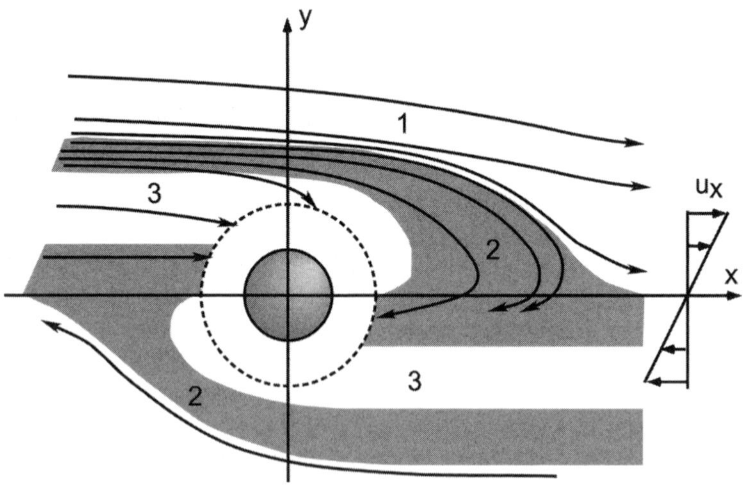

Figure 12.13. Trajectories in the $z = 0$ plane for two spheres in a simple shear flow, with $N_r = 0$ and $N_f = 1$: (1) open trajectories extending from upstream to downstream infinity; (2) trajectories leading to collision after crossing the plane of shear ($y = 0$) and reversing direction; (3) trajectories leading to collision prior to crossing the plane of shear (Feke and Schowalter, 1983).

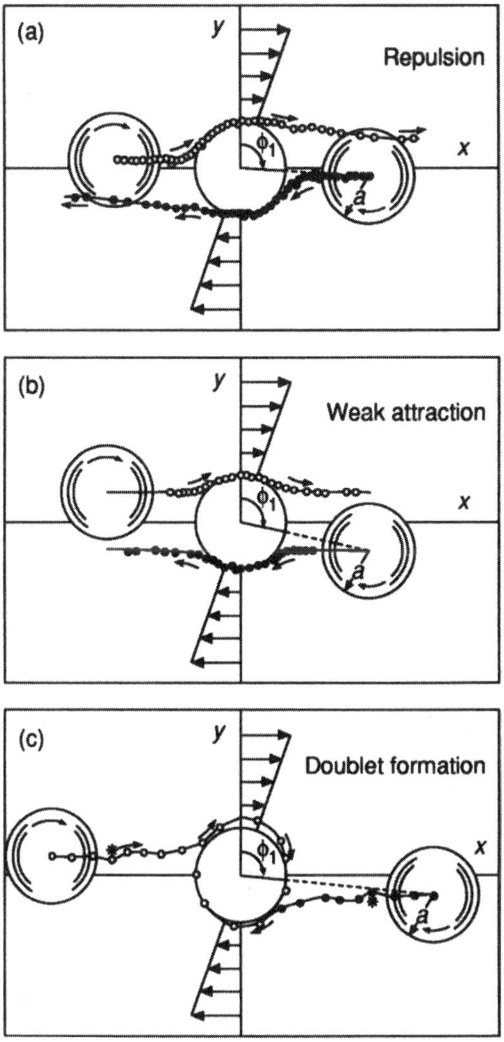

Figure 12.14. Trajectories of shear induced collisions of 2.6 μm PS latex spheres in 50% aqueous glycerol showing the projection on the xy-plane of the paths of the sphere centers from the midpoint between them. At the center is the exclusion sphere which cannot be penetrated when the collision occurs in the xy-plane (Takamura *et al.*, 1981).

with their centers $\pm 0.15a$ off the x-axis. However, after they encounter each other, they recede with their centers $\pm 0.8a$ away from the x-axis.

Figure 12.14(b) shows the case of a weak attraction upon increasing the KCl electrolyte concentration to 10 mM. The paths of approach and recession of the two spheres become almost a mirror image of each other. By further increasing the electrolyte concentration to 100 mM, the net force becomes attractive and a doublet is formed upon the spheres collision as is shown in Figure 12.14(c).

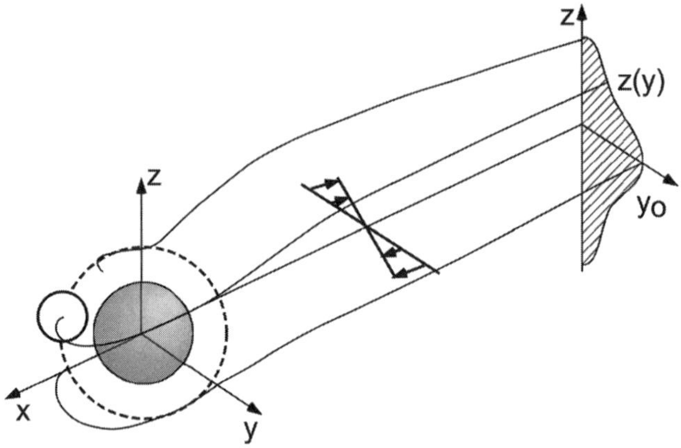

Figure 12.15. Trajectory of real systems (Higashitani *et al.*, 1982).

To evaluate the coagulation rate, it is necessary to know the limiting trajectory within which all particles collide. A limiting trajectory for the case of two equal-sized spheres is defined as that which terminates at $\phi = 0$ with $r = 2a$ (see Figure 12.11). The area occupied by the limiting trajectories is usually evaluated by retracing backward in time the path of the particles. The limiting interception area is shown in Figure 12.15 by the shaded area. The corresponding limiting interception area for Smoluchowski's analysis is shown for comparison in Figure 12.16.

The rate of coagulation, J, between particles of equal-sized spheres per unit volume is given by counting the number of particles crossing the interception area, where

$$J = 4n^2\dot{\gamma}\int_0^{y_0} yz(y)dy \tag{12.102}$$

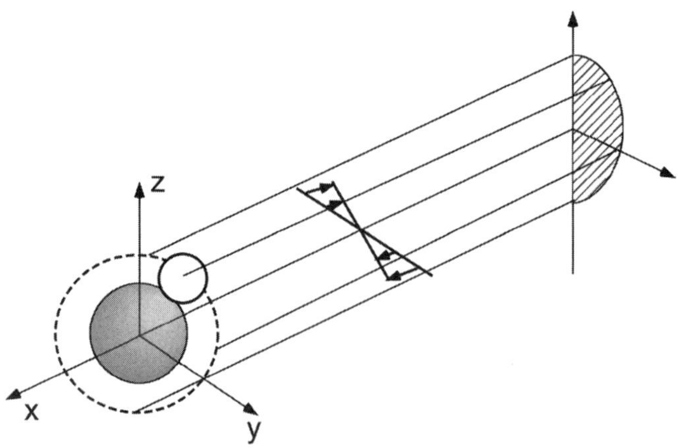

Figure 12.16. Trajectory of Smoluchowski model (Higashitani *et al.*, 1982).

where $z = z(y)$ is the function defining the boundary of the interception area and y_0 is the y value of the boundary at $z = 0$. Recalling Smoluchowski's collision rate $J_{0,sh}$ defined as

$$J_{0,sh} = \left(\frac{32}{3}\right) a^3 \dot{\gamma} n^2 \tag{12.103}$$

it is possible to define a stability ratio W as

$$W = \frac{J_{0,sh}}{J} \tag{12.104}$$

Large W values indicate a stable colloidal system. In some literature, the inverse of W is used, and W^{-1} is called the shear coagulation coefficient. Figure 12.17 shows the stability ratio W as a function of N_f for the case of $N_r = 0$.

At small N_f values, W becomes less than unity, indicating a strong attractive London–van der Waals force. However, at large value of N_f, where the London–van der Waals force becomes weaker relative to the hydrodynamic force, W increases rapidly indicating a high hydrodynamic resistance as compared to London–van der Waals forces for the approaching particles and a reduction in J.

The preceding interpretation of Figure 12.17 is to keep the hydrodynamic potential $6\pi a^3 \dot{\gamma} \mu$ constant (see Eq. 12.74) and the change in N_f is due to the Hamaker constant. The variation in A_H reflects changes in the attractive London–van der Waals forces. One can also use Figure 12.17 to examine the effect of increasing the flow strength $\dot{\gamma}$. Here A_H is kept constant. Recognizing that $\dot{\gamma}$ appears both in W and in N_f, a change in N_f from, say 10^2 to 10^5 gives a corresponding change in W from ~5 to ~20. This

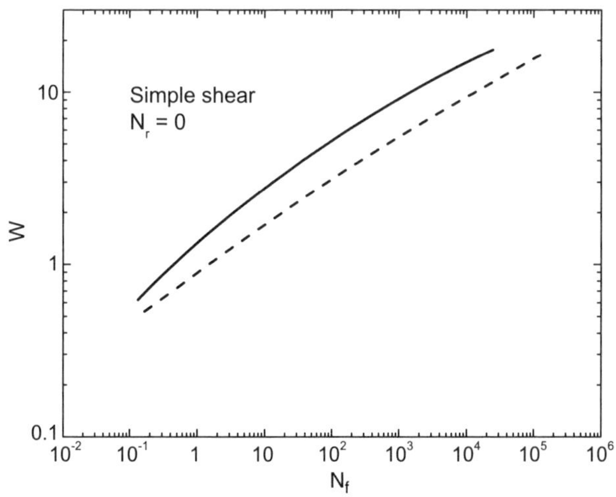

Figure 12.17. Comparison of the stability ratios, $W = 32a^3\dot{\gamma}n^2/(3J)$, for flow-induced coagulation for $Pe \rightarrow \infty$ using unretarded potential. Dashed line: trajectory analysis of Zeichner and Schowalter (1977). Solid line: Feke and Schowalter (1983).

means that J at $N_f = 10^5$ is ~250 times that at $N_f = 10^2$. Consequently, increasing $\dot{\gamma}$ by a factor of 10^3 has a profound effect on increasing the collision rate per unit volume.

Now let us study the effect of viscosity by keeping other parameters constant. N_f contains the viscosity μ and W does not implicitly contain μ. Increasing N_f from 10^2 to 10^5 would give a corresponding change in J of $5/20 = 0.25$. This means that increasing the fluid viscosity by a factor of 10^3 only reduces J by a factor of 4.

The integration of the trajectory Eqs. (12.97) to (12.99) for different N_r and N_f produces stability diagrams (van de Ven and Mason, 1976; Zeichner and Schowalter, 1977). A typical stability plot is shown in Figure 12.18. It should be recalled that for a given fluid viscosity, particle radius, medium dielectric constant, and Hamaker constant, one obtains $N_r \propto \psi_s^2 \propto$ (repulsion force) and $N_f \propto \dot{\gamma} \propto$ (flow strength). Consequently, $\sqrt{N_r}$ represents changes in the surface potential ψ_s and N_f represents the strength of the flow field. From an experimenter's point of view, ψ_s is controlled by the ionic strength and pH of the electrolyte solution.

The stability diagram shown in Figure 12.18 is divided into three regions. In region 1, coagulation occurs in the primary minimum. In region 2, coagulation occurs in the secondary minimum, and finally, in region 3, no coagulation occurs (Zeichner and Schowalter, 1977). For example, if one is to start with a sol (a colloidal suspension) under no shear containing spherical particles with, say, $N_r = 20$, then upon subjecting the sol to increasing shear, doublets coagulating in the primary minimum would be formed. Increasing N_f beyond 5×10^3 would cause the doublet to deflocculate. For $N_r = 100$, if a sol is sheared then doublets would form at the secondary minimum. A further increase in the rate of shear can lead to deflocculation of the doublets out of

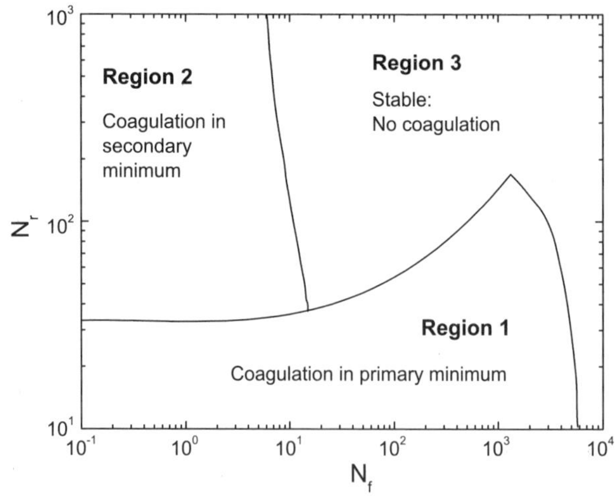

Figure 12.18. Stability plane for shear flow (Zeichner and Schowalter, 1977). Region 1 – primary minimum coagulation; Region 2 – secondary minimum coagulation; Region 3 – stable. (Calculated for unretarded attraction, constant surface potential, $\kappa a = 100$, and $d_{min}/a = 2.004$.)

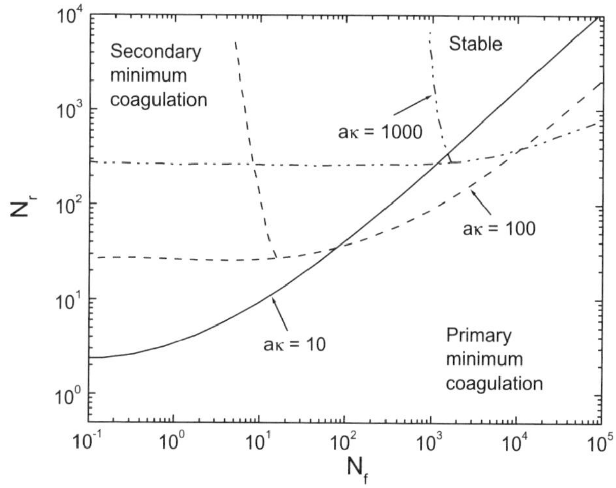

Figure 12.19. Stability diagram for doublet formation in a simple shear flow at $Pe \to \infty$, with a constant-potential boundary condition and the non-retarded Hamaker form for the attraction force (Zeichner and Schowalter, 1977).

the secondary minimum. Increasing the shear rate further would cause the particles to coagulate into the primary minimum. At still higher shear rates, doublets can deflocculate out of the primary minimum. If one is to decrease the shear rate, the dispersed sol will again coagulate into the primary minimum. Further lowering of the shear rate will not, however, cause deflocculation or a shift to secondary minimum flocculation.

The boundaries of the various regions shift with κa as shown in Figure 12.19. This type of a stability diagram is very important in the formation of emulsions and in the process of demulsification under shear. In all cases discussed above, the Brownian motion was not considered and we dealt with cases corresponding to large *Pe*.

Figure 12.18 is useful in interpreting many industrial processes where colloidal systems are of concern. For example, Yan and Masliyah (1993) found that for solids stabilized oil-in-water emulsions, it is possible to demulsify such emulsions by the addition of fresh oil under mixing action. The mixing action breaks down the added fresh oil into large oil droplets which act as scavengers to the solids-coated oil emulsion droplets. Here coagulation occurs in the primary minimum and hence scavenging takes place. In this process, the mixing action, *i.e.*, the shear rate, is very important. Very high mixing rates were found to be detrimental to the scavenging process because the added fresh oil would form stable emulsion at large N_f values.

Another example is in the processing of oil sands and the recovery of bitumen. Oil sands containing bitumen are diluted with hot water in the presence of NaOH and are slurried in a rotational conditioning drum. After the liberation of the bitumen from the sand grains, it becomes desirable to avoid the coagulation of the liberated bitumen with the fines (or clays) in the slurry. Proper N_r and N_f values should be maintained so that the slurrying operation occurs in region 3. This depends on the amount of NaOH added and the degree of mixing in the conditioning equipment.

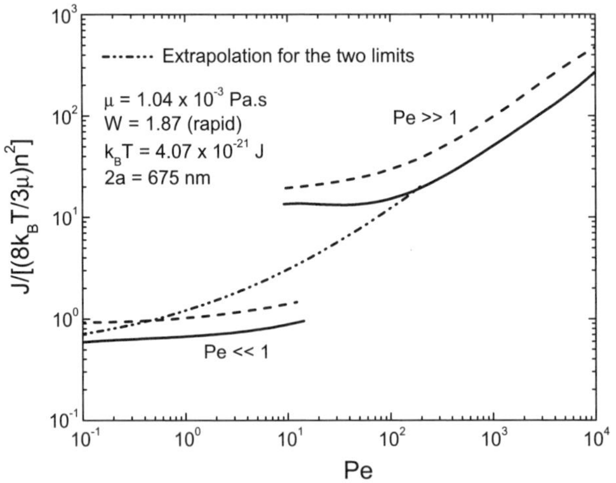

Figure 12.20. Rate of doublet formation for spheres as a function of *Pe*, showing the results from the perturbation expansions in the $Pe \ll 1$ and $Pe \gg 1$ limits for $A_H/k_BT = 2$ (solid lines) and 20 (dashed lines). These results were obtained in absence of any repulsive force ($N_r = 0$) (Feke and Schowalter, 1985). The Peclet number in this figure is given as $3\pi a^2\dot{\gamma}\mu/k_BT$.

For cases of intermediate *Pe* values, the combined effects of the Brownian diffusion on shear-induced coagulation of colloidal dispersions are dealt with by Feke and Schowalter (1985). The protocol is to combine the analysis made for the Brownian and shear coagulation. Figure 12.20 shows the rate of doublet formation normalized with $J_{o,Br}$. Two regions are given, one for $Pe \ll 1$ and $Pe \gg 1$. No solution was provided for the intermediate region. Figure 12.20 indicates that in the region of $Pe \ll 1$, increasing $\dot{\gamma}$ (all other parameters being constant) has little effect on the collision rate per unit volume J. For the case of $Pe \gg 1$, J increases rapidly with changes in the flow strength.

The analysis of the preceding sections clearly indicates the complexity of the particle-particle coagulation process where the Brownian motion, hydrodynamic forces and interaction forces due to electrostatic and dispersion can play a major role. Experimental studies tend to confirm the theoretical analysis for rapid coagulation. However, deviation from theory occurs for slow coagulation where appreciable repulsive forces are present (Zeichner and Schowalter, 1979; Kihira *et al.*, 1992).

12.7 NOMENCLATURE

$\mathcal{A}, \mathcal{B}, \mathcal{C}$	flow functions, dimensionless
a, a_1, a_2, a_3	particle radius, m
A_H	Hamaker constant, J
C_A	dimensionless parameter describing the relative importance of dispersion and hydrodynamic potentials (for forces), $A_H/36\pi\mu a^3\dot{\gamma}$

C_R	dimensionless parameter describing the relative importance of electrostatic and hydrodynamic forces, $2\epsilon\psi_s^2/3a^2\mu\dot{\gamma}$
$d_\parallel$	correction factor for the diffusion coefficient for motion parallel to the surface of the spheres, dimensionless
$d_\perp$	correction factor for the diffusion coefficient for motion normal to the surface of the spheres, dimensionless
D	particle diffusion coefficient, m^2/s
$\bar{\bar{D}}$	particle diffusion coefficient tensor, m^2/s
D_∞	particle diffusion coefficient at infinite dilution, m^2/s
D_i, D_j	diffusion coefficient of particles i and j, respectively, m^2/s
f	friction coefficient, Eq. (12.7), Ns/m
$f_A(R)$	dimensionless dispersion potential
$f_R(R)$	dimensionless electrostatic potential
$f(x)$	dimensionless function for the dispersion potential, Eq. (12.62)
$F(r)$	interaction force function, N
$g(X, \kappa a)$	dimensionless function for the electrostatic potential
h	dimensionless gap between two surfaces, $h = R - 2$
i, j, k	particle species
J_{ij}	collision rate (frequency) per unit volume, $m^{-3}s^{-1}$
$J_{0,Br}$	Smoluchowski collision rate per unit volume due to Brownian diffusion, $8k_BTn^2/3\mu$, $m^{-3}s^{-1}$
$J_{0,sh}$	Smoluchowski collision rate per unit volume due to simple shear, $(32/3)a^3\dot{\gamma}n^2$, $m^{-3}s^{-1}$
$\mathbf{j}_D^{**}$	diffusive flux based on number concentration, $(m^{-2}s^{-1})$
k_B	Boltzmann constant, J/K
K	function defined by Eq. (12.38), m^3/s
m	mass of a particle, kg
n	number of particle per unit volume (number concentration), m^{-3}
n_∞	number concentration of particle at time zero, m^{-3}
n_i, n_j	number concentration of particle species i and j, respectively, m^{-3}
$n_{i\infty}, n_{j\infty}$	number concentration of particle species i and j in the undisturbed state, respectively, m^{-3}
N_A	dimensionless parameter describing the relative importance of the dispersion potential and Brownian motion, A_H/k_BT
N_f	dimensionless parameter describing the relative importance of the hydrodynamic and dispersion potentials (or forces), $6\pi a^3\dot{\gamma}\mu/A_H$
N_i	total number of released particles per unit area at time zero and $x = 0$, m^{-2}
N_0	value of N_{tot} at time zero, m^{-3}
N_r	dimensionless parameter describing the relative importance of electrostatic and dispersion potentials (or forces), $4\pi\epsilon a\psi_s^2/A_H$
N_R	dimensionless parameter describing the relative importance of the electrostatic potential and Brownian motion, $4\pi\epsilon a\psi_s^2/k_BT$

N_{tot}	total number of particles per unit volume (number of concentrations), m^{-3}
Pe	Peclet number, a measure of convection to diffusional transport, dimensionless
r	radial position; dimensionless center-to-center distance between two spheres, m
R	dimensionless center-to-center distance between two spheres, r/a
t	time, s
t_{Br}	characteristic half-time for coagulation due to Brownian diffusion, $\pi\mu a^3/\alpha_p k_B T$, s
t_{sh}	characteristic half-time for coagulation due to shear, $\pi/4\dot{\gamma}\alpha_p$, s
T	absolute temperature, K
u_x	fluid velocity in x-direction, m/s
U_x	dimensionless fluid velocity $u_x/(a\dot{\gamma})$
U	interaction potential, J
U_A	dispersion potential, normally assumed to be attractive, J
U_R	electrostatic potential, normally assumed to be repulsive, J
v	particle velocity, m/s
v_i, v_j, v_k	volume of particle species i, j and k, respectively, m^3
W	stability ratio, dimensionless
x	Cartesian coordinate, m
X	dimensionless distance or coordinate, x/a

Greek Symbols

α_p	volume fraction of particles, dimensionless
β	function relating t , Eq. (12.8); collision frequency function, m^3/s
β_{ij}	collision frequency function pertaining to particles i and j, m^3/s
$\dot{\gamma}$	shear rate, s^{-1}
ϵ	dielectric permittivity of a medium, C/Vm
θ	spherical coordinate
κ	inverse Debye length, m^{-1}
μ	fluid viscosity, Pa s
τ	dimensionless time, $t\dot{\gamma}$
ϕ	spherical coordinate
ψ_s	surface potential, V

12.8 PROBLEMS

12.1. We have discussed Brownian coagulation under a field force in terms of the stability ratio, W. Let us investigate the stability ratio, defined by Eq. (12.57), where van der Waals and repulsive forces are considered for the case of mono-sized particles and variable diffusion coefficient. The van der Waals interaction potential can be taken from Eq. (12.59) and the electrostatic potential from

Eq. (5.164) with $a_1 = a_2 = a$. Assume a solution temperature of 20°C and a Hamaker constant of 10^{-21} J. Evaluate the stability ratio and comment on the sensitivity of W to changes in the system properties for the following cases:

Case 1: Uncharged spheres in a mono-sized particle dispersion.
Case 2: Particle surface zeta potential of 25 mV in a 0.001 M (2 : 1) electrolyte solution. Calculate the stability ratio for three different particle sizes in the monodisperse suspension, namely $a = 0.1$, 1, and 10 μm.
Case 3: Particle radius, $a = 1.0$ μm in a 0.001 M (2 : 1) electrolyte solution. Calculate W for particle surface zeta potentials of 10 and 50 mV. Also, plot the stability ratio, W against zeta potential for the range of 5 to 50 mV.
Case 4: Particle radius of 1.0 μm and particle surface zeta potential of 25 mV. Calculate the stability ratio in a 0.001 M electrolyte solution for a (1 : 1), (2 : 1), and (3 : 1) electrolyte.

12.2. We have discussed coagulation due to Brownian motion and due to shear using the Smoluchowski approach. Assume additivity of the coagulation rates due to these two mechanisms (see Eqs. 12.40 and 12.83). Show that the rate of change of total particle concentration is given by

$$\frac{dN_{tot}}{dt} = -\frac{4k_BT}{3\mu}N_{tot}^2 - \frac{4\dot{\gamma}\alpha_p}{\pi}N_{tot}$$

where N_{tot} is the total number of particles per unit volume, m^{-3}. The other symbols have their usual definition.

(a) Show that for an initial particle concentration, $N_{tot}(0) = N_0$, the solution to the above coagulation rate equation is given by

$$\frac{N_{tot}}{N_0} = \frac{R\exp(-4\dot{\gamma}\alpha_p t/\pi)}{1 + R - \exp(-4\dot{\gamma}\alpha_p t/\pi)} \tag{12.105}$$

where

$$R = \frac{3\dot{\gamma}\alpha_p\mu}{\pi k_B T N_0}$$

The parameter R can be thought of as being a measure of shear to Brownian motion.

(b) Show that in the limiting case of $\dot{\gamma} \to 0$ (which makes $R \to 0$), Eq. (12.105) degenerates to Eq. (12.41) for the case of Brownian coagulation.

(c) Show that in the limiting case of $\dot{\gamma} \to \infty$, making $R \to \infty$, Eq. (12.105) becomes identical to Eq. (12.84) for the case of coagulation due to shear.

(d) Plot the variation of N_{tot}/N_0 with time for $\mu = 0.001$ Pa s, $T = 300$ K, $N_0 = 10^{14}$ m^{-3} and $\alpha_p = 0.1$. Use $\dot{\gamma} = 0.1, 1, 3, 10$, and 100 s^{-1}. Comment on your plot.

12.9 REFERENCES

Arp, P. A., and Mason, S. G., The kinetics of flowing dispersion. VIII. Doublets of rigid spheres (theoretical), *J. Colloid Interface Sci.*, **61**, 21–43, (1977).

Batchelor, G. K., and Green, J. T., The hydrodynamic interaction of two small freely-moving spheres in a linear flow field, *J. Fluid Mech.*, **56**, 275–400, (1972).

Brenner, H., The slow motion of a sphere through a viscous fluid towards a plane surface, *Chem. Eng. Sci.*, **16**, 242–251, (1961).

Cox, R. G., Zia, I. Y. Z., and Mason, S. G., Particle motions in sheared suspensions. XXV. Streamlines around cylinders and spheres, *J. Colloid Interface Sci.*, **27**, 7–18, (1941).

Derjaguin, B. V., and Landau, L., Theory of the stability of strongly charged lyophobic sols and the adhesion of strongly charged particles in solutions of electrolytes, *Acta Physicochim. URSS*, **14**, 633–662, (1941).

Derjaguin, B. V., and Muller, V. M., Slow coagulation of hydrophobic colloids, *Dokl. Akad. Nauk. SSSR*, **176**, 738–741, (1967).

Feke, D. L., and Schowalter, W. R., The effect of Brownian diffusion on shear-induced coagulation of colloidal dispersion, *J. Fluid Mech.*, **133**, 17–35, (1983).

Feke, D. L., and Schowalter, W. R., The influence of Brownian diffusion on binary flow-induced collision rates in colloidal dispersion, *J. Colloid Interface Sci.*, **106**, 203–214, (1985).

Friedlander, S. K., *Smoke, Dust and Haze: Fundamentals of Aerosol Dynamics*, 2nd ed., Oxford University Press, New York, (2000).

Fuchs, N., Über die stabilität und aufladung der aerosole, *Z. Physik.*, **89**, 736–743, (1934).

Kao, S. V., Cox, R. G., and Mason, S. G., Streamlines around single spheres and trajectories of pairs of spheres in two-dimensional creeping flow, *Chem. Eng. Sci.*, **32**, 1505–1515, (1977).

Kihira, H., Ryde, N., and Matijevic, E., Kinetics of heterocoagulation i. a comparison of theory and experiment, *Colloids and Surfaces*, **64**, 317–324, (1992).

Kim, S., and Mifflin, R. T., The resistance and mobility functions of two equal spheres in low Reynolds number flow, *Phys. Fluids*, **28**, 2033–2045, (1985).

Higashitani, K., and Matsuno, Y., Rapid Brownian coagulation of colloidal dispersions, *J. Chem. Eng. Japan*, **12**, 460–465, (1979).

Higashitani, K., Ogawa, R., Hosokawa, G., and Matsuno, Y., *J. Chem. Eng. Japan*, **15**, 299–304, (1982).

Honig, E. P., Roebersen, G. J., and Wiersema, P. H., Effect of hydrodynamic interaction on the coagulation rate of hydrophobic colloids, *J. Colloid Interface Sci.*, **36**, 97–109, (1971).

Lichtenbelt, J. W. Th., Pathmanamoharan, C., and Wiersema, P. H., Rapid coagulation of polystyrene latex in a stopped flow spectrophotometer, *J. Colloid Interface Sci.*, **49**, 281–285, (1974).

Lin, C. J., Lee, K. J., and Sather, N. F., Slow motion of two spheres in a shear field, *J. Fluid Mech.*, **43**, 35–47, (1970).

Lips, A., and Willis, W. E., Low angle scattering technique for the study of coagulation, *J. Chem. Soc. Faraday Trans. I*, **69**, 1226–1236, (1973).

Mathews, B. A., and Rhodes, C. T., Studies of the coagulation kinetics of mixed suspensions, *J. Colloid Interface Sci.*, **32**, 332–338, (1970).

Probstein, R. F., *Physicochemical Hydrodynamics, An Introduction*, 2nd ed., Wiley Interscience, New York, (2003).

Russel, W. B., Saville, D. A., and Schowalter, W. R., *Colloidal Dispersions*, Cambridge University Press, Cambridge, (1989).

Smoluchowski, M. von, Versuch einer mathematischen theorie der koagulationskinetic kolloider Lösungen, *Z. Phys. Chem.*, **92**, 129–168, (1917).

Spielman, L. A., Viscous interactions in Brownian coagulation, *J. Colloid Interface Sci.*, **33**, 562–571, (1970).

Spielman, L. A., Particle capture from low-speed laminar flows, *Ann. Rev. Fluid Mech.*, **9**, 297–319, (1977).

Takamura, K., Goldsmith, H. L., and Mason, S. G., The microrheology of colloidal dispersions xii. trajectories of orthokinetic pair-collisions of latex spheres in a simple electrolyte, *J. Colloid Interface Sci.*, **82**, 175–189, (1981).

van de Ven, T. G. M., *Colloidal Hydrodynamics*, Academic Press, London, (1989).

van de Ven, T. G. M., and Mason, S. G., The microrheology of colloidal dispersion. IV. Pairs of interacting spheres in a shear flow, *J. Colloid Interface Sci.*, **57**, 505–516, (1976).

Verwey, E. J. W., and Overbeek, J. Th. G., *Theory of the Stability of Lyophobic Colloids*, Elsevier, Amsterdam, (1948).

Yan, Y., and Masliyah, J. H., Solids-stabilized oil-in-water emulsion: scavenging of emulsion droplets by fresh oil addition, *Colloids Surf.*, **75**, 123–132, (1993).

Zeichner, G. R., and Schowalter, W. R., Use of trajectory analysis to study stability of colloidal dispersions in flow fields, *AIChE J.*, **23**, 243–254, (1977).

Zeichner, G. R., and Schowalter, W. R., Effects of hydrodynamic and colloidal forces on the coagulation of dispersions, *J. Colloid Interface Sci.*, **71**, 237–253, (1979).

CHAPTER 13

DEPOSITION OF COLLOIDAL PARTICLES

13.1 INTRODUCTION

In Chapter 12 we discussed particle–particle interaction under Brownian motion and field forces. A stability ratio W was defined to assess the stability of a colloidal system. The cases considered were concerned with particles of similar size and consequently the analysis was of interest in studying the stability of colloidal systems under Brownian motion, shear and interparticle forces such as the London–van der Waals attractive and electrostatic repulsive forces.

In this chapter we shall deal with the interactions between a small particle and a large surface, generally referred to as the collector. The small particle is of colloidal size that is much smaller than the collector with which it is interacting. The large surface can be a cylindrical fiber in a filter mat, a granular particle in a packed bed or simply a surface surrounding the colloidal dispersion, *e.g.*, a container's wall. Various deposition configurations are shown in Figure 13.1. We shall deal only with the process of deposition rather than with the process of detachment. The latter case is more complicated to analyze as the depth of the energy minimum in which a particle is captured is usually not known. However, the study of detachment itself can provide information regarding the strength of the forces between a particle and a surface (van de Ven, 1989; Liu *et al.*, 2004).

There are two approaches to the study of particle deposition. The first method is the Eulerian approach where the distribution of particles is evaluated in space. In the absence of colloidal forces and for infinitesimal particle size, the Eulerian approach

Electrokinetic and Colloid Transport Phenomena, by Jacob H. Masliyah and Subir Bhattacharjee
Copyright © 2006 John Wiley & Sons, Inc.

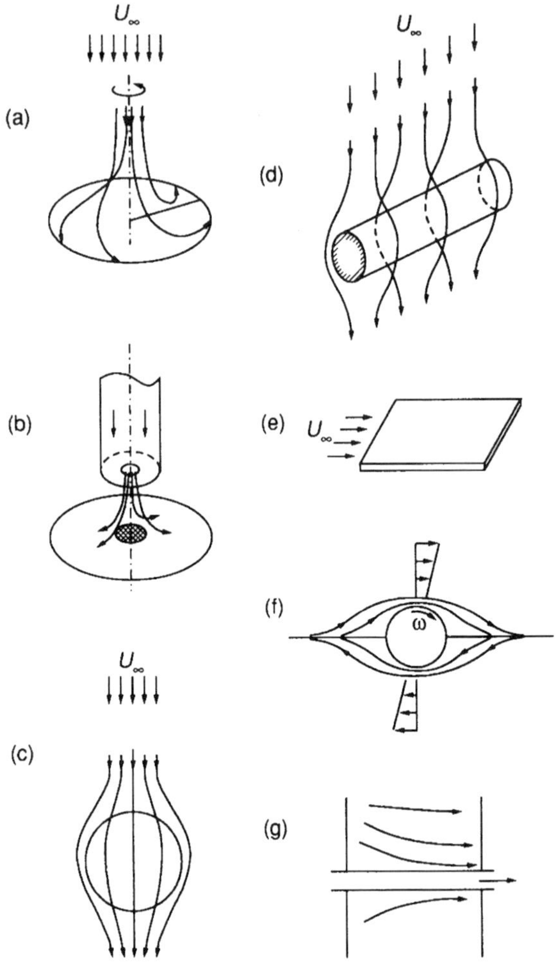

Figure 13.1. Various deposition geometries: (a) rotating disc; (b) impinging jet; (c) flow over a spherical collector; (d) cross flow over a cylinder; (e) flow along a flat plate; (f) sphere in shear flow; (g) flow parallel to a cylinder (Adamczyk, 1989a,b).

becomes that of the study of heat or mass transfer from submerged objects in a flowing stream. To this end, the generalized convection-diffusion equation is solved for the diffusing colloidal particles subject of the appropriate boundary conditions. The second method is the Lagrangian approach. Here attention is focused on a single particle trajectory which is described by Newton's laws of motion. The particle path is followed and the particle trajectories determine the collision or capture efficiency between the colloidal particle and the larger collector particle. The forces associated with Newton's law are deterministic in nature and consequently most of the literature studies employing the Lagrangian approach usually do not consider the Brownian motion and, hence, the diffusion of the particles. However, it is possible to include the Brownian motion in the analysis and both approaches should, in principle, yield similar results (van de Ven, 1989).

13.2 CLASSICAL DEPOSITION MECHANISMS

Deposition of particles onto surfaces occurs due to several interparticle forces. However, in limiting cases it is possible to distinguish between the various modes of particle deposition. The various modes of deposition can be classified as follows:

13.2.1 Brownian Diffusion: Classical Convection–Diffusion Transport

Submicron particles undergo Brownian motion, which enhances their deposition during flow past the surface of a collector. The governing transport equation is given by the convection–diffusion equation as

$$\mathbf{u} \cdot \nabla c = D_{\infty} \nabla^2 c \tag{13.1}$$

where $\mathbf{u}$ is the fluid velocity, c is the particle concentration and D_{∞} is the particle diffusion coefficient given by the Stokes–Einstein equation as

$$D_{\infty} = \frac{k_B T}{6\pi \mu a} \tag{13.2}$$

This analysis treats the particles as diffusing non-interacting points with D_{∞} being a constant. The particle hydrodynamic velocity is the same as the fluid velocity. The solution of the Navier–Stokes equation provides the fluid velocity $\mathbf{u}$.

The mass transfer rate is related to the flow Reynolds and Schmidt numbers. In the limit of low Reynolds number (creeping) flow, the mass transfer is related to the flow Peclet number, Pe, defined as LU_{∞}/D_{∞} where L and U_{∞} represent a characteristic length and the undisturbed fluid velocity, respectively. In this context, L represents the collector length scale.[1] The Peclet number is obtained as the product of the Reynolds number ($Re = LU_{\infty}\rho/\mu$) and the Schmidt number ($Sc = \mu/(\rho D_{\infty})$), where ρ is the fluid density and μ is the fluid dynamic viscosity. The Peclet number represents the ratio of the convective force to the diffusive (Brownian) force. In general, for $Pe \ll 1$, the transport is dominated by diffusion, while for $Pe \gg 1$, the transport is dominated by convection. For small Peclet numbers, $Pe \ll 1$, the dimensionless mass transfer group as presented by the Sherwood number is directly proportional to Pe. On the other hand, when $Pe \gg 1$, the Sherwood number becomes proportional to $Pe^{1/3}$.

13.2.2 Interception Deposition

Capture by interception assumes that the particles have a finite size and they are non-interacting and non-diffusing. The center of a particle follows exactly the undisturbed

[1] In Chapter 12, the characteristic length scale was that of the colloidal particle radius, a.

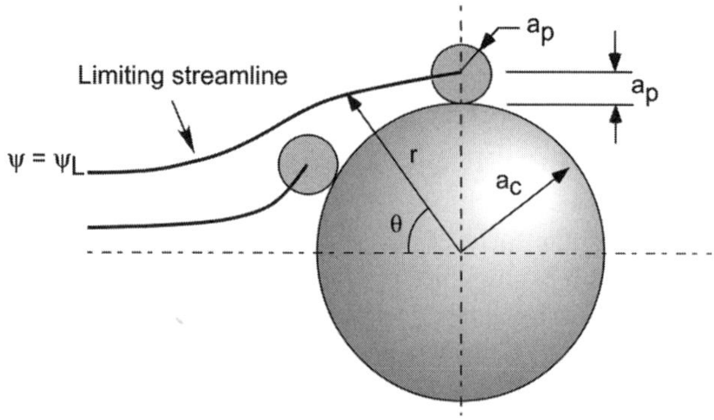

Figure 13.2. Capture due to interception by a spherical collector.

fluid streamline past the collector. When the particle touches the collector, capture or deposition takes place. Capture by pure interception ignores the increased hydrodynamic resistance between the particles and the collector upon approach. Figure 13.2 shows capture by interception of a spherical particle by a spherical collector. Particles located above the limiting trajectory or streamline, ψ_L, will not intercept the collector. Particles located below the limiting trajectory will intercept the collector.

As the particles follow the flow streamline, it is possible to evaluate the flux of particles that intercept a collector from the knowledge of the flow stream function. For the case of an isolated spherical collector, the flow stream function, under creeping flow conditions, is given by

$$\psi = \frac{1}{2}U_\infty\left(r^2 - \frac{3}{2}a_c r + \frac{a_c^3}{2r}\right)\sin^2\theta \tag{13.3}$$

The spherical coordinates are shown in Figure 13.2 together with the streamlines which are the loci of a constant ψ-value. The volumetric flow rate between any streamline and the collector axis is given, by definition, as $2\pi\psi$. A particle is considered to be captured or intercepted by the collector when its center is within one particle radius from the collector surface. Consequently, all particles within such a limiting streamline are captured. The limiting streamline, Ψ_L, for the special case of interception deposition is given by setting $r = a_p + a_c$ and $\theta = \pi/2$; hence

$$\psi_L = \frac{1}{2}U_\infty\left[(a_p + a_c)^2 - \frac{3}{2}a_c(a_p + a_c) + \frac{a_c^3}{2}\left(\frac{1}{a_p + a_c}\right)\right] \tag{13.4}$$

The total flux of the captured particles is given by $2\pi\psi_L n_\infty$. Capture efficiency is defined by the ratio of the actual captured particles to the idealized capture given by the area swept by the collector $\pi a_c^2 U_\infty n_\infty$. The capture efficiency is then given by

$$\eta = \frac{2\pi\psi_L n_\infty}{\pi a_c^2 U_\infty n_\infty} = \frac{2\psi_L}{a_c^2 U_\infty} \tag{13.5}$$

Making use of Eq. (13.4), the interception capture efficiency becomes

$$\eta_I = \frac{1}{a_c^2}\left[(a_p + a_c)^2 - \frac{3}{2}a_c(a_p + a_c) + \frac{a_c^3}{2}\left(\frac{1}{a_p + a_c}\right)\right] \tag{13.6}$$

For the case of $a_p/a_c \ll 1$, the above expression reduces to

$$\eta_I = \frac{3}{2}\left(\frac{a_p}{a_c}\right)^2 \tag{13.7}$$

The interception capture efficiency η_I is normally used as a means to assess other modes of capture. For the case of flows with a high Reynolds number, the interception capture efficiency due to a spherical collector was given by Weber and Paddock (1983) as

$$\eta_I = \frac{3}{2}\left(\frac{a_p}{a_c}\right)^2\left[1 + \frac{0.375Re}{1 + 0.367Re^{0.56}}\right] \quad \text{for } 0 < Re \leq 150 \tag{13.8}$$

where $Re = U_\infty a_c \rho/\mu$.

For the case of creeping flow ($Re \to 0$) in a packed bed of spherical collectors, the interception capture efficiency of a single collector is modified as

$$\eta_I = \frac{3}{2}\left(\frac{a_p}{a_c}\right)^2 A_{sph} \tag{13.9}$$

where A_{sph} is a dimensionless parameter that expresses the modification of the stream function due to the presence of other collector particles (Spielman, 1977; Neale and Masliyah, 1975). This type of approach has been quite successful in the modeling of capture by pure interception and under colloidal forces in a packed bed or a filter mat. For Happel's cell model simulating flow through a packed bed of spheres, the dimensionless function A_{sph} is provided by (Happel, 1958; Spielman and Fitzpatrick, 1973)

$$A_{sph} = \frac{2(1 - \alpha_c^{5/3})}{2 - 3\alpha_c^{1/3} + 3\alpha_c^{5/3} - 2\alpha_c^2} \tag{13.10}$$

where α_c is the collector volume fraction. Note when $\alpha_c \to 0$, A_{sph} becomes unity.

The collection efficiency for a single collector can be applied to evaluate the decrease in the particle number concentration through a packed bed or a filter mat. Assume a packed bed having n_c spherical collectors per unit volume of the bed. Consider an element of thickness dx within the packed bed. A particle number balance on the differential thickness dx provides

$$[U_\infty n A_b] - \left[U_\infty n A_b + U_\infty A_b \frac{dn}{dx}dx\right] - (n_c A_b dx)(\pi a_c^2 U_\infty n)\eta = 0 \tag{13.11}$$

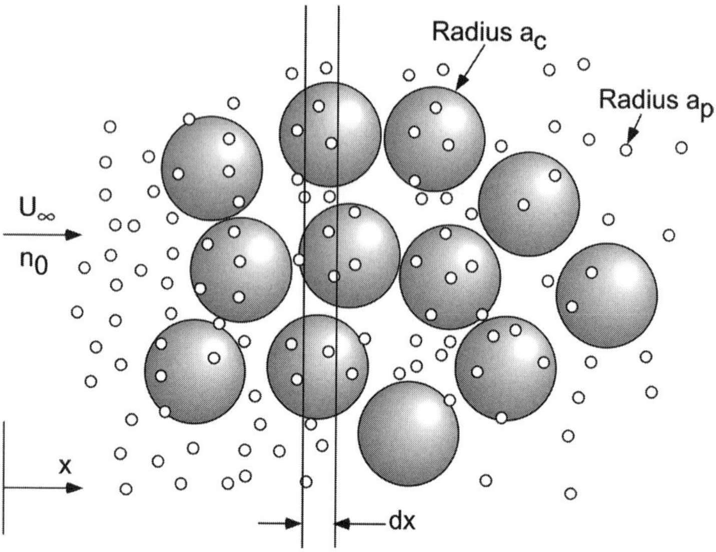

Figure 13.3. Particle deposition in a packed bed of spherical collectors.

where U_∞ is the fluid velocity away from the collector surface, A_b is the bed cross-sectional area and n is the particle number concentration at distance x along the bed. The differential element is shown is Figure 13.3. The particle balance equation (13.11) reduces to

$$\frac{dn}{dx} = -\pi a_c^2 n_c \eta n \tag{13.12}$$

The number of collector grains per unit volume, n_c, is given by

$$n_c = \frac{3\alpha_c}{4\pi a_c^3} \tag{13.13}$$

Combining Eqs. (13.12) and (13.13) leads to

$$\frac{dn}{dx} = -\left(\frac{3\alpha_c \eta}{4a_c}\right) n \tag{13.14}$$

For convenience, a filter coefficient is normally defined as

$$\lambda = \frac{3\alpha_c \eta}{4a_c} \tag{13.15}$$

Making use of the filter coefficient definition, we obtain

$$\frac{dn}{dx} = -\lambda n \tag{13.16}$$

For an incoming suspended particle number concentration of n_o (at $x = 0$), Eq. (13.16) can be integrated to give

$$n = n_o \exp(-\lambda x) \tag{13.17}$$

Equation (13.17) indicates that the decay in the number concentration is exponential and that the filter coefficient λ can be considered as a characteristic penetration depth. The collision efficiency η can be determined for cases other than pure interception as will be shown in later sections.

13.2.3 Inertial Deposition

When suspended non-interacting particles do not follow the flow streamlines due to their inertia, their capture by a collector is termed inertial deposition. For particles of less that 10 μm in liquids, inertial deposition is not important. However, inertial deposition in gases can be very significant (Friedlander, 2000). Figure 13.4 shows inertial deposition of a spherical particle onto a cylindrical collector.

Newton's second law can be used to write a force balance for a particle in a flow field in the absence of gravity and for $\rho_p \gg \rho$. It is given as

$$m_p \frac{d^2\mathbf{x}}{dt^2} = 6\pi \mu a_p \left(\mathbf{u} - \frac{d\mathbf{x}}{dt}\right) \tag{13.18}$$

where m_p is the mass of the particles. In such a formulation it is assumed that the particle is much smaller than the collector and at sufficiently low concentration such that the flow field about the collector is not altered by the presence of the particles (Clift *et al.*, 1978; Friedlander, 2000). The term on the left side of Eq. (13.18) is the particle acceleration; the term on the right side is the drag on the particle. Here Stokes drag is used because the particle Reynolds number is assumed to be small. The term $\mathbf{u}$ is the fluid velocity vector and $\mathbf{x}$ is the particle position vector. Procedures to solve Eq. (13.18) are given by Michael and Norey (1969), Griffin and Meisen (1973), and Masliyah and Duff (1975).

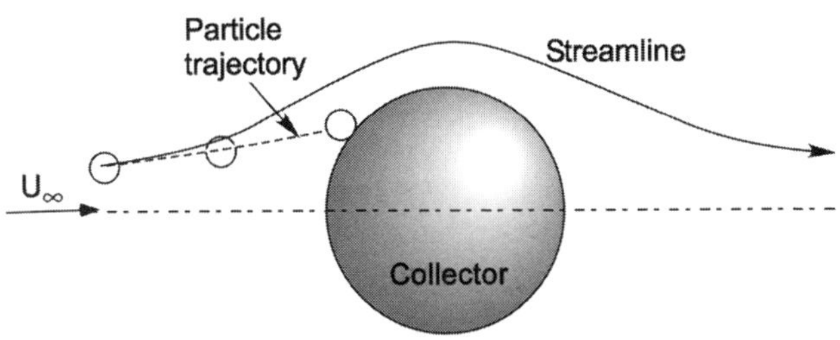

Figure 13.4. Inertial impaction of a spherical particle on a cylindrical fibre in the absence of gravity and for $\rho_p \gg \rho$.

Non-dimensionalization of Eq. (13.18) leads to

$$St\frac{d^2\mathbf{X}}{d\tau^2} = \mathbf{u}^* - \frac{d\mathbf{X}}{d\tau} \tag{13.19}$$

where

$$\tau = \frac{U_\infty t}{a_c} \tag{13.20}$$

$$\mathbf{X} = \frac{\mathbf{x}}{a_c} \tag{13.21}$$

$$\mathbf{u}^* = \frac{\mathbf{u}}{U_\infty} \tag{13.22}$$

and

$$St = \frac{2}{9}\left(\frac{a_p}{a_c}\right)^2\left(\frac{\rho_p}{\rho}\right)\left(\frac{\rho a_c U_\infty}{\mu}\right) \tag{13.23}$$

Here ρ and ρ_p are the fluid and particle densities, respectively. St is the Stokes number which is a measure of the particle inertia with respect to viscous forces. Variation of the collection efficiency η with the Stokes number is shown in Figure 13.5. The flow field used is that of a potential flow over a sphere given by

$$\psi = \frac{1}{2}U_\infty r^2\left[1 - \left(\frac{a_c}{r}\right)^3\right]\sin^2\theta$$

The collection efficiency increases with increasing Stokes number where the particle inertia becomes important.

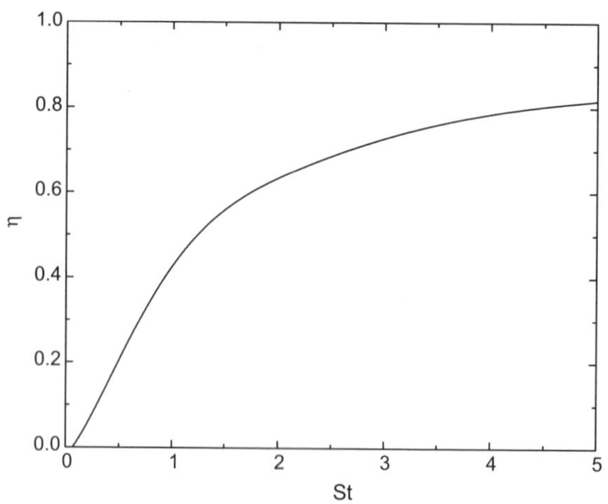

Figure 13.5. Deposition efficiency due to inertial capture by a spherical collector using potential flow velocity (Michael and Norey, 1969).

13.3 EULERIAN APPROACH

Capture of colloidal particles from a flowing suspension of particles by a stationary collector surface can occur due to the combination of Brownian motion, hydrodynamic interaction, and the presence of other forces, such as gravitational, London–van der Waals, and electric double layer forces. In this section, we present the methodology of analyzing particle capture or deposition employing the Eulerian approach. The general mathematical construct of the Eulerian approach is first presented, followed by consideration of specific geometries, and the different type of forces.

The generalized picture of particle deposition on a stationary collector surface involves convective transport of particles with the fluid flow toward the collector, diffusive transport of the particles from regions of high to low concentrations, and migration due to other forces like gravity or colloidal forces. The generalized convection–diffusion equation which governs particle transfer due to a flow field characterized by a velocity vector $\mathbf{u}$ is given by

$$\frac{\partial n}{\partial t} + \nabla \cdot \mathbf{j} = Q \tag{13.24}$$

where

$$\mathbf{j} = \mathbf{v}n - \bar{\bar{D}} \cdot \nabla n + \frac{n}{k_B T} \bar{\bar{D}} \cdot \mathbf{F} \tag{13.25}$$

where $\mathbf{j}$ is the flux of the particles (particles per unit time and unit area), n is the local particle number concentration, $\bar{\bar{D}}$ is the diffusion coefficient tensor, $\mathbf{F}$ is the force experienced by the particle, $\mathbf{v}$ is the particle hydrodynamic velocity, *i.e.*, the particle velocity that is induced by the fluid motion in the absence of all other forces such as Brownian or colloidal forces, T is the absolute temperature, k_B is the Boltzmann constant and Q is a source term.

It is noted that in Eq. (13.25), the hydrodynamic particle velocity, $\mathbf{v}$, is used instead of the fluid velocity, $\mathbf{u}$, as was discussed in Chapter 6, Section 6.3. In this chapter, we are discussing large particles, which cannot be treated as point masses. Consequently, when such finite sized particles are suspended in a fluid with a spatially non-uniform velocity field, different parts of the particle are subjected to fluid velocities of different intensities. As a result, the net hydrodynamic particle velocity determined at the particle center of mass becomes different from the fluid velocity. Particularly, such differences between the hydrodynamic particle velocity and the fluid velocity become pronounced for flows close to a stationary surface. It is for this reason that the fluid velocity, $\mathbf{u}$, is replaced by the particle hydrodynamic velocity, $\mathbf{v}$. It is this hydrodynamic particle velocity that needs to be adjusted by the diffusional fluxes and other external forces to arrive at the particle flux $\mathbf{j}$.

The reason for the introduction of the diffusion coefficient tensor rather than simply using a scalar quantity is that the particle diffusion coefficient is a function of the particle position relative to a surface. Once again, this artifact of incorporating particle hydrodynamic interactions was briefly discussed in Section 6.3. This was illustrated by Honig *et al.* (1971), who gave expressions for the diffusion coefficient variation with gap width between two spherical particles. For the case of a spherical particle (1) diffusing near another spherical particle (2), one can write the diffusion coefficient tensor as

$$\bar{\bar{D}} = (D_1 + D_2)_\infty \begin{pmatrix} d_\parallel & 0 \\ 0 & d_\perp \end{pmatrix} \tag{13.26}$$

where $(D_1 + D_2)_\infty$ is the diffusion coefficient of the diffusing spherical particles in an infinite medium given by the Stokes–Einstein equation. The term $d_\parallel$ and $d_\perp$ are correction factors due to the finite gap width between the two spheres. They are both functions of the ratio of the spheres radii and the separation gap width. For the case of equal-sized spheres, $d_\perp = 1/G(h)$ as was given by Eq. (12.54). For the case of a spherical colloidal particle of radius a and a very large collector grain, which is treated as a flat surface, Eq. (13.26) becomes

$$\bar{\bar{D}} = D_\infty \begin{pmatrix} d_\parallel & 0 \\ 0 & d_\perp \end{pmatrix} \tag{13.27}$$

It should be noted, however, that $d_\parallel$ and $d_\perp$ of Eq. (13.27) are different from those of Eq. (13.26) because $d_\parallel$ and $d_\perp$ are functions of the sphere size ratio. Both $d_\parallel$ and $d_\perp$ approach zero as the gap width $h \to 0$. For small h

$$d_\perp = 2h \quad \text{for } a_1 = a_2$$

and

$$d_\perp = h \quad \text{for } a_2 \to \infty$$

The correction function $d_\parallel$ is proportional to $1/|\ln h|$ for $h \to 0$. As the dimensionless gap width becomes large, $d_\parallel \to 1$ and $d_\perp \to 1$ (Batchelor, 1976; Adler, 1981; Jeffrey and Onishi, 1984; van de Ven, 1989). Note that is some publications the definition of the terms $d_\parallel$ and $d_\perp$ or their equivalent are interchanged.

The velocity of the fluid $\mathbf{u}$ is usually evaluated from the solution of the Navier–Stokes equation in the absence of the dispersed particles. To this end, the analysis becomes limited to a very diluted dispersed phase. The required hydrodynamic velocity of the particle $\mathbf{v}$ is related to the fluid velocity $\mathbf{u}$ by making corrections due to the presence of the collector surface.

For a deposition problem involving a specific collector geometry, the forces and the flow configuration are first identified. The particle flux $\mathbf{j}$ is then formulated in terms of the convection, diffusion, and migration contributions in Eq. (13.25). Finally, solution to Eq. (13.24) is obtained employing appropriate boundary conditions at the collector surface. In the following subsections, various situations will be presented to illustrate the use to the Eulerian approach.

13.3.1 Deposition Due to Brownian Diffusion Without External Forces: Spherical Collector

Here we shall deal with the classical case of deposition onto a spherical collector due to Brownian diffusion alone in the absence of any other forces. The diffusing particles are considered to be of submicron size (strictly speaking, point masses) with their hydrodynamic velocity identical to the fluid velocity ($\mathbf{v} = \mathbf{u}$). In this case, we ignore any hydrodynamic interactions. With the radius a_p being very small (point particles), the colloidal forces are taken to be zero and the diffusion coefficient is constant. For steady-state conditions, with $Q = \mathbf{F} = 0$ and for a constant diffusion coefficient, D_∞, combining Eqs. (13.24) and (13.25) leads to

$$\nabla\cdot(n\mathbf{u}-D_\infty\nabla n) = 0 \tag{13.28}$$

Recognizing that $\nabla \cdot \mathbf{u} = 0$ for an incompressible fluid, the above equation becomes

$$\mathbf{u}\cdot\nabla n - D_\infty\nabla^2 n = 0 \tag{13.29}$$

Equation (13.29) in spherical coordinates becomes

$$u_r\frac{\partial n}{\partial r} + \frac{u_\theta}{r}\frac{\partial n}{\partial \theta} = D_\infty\left[\frac{1}{r^2}\frac{\partial}{\partial r}\left(r^2\frac{\partial n}{\partial r}\right) + \frac{1}{r^2\sin\theta}\frac{\partial}{\partial\theta}\left(\sin\theta\frac{\partial n}{\partial\theta}\right)\right] \tag{13.30}$$

For Stokes flow about an isolated sphere (in the limit of a very small Reynolds number), the flow velocities in the radial and angular-directions are (Bird *et al.*, 2002):

$$u_r = -U_\infty\cos\theta\left[1 - \frac{3}{2}\frac{a_c}{r} + \frac{1}{2}\frac{a_c^3}{r^3}\right] \tag{13.31}$$

and

$$u_\theta = U_\infty\sin\theta\left[1 - \frac{3}{4}\frac{a_c}{r} - \frac{1}{4}\frac{a_c^3}{r^3}\right] \tag{13.32}$$

For large Peclet numbers, the variation in the concentration is much higher in the r-direction that in the θ-direction and one can rewrite Eq. (13.30) as

$$u_r\frac{\partial n}{\partial r} + \frac{u_\theta}{r}\frac{\partial n}{\partial\theta} = D_\infty\left[\frac{\partial^2 n}{\partial r^2} + \frac{2}{r}\frac{\partial n}{\partial r}\right] \tag{13.33}$$

The flow and the coordinate system is shown in Figure 13.6. Assuming a perfect sink at the spherical collector surface, the boundary conditions become

$$\text{symmetry at}\quad \theta = 0, \pi \quad \text{for all } r \tag{13.34}$$

$$n = 0 \quad \text{at } r = a_c \quad \text{for all } \theta \tag{13.35}$$

and

$$n \to n_\infty \quad \text{as } r \to \infty \quad \text{for all } \theta \tag{13.36}$$

Here n_∞ is the bulk value of n at a large distance from the spherical collector. A solution to Eq. (13.33) is given by Masliyah and Epstein (1973) and other solutions

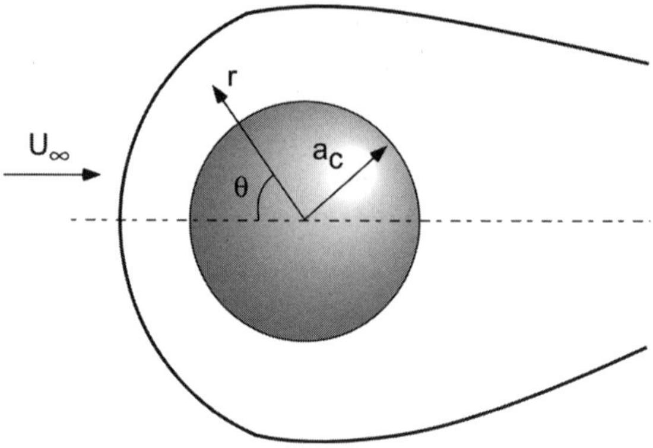

Figure 13.6. Diffusion boundary layer around a spherical collector.

are cited by Clift *et al.* (1978). However, simplification in the solution can be made by assuming that the diffusion process is limited to a region very close to the surface and hence the velocity profile needs to be true only close to the spherical collector. Consequently, the velocity components are approximated by

$$u_r = -\frac{3}{2}\left(\frac{y}{a_c}\right)^2 U_\infty \cos\theta \tag{13.37}$$

and

$$u_\theta = \frac{3}{2}\left(\frac{y}{a_c}\right) U_\infty \sin\theta \tag{13.38}$$

where

$$\frac{r}{a_c} = 1 + \frac{y}{a_c} \tag{13.39}$$

The above velocity expressions are valid close to the collector surface.

Levich (1962) obtained a similar solutions to Eqs. (13.33) to (13.39). At the collector surface, the flux component normal to the collector surface, $j_\perp$, is given by

$$j_\perp = -j_r = D\left.\frac{\partial n}{\partial r}\right|_{surface} \tag{13.40}$$

in $m^{-2}s^{-1}$, and the total rate of particles diffusing to the surface is given by

$$\int_S j_\perp dS = 2\pi a_c^2 \int_0^\pi j_\perp \sin\theta d\theta \tag{13.41}$$

where $j_\perp$ is the flux normal to the collector surface and S is the spherical collector surface area.

The local Sherwood number characterizing the mass transfer is defined as

$$Sh(\theta) = \frac{a_c j_\perp}{D_\infty n_\infty} \tag{13.42}$$

where n_∞ is the particle number concentration at a large distance from the surface of the spherical collector. The average Sherwood number is defined as

$$Sh = \frac{1}{S} \int_S Sh(\theta)\, dS \tag{13.43}$$

The solution of Eq. (13.33) with the velocities defined by Eqs. (13.37) through (13.39) subject to the boundary conditions of Eqs. (13.34) through (13.36) provides the dimensionless mass transfer number (Clift *et al.*, 1978):

$$Sh = 0.624(Re\, Sc)^{1/3} \quad \text{for } ReSc \gg 1 \tag{13.44}$$

where the Schmidt number (ratio of momentum transfer to molecular diffusion) is given by

$$Sc = \frac{\mu}{\rho D_\infty} \tag{13.45}$$

and the Reynolds number (ratio of inertial to viscous forces) is given by

$$Re = \frac{a_c \rho U_\infty}{\mu} \tag{13.46}$$

Equation (13.44) is usually referred to as the *Lighthill–Levich equation* or simply the *Levich equation*.[2] The product of Re and Sc is normally called the Peclet number and it is given by

$$Pe = ReSc = \frac{a_c U_\infty}{D_\infty} \tag{13.47}$$

Acrivos and Goddard (1965) gave a first-order correction to Eq. (13.44) as

$$Sh = 0.624(Pe)^{1/3} + 0.46 \quad Pe \gg 1 \tag{13.48}$$

For the case of small Peclet numbers, solution of the complete convection-diffusion equation for the case of creeping flow is given by Acrivos and Taylor (1962) as

$$Sh = 1 + \frac{Pe}{2} \quad Pe \ll 1 \tag{13.49}$$

[2]The original derivation of Levich for $Pe \to \infty$ was given as $Sh = (7.98/4\pi)Pe^{1/3}$, where the coefficient becomes 0.635. This value of the coefficient is slightly higher than in the more recent analysis, Clift *et al.* (1978).

Equations (13.44), (13.48), and (13.49) were derived using creeping flow velocities ($Re \to 0$) about the spherical collector and the equations are identical to those given for classical mass transfer from spheres. Equation (13.49) states that for pure diffusion, the Sherwood number is unity. In the definition of both Peclet and Sherwood numbers, the radius of the diffusing particle is used as the characteristic length.

13.3.2 Deposition due to Brownian Diffusion with External Forces: Stagnation Flow

Convenient techniques that are often used in experimental and theoretical studies for deposition of colloidal particles are those of rotating disks (Marshall and Kitchener, 1966; Hull and Ketchner, 1969; Prieve and Lin, 1980) and of stagnation flows due to impinging jets (Dabros and van de Ven, 1983; Boluk and van de Ven, 1989; Adamczyk *et al.*, 1989). Other geometries are discussed by Adamczyk (1989a,b) and Elimelech *et al.* (1995).

In the previous section, diffusion due to Brownian motion was considered with the diffusing particles taken as "point" particles. In this section we shall discuss in detail the Eulerian approach to the study of finite-sized particle deposition in a stagnation flow due to an impinging jet. We shall take into account the relationship between the fluid and particle velocities, a variable diffusion coefficient, and interparticle forces.

Consider a fluid containing colloidal particles impinging on a flat plate as shown in Figure 13.7. Due to the axisymmetry of the flow, the velocity components in the angular direction are zero and the velocity field is given by the radial and the normal components, namely u_r and u_z, respectively. As the Schmidt number of the colloidal system exceeds unity by one or two orders of magnitude, the diffusion boundary layer

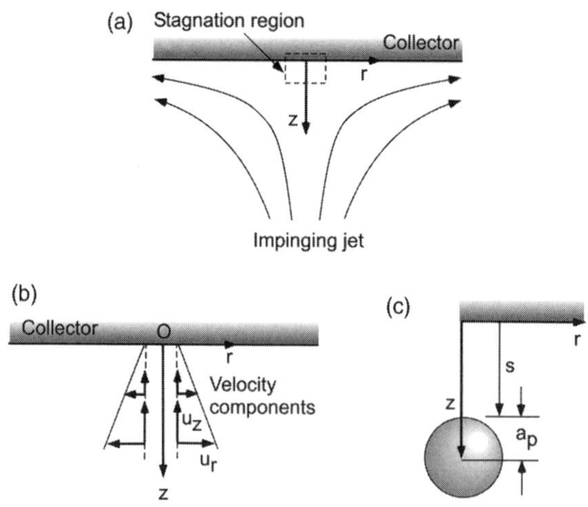

Figure 13.7. Impinging jet geometry. (a) Overall flow geometry. (b) Velocity profile at the stagnation region shown by the dashed box in (a). O is the stagnation point where the velocity is zero. (c) Position of a colloidal particle with respect to the impinging surface.

is much thinner than the hydrodynamic boundary layer. It is for this reason that the flow velocities need to be characterized close to the impingement region only. For flow near the stagnation region, the velocities of the fluid are given by

$$u_r = \alpha z r \tag{13.50}$$

and

$$u_z = -\alpha z^2 \tag{13.51}$$

The parameter α characterizes the intensity of the flow (Dabros and van de Ven, 1983). Here it is assumed that the presence of the particles does not alter the flow field.

The movement of a particle entrained in the stagnation region of the flow is decomposed into normal (z-direction) and tangential (r-direction) components. The colloidal particle Reynolds number is assumed to be sufficiently small such that Stokes flow holds. Equation (13.25) can be written as

$$j_r = v_r n - \left(D_{rr}\frac{\partial n}{\partial r} + D_{rz}\frac{\partial n}{\partial z} \right) + \frac{n}{k_B T}\left[D_{rr}F_r + D_{rz}F_z \right] \tag{13.52}$$

and

$$j_z = v_z n - \left(D_{zr}\frac{\partial n}{\partial r} + D_{zz}\frac{\partial n}{\partial z} \right) + \frac{n}{k_B T}[D_{zr}F_r + D_{zz}F_z] \tag{13.53}$$

Note that in the above expressions, the product of the diffusion coefficient and the force components represents a product of the diffusion coefficient tensor, $\bar{\bar{D}}$, with the force vector, **F**. The dot product of a tensor $\bar{\bar{T}}$ with a vector **A** is given by

$$\begin{aligned} \bar{\bar{T}} \cdot \mathbf{A} = {} & (T_{xx}A_x + T_{xy}A_y + T_{xz}A_z)\mathbf{i}_x \\ & + (T_{yx}A_x + T_{yy}A_y + T_{yz}A_z)\mathbf{i}_y \\ & + (T_{zx}A_x + T_{zy}A_y + T_{zz}A_z)\mathbf{i}_z. \end{aligned}$$

Recognizing from Eq. (13.27) that $D_{zr} = D_{rz} = 0$, we can write

$$j_r = v_r n - D_{rr}\frac{\partial n}{\partial r} + \frac{n}{k_B T}D_{rr}F_r \tag{13.54}$$

and

$$j_z = v_z n - D_{zz}\frac{\partial n}{\partial z} + \frac{n}{k_B T}D_{zz}F_z \tag{13.55}$$

where j_r and j_z are the fluxes in r- and z-directions, respectively.

Following Spielman and Fitzpatrick (1973), Prieve and Lin (1980), Dabros and van de Ven (1983), and van de Ven (1989), the relationships between the fluid and

TABLE 13.1. Asymptotic Expressions for the Universal Hydrodynamic Functions for a Spherical Particle Near a Flat Solid Surface (Spielman and Fitzpatrick, 1973).

Function	$h \to 0$	$H \to \infty$
$f_1(h)$	h	$1 - \dfrac{9}{8(h+1)}$
$f_2(h)$	3.23	$\left[1 - \dfrac{9}{8(h+1)}\right]^{-1}$
$f_3(h)$	$\dfrac{0.7431}{0.63736 - 0.2\ln h}$	$1 - \dfrac{5}{16(h+1)^3}$

particle velocities and the local particle diffusion coefficients are given by

$$v_r = u_r f_3(h) \tag{13.56}$$

$$v_z = u_z f_1(h) f_2(h) \tag{13.57}$$

$$D_{rr} = D_\infty d_\parallel = D_\infty f_4(h) \tag{13.58}$$

and

$$D_{zz} = D_\infty d_\perp = D_\infty f_1(h) \tag{13.59}$$

where f_1 to f_4 are the universal hydrodynamic (correction) functions relating the deviation from Stokes flow and the Stokes–Einstein relationship due to the presence of the collector wall. f_1 is given by Brenner (1961), f_2 is given by Goren (1970) and Goren and O'Neill (1971), and f_3 and f_4 are given by Goldman *et al.* (1967a,b). Asymptotic behaviors of f are listed in Table 13.1. Curve-fit expressions for f are given in Table 13.2. Figure 13.8 shows the variation of the functions with the dimensionless gap h, where

$$\frac{z}{a_p} = h + 1 \tag{13.60}$$

It should be noted that the correction factors f_1 to f_4 and $d_\parallel$ and $d_\perp$ are for the geometry of a sphere and a solid plane as applicable to the problem at hand.

Making use of Eqs. (13.56) to (13.59), the flux Eqs. (13.54) and (13.55) become

$$j_r = u_r f_3 n - D_\infty f_4 \frac{\partial n}{\partial r} + \frac{n}{k_B T} D_\infty f_4 F_r \tag{13.61}$$

TABLE 13.2. Curve Fit for the Universal Functions f_i.

i	a	b	c	d	e
1	0.9267	−0.3990	0.1487	−0.601	1.202
2	0.5695	1.355	1.36	0.875	0.525
3	0.15	−0.375	3.906	−0.625	3.105
4	1.26	−2.676	0.3581	1.999	0.232

$f_i = 1.0 + b_i \exp(-c_i h) + d_i \exp(-e_i h^{a_i})$

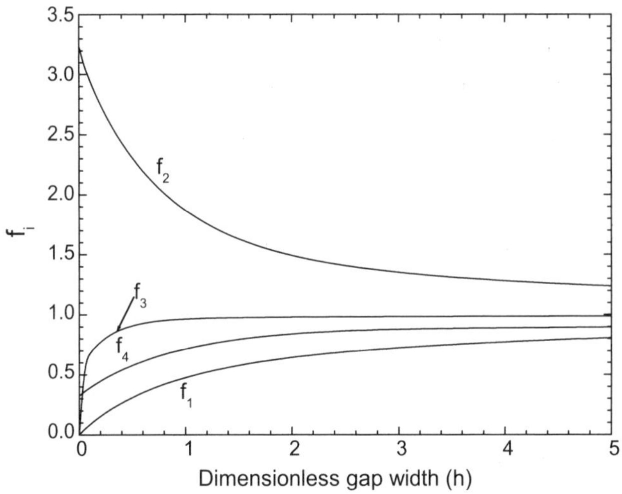

Figure 13.8. Universal hydrodynamic correction functions for diffusion and particle motion near a plane for the case of a sphere.

and

$$j_z = u_z f_1 f_2 n - D_\infty f_1 \frac{\partial n}{\partial z} + \frac{n D_\infty}{k_B T} f_1 F_z \tag{13.62}$$

where D_∞ is given by the Stokes–Einstein equation.

The z-direction component of the force F_z is composed of gravitational force, London–van der Waals force, and electrostatic (electric double layer) force. F_z is given by

$$F_z = F_{gz} + F_A + F_R \tag{13.63}$$

Here it is assumed that the colloidal forces act normal to the surface of the collector. The r-direction component of the force is due to gravity alone:

$$F_r = F_{gr} \tag{13.64}$$

For a horizontal deposition plate, F_r is zero.

For steady-state conditions and a zero source term, Eq. (13.24) in cylindrical coordinates becomes

$$\frac{1}{r}\frac{\partial}{\partial r}(r j_r) + \frac{\partial j_z}{\partial z} = 0 \tag{13.65}$$

From Eqs. (13.61) and (13.62), the flux conservation Eq. (13.65) becomes

$$\frac{1}{r}\frac{\partial}{\partial r}\left[r\left(u_r f_3 n - D_\infty f_4 \frac{\partial n}{\partial r} + \frac{n}{k_B T} D_\infty f_4 F_r\right)\right] + \frac{\partial}{\partial z}\left[u_z f_1 f_2 n - D_\infty f_1 \frac{\partial n}{\partial z} + \frac{n}{k_B T} D_\infty f_1 F_z\right] = 0 \tag{13.66}$$

Equation (13.66) is the convection-diffusion equation for the transport process in the presence of external forces. This equation was derived for a stagnation flow configuration and is also applicable to other axisymmetric deposition configuration such as a rotating disc.

The elements that make up the forces F_r and F_z are discussed below.

The gravitational force is given by

$$F_{gr} = \frac{4\pi}{3} a_p^3 \Delta\rho g \sin\theta \tag{13.67}$$

and

$$F_{gz} = \frac{4\pi}{3} a_p^3 \Delta\rho g \cos\theta \tag{13.68}$$

where $\Delta\rho$ is the density difference between the particle and the fluid. The inclination angle θ is between the z-axis and the vertical direction.

The attractive interaction force, F_A, is typically represented by an appropriate expression for the van der Waals forces (see Chapter 11). A convenient expression for the retarded London–van der Waals dispersion force between a spherical particle and a planar collector surface is given by (Suzuki *et al.*, 1989)

$$F_A = -\frac{A_H \bar{\lambda}(\bar{\lambda} + 22.232h)}{6a_p h^2(\bar{\lambda} + 11.116h)^2} \tag{13.69}$$

where A_H is the effective Hamaker constant for interaction between the particle and the collector in the suspending medium ($= A_{123}$) and $\bar{\lambda} = \lambda/a_p$ is a dimensionless parameter accounting for electromagnetic retardation, with λ being the London wavelength. Usually, the value of the London wavelength is taken as 10^{-7} m (100 nm).

The repulsive interaction force, F_R, between the particle and the collector predominantly arises due to the electrical double layer interactions. For spherical particles and planar collectors, the electrical double layer force expression is given by (Hogg *et al.*, 1966; Usui, 1973)

$$F_R = 4\pi\epsilon\kappa a_p \zeta_p \zeta_c \left[\frac{\exp(-\kappa a_p h)}{1 \pm \exp(-\kappa a_p h)} \mp \frac{(\zeta_c - \zeta_p)^2}{2\zeta_c \zeta_p} \frac{\exp(-2\kappa a_p h)}{\{1 - \exp(-2\kappa a_p h)\}} \right] \tag{13.70}$$

The upper and lower signs in Eq. (13.70) correspond to constant surface potential and constant surface charge, respectively.

The following dimensionless groups are defined as:

$$\textit{Gravitational parameter, } Gr = \frac{\text{gravitational force}}{\text{Brownian motion force}}$$

$$Gr = \frac{2}{9} \frac{\Delta\rho g a_p^3}{\mu D_\infty} \tag{13.71}$$

where $D_\infty = k_B T/(6\pi\mu a_p)$.

$$\textit{Double layer parameter, } Dl = \frac{\text{electrostatic force}}{\text{Brownian motion force}}$$

$$Dl = \frac{4\pi \epsilon a_p \zeta_p \zeta_c}{k_B T} \tag{13.72}$$

Double layer asymmetry parameter, Da, referring to the asymmetry of the double layers for different surface potentials on the interacting surfaces, defined as

$$Da = \frac{(\zeta_c - \zeta_p)^2}{2\zeta_c \zeta_p} \tag{13.73}$$

For $\zeta_c = \zeta_p$, Da becomes zero.

Dimensionless inverse Debye length, κa_p, given by

$$\kappa a_p = \left(\frac{2000 z^2 e^2 N_A M}{\epsilon k_B T} \right)^{1/2} a_p \tag{13.74}$$

where N_A is the Avogadro number, and M is the molar concentration of the symmetric $(z : z)$ electrolyte.

Normalizing the forces by $(k_B T/a_p)$, neglecting radial derivatives, and setting $\theta = 0$, Eq. (13.66) becomes

$$\frac{d}{dh}\left[\frac{1}{2} Pe f_1 f_2 n (1+h)^2 + f_1 \frac{dn}{dh} - n f_1 F_z^* \right] - Pe f_3 (1+h) n = 0 \tag{13.75}$$

where

$$F_z^* = -Ad \frac{\bar{\lambda}(\bar{\lambda} + 22.232h)}{h^2(\lambda + 11.116h)^2} + (Dl)(\kappa a_p) \left[\frac{\exp(-\kappa a_p h)}{1 \pm \exp(-\kappa a_p h)} \right.$$
$$\left. \mp Da \frac{\exp(-2\kappa a_p h)}{\{1 - \exp(-2\kappa a_p h)\}} \right] + Gr \tag{13.76}$$

In Eqs. (13.75) and (13.76)

$$Pe = \frac{2a_p^3 \alpha}{D_\infty} \tag{13.77}$$

and

$$Ad = \frac{A_H}{6 k_B T} \tag{13.78}$$

Ad is a measure of the relative strengths of London–van der Waals force to the Brownian motion force. It is usually referred to as the adhesion number or parameter.

The parameters that influence particle deposition on smooth surfaces are κa, $\bar{\lambda}$, Ad, Dl, Da, Gr, and Pe.

Although the above analysis was derived for stagnation flow, it is valid for any flow configuration having velocities of the form given by Eqs. (13.50) and (13.51). Therefore by using an appropriate definition for the Peclet number, Pe, the analysis for various flow geometries becomes similar. For the case of an impinging jet, the flow strength α is given by

$$\alpha = \bar{\alpha}(Re)\frac{Re\, \nu}{R_{jet}^3}$$

where $\bar{\alpha}(Re) = \beta Re^{1/2}$, and β is of order unity (Dabros and van de Ven, 1987). R_{jet} is the jet tube radius and $\nu = \mu/\rho$ is the kinematic viscosity. The Reynolds number Re is given by

$$Re = \frac{U_\infty R_{jet}}{\nu} \tag{13.79}$$

where U_∞ is the average exit velocity of the jet. Various definitions of Peclet number for other geometries are given by van de Ven (1989).

The boundary conditions of Eq. (13.75) are

$$n \to n_\infty \quad \text{as } h \to \infty \tag{13.80}$$

and

$$n \to 0 \quad \text{at } h = \delta_m/a_p \tag{13.81}$$

where δ_m is generally identified with the distance of primary energy minimum. The solution of Eq. (13.75) subject to Eqs. (13.80) and (13.81) can be expected to yield an upper limit to the flux towards the collector for a given energy of interaction between the particle and surface and for given hydrodynamic conditions (Dabros and van de Ven 1983, 1987). The boundary condition [Eq. (13.81)] assumes that all particles arriving close to the collector surface are irreversibly captured in an infinite-depth energy sink and disappear from the system (Adamczyk, 1989a). It is often referred to as the perfect sink boundary condition. In most studies, δ_m/a_p is taken as a fixed small value, say 10^{-3} (Prieve and Lin, 1980), since setting $\delta_m = 0$ in conjunction with any of the Hamaker type expressions for the van der Waals interactions will lead to a divergence of the interaction forces to $-\infty$ at contact.

The local Sherwood number is defined as

$$Sh_l = -j_{zw}\frac{a_p}{D_\infty n_\infty} = j_\perp \frac{a_p}{D_\infty n_\infty}$$

where j_{zw} is the flux in the z-direction at the surface of the collector (i.e., at $z = a + \delta_m$). The average Sherwood number is given by

$$Sh = \frac{1}{S}\int_S Sh_l dS \tag{13.82}$$

where Sh is the average Sherwood number and S is the deposition surface area. j_{zw} is evaluated using Eq. (13.62) once the distribution on n is known. Numerical techniques are normally employed in solving Eq. (13.75), thus providing the variation of n with h.

Transport equations similar to Eq. (13.75) are given by Prieve and Ruckenstein (1974) for the case of a spherical collector; Prieve and Lin (1980), Rajagopalan and Kim (1981), and Clint *et al.* (1973) for the case of a rotating disc; Dabros and van de Ven (1983, 1987) and van de Ven (1989) for the case of an impinging jet; Adamczyk and van de Ven (1981a) for the case of parallel plates channel; and Adamczyk and van de Ven (1981b) for the case of cylindrical collectors. General discussions on the kinetics of particle accumulation on collectors is given by Adamczyk *et al.* (1984), Adamczyk and van de Ven (1984), and Adamczyk *et al.* (1992).

Solution of the transport equations such as Eq. (13.75) provides values for the Sherwood number Sh and a knowledge of how the various parameters, *e.g.*, Pe, Dl, Da, and Ad affect mass transport.

In this section, detailed analysis was given for the impinging jet geometry. Mass transfer analysis, *i.e.*, particle deposition for other geometries, is very similar. The effect of the various parameters on the dimensionless mass transfer Sh will be given below. Table 13.3 gives the definitions of the Reynolds, Peclet, and Sherwood numbers for different geometries employed in the literature. van de Ven (1989) provides tabulation of the velocity components near the surface of the various geometries.

For the case of Brownian particles with negligible London–van der Waals and electric double-layer forces, the stagnation flow Sherwood number is given by the Levich-type equation as

$$Sh = 0.616 Pe^{1/3}$$

for $Pe \gg 1$ and $Dl = Ad = Gr = 0$, where Pe and Sh are defined by Eqs. (13.77) and (13.82), respectively. Sherwood number variation is similar to that obtained for mass transfer of point particles towards a spherical collector as given by Eq. (13.44).

TABLE 13.3. Definition of Reynolds, Peclet, and Sherwood Numbers for Various Deposition Geometries.

Collector Geometry	Reynolds Number	Peclet Number	Sherwood Number
Spherical collector	$\dfrac{a_c \rho U_\infty}{\mu}$	$\dfrac{a_c U_\infty}{D_\infty}$	$\dfrac{1}{S}\int_S \left(\dfrac{a_c j_\perp}{D_\infty n_\infty}\right) dS$
Cylindrical collector	$\dfrac{a_c \rho U_\infty}{\mu}$	$\dfrac{a_c U_\infty}{D_\infty}$	$\dfrac{1}{S}\int_S \left(\dfrac{a_c j_\perp}{D_\infty n_\infty}\right) dS$
Impinging jet	$\dfrac{R_{jet} \rho U_\infty}{\mu}$	$\dfrac{2a_p^3 \alpha}{D_\infty}$	$\dfrac{1}{S}\int_S \left(\dfrac{a_p j_\perp}{D_\infty n_\infty}\right) dS$
Rotating disc	$a_p^3 \left(\dfrac{\omega}{\upsilon}\right)^{3/2}$	$\dfrac{1.02\omega^{3/2} a_p^3}{\upsilon^{1/2} D_\infty}$	$\dfrac{1}{S}\int_S \left(\dfrac{a_p j_\perp}{D_\infty n_\infty}\right) dS$

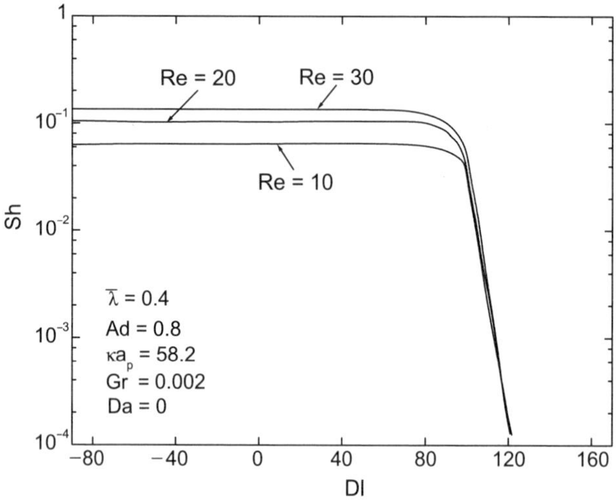

Figure 13.9. Calculated values of the Sherwood number for stagnation flow as a function of the double layer parameter Dl (Dabros and van de Ven, 1983).

Dabros and van de Ven (1983) gave the variation of Sherwood number, Sh, with the double-layer parameter, Dl, for various Reynolds numbers, Re. A negative Dl number means that the particle and the collector have opposite signs for their surface (zeta) potentials. It is clear from Figure 13.9 that a critical Dl value exists. Beyond the critical Dl value the deposition rate is suddenly reduced by several orders of magnitude. By increasing Dl, the energy barrier increases and the height of the energy barrier becomes the rate-determining factor. This type of sudden drop in Sh with Dl is characteristic of all collectors and is also observed during coagulation of colloidal particles (van de Ven, 1989). Figure 13.9 also shows that changing the Reynolds number from 10 to 30 has the effect of increasing Sh when deposition occurs, but has an insignificant effect when the deposition rate is small. Other examples for the effect of Dl are given by Adamczyk (1989a,b).

Figure 13.10 shows the dependence of the Sherwood number, Sh, on Reynolds number, Re, for various values of the double layer parameter, Dl, for $\bar{\lambda} = 0.4$, $Ad = 0.8$, $\kappa a_p = 58.2$, $Gr = 0.002$, and $Da = 0$ for stagnation flow collector. When $Dl = 0$, where no electrostatic repulsion exists, the flux of the particles, as represented by the Sherwood number, to the collector is high and depends on the Reynolds number, unlike the case when a high energy barrier exists for $Dl > 0$. When $Dl = 110$, the Sherwood number becomes quite small for all Reynolds numbers tested (Dabros and van de Ven, 1983).

13.3.3 Deposition due to Brownian Diffusion with External Forces: Spherical Collectors

Deposition analysis due to Brownian diffusion with external forces for spherical collectors was made by Prieve and Ruckenstein (1974). In their study, Prieve and

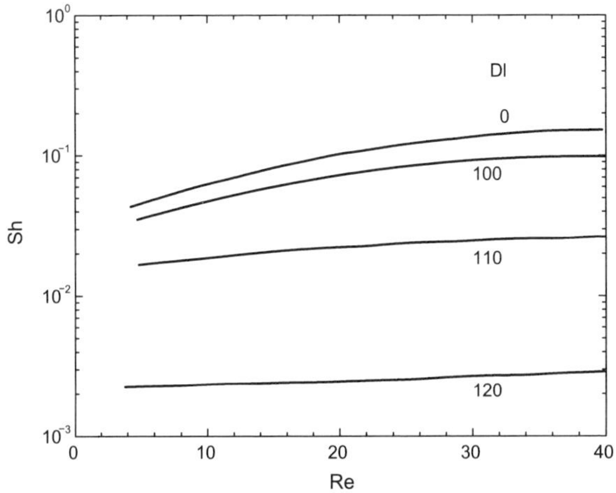

Figure 13.10. Theoretical values of the Sherwood number for stagnation flow as a function of the jet Reynolds number for $Ad = 0.8$, $\bar{\lambda} = 0.4$, $Gr = 0.002$, and $\kappa a_p = 58.2$ (Dabros and van de Ven, 1983).

Ruckenstein considered a packed bed made up of spherical collectors. They used the Spielman and Fitzpatrick (1973) approximation of Happel's cell model to describe the flow in a packed bed of spherical particles (see Figure 13.6 for coordinate system).

The exact Happel's cell model that provides the stream function of the flow field around a spherical collector embedded in the packed bed is given by

$$\psi = \frac{1}{2}U_\infty \left[\frac{A}{(r/a_c)} + B\left(\frac{r}{a_c}\right) + C\left(\frac{r}{a_c}\right)^2 + D\left(\frac{r}{a_c}\right)^4 \right] a_c^2 \sin^2\theta \tag{13.83}$$

where

$$A = 1/w \tag{13.84-a}$$

$$B = -(3 + 2\alpha_c^{5/3})/w \tag{13.84-b}$$

$$C = (2 + 2\alpha_c^{5/3})/w \tag{13.84-c}$$

$$D = -\alpha_c^{5/3}/w \tag{13.84-d}$$

and

$$w = (2 - 3\alpha_c^{1/3} + 3\alpha_c^{5/3} - 2\alpha_c^2) \tag{13.84-e}$$

Here α_c is the volume fraction of solids in the packed bed, U_∞ is the approach fluid velocity, and a_c is the collector radius.

The fluid velocity components are given by

$$u_r = -\frac{1}{r^2 \sin\theta} \frac{\partial \psi}{\partial \theta} \tag{13.85-a}$$

and

$$u_\theta = \frac{1}{r \sin\theta} \frac{\partial \psi}{\partial r} \tag{13.85-b}$$

To simplify the analysis for flow in a packed bed of spheres, Spielman and Fitzpatrick (1973) introduced a flow model parameter, A_{sph}, for the case of spherical collectors in a packed bed defined as

$$A_{sph} = \frac{2(1-\alpha_c^{5/3})}{2 - 3\alpha_c^{1/3} + 3\alpha_c^{5/3} - 2\alpha_c^2} \tag{13.86}$$

together with an approximate expression for the stream function that is valid near the collector. The approximate expression for the stream function is given by

$$\psi = \frac{3}{4} A_{sph} U_\infty \, [r - a_c]^2 \sin^2\theta \tag{13.87}$$

The factor A_{sph} accounts for the influence of neighboring spheres on the fluid flow field. The expression for A_{sph} would vary according to the model used to describe the flow over a bed of spherical collectors (Tien, 1989).

Using an approach similar to that presented in the previous section for stagnation flow, Prieve and Ruckenstein (1974) solved the flux equations equivalent to that given by Eq. (13.75) in the presence of gravity and van der Waals forces. Electrostatic forces were not included in their analysis. As their study included packed bed of spheres, the Peclet number can be conveniently defined as[3]

$$Pe = \frac{a_c U_\infty A_{sph}}{D_\infty} \tag{13.88}$$

to allow for the effects of neighboring spheres. The Sherwood number is given by Eq. (13.42). Prieve and Ruckenstein (1974) used in their analysis fluid velocities given by (see Figure 13.6 for coordinate system)

$$u_r = -A_{sph} U_\infty \left[1 - \frac{3}{2}(a_c/r) + \frac{1}{2}(a_c/r)^3\right] \cos\theta \tag{13.89-a}$$

and

$$u_\theta = A_{sph} U_\infty \left[1 - \frac{3}{4}(a_c/r) - \frac{1}{4}(a_c/r)^3\right] \sin\theta \tag{13.89-b}$$

[3]In the original paper by Prieve and Ruckenstein, (1974) Sherwood and Peclet numbers were based on the collector diameter, a_c.

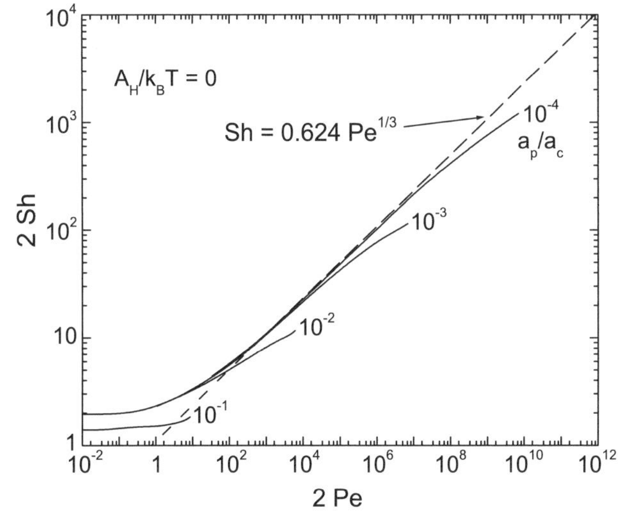

Figure 13.11. Sherwood numbers computed for the convection–diffusion of particles of finite size to the surface of a spherical collector by neglecting interaction forces. The dashed line is for the Levich equation which is valid when a diffusion boundary layer exists and the particles are infinitesimal (Prieve and Ruckenstein, 1974).

For a spherical collector, in the absence of external forces, the effect of a finite particle size on the mass transfer (deposition) is shown in Figure 13.11. Here the Sherwood number is given for the transport of particles of finite size to the surface of a spherical collector. The Levich-type equation, Eq. (13.44), which is valid for $a_p/a_c \to 0$ and when a diffusion boundary layer exists ($Pe \gg 1$), is plotted for comparison. Deviation from the Levich equation occurs when the diffusion boundary layer thickness is not small compared to the collector radius. This occurs at low values of Pe. Also deviation from the Levich equation becomes apparent when the particle size becomes large as compared with the diffusion boundary layer thickness. This situation occurs at high values of Pe.

For the case of a spherical collector under no-flow conditions, *i.e.*, $Pe = 0$, the effect of London–van der Waals forces on the Sherwood number is shown in Figure 13.12 for finite size particles. It is clear from Figure 13.12 that when a_c/a_p becomes large, $Sh \to 1$ irrespective of the values of A_H/k_BT, which are a measure of the attractive dispersion force. However, for smaller values of a_c/a_p, the dispersion force has a large influence on the Sherwood number which is a measure of the deposition (Prieve and Ruckenstein, 1974). The limit of $Sh \to 1$ as $Pe \to 0$ is that of pure diffusion of infinitesimal particles.

For the case of a spherical collector, the effect of the London–van der Waals forces on the deposition in the absence of other forces was studied by Prieve and Ruckenstein (1974). They evaluated the Sherwood number as a function of the Peclet number for different values of a_p/a_c. Figure 13.13 shows the case of $a_p/a_c = 10^{-4}$. The Levich Eq. (13.44) is shown for comparison. At low values

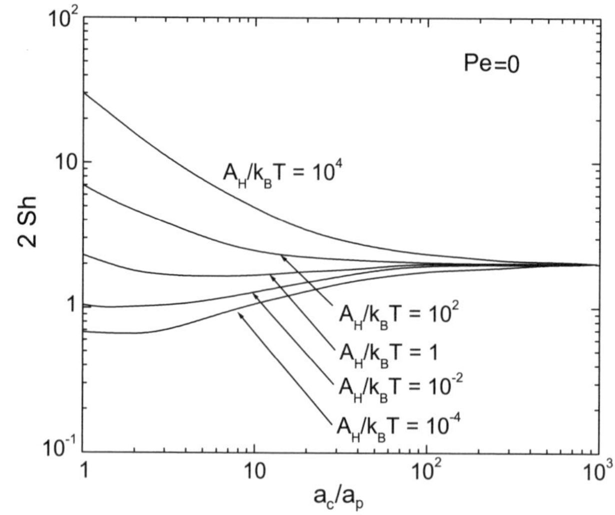

Figure 13.12. Sherwood numbers computed for the transport of finite sized particles through a stagnant fluid to a spherical collector under the action of diffusion and van der Waals forces (Prieve and Ruckenstein, 1974).

of Peclet number, Sh becomes insensitive to A_H/k_BT. This is because the deposition is diffusion-controlled. At higher values of Pe, the Sherwood number becomes more dependent on A_H/k_BT. This is due to the fact that the diffusion boundary layer becomes very thin and offers little resistance to mass transfer, consequently the

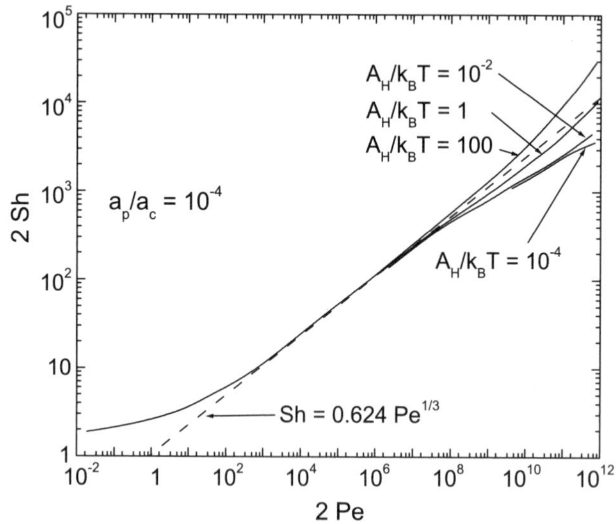

Figure 13.13. Sherwood numbers computed for the transport of finite particles to a spherical collector under the combined influence of diffusion and van der Waals forces (Prieve and Ruckenstein, 1974).

dispersion force plays a more important role in the mass transfer process. The reasons for the deviation from the Levich equation at very small and large Pe are similar to those advanced for Figure 13.11.

Plots for the case of a rotating disc are given by Prieve and Lin (1980). For large values of Pe and $(A_H/k_BT)/Pe$, the deposition Sherwood number in the absence of electrostatic and gravity forces is given by

$$Sh = 0.55\left(\frac{A_H}{k_BT}\right)^{1/3} Pe^{2/3}$$

where the terms Sh and Pe are defined in Table 13.3.

Prieve and Ruckenstein (1974) gave an approximate plot indicating the regions where dispersion or convection-diffusion is rate controlling for the case of a spherical collector. Figure 13.14 identifies the regions in which either London–van der Waals or diffusion may be neglected when calculating the deposition rate. When the diffusion boundary layer is thin, the dispersion forces control the deposition rate. On the other hand, when the diffusion boundary layer is thick, the convection–diffusion process controls the deposition rate.

For a spherical collector, the effect of gravitational forces upon the rate of deposition is shown in Figure 13.15 in the absence of an electric double-layer force. The gravitational field is in the main flow direction. The effect of gravitational force is substantial at lower values of Pe. The gravitational force is characterized by the gravitational parameter which is defined as the ratio between the gravitational force to

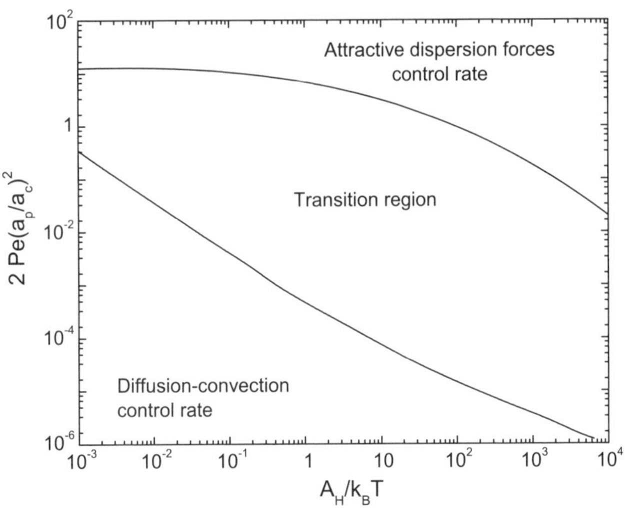

Figure 13.14. Regions of capture mechanism and controlling rate for the case of a spherical collector for $a_c/a_p > 10^3$ in the absence of electrostatic double layer forces (Prieve and Ruckenstein, 1974).

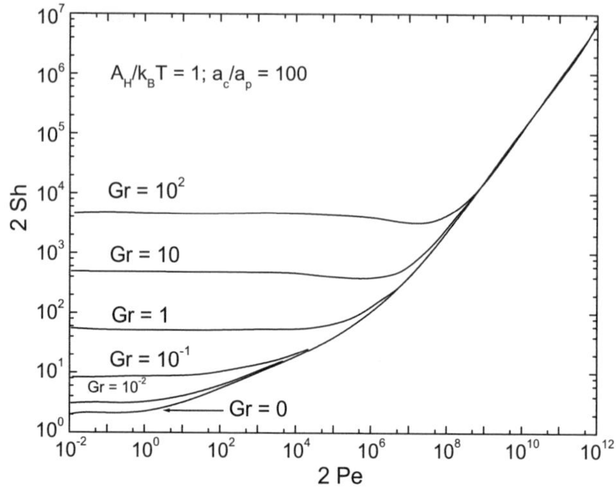

Figure 13.15. Effect of gravitational forces upon the rate of deposition for a given radius ration for the case of spherical collectors in the absence of electrostatic force (Prieve and Ruckenstein, 1974).

that of thermal Brownian force, where

$$Gr = \frac{4\pi}{3}\frac{a_p^3 \Delta\rho g}{k_B T/a_p} = \frac{4\pi a_p^4 \Delta\rho g}{3k_B T}$$

Here, $\Delta\rho$ is the density difference between the particle and the fluid.

In the limit of $Gr \gg 1$, the rate of particle deposition becomes $\pi a_c^2 v_t n_\infty$, particles/unit time where v_t is the Stokes terminal velocity of the particle given by

$$v_t = \frac{2g\Delta\rho a_p^2}{9\mu} \tag{13.90}$$

The deposition Sherwood number is then given by

$$Sh = \left(\frac{\pi a_c^2 v_t n_\infty}{4\pi a_c^2}\right)\frac{a_c}{D_\infty n_\infty} \tag{13.91}$$

leading to

$$Sh = \frac{1}{4}\left(\frac{a_c}{a_p}\right) Gr \tag{13.92}$$

The flat regions of the curves of Figure 13.15 are given by Eq. (13.92).

The question of whether adding the individual process rates produces the total effective mass transfer rate was addressed by Prieve and Ruckenstein (1974). They showed that there is good agreement between the exact solution of the transport equation and that obtained from the summation of the individual contributions to

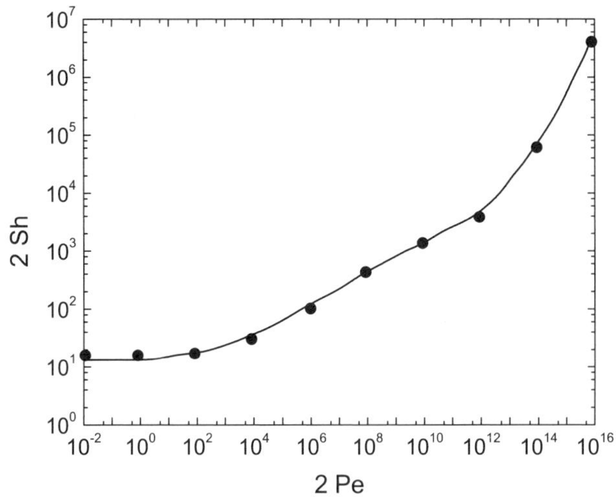

Figure 13.16. Adequacy of the additivity rule for $a_c/a_p = 10^4$ and $A_H/k_BT = 10^4$. The curve represents results obtained by summing individual contributions from gravitational forces, Brownian motion, and van der Waals force. Filled circles were obtained from simultaneously considering all effects (Prieve and Ruckenstein, 1974).

the deposition rate from each mechanism. For the case of a spherical collector with $a_c/a_p = 10^4$, and $A_H/k_BT = 10^4$, individual contributions due to gravity, Brownian motion (diffusion) and dispersion forces were summed and compared with the exact solution that considers these effects simultaneously. The comparison between the two approaches is shown in Figure 13.16. The solid circles are given by the exact solution and the solid line is from the addition of individual contributions. The agreement between the two approaches is fairly good. However, there is no theoretical basis for concluding that the concept of additivity should hold under other physical parameters. It should be recognized that the comparison is made for $A_H/k_BT = 10^4$ which gives a rather large value for A_H.

13.4 LAGRANGIAN APPROACH

In the previous section, we dealt with the deposition of particles having a finite Peclet number by solving the generalized convection-diffusion equation. Here we shall deal with particle capture (or deposition) in the absence of Brownian motion and hence diffusion, *i.e.*, $Pe \to \infty$. The special case of particle capture by a spherical collector will be treated in detail.

13.4.1 Particle Collisions on a Spherical Collector: With the Presence of External Forces

The case of particle capture by a spherical collector is of great practical importance in a variety of industrial applications, such as, in flotation columns where air bubbles are used as collectors of small mineral particles, in filtration using a packed bed, in analysis of colloid transport in groundwater, and in chromatographic separations.

In the Lagrangian approach, the flow field about the collector is assumed to be undisturbed by the presence of the colloidal particles. The fluid flow near the collector is deduced from the general expression of the flow and it is decomposed into a normal component and a tangential component with respect to the collector surface. A force balance on the colloidal particle is made in the directions normal and tangential to the collector surface. From the force balance equations, the angular variation of the gap between the colloidal particle and the surface is obtained and the trajectory equations provide the basis for the evaluation of deposition efficiency.

When the spherical collector is small, it is possible to neglect the fluid inertia and the flow field is given by the Stokes solution. The undisturbed axisymmetric flow field close to a spherical collector is approximated by

$$\psi = \frac{3}{4} A_{sph} U_{\infty} (r - a_c)^2 \sin^2\theta \tag{13.93}$$

where ψ is the stream function, U_{∞} is the undisturbed fluid velocity, a_c is the collector radius, r is the radial coordinate and A_{sph} is a dimensionless parameter characterizing the flow model used to account for the presence of other collectors. For an isolated spherical collector, $A_{sph} = 1$. For a packed bed of spherical particles, A_{sph} is given using Happel's (1958) model by Spielman and Fitzpatrick (1973) as

$$A_{sph} = \frac{2(1 - \alpha_c^{5/3})}{2 - 3\alpha_c^{1/3} + 3\alpha_c^{5/3} - 2\alpha_c^2}$$

where α_c is the collector grain volume fraction in the packed bed. Figure 13.17 shows the flow geometry together with the collector and the colloidal particle.

The fluid velocity components are related to the stream function as

$$u_r = -\frac{1}{r^2 \sin\theta} \frac{\partial \psi}{\partial \theta} \tag{13.94}$$

and

$$u_\theta = \frac{1}{r \sin\theta} \frac{\partial \psi}{\partial r} \tag{13.95}$$

Following the approach of Spielman and FitzPatrick (1973), it is possible to define a system of local cylindrical coordinates ϖ and z whose origin is on the collector surface with $r = a_c$ and θ. Here θ is the angle corresponding to the center position of the colloidal particle to be captured. With this coordinate system, the origin of the coordinates changes position as the entrained particle moves around the collector. The fluid flow near the collector in the new cylindrical coordinates is given by

$$\mathbf{u} = \mathbf{u}_{st} + \mathbf{u}_{sh} \tag{13.96}$$

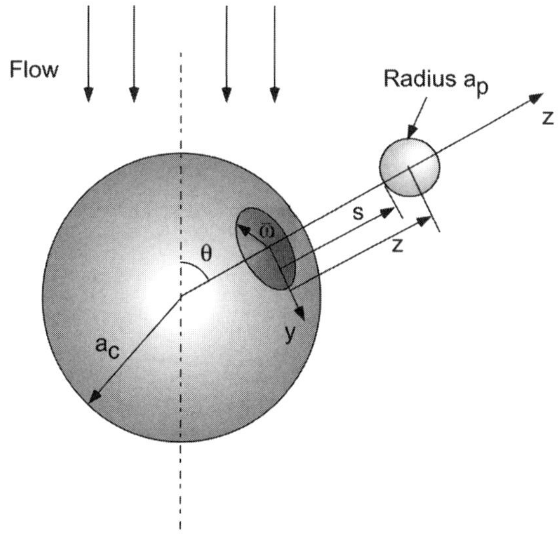

Figure 13.17. Flow geometry of a colloidal particle in the proximity of a larger spherical collector.

where

$$\mathbf{u}_{st} = \frac{3A_{sph}U_\infty \cos\theta}{2a_c^2}[\varpi z\mathbf{i}_\varpi - z^2\mathbf{i}_z] \tag{13.97}$$

and

$$\mathbf{u}_{sh} = \frac{3A_{sph}U_\infty \sin\theta}{2a_c} z\mathbf{i}_y \tag{13.98}$$

where $\mathbf{i}_y$, $\mathbf{i}_\varpi$, $\mathbf{i}_z$ are unit vectors in the y, ϖ, and z-directions, respectively.

The velocity vector $\mathbf{u}_{st}$ represents an axisymmetric stagnation flow at $\varpi = 0$, whereas $\mathbf{u}_{sh}$ represents a shear flow parallel to the collector surface. Locally, one can think of the entrained colloidal particle as being subjected to a stagnation flow and a shear flow as described by Eqs. (13.97) and (13.98). Both $\mathbf{u}_{st}$ and $\mathbf{u}_{sh}$ represent the undisturbed fluid flow velocities about the collector sphere in the absence of the colloidal particle. This decomposition of the flow field is depicted in Figure 13.18. In part (a) of Figure 13.18, the overall geometrical representation of the variation in the deposition behavior from a completely stagnation point flow regime (when $\theta = 0$ to a completely parallel flow regime (for $\theta = 90°$) is depicted. At intermediate angular positions, the deposition is governed by both stagnation and parallel flow. The stagnation flow and the shear flow components are shown in Figure 13.18(b) and (c), respectively.

The entrained particle velocity is decomposed separately into flow fields corresponding to its normal and tangential velocities. This is allowed as the flow equation is linear and the method of superposition can be used (Happel and Brenner, 1965).

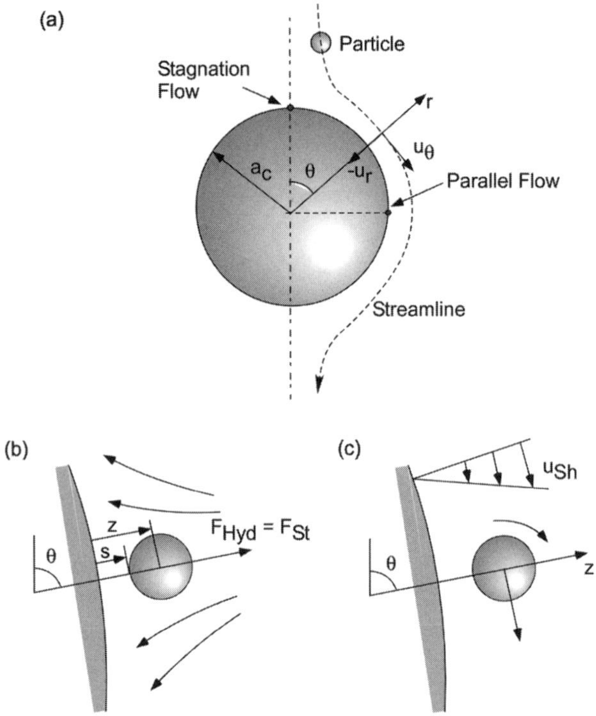

Figure 13.18. Flow geometry of a particle in the proximity of a larger spherical collector. (a) Overall geometry. The radial and angular velocity components of the fluid, u_r and u_θ, respectively, are shown for the spherical coordinate system with the collector center as the origin. This velocity field can be represented as linear superposition of two types of flow, namely, stagnation flow shown in part (b) and shear flow shown in (c).

The particle normal velocity ds/dt under the influence of a total normal force F_n is given by

$$\frac{ds}{dt} = \frac{F_n f_1}{6\pi \mu a_p} \tag{13.99}$$

where F_n is the total force acting on the entrained particle normal to the collector surface. Here f_1 is the hydrodynamic universal correction factor given by Brenner (1961). It relates to the additional retardation in the translation velocity of the colloidal particle in a stagnant fluid towards a flat surface under the influence of a force F_n. Here s is the dimensional separation distance given by

$$s = z - a_p \tag{13.100}$$

The total force acting on the entrained particle normal to the collector surface is the sum of the London–van der Waals, electric double layer, gravitational, and hydrodynamic forces. The normal force F_n is given by

$$F_n = F_A + F_R + F_{gz} + F_{Hyd} \tag{13.101}$$

The London–van der Waals force between a spherical particle and a plane is given by Eq. (13.69), the corresponding electrical double-layer force is given by Eq. (13.70), and the gravitational force is given by

$$F_{gz} = -\frac{4\pi}{3} a_p^3 \Delta\rho g \cos\theta \tag{13.102}$$

The direction of the gravity force is assumed to coincide with the main direction of the flow. The hydrodynamic force is due to the fluid stagnation flow and it is given by

$$F_{Hyd} = \left[-\frac{3(s+a_p)^2 A_{sph} U_\infty \cos\theta}{2a_c^2} \right] (6\pi\mu a_p) f_2 \tag{13.103}$$

where the term in the first square brackets of Eq. (13.103) represents the z-directed fluid velocity as given by Eq. (13.97).

To simplify the analysis we shall use the non-retarded expression for the London–van der Waals potential and assume the collector and the particle to have the same value for the zeta potential. This will allow the dropping of the second term in Eq. (13.70). Letting $h = s/a_p$, Eq. (13.99) when combined with the constitutive expressions for the normal force components, yields

$$\begin{aligned} \frac{dh}{dt}\left(\frac{6\pi\mu a_p^2}{f_1}\right) = & -\frac{2A_H}{3a_p(h+2)^2 h^2} \\ & + 4\pi\epsilon\kappa a_p \zeta_p \zeta_c \left[\frac{\exp(-\kappa a_p h)}{1 \pm \exp(-\kappa a_p h)}\right] - \frac{4\pi}{3} a_p^3 \Delta\rho g \cos\theta \\ & - \left(6\pi\mu a_p^3 f_2\right) \left[\frac{3(h+1)^2 A_{sph} U_\infty \cos\theta}{2a_c^2}\right] \end{aligned} \tag{13.104}$$

The above equation relates the variation of the dimensionless separation gap, h, with time to the various flow conditions and physical properties of the system. The separation gap is normalized by the spherical particle radius, a_p.

The motion of the entrained colloidal particle parallel to the collector (y- direction) is composed of two parts. The first part is the free rotation and translation due to the fluid shear velocity $\mathbf{u}_{sh}$. The particle velocity due to the fluid shear velocity is given by

$$v_{py1} = \left[\frac{3A_{sph} U_\infty (s+a_p) \sin\theta}{2a_c}\right] f_3 \tag{13.105}$$

The quantity in the square brackets of Eq. (13.105) is the shear velocity given by Eq. (13.98). The second part of the y-direction velocity of the particle corresponds to the free rotation and translation of the particle under the external force of gravity and it is given by

$$v_{py2} \simeq \left[\frac{f_4}{6\pi\mu a_p}\right]\left[\frac{4\pi a_p^3 \Delta\rho g \sin\theta}{3}\right] \tag{13.106}$$

The functions f_3 and f_4 are given by Goldman *et al.* (1967a,b).

The resultant particle velocity in the y-direction is given by combining Eqs. (13.105) and (13.106) leading to

$$a_c \frac{d\theta}{dt} = \frac{3a_p A_{sph} U_\infty (h+1)\sin\theta}{2a_c} f_3 + \frac{2a_p^2 \Delta\rho g \sin\theta}{9\mu} f_4 \tag{13.107}$$

The trajectory equation for the entrained particle can be obtained by eliminating the time from Eqs. (13.104) and (13.107). The trajectory equation is given by

$$\begin{aligned}\left[\frac{3}{2}(h+1)f_3 + N_G\left(\frac{a_p}{a_c}\right)f_4\right]\frac{\sin\theta}{f_1}\frac{dh}{d\theta} &= -\frac{N_A}{h^2(h+2)^2} \\ &\quad + N_A N_\zeta \kappa a_p \left[\frac{\exp(-\kappa a_p h)}{1 \pm \exp(-\kappa a_p h)}\right] \\ &\quad - \left[\frac{3}{2}f_2(h+1)^2 + N_G\right]\cos\theta\end{aligned} \tag{13.108}$$

where N_A is the attraction number defined as

$$N_A = \frac{A_H a_c^2}{9\pi\mu a_p^4 U_\infty A_{sph}} \tag{13.109}$$

N_ζ is the electrostatic (repulsion) number defined as

$$N_\zeta = \frac{6\pi\epsilon\zeta_c\zeta_p a_p}{A_H} \tag{13.110}$$

and N_G is a dimensionless gravity number defined as

$$N_G = \frac{2\Delta\rho g a_c^2}{9\mu U_\infty A_{sph}} \tag{13.111}$$

In the previous section when dealing with the deposition rate due to an impinging jet, the Brownian motion was taken into consideration. Consequently, the gravitational parameter, Gr, double layer parameter, Dl, and adhesion parameter, Ad, were defined relative to the Brownian motion thermal energy $k_B T$. As the diffusion mechanism is not considered here, the gravitational N_G, electrostatic number N_ζ, and the attraction number N_A, cannot be defined relative to $k_B T$. Consequently, even though these dimensionless numbers represent the strength of the forces to which they are named, their bases are different.

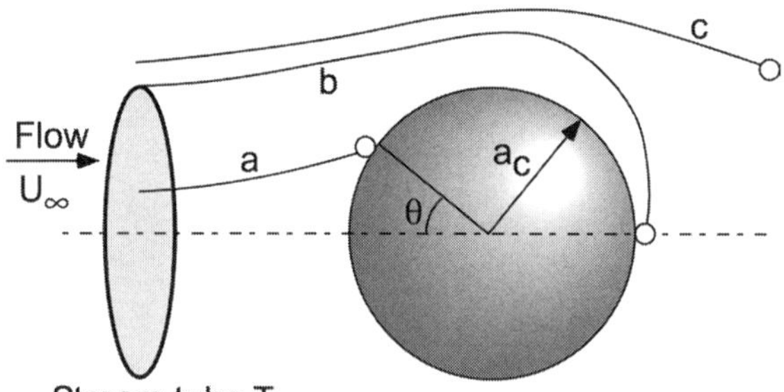

Figure 13.19. Paths of a colloidal particle: trajectory a leads to a collision, trajectory b is the limiting trajectory, and trajectory c leads to no collision.

The electrostatic repulsion number N_ζ represents the ratio of electrostatic repulsive to London–van der Waals attractive forces. The gravitational number N_G represents the ratio of gravitational force to the viscous force modified by $(a_c/a_p)^2$. Similarly, the attraction number N_A represents the ratio of the dispersion attraction force to the viscous force modified by $(a_c/a_p)^2$.

Spielman and FitzPatrick (1973) in their analysis of colloidal deposition on spherical collectors neglected the term $N_G(a_p/a_c)$, which appears on the right hand side of Eq. (13.108) and further assumed negligible electrostatic repulsion forces, *i.e.*, $N_\zeta = 0$. On the other hand, Spielman and Cukor (1973) assumed $N_G = 0$ but incorporated electric double-layer repulsion in their analysis.

Figure 13.19 shows the paths of a colloidal particle near a spherical collector. Three types of particle trajectories are shown. Particle trajectory a leads to a collision. Trajectory b is the limiting trajectory separating a collision trajectory from a non-collision trajectory. Trajectory c is a non-collision trajectory. All particles within the stream tube T would lead to a collision. The capture efficiency is given by Eq. (13.5) as

$$\eta = \frac{2\psi_L}{a_c^2 U_\infty} \tag{13.112}$$

For the case of $(a_c/a_p) \ll 1$, the stream function for the flow over a spherical collector is approximated by

$$\psi = \frac{3}{4} A_{sph} a_c^2 U_\infty \left(\frac{a_p}{a_c}\right)^2 (h+1)^2 \sin^2\theta$$

and the collision efficiency becomes

$$\eta = \frac{3A_{sph}}{2}\left(\frac{a_p}{a_c}\right)^2 \lim_{\theta\to 0}\left[(h+1)^2 \sin^2\theta\right] \tag{13.113}$$

Solution of Eq. (13.108) by backward integration from $\theta = \pi$ and $h = 0$ to large values of h with $\theta \to 0$ would then, in principle, determine the limit of $(h+1)^2 \sin^2\theta$.

Hence, the collision efficiency can be determined. In practice, however, there is an infinite attraction ($N_A \neq 0$) at $\theta = \pi$ and $h = 0$. Spielman and FitzPatrick (1973) discussed the details of the initial condition necessary in order to be able to proceed with backward integration of Eq. (13.108).

It should be recalled here that capture by pure interception for the case of a single isolated spherical collector is given by Eq. (13.7). For a collector within a packed bed of spheres, the corresponding capture efficiency is

$$\eta_I = \frac{3}{2} A_{sph} \left(\frac{a_p}{a_c} \right)^2$$

where, for an isolated spherical particle, $A_{sph} = 1$.

In the limit of a very large attraction number N_A and in the absence of gravitational and electrostatic repulsion forces, Spielman and Goren (1970, 1971) and Goren and Fitzpatrick (1973) used limiting expressions for the universal hydrodynamic functions and obtained an analytical expression for the collision efficiency for a spherical collector as

$$\eta = \left[\frac{3}{2} A_{sph} \left(\frac{a_p}{a_c} \right)^2 \right] \left(\frac{4}{3} \right) \left[\frac{9}{5} N_A \right]^{1/3} \tag{13.114}$$

for $N_A \gg 1$, with $N_G = N_\zeta = 0$.

The term appearing in the first square brackets of Eq. (13.114) is the pure interception capture efficiency for a spherical collector. Consequently, Eq. (13.114) can be written as

$$\frac{\eta}{\eta_I} = \frac{4}{3} \left[\frac{9}{5} N_A \right]^{1/3} \tag{13.115}$$

Clearly, for $N_A \gg 1$, η/η_I becomes larger than unity. The enhancement in the collection efficiency is due to the presence of the London–van der Waals attractive force.

Similarly, Spielman and Goren (1970) obtained a limiting expression for the case of a cylindrical collector:

$$\eta = \left[2A_{cyl} \left(\frac{a_p}{a_c} \right)^2 \right] \left(\frac{3\pi}{4} N_A \right)^{1/3} \tag{13.116}$$

for $N_A \gg 1$, $N_G = N_\zeta = 0$. Here the term in the square brackets is the pure interception efficiency for a cylindrical collector. The hydrodynamic flow parameter A_{cyl} is given by

$$A_{cyl} = \frac{1}{2} \left[2.0 - \ln \left(\frac{2\rho a_c U_\infty}{\mu} \right) \right]^{-1} \tag{13.117}$$

for an isolated cylinder (Lamb's solution). For a fibre mat, the corresponding expression for A_{cyl} is (Happel, 1959; Guzy *et al.*, 1983; Adamczyk and van de Ven, 1981b)

$$A_{cyl} = -\left[\ln \alpha_c + \frac{1}{(1+\alpha_c^2)}\right]^{-1} \quad \text{Happel's model} \tag{13.118}$$

where α_c is the volume fraction of the cylindrical collectors in the fibre mat. The stream function for flow past a cylinder for the case of $a_p \ll a_c$ is defined as

$$\psi = 2A_{cyl}U_\infty (r - a_c)^2 \frac{\sin\theta}{a_c} \tag{13.119}$$

Equation (13.119) is valid for $Re \leq 0.5$. For the case of a cylindrical collector, a_c becomes the radius of the cylinder. The attraction number for the case of the cylinder becomes

$$N_A = \frac{A_H a_c^2}{9\pi \mu a_p^4 U_\infty A_{cyl}}$$

and the ratio of the deposition efficiency to that of the interception efficiency for $N_A \gg 1$ becomes for the case of a cylindrical collector

$$\frac{\eta}{\eta_I} = \left[\frac{3\pi}{4} N_A\right]^{1/3} \tag{13.120}$$

Once again, η/η_I becomes larger than unity for $N_A \gg 1$.

For a spherical collector with $N_\zeta = 0$, the variation of the ratio of the capture efficiency to that due to pure interception as obtained by the numerical solution of Eq. (13.108) is shown in Figure 13.20. As would be expected, for a given value of the gravitational group N_G, the capture efficiency ratio increases with the attraction number N_A. For a fixed N_A, the capture efficiency ratio increases with increasing gravitational number N_G. The asymptotic solution given by Eq. (13.115) for $N_\zeta = N_G = 0$ is also shown in Figure 13.20. Good agreement between the numerical and the asymptotic solution is evident for $N_A \gg 1$. Such a good agreement indicates that the separation distance s does not need to be very large compared to a_p before the hydrodynamic interactions become weak (Probstein, 2003).

For a cylindrical collector, Figure 13.21 shows the variation of the ratio of the capture efficiency to that of pure interception computed using an equation corresponding to Eq. (13.108) (Spielman and FitzPatrick, 1973). The asymptotic solution given by Eq. (13.116) for $N_A \gg 1$, $N_G = N_\zeta = 0$ is also shown on Figure 13.21. Good agreement is once again present for $N_A > 1$ between the two solutions.

In the absence of gravitational force $N_G = 0$, Spielman and Cukor (1973) numerically solved the trajectory equation for the case of a spherical collector for both constant potential and constant charge cases. Here the solution is governed by the scaled double layer thickness κa_p, the attraction (adhesion) number

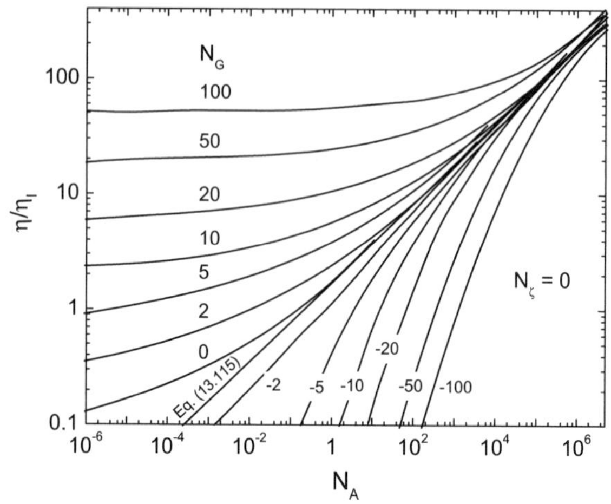

Figure 13.20. Variation of normalized capture efficiency for a spherical collector with attraction number N_A (Spielman and Fitzpatrick, 1973).

N_A, and the repulsion number N_ζ. The analysis is made using an unretarded dispersion potential. The expression for the repulsive double layer is restricted to $|\zeta_c \sim \zeta_p| < 0.05$ V and $\kappa a_p > 10$. A typical potential energy of interaction is shown in Figure 13.22.

Depending on the values of κa_p, N_A, and N_ζ, four possible modes of capture are predicted. For the case of $\kappa a_p \gg 1$ and constant charge, Figure 13.23 shows the four modes of capture. In zone I, where no capture is possible, the particles are unable to surmount the repulsive energy barrier as shown in Figure 13.22. Here all particles

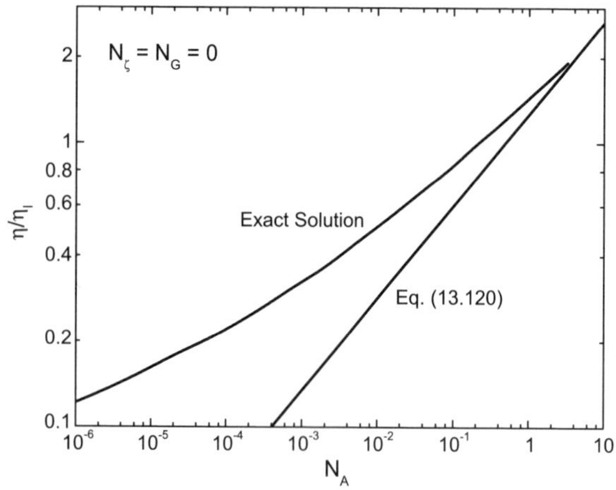

Figure 13.21. Variation of normalized capture efficiency with attraction number N_A for capture of neutrally buoyant particles by a cylinder. Hydrodynamic interactions and van der Waals attraction are incorporated (Spielman and Fitzpatrick, 1973).

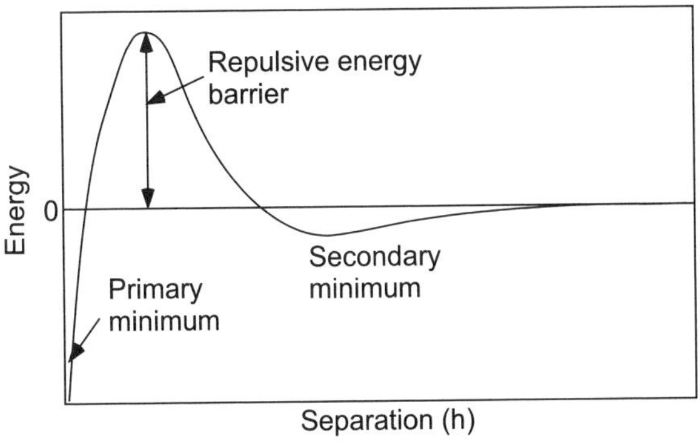

Figure 13.22. Variation of combined London–van der Waals attraction and electric double layer repulsion potentials with separation distance between a colloidal particle and a collector.

escape capture. In zone II, capture occurs at the primary minimum. In this region, $N_\zeta/\kappa a_p$ is small indicating a weak repulsive energy barrier. In zone III, capture occurs at the secondary minimum. Here, no approaching particles are carried over the repulsive energy barrier. The secondary minimum is sufficiently deep to prevent some particles from escaping. In zone IV, a small region is present where combined capture by the primary and secondary minima occurs (Spielman, 1977).

The effect of the dimensionless inverse Debye length κa_p on the location of the four zones is shown in Figure 13.24 for the case of constant surface charge and in Figure 13.25 for the case of constant surface potential.

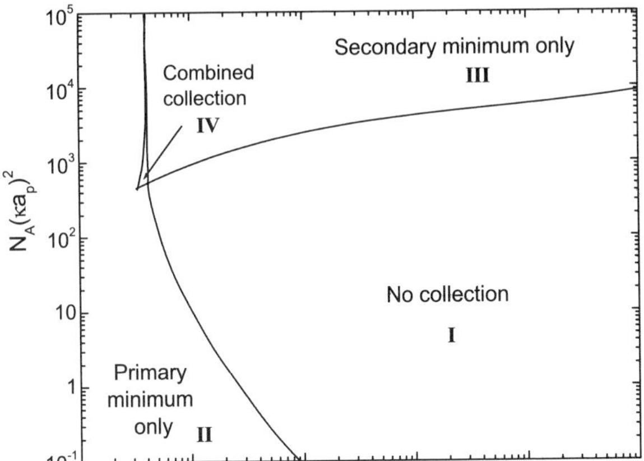

Figure 13.23. Modes of capture of non-diffusing particles by a spherical collector at constant surface charge with $\kappa a_p \gg 1$. Hydrodynamic interactions, London–van der Waals attraction, and double layer repulsion are incorporated with $N_G = 0$ (Spielman and Cukor, 1973).

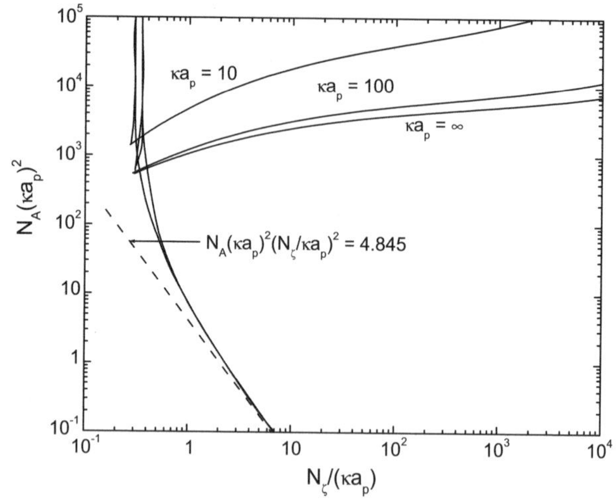

Figure 13.24. Regions of capture modes for non-diffusing particles by a spherical collector at constant surface charge for selected values of κa_p for $N_G = 0$ (Spielman and Cukor, 1973).

Figure 13.26 illustrates the variation of the capture efficiency (normalized with pure interception capture efficiency) with the attraction number at fixed repulsion numbers. Recall that the repulsion number N_ζ represents the ratio of repulsive forces to attractive forces. In other words, for a constant non-zero value of N_ζ, the ratio of the repulsive to the attractive forces is fixed. Consequently, for a given system, where the physical properties are held constant, a change in N_A is derived from changes in the flow velocity U_∞. At $N_\zeta = 0$, the capture efficiency increases with N_A as

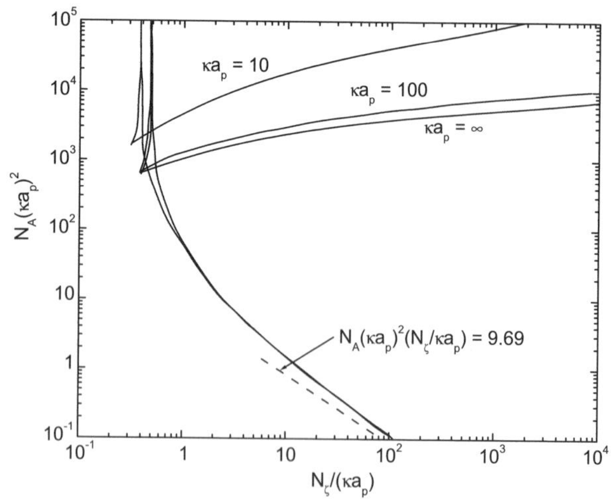

Figure 13.25. Regions of capture modes for non-diffusing particles by a spherical collector at constant potential for selected values of κa_p for $N_G = 0$ (Spielman and Cukor, 1973).

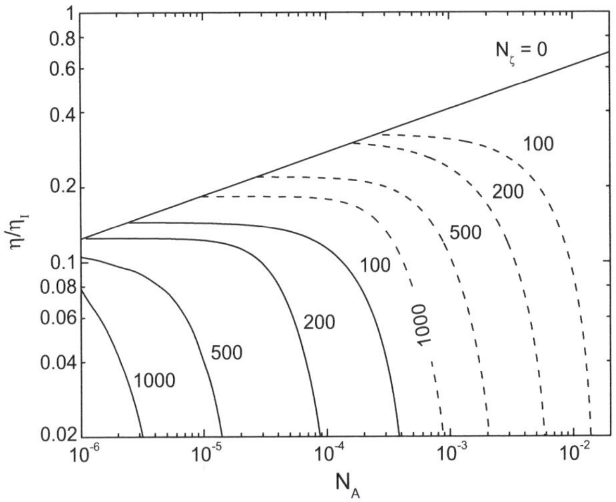

Figure 13.26. Normalized efficiency *vs.* attraction number at fixed selected repulsion numbers and $\kappa a_p = 10$ for the spherical collector: —— constant charge; – – – constant potential (Spielman and Cukor, 1973).

a consequence of increasing the attractive forces. A high capture efficiency can be achieved either by increasing the value of the Hamaker constant or decreasing the flow velocity. All the curves for non-zero N_ζ values give a lower capture efficiency. This is to be expected due to the presence of a repulsive force. Consider now a given value of N_ζ, say $N_\zeta = 200$ for the constant charge case. As stated earlier, a change in N_A can be attributed to a change in the fluid velocity which may be capable of propelling a particle over the repulsive energy barrier to be captured at the collector surface. Indeed, for $N_\zeta = 200$, the capture efficiency is slightly below that for $N_\zeta = 0$ at $N_A < 10^{-5}$. However, as N_A increases, it leads to a weaker flow which may not be able to push particles over the repulsive energy barrier. Consequently, increasing N_A leads to a lower capture and, finally, when N_A is sufficiently large, no capture occurs.

13.5 DEPOSITION EFFICIENCY AND SHERWOOD NUMBER

The deposition rates were reported in terms of Sherwood numbers when the analysis was made using the Eulerian approach. However, capture or deposition efficiency, η, was used when dealing with the Lagrangian approach. The Sherwood numbers for the case of the spherical collector reported in Section 13.3 can be cast in terms of the capture efficiency. The establishment of a relationship between the capture efficiency and the Sherwood number for the case of a spherical collector is given below.

Recall that the average Sherwood number was defined (see Table 13.3) as

$$Sh = \frac{1}{S}\int_S \left(\frac{j_\perp a_c}{D_\infty n_\infty}\right) dS = \frac{a_c}{S D_\infty n_\infty}\int_S j_\perp dS \tag{13.121}$$

The integral $\int_S j_\perp dS$ is given by

$$\int_S j_\perp dS = 2\pi \psi_L n_\infty \tag{13.122}$$

where $2\pi \psi_L n_\infty$ represents the total flux of the particles. Here ψ_L represents the limiting stream function for a given capture mechanism.

Combining Eqs. (13.121) and (13.122) leads to

$$Sh = \left(\frac{a_c}{SD_\infty n_\infty} \right) (2\pi \psi_L n_\infty) \tag{13.123}$$

From the definition of the capture efficiency for the case of a spherical collector as given by Eq. (13.5) and by setting $S = 4\pi a_c^2$, Eq. (13.123) becomes

$$Sh = \left(\frac{a_c}{4\pi a_c^2 D_\infty n_\infty} \right) (\eta \pi a_c^2 U_\infty n_\infty)$$

leading to

$$Sh = \frac{\eta Pe}{4} \tag{13.124}$$

where $Pe = a_c U_\infty / D_\infty$ for a spherical collector. Equation (13.124) relates the Sherwood number of a spherical collector to its capture efficiency.

The equivalent expression for the case of a cylindrical collector is given as

$$Sh = \frac{\eta Pe}{\pi}$$

Using Eqs. (13.124), (13.44), and (13.47), one can obtain an expression for the deposition efficiency for purely diffusive (Brownian) transport, η_D, around an isolated spherical collector as

$$\eta_D = 4Sh/Pe = 4[0.624 Pe^{-2/3}] \tag{13.125}$$

When the deposition is dominated by gravitational force, using Eqs. (13.92), (13.124), and the Stokes–Einstein equation, (12.13), together with Eq. (12.7) we can define a deposition efficiency due to gravity, η_G, as

$$\eta_G = \frac{2}{9} \frac{\Delta\rho g a_p^2}{\mu U_\infty} \tag{13.126}$$

Finally, the deposition efficiency due to interception is given by Eq. (13.7)

$$\eta_I = \frac{3}{2} \left(\frac{a_p}{a_c} \right)^2 \tag{13.127}$$

An overall deposition efficiency, η_0, can be obtained by adding the deposition efficiencies due to diffusion, gravity and interception

$$\eta_0 = \eta_D + \eta_G + \eta_I \tag{13.128}$$

Such correlations were developed for filtration by Yao *et al.* (1971), who based their work on earlier developments in aerosol filtration (Friedlander, 1958; Fuchs, 1964). Subsequently, correlations for the deposition efficiency were modified to include the effects of the van der Waals attraction (Rajagopalan and Tien, 1976). A more complete description of these correlations can be found in Elimelech *et al.* (1995). One should note, however, that the simple additivity of the different mechanisms for deposition is an approximation, and although it is found to be quite accurate for most applications, such additivity of different mechanisms may not be fundamentally sound. Furthermore, these correlations do not account for the influence of electric double layer interactions on the deposition efficiency. Thus, the numerical methodologies outlined in the previous sections are preferable when more accurate results are desired.

The correlation given by Eq. (13.128) predicts the dependence of the deposition efficiency on the particle size, and provides a means for assessing the dominant mechanisms for the deposition process. Figure 13.27 depicts the variation of the overall deposition efficiency with particle radius obtained using Eq. (13.128), as well as the individual contributions due to diffusion, gravity, and interception. It is interesting to note that there is a minimum in the deposition efficiency observed for a

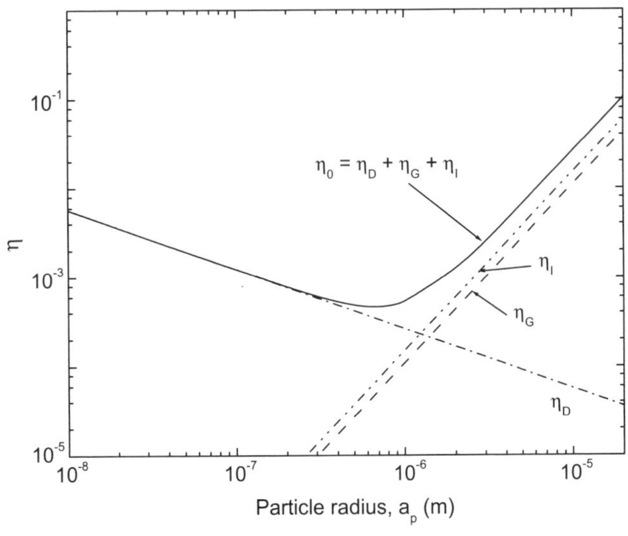

Figure 13.27. Dependence of the deposition efficiency on particle radius. The solid line shows the overall deposition efficiency, while the other lines depict the contributions due to diffusion, interception, and gravity. $U_\infty = 2.0 \times 10^{-3}$ m/s, $a_c = 0.1$ mm; $\rho_p = 1100$ kg/m^3; $\rho = 1000$ kg/m^3 and $T = 298$ K.

particle radius of about 0.6–0.7 μm. Particles smaller than *ca.* 0.2 μm exhibit purely Brownian (diffusion controlled) deposition, while for $a_p > 1.5$ μm, the deposition process is governed solely by interception and gravity. The particle size dependence of η_0 shown in Figure 13.27 is only applicable in absence of electric double layer interactions. In presence of other interactions, the deposition efficiency changes quite markedly. However, the presence of a minimum deposition efficiency corresponding to a particle radius of 0.5–1.0 μm is almost universally observed. This size dependence is of immense importance in filtration, since the behavior shown in Figure 13.27 indicates that the filtration will be least efficient (capture efficiency will be lowest) for particle radii in the range of 0.5 to 1.0 μm.

13.6 EXPERIMENTAL VERIFICATIONS

In the previous sections we discussed the theoretical development of the deposition of colloidal particles on a surface. The mass transfer rate was given by dimensionless groups, either in terms of a collection efficiency η or in terms of a Sherwood number Sh. For the case of a spherical collector, the relationship between η and Sh was derived and it is given by Eq. (13.124).

For the case of deposition in a stagnation flow, spinning disc, and channel flow, the theoretical descriptions of the flow field are well established. On the other hand, for the case of flow in a packed bed or in a fibre mat, the flow field is deduced using approximate flow models. Consequently, any disagreement between the experimental and theoretical deposition rates in the case of a well-defined flow geometry can only be attributed to the manner by which interparticle forces are modelled. Hence, experimental data from well-defined flow geometries are valuable for establishing the validity of the DLVO theory and deposition models.

Hull and Kitchener (1969) studied the deposition of polystyrene particles ($a_p = 0.154 \times 10^{-6}$ m) onto a rotating disc under laminar flow conditions. In their first set of deposition experiments, the surface of the rotating disc was covered with polyvinylpyridine copolymer film giving a zeta potential of 0.072 V. The polystyrene particles zeta potential was −0.070 V. For this case of opposite signs in the zeta potential, the experimentally measured deposition rate was very close to that given by the Levich theory, where

$$j_{\perp} = 0.62 D_{\infty}^{2/3} \left(\frac{\rho}{\mu}\right)^{1/6} \omega^{1/2} n_{\infty} \tag{13.129}$$

The Levich equation represents the case where the deposition process is purely convection-diffusion (mass-transfer) controlled in the absence of interparticle forces. The agreement of the experimental results with the Levich theory is not surprising as, for the system studied, the adhesion number Ad is of order 10^{-1} and the double-layer Dl is negative. For the case of a stagnation flow, Figure 13.9 showed that when Dl is small or negative, the Sherwood number for a given flow Reynolds number is not sensitive to variations in Dl. Equation (13.129) gives

$$\frac{j_{\perp}}{n_{\infty}\omega^{1/2}} = 0.62 D_{\infty}^{2/3} \left(\frac{\rho}{\mu}\right)^{1/6} \tag{13.130}$$

The right side of Eq. (13.130) is a constant for a given system. Making use of the definitions given in Table 13.3, Eq. (13.130) leads to

$$\frac{Sh}{Re^{1/3}} = 0.62\, Sc^{1/3} \tag{13.131}$$

In this case, the local and the average Sherwood numbers are the same.

For negative and small values of Dl, the curves of Figure 13.9 conform to the relationship given by Eq. (13.131) and in turn to Eq. (13.130). According to Hull and Kitchener (1969) experimental measurements for the case of opposite signs in zeta potential, the particles flux is given as

$$\frac{j_{\perp}}{n_{\infty}\omega^{1/2}} = 7.59 \times 10^{-8}$$

(in $\mathrm{m/s^{1/2}}$) whereas the Levich theory gives $7.66 \times 10^{-8}\ \mathrm{m/s^{1/2}}$. The agreement between the experimental measurements and the Levich theory is excellent.

Hull and Kitchener carried out further experimental tests where the zeta potential of the surface of the rotating disc is rendered negative by coating the disc surface with Formvar. Here both the particles and the disc surface have the same signs for the zeta potential. In this case, there is an electrostatic repulsion between the polystyrene particles and the disc surface. The experimental results showed poor agreement with the deposition theory which takes into account the dispersion and electrostatic forces.

Clint *et al.* (1973) carried out similar experiments to those conducted by Hull and Kitchener. Clint *et al.* regulated the zeta potential of the polystyrene particles ($a_p = 1 - 0.214 \times 10^{-6}$ m) and the rotating disc surface by the addition of $Ba(NO_3)_2$. The zeta potential of the particles varied from about −0.025 to −0.012 V and the deposition surface zeta potential from −0.005 to −0.007 V by changing the electrolyte concentration from 5 to 20 $\mathrm{mol/m^3}$, respectively.

When Clint *et al.* (1973) accounted for the formation of doublets and triplets within the bulk suspension for the case of high electrolyte concentration, they were able to obtain reasonable agreement with the deposition theory where dispersion and electrostatic forces are accounted for. However, poor agreement with theory was obtained at low electrolyte concentrations where the repulsive forces are strongest. Clint *et al.* pointed out that the deposition rate was very sensitive to the value used for the zeta potential of the deposition surface. Changing ζ_c from −0.0061 to −0.0062 V had the effect of changing the deposition rate by 16%. The change in ζ_c corresponds to 0.1 mV which is well within experimental error for a zeta potential measurement.

Bowen and Epstein (1979) carried out deposition experiments from a flowing suspension of silica particles in a parallel-plates channel. The deposition corresponding to their experimental Run II-2 is shown in Figure 13.28. Table 13.4 gives details

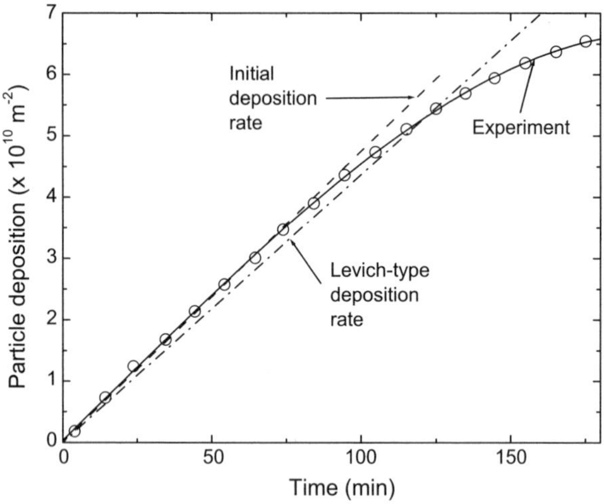

Figure 13.28. Deposition onto a positive 2VP/S substrate in a parallel plate channel. Run II-2 (Bowen, 1978).

of their experimental run. For this test, the silica particles and the deposition surface have opposite signs for the zeta potential. It is clear that the initial variation of the concentration of deposited particles with time is linear and the deposition rate declines with time. The experimental initial deposition rate was found to be $7.87 \times 10^6\ \mathrm{m^{-2}s^{-1}}$. The theoretical value as calculated by Adamczyk and van de Ven (1981b) is given as $7.63 \times 10^6\ \mathrm{m^{-2}s^{-1}}$. Again excellent agreement between the experimental and theoretical deposition rates exists. As in the case of the rotating disc deposition

TABLE 13.4. Physical Data for Run II-2 of Bowen (1978) and Bowen and Epstein (1979).

Silica particle radius, a_p	0.324×10^{-6} m
Zeta potential of particle, ζ_p	-0.075 V
Zeta potential of deposition surface, ζ_c	0.015 V
Temperature, T	298 K
Axial distance, x	0.125 m
Channel half-width, b	4.26×10^{-4} m
Particle number concentration, n_∞	0.795×10^{14} m^{-3}
Average axial velocity, V_m	0.149 m/s
Counterion concentration	6.8×10^{-6} mol/L
Hamaker constant (assumed), A_H	1×10^{-21} J
Diffusion coefficient, D_∞	7.53×10^{-13} m^2/s
κa_p	2.77
$Dl = 4\pi\epsilon a_p\zeta_p\zeta_c/k_BT$	-780
$Ad = A_H/6k_BT$	0.04
$Pe = 3V_m a_p^3/2b^2 D_\infty$	0.056
Channel Reynolds number, $4V_m b\rho/\mu$	284

measurements, for the case of opposite signs for the zeta potential, the deposition rate is good in agreement with Levich-type analysis. For the case of slit channel flow, the theoretical analysis was conducted by Levesque (1928) (see Adamczyk and van de Ven, 1981a,b; Bowen *et al.*, 1976). The Levesque-type local deposition rate in a channel is given by

$$j_{\perp} = \frac{n_{\infty} D_{\infty}}{b\Gamma(4/3)} \left(\frac{2}{9\gamma}\right)^{1/3} \tag{13.132}$$

where

$$\gamma = \frac{2D_{\infty}x}{3V_m b^2} \tag{13.133}$$

and

$$D_{\infty} = \frac{k_B T}{6\pi \mu a_p} \tag{13.134}$$

Here V_m is the average velocity in the channel, b is the channel half-width, x is the axial downstream distance at which the deposition measurements are taken, and $\Gamma(4/3)$ is the Gamma function given as 0.893.

The Levich deposition rate is given as 7.22×10^6 m^{-2} s^{-1} which agrees quite well with the experimentally determined value. In dimensionless form, Eq. (13.132) is given by

$$Sh_l = \frac{a_p j_{\perp}}{n_{\infty} D_{\infty}} = 0.678 \left(\frac{Pe}{\bar{x}}\right)^{1/3} \tag{13.135}$$

where

$$\bar{x} = \frac{x}{b} \tag{13.136}$$

The Peclet number is given after Adamczyk and van de Ven (1981a) as

$$Pe = \frac{3V_m a_p^3}{2b^2 D_{\infty}} \tag{13.137}$$

Similar to the previous studies with a rotating disc, poor agreement for deposition was obtained when the deposition channel wall had a zeta potential ζ_c of the same sign as the colloidal particles. ζ_c varied from -0.009 V to -0.046 V. The theoretical analysis predicted no deposition whereas experimentally significant deposition was observed.

In summary, when the electrostatic forces are attractive, good agreement was observed between the theoretical analysis and the experimental measurements. Hence, the mass transfer is not much affected by either the attractive dispersion or the attractive electrostatic forces. In most cases, Levich-type analysis predicts the deposition rate quite adequately. However, when strong repulsive forces due to the electric double layer are present, large deviations between the theoretical analysis and the experimental measurements occur. There are many possible explanations for these discrepancies. As was indicated earlier, when electrostatic repulsive forces are present, the deposition rates become very sensitive to the exact value of the zeta potential used. Moreover, unlike an idealized surface, real surfaces have small-scale roughness elements and a non-uniform distribution of charge sites on the surface. The interaction between a given particle and a substrate is strongly dependent upon the nature of the local areas coming into contact (Bowen and Epstein, 1979). Kihira *et al.* (1992) were able to develop a mathematical model that accounts for the discreteness of the surface charge where it was assumed that it followed a Gaussian distribution with an assigned mean and a standard deviation. The mean Stern potential is assigned to the surface zeta potential. The standard deviation becomes a measure of the heterogeneity of the surface. By adjusting the standard deviation of the surface potential, they were able to match their experimental coagulation stability ratio with their theoretical model. To this end, a new theory of electric double layer interactions is required that can account for heterogeneities of the geometry and the charge distribution as applied to real surfaces.

Sanders *et al.* (2003) conducted stagnation point flow deposition experiments with silane-treated (hydrophobic) silica particles ($a_p = 0.47$ μm) onto silane-treated (hydrophobic) collectors at a pH of 4 for different electrolyte (NaCl) concentrations. As shown in Figure 13.29, the experimental deposition data lie much above the DLVO predictions calculated for 0.1 M NaCl solution. The particle deposition is nearly independent of NaCl concentration and appears to depend predominantly upon the flow rate as characterized by the jet Reynolds number, see Table 13.3. Their data suggest that an additional attractive particle-collector interaction that is not

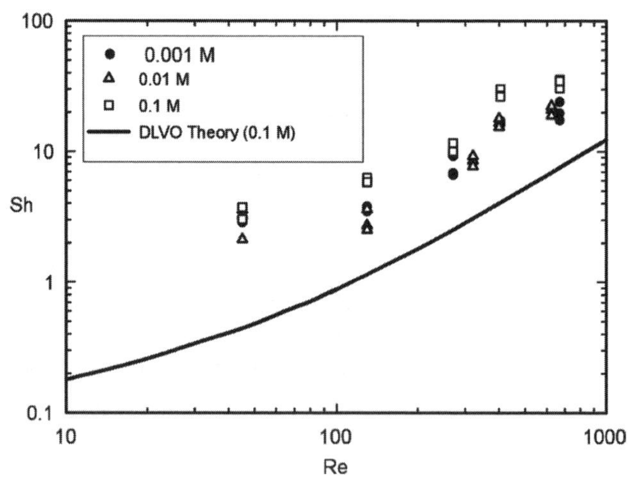

Figure 13.29. Dimensionless mass transfer (expressed as Sherwood number) as a function of NaCl concentration and Reynolds number for silane-treated (hydrophobic) silica particle deposition onto a silane-treated (hydrophobic) collector: $a_p = 47$ μm; pH = 4; $Ad = 1.75$; $Gr = 0.187$. Solid line shows mass transfer rates predicted using DLVO theory for 0.1 M NaCl. The zeta potential of silane-treated silica are −38, −21, and −11 mV at 0.001, 0.01, and 0.1 M NaCl, respectively (Sanders *et al.*, 2003).

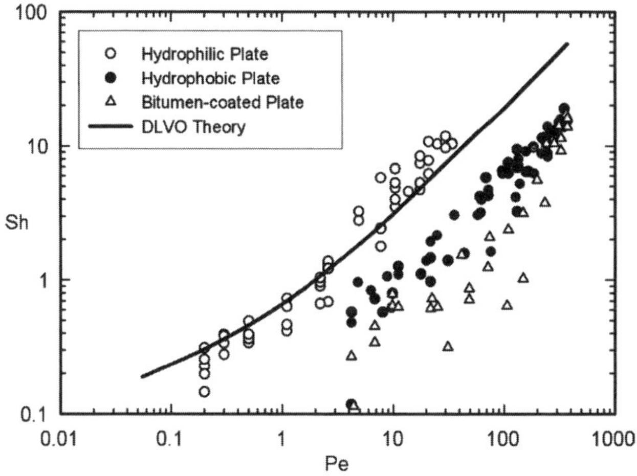

Figure 13.30. Effect of collector surface properties on the rate of bitumen droplet deposition due to jet impingement (Sanders *et al.*, 2003).

included in the DLVO analysis, exists. This observation is in agreement with other studies of hydrophobic interactions (Israelachvili, 1992; Israelachvili and Pashley, 1984; Rabinovich and Yoon, 1994; Song, *et al.*, 2000) and is also in agreement with more specific studies of hydrophobic silica particles (Xu and Yoon, 1989, 1990).

Sanders *et al.* (1995, 2003) also conducted deposition experiments using bitumen droplets on silane-treated, non-treated, and bitumen coated impingement plates. They found excellent agreement with the DLVO theory while using the non-treated hydrophilic impingement plate. However, a lower deposition rate than theoretically predicted by the DLVO theory was found for the silane-treated hydrophobic plate and bitumen-coated plate. Their findings are shown in Figure 13.30. It is clear that in deposition experiments, not only the surface charge can play a role in modifying the deposition rate, but also the nature of the surface hydrophobicity as well as any additional steric interaction.

For the case of deposition in packed beds, FitzPatrick and Spielman (1973) conducted a very extensive experimental study for the deposition of latex particles in beds of glass spheres. The colloidal particles had a wide range of size with a_p being in the range of 0.36–11 $\times$ 10^{-3} μm and the collector size a_c being in the range of 0.05–2 mm. For the case of a negligible double-layer repulsion, *i.e.*, small N_ζ numbers and negligible gravity numbers N_G, they plotted the dimensionless filter coefficient λ/λ_I versus the attraction number N_A as shown in Figure 13.31. The filter coefficient λ is defined by Eq. (13.15) and the pure interception filter coefficient λ_I is given by

$$\lambda_I = \frac{9\alpha_c A_{sph} a_p^2}{8a_c^3} \tag{13.138}$$

The flow model correction factor A_{sph} used in the attraction number N_A is due to Happel (1958) and it is given by Eq. (13.10). The solid line on Figure 13.31 is from the

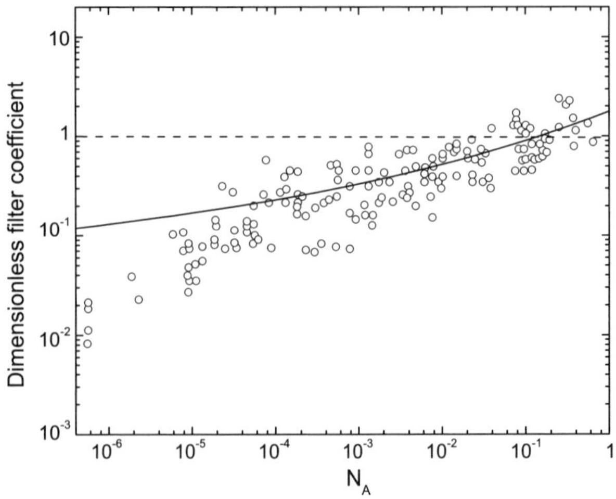

Figure 13.31. Variation of the dimensionless filter coefficient λ/λ_I with the attraction number for negligible repulsion and sedimentation. Solid line: Theoretical analysis with $N_\zeta = N_G = 0$ (adapted from Fitzpatrick and Spielman, 1973). The dimensionless filter coefficient is given by $\lambda/\lambda_I = 8a_c^3\lambda/(9\alpha_c A_{sph}a_p^2)$ and the attraction number is given by $N_A = A_H a_c^2/(9\pi\mu a_p^4 U_\infty A_{sph})$.

solution of Eq. (13.108). Good agreement between the experimental and theoretical deposition results is evident even though to some extent the theoretical analysis is dependent on the choice of the flow model A_{sph}.

When repulsion forces are not negligible, considerable scatter in η/η_I versus N_A plot was observed and the data could not be correlated with the repulsion number N_ζ.

The effect of sedimentation on λ/λ_I is shown in Figure 13.32 for $N_G = 4.3$, 12.9, and 43.2. Curves for $N_G = 0$ are also shown for comparison purposes. The solid curves are due to theory, Eq. (13.108). It is clear that there is good agreement between the theory and the experimental data. The dashed lines give the dimensionless filter coefficient for the combined interception and gravity. They are given by

$$\frac{\lambda}{\lambda_I} = 1 + \frac{2}{3}N_G \quad \text{for } N_G > -3/2 \tag{13.139}$$

The development of the above equation does not include the hydrodynamic interaction between a colloidal particle sand the collector surface. At higher N_G numbers, the exact solution and that given by Eq. (13.139) become closer.

Vaidyanathan and Tien (1988) re-examined FitzPatrick's (1972) data by incorporating non-uniform potentials in the trajectory analysis. The average collection efficiency is defined as

$$\bar{\eta} = \int_{-\infty}^{\infty}\int_{-\infty}^{\infty} f_p(\zeta_p)\, f_c(\zeta_c)\, \eta(\zeta_p, \zeta_c)\, d\zeta_p\, d\zeta_c \tag{13.140}$$

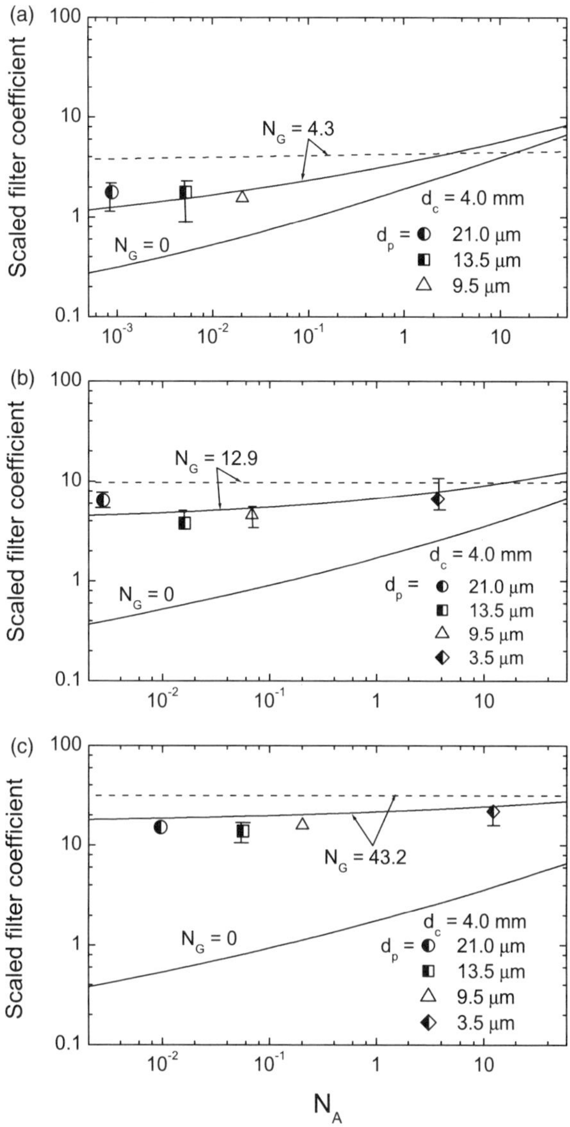

Figure 13.32. Variation of the dimensionless filter coefficient with attraction number for different gravity numbers. Solid line: theoretical analysis; dashed line: Eq. (13.139) (Spielman and Fitzpatrick, 1973). The dimensionless filter coefficient is given by $\lambda/\lambda_I = 8a_c^3\lambda/(9\alpha_c A_{sph}a_p^2)$ and the attraction number is given by $N_A = A_H a_c^2/(9\pi\mu a_p^4 U_\infty A_{sph})$.

where $f_c(\zeta_c)$ and $f_p(\zeta_p)$ are probability density functions of the collector and particle zeta potential, respectively. $\eta(\zeta_p, \zeta_c)$ is the collection efficiency when the collector has a zeta potential ζ_c and the particle ζ_p. Figure 13.33 shows the filter coefficient variation with the ionic strength. The collector and particle zeta potentials are included

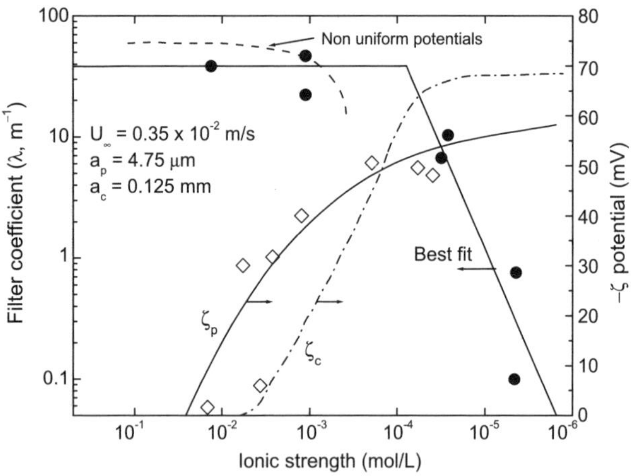

Figure 13.33. Filter coefficient variation with electrolyte ionic strength (Fitzpatrick and Spielman, 1973). Non-uniform potentials type solution from Vaidyanathan and Tien (1988).

in the plot. The agreement between the prediction of Vaidyanathan and Tien with the experimental data of FitzPatrick (1972) for the filter coefficient λ is found to be reasonably good at the onset of unfavorable surface interactions but poor as the ionic concentrations decline further.

Guzy *et al.* (1983) showed that particle retention on unconsolidated fibrous porous media depends on the flow model used in estimating the flow field within the porous bed. The various flow models, *e.g.*, the Happel, the Kuwabara, and the Brinkman cell models, lead to small but very significant differences in the streamlines near the collector surface. The differences are most pronounced for lower porosity beds of filters.

The sensitivity of the collision efficiency to the flow model and electric double-layer thickness was illustrated by Guzy *et al.* (1983). Figure 13.34 shows the sum of the London–van her Waals attractive force F_A and the electric double layer repulsive force F_R for $Dl = 100$, $Ad = 1.0$, and $\bar{\lambda} = 0.4$ with $Da = Gr = 0$. Plots for various κa_p ranging from 2 to 80 are shown. The total interaction force has a repulsive peak corresponding to all values of κa_p shown except for $\kappa a_p = 80$. As κa_p is increased, the location of the repulsive peak shifts closer to the collector surface ($h \to 0$). The magnitude of the peak first increases with increasing κa_p, and then decreases for $\kappa a_p > 30$. It is of interest to compare the force profiles corresponding to $\kappa a_p = 5$ and 50 (also $\kappa a_p = 2$ and 57), since the maxima of the force curves are quite similar ($F_z^* \sim 140$ for $\kappa a_p = 5$ and 50, while $F_z^* \sim 70$ for $\kappa a_p = 2$ and 57). However, the maximum values occur at different separation gaps, h. Due to the lubrication effects, the "attractive" hydrodynamic force is weaker near the collector surface than farther out. Thus, for an expanded double layer, $\kappa a_p = 2$, the hydrodynamic force is sufficient to move the particle through the energy barrier as opposed to the case of $\kappa a_p = 57$. A small change in the value of the normal velocity near the collector surface (*i.e.*, the position of the streamline) can make a difference as to whether capture will occur

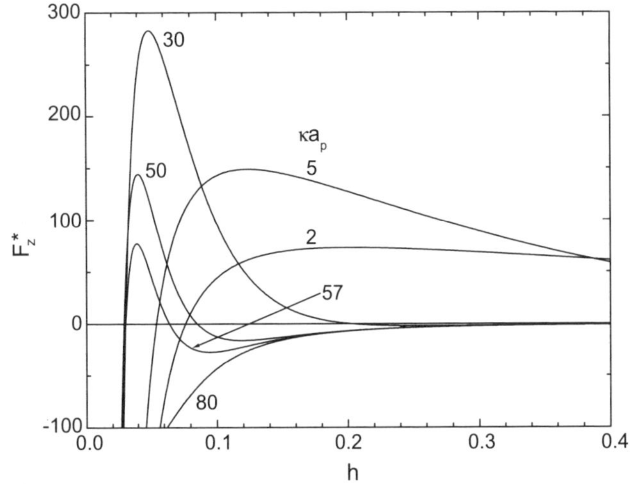

Figure 13.34. Variation of the sum of the electric double layer and London–van der Waals dimensionless forces as given by Eq. (13.76) with dimensionless gap length.

or not. This is why an accurate prediction of the flow field is crucial in determination of the deposition efficiency.

It is clear from the comparison between the experimental measurements and the theoretical analysis that there is much room for improvement in the DLVO theory as applied to real surfaces and in the flow models for deposition in filter beds (Hirtzel and Rajagopalan, 1985). This is particularly important for repulsive electric double layer interactions and surface hydrophobicity, for which case, a consistent deviation between the theory and the experiments have been reported. For a comprehensive treatise on granular filtration and deposition of particles, the reader is referred to Tien (1989) and Elimelech *et al.* (1995).

13.7 APPLICATION OF DEPOSITION THEORY

Theoretical analysis of particle deposition is a mature subject, and the foregoing discussion in this chapter provides a summary of the fundamental aspects of the theory. The primary driving factors for studies on particle deposition are the immense importance of the subject in a variety of industrial and environmental applications. Although the basic modeling approach still remains firmly rooted in the Eulerian or trajectory analysis, several modifications in the theoretical construct are needed before one can apply these models to predict deposition processes in real systems. Here we summarize some of the key directions.

13.7.1 Deposition in Porous Media

Hydrodynamic Dispersion in Porous Media The Eulerian approach presented earlier considered diffusion of the particles. However, we noted in Chapter 8 that hydrodynamic dispersion in narrow channels can cause considerable variations in the solute (particle) distribution. A straightforward modification of the particle transport equation in porous beds of collectors involves replacing the diffusion coefficient with the hydrodynamic dispersion coefficient. The hydrodynamic dispersion coefficient, D_h, for a one dimensional flow is given by

$$D_h = D_\infty + \alpha_L V_p \tag{13.141}$$

where D_∞ is the Stokes–Einstein diffusivity of the particle, α_L is the longitudinal dispersivity parameter, and V_p is the interstitial velocity, which is the average velocity of the fluid through the packed bed of collectors.

Long Term Deposition Behavior The deposition efficiencies or Sherwood numbers discussed in the previous sections are all based on the assumption that the deposited particles do not accumulate on the collector, and that the collector surface is clean. Such an assumption applies only during the initial stages of deposition onto a clean collector. Consequently, the deposition efficiency provided by the theoretical analysis discussed earlier is applicable to the initial deposition behavior, and is generally termed as the initial deposition efficiency. In reality, as the collector surface becomes covered with deposited particles, its capture efficiency changes with the extent of particle coverage. In context of filtration processes, such a transient behavior is more pertinent. There can be two outcomes of the transient variations of the deposition efficiency: When the deposited particles hinder further deposition, the process is referred to as blocking, while if the deposited particles enhance further deposition, the process is called ripening.

The transient deposition behavior is typically observed in deep bed filtration and in chromatographic separations. Consider a packed column of collectors, through which a particle suspension is continuously injected. Assume the inlet particle concentration to be C_0. The effluent particle concentration, C_e, exhibits variations with time in such a system. The transient variation of the concentration ratio C_e/C_0 is generally referred to as particle breakthrough curves and is depicted in Figure 13.35. The initial stages of the particle breakthrough curves can be represented in terms of the deposition theories applicable to clean beds. However, modeling the long term behavior of these breakthrough curves owing to blocking or ripening necessitates additional insights into the role played by previously deposited particles in modifying the collector efficiency.

The basic physics of blocking involves reduction in the available surface sites on the collector grains due to the deposited particles. Considering the fact that similarly charged particles will electrostatically repel each other, it is discernable that blocking is engendered by (i) physically reducing the available collector surface area (steric hindrance), and (ii) by rendering the collector surface less attractive or less favorable (electrostatic hindrance). Ripening, on the other hand, is predominantly caused by increase in the net capture cross section of the collectors, resulting in a greater interception of flowing particles.

The dynamics of collector surface blocking by deposited particles has received significant attention over the past several decades, and one of the most prominent

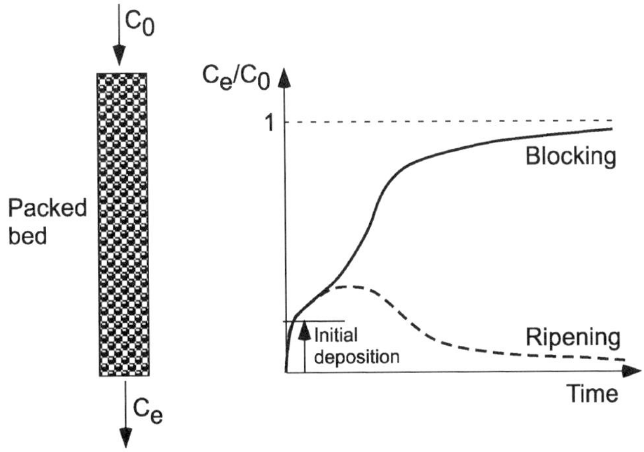

Figure 13.35. Schematic representation of the variation of particle breakthrough curves with time during blocking and ripening.

developments in this regard is the random sequential adsorption (RSA) model (Feder, 1980; Schaaf and Talbot, 1989; Adamczyk *et al.*, 1999; Ko *et al.*, 2000). The RSA model provides a relationship between the deposition efficiency and the extent of collector surface coverage by deposited particles in the form of a chemical kinetic equation. Solving the particle transport model in conjunction with the RSA based kinetic equation provides a reasonable mathematical construct for analyzing long term deposition behavior (Ko *et al.*, 2000).

Release of Deposited Particles Although release of deposited particles has received considerably less attention, it is observed in nature, and has tremendous consequences. Colloid mobilization in groundwater is a consequence of release of deposited particles from the soil matrix. Due to chemical variations in groundwater (such as a sudden change of salt concentration or pH following infiltration of fresh rain water), the electrostatic interactions between the particles and collectors can be dramatically modified. Such a modification can result in release and mobilization of colloidal particles. Empirical approaches based on kinetic expressions are available to assess the extent of release in porous media (see, for instance, Bhattacharjee *et al.*, 2002).

Chemical Heterogeneity of the Collector Bed As discussed earlier, chemical heterogeneity of collector beds can have a profound influence on the deposition behavior. Incorporation of such effects in models for predicting the performance of packed bed filters can be quite challenging if one has to consider the exact heterogeneous nature of the collectors. An effective means for modeling such heterogeneities in a simplified manner for packed beds is afforded by the two-site patchwise charge heterogeneity model (Johnson *et al.*, 1996). This model assumes that the total deposition surface area of the collectors in a packed bed is composed of a favorable fraction, $\lambda_f = A_f/A_c$ and an unfavorable fraction, $(1 - \lambda_f)$, where A_f is the favorable (attractive) surface area of the collectors, and A_c is the total collector surface area.

13.7.2 Colloid Transport Models in Porous Media

The different mechanisms affecting the long term deposition process discussed above can be simultaneously incorporated in a theoretical formulation to assess the particle transport and breakthrough behaviors through porous media. Referring back to Figure 13.3, the general equation for particle transport in presence of hydrodynamic dispersion and blocking effects can be written (for one-dimensional transport through a packed bed of collectors) as (Ko *et al.*, 2000)

$$\frac{\partial c}{\partial t} = D_h \frac{\partial^2 c}{\partial x^2} - V_p \frac{\partial c}{\partial x} - \frac{f}{\pi a_p^2} \frac{\partial \theta}{\partial t} \tag{13.142}$$

Here, c is the local particle concentration in the flowing suspension at the location x, D_h is the hydrodynamic dispersion coefficient, V_p is the interstitial (pore) velocity, which can be related to the external (superficial) velocity, U_∞, of the injected suspension as $V_p = U_\infty/(1 - \alpha_c)$. The term f is the collector surface area per unit length of the packed bed, a_p is the depositing particle radius, and θ is the fraction of the collector surface covered by deposited particles at a given instant. For irreversible monolayer deposition, the fractional surface coverage is given by a kinetic expression of the form

$$\frac{\partial \theta}{\partial t} = \pi a_p^2 k_{dep} B(\theta) c \tag{13.143}$$

Equation (13.143) is a kinetic expression that is applied as a source/sink term in the convection–dispersion equation (13.142). Here, k_{dep} is the deposition rate constant, which for spherical collectors, can be related to the collector deposition efficiency through

$$k_{dep} = \frac{\eta U_\infty}{4} \tag{13.144}$$

The term $B(\theta)$ in Eq. (13.143) is the dynamic blocking function or the available surface function, which relates the modification of the deposition kinetics with the fractional surface coverage. One can express the dynamic blocking function as a virial of the form

$$B(\theta) = 1 + a_1\theta + a_2\theta^2 + a_3\theta^3 + \cdots \tag{13.145}$$

where the coefficients a_1, a_2, a_3, *etc.*, are dictated by different models of adsorption. For simple Langmuirian adsorption, the blocking function is given by $B(\theta) = 1 - \theta$. The Langmuirian blocking model is applicable to infinitesimal particles. For finite sized particles, one needs to consider the steric effects due to the occluded area by each particle. For hard spherical particles depositing on a planar collector, the random sequential adsorption (RSA) model provides the following values of the coefficients (Schaaf and Talbot, 1989):

$$a_1 = -4; \quad a_2 = \frac{6\sqrt{3}}{\pi}; \quad \text{and} \quad a_3 = \left(\frac{40}{\pi\sqrt{3}} - \frac{176}{3\pi^2}\right)$$

It is interesting to note from the first virial coefficient that a hard sphere occupying a surface will occlude (prevent other particles to deposit on) four times its own projected area due to its finite size. The hard-sphere RSA model also provides insight about the maximum surface coverage on a collector beyond which no more particles can deposit. This maximum coverage is given by $\theta_\infty = 0.546$, which implies that if 54.6% of a collector surface is covered by hard spheres of finite dimensions, it can no longer accommodate any more particles.

The RSA model has been explored in context of particle deposition studies to investigate how electric double layer repulsion between the adsorbed and depositing particles as well as hydrodynamic interactions modify the virial coefficients a_1 to a_3 in Eq. (13.145). Extensive discussions of these effects are available elsewhere (Adamczyk *et al.*, 1994; Adamczyk *et al.*, 1999; Ko *et al.*, 2000). Here we provide a qualitative summary of the pertinent effects. Consider the deposition process on a planar collector for three possible situations, as shown in Figure 13.36. In each case, we consider the approach of a suspended particle near an already deposited particle on the planar surface. First, when one considers deposition of hard spheres in a quiescent fluid (case a), each sphere will block four times its projected area of the surface. When the particles are charged, as in case b, the projected area of the surface blocked will be larger, owing to the presence of the electric double layer around the particle. The double layer will effectively increase the deposited particle size, and consequently, the overall blocked area. Finally, in case (c), we observe the coupled influence of the electric double layer and a tangential flow field on the blocked area. Here, the combined influence of the electrostatic and hydrodynamic interactions results in an

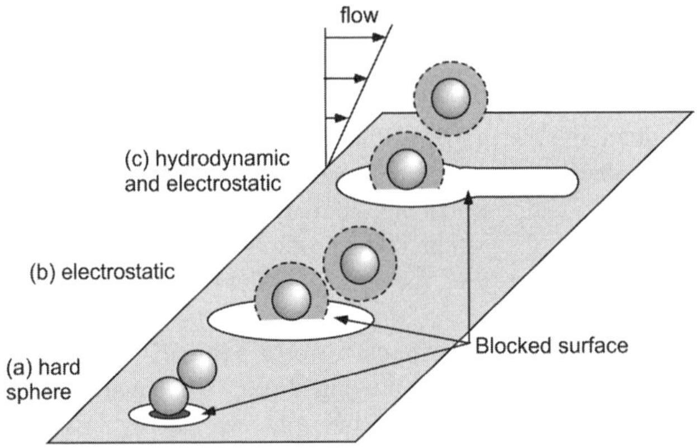

Figure 13.36. Schematic representation of influence of electric double layer and hydrodynamic interactions on the dynamic blocking function. (a) Hard spheres in a quiescent medium; (b) charged hard spheres in a quiescent medium; (c) charged hard spheres in a tangentially flowing suspension. The regions on the planar substrate marked as blocked surfaces are regions where other particles cannot deposit. The shapes of the blocked regions vary owing to the different interactions.

elongated "shadow region" behind each deposited particle. It is therefore evident that the particle-particle electrostatic and hydrodynamic interactions can significantly modify the long term deposition kinetics as well as the maximum surface coverage of the collector.

Incorporation of particle release and chemical heterogeneity on the collector surface in the filtration models involves further modification of the source/sink term of the convection-dispersion equation. A general format for including these effects is to consider the deposition on two types of surface sites, namely, favorable sites given by the fraction θ_f and an unfavorable fraction, θ_u. The total fractional coverage of the collector at any given instant can then be represented in light of a patchwise heterogeneity model as

$$\theta = \lambda_f \theta_f + (1 - \lambda_f)\theta_u \tag{13.146}$$

The corresponding rate expression for the patchwise deposition model can be written as (Chen *et al.*, 2001; Bhattacharjee *et al.*, 2002)

$$\frac{\partial \theta}{\partial t} = \lambda_f \frac{\partial \theta_f}{\partial t} + (1 - \lambda_f)\frac{\partial \theta_u}{\partial t} \tag{13.147}$$

For each type of site, the deposition and release kinetics can be modeled as

$$\frac{\partial \theta_f}{\partial t} = \pi a_p^2 k_f B(\theta_f)c - k_{release,f}\,\theta_f \tag{13.148-a}$$

$$\frac{\partial \theta_u}{\partial t} = \pi a_p^2 k_u B(\theta_u)c - k_{release,u}\,\theta_u \tag{13.148-b}$$

In the above equations, k_f and k_u are the deposition rate constants on the favorable and unfavorable surface sites, respectively. Similarly, $k_{release,f}$ and $k_{release,u}$ represent the release rate coefficients from the favorable and unfavorable sites, respectively. While the deposition rate constants can be evaluated theoretically using the deposition efficiency, no theoretical model is available for the release rate constants.

Once the transport model is formulated for a particular deposition problem, it can be solved for a given system employing appropriate boundary conditions. The application of the one dimensional and two dimensional versions of the model in assessing colloid transport in porous media and virus transport in groundwater have been discussed by Ko *et al.* (2000); Bhattacharjee *et al.* (2002); and Loveland *et al.* (2003). Such models can either be used as predictive tools for assessing particle deposition dynamics in packed media, or as tools for solving inverse problems for estimation of the deposition and release rate constants from experimental observations of colloid transport in porous media and aquifer beds.

13.8 SUMMARY OF DIMENSIONLESS GROUPS

13.8.1 Dimensionless Groups in the Flux Equation

Double-layer parameter I, Dl: Measure of electrostatic to Brownian motion forces

$$Dl = \frac{4\pi\epsilon a_p \zeta_p \zeta_c}{k_B T}$$

Double-layer parameter II, Da: Measure of asymmetry of the electrical double layer

$$Da = \frac{(\zeta_c - \zeta_p)^2}{2\zeta_c \zeta_p}$$

Adhesion parameter, Ad: Measure of London–van der Waals to Brownian motion forces

$$Ad = \frac{A_H}{6k_B T}$$

Retardation parameter, $\bar{\lambda}$: Measure of retardation due to the finite speed of light

$$\bar{\lambda} = \lambda / a_p \quad \text{where } \lambda = 10^{-7} m$$

Gravitational parameter, Gr: Measure of gravitational to Brownian motion forces

$$Gr = \frac{2}{9} \frac{\Delta\rho g a_p^3}{\mu D_\infty}$$

Double-layer thickness parameter, κa_p: Measure of electric double-layer thickness

$$\kappa a_p = \left(\frac{2e^2 z^2 n_\infty}{\epsilon k_B T} \right)^{1/2} a_p = \left(\frac{2000 e^2 z^2 N_A M}{\epsilon k_B T} \right)^{1/2} a_p$$

Table 13.3 gives the definitions of the Reynolds, the Peclet, and the Sherwood numbers for the various geometries. For the case of a cylinder, the dimensionless groups in Table 13.3 are the same as for the case of a spherical collector with a_c being the cylindrical collector radius.

13.8.2 Dimensionless Groups in the Trajectory Equation

Electrostatic number, N_ζ: Measure of electrostatic repulsion to London–van der Waals forces

$$N_\zeta = \frac{6\pi\epsilon \zeta_c \zeta_p a_p}{A_H}$$

Attraction number, N_A: Measure of London–van der Waals to viscous forces

$$N_A = \frac{A_H a_c^2}{9\pi \mu a_p^4 U_\infty A_c}$$

where the collector area, A_c is given by

$$A_c \equiv A_{cyl} \text{ for a cylindrical collector}$$

and

$$A_c \equiv A_{sph} \text{ for a spherical collector}$$

Gravity number, N_G: Measure of gravitational to viscous forces

$$N_G = \frac{2\Delta\rho g a_c^2}{9\mu U_\infty A_c}$$

Double-layer thickness parameter, κa_p: Measure of electric double-layer thickness

$$\kappa a_p = \left(\frac{2e^2 z^2 n_\infty}{\epsilon k_B T}\right)^{1/2} a_p = \left(\frac{2000 e^2 z^2 N_A M}{\epsilon k_B T}\right)^{1/2} a_p$$

13.9 NOMENCLATURE

a	spherical particle radius, m
a_1, a_2, a_3	coefficients of the dynamic blocking function
a_c	collector radius (spherical or cylindrical), m
a_p	colloidal particle radius, m
A_H	Hamaker constant, J
Ad	adhesion parameter, $A_H/6k_BT$
A_b	empty bed cross-sectional area, m^2
A_c	dimensionless flow parameter ($= A_{cyl}$ or A_{sph})
A_{cyl}	dimensionless flow parameter accounting for the presence of cylindrical particle in a fibre mat, Eqs. (13.117) and (13.118)
A_{sph}	dimensionless flow parameter accounting for the presence of spherical particle in a suspension, Eq. (13.10)
$B(\theta)$	dynamic blocking function or available surface function
c	particle concentration, m^{-3} or mol/m^3
d_c	spherical collector diameter
d_p	colloidal particle diameter
$d_\parallel$	correction factor for diffusion coefficient parallel to surfaces, dimensionless
$d_\perp$	correction factor for diffusion coefficient normal to surfaces, dimensionless

Da	double layer parameter due to potential asymmetry, $(\zeta_c - \zeta_p)^2/2\zeta_c\zeta_p$
D_l	double layer parameter, $4\pi\epsilon a_p\zeta_c\zeta_p/k_BT$
D_{rr}	diffusion coefficient correction for parallel to surfaces, m^2/s
D_{rz}	diffusion coefficient correction for normal and tangential movements, m^2/s
D_{zz}	diffusion coefficient correction for movements normal to surfaces, m^2/s
$D_{1\infty}$, $D_{2\infty}$	particle diffusion coefficient in infinite dilution, m^2/s, given by $k_BT/6\pi\mu a$
$\bar{\bar{D}}$	diffusion coefficient tensor, m^2/s
f_1, f_2, f_3, f_4	universal hydrodynamic correction functions (Tables 13.1 and 13.2), dimensionless
F_r	force in r-direction, N
F_z	force in z-direction, N
F_z^*	dimensionless interparticle z-directed force
F_{hyd}	hydrodynamic forces, N
F_{gr}	gravitational force in r-direction, N
F_{gz}	gravitational force in z-direction, N
F_A	dispersion force (attractive), N
F_R	electrostatic (repulsion) force, N
$\mathbf{F}$	fluctuating force vector, N
g	gravitational acceleration, m/s^2
Gr	gravitational parameter, $2\Delta\rho g a_p^3/9\mu D_\infty$
h	dimensionless gap width, s/a_p
$\mathbf{i}$	unit vector
$\mathbf{j}$	particles flux vector, $m^{-2}s^{-1}$
j_r	particles flux in r-direction, $m^{-2}s^{-1}$
$j_\perp$	particles flux normal to the collector surface, $m^{-2}s^{-1}$
k_B	Boltzmann constant, J/K
k_{dep}	particle deposition rate constant, m/s
k_f, k_u	favorable and unfavorable deposition rate constants, m/s
L	characteristic length, m
m_p	particle buoyant mass, kg
n	local number concentration, m^{-3}
n_c	number of collectors per unit volume, m^{-3}
n_o	number concentration of particles at bed inlet, m^{-3}
n_∞	number concentration of particles far from a collector, m^{-3}
N_A	attraction number, $A_H a_c^2/9\pi\mu a_p^4 U_\infty A_{sph}$
N_G	gravity number, $2\Delta\rho g a_c^2/9\mu U_\infty A_{sph}$
N_ζ	electrostatic (repulsion) number, $6\pi\epsilon\zeta_c\zeta_p a_p/A_H$
Pe	Peclet number, defined differently for the various geometries
Q	source term, $m^{-3}s^{-1}$
r	radial coordinate, m
Re	Reynolds number (see Table 13.3)

s	dimensional gap between two surfaces, $s = a_p h$
S	collector surface area, m^2
Sc	Schmidt number, $(\mu/\rho)/D_\infty$
Sh	average Sherwood number (see Table 13.3)
$Sh(\theta)$, Sh_l	local Sherwood number
St	Stokes number, Eq. (13.23)
t	time, s
T	absolute temperature, K
u_r	radial fluid velocity, m/s
u_z	normal fluid velocity, m/s
$\mathbf{u}$	fluid velocity vector, m/s
$\mathbf{u}^*$	dimensionless fluid velocity vector
$\mathbf{u}_{st}$	stagnation fluid velocity vector, m/s
$\mathbf{u}_{sh}$	shear fluid velocity vector, m/s
$\mathbf{u}_\infty$	characteristic velocity; fluid velocity far from a collector; superficial velocity in a packed bed, m/s
v_r	radial component of particle velocity, m/s
v_t	terminal velocity of a free settling spherical particle, m/s
v_z	normal component of particle velocity, m/s
$\mathbf{v}$	particle velocity vector, m/s
v_m	average velocity in a channel, m/s
x	Cartesian coordinate, distance along a channel, m
$\mathbf{x}$	direction vector, m
$\mathbf{X}$	dimensionless direction vector, Eq. (13.21)

Greek Symbols

α	impinging jet flow coefficient, $m^{-1}s^{-1}$
α_c	volume fraction of collector particles
$\bar{\alpha}$	dimensionless impinging jet flow coefficient
$\Delta\rho$	density difference between particle and suspending fluid, kg/m^3
ϵ	permittivity of liquid, C/Vm
ζ_c	collector zeta potential, V
ζ_p	colloidal particle zeta potential, V
η	collection efficiency
η_I	pure interception collection efficiency
θ	angular coordinate
$\theta, \theta_f, \theta_u$	fractional surface coverage
κ	inverse Debye length, m^{-1}
λ	filter coefficient, Eq. (13.15), m^{-1}
	London wavelength, of order of 10^{-7} m
$\bar{\lambda}$	dimensionless London wavelength, λ/a_p
μ	fluid viscosity, Pa s
ν	kinematic viscosity, m^2/s
ρ	fluid density, kg/m^3

ρ_p	particle density, kg/m^3
ψ	stream function, m^3/s for flow over a spherical particle; m^2/s for flow over a cylinder
ψ_L	limiting stream function defining a particular collection mode
ω	angular velocity of a spinning disc, radians/s
ϖ	local radial coordinate, m
∇	del operator, m^{-1}
∇^2	Laplacian operator, m^{-2}

13.10 PROBLEMS

13.1. A spherical particle of radius a contained in a large vessel is released from a gap height of h_0. The vessel contains a liquid of viscosity μ. The density difference between heavy particle and the liquid is $\Delta\rho$. Gravity acts downwards. The geometry for the approach of the particle to the bottom surface of the vessel, which is assumed planar, is shown in Figure 13.37.

(a) Assume that gravity is the sole force acting on the particle. Show that under pseudo steady-state conditions

$$\frac{dh}{dt} = -\frac{2a^2\Delta\rho g}{9\mu} f_1$$

where g is the gravitational constant and f_1 is the universal hydrodynamic correction function that accounts for the presence of the vessel bottom surface.

(b) Let the van der Waals force be included in the force balance. Show that

$$\frac{dh}{dt} = -\left[\frac{2a^2\Delta\rho g}{9\mu} + \frac{A_H}{36\pi\mu h^2}\right] f_1$$

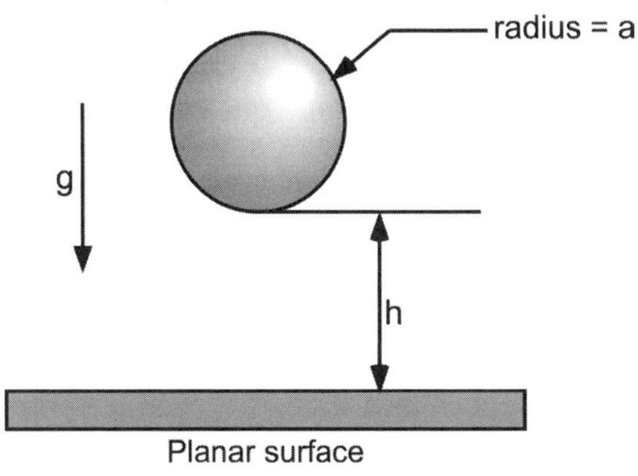

Figure 13.37. Schematic depiction of a particle approaching a planar surface of a vessel.

where A_H is the Hamaker constant and the van der Waals interaction energy between a sphere and a flat surface is given by $-A_H a/(6h)$, see Table 11.2.

(c) Let the dimensionless time, τ, be defined as

$$\tau = \frac{2\Delta\rho g a}{9\mu} t$$

and dimensionless length, H, as

$$H = \frac{h}{a}$$

Show that

$$\frac{dH}{d\tau} = -\left[1 + \frac{G}{H^2}\right] f_1$$

where

$$G = \frac{A_H}{8\pi a^4 \Delta\rho g}$$

Here, G is a dimensionless group that accounts for the ratio of attractive to gravitational forces.

(d) The hydrodynamic correction function, for convenience, can be taken as

$$f_1 = \frac{H}{H+1}$$

Show that

$$\tau = -\int_{H_0}^{H} \frac{H(H+1)}{H^2 + G} dH$$

where $H_0 = h_0/a$ at time zero.

(e) Let $a = 10^{-5}$ m, $\Delta\rho = 1000\,\text{kg/m}^3$, $\mu = 10^{-3}$ Pa.s, and $H_0 = 5$. Consider four values of the Hamaker constant, namely, $A_H = 0$, 10^{-21}, 10^{-20}, and 10^{-19} J. For each A_H value, plot the variation of the scaled separation H with τ. Comment on your solution. What is your observation for $G = 0$?

13.2. **(a)** Derive the equivalent expression for $dH/d\tau$ in the presence of gravitational, van der Waals, and electrostatic repulsion forces in a manner similar to part (c) of Problem 13.1. Assume the zeta potential of the particle is the same as the surface of the vessel. What is the additional dimensionless group and how would you characterize it?

(b) Plot the variation of the dimensionless separation gap H with time τ for zeta potentials of 50 mV on the sphere and the vessel surfaces for water at 20°C containing 0.001 M NaCl. Assume $A_H = 10^{-20}$ J. Compare with the case of zero repulsive force. All other physical data are the same as in Problem 13.1.

13.11 REFERENCES

Acrivos, A., and Goddard, J. D., Asymptotic expansions for laminar forced-convection heat and mass transfer, *J. Fluid Mech.*, **23**, 273–291, (1965).

Acrivos, A., and Taylor, T., Heat and mass transfer from single spheres in stokes flow, *Phys. Fluids*, **5**, 387–394, (1962).

Adamczyk, Z., Particle transfer and deposition from flowing colloid suspensions, *Colloids and Surfaces*, **35**, 283–308, (1989a).

Adamczyk, Z., Particle deposition from flowing suspensions, *Colloids and Surfaces*, **39**, 1–37, (1989b).

Adamczyk, Z., Dabros, T., Czamecki, J., and van de Ven, T. G. M., Kinetics of particle accumulation at collectors surfaces. II. Exact numerical solutions, *J. Colloid Interface Sci.*, **97**, 91–104, (1984).

Adamczyk, Z., Siwek, B., and Zembala, M., Reversible and irreversible adsorption of particles on homogeneous surfaces, *Colloids Surf.*, **62**, 119–130, (1992).

Adamczyk, Z., Siwek, B., Zembala, M., and Warszynski, P., Enhanced deposition of particles under attractive double-layer forces, *J. Colloid Interface Sci.*, **130**, 578–587, (1989).

Adamczyk, Z., Siwek, B., Zembala, M., and Belouschek, P., Kinetics of localized adsorption of colloid particles, *Adv. Colloid Interface Sci.*, **48**, 151–280, (1994).

Adamczyk, Z., and van de Ven, T. G. M., Deposition of particle under external forces in laminar flow through parallel-plate and cylindrical channels, *J. Colloid Interface Sci.*, **80**, 340–357, (1981a).

Adamczyk, Z., and van de Ven, T. G. M., Deposition of Brownian particles onto cylindrical collectors, *J. Colloid Interface Sci.*, **84**, 497–518, (1981b).

Adamczyk, Z., and Van de Ven, T. G. M., Kinetics of particle accumulation at collector surfaces. II. Approximate analytical solutions, *J. Colloid Interface Sci.*, **97**, 68–90, (1984).

Adamczyk, Z., Senger, B., Voegel, J. C., and Schaaf, P., Irreversible adsorption/desorption kinetics: A generalized approach, *J. Chem. Phys.*, **110**, 3118–3128, (1999).

Adler, P. M., Interaction of unequal spheres I. Hydrodynamic interaction: colloidal forces, II. Conducting spheres, III. Experimental, *J. Colloid Interface Sci.*, **84**, 461–474, 475–488, 489–496, (1981).

Batchelor, G. K., Brownian diffusion of particles with hydrodynamic interaction, *J. Fluid Mech.*, **74**, 1–29, (1976).

Bhattacharjee, S., Ryan, J. N., and Elimelech, M., Virus transport in physically and geochemically heterogeneous porous media, *J. Contaminant Hydrology*, **57**, 161–187, (2002).

Bird, R. B., Stewart, W. E., and Lightfoot, E. N., *Transport Phenomena*, Wiley, New York, (1960).

Boluk, M. Y., and van de Ven, T. G. M., Kinetics of electrostatically controlled deposition of colloidal particles on solid surfaces in stagnation point flow, *PCH Physicochem. Hydrodyn.*, **11**, 113–127, (1989).

Bowen, B. D., *Fine particle deposition in smooth channels*, PhD Dissertation, University of British Columbia, Vancouver, British Columbia, Canada, (1978).

Bowen, B. D., and Epstein, N., Fine particle deposition in smooth parallel-plate channels, *J. Colloid Interface Sci.*, **72**, 81–97, (1979).

Bowen, B. D., Levine, S., and Epstein, N., Fine particle deposition in laminar flow through parallel-plate and cylindrical channels, *J. Colloid Interface Sci.*, **55**, 275–290, (1976).

Brenner, H., The slow motion of a sphere through a viscous fluid towards a plane surface, *Chem. Eng. Sci.*, 16, 242–251, (1961).

Chen, J. Y., Ko, C.-H., Bhattacharjee, S., and Elimelech, M., Role of spatial distribution of porous medium surface charge heterogeneity in colloid transport, *Colloids Surf. A*, **191**, 3–15, (2001).

Clift, R., Grace, J. R., and Weber, M. E., *Bubbles, Drops and Particles*, Academic Press, London, (1978).

Clint, G. E., Clint, J. G., Corkill, J. M., and Walker, T., Deposition of latex particle on a planar surface, *J. Colloid Interface Sci.*, **44**, 121–132, (1973).

Dabros, T., and van de Ven, T. G. M., A direct method for studying particle deposition onto solid surfaces, *Colloid Polym. Sci.*, **261**, 694–707, (1983).

Dabros, T., and van de Ven, T. G. M., Deposition of latex particles on glass surfaces in an impinging jet, *Physicochem. Hydrodyn.*, **8**, 161–172, (1987).

Elimelech, M., Gregory, J., Zia, X., and Williams, R. A., *Particle Deposition and Aggregation: Measurement, Modelling, and Simulation*, Butterworth, London, (1995).

Feder, J., Random sequential adsorption, *J. Theor. Biol.*, **87**, 237–254, (1980).

Fitzpatrick, J. A., *Mechanisms for particle capture in water filtration*, PhD Dissertation, Division of Engineering and Applied Physics, Harvard University, Cambridge, MA, (1972).

Fitzpatrick, J. A., and Spielman, L. A., Filtration of aqueous latex suspensions through beds of glass spheres, *J. Colloids Interface Sci.*, **43**, 350–369, (1973).

Friedlander, S. K., Theory of aerosol filtration, *Ind. Eng. Chem.*, **50**, 1161–1164, (1958).

Friedlander, S. K., *Smoke, Dust and Haze: Fundamentals of Aerosol Dynamics*, 2nd ed., Oxford University Press, New York, (2000).

Fuchs, N. A., *The Mechanics of Aerosols*, Dover, New York, (1964).

Goldman, A. J., Cox, R. G., and Brenner, H., Slow viscous motion of a sphere parallel to a plane wall: I. Motion through a quiescent fluid, *Chem. Eng. Sci.*, **22**, 637–651, (1967a).

Goldman, A. J., Cox, R. G., and Brenner, M., Slow viscous motion of a sphere parallel to a plane wall: II. Couette flow, *Chem. Eng. Sci.*, **22**, 653–660, (1967b).

Goren, S. L., The normal force exerted by creeping flow on a small sphere touching a plane, *J. Fluid Mech.*, **41**, 619–625, (1970).

Goren, S. L., and O'Neill, M. E., On the hydrodynamic resistance to a particle of a dilute suspension when in the neighborhood of a large obstacle, *Chem. Eng. Sci.*, **26**, 325–338, (1971).

Griffin, F. O., and Meisen, A., Impaction of spherical particles on cylinders at moderate Reynolds numbers, *Chem. Eng. Sci.*, **28**, 2155–2164, (1973).

Guzy, C. J., Bonano, E. J., and Davis, E. J., The analysis of flow and colloidal particle retention in fibrous porous media, *J. Colloid Interface Sci.*, **95**, 523–543, (1983).

Happel, J., Viscous flow in multiparticle systems, *AIChE J.*, **4**, 197–201, (1958).

Happel, J., Viscous flow relative to arrays of of cylinders, *AIChE J.*, **5**, 174–177, (1959).

Happel, J., and Brenner, H., *Low Reynolds Number Hydrodynamics*, Prentice-Hall, Englewood Cliffs, NJ, (1965).

Hirtzel, C. S., and Rajagopalan, R., *Colloidal Phenomena, Advanced Topics*, Noyes, NJ, (1985).

Hogg, R., Healy, T. W., and Fuerstenau, D. E., Mutual coagulation of colloidal dispersions, *Trans. Faraday Soc.*, **63**, 1638–1651, (1966).

Honig, E. P., Roeberson, G. J., and Wiersema, P. H., Effect of hydrodynamic interaction on the coagulation rate of hydrophobic colloids, *J. Colloid Interface Sci.*, **36**, 97–109, (1971).

Hull, M., and Kitchener, J. A., Interaction of spherical colloidal particles with planar surfaces, *Trans. Faraday Soc.*, **65**, 3093–3104, (1969).

Israelachvili, J. N., *Intermolecular and Surface Forces*, 2nd ed., Academic Press, London, (1992).

Israelachvili, J. N., and Pashley, R. M., Measurement of the hydrophobic interaction between two hydrophobic surfaces in aqueous electrolyte solutions, *J. Colloid Interface Sci.*, **98**, 500–514, (1984).

Jeffrey, D. J., and Onishi, Y., Calculations of the resistance and mobility functions for two unequal rigid spheres in low-Reynolds number flow, *J. Fluid Mech.*, **139**, 261–290, (1984).

Johnson, P. R., Sun, N., and Elimelech, M., Colloid transport in geochemically heterogeneous porous media: Modeling and measurements, *Environ. Sci. Technol.*, **30**, 3284–3293, (1996).

Kihira, H., Ryde, N., and Matijevic, E., Kinetics of heterocoagulation. Part 2: The effect of discreteness of surface charge, *J. Chem. Soc. Faraday Trans.*, **88**, 2379–2386, (1992).

Ko, C.-H., Bhattacharjee, S., and Elimelech, M., Coupled model of colloidal and hydrodynamic interactions on the RSA dynamic blocking function for particle deposition onto packed spherical collectors, *J. Colloid Interface Sci.*, **229**, 554–567, (2000).

Levesque, M. , *Ann. Mines*, **13**, 201, (1928).

Levich, V. G., *Physicochemical Hydrodynamics*, Prentice-Hall, Englewood Cliffs, NJ, (1962).

Liu, J., Xu, Z., and Masliyah, J. H., Role of fine clays in bitumen extraction from oil sands, *AIChE J.*, **50**, 1917–1927, (2004).

Loveland, J. P., Bhattacharjee, S., Ryan, J. N., and Elimelech, M., Colloid transport in a geochemically heterogeneous porous medium: Aquifer tank experiment and modeling, *J. Contaminant Hydrology*, **65**, 161–182, (2003).

Marshall, J. K., and Kitchener, J. A., The deposition of colloidal particles on smooth solids, *J. Colloid Interface Sci.*, **22**, 342–351, (1966).

Masliyah, J. H., and Duff, A., Impingement of spherical particles on elliptical cylinder, *Aerosol Sci.*, **6**, 31–43, (1975).

Masliyah, J. H., and Epstein, N., Numerical solution of heat and mass transfer from spheroids in steady axisymmetric flow, *Prog. Heat Mass Transf.*, **6**, 613–632, (1973).

Michael, D. M., and Norey, P. W., Particle collision efficiencies for a sphere, *J. Fluid Mech.*, **37**, 567–575, (1969).

Neale, G., and Masliyah, J. H., Flow perpendicular to mats of randomly arranged cylindrical fibers (Importance of cell models), *AIChE J.*, **21**, 805–807, (1975).

Prieve, D. C., and Lin, M. M. J., Adsorption of Brownian hydrosols onto rotating disc aided by uniform applied force, *J. Colloid Interface Sci.*, **76**, 32–47, (1980).

Prieve, D. C., and Ruckenstein, E., Effect of London forces upon the rate of deposition of Brownian particles, *AIChE J*, **20**, 1178–1187, (1974).

Probstein, R. F., *Physicochemical Hydrodynamics, An Introduction*, 2nd ed., Wiley Interscience, New York, (2003).

Rabinovich, Y. I., and Yoon, R.-H., Use of atomic force microscope for the measurements of hydrophobic forces between silanated silica plate and glass sphere, *Langmuir*, **10**, 1903–1909, (1994).

Rajagopalan, R., and Kim, J. S., Adsorption of Brownian Particles in the presence of potential barriers: Effect of different modes of double-layer interaction, *J. Colloid Interface Sci.*, **83**, 428–448, (1981).

Rajagopalan, R., and Tien, C., Trajectory analysis of deep bed filtration with sphere-in-cell porous media model, *AIChE J.*, **22**, 523–533, (1976).

Sanders, R. S., Chow, R. S., and Masliyah, J. H., Deposition of bitumen and asphaltene stabilized emulsions in an impinging jet cell, *J. Colloid Interface Sci.*, **174**, 230–245, (1995).

Sanders, R. S., Chow, R. S., and Masliyah, J. H., Hydrophobic interactions in silane treated silica suspensions and bitumen emulsions, *Can. J. Chem. Eng.*, **81**, 43–52, (2003).

Schaaf, P., and Talbot, J., Kinetics of random sequential adsorption, *Phys. Rev. Lett.*, **62**, 175–178, (1989).

Song, S., Lopez-Valdivieso, A., Reyes-Bahena, J. L., Bermejo-Perez, H. I., and Trass, O., Hydrophobic flocculation of galena fines in aqueous suspensions, *J. Colloid Interface Sci.*, **227**, 272–281, (2000).

Spielman, L. A., Particle capture from low-speed laminar flows, *Ann. Rev. Fluid Mech.*, **9**, 297–319, (1977).

Spielman, L. A., and Cukor, P. M., Deposition of non-Brownian particles under colloidal forces, *J. Colloid Interface Sci.*, **43**, 51–65, (1973).

Spielman, L. A., and Fitzpatrick, J. A., Theory for particle collection under London and gravity forces, *J. Colloid Interface Sci.*, **42**, 607–623, (1973).

Spielman, L. A., and Goren, S. L., Capture of small particles by london forces from low-speed liquid flows, *Environ. Sci. and Tech.*, **4**, 135–140 (see corrections in **5**, 254, 1971), (1970).

Suzuki, A., Ho, N. F. H., and Higuchi, W. I., Predictions of the particle size distribution changes in emulsions and suspensions by digital computation, *J. Colloid Interface Sci.*, **29**, 552–564, (1969).

Tien, C., *Granular Filtration of Aerosols and Hydrosols*, Butterworths, Boston, (1989).

Usui, S., Interaction of electrical double layers at constant surface charge, *J. Colloid Interface Sci.*, **44**, 107, (1973).

van de Ven, T. G. M., *Colloidal Hydrodynamics*, Academic Press, London, (1989).

Vaidyanathan, R., and Tien, C., Hydrosol deposition in granular beds, *Chem. Eng. Sci.*, **43**, 289–302, (1988).

Weber, M. E., and Paddock, D., Interceptional and gravitational collision efficiencies for single collectors at intermediate Reynolds numbers, *J. Colloid Interface Sci.*, **94**, 328–335, (1983).

Xu, Z., and Yoon, R.-H., The role of hydrophobic interactions in coagulation, *J. Colloid Interface Sci.*, **132**, 532–541, (1989).

Xu, Z., and Yoon, R.-H., A study of hydrophobic coagulation, *J. Colloid Interface Sci.*, **134**, 427–434, (1990).

Yao, K. M., Habibian, M. T., and O'Melia, C. R., Water and wastewater filtration: Concepts and applications, *Environ. Sci. Technol.*, **5**, 1105–1112, (1971).

CHAPTER 14

NUMERICAL SIMULATION OF ELECTROKINETIC PHENOMENA

Application of numerical solution techniques often becomes the only recourse when one intends to employ the electrokinetic models discussed in the earlier chapters to complex systems. It was noted in the earlier chapters, that the closed form mathematical expressions describing various electrokinetic phenomena were derived for low surface potentials and highly simplified geometries using linear perturbation approximations. Such approximations and simplifications may potentially restrict the application of analytical results when addressing more complicated systems of practical interest. For instance, the electric potential and free charge distribution used in the treatment of electroosmosis and electrophoresis were obtained from a solution of the linearized Poisson–Boltzmann equation, which is strictly valid for low surface potentials, namely $|\psi_s| < 25$ mV. For higher surface potentials, one must solve the full non-linear Poisson–Boltzmann equation. This becomes a difficult task if one attempts analytic routes, particularly for complex geometries. Furthermore, analytic solutions of the non-linear Poisson–Boltzmann equation are only feasible for symmetric electrolytes.

Modeling electrokinetic transport phenomena requires simultaneous solution of at least three coupled physical models, namely, the governing equations for electrostatics (Poisson equation), fluid flow (Navier–Stokes and continuity equations), and transport of ions (Nernst–Planck equations). For a multicomponent system in a three-dimensional flow, this translates into solving n coupled Nernst–Planck equations for the n ionic species, the Poisson equation, three equations for momentum conservation along the three coordinate directions,and the fluid continuity equation. This

---

Electrokinetic and Colloid Transport Phenomena, by Jacob H. Masliyah and Subir Bhattacharjee
Copyright © 2006 John Wiley & Sons, Inc.

implies that, in general $n + 5$ partial differential equations must be solved simultaneously for an electrokinetic problem involving n ionic species. The equations are coupled as follows: The Poisson equation relates the electric potential distribution to the volumetric charge density, the latter being related to the distribution of the ions. The ion distributions are provided by the Nernst–Planck equations. The Nernst–Planck equation requires the fluid velocity components to obtain the convective flux, as well as the electric field to obtain the migration flux. The velocity components are provided by the Navier–Stokes (and continuity) equations, while the electric field is provided by the solution of the Poisson equation. Finally, the solution of the Navier–Stokes equations involves consideration of the electrical body force, which is determined from the Poisson and the Nernst–Planck equations. Even for relatively simple geometries, the task of solving all these equations simultaneously is quite formidable.

In this chapter, we will discuss the key issues involved in attempting numerical solutions for electrokinetic problems. The objective of this chapter is not to expound on basic numerical techniques. One should refer to appropriate textbooks on numerical methods for that purpose. Here, we specifically illustrate, through some examples, how one can implement a numerical scheme to solve typical electrokinetic transport problems, which are cast as a set of coupled partial differential equations. Numerous sophisticated techniques exist for solution of partial differential equations numerically. In this chapter, we use a fairly generic approach, namely, finite element analysis, for the solution of the governing electrokinetic equations. This, of course, does not imply that finite element technique is the only approach to obtain a viable solution of electrokinetic problems. Nearly any technique for solving coupled partial differential equations, like finite difference, finite volume, boundary element, method of lines, and spectral techniques, to name a few, can be used to solve such problems numerically. The sole advantage of the finite element technique seems to be its ability to provide a solution for arbitrary geometries and computational domains. Furthermore, since numerous commercial software implementing the finite element technique are readily available, it was felt that relegating the numerical task to a robust software will allow one to focus on the problem formulation, which is our key objective in this chapter.

14.1 TOOLS AND METHODS FOR COMPUTER BASED SIMULATIONS

The models for electrokinetic transport are continuum models, which yield partial differential equations. Combined, the Poisson, the Navier–Stokes, and the Nernst–Planck equations span the entire spectrum of generic classification of partial differential equations (PDE), representing parabolic, elliptic, and hyperbolic equations. The Poisson equation is an example of elliptic PDEs. The Nernst–Planck equations are generally parabolic, but depending on parameter values, may sometimes behave as hyperbolic equations. The Navier–Stokes equations are elliptic for quasi-steady creeping flow regimes (no inertia term), but the general Navier–Stokes equation can assume all three forms of PDEs depending on parameter values. An essential component for attempting to solve these equations simultaneously is to have a numerical technique for solving PDEs that is generic enough to solve elliptic, parabolic, and hyperbolic equations. Furthermore, most of these equations are non-linear, which necessitates a robust equation solver that can handle non-linearities. Adding to this list, the requirement of handling arbitrary geometries and computational domains leaves very few options other than to use techniques like finite difference and finite element analysis. In this chapter, we will use the finite element technique owing to the fact that most finite element solvers are quite generic in nature to accommodate diverse types of equations, and allow handling of different geometries quite readily. Several modeling software based on finite element technique have become available over the past decade, including ANSYS, Coventor, FEMLab, Fluent, ALGOR, *etc*. These software generally allow (to varying degrees of sophistication) what is commonly referred to as *multiphysics modeling*, and generally provide tools and algorithms to deal with electrokinetic problems associated with microfluidic simulations. The computational results presented in this chapter were obtained using one such commercially available software, FEMLab. Relegating the numerical aspects of the finite element implementation to a commercial software allows us to focus on the specific issues related to formulation of the electrokinetic problems to render them amenable to numerical solution. This brings us to an important fact one should keep in mind when using numerical techniques.

The key to successfully implementing a numerical methodology is to realize at the outset that numerical methods are approximate, and generally work within a range of real numbers roughly spanning eight orders of magnitude (in single precision). As an example, if one is to solve a partial differential equation with a dependent variable that ranges from 1 to 0, the numerical solution will only provide reasonably accurate predictions of the variable over a range spanning 1 to 10^{-8}. If instead, the dependent variable ranges from 10^5 to 0, perhaps the smallest value that will be accurately predicted by the numerical solution is 10^{-3}. The problem is exacerbated when we deal with coupled partial differential equations. In this case, one needs to carefully scale all the dependent (and independent) variables such that they are of comparable magnitudes. In this context, the most important tools for implementation of a numerical solution are still pen and paper. One should perhaps spend ample time in properly setting up the model problem such that it can be solved readily using a standard numerical technique. This can be achieved by careful non-dimensionalization and scaling of the governing equations, appropriate simplification of the geometry, careful choice of the parameter phase space, use of symmetry, coordinate transformation, and a variety of other techniques. None of these initial steps require a computer or any software. Nevertheless, spending time on these aspects is important if one desires to obtain a quick and sufficiently accurate numerical solution for any mathematical model.

There are several challenging issues specific to electrokinetic problems that need consideration during numerical simulations. The foremost of these is the frequent use of boundary conditions at infinity when defining an electrokinetic problem. Consider for instance, solution of the Poisson–Boltzmann equation for the electric potential distribution in an electrolyte near an infinite planar charged surface (see Section 5.2). The boundary conditions used for this problem are:

1. at the surface of the plate, the electric potential is equal to the surface potential of the plate, and,
2. at an infinite distance away from the plate, the surface potential and the electric field in the direction normal to the plate are zero.

The second boundary condition at infinity poses a serious problem in numerical computations. One cannot readily implement infinity in a numerical solution technique. Consequently, one must implement an alternative approach to define the boundary conditions in a numerical problem. A coordinate transformation is the commonly adopted approach, whereby the infinity is somehow converted to a finite number. This type of transformation will render such problems amenable to a numerical treatment. It is also possible to artificially implement the outer boundary (corresponding to infinity) at a sufficiently large distance away from the plate, such that the electric field and potential can be assumed to be infinitesimally small at this location. This type of artificial imposition of a boundary condition at infinity cannot be considered a sound mathematical practice, since one cannot *a priori* tell whether a given distance is sufficiently far away to actually validate the assumption of infinitesimal potential or field at this location. Therefore, whenever such artificial boundary conditions are used, the validity of the methodology should be verified by placing the boundary further and further away and comparing the solutions obtained at these different locations.

A second issue arising in numerical solution of electrokinetic problems stems from the high non-linearity of the Poisson equation. Consider the Poisson–Boltzmann equation, for example. We observed in the earlier chapters that for high electrolyte concentrations, when the Debye screening length, κ^{-1}, is small, the analytical solution of the Poisson–Boltzmann equation is quite straightforward. This is not the case for numerical solutions. At high electrolyte concentrations, due to small Debye screening lengths, the electrical potential decays so rapidly near the charged surface, that many standard numerical differential equation solvers fail to capture the sharply decaying profile of the electric potential. This gives rise to convergence problems, or results that are grossly erroneous. Contrary to what was observed in case of most analytic solutions of electrostatic double layer interactions, numerical approaches work quite well when we consider the solution of the Poisson–Boltzmann equation at lower electrolyte concentrations or larger values of κ^{-1}. There are several other issues of the above nature that one needs to consider during a numerical solution of an electrokinetic problem. We will discuss some of these in the following sections.

In this chapter, the application of numerical techniques will be discussed in context of three problems. First, a solution of the Poisson–Boltzmann equation will be obtained in a cylindrical capillary, which contains two colloidal particles with their centers lying on the capillary axis. This is a highly non-linear problem, which involves solution of an elliptic partial differential equation for a single dependent variable, namely, the electric potential. In context of this problem, several issues regarding numerical implementation of finite element approximation will be discussed. The second problem to be discussed involves electrokinetic flow of an electrolyte solution in a charged cylindrical capillary under the combined influence of pressure and electric potential gradient. This will serve as an example of typical multiphysics problems associated with electrokinetic transport phenomena. We will solve both steady-state and time dependent versions of the governing equations. A specific case of the problem involving solely pressure driven flow will provide a direct measure of the streaming potential developed across the capillary. Finally, the third problem we consider in this chapter involves simulation of the electrophoretic mobility of a spherical particle under the influence of an externally imposed electric potential gradient. All three problems will be solved using the general forms of the governing equations obtained from the continuum models of electrostatics and electrokinetics. These problems are discussed with the goal of comparing the numerical results with various analytical results derived in the previous chapters. The comparisons will, on one hand, provide a measure of the accuracy of the numerical calculations. On the other hand, the comparisons will show that although approximate, some of the analytical results derived in the previous chapters are remarkably accurate.

14.2 NUMERICAL SOLUTION OF THE POISSON–BOLTZMANN EQUATION

The electric double layer (EDL) interaction energy or force between two colloidal particles is of critical importance in prediction of colloidal stability as noted in Chapter 11. The EDL force is susceptible to tremendous variations with physico-chemical conditions like the electrolyte concentration, the charging behavior of the particles (constant potential or constant charge), the dielectric constant of the solvent, as well as the shape of the particles. In many problems, for instance, particle transport in porous media, membrane filtration, chromatographic separations, and capillary electrophoresis, the colloidal particles are usually confined in narrow electrolyte filled capillaries. For modeling purposes, the capillaries are often assumed to be straight circular cylinders. Evaluating the EDL interaction force on a single colloidal particle trapped inside a narrow capillary is a complex task. An analytic solution of the problem exists for low potentials on the capillary wall and particle surfaces, which allows linearization of the governing Poisson–Boltzmann equation (Smith III and Deen, 1980). This solution is valid for symmetric electrolytes. Analytical solution for this problem does not exist for higher potentials on the interacting surfaces of the capillary and the particle, where the complete (non-linear) Poisson–Boltzmann equation needs to be solved. The problem becomes far more complex when we consider two particles trapped inside the capillary. There is no analytic solution of the Poisson–Boltzmann equation, even for a symmetric electrolyte, for this problem. Numerical solutions for this problem, however, have been reported (Bowen and Sharif, 1998; Gray *et al.*, 1999).

The calculation of the EDL interaction force between two spherical colloidal particles lying on the axis of a straight cylindrical capillary, with their surfaces separated by a distance h, forms an interesting case study in numerical analysis. The geometry

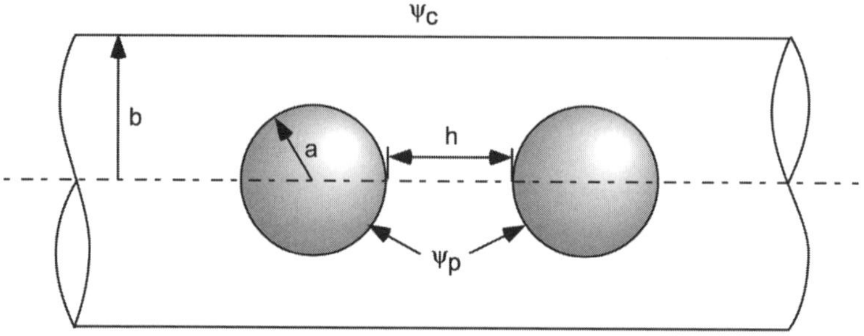

Figure 14.1. Schematic representation of two spherical colloidal particles interacting within a straight cylindrical capillary. The particles lie on the axis of the cylinder, and are separated by a distance h.

is schematically depicted in Figure 14.1. Here, two charged particles of radius a with surface potential (zeta potential) ψ_p are suspended in an electrolyte solution confined within an infinitely long capillary of radius b. The capillary wall has a surface potential of ψ_c. This problem represents a class of many-body (three-body to be specific) electrostatic interactions, where the interaction energy between the two spheres is influenced by the presence of the charged capillary wall. The first two numerical studies of this problem (Bowen and Sharif, 1998, 1999; Gray *et al.*, 1999) appeared in the late nineties, both employing the finite element technique. Among these, the numerical results of Bowen and Sharif (1998) incorrectly showed that there will be an attractive force between two similarly charged spherical particles at large separation distances when the particles are trapped inside a cylindrical capillary. Their study was spurred by the experiments of Larsen and Grier (1997), which indicated the presence of a long-range attraction between two similarly charged colloidal particles. The numerical predictions of Gray *et al.* (1999), however, did not show the presence of any attractive force between the spherical particles, although they employed the exact problem formulation used by Bowen and Sharif (1998). Later, it was theoretically demonstrated that the long-range attraction observed in the experiments of Larsen and Grier (1997) can not be replicated in the framework of the Poisson–Boltzmann equation (Sader and Chan, 1999; Sader and Chan, 2002).

We choose this problem as our model to demonstrate the application of finite element techniques for solution of the Poisson–Boltzmann equation. Apart from revisiting a topic of considerable recent interest, and addressing a geometrical system that is relevant to electrokinetic transport phenomena, the model also allows numerical evaluation of the interaction force between two spherical particles as a limiting case. This limiting case is obtained when the capillary radius, b, is considerably larger than the radius of the spheres, a. A variety of numerical solution schemes for the Poisson–Boltzmann equation to evaluate the interaction force between two spheres immersed in an infinitely large electrolyte bath are available (Glendinning and Russel, 1983; Carnie *et al.*, 1994; Stankovich and Carnie, 1996; Warszynski and Adamczyk, 1997). Amongst these, the solution of Glendinning and Russel (1983) is based on a multipole expansion technique, and applies to the linearized Poisson–Boltzmann equation for two equal spheres in a symmetric electrolyte. The remaining studies are based on a finite difference approach, employing a bispherical coordinate system to tackle the boundary condition at infinity. Use of finite element and boundary element techniques for solving the problem have also been reported (Bowen and Sharif, 1997; Grant and Saville, 1995). Incidentally, the boundary element technique can only be applied to solve the linearized Poisson–Boltzmann equation.

For the geometry of the present problem, the cylindrical coordinate system (r, θ, x) appears to be the most convenient, since it allows the utilization of axial symmetry (invariance of the parameters with respect to the angular coordinate, θ) to render the three dimensional geometry using a two dimensional model. The typical steps of a numerical solution procedure based on finite element analysis involve: (i) model formulation, (ii) design of the computational geometry, (iii) mesh generation, (iv) solution of the governing equations, (v) postprocessing, and (vi) model validation. In the following, these steps are described, focussing on how the approach is employed for the solution of the Poisson–Boltzmann equation.

14.2.1 Problem Formulation

For a symmetric ($z : z$) electrolyte, the Poisson–Boltzmann equation can be written as

$$\nabla^2 \Psi = \kappa^2 \sinh(\Psi) \tag{14.1}$$

where Ψ $(= ze\psi/k_B T)$ is the scaled electric potential, z is the ionic valence, e is the magnitude of the electronic charge, ψ is the electrical potential (V). The parameter κ in Eq. (14.1) is the inverse Debye length, given by

$$\kappa = \sqrt{\frac{2z^2 e^2 n_\infty}{\epsilon k_B T}} \tag{14.2}$$

where n_∞ is the bulk electrolyte concentration (m^{-3}) and ϵ is the dielectric permittivity of the electrolyte solution ($\mathrm{C}^2/\mathrm{Nm}^2$).

The Poisson–Boltzmann equation (14.1) is applied to the electrolyte filled region in the capillary (Figure 14.1). When the particle centers reside on the axis of the cylindrical capillary, one can employ axial symmetry to represent the problem in a two-dimensional axisymmetric cylindrical coordinate system (r, x) as shown in Figure 14.2. The explicit form of Eq. (14.1) in axisymmetric cylindrical coordinates is

$$\frac{\partial^2 \Psi}{\partial r^2} + \frac{1}{r}\frac{\partial \Psi}{\partial r} + \frac{\partial^2 \Psi}{\partial x^2} = \kappa^2 \sinh(\Psi) \tag{14.3}$$

Note that use of axial symmetry to render the problem two dimensional is only possible when the particle centers are on the axis of the capillary. If, further, one scales the coordinates r and x with respect to the Debye screening length, κ^{-1}, such that $\bar{r} = \kappa r$

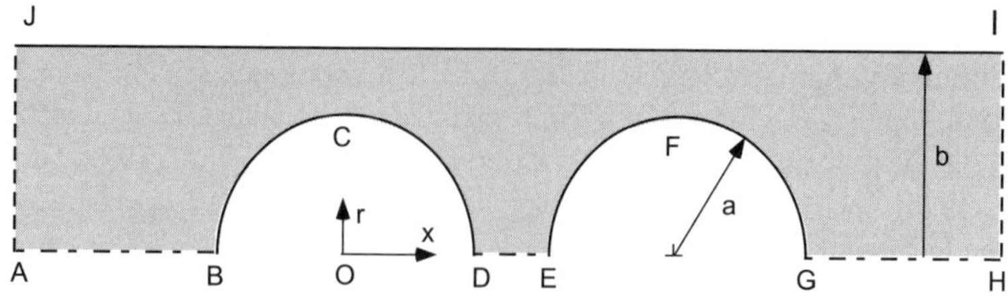

Figure 14.2. Schematic representation of the geometrical domain used in the finite element analysis. The point O, located at the center of the sphere BCD, is the origin of the cylindrical coordinate system.

and $\bar{x} = \kappa x$,[1] Eq. (14.3) becomes

$$\frac{\partial^2 \Psi}{\partial \bar{r}^2} + \frac{1}{\bar{r}}\frac{\partial \Psi}{\partial \bar{r}} + \frac{\partial^2 \Psi}{\partial \bar{x}^2} = \sinh(\Psi) \tag{14.4}$$

Equation (14.4) is the non-dimensional form of the Poisson–Boltzmann equation written in axisymmetric cylindrical coordinates.

We now define the boundary conditions for this equation. Referring to Figure 14.2, and assuming constant surface potentials on the spheres and the cylindrical capillary, we can write

$$\Psi = \Psi_p \quad \text{for } \partial\Omega \in \text{ BCD and EFG} \tag{14.5-a}$$

$$\Psi = \Psi_c \quad \text{for } \partial\Omega \in \text{ IJ} \tag{14.5-b}$$

The above conditions specify the surface potentials on the spheres and the capillary wall, respectively. Here, the potential on both spheres are assumed to be equal (Ψ_p), while the potential on the capillary wall, Ψ_c, may be different from the particle surface potentials.

On the axis of the cylindrical capillary, namely, segments AB, DE, and GH (Figure 14.2), one can employ the symmetry condition, stated as

$$\mathbf{n} \cdot \bar{\nabla}\Psi = 0 \quad \text{for } \partial\Omega \in \text{ AB, DE, and GH} \tag{14.5-c}$$

In the above equation, **n** represents a unit normal to the surface. In the present system, the axial symmetry simply means that at $r = 0$, $\partial\Psi/\partial\bar{r} = 0$.

We are now left only with two line segments, namely, AJ and HI, where the boundary conditions need to be defined. Unfortunately, there are no well-defined boundary

[1] In this case, the gradient operator, ∇, can be scaled as

$$\bar{\nabla} = \left(\frac{\partial}{\partial \bar{r}}\mathbf{i}_{\bar{r}} + \frac{\partial}{\partial \bar{x}}\mathbf{i}_{\bar{x}}\right) = \kappa^{-1}\nabla$$

where $\mathbf{i}_{\bar{r}}$ and $\mathbf{i}_{\bar{x}}$ represent unit vectors along the scaled coordinates, $\bar{r}$ and $\bar{x}$, respectively.

conditions for these two segments unless these are located at infinity. In a numerical solution, these lines cannot be assumed to be located at infinite distances from the origin. Let us first consider what boundary conditions can be applied on segments AJ and HI, if there are no particles in the capillary. In this case, if the capillary is assumed to be infinitely long, then the electric field component along the axial direction will be zero on AJ and HI. In other words, $\partial\Psi/\partial\bar{x} = 0$ on these two segments. Thus, one can assume that if AJ and HI are sufficiently far away from the particles, the electric field at these segments will have negligible influence of the particles. In other words, placing AJ and HI sufficiently far away from the particles allows the application of a symmetry boundary condition of the form

$$\mathbf{n} \cdot \bar{\nabla}\Psi = 0 \quad \text{for } \partial\Omega \in \text{ AJ and HI} \tag{14.5-d}$$

This condition is not exactly a true boundary condition. The validity of this artificial and approximate boundary condition must be verified once the numerical solution is obtained. However, it is discernable that as AJ and HI are moved further away from the particles, the accuracy of this boundary condition will improve.

Equations (14.5-a)–(14.5-d) represent the boundary conditions for the Poisson–Boltzmann equation applied to the geometry of Figure 14.2. If, instead of constant surface potential, constant charge density conditions are applied on the particles and the capillary wall, Eqs. (14.5-a) and (14.5-b) are replaced by constant surface charge boundary conditions, given by

$$-\mathbf{n} \cdot \bar{\nabla}\Psi = \frac{ze}{\kappa\epsilon k_B T} q_p = \sigma_p \quad \text{for } \partial\Omega \in \text{ BCD and EFG} \tag{14.6-a}$$

and

$$-\mathbf{n} \cdot \bar{\nabla}\Psi = \frac{ze}{\kappa\epsilon k_B T} q_c = \sigma_c \quad \text{for } \partial\Omega \in \text{ IJ} \tag{14.6-b}$$

Here, q_p and q_c are the charge densities (C/m^2) on the surfaces of the particles and the capillary wall, respectively, while σ_p and σ_c are the corresponding non-dimensional surface charge densities.

Equation (14.4), along with Eqs. (14.5-a)–(14.5-d) can be solved using an appropriate numerical technique. This will provide the results for constant potential surfaces. Replacing Eqs. (14.5-a) and (14.5-b) with (14.6-a) and (14.6-b), will yield the corresponding results for constant surface charge boundary conditions. The above formulation can be considered as fairly general, and should provide insight into the behavior of EDL interactions for a wide range of conditions. One should note a few simplifications made in the formulation of the problem. First, utilization of axial symmetry reduces the dimensionality of the problem (from a three dimensional domain to an axisymmetric two dimensional plane). Secondly, scaling the length with respect to the screening length of the EDL reduces the number of independent variables – one can now simply perform the calculations for different selected values of the parameter κa, where a is the particle radius, without explicitly having to consider κ^{-1} and a separately. Furthermore, such scaling makes the implementation of the numerical procedure more facile, since one does not need to deal with real length scales of the system anymore.

14.2.2 Finite Element Formulation

The finite element formulation employs the so-called general form of a partial differential equation, written in rectangular cartesian coordinates (x, y, z), given by

$$\frac{\partial \phi}{\partial t} + \nabla \cdot \Gamma = R \tag{14.7}$$

where ϕ is the dependent variable, t is time, ∇ is the gradient operator in cartesian coordinates, Γ represents the spatial gradient of the dependent variable, and R is the source term or a forcing function. For a time-independent problem, Eq. (14.7) becomes

$$\nabla \cdot \Gamma = R \tag{14.8}$$

The function Γ is written in a two-dimensional cartesian coordinate system (x, y) as

$$\Gamma = \nabla \phi = \frac{\partial \phi}{\partial x}\mathbf{i}_x + \frac{\partial \phi}{\partial y}\mathbf{i}_y \tag{14.9}$$

where $\mathbf{i}_x$ and $\mathbf{i}_y$ are unit vectors along the x and y coordinate directions, respectively, and the gradient operator in the (x, y) coordinate system is

$$\nabla = \frac{\partial}{\partial x}\mathbf{i}_x + \frac{\partial}{\partial y}\mathbf{i}_y \tag{14.10}$$

Using Eqs. (14.9) and (14.10), Eq. (14.8) can be written as

$$\left(\frac{\partial}{\partial x}\mathbf{i}_x + \frac{\partial}{\partial y}\mathbf{i}_y\right) \cdot \left(\frac{\partial \phi}{\partial x}\mathbf{i}_x + \frac{\partial \phi}{\partial y}\mathbf{i}_y\right) = \frac{\partial^2 \phi}{\partial x^2} + \frac{\partial^2 \phi}{\partial y^2} = R \tag{14.11}$$

The forcing function, R, can be any function of ϕ, $\nabla\phi$, x, and y, and is obtained from the governing physics of the problem.

Typically, most finite element codes are developed to solve a partial differential equation in the general form written in cartesian coordinates represented by Eq. (14.7) or Eq. (14.8). Employing such a cartesian computational domain simplifies conversion of the partial differential equations from the general form to the so-called "weak form," which is often a necessary step in implementation of finite element techniques (Zienkiewicz and Taylor, 1989). Consequently, when one does not develop the entire finite element code for solving a problem in a specific coordinate system, but relies on either standard software, or uses a "black-box" code, it is perhaps safer to recast any given model into the general form and solve the problem in a cartesian-type coordinate frame. We now investigate how this can be achieved for our model problem, which was developed in an axisymmetric cylindrical coordinate system.

Equation (14.4), which represents the Poisson–Boltzmann equation in a cylindrical coordinate system, can be rearranged as

$$\frac{\partial^2 \Psi}{\partial \bar{r}^2} + \frac{\partial^2 \Psi}{\partial \bar{x}^2} = \sinh(\Psi) - \frac{1}{\bar{r}}\frac{\partial \Psi}{\partial \bar{r}} \tag{14.12}$$

It is immediately evident from Eqs. (14.11) and (14.12), that the left hand side of Eq. (14.12) can be treated as the Laplacian operator written in cartesian coordinates. In this case, one obtains a modified form of the forcing function, given by

$$R = \sinh(\Psi) - \frac{1}{\bar{r}}\frac{\partial \Psi}{\partial \bar{r}} \tag{14.13}$$

The form of Eq. (14.13) implies that the term $(\partial \Psi/\partial \bar{r})/\bar{r}$ arising from the Laplacian operator in the cylindrical coordinate system can be added to the forcing function term, R. This rearrangement renders the resulting partial differential equation to behave like one written in a cartesian coordinate system. Note that the solution of Eq. (14.12) will still provide the variation of the dependent variable Ψ with the scaled radial ($\bar{r}$) and axial ($\bar{x}$) coordinates in the cylindrical coordinate system. However, the numerical code will "see" the governing equation as one written in a cartesian coordinate system, with a slightly different forcing function. This technique allows a computer code written for solving a partial differential equation in the cartesian coordinate system the flexibility of tackling other types of coordinate systems, as long as it can handle different definitions of the forcing function R.

Writing the non-dimensional Poisson–Boltzmann equation in the form of Eq. (14.12) allows its solution in many standard finite element programs. The finite element method approximates the spatial variation of the dependent variable (ϕ) as a combination of piecewise polynomial shape functions. The shape function is defined such that it has a value of 1 at a given node in an element and smoothly approaches a value of zero at all other nodes of the element. The sum of these shape functions, multiplied with appropriate weights, over all the nodes constituting the element provides the approximation for the dependent variable. Denoting the basis functions as ξ_i and the weighting variable (also known as the degree of freedom) as U_i, the solution can be written for an element as

$$\phi \simeq \sum_i U_i \xi_i \tag{14.14}$$

where the index i refers to the nodes in an element. The degrees of freedom are the desired solution variables in a finite element solution. In other words, one attempts to obtain the appropriate values of U_i that allow the constructed solution based on the summation of the basis functions ξ_i to match the actual solution of the governing partial differential equation.

14.2.3 Mesh Generation

The implementation of a finite element solution starts with the definition of a shape function and an element type. Typically, one discretizes a two dimensional computational domain into triangular or quadrilateral elements. In the present formulation, triangular elements were employed with quadratic Lagrangian shape functions. The quadratic triangular elements typically have six nodes, three located at the vertices, and three at the mid-points of the triangle edges. The computational domain is discretized into non-uniform triangular elements, with smaller elements placed near the curved boundaries to ensure a more refined spatial discretization as shown in Figure 14.3.

The mesh largely dictates the accuracy of the finite element solution. The mesh should adequately resolve the geometry, be sufficiently refined in regions where large gradients of the dependent variable exist, and should be of good "quality". One needs to place a large number of elements in the regions where the boundaries are curved to capture the curvature adequately, such that geometrical errors arising from discretization is minimized. One also needs to have a refined mesh in regions where the dependent variable changes sharply (the spatial gradients of ϕ are large). Finally, the element quality is another important criterion in dictating the accuracy of the solution. The quality of an element, in simple terms, is a measure of how comparable the lengths of the three sides of a triangular element are; the similar the lengths are, the better the quality of the element. The degrees of freedom indicate the size of the matrix equation required in the solution for a given mesh. It depends on the number of elements, number of nodes in the element, number of governing equations, and the boundary conditions. The degrees of freedom dictate the size of the computational problem. The efficient solution of large matrix equations with minimal memory requirement constitutes the main algorithmic finesse of finite element solvers.

For many problems, it is difficult to assess *a priori* which regions of the computational domain will have large spatial gradients, and hence, will require a refined mesh. Accordingly, it is common to employ mesh refinement strategies to converge to an accurate solution. Normally, most numerical calculations involve setting up an error criterion, and a tolerance level for that error. Then, the solution procedure is repeated with finer and finer meshes, until the tolerance is reached. Refinement of the mesh can

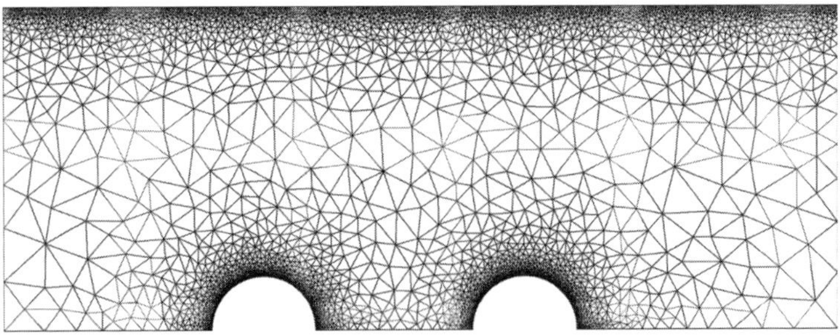

Figure 14.3. A typical initial finite element mesh consisting of quadratic triangular elements. The mesh has 10,299 elements and 21,416 degrees of freedom, with a minimum element quality of 0.53. Note that the elements are more refined near the curved surfaces of the particles and on the surface of the capillary wall. This is done to ensure that the curved surfaces and the regions where the potential decays more rapidly are adequately resolved.

be performed by simply making every element smaller by a preset factor, a technique sometimes referred to as uniform refinement. However, this is computationally inefficient, and is not usually done in a commercial solver. These solvers frequently use adaptive refinement, where the mesh is refined locally in the regions where the spatial gradients are large. This type of adaptive mesh refinement is important in solution of Poisson–Boltzmann equation, since the spatial gradients of the electric potential are large in the vicinity of the charged surfaces, and refined meshes are often necessary in these regions.

During refinement of mesh to obtain a more accurate solution, one does not have an absolute measure of the accuracy of the solution. The adaptation is usually done by comparing the results of two consecutive solutions of the same problem with two levels of mesh refinement. The difference between these two consecutive results then forms the basis for further mesh refinement. In the present problem, one of our goals is to obtain the net electric double layer force experienced by one of the particles in the capillary. Accordingly, we use this quantity as a metric for the mesh refinement. The calculation of the force will be discussed shortly. Table 14.1 depicts the manner in which the electric double layer force converges as the mesh is refined. The forces were obtained for a scaled separation $\kappa h = 0.4$ and scaled particle size $\kappa a = 1.0$, with a very large capillary radius $\kappa b \simeq 10$. The results for both uniform and adaptive refinement are shown in Table 14.1. It is evident that the predicted force converges quite rapidly to a mesh independent value with adaptive mesh refinement as compared to uniform refinement, and requires considerably fewer number of elements in attaining the accuracy. This superiority of adaptive mesh refinement is ascribed to the ability of the technique to selectively place refined elements in the regions of the computational domain where more accuracy is required.

14.2.4 Solution Methodology

Once the governing equations have been defined with appropriate boundary conditions, and the initial mesh has been generated, the weak form of the equations is written for each element. Incorporating information regarding how the various nodes

TABLE 14.1. Scaled Electric Double Layer Interaction Forces (see text for details) Obtained from Finite Element Calculations Employing Uniform and Adaptive Mesh Refinement.

	Uniform Refinement		Adaptive Refinement	
Step Number	No. of Elements	Force	No. of Elements	Force
1	212	23.085	212	23.085
2	848	25.662	505	25.178
3	3392	26.581	1153	26.286
4	13568	26.864	2506	26.682
5	54272	26.963	5496	26.903
6	—	—	11421	26.910

in adjacent elements are connected, a global set of algebraic equations is then developed. These coupled algebraic equations have the degrees of freedom (DOF), or the weighting parameters in Eq. (14.14), as unknowns. The degrees of freedom represent the number of coupled algebraic equations requiring simultaneous solution. Frequently this is a large number, and the resulting coefficient matrices are quite large. Furthermore, unlike finite difference techniques, these matrices may not be very ordered. In most finite difference formulations, the non-zero elements of the coefficient matrix are packed close to the diagonal elements. This is generally not the case for finite element problems. Finally, for non-linear problems, the matrix equations represent the linearized forms of the governing equation, and must be solved iteratively. Consequently, one has to solve large matrix equations employing specialized types of solvers during finite element analysis. This is one of the most memory intensive and computationally burdensome aspects of finite element calculations. This is also the reason why commercial solvers are so popular in finite element analysis. Most commercial finite element packages incorporate a robust matrix solver. These solvers can be of two types, namely, direct solvers and iterative solvers. Direct solvers are predominantly based on implementations of the Gaussian elimination procedure. Iterative solvers coupled with appropriate pre-conditioners are often less memory intensive than direct solvers. However, for highly non-linear problems, these techniques require a good initial guess for the solution. There are different algorithms for iterative solution, for instance, techniques based on Broyden search, generalized minimum residual (GMRES), and quasi-minimal residual (QMR).

For the present problem, a direct solver was used to obtain the results. Starting from an initial mesh, the problem was solved incorporating adaptive mesh refinement. The tolerance in the global error was set to 10^{-6}. A maximum of 10 iterations in mesh adaptation was allowed. Finally, a limit of 30,000 for the maximum number of elements was imposed. Whenever one of the above criterion was met, the solution was terminated. The entire computation was performed in FEMLab.

As mentioned earlier, the boundary conditions on segments AJ and HI (Figure 14.2) are artificial conditions, which may be valid only when these boundaries are located sufficiently far away from the particles. To ensure that the influence of these boundaries are negligible on the obtained solution, the simulations were performed by setting the length of AB and GH (Figure 14.2) to different values ranging from 1 to 10 times the particle radius. The resulting force calculations indicate that the artificial boundary conditions have negligible effect on the solution (computed force) when the lengths of the segments AB and GH are set to values $\geq 2\kappa a$. Accordingly, in all subsequent calculations, the boundaries AJ and HI were located at a distance 2.5 to 5 times the particle radius from the nearest particle surfaces.

14.2.5 Postprocessing: Calculation of the EDL Force

The finite element solution provides the electric potential distribution over the computational domain. From this, the electric field components can be evaluated. One should note that to evaluate the EDL force on a particle, the components of the electric field (which constitute the Maxwell stress) must be obtained accurately from the solution of the Poisson–Boltzmann equation. Accordingly, the solution should provide highly accurate estimates of the electric field or the gradient of the potential. In the present case, a fairly accurate solution for the potential distribution could be obtained even using linear elements. However, since our focus was to obtain an accurate measure of the electric field, quadratic elements were employed in the solution. Taking this to a higher order accuracy, one could even have employed cubic elements, but higher order elements lead to more involved computation, and sometimes may even lead to an unstable solution. Decisions of this type are quite important in application of a numerical technique, and sound judgement in this regard is often acquired through experience.

The force acting on a particle can be obtained by integrating the total stress tensor over the surface of the particle. Referring back to Chapter 5 (Section 5.4, see Example 5.8), the stress tensor at a given point on the particle surface is

$$\overline{\overline{\mathbf{T}}} = -p\overline{\overline{\mathbf{I}}} + \overline{\overline{T}}_e = -\left(p + \frac{1}{2}\epsilon\mathbf{E}\cdot\mathbf{E}\right)\overline{\overline{\mathbf{I}}} + \epsilon\mathbf{E}\mathbf{E} \tag{14.15}$$

Here $p\overline{\overline{\mathbf{I}}}$ is the hydrostatic (or osmotic) stress, $\overline{\overline{T}}_e$ is the Maxwell stress tensor, Eq. (3.156), $\mathbf{E}$ $(= -\nabla\psi)$ is the electric field, and $\overline{\overline{\mathbf{I}}}$ is a unit tensor. Integrating the stress tensor over the surface of a spherical particle yields the force

$$\mathbf{F} = \int_S \overline{\overline{\mathbf{T}}}\cdot\mathbf{n}dS = \int_S \left[-\left(p + \frac{1}{2}\epsilon\mathbf{E}\cdot\mathbf{E}\right)\overline{\overline{\mathbf{I}}} + \epsilon\mathbf{E}\mathbf{E}\right]\cdot\mathbf{n}dS \tag{14.16}$$

Here, $\mathbf{n}$ is a unit outward normal to the particle surface. The integration should be performed over the particle surface S. It should be noted that the above expression for force is independent of the choice of the surface over which the integration is performed. In other words, the force can be computed by integrating the stress tensor over either particle surfaces BCD or EFG in Figure 14.2.

Let us explicitly formulate the force for the present problem. From Eq. (5.105), it follows that for a symmetric $(z:z)$ electrolyte, the osmotic (hydrostatic) stress contribution is given by

$$p = 2n_\infty k_B T\left[\cosh(\Psi) - 1\right] \tag{14.17}$$

where n_∞ is the bulk electrolyte concentration. Recalling from Chapter 5 that

$$\frac{n_\infty k_B T}{\kappa^2} = \frac{\epsilon}{2}\left(\frac{k_B T}{ze}\right)^2$$

one can write the osmotic stress as

$$p = -\kappa^2\epsilon\left(\frac{k_B T}{ze}\right)^2\left[1 - \cosh(\Psi)\right] \tag{14.18}$$

It should be noted that $\kappa^2\epsilon(k_BT/ze)^2$ has units of force per unit area (N/m^2).

The Maxwell stress for the two dimensional system is given explicitly in a matrix form by

$$
\begin{aligned}
\bar{\bar{T}}_e &= -\left(\frac{1}{2}\epsilon\mathbf{E}\cdot\mathbf{E}\right)\bar{\bar{\mathbf{I}}} + \epsilon\mathbf{E}\mathbf{E} \\
&= -\frac{1}{2}\epsilon\left[(\partial\psi/\partial r)^2 + (\partial\psi/\partial x)^2\right]\begin{bmatrix}1 & 0\\ 0 & 1\end{bmatrix} \\
&\quad + \epsilon\begin{bmatrix}(\partial\psi/\partial r)^2 & (\partial\psi/\partial r)(\partial\psi/\partial x)\\ (\partial\psi/\partial r)(\partial\psi/\partial x) & (\partial\psi/\partial x)^2\end{bmatrix}
\end{aligned}
$$

This expression can be written in terms of the non-dimensional potential ($\Psi = ze\psi/k_BT$) and the scaled coordinates ($\bar{r} = \kappa r$ and $\bar{x} = \kappa x$) as

$$
\begin{aligned}
\bar{\bar{T}}_e = \kappa^2\epsilon\left(\frac{k_BT}{ze}\right)^2 &\left\{-\frac{1}{2}\left[\left(\frac{\partial\Psi}{\partial\bar{r}}\right)^2 + \left(\frac{\partial\Psi}{\partial\bar{x}}\right)^2\right]\begin{bmatrix}1 & 0\\ 0 & 1\end{bmatrix}\right. \\
&\left. + \begin{bmatrix}(\partial\Psi/\partial\bar{r})^2 & (\partial\Psi/\partial\bar{r})(\partial\Psi/\partial\bar{x})\\ (\partial\Psi/\partial\bar{r})(\partial\Psi/\partial\bar{x}) & (\partial\Psi/\partial\bar{x})^2\end{bmatrix}\right\}
\end{aligned}
$$

Combining the osmotic and Maxwell stresses, and simplifying the resulting expression, the total stress tensor, Eq. (14.15), can be written explicitly in a matrix form as

$$
\bar{\bar{\mathbf{T}}} = \kappa^2\epsilon\left(\frac{k_BT}{ze}\right)^2\begin{bmatrix}(1-\cosh\Psi) + \frac{1}{2}\left(\Psi_{\bar{r}}^2 - \Psi_{\bar{x}}^2\right) & \Psi_{\bar{r}}\Psi_{\bar{x}}\\ \Psi_{\bar{r}}\Psi_{\bar{x}} & (1-\cosh\Psi) + \frac{1}{2}(\Psi_{\bar{x}}^2 - \Psi_{\bar{r}}^2)\end{bmatrix} \tag{14.19}
$$

where the terms $\Psi_{\bar{r}}(= \partial\Psi/\partial\bar{r})$ and $\Psi_{\bar{x}}(= \partial\Psi/\partial\bar{x})$ are shorthand notations for the potential gradients (or electric field components) along $\bar{r}$ and $\bar{x}$, respectively.

For the spherical particle BCD in Figure 14.2, which is centered at the origin of the coordinate system, the unit outward surface normal can be written as

$$
\begin{aligned}
\mathbf{n} &= n_{\bar{r}}\mathbf{i}_{\bar{r}} + n_{\bar{x}}\mathbf{i}_{\bar{x}} \\
&= \frac{\bar{r}}{\sqrt{\bar{r}^2+\bar{x}^2}}\mathbf{i}_{\bar{r}} + \frac{\bar{x}}{\sqrt{\bar{r}^2+\bar{x}^2}}\mathbf{i}_{\bar{x}}
\end{aligned} \tag{14.20}
$$

Using the expressions of the total stress tensor, Eq. (14.19) and the unit surface normal, Eq. (14.20) in Eq. (14.16), one obtains

$$
\mathbf{F} = \kappa^2\epsilon\left(\frac{k_BT}{ze}\right)^2\int_S\begin{bmatrix}\left[1-\cosh\Psi + \frac{1}{2}\left(\Psi_{\bar{r}}^2 - \Psi_{\bar{x}}^2\right)\right]n_{\bar{r}} + \Psi_{\bar{r}}\Psi_{\bar{x}}n_{\bar{x}}\\ \Psi_{\bar{r}}\Psi_{\bar{x}}n_{\bar{r}} + \left[1-\cosh\Psi + \frac{1}{2}(\Psi_{\bar{x}}^2 - \Psi_{\bar{r}}^2)\right]n_{\bar{x}}\end{bmatrix}dS \tag{14.21}
$$

Note that the term within the integral represents a vector quantity, with the upper line indicating the force component acting along the $\bar{r}$ coordinate, and the lower line representing the force component along $\bar{x}$.

Since we are interested in the component of the force acting along the axis of the cylinder ($\bar{x}$), we obtain this force as

$$F_{\bar{x}} = \mathbf{F} \cdot \mathbf{i}_{\bar{x}} = \kappa^2 \epsilon \left(\frac{k_B T}{ze} \right)^2 \int_S \left\{ \Psi_{\bar{r}} \Psi_{\bar{x}} n_{\bar{r}} + \left[1 - \cosh \Psi + \frac{1}{2} (\Psi_{\bar{x}}^2 - \Psi_{\bar{r}}^2) \right] n_{\bar{x}} \right\} dS \tag{14.22}$$

Now, for the spherical particle denoted by the arc BCD, the area elements can be defined in the present coordinate system as $dS = 2\pi \bar{r} d\bar{x}/\kappa^2$. Substituting this in Eq. (14.22), one obtains,

$$F_{\bar{x}} = 2\pi \epsilon \left(\frac{k_B T}{ze} \right)^2 \int_{\mathrm{BCD}} \left\{ \Psi_{\bar{r}} \Psi_{\bar{x}} n_{\bar{r}} + \left[1 - \cosh \Psi + \frac{1}{2} (\Psi_{\bar{x}}^2 - \Psi_{\bar{r}}^2) \right] n_{\bar{x}} \right\} \bar{r} d\bar{x} \tag{14.23}$$

Equation (14.23) represents the net interaction force acting along the axial coordinate ($\bar{x}$) on the particle BCD in Figure 14.2. Referring to Figure 14.2, the force will be acting toward the left (along the negative x direction) if it is repulsive, while it will be acting toward the right (along positive x direction) if there is attraction between the two spherical particles. We note that representing the force in the form of Eq. (14.23) allows the use of the scaled potentials and their spatial gradients in the same form as employed during the finite element solution. Finally, the force can be expressed in a non-dimensional form as

$$f_{\bar{x}} = \frac{F_{\bar{x}}}{\epsilon} \left(\frac{ze}{k_B T} \right)^2 \tag{14.24}$$

In the remainder of the discussion in this section, we will use this non-dimensionalized form of the force. It should be noted that one can obtain an analogous expression for the force on the particle EFG. While the expression for the force will be somewhat different from Eq. (14.23), the magnitude of the force calculated on either BCD or EFG will be same. This is a very pertinent self-consistency check of the computation, and should be carried out to ensure the accuracy of the numerical program.

The integration of the stress tensor according to Eq. (14.23) can be conducted employing the built-in postprocessing unit of the finite element software. The boundary integration procedure adopted for this purpose employs Gaussian quadrature, which provides a highly accurate result with few integration points.

14.2.6 Validation of Numerical Results

A validation of the finite element solution against existing results can be obtained for the case of two spherical particles interacting in a large electrolyte reservoir. The interaction between two constant potential or constant charge spheres is a well-studied problem, and numerous analytical and numerical solutions exist for this problem. In the present formulation, the problem of two particles interacting inside a cylindrical capillary can be modified to obtain a solution for the two interacting spheres. This can be performed by making the capillary radius b considerably larger than the particle radii, a, and making the capillary surface neutral. The capillary surface can be made neutral by simply assigning a boundary condition of the form

$$\mathbf{n} \cdot \bar{\nabla}\Psi = 0 \quad \text{on IJ} \tag{14.25}$$

One should, however, note that this artificial condition can only be applied when $b \gg a$. Employing these conditions, one can obtain the interaction force between two spheres separated by a distance h without having to re-formulate the problem.

The electrostatic force on a particle determined from the finite element analysis for this limiting case can be compared with independent estimates of the force available in literature. A numerical solution procedure for the Poisson–Boltzmann equation was developed by Carnie *et al.* (1994), which utilized a bispherical coordinate system. They employed a cubic Hermite collocation based technique for discretization of the governing equation, followed by an iterative solution of the resulting non-linear system of equations employing a Newton–Raphson technique to obtain the interaction force. This numerical solution provides very accurate estimates of the electrostatic interaction force between two spherical particles. The solution is applicable for any surface potential on the particles, and for a wide range of separation distances and the parameter κa. One should recall from our earlier discussions that the numerical solutions tend to falter at large values of the parameter κa. However, these solutions are accurate when $\kappa a \leq 10$.

Exact analytical expressions for the sphere-sphere interaction force based on a solution of the non-linear Poisson–Boltzmann equation are not available. There are several results based on the solution of the linearized Poisson–Boltzmann equation. These range from the simple result of Hogg *et al.* (1966), which is based on Derjaguin's approximation, to some complex series solutions based on a multipole expansion technique (Glendinning and Russel, 1983).

Here, we will compare the numerical results obtained from the finite element calculations with the results of Carnie *et al.* (1994), as well as, with the simple analytic result of Hogg *et al.* (1966). Figure 14.4 shows a comparison between the scaled interaction force, f_z, obtained from the finite element calculations, and those obtained by applying the Hermite collocation solution of Carnie *et al.* (1994). The agreement between the two solutions is excellent, given that they were obtained using completely different coordinate systems, and two different numerical techniques. However, the good agreement is expected, since we are essentially solving the same physical problem. What makes the agreement remarkable is that in the finite element calculations, the cylinder wall radius was chosen to be only five times the particle radius, whereas the solution in the bispherical coordinate system employs the true boundary condition at infinity. There is another difference between the two calculations. The interaction force on a sphere was computed in the finite element calculations using an integration of the stress tensor over the particle surface. In the Hermite collocation solution, the interaction force was computed by integrating the stress tensor at the mid-plane between the two particles. Integration of the stress tensor over the mid-plane makes considerably more sense when one employs the bispherical coordinate system. However, this is inadequate for the geometry employed in the finite element calculations.

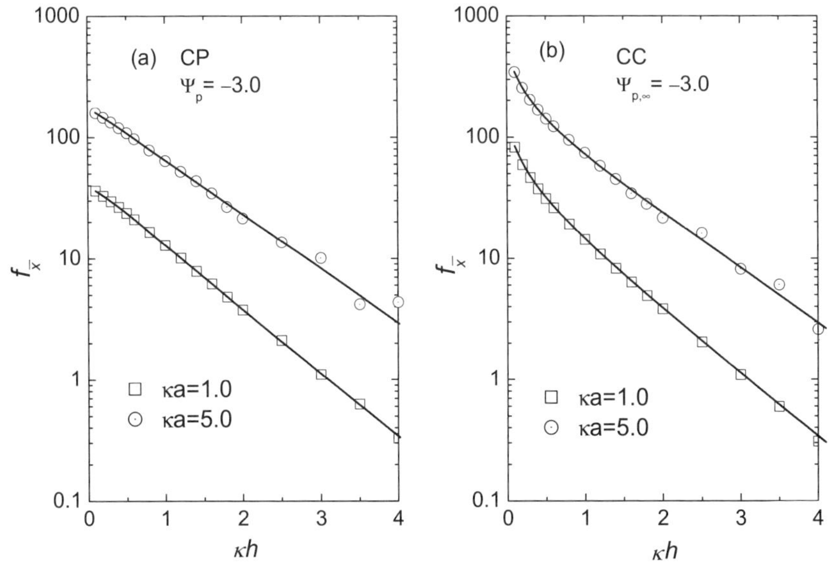

Figure 14.4. Comparison of the electric double layer force between two spherical particles obtained using the finite element analysis (symbols) with the calculations of Carnie *et al.* (1994) (lines) for $b \gg a$ and a neutral capillary surface. (a) Results for constant surface potential (CP) on the particles. (b) Results for constant surface charge density (CC) on the particles. In case (b) the surface charge density on an isolated particle was computed from the given value of $\Psi_{p,\infty}$. Details of this comparison is available in Das *et al.* (2003).

The agreement between the forces calculated on two surfaces clearly indicates that integration of the stress tensor over any closed surface enclosing a sphere will provide the total interaction force. In case of the bispherical coordinate system, the integration over the mid-plane actually represents an infinite surface enclosing one of the spheres.

The comparison of the forces obtained from the two procedures indicates the ability of finite element approximation to provide a reasonable estimate of the EDL interaction forces. This, however, comes at the expense of fairly involved computation. It should be noted that the good agreement between the two force estimates could not be attained without using adaptive mesh refinement, employing a large number of elements (generally 20,000 to 30,000), and carefully choosing the convergence criterion. Furthermore, one should note that all the boundary conditions employed in the finite element model for the sphere-sphere interaction, except for those on the particle surfaces, are artificial. If one plans to solve the Poisson–Boltzmann equation to obtain the interaction force between two spheres in an infinitely large domain, use of bispherical coordinates is definitely the most appropriate route. Nevertheless, the accuracy of the finite element calculations for this limiting case does validate the correctness of the numerical solution procedure, and provides confidence regarding its applicability in cases when the effect of the charged capillary wall on the interaction force is calculated.

In Figure 14.5, the scaled interaction force predicted by the finite element calculations are compared with the scaled EDL force computed by the expression given by Hogg *et al.* (1966), often referred to as the Hogg–Healy–Fuerstenau (HHF) expression. The HHF expression for the interaction force between two identical spheres is given by Eq. (13.70) for constant surface potential on the spheres. Note that for identical surface potentials, the double layer asymmetry parameter, Da, (*cf.*, Eq. 13.73) will be zero, resulting in a somewhat simplified form of the equation. One should note that the expression of Hogg *et al.* (1966) is applicable only when the surface potentials on the particles are small (strictly speaking, $ze\psi_p/k_BT \ll 1$), and for large particles ($\kappa a > 10$). The above limitations of the HHF result stems from the use of linearized Poisson–Boltzmann equation and Derjaguin's approximation. Nevertheless, this expression has been used to obtain the interaction forces between particles that have much larger surface potentials, and sometimes even for $\kappa a \simeq 2$). The comparison of the two calculations in Figure 14.5 (left) indeed shows that at low surface potentials, $\Psi_p = -1$, the analytical HHF result provides fairly accurate estimates of the interaction force for $\kappa a \simeq 5$. The deviation of the HHF result from the finite element predictions increases at lower values of κa. In these cases, the HHF expression underpredicts the interaction force at small separations, but overpredicts the interaction force at large separations. Figure 14.5 (right) shows the comparison between the analytic and numerical results for $\kappa a = 5$ corresponding to different values of surface potentials. At higher surface potentials, the analytical result deviates from the numerical predictions of the interaction force considerably. The deviations indicate the extent of error introduced in the force estimates due to linearization of the

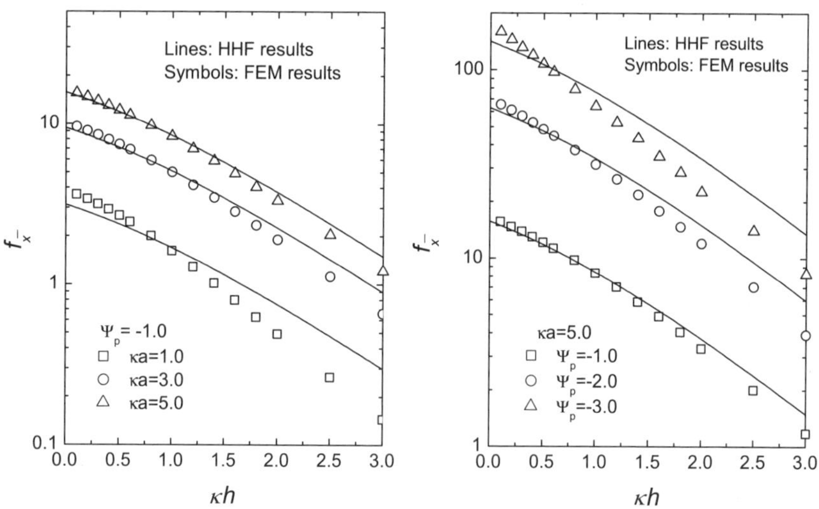

Figure 14.5. Comparison of the electric double layer force between two spherical particles obtained using the finite element analysis (Symbols) with the analytic solution of Hogg *et al.* (1966) (lines). Left: Comparison for different κa for low surface potential, $\Psi_p = -1$. Right: Comparison for $\kappa a = 5$ and different values of surface potential.

Poisson–Boltzmann equation, and the use of Derjaguin's approximation embedded in the HHF result.

The above comparisons of the finite element estimates of the interaction force serve as validation of the numerical code against known solutions of the Poisson–Boltzmann equation. This validation procedure is essential in any numerical calculation. Once the numerical methodology is validated, it can be employed to explore the dependence of the interactions on different parameters.

14.2.7 EDL Interaction Force on Particles in a Charged Capillary

The interaction force between two spherical particles can be considerably modified when the particles are trapped in a capillary of comparable radius. Detailed analysis of the interaction forces experienced by the spherical particles in a capillary with different surface potentials has been recently reported (Das *et al.*, 2003). Figure 14.6 depicts the variation of the interaction force with separation distance between the particles for different scaled surface potentials, Ψ_c, on the capillary wall. The scaled particle surface potential is fixed at $\Psi_p = -3$, and the aspect ratio, $\lambda = b/a = 1.2$ for all the calculations. In this figure, the symbols represent the scaled interaction force between the spheres in an unbounded electrolyte solution. When the capillary radius is comparable to the particle radius and the capillary surface potential has the same sign and magnitude as the particle surface potentials, there is a significant reduction in the electric double layer repulsion between the particles. In no case, however, an

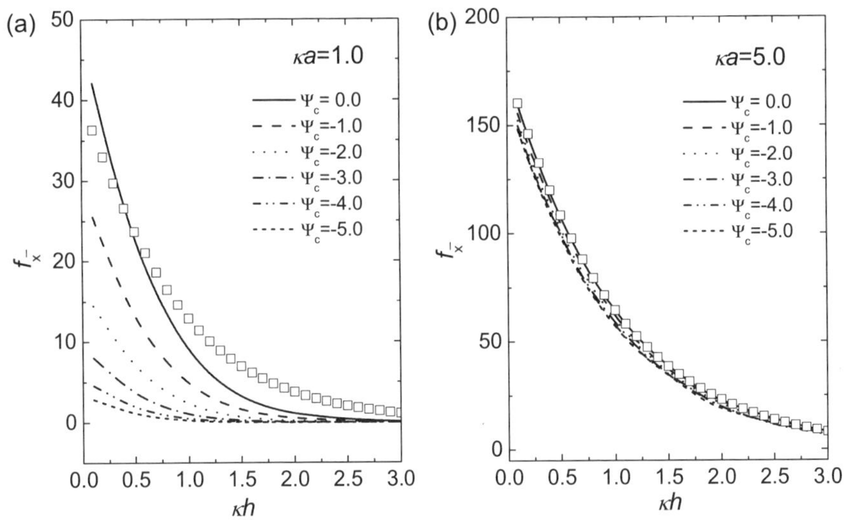

Figure 14.6. Scaled electric double layer interaction force, $f_{\bar{x}}$, between two spherical particles inside a cylindrical charged capillary (lines). In all cases, the particles have a constant scaled surface potential, $\Psi_p = -3$. The capillary surface potential, Ψ_c, is varied to study its influence on the particle-particle interaction force. The ratio of the capillary radius to the particle radius, $\lambda = b/a$, is fixed at 1.2. The symbols represent the interaction force between the spheres in an infinite reservoir in absence of the charged capillary wall (Das *et al.*, 2003).

attractive EDL force is observed for such a geometry, as long as the capillary wall remains smooth. The influence of the capillary wall on the interaction force is most pronounced for small values of the parameter $\kappa a \simeq 1$ [Figure 14.6(a)]. For larger values of this parameter ($\kappa a = 5$), there is virtually no influence of the capillary wall on the interaction force between the particles [Figure 14.6(b)]. Das *et al.* (2003) also observed that the EDL repulsion between the particles is enhanced when the capillary wall surface potential has an opposite sign to the particle surface potentials.

Similar calculations can also be performed by employing constant charge boundary conditions on the surface of the particles and the capillary wall. Although presence of the capillary wall has some influence on the interaction force between constant charge particles, the modification of the force is not as significant as in the case of constant potential particles. Simulations for the interaction between particles and capillary walls having mixed boundary conditions (constant charge on one particle while constant potential on the other) can also be performed. The presence of roughness on the capillary wall considerably influences the interaction force between the particles. Simulations of these different scenarios have also been reported (Das and Bhattacharjee, 2004, 2005). The presence of roughness on the capillary wall modifies the electric field distribution at different locations of the capillary wall surface. Such local modifications can give rise to different forces on each particle. In such situations, one may even observe an attractive force between two similarly charged particles.

An interesting observation regarding the interaction force components, namely the osmotic stress and the electrical (Maxwell stress) is that for a constant potential particle, the osmotic stress is equal at all locations on the surface (*cf.*, Eq. 14.17). Consequently, integration of this stress component over the closed surface of the particle makes the osmotic pressure contribution to the total interaction force zero. Therefore, the entire force modification observed for constant potential particles is due to modifications in the Maxwell stresses. For constant charge particles, however, the surface potential on the particle will be different at different locations. Consequently, the integration of the osmotic stresses over the particle surface will result in a finite osmotic pressure force.

The analysis of the Poisson–Boltzmann equation described in this section should not be considered a more accurate representation of the actual physics of electrostatic interactions. All the approximations inherent in the development of the Poisson–Boltzmann equation are still present in the numerical results, and hence, the influence of the various assumptions, for instance, the point charge ions, continuum solvent, *etc.*, on the result need to be assessed. Noting that the solution was obtained for small values of the parameter $\kappa a \simeq 1$, this would imply that for aqueous systems, the capillary radius is of the order of ten nanometers for salt concentrations of about 10^{-4} M. This may seem to be too small a pore radius to allow a continuum analysis. However, small κa values can also be achieved with fairly large particle radii in solvents with lower dielectric constants than water. The developed numerical solution is therefore also of interest when dealing with non-aqueous, low dielectric constant solvents. For such systems, it is evident from the above results that the electric double layer force between two spherical colloidal particles can be considerably modified due to the presence of the capillary wall.

In this section, the numerical procedure for solving the Poisson–Boltzmann equation was presented in considerable detail. In particular, the rendition of the physical problem to a finite element formulation was elaborated in some detail. In summary, the Poisson–Boltzmann equation represents a single stationary (time independent) partial differential equation. From a numerical standpoint, solution of such an equation should be fairly tractable. However, the correct accounting of the physics in the mathematical formulation required considerable attention to detail, particularly concerning the placement of boundary conditions. In the following sections, we will address two problems that are more complicated from a numerical standpoint.

14.3 FLOW OF ELECTROLYTE IN A CHARGED CYLINDRICAL CAPILLARY

In this section, we will discuss numerical simulations of the combined pressure and electric potential gradient driven electrokinetic flow of an electrolyte solution through a cylindrical capillary microchannel having a charged wall. This problem was discussed in Chapter 8 as electrokinetic flow in capillary microchannels.

Electrokinetic flow is of significant interest in microfluidic devices (Li, 2004). During pressure driven flows through microchannels, a streaming potential is developed across the capillary due to the flow of the ions relative to the stationary charged walls of the channel. The transport of an electrolyte solution through a capillary microchannel can be treated as either a steady-state or a transient problem. The model requires a coupled solution of three sets of governing equations representing the coupled effects of three physical processes. First, the distribution of the electric potential in the system is governed by the Poisson equation. Secondly, the flow of the electrolyte solution is dictated by the fluid continuity and Navier–Stokes equations. Finally, the transport and distribution of the ions are governed by the Nernst–Planck equations. These three types of governing equations must be solved simultaneously to obtain the electric potential distributions, the flow velocities and fluxes, and the ion concentration distributions in the system.

The theory of electrokinetic transport, particularly analyses pertaining to the calculation of streaming potential, has been thoroughly developed over the past century, with numerous studies exploring different aspects of the governing transport phenomena (Helmholtz, 1879; Smoluchowski, 1903; Lewis, 1960; Osterle, 1964; Burgreen and Nakache, 1964; Rice and Whitehead, 1965; Morrison, 1969; Levine *et al.*, 1975; Yang *et al.*, 2001; Keh and Tseng, 2001; Daiguji *et al.*, 2004). Most of these studies are based on a steady-state analysis of the governing electrochemical transport equations applied to microchannels of infinite length. These studies, apart from those of Morrison (1969) and Keh and Tseng (2001), do not address issues associated with the transient development of the electric potential across the microchannel. Furthermore, barring a handful of studies (Yang *et al.*, 2001; Daiguji *et al.*, 2004) the entry effects, or the effects of charged walls of the reservoirs connecting the capillary are rarely considered.

Electrokinetic flow in narrow capillary channels has also received considerable attention in the context of transport of electrolytes through reverse osmosis and ultrafiltration membranes (Gross and Osterle, 1968; Jacazio *et al.*, 1972; Anderson and Malone, 1974; Sasidhar and Ruckenstein, 1981, 1982; Deen, 1987; Cwirko and Carbonell, 1989). It is well known that transport of an electrolyte through membrane pores results in a rejection of the ions, yielding a lower electrolyte concentration in the solution emerging from the pore. The space charge model was an outcome of the initial attempts at analyzing the transport of ions through narrow porous media. More recently, a considerable body of studies employing the so-called one-dimensional extended Nernst–Planck equations (Smit, 1989; Basu and Sharma, 1997; Hall *et al.*, 1997; Bhattacharjee *et al.*, 2001) have been used to study ion transport and rejection by membrane pores. Once again, most of these modeling approaches are based on steady-state analysis.

Electrokinetic flow in microchannels and transport through membrane pores are governed by the same fundamental equations, namely, the Navier–Stokes equations for the fluid flow, the Nernst–Planck equations for the ion transport, and the Poisson equation for the electric potential and field distributions. Both types of transport problems essentially attempt to emulate a physical picture shown in Figure 14.7, where a capillary connects two reservoirs containing bulk electrolytes. Note that in Figure 14.7, the outer boundaries of the reservoirs (shown by dashed lines) simply represent locations in the reservoir that are sufficiently far away from the capillary to allow the assumption of bulk conditions like electroneutrality to hold. Depending on the

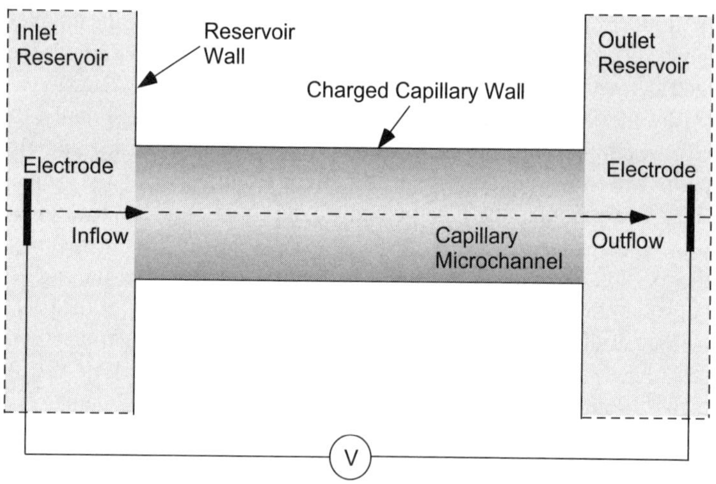

Figure 14.7. Schematic diagram of a capillary microchannel connecting two reservoirs. When a pressure gradient is set up between the two reservoirs, a potential difference can be recorded between the electrodes placed in the reservoirs. The potential difference under steady-state flow conditions is referred to as streaming potential. Alternately, setting up a potential difference between the electrodes will induce an electroosmotic flow through the capillary in absence of any applied pressure gradient.

nature of the problem, for instance, whether one intends to determine the streaming potential during electrokinetic flow, or assess salt rejection during membrane transport, the conditions in the outlet reservoir are stated differently. In streaming potential analysis, the electrolyte concentrations in the inlet and outlet reservoirs are held fixed at same values. In problems dealing with prediction of salt rejection by capillary microchannels, the electrolyte concentration in the outlet reservoir is not known *a priori*. Consequently, one needs to specify an alternate set of conditions for this reservoir.

The analytical models for electroosmotic flow in Chapter 8 were developed without explicitly considering the two reservoirs at the two ends of the capillary microchannel. The analytical model simply applied to a section of an infinitely long capillary, and neglected any effects of the entry and exit flows at the capillary ends. The fluid flow inside the capillary was assumed to be strictly one dimensional. Adopting a suitable numerical procedure to solve the governing equations for the geometry of Figure 14.7 can lead to an assessment of the deviations in streaming potential engendered by the approximations inherent in the analytical models. In particular, a numerical procedure for solving the governing equations can provide detailed information regarding the ion concentration distributions in the channel and the reservoirs, the velocity distribution, and the electric potential distribution.

Let us first summarize the implications of the different approximations inherent in the analytical models for electrokinetic flow. Typically, these models assume an infinitely long capillary, and ignore any axial (along the length of the capillary) variation of ion concentrations (Rice and Whitehead, 1965). Such assumptions allow the use of an equilibrium radial ion concentration distribution in the capillary, which is obtained from the solution of the one-dimensional Poisson–Boltzmann equation. Most analytical solutions, and sometimes numerical simulations, of electrokinetic transport processes are obtained employing the undisturbed ion concentration profiles based on the Poisson–Boltzmann equation (see, for instance, the theoretical development in Chapter 8 for electroosmotic flow in capillary microchannels). The obvious limitation of such an approach is its inability to predict any axial variation of ion concentrations, and hence, any rejection of ions by the capillary. In contrast, the extended Nernst–Planck models applied to study ion transport through membrane pores predict a substantial axial variation of the ion concentrations in the capillary, and hence, salt rejection.

The steady state analytical models also fail to provide a clear assessment of the time scales required for development of the steady-state streaming potential. According to some electrokinetic theories (Morrison, 1969; Keh and Tseng, 2001), the steady state streaming potential is attained within the time scale of the hydrodynamic relaxation. In contrast, the extended Nernst–Planck approach applied to pore transport problems predicts establishment of the steady state over the time scale of diffusion of the ions. The diffusion time scale is several orders of magnitude slower than the hydrodynamic time scale. This leads to the question as to what is the true time scale for attainment of steady state during electrokinetic flow through narrow capillaries?

The above mentioned apparent discrepancies between the electrokinetic flow and membrane transport models can be systematically resolved through a coupled solution of the transient electrochemical transport equations for the geometry depicted in Figure 14.7. More specifically, the geometry obviates the requirement of specifying the boundary conditions directly at the capillary entrance and exit. Furthermore, a transient solution of the governing equations can provide considerable insight regarding the time scales of the development of the steady state flow characteristics. Using such a geometry, Mansouri *et al.* (2005) studied the ion transport behavior in a finite length capillary, with different types of boundary conditions imposed at the inlet and outlet reservoirs. Their analysis was primarily restricted to the flows engendered by an applied pressure gradient along the capillary. Here we generalize the model to accommodate simultaneous application of the pressure and electric potential gradients along the capillary. Accordingly, solutions for the general electroosmotic flow problem occurring under the combined influence of pressure and electric potential gradients across the capillary channel will also be discussed in this chapter. Thus, within the scope of the same geometrical description and the same governing equations, we address three separate case studies pertaining to electrokinetic flow. These are:

Case 1: Prediction of streaming potential developed at steady-state during pressure driven flow through a charged capillary microchannel connecting two reservoirs. In this case, the two reservoirs contain electroneutral electrolyte solutions of identical concentrations.

Case 2: Prediction of the transient evolution of steady-state electric potential and ion concentration distributions in a capillary microchannel during pressure driven flow. In this case, the electrolyte concentration in the outlet reservoir is unknown, thus emulating the problem of ion rejection through a membrane pore.

Case 3: Prediction of the steady-state fluid axial velocity in a capillary microchannel under the coupled influence of pressure and electric potential gradient.

14.3.1 Problem Formulation

The computational geometry and the governing equations common to all three case studies mentioned above are presented here. The numerical solution procedure for the three case studies primarily differ in the manner in which the boundary conditions are specified. Accordingly, we will first present all aspects of the numerical solution procedure that are common to all three cases, following which, we will separately discuss the boundary conditions of the three case studies.

The model geometry represents a cylindrical capillary connected to two reservoirs, as shown in Figure 14.8. The entrance and exit reservoirs are modeled as cylindrical manifolds of radius b, which is much larger than the capillary radius, a. Utilizing the axial symmetry of the geometry, all the governing equations will be written in axisymmetric cylindrical coordinates (r, x) to simplify the model to two dimensions. The axisymmetric model geometry is represented by the region bounded by the outer boundary ABCDEF, and the line of axial symmetry, PQ. The dashed boundaries AP, AB, EF, and FQ represent the regions in the reservoirs that are unaffected by the capillary. These boundaries are either represented through bulk conditions or appropriate symmetry conditions. The distances of these boundaries from the entrance and exit

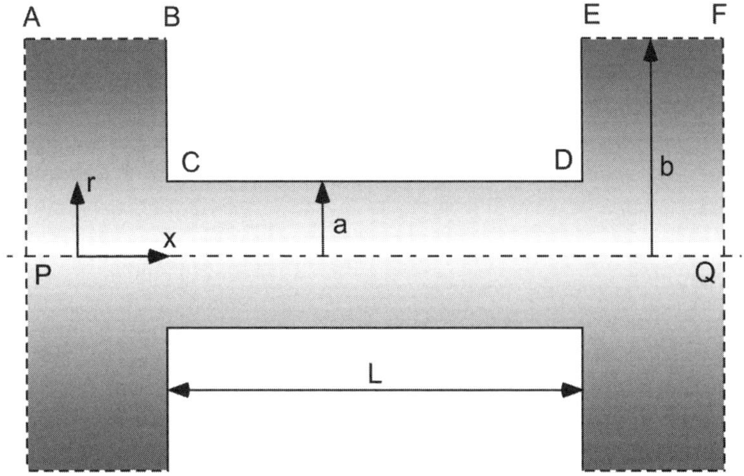

Figure 14.8. Computational geometry for modeling electrokinetic flow of an electrolyte in a cylindrical capillary.

regions of the capillary should be sufficiently large to ensure that the electrochemical properties at these locations are not influenced by the capillary. The model geometry also incorporates the walls of the inlet and outlet reservoirs (BC and DE). One can assign appropriate electric potentials or charge densities on these walls or simply render them as electrically neutral surfaces. The wall of the capillary (CD) can bear a constant electric potential or a constant charge density. All the solid surfaces are assigned no slip boundary condition, and are assumed to be impermeable to the ions.

In all the three case studies, we consider the transport of a symmetric $(z : z)$ binary electrolyte solution of a specified bulk molar concentration, c_∞. In Cases 1 and 2, the flow is driven solely by an externally applied pressure gradient. In Case 3, an external electric field is also applied in the axial direction, causing a flow influenced by the combined effects of pressure and electric potential gradients. Various dimensions and parameter values pertaining to the model are shown in Table 14.2. The ionic diffusivity is assumed to be equal for all ions. The ratio of the radii b to a is referred to as the capillary expansion factor, and is assigned a value of 5 in all subsequent discussion. The capillary length is chosen to be larger than the capillary radius by a factor of 10, so that the entrance and exit effects on the overall transport phenomenon is small.

The electric potential distribution in the electrolyte solution is related to the volumetric free charge density by Poisson's equation:

$$\nabla^2 \psi = -\frac{\rho_f}{\epsilon} \tag{14.26}$$

Here ϵ is the permittivity of the fluid, ψ is the prevailing electric potential, and ρ_f is the free charge density, given by the equation

$$\rho_f = \sum_i z_i e n_i \tag{14.27}$$

TABLE 14.2. Parameter Values Employed in the Model for Flow of an Electrolyte in a Charged Capillary.

Parameter	Value/Range
Solvent permittivity, ϵ	$78.54 \times 8.854 \times 10^{-12}$ C^2/Nm^2
Capillary wall potential, ψ_c	0 to -75 mV
Bulk electrolyte concentration, c_∞	10^{-5} M
Inverse Debye length, κ	1.038×10^7 m^{-1}
Ion valence, z_i	1
Ion diffusivity, D	1×10^{-9} m^2/s
Temperature	298 K
Fluid density, ρ	1000 kg/m^3
Fluid viscosity, μ	0.001 N.s/m^2
Axial pressure gradient, $p_x = -\partial p/\partial x$	10^7 Pa.m^{-1}
Capillary radius, a	κ^{-1}–$10\kappa^{-1}$ m
Capillary expansion factor, b/a	5
Capillary length	$10\kappa^{-1}$ m

where z_i and n_i are the valence and number concentration of the ionic species, respectively.

The movement of the ions in the system is governed by the Nernst–Planck equations. For each ionic species, the Nernst–Planck equation can be written as

$$\frac{\partial n_i}{\partial t} = -\nabla \cdot \mathbf{j}_i^{**} = -\nabla \cdot \left[n_i \mathbf{u} - D \nabla n_i - \frac{z_i e n_i D}{k_B T} \nabla \psi \right] \tag{14.28}$$

where $\mathbf{j}_i^{**}$ is the ionic number flux (m^{-2}s^{-1}), $\mathbf{u}$ is the fluid velocity vector, and D is the diffusion coefficient. If we consider only the steady-state solution, Eq. (14.28) becomes

$$\nabla \cdot \left[n_i \mathbf{u} - D \nabla n_i - \frac{z_i e n_i D}{k_B T} \nabla \psi \right] = 0 \tag{14.29}$$

To determine the velocity field, the Navier–Stokes equation is used. The Navier–Stokes equation, including a volume force due to an electric field, is given by

$$\rho \frac{\partial \mathbf{u}}{\partial t} + \rho \mathbf{u} \cdot \nabla \mathbf{u} = -\nabla p + \mu \nabla^2 \mathbf{u} + \rho \mathbf{g} - \rho_f \nabla \psi \tag{14.30}$$

Here ρ and μ are the fluid density and viscosity, respectively, and p is the pressure. Gravitational force can be neglected due to the small scale of the problem. The electrical body force per unit volume is given by the product of the charge density ρ_f and the local electric field $\mathbf{E} = -\nabla \psi$. When the steady-state solution of the problem is considered, removing the time dependent term from Eq. (14.30) yields,

$$\rho \mathbf{u} \cdot \nabla \mathbf{u} = -\nabla p + \mu \nabla^2 \mathbf{u} + \rho \mathbf{g} - \rho_f \nabla \psi \tag{14.31}$$

TABLE 14.3. Dimensionless Variables Employed to Scale the Governing Equations.

Dimensionless Variable	Expression
Radial coordinate, $\bar{r}$	κr
Axial coordinate, $\bar{x}$	κx
Time, τ	$\kappa^2 D t$
Velocity, $\mathbf{U}$	$\mathbf{u}/\kappa D$
Ion concentration, N_i	n_i/n_∞
Electric potential, Ψ	$\dfrac{ze\psi}{k_B T}$
Electric field, $\bar{E}$	$\dfrac{ze}{k_B T \kappa} E$
Surface charge density, σ_c	$\dfrac{ze}{\epsilon k_B T \kappa} q_c$
Fluid density, $\bar{\rho}$	$\dfrac{z^2 e^2 D^2}{\epsilon k_B^2 T^2} \rho$
Viscosity, $\bar{\mu}$	$\dfrac{z^2 e^2 D}{\epsilon k_B^2 T^2} \mu$
Pressure, P	$\dfrac{z^2 e^2}{\epsilon k_B^2 T^2 \kappa^2} p$

For an incompressible fluid, the Navier–Stokes equation is solved along with the continuity equation given by

$$\nabla \cdot \mathbf{u} = 0 \tag{14.32}$$

In order to obtain the governing equations in dimensionless form, we will consider the case of a symmetric electrolyte, and scale all dimensions with respect to the inverse Debye length

$$\kappa = \left(\frac{2 n_\infty z^2 e^2}{\epsilon k_B T} \right)^{1/2} \tag{14.33}$$

where n_∞ is the bulk electrolyte concentration. By introducing the dimensionless variables given in Table 14.3, the governing equations can now be stated in their scaled forms.

The Poisson equation in the non-dimensional form is given by

$$\bar{\nabla}^2 \Psi = -\left(\frac{N_p - N_n}{2} \right) \tag{14.34}$$

where N_p and N_n are the dimensionless concentrations of the positively and negatively charged species, respectively. Here, the scaled gradient operator $\bar{\nabla}$ is defined as

$$\bar{\nabla} = \kappa^{-1}\nabla = \left(\frac{\partial}{\partial \bar{r}}\mathbf{i}_{\bar{r}} + \frac{\partial}{\partial \bar{x}}\mathbf{i}_{\bar{x}}\right) \tag{14.35}$$

where $\mathbf{i}_{\bar{r}}$ and $\mathbf{i}_{\bar{x}}$ are the unit vectors along the scaled radial and axial coordinates, respectively.

Scaling the Nernst–Planck equation gives the following equations for the transport of positive and negative ions, respectively,

$$\frac{\partial N_p}{\partial \tau} = -\bar{\nabla} \cdot [N_p\mathbf{U} - \bar{\nabla}N_p - N_p\bar{\nabla}\Psi] \tag{14.36-a}$$

$$\frac{\partial N_n}{\partial \tau} = -\bar{\nabla} \cdot [N_n\mathbf{U} - \bar{\nabla}N_n + N_n\bar{\nabla}\Psi] \tag{14.36-b}$$

Finally, the Navier–Stokes in absence of gravity and continuity equations are also scaled to obtain

$$\bar{\rho}\frac{\partial \mathbf{U}}{\partial \tau} + \bar{\rho}\mathbf{U} \cdot \bar{\nabla}\mathbf{U} = -\bar{\nabla}P + \bar{\mu}\bar{\nabla}^2\mathbf{U} - \left(\frac{N_p - N_n}{2}\right)\bar{\nabla}\Psi \tag{14.37}$$

and

$$\bar{\nabla} \cdot \mathbf{U} = 0 \tag{14.38}$$

Note that in the scaled form of the Navier–Stokes equation, the gravitational body force term was dropped.

Equations (14.34) to (14.38) represent the dimensionless forms of the governing equations for the general electrokinetic flow problem pertaining to all three case studies that will be conducted in this section. These equations were written retaining the time dependent terms. Dropping the time derivative terms from the Nernst–Planck and Navier–Stokes equations will provide the corresponding steady-state transport model. Although these equations seem to be quite general, they are based on several assumptions, for instance, symmetric electrolyte ($z_p = -z_n = z$), same diffusivity for all the ionic species, fixed solution density, and constant viscosity. Furthermore, the diffusivities of the ions will generally be much smaller in narrow capillaries than the corresponding values of the parameter in bulk solutions. Most of these assumptions can be generalized. However, to render the numerical model identical to the analytical problems described in Chapter 8, these assumptions were retained in the governing equations. It should be noted that all the linear dimensions in the governing equations were scaled with respect to the screening length (Debye length) of the electric double layer. This was done deliberately to explore the electroosmotic phenomena in capillaries that are comparable in dimensions to the screening length of the electric double layer. As mentioned earlier in this chapter, most analytical theories of electroosmotic flow were developed for cases when the channel dimensions are significantly larger than the Debye length. Typically, for aqueous systems containing a monovalent salt at a reasonable concentration flowing through a large radius capillary, the parameter κa, where a is the channel radius, is often >100. Attempting a numerical solution for such large values of the parameter κa should not, in principle, provide any further insight compared to analytic solutions, such as the Helmholtz–Smoluchowski result. In this context, application of numerical techniques can be justified when addressing electroosmotic flows in narrow capillaries, where the analytic expressions derived for the limiting cases of large $\kappa a \gg 1$ might provide inaccurate results.

14.3.2 Mesh Generation and Numerical Solution

The mesh generation and the numerical solution procedure are similar for all the case studies. Hence, we present these components of the numerical modeling procedure prior to the individual description of these case studies. The computational domain bounded by ABCDEFQP in Figure 14.8 was discretized into quadratic triangular elements. For the present problem, there are three governing models dictated by three sets of equations. A complete adaptive mesh refinement scheme that allows minimization of numerical errors for all three sets of governing equations is quite difficult and memory intensive. Consequently, we solve the problem with a predefined mesh containing at least 10,000 elements. A typical mesh is depicted in Figure 14.9. The mesh consists of about 25,000 elements, with smaller elements placed along the capillary and reservoir walls, as well as the line of axial symmetry. Since adaptive mesh refinement could not be used in the computations, the problem was solved with meshes having different numbers of elements, and the solutions obtained for different mesh sizes were compared to ensure that mesh independent results were obtained. Most of the simulation results became mesh independent when the number of elements was about 15,000.

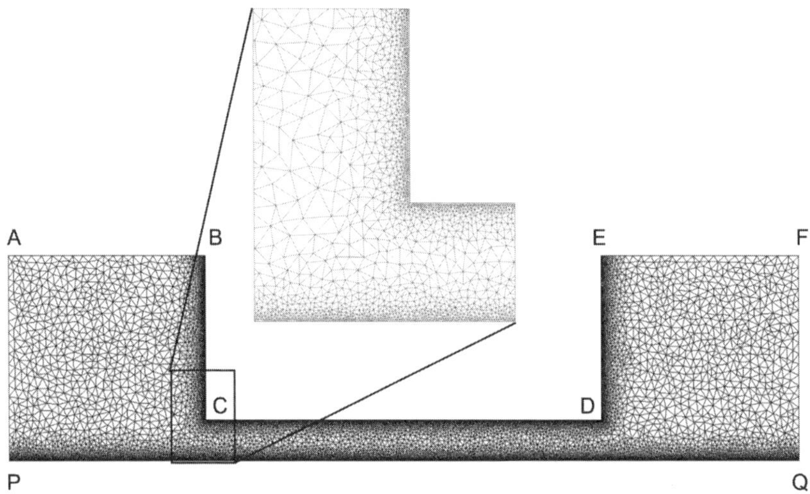

Figure 14.9. Finite element mesh used to solve the electrokinetic flow problem. A close-up view of the mesh is shown over the boxed region of the computational domain.

The finite element software FEMLab provides a simple and intuitive interface for modeling the above multiphysics problem. The software provides individual modules for solving the Poisson, Navier–Stokes, and Nernst–Planck Equations. A separate problem module is set up for each governing physics. One selects the governing equation to be solved for the geometry, defines the parameters of these equations, along with the appropriate initial and boundary conditions, and ensures that the coupled variables are denoted similarly in different equation modules. For instance, one should use the same parameter name, Ψ, for the scaled electric potential in the Poisson, Nernst–Planck, and the Navier–Stokes modules. Similarly, the concentration of the positive and negative ions are denoted by N_p and N_n, respectively, in all the three equation modules. Availability of these predefined modules renders the setting up of the multiphysics problem quite facile. Note that in this case, we are not explicitly required to recast every governing equation to the general PDE form, as was done in Section 14.2. Instead, the selected module for the governing equation in FEMLab reads in the parameters for the actual equation, and performs the conversion of the governing equation to the required general or weak forms internally.

Most of the numerical methodology validation, and subsequent calculations were done using a value of $\kappa a = 5$, which means that the capillary radius is five times the Debye length. The radius of the inlet and outlet manifolds, b, was five times the capillary radius. The scaled length of the capillary (CD) was 50, giving a length to diameter ratio of 5. Finally, the scaled length of the inlet and outlet reservoirs (AB and EF) was 25, yielding a total length of the capillary and reservoirs (PQ) of 100.

The program was set up to solve either the steady-state or the transient equations. The transient governing equations were solved from an initially no flow condition until the steady-state flow conditions were attained after imposition of an axial pressure and/or an applied axial electric potential gradient. The numerical scheme for the solution of the transient equations for simulating the flow under an applied pressure gradient is depicted in Figure 14.10. After defining the problem parameters, the geometry, and creation of the mesh, we first solve the Nernst–Planck and Poisson equations in absence of any fluid flow. This yields the Poisson–Boltzmann stationary ion and electric potential distribution over the computational domain. Once this initial distribution is calculated, a pressure gradient is employed along the capillary, and the transient governing equations are solved iteratively starting from the initial quiescent solution. Using the Poisson–Boltzmann ion and concentration distributions, the free charge density and electric field terms appearing in the body force of the Navier–Stokes equation are first determined. The velocity field is then obtained by solving the Navier–Stokes equation. The velocity field is substituted in the Nernst–Planck equations to obtain an updated estimate of the ion concentration distributions in presence of convection. Substituting these ion concentration distributions in the Poisson equation then provides the electric potential distribution at the next time step. The updated electric potential and charge density distributions are then substituted back in the Navier–Stokes equations to provide the new velocity field. This iterative solution procedure is continued until the solution at two consecutive time steps converges to within a preset tolerance. The solution procedure was implemented on the software FEMLab employing its in-built solver manager. The solver manager runs the solution

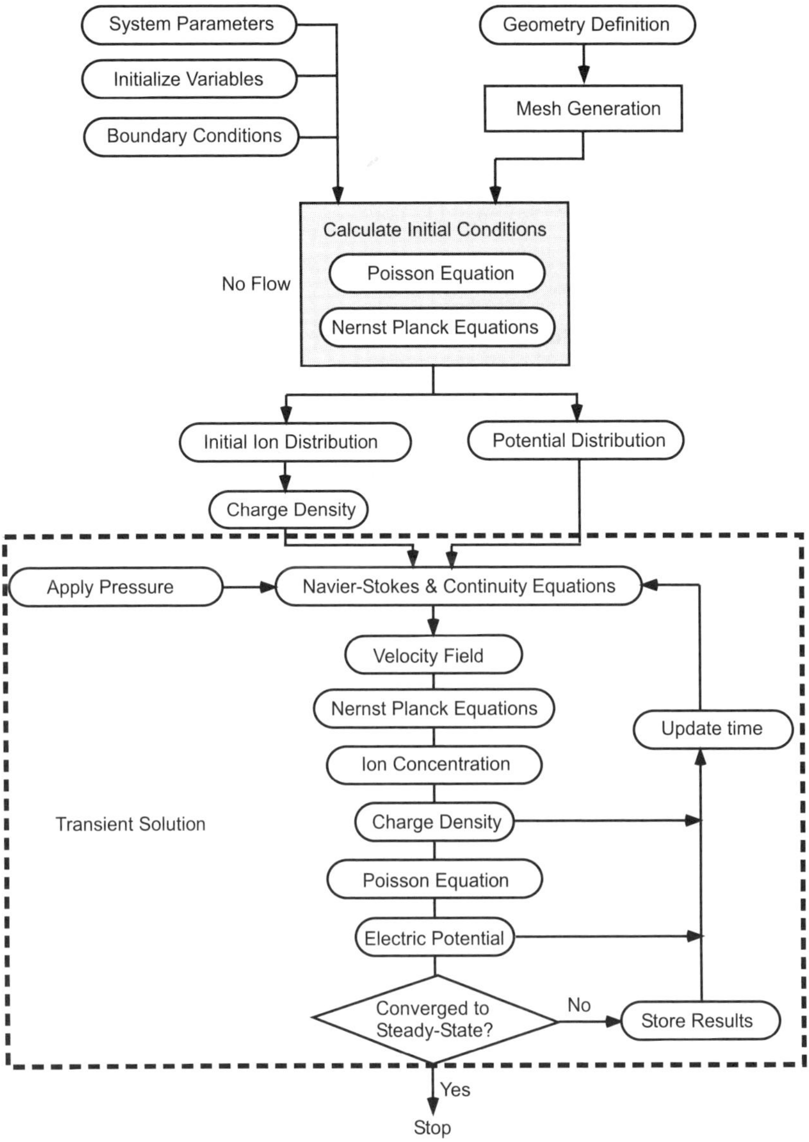

Figure 14.10. Flowchart depicting the finite element solution methodology for the streaming potential development problem. The region demarcated within the dashed rectangle represents the iterative solution procedure for the transient problem.

procedure using a script which automates the sequential solution of the three sets of governing equations.

The solution procedure depicted in Figure 14.10 addresses the development of the flow under the sole influence of an applied pressure gradient (which is pertinent to the first two case studies). To solve the complete electroosmotic flow problem in presence of both an applied pressure gradient and an external electric field (Case 3), one only needs to modify the electrical boundary condition (for the Poisson equation) at the flow outlet, where an external electric field E_0, is imposed as a boundary condition. Even though the governing equations, the computational domain and mesh, as well as the general solution methodology remain unchanged in all subsequent analysis, application of different boundary conditions alter the overall problem formulation to reflect the three case studies we intend to conduct. We will discuss these three cases separately in the following.

14.3.3 Case 1: Streaming Potential Across a Capillary Microchannel

In this case study, we simulate the steady-state streaming potential developed between two reservoirs connected by a narrow capillary microchannel. This case was treated analytically in Chapter 8. Here we describe how one can perform the numerical simulations to obtain results that emulate those predicted by the analytical approach.

14.3.3.1 Boundary Conditions To define the boundary conditions for the steady-state transport problem, we focus on the computational domain in the two dimensional cylindrical coordinate system (r, x) as shown in Figure 14.11. Several of the boundary conditions remain unaltered in all three case studies. For instance, PQ is the line of axial symmetry (axis of the capillary microchannel). Accordingly, all the governing equations will have the axial symmetry condition on the segment PQ for all three case studies. The boundary conditions on the segments AB and EF will also be similar in all three problems. We will only change the conditions at the charged walls (BC, CD, and DE) as well as on the line segments AP and FQ to differentiate between the problems we are solving.

The appropriate boundary conditions applicable for each governing equation that are pertinent to the streaming potential analysis are described in Table 14.4. As stated earlier, axial symmetry exists for all the governing equations on segment PQ. The boundary conditions on the segments AB and EF are defined assuming these surfaces to be in the bulk electrolyte reservoirs. Accordingly, for the Poisson equation, we will

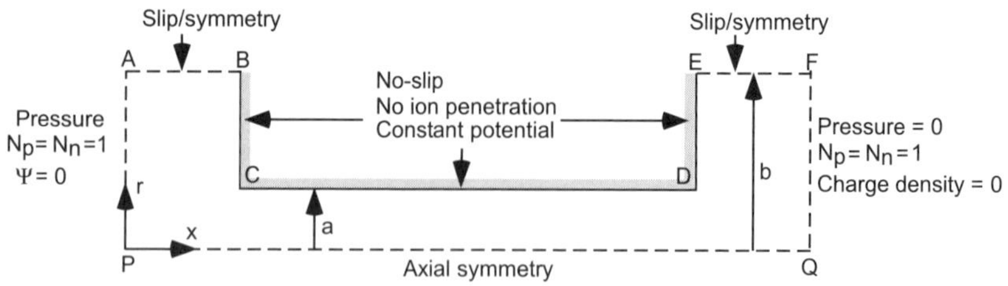

Figure 14.11. Schematic representation of the boundary conditions for the Navier–Stokes, Nernst–Planck, and Poisson equations used in the simulation of streaming potential (Case 1). On the symmetry planes AB and EF, the ion concentration gradients and potential gradients normal to the planes are assumed to be zero.

TABLE 14.4. Boundary Conditions of the Governing Equations for the Streaming Potential Problem (Case 1).

Boundary	Poisson Equation	Nernst–Planck Equations	Navier–Stokes & Continuity Eqs.
PQ	Axial symmetry	Axial symmetry	Axial symmetry
AB, EF	Symmetry[1]	Zero normal flux	Slip
BC, DE	Potential = 0	Zero normal flux	No slip
CD	Constant potential ($\Psi = \Psi_c$, specified)	Zero normal flux	No slip
AP	Potential = 0	Bulk ion conc. ($N_p = N_n = 1$)	Normal pressure, P_0 (specified)
FQ	Symmetry	Bulk ion conc. ($N_p = N_n = 1$)	Normal pressure ($P = 0$)

[1]Symmetry refers to the condition when no charge is developed at the boundary ($\mathbf{n} \cdot \nabla\Psi = 0$).

use the symmetry condition, $\mathbf{n} \cdot \bar{\nabla}\Psi = 0$. For the Nernst–Planck equations, we will assume no concentration gradient normal to these interfaces, $\mathbf{n} \cdot \bar{\nabla} N_i = 0$. Finally, we will use the slip boundary condition for the Navier–Stokes equation. Note that for these boundary conditions to hold, the segments AB and EF should be located sufficiently far away from the capillary entrance.

At the inlet plane (AP), we set the concentration of the ions to be equal ($N_p = N_n = 1$) for the Nernst–Planck equations, the potential to zero ($\Psi = 0$) for the Poisson equation, and the normal pressure to a predetermined value, P_0, calculated from the user-defined axial pressure gradient, for the Navier–Stokes equation.

For the evaluation of the streaming potential, the exit reservoir should contain the same electrolyte concentration as the inlet reservoir. Accordingly, we set the conditions at the exit boundary (FQ) as follows. For the Poisson equation, we use

$$\mathbf{n} \cdot \bar{\nabla}\Psi = 0 \tag{14.39}$$

while for the Nernst–Planck equations, we set the concentrations of ions to be equal to the inlet reservoir concentrations,

$$N_p = N_n = 1 \tag{14.40}$$

For the Navier–Stokes equation, we set the normal pressure to zero at this boundary.

The boundary conditions on BC, CD, and DE pertaining to the Nernst–Planck equations are those of no net normal flux of ions (no ion penetration)

$$\mathbf{n} \cdot \mathbf{j}_i^{**} = 0 \tag{14.41}$$

where $\mathbf{j}_i^{**}$ is the number flux of the ith ionic species. For the Navier–Stokes equation, the appropriate condition on these boundaries is that of no slip ($\mathbf{u} = 0$). The boundary condition on the capillary wall (CD) for the Poisson equation is set as a constant surface potential condition

$$\psi = \psi_c \quad \text{or} \quad \Psi = \Psi_c \tag{14.42}$$

On the walls BC and DE, the potential is set to zero.

The streaming potential problem requires specification of the scaled pressure, P_0, at the channel inlet (AP). The specification of the pressure requires some elaboration. In the subsequent discussion, we will generally refer to an applied axial pressure gradient, p_x, which is defined as

$$p_x = -\frac{\partial p}{\partial x} \tag{14.43}$$

In all the simulations, we fix the pressure at the exit plane FQ as 0. The axial pressure gradient used in these simulations is chosen as 10^7 Pa/m. Assuming this pressure gradient to be constant between the inlet plane AP and the exit plane FQ, we can write

$$p_x = 10^7 = \frac{p_0 - 0}{L} \tag{14.44}$$

where p_0 is the pressure at plane AP, and L is the overall length of the channel (PQ). From the above relationship, Eq. (14.44), we obtain the pressure at the inlet section

$$p_0 = p_x L = p_x \frac{\bar{L}}{\kappa} \tag{14.45}$$

where the scaled length $\bar{L} = 100$. The scaled form of this pressure, P_0, is specified as a boundary condition for the Navier–Stokes equation at the inlet section (AP) in Table 14.4. Note that the use of the linear pressure gradient to specify a pressure at the channel inlet section AP is completely arbitrary. As we will observe shortly, the actual axial pressure gradient will vary at different locations in the simulations. The procedure simply allows one to specify a realistic pressure drop across the capillary and a means to provide an appropriate value of P_0 at the inlet section.

14.3.3.2 Comparison of Numerical and Analytical Results Solving the steady-state governing equations with the above set of boundary conditions will lead to the prediction of the streaming potential developed between the two reservoirs connected by the capillary microchannel. The parameters that need to be specified are the capillary radius, the bulk electrolyte concentration, the surface potential at the capillary wall, and the pressure at the reservoir inlet plane (AP). For the present simulations, the capillary wall surface potential was set to $\Psi_c = -1$.

The accuracy of the numerical results can be compared against the analytical results of electroosmotic flow in capillary microchannels described in Chapter 8. The analytical approach described in Chapter 8 employs several assumptions. These include application of linearized Poisson–Boltzmann equation, neglecting axial concentration gradients, and ignoring the entrance/exit effects. In the numerical model, none of these assumptions were made. Consequently, comparison of the numerical and analytical results is not straightforward. To perform an accurate comparison of the numerical and analytical results, we first need to assess the behavior of some of the key variables along the capillary as obtained from the numerical solution. Figure 14.12(a) depicts the variations of two important parameters, namely, the electric potential and the pressure along the axis of the cylindrical capillary obtained from the steady-state numerical calculations. The corresponding variations of the electric field component along the axial direction (along $\bar{x}$), as well as the axial pressure gradient are shown in Figure 14.12(b). It is evident that the electric potential and the pressure are virtually constant in the two reservoirs flanking the capillary. Within the capillary, the electric potential shows a non-linear increase with axial position, while the pressure shows a nearly linear variation with axial position. Figure 14.12(b) clearly shows that the axial component of the electric field varies perceptibly along the length of the capillary. The axial pressure gradient within the capillary, on the other hand, shows a more modest variation with axial location. All these quantities undergo marked variations near the capillary entrance and exit regions. It is therefore clearly discernable that a comparison of the results obtained from the detailed numerical calculations with the approximate analytical results is somewhat convoluted.

Nevertheless, if we focus our attention to the mid-section of the capillary (corresponding to $\bar{x} = 50$), which is sufficiently removed from the channel entrance and exit regions, it is likely that the local conditions at this section from our numerical results will closely emulate those existing in an infinitely long capillary. Furthermore, if the capillary wall surface potential is small, one might expect that the linearization of the Poisson equation in the analytical model will be reasonably accurate. Therefore, we first compare the axial velocity profile obtained from the numerical simulation at the capillary mid-section, $\bar{x} = 50$, with the corresponding analytical expression of the velocity profile, Eq. (8.108), which is given by

$$u_x(r) \equiv u_x = \frac{a^2 p_x}{4\mu}\left[1 - \left(\frac{r}{a}\right)^2\right] - \frac{\epsilon\psi_c}{\mu}\left[1 - \frac{I_o(\kappa r)}{I_o(\kappa a)}\right] E_x \tag{14.46}$$

Here, ψ_c is the surface potential of the capillary wall. Note that to compare the numerical and the analytical velocity profiles, one will need to employ values of $p_x = -\partial p/\partial x$ and E_x in Eq. (14.46) that are identical to the values of these parameters at the mid-plane of the capillary ($\bar{x} = 50$) obtained from the numerical solution (Figure 14.12).

To perform the comparison of the velocity profiles, we conducted the numerical simulation for $\kappa a = 1, 5$, and 9 using a constant surface potential of $\Psi_c = ze\psi_c/k_BT = -1$ at the capillary wall. The numerical simulations were conducted with an axial pressure gradient of $p_x = 10^7$ Pa/m between the reservoirs.[2] From the steady-state numerical solution of the governing equations, the values of p_x and E_x at $\bar{x} = 50$ and $\bar{r} = 0$ were determined. These values were then substituted in Eq. (14.46) to calculate the analytical velocity profile. The comparison of the numerical and

[2]This is, however, not the actual pressure gradient at the capillary mid-section, as is evident from Figure 14.12.

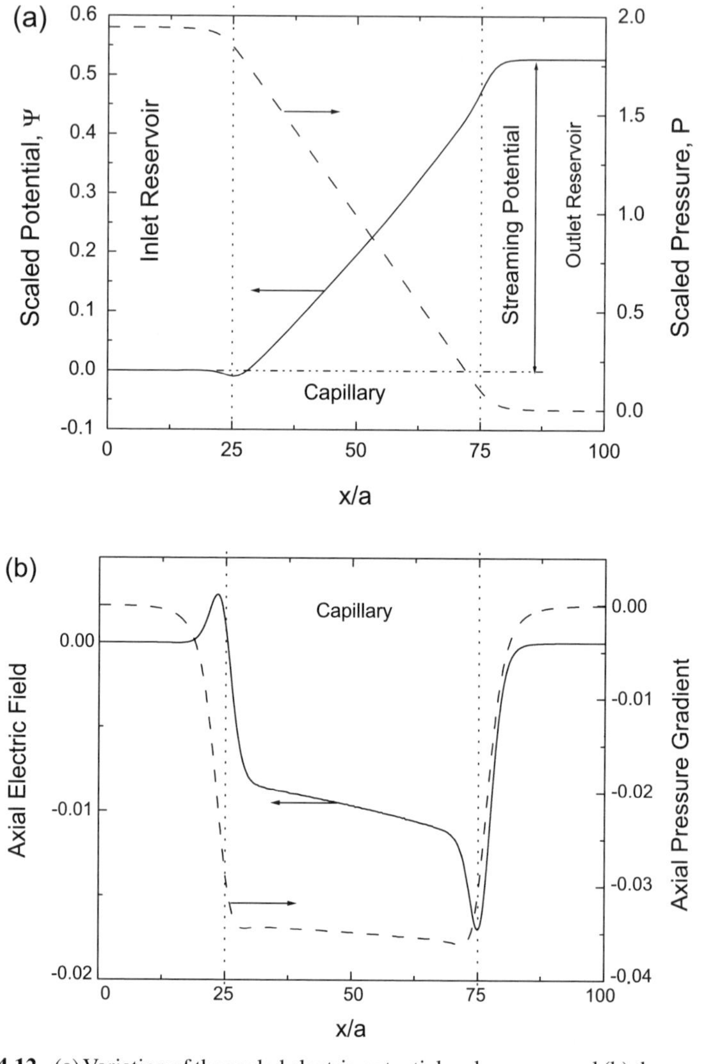

Figure 14.12. (a) Variation of the scaled electric potential and pressure and (b) the corresponding dimensionless electric field ($\bar{E}$) and axial pressure gradient ($\partial P/\partial \bar{x}$) along the axis of the capillary and the inlet and outlet reservoirs obtained from the steady-state numerical solution of the governing electrochemical transport equations. Solid lines represent the electric potential and field, while the dashed lines represent the pressure and pressure gradient. The simulations were performed for a scaled surface potential of $\Psi_c = -1$ on the capillary wall, an applied pressure gradient of 10^7 Pa/m between the reservoirs, and a scaled capillary radius of $\kappa a = 5$. The vertical dotted lines in each figure represent the entrance and exit planes of the capillary microchannel.

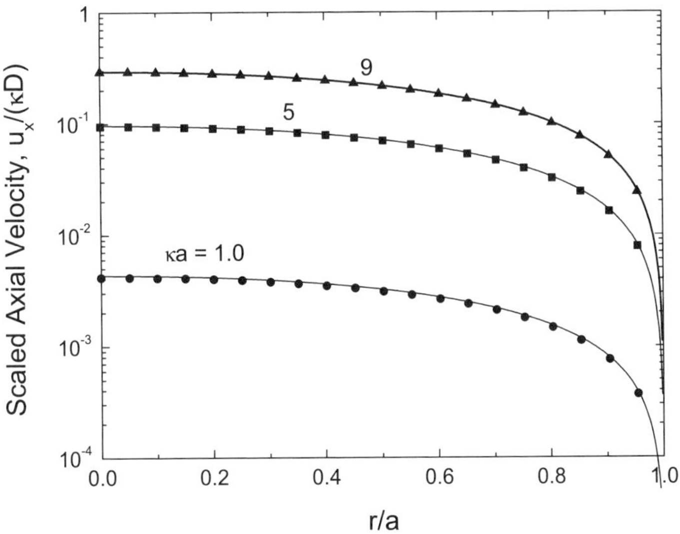

Figure 14.13. Comparison of the steady-state axial velocity profiles at the capillary mid-section obtained from the numerical solution with the corresponding predictions of the analytical expression, Eq. (14.46). Symbols are numerical predictions, while lines represent analytical results. The numerical and analytical velocities were compared using the *local* pressure and electric potential gradients existing at the capillary mid-section, $\bar{x} = 50$, obtained from the numerical solution.

analytical velocity profiles is shown in Figure 14.13. It is evident from the figure that the velocity profile obtained numerically at the mid-section of the channel ($\bar{x} = 50$) is in remarkably good agreement with the velocity profile obtained for an infinitely long cylindrical capillary using the analytical procedure.

It is important to note that the comparison of the numerical and analytical velocity profiles were performed using the values of the local axial pressure gradient and electric field that exists at the mid-section of the capillary (at $\bar{r} = 0$ and $\bar{x} = 50$). The axial pressure gradient at this position is different from the overall pressure gradient between the reservoirs employed to obtain the numerical solution. Furthermore, the electric field is different at different locations in the capillary. In this context, it now becomes pertinent to verify whether the numerical prediction of the streaming potential, as depicted in Figure 14.12(a), compares with the corresponding analytical prediction of the streaming potential. The numerical prediction of the streaming potential is simply the potential difference observed between the inlet and outlet reservoirs in Figure 14.12(a). The analytical results of Chapter 8 do not directly yield a potential difference between the two reservoirs connected by the capillary. Instead, it provides the following relationship for streaming potential under zero current condition (*cf.*, Eq. 8.143)

$$\left(\frac{E_x}{p_x}\right)_{I=0} = \frac{\epsilon\psi_c}{\mu\sigma^\infty}\left(1 - \frac{2I_1(\kappa a)}{(\kappa a)I_0(\kappa a)}\right) f(\kappa a, \beta, F_{cc}) \tag{14.47}$$

where $\beta = \epsilon^2\psi_c^2\kappa^2/\mu\sigma^\infty$, and

$$f(\kappa a, \beta, F_{cc}) = \frac{1}{F_{cc}\left\{1 - \beta\left[1 - 2A_1/(\kappa a) - A_1^2\right]/F_{cc}\right\}}$$

Here, F_{cc} is a correction term for the conduction current, and $A_1 = I_1(\kappa a)/I_0(\kappa a)$. Recall the earlier discussion in Chapter 8 that the function $f(\kappa a, \beta, F_{cc})$ has a sizeable contribution to the streaming potential only for large values of ψ_c. Equation (14.47) only provides the ratio of the electric field and the pressure gradient that should exist within the capillary at steady state for the streaming potential condition. It is therefore instructive to explore how the potential difference obtained across the capillary in the numerical calculations as shown in Figure 14.12(a) relates to Eq. (14.47).

To calculate a potential difference over a given length of the capillary using the analytical approach, the right hand side of Eq. (14.47) was evaluated employing identical capillary radius and other conditions applied to our numerical simulations. In the analytical calculations, the value of F_{cc} was set to 1.0. From the numerical simulation, the pressure difference, Δp, over an axial distance L between a location in the inlet reservoir and another location in the outlet reservoir was computed. These locations were selected such that they were equidistant from the capillary mid-section. Dividing the pressure difference by L provided the axial pressure gradient, $p_x(= \Delta p/L)$. Using the pressure gradient in Eq. (14.47), the axial component of the electric field, E_x, was determined. This value of the electric field was then employed to calculate the potential difference over the length L using

$$\Delta\psi = E_x L$$

The potential difference thus calculated using the analytical expression, Eq. (14.47), was compared with the potential difference between identical locations in the two reservoirs obtained from the numerical simulations. Figure 14.14 shows the comparison of the two potential differences for different values of scaled capillary radius κa. The comparison is based on the parameters evaluated along the axis of the capillary. Under these conditions, it is evident from Figure 14.14 that the numerically and analytically evaluated potential differences across the capillary are in remarkably good agreement. Furthermore, it is interesting to note that the choice of L does not affect the predicted streaming potential, as long as the two locations in the inlet and outlet reservoirs are equidistant from the capillary mid-plane ($\bar{x} = 50$). The numerical results slightly overpredict the potential difference for lower values of $\kappa a < 5$. However, for $\kappa a \geq 5$ the two predictions are virtually identical.

The excellent agreement between the numerical and analytical results not only serves as a validation of the numerical model, but also provides considerable insight regarding the development of streaming potential across charged capillaries. First, the comparison of the two solutions were performed on the basis of conditions, particularly the axial pressure gradient and electric field, determined at the mid-point of the capillary ($\bar{r} = 0, \bar{x} = 50$). Secondly, the streaming potential in the analytical solution was based solely on the streaming and conduction current, whereas in the numerical

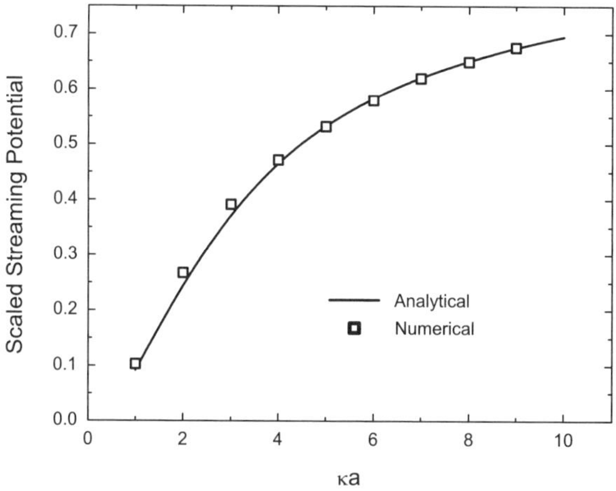

Figure 14.14. Comparison of the numerical and analytical predictions of the streaming potential for different values of scaled capillary radius.

calculations, it was obtained considering all three modes of ion transport (convection, diffusion, and migration). These comparisons shed some insight regarding how the streaming potential from the analytical solution can be interpreted for a more realistic geometry of a finite length capillary connecting two reservoirs. It is important to note that the streaming potential is measured between the two reservoirs connected by the capillary, and not within the capillary.

14.3.4 Case 2: Transient Analysis of Electrolyte Transport in a Capillary Microchannel

We turn our attention now to a slightly different problem, which involves transport of an electrolyte solution through a capillary microchannel representing a membrane pore. In this case, unlike the streaming potential simulations, the ion concentrations in the outlet reservoir at steady-state are not known *a priori*. Furthermore, instead of a steady-state solution, we solve the time dependent governing equations to simultaneously observe how the steady-state electric potential and the ion concentration distributions are established upon imposition of an axial pressure gradient. The analysis of the transient development of the streaming potential condition across a finite length capillary was recently addressed by Mansouri *et al.* (2005), who used a similar numerical approach and a computational geometry as described here. They studied the transient distributions of the electric potential, field, and ion concentrations in the system, and assessed the different time scales required for the development of the steady-state potential and the ion concentration fields in the system.

14.3.4.1 Boundary and Initial Conditions We conduct the simulations for a capillary radius of $\kappa a = 5$, and a constant surface charge density on the capillary wall.

All geometrical and parametric specifications are identical to the previous case study. The only difference in the present model pertains to specification of an alternate set of boundary conditions that emulate the membrane transport process. The boundary and initial conditions for this transient problem are stated in Table 14.5.

For this problem, we demonstrate the use of constant surface charge density boundary conditions on the charged walls of the capillary, as well as the two reservoirs (BC, CD, and DE). Accordingly, on these boundaries, we use the condition

$$-\mathbf{n}\cdot[\epsilon\nabla\psi] = q_c \tag{14.48}$$

where q_c (C/m^2) is the surface charge density on the walls. This quantity is provided as an input to the problem. The non-dimensional form of the surface charge density is

$$\sigma_c = \frac{ze}{\epsilon k_B T\kappa} q_c$$

Here we intend to emulate the transport through a membrane pore, and hence, cannot *a priori* provide the electrolyte concentration in the exit reservoir. Consequently, we specify the convective flux condition at FQ for the Nernst–Planck equations, which is written as

$$\mathbf{j}_i^{**} = n_i\mathbf{u} \tag{14.49}$$

TABLE 14.5. Boundary and Initial Conditions of the Governing Equations for Transient Electrokinetic Flow in a Capillary Microchannel (Case 2).

Boundary	Poisson Equation	Nernst–Planck Equations	Navier–Stokes & Continuity Equations
PQ	Axial symmetry	Axial symmetry	Axial symmetry
AB, EF	Symmetry	Zero normal flux	Slip/symmetry
BC, DE	Constant charge density, q_c (specified)	Zero normal flux	No slip
CD	Constant charge density, q_c (specified)	Zero normal flux	No slip
AP	Zero potential	Bulk ion conc. ($N_p = N_n = 1$)	Normal pressure, P_0 (specified)
FQ	Symmetry	Convective flux ($\bar{\nabla} N_i = 0$)	Normal pressure ($P = 0$)
Initial condition:			
AP	—	—	Normal pressure $P = 0$

In other words, the diffusional or migration contributions to the ionic flux are assumed to be zero, which implies

$$\bar{\nabla} N_i = 0 \tag{14.50}$$

and requires setting the boundary condition for the Poisson equation as

$$\mathbf{n} \cdot \bar{\nabla} \Psi = 0 \tag{14.51}$$

In this case, the reservoir at the capillary outlet will have a lower electrolyte concentration than the feed side reservoir. Although the individual ion concentrations in the outlet reservoir will be lower than the feed reservoir, the electrolyte solution will be electroneutral in this outlet reservoir.

All other boundary conditions for the present model are identical to that of the streaming potential simulation case discussed previously.

Finally, one needs to provide the initial conditions (at $t = 0$) for this transient problem. To formulate the initial conditions for the problem, we consider an initial state where the electrolyte solution is in equilibrium with no externally imposed pressure gradient. In this case, the ion distributions are set up in response to the charged walls of the capillary and reservoirs according to Boltzmann distribution. Thus, initially, we calculate the electrolyte concentration in the computational domain from the Poisson and Nernst–Planck equations in absence of any convection ($\mathbf{u} = 0$). Under no-flow conditions, these equations simply become the Poisson–Boltzmann equation. This equation can be solved subject to the conditions of a specified bulk electrolyte concentration at the inflow and exit boundaries (AP and FQ, respectively), and a specified potential on the solid walls (in these simulations, $\Psi_c = -1$). The no flow situation is modeled by setting the pressure to zero at AP ($P_0 = 0$). The ion concentration and electric potential distributions thus obtained can serve as the initial conditions for the transient problem.

Once the stationary potential distribution is calculated, one can extract the scaled surface charge density, σ_c, at the capillary and reservoir walls from this solution. This scaled surface charge density is then used in the transient simulations as a boundary condition. The scaled charge density used in the transient calculations was $\sigma_c = -1.04$.

The transient problem tracks the modification of the initial stationary distributions of ion concentrations and electric potential due to the fluid motion (convection) once a pressure gradient is applied across the capillary microchannel. The transient problem is solved according to the flowchart shown in Figure 14.10.

14.3.4.2 Transient Simulation Results Figure 14.15 depicts the development of the electric potential and the ion concentration distributions along the capillary axis obtained from the transient numerical simulations immediately after imposing an axial pressure gradient on the system. In Figure 14.15, the axial positions $25 \leq \kappa x \leq 75$ correspond to the capillary, while the regions $\kappa x < 25$ and $\kappa x > 75$ represent the inlet and outlet reservoirs, respectively. The parameter $\tau = \kappa^2 D t$ corresponding to each solid line represents the scaled time (see Table 14.3) at which the profile was

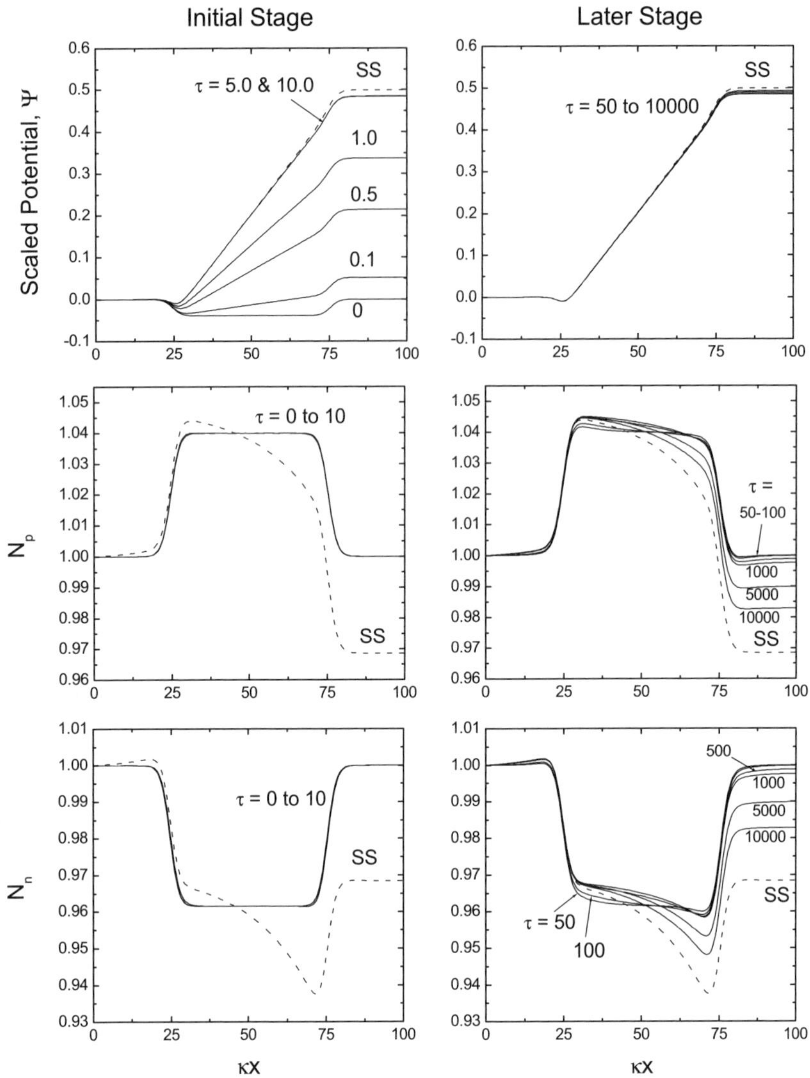

Figure 14.15. Transient development of scaled electric potential, Ψ, and the positive (N_p) and negative (N_n) ion concentration profiles along the capillary axis. The scaled capillary radius is $\kappa a = 5$. The figures on the left hand column show the developing potential and ion concentration profiles immediately after application of the pressure gradient (scaled time, τ varying from 0 to 10), while the right hand side column depicts the corresponding profiles at the later stages of the transient simulations ($\tau = 50$ to steady-state, denoted by SS) (Mansouri *et al.*, 2005).

determined. The dashed lines marked "SS" indicate the steady-state ($\tau \to \infty$) results. The graphs on the left hand column depict the development of the potential and ion concentration profiles during the initial period ($\tau \leq 10$) after the pressure gradient is applied. The graphs on the right hand column show the corresponding profiles at later stages of the transient simulation ($\tau \geq 50$). For the simulation parameters used in this study, a scaled time of $\tau = 1$ corresponds to a real time of $t = 10$ microseconds. The remarkable feature of these simulations is the rapid buildup of the potential difference between the two reservoirs after imposition of the pressure gradient. Within $\tau = 10$, the potential distribution attains values within 5% of the steady-state potential distribution. In contrast, the ion concentration distributions are not affected during these initial times. Both the positive and negative ion distributions remain unchanged from the initial Poisson–Boltzmann distribution during this period. At later stages of the transient simulation, the potential distribution does not undergo any significant change. The ion concentration distributions, however, undergo substantial modification during this long term relaxation of the system toward a steady-state.

In the above simulations, the convective flux boundary condition was used for the Nernst–Planck equations at the exit plane (FQ). Accordingly, the ion concentration profiles obtained at longer durations show a distinct difference between the ion concentrations in the inlet and outlet reservoirs. At steady state, the concentrations of both the co- and counterions (N_n and N_p, respectively) in the outlet reservoir are lower than the corresponding concentrations in the inlet reservoir. This difference is attributed to the ion rejection by the capillary. The interesting feature of the results is that both the co- and counterions have the same concentration in the outlet reservoir, indicating that this reservoir contains an electroneutral electrolyte solution. Thus, although a charged capillary will reject co-ions, the overall coupled transport of the ions through the capillary self-consistently ensures that the permeate solution is electroneutral, and there is no current transport through the capillary at steady-state.

It is evident from these simulations that the steady-state electric potential and ion concentration distributions evolve over largely disparate time scales. The development of the steady-state potential is almost instantaneous, while the equilibrium ion concentrations are established over a much longer time period. For more detailed discussions on these transient simulations, one may refer to Mansouri *et al.* (2005).

14.3.5 Case 3: Electroosmotic Flow due to Axial Pressure and Electric Potential Gradients

We now focus our attention to the flow of an electrolyte solution through a capillary microchannel engendered by the combined influence of axial pressure and electric potential gradients. This general case of electroosmotic flow will be treated as a variant of the steady-state problem discussed in conjunction with the prediction of streaming potential (Case 1). In the present problem, we will simply demonstrate how application of an external electric field at the exit plane of the system in addition to an applied axial pressure gradient modifies the fluid velocity distribution within the capillary microchannel.

14.3.5.1 Boundary Conditions Since this case involves simultaneous application of a pressure and an electric potential gradient, one can retain the pressure boundary conditions on segments AP and FQ used in Case 1. To consider the applied electric field, one can either apply an electric potential, ψ_0, at the exit plane FQ, or specify the electric field, E_0, at this plane. Here, we demonstrate how the electric field is imposed at the exit plane through the boundary condition

$$\mathbf{n} \cdot \bar{\nabla}\Psi = -\bar{E}_0 \tag{14.52}$$

on FQ. Here $\bar{E}_0$ is the scaled external electric field, provided as an input to the problem. This boundary condition is specified in the FEMLab program using the displacement vector, **D**. In terms of the non-dimensionalization used in developing the model, one can write $\bar{\mathbf{D}} = \bar{\mathbf{E}}$. In non-dimensional form, the displacement boundary condition at FQ is specified as

$$\bar{\mathbf{D}}|_{\text{FQ}} = \bar{\mathbf{E}}|_{\text{FQ}} = \bar{E}_0 \mathbf{i}_{\bar{x}} \tag{14.53}$$

where $\bar{E}_0$ is the specified scaled axial electric field. The radial component of the displacement vector is specified as zero at the boundary FQ. The scaled axial electric field is related to the external electric field specified on the boundary FQ, E_0, (V/m) through

$$\bar{E}_0 = \frac{ze}{k_B T \kappa} E_0$$

In terms of the parameters used in the present model, for an electrolyte concentration of $c_\infty = 10^{-5}$M, when $\kappa = 1.038 \times 10^7\ \text{m}^{-1}$, a scaled axial field of 0.01 will correspond to an applied electric field of 2.67×10^3 V/m.

The electric potential on plane AP is set to zero. On the charged walls of the capillary and the reservoirs (BC, CD, and DE), we employ the constant surface charge density condition, where the charge density is specified as a problem input. All other boundary conditions for this problem are identical to those specified in context of Case 1. Table 14.6 describes these boundary conditions.

For the problem, the required input parameters are the applied pressure at the inlet (AP), the electric field at the exit plane (FQ), and the charge densities on the walls (BC, CD, and DE). As in case 2, the scaled charge density at the solid walls (BC, CD, DE) was set to $\sigma_c = -1.04$, which roughly corresponds to a scaled surface potential of $\Psi_c = -1$ in absence of any flow.

14.3.5.2 Velocity Profiles in Electroosmotic Flow When discussing electrokinetic flow through a capillary microchannel in Chapter 8, it was shown within the context of the linearized theory (based on the linearized Poisson–Boltzmann equation) that the overall electrokinetic flow in presence of both axial pressure and electric potential gradients is a simple superposition of purely pressure driven flow and purely electroosmotic flow. Here, we first test whether such an assumption is valid when the complete problem formulation is employed. The complete problem formulation in the present numerical simulation does not involve linearization of the governing equations, and considers a finite length capillary. Furthermore, with a constant surface

TABLE 14.6. Boundary Conditions of the Governing Equations for Electroosmotic Flow in Presence of Combined Axial Pressure and Electric Potential Gradients (Case 3).

Boundary	Poisson Equation	Nernst–Planck Equations	Navier–Stokes & Continuity Eqs.
PQ	Axial symmetry	Axial symmetry	Axial symmetry
AB, EF	Symmetry	Zero normal flux	Slip/symmetry
BC, DE	Constant charge density, q_c (specified)	Zero normal flux	No slip
CD	Constant charge density, q_c (specified)	Zero normal flux	No slip
AP	Potential = 0	Bulk ion conc. ($N_p = N_n = 1$)	Normal pressure, P_0 (specified)
FQ	Electric field, $\bar{E}_0$ (specified)	Bulk ion conc. ($N_p = N_n = 1$)	Normal pressure ($P = 0$)

charge density boundary condition, the surface potential on the capillary wall can vary axially.

Figure 14.16 depicts the scaled axial velocity profile, $u_x/(\kappa D)$, at the mid-section of the capillary microchannel obtained under different combinations of the axial pressure gradient, $p_x = -\partial p/\partial x$, and scaled axial electric field, $\bar{E}_0$, applied at the plane FQ. The velocity profile shown by the dashed line for purely pressure driven flow (with $p_x = 10^7$ Pa/m) was obtained by numerically solving the Navier–Stokes equation for an uncharged capillary. No electrokinetic effects were considered in obtaining this velocity distribution. The velocity profile for a purely electroosmotic flow ($p_x = 0$ and $\bar{E}_0 = 0.01$) was obtained by solving the complete electrokinetic transport model numerically, after setting the axial pressure gradient to zero (using $P = 0$ at AP). In this case, the electric field was directed along the the positive x-coordinate. For the purely pressure driven flow, the velocity profile is parabolic, while for the electroosmotic flow, the velocity distribution has a flatter profile near the channel axis. The symbols in Figure 14.16 were obtained by adding the two above results for purely pressure driven and electroosmotic flows. Finally, Figure 14.16 also shows the velocity profile when both the axial pressure gradient and the electric field are simultaneously applied to the electrolyte solution, as shown by the solid line (with $p_x = 10^7$ Pa/m and $\bar{E}_0 = 0.01$). The resulting velocity profile obtained from the numerical solution (solid line) has a slight deviation from the corresponding velocity distribution obtained by linear superposition of the pressure driven and electroosmotic velocity distributions (symbols). The maximum deviation between these two results is about 2%. While this difference definitely indicates that the linear superposition of the pressure driven and electroosmotic velocities is not exact, it is also evident that such a superposition is a fairly good approximation.

The slight difference between the two results originates from the fact that the local electric fields and pressure gradients at the capillary mid-section are quite different in the case of purely pressure driven flow, purely electroosmotic flow, and electrokinetic

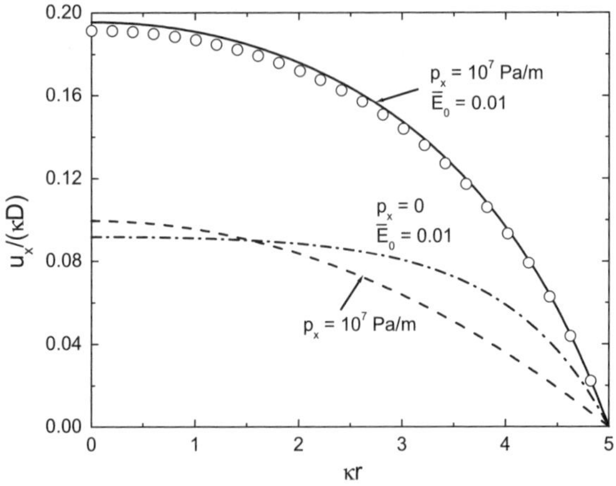

Figure 14.16. Variation of the steady-state axial velocity, $u_x/(\kappa D)$, with scaled radial position at the capillary mid-section ($\bar{x} = 50$). Solid line: numerical simulation of electrokinetic flow under the combined influence of pressure ($p_x = 10^7$ Pa/m) and an applied electric field ($\bar{E}_0 = 0.01$). Dashed line: purely pressure driven flow (no electrokinetic effect, $\bar{E}_0 = 0$). Dash-dotted line: purely electroosmotic flow ($\bar{E}_0 = 0.01$, $p_x = 0$). Symbols: summation of purely pressure driven and electroosmotic flows (dashed and dash-dotted lines). The scaled capillary radius is 5, and the scaled capillary wall surface charge density is $\sigma_c = -1.04$.

flow under the combined influence of pressure and electric field. To demonstrate this, we plot the axial electric fields and the axial pressure gradients along the capillary axis in Figure 14.17 corresponding to the different flow situations depicted in Figure 14.16. For a purely pressure driven flow in absence of electrokinetic effects, the electric field is zero. The scaled axial electric fields obtained for the purely electroosmotic flow (dash-dotted line) and the combined pressure and electric field driven electrokinetic flow (solid line) are slightly different at the mid-section of the capillary, as seen from Figure 14.17(a). It is also important to note that the field inside the capillary microchannel is substantially different from the applied field, $\bar{E}_0 = 0.01$. The axial pressure gradients at the capillary mid-section corresponding to purely pressure driven flow (dashed line), purely electroosmotic flow (dash-dotted line), and the combined electrokinetic flow (solid line) are quite different, as is evident from Figure 14.17(b). One may note that there exists an adverse axial pressure gradient within the capillary (dash-dotted line) during purely electroosmotic flow. For the purely pressure driven flow, the axial pressure gradient within the capillary is constant. It is discernable from Figure 14.17 that the three flow scenarios considered here result in different axial electric fields and axial pressure gradients at the channel mid-section. In this context, it is indeed remarkable that despite these different driving forces, the results from superposition of pressure driven and electroosmotic velocities are in such close proximity to the complete electrokinetic velocity profile in presence of combined axial pressure gradient and electric field.

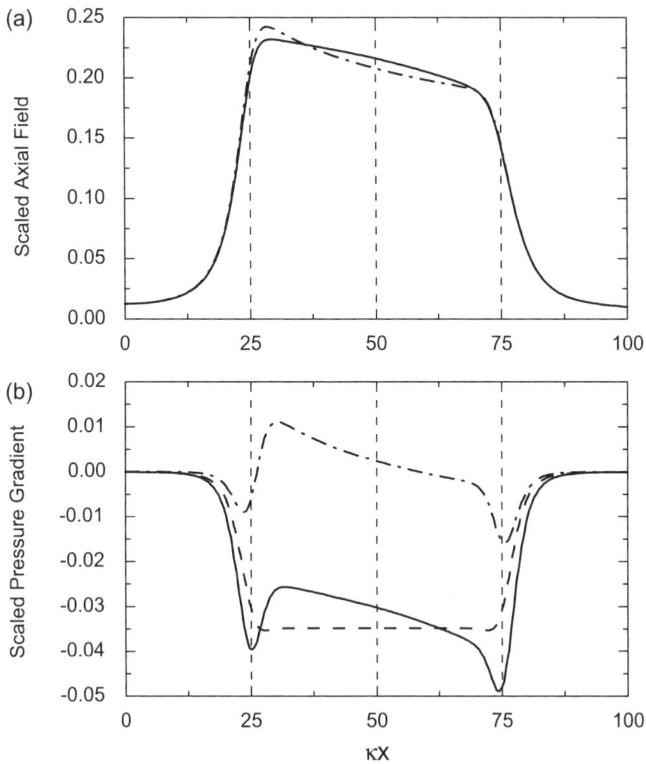

Figure 14.17. Axial variation of the electric field, $\bar{E}_x$, and the pressure gradient, $\partial P/\partial \bar{x}$, along the capillary axis ($\bar{r} = 0$) corresponding to the results shown in Figure 14.16. The vertical dashed lines represent the capillary entrance, mid-section, and the exit. The different line types correspond to the line types shown in Figure 14.16.

The simulation results for the axial velocity profile in the capillary under the influence of oppositely directed pressure and electric field gradients are presented next. The simulations were performed by imposing external electric fields of different magnitudes acting along the capillary axis and observing the steady-state velocity profile of the electrolyte at the channel mid-section. The scaled electric field, denoted by $\bar{E}_0$, and applied at the plane FQ, was varied from zero to -0.05. When the external field is zero, we have the situation of pressure driven electrokinetic flow, with the pressure gradient directed along the positive x coordinate. This velocity profile is depicted by the topmost curve of Figure 14.18. Note that the velocity profile obtained for this case will be slightly different from the purely pressure driven velocity distribution shown in Figure 14.16. This is owing to the fact that in the present case, which considers the complete electrokinetic transport, the electric field generated by the streaming potential will counter the pressure driving force and slightly diminish the velocity.

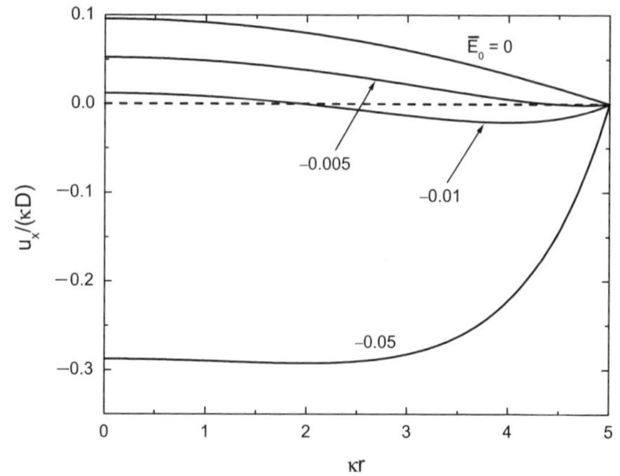

Figure 14.18. Variation of the steady-state axial velocity with radial position at the capillary mid-section under the combined influence of oppositely directed pressure gradient and electric field ($\bar{E}_0$). The axial pressure gradient used in all cases was $p_x = 10^7$ Pa/m. The scaled capillary radius is 5, and the scaled capillary wall surface charge density is $\sigma_c = -1.04$.

The electric field, $\bar{E}_0$, is now applied in a direction opposite to the axial pressure gradient (along negative x direction). The ensuing steady-state velocity profiles are depicted in Figure 14.18. Positive values of the scaled axial velocity represent the flow along the positive axial direction (along the direction of the pressure gradient, p_x), while negative values indicate a reversal of the flow.

The simulations show that the parabolic velocity profile observed under the influence of the pressure gradient is modified upon application of a counteracting externally imposed electric field. For small electric fields, the velocity decreases slightly from the purely pressure driven flow. The velocity near the capillary wall reverses direction upon application of the electric field. For the scaled field of -0.01, we observe that the fluid near the capillary wall flows in a direction opposite to the flow in the core of the capillary microchannel. In this case, the velocity is zero at two radial locations, namely, at the capillary wall (no-slip), and at another location ($\bar{r} \simeq 2$) between the capillary axis and the wall. This second location is the stationary plane where the fluid does not undergo any axial motion. Between this stationary plane and the capillary wall, there exists a location (corresponding to the maximum negative velocity), where the shear stresses will vanish. Further increasing the electric field ($\bar{E}_0 = -0.05$) completely reverses the flow direction.

The different types of simulation results presented above indicate the general applicability of the numerical simulation procedure to address a wide range of problems associated with electrokinetic transport of an electrolyte through a charged capillary microchannel. The advantage of such a numerical simulation code is the flexibility of modifying the geometry, surface properties, and boundary conditions, leading to different applications. Although a symmetric electrolyte with identical diffusivities of different ionic species was considered in the foregoing simulations, it is fairly straightforward to incorporate different diffusivities of different ionic species and asymmetric electrolytes in the current model. Furthermore, multi-component electrolytes (with more than two ionic species) can be considered through inclusion of additional Nernst–Planck equations in the model. Although the initial development and validation of a numerical methodology can be somewhat tedious, a sufficiently general numerical code can eventually perform simulation tasks for more realistic situations compared to what can be addressed through approximate analytic methods of solving the governing electrokinetic transport equations.

14.4 ANALYSIS OF ELECTROPHORETIC MOBILITY

The electrophoretic mobility of a single spherical particle in a large reservoir containing an electrolyte solution was discussed extensively in Chapter 9. The classical approach of Henry (1931) was employed to determine the mobility for low surface (zeta) potentials on the particle. Furthermore, it was stated that rigorous solutions accounting for the convective relaxation effects, such as the perturbation solution of O'Brien and White (1978), leads to a deviation of the electrophoretic mobility from the values predicted by Henry's solution at higher values of particle surface potential. Of particular interest in this context is the fact that for large values of the scaled particle size, $\kappa a_p > 3$, the plot of electrophoretic mobility, η, *versus* zeta potential attains a maximum between $4 < |ze\zeta_p/k_BT| < 7$, where ζ_p is the particle zeta potential. In this section, we aim to reproduce these results for selected values of the parameters κa_p and the scaled particle surface potential, $ze\zeta_p/k_BT$, using a finite element solution of the governing electrochemical transport equations representing the electrophoresis of a spherical particle of radius a_p.

As discussed in Chapter 9, the general governing equations for electrophoresis are the Navier–Stokes and continuity equations, the Poisson equation, and the Nernst–Planck equations. These equations need to be solved simultaneously for a charged particle moving under the influence of an applied electric field in an electrolyte solution, which is stationary at an infinite distance from the particle.

Direct numerical simulations of electrophoretic mobility of a single spherical particle in an unbounded fluid are relatively scarce. The perturbation techniques of Wiersema *et al.* (1966) and O'Brien and White (1978) are still considered to be the most robust methodologies for obtaining a solution of the problem. Several numerical methodologies for calculation of the electrophoretic mobility have subsequently been developed, some of which essentially follow the approach of Teubner (1982). Teubner's analysis starts from the observation that amongst the equations governing electrophoresis, namely, the Stokes and continuity equations (incorporating electrical body forces) for fluid velocity, the Poisson equation, and the Nernst–Planck equations, only the equations governing fluid velocity are linear. The Nernst–Planck equations remain non-linear even for small applied electric fields. He then proceeded to develop an elegant methodology for circumventing the difficulties associated with the nonlinearities in the Nernst–Planck equations. Based on this methodology, Shugai *et al.*

(1997) and Shugai and Carnie (1999) numerically calculated the electrophoretic mobilities of two spherical particles, and for spherical particles in bounded flows. Previously, Keh and Chen (1989a,b) also computed the mobilities for two particles using a bispherical coordinate system. More recently, numerical simulations of electrophoresis of charged particles in a straight cylindrical capillary were reported by Hsu and Hung (2003), as well as Hsu *et al.* (2004). A direct simulation of electrophoresis of two particles in a rectangular microchannel has also been reported by Ye and Li (2004).

Although most of the recent numerical studies mentioned above appear to address electrophoresis in complex geometries, such as, particle motion in bounded flows, or multiparticle systems, there are several simplifications inherent in many of the studies. Most of these simplifications are related to avoiding direct solution of the Nernst–Planck equations governing the ionic movement in the electrolyte solution. The approach of O'Brien and White (1978) is based on perturbation analysis. The methodologies based on Teubner's analysis (Shugai *et al.*, 1997; Shugai and Carnie, 1999) essentially rely on circumventing the direct solution of the Nernst–Planck equations. Many numerical solutions are based on the assumption of Boltzmann distribution of ions (Hsu and Hung, 2003; Hsu *et al.*, 2004). These techniques are similar in scope as the method of Henry (1931), which neglects convective relaxation effects altogether. Finally, there are approaches which do not explicitly consider the electric double layer in the analysis of the motion, but employ a slip velocity at an arbitrary slip plane surrounding the particle to solve the Navier–Stokes equations for fluid flow (in absence of electrical body forces) (Ye and Li, 2004).

The governing equations for electrophoresis might seem fairly tractable, particularly after considering the multiphysics formulation of electroosmosis in the previous section. However, two factors complicate the problem dramatically. The first is associated with the severe non-linearity of the Nernst–Planck equations. As was evident from the foregoing discussions, most numerical treatments of electrophoresis tend to bypass the direct solution of these equations. The second issue is related to properly interpreting the force balance on the particle that leads to the steady-state translational velocity of the particle in an applied field. Anderson (1989) discussed this matter elegantly.

Consider the scenario depicted in Figure 14.19. When a negatively charged particle is immersed in an electrolyte solution, the counterions form a positively charged double layer sheath around the particle. Combined, the particle and the sheath form an electroneutral body. The question then is, why will this electroneutral object (particle plus the oppositely charged sheath) move under the influence of an applied electric field? The answer to this question lies in the fact that the positively charged double layer sheath is mobile. In other words, the ionic sheath surrounding the particle can move under the influence of the external electric field. The resulting electroosmotic fluid flow occurs in a direction that is opposite to the direction of the electrical force acting on the particle. The velocity field in this sheath determines the velocity in the fluid outside the charged sheath (the external electroneutral fluid). Consequently, the outer boundary of the charged sheath (denoted as the slip plane) will have a velocity that differs from the actual particle velocity. Thus, if one can calculate the velocity

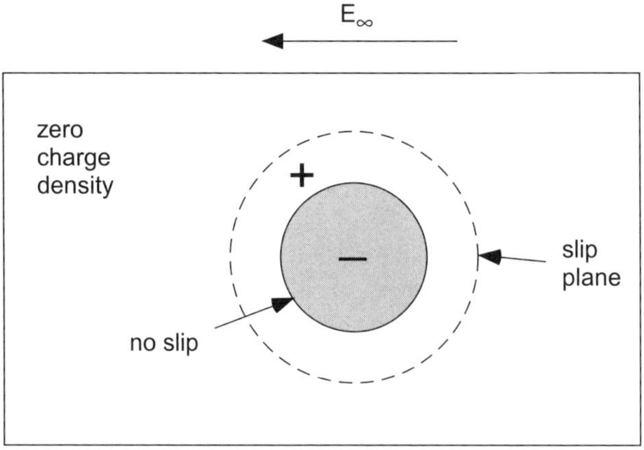

Figure 14.19. Schematic representation of electrophoresis of a charged spherical particle (shown in gray) in an electrolyte solution under the influence of a uniform external electric field, E_{∞}. The negatively charged particle and the positively charged electric double layer sheath surrounding it (dashed circle) is electrically neutral. The fluid outside the dashed line denoted as the slip plane is also electrically neutral.

at the slip plane relative to a stationary particle, the electrophoretic mobility of the particle can be inferred from this slip plane velocity.

The above picture is considerably different from the case of electrophoretic motion of a charged particle immersed in a charge free dielectric liquid. Here, the fluid has no free charge (ions) to "neutralize" the charge of the particle. Furthermore, due to absence of free charge, there is no electric double layer sheath or an electroosmotic flow. Hence, the steady-state particle velocity in this case arises simply from the electrical force acting directly on the charged particle due to the external field and the counteracting fluid drag force. Note that this fluid drag force is simply the force that even an uncharged particle moving through the viscous fluid will experience. In context of Figure 14.19, for such types of motion, the slip plane moves to infinity.

For any finite electrolyte concentration, when a double layer sheath is formed around the charged particle, the net particle motion arises from a combination of three forces. These are, (i) the electrical force acting on the particle due to the imposed field, (ii) the fluid drag force acting on the particle in absence of any charge effects when the particle translates under the influence of the electrical force, and (iii) the excess drag force on the particle due to the counteracting electroosmotic flow (Teubner, 1982; Shugai and Carnie, 1999). The overall steady-state velocity of the particle is governed by a balance of the above three forces.

If one is to model the complete transient development of the electrophoretic motion of the particle, a highly sophisticated numerical method is needed, which tracks the movement of the particle in the electric field. Such particle tracking algorithms are becoming increasingly accessible in the field of computational fluid dynamics (Ye and Li, 2004). However, implementation of such an algorithm in a complete electrokinetic model is still somewhat complicated. In particular, the software we use for the simulations in this book (FEMLab) does not at present have a robust implementation of a particle tracking algorithm for a multiphysics problem. Typically, finite element methods allow tracking of a moving particle by moving the nodal points of a mesh, a technique akin to the Lagrangian approach. However, in such techniques, the mesh can become considerably distorted (the quality of the elements can become extremely poor) over successive time steps. Thus, one needs to have a methodology for restoring the element quality after several moves of the particle. The implementation of this Lagrangian particle tracking method is fairly straightforward in context of a fluid dynamics problem. However, when one considers the coupled solution of Poisson and Nernst–Planck equations along with the Navier–Stokes equation, the problem becomes even more complicated from a numerical perspective.

To avoid the above-mentioned difficulty, we will present here the simulation of the steady-state electrophoretic motion of a particle. The approach we discuss does not require the particle position to be updated. Instead, we fix the coordinate system at the center of the particle, assuming it to be stationary. In this particle fixed reference frame, the fluid surrounding the particle will appear to be moving in a direction opposite to the particle. One can then infer the particle velocity with respect to a stationary observer using the fluid velocity sufficiently far from the particle. The simulations only provide the electroosmotic velocity for finite locations of the slip plane surrounding the particle, and hence, are not applicable directly to obtain the electrophoretic mobility of particles in pure dielectric fluids. Notwithstanding these limitations, the numerical methodology presented here is adequate to provide comparisons of the electrophoretic mobility with some of the analytical and numerical results discussed in Chapter 9.

14.4.1 Problem Formulation

Consider a spherical particle of radius a_p suspended in a large volume of electrolyte as shown in Figure 14.20(a). Upon application of an electric field, E_∞, the negatively charged particle starts to translate toward the positive electrode. The steady state velocity of the particle, u_p, is the unknown we wish to determine from the numerical simulation. For the numerical calculations, the geometry is represented in cylindrical coordinates, with the particle center located at the origin (O), Figure 14.20(b). In this particle fixed reference frame, with the electric field aligned along the x axis, we can write the three dimensional problem as an axi-symmetric two dimensional model in cylindrical coordinates (r,x). In the particle fixed reference frame, the fluid flows in the direction opposite to the direction of particle motion. The far-field fluid velocity obtained from the solution of the governing equations is used to calculate the particle electrophoretic mobility.

The computational domain used for the simulations is represented by the shaded region in Figure 14.20(b) which is obtained by the intersection of the rectangular domain ABQP and the semi-circle CDE. We will only restrict our attention to solving the electrochemical transport equations in the shaded region (within the electrolyte solution). For a more general solution, one can additionally solve a Laplace equation for the electric potential distribution within the charged particle (inside the domain CDE). However, we will simply assume the particle to have a constant

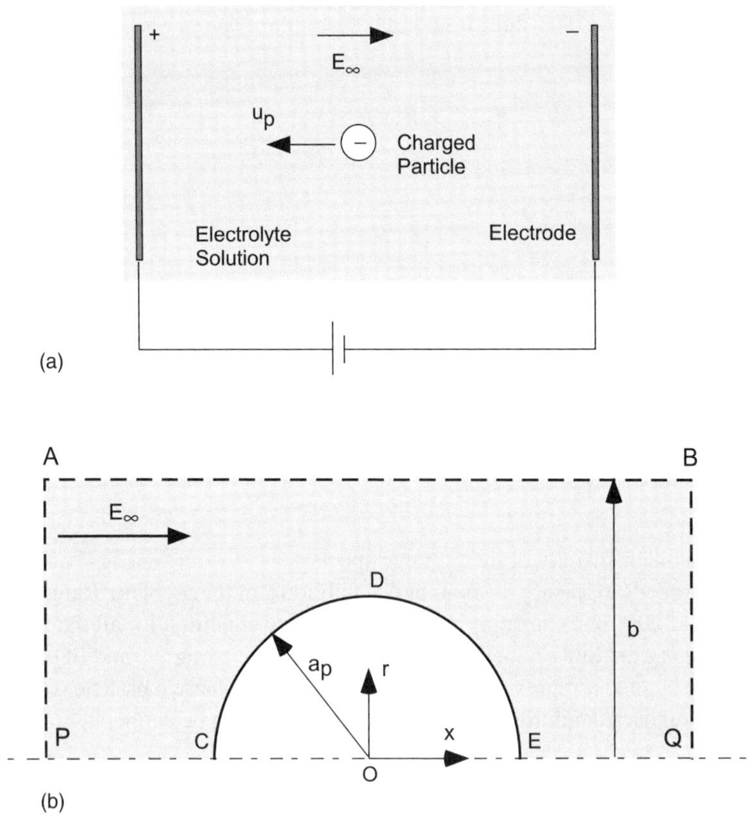

Figure 14.20. Schematic representation of (a) electrophoresis of a charged spherical particle in an electrolyte solution under the influence of a uniform external electric field, and (b) the axi-symmetric cylindrical computational geometry used for analysis of the electrophoretic mobility. The origin O is set at the center of the particle.

potential or charge density over its surface, and avoid solving the additional Laplace equation within the particle. This is generally justified when the particle dielectric permittivity is much smaller than the dielectric permittivity of the surrounding fluid.

The equations governing the problem were described in Section 9.2. Here we briefly reiterate the steady-state forms of these governing equations and pertinent boundary conditions. The electric potential distribution in the electrolyte solution surrounding the particle is given by the Poisson equation

$$\epsilon \nabla^2 \psi = -\rho_f = -\sum_i z_i e n_i \tag{14.54}$$

where ϵ is the dielectric permittivity of the fluid, ρ_f is the volumetric free charge density, z_i and n_i are the valency and the number concentration of the ith ionic species, respectively. The fluid flow is given by the Navier–Stokes and continuity equations,

$$\rho \mathbf{u} \cdot \nabla \mathbf{u} = -\nabla p + \mu \nabla^2 \mathbf{u} + \rho \mathbf{g} + \rho_f \mathbf{E} \tag{14.55}$$

and

$$\nabla \cdot \mathbf{u} = 0 \tag{14.56}$$

where $\mathbf{u}$ is the fluid velocity vector, with the velocity components along the r and x directions represented by u and v, respectively, ρ is the fluid density, μ is the viscosity, $\mathbf{g}$ is the gravitational acceleration, and $\mathbf{E} = -\nabla\psi$ is the electric field. Finally, the ion concentration distributions are given in terms of the Nernst–Planck equations,

$$\nabla \cdot \mathbf{j}_i^{**} = \nabla \cdot \left[n_i \mathbf{u} - D_i \nabla n_i - \frac{z_i e D_i n_i}{k_B T} \nabla \psi \right] = 0 \tag{14.57}$$

where D_i is the diffusivity of the ith ionic species.

One now needs to specify the boundary conditions for the problem. Referring back to Figure 14.20(b), one can apply the axial symmetry condition for all the governing equations along the line segment PQ (more specifically, along PC and EQ). We next focus on the boundary represented by the surface of the spherical particle (CDE). The electrical boundary condition for the Poisson equation can be defined as

$$-\epsilon \, \mathbf{n} \cdot \nabla \psi = q_p \tag{14.58}$$

for a constant surface charge density, q_p, on the particle. The requirement for the electrophoresis problem is that the particle charge density is not modified upon imposition of the external electric field. To obtain the charge density, one can initially solve the Poisson and Nernst–Planck equations in absence of any external field (and any fluid flow) using a constant surface potential at the particle. In this case, the boundary condition on CDE is set as

$$\psi = \zeta_p \tag{14.59}$$

where ζ_p is the specified zeta potential on the particle surface. From the solution of the Poisson and Nernst–Planck equations (with no fluid flow), one can then calculate the surface charge density, q_p, on the particle corresponding to the specified surface potential. Following this, during the solution of the complete problem in presence of an external field, one reverts to the constant surface charge boundary condition, Eq. (14.58), where the calculated value of the surface charge density is used.

Assuming the spherical particle to be fixed in space, the no slip condition for the Navier–Stokes equation on the boundary CDE becomes

$$\mathbf{u} = 0 \tag{14.60}$$

Finally, for the Nernst–Planck equations, we impose the condition that no electrolyte ions penetrate the sphere surface

$$\left[D_i \nabla n_i + \frac{z_i e n_i D_i}{k_B T} \nabla \psi\right] \cdot \mathbf{n} = 0 \quad \text{and} \quad \mathbf{u} \cdot \mathbf{n} = 0 \text{ on CDE} \tag{14.61}$$

In the above equations, $\mathbf{n}$ is a unit outward normal vector (pointing toward the fluid) on the sphere surface.

We now turn our attention to the boundary conditions far away from the sphere, on the outer boundaries of the computational domain, denoted by the line segments PA, AB, and BQ. The electrical boundary condition for the Poisson equation on these three boundaries should be

$$\nabla \psi = -\mathbf{E}_\infty \tag{14.62}$$

where $\mathbf{E}_\infty = E_\infty \mathbf{i}_x$ is the applied electric field acting along the positive x direction. However, implementing this Neumann condition on all the outer boundaries will not yield a solution for the governing equations. To obtain a solution, one needs to specify the electric potential, Ψ, at least on one of these external boundaries. Therefore, we implement the electrical boundary condition on these outer boundaries as follows. At the outset, we assume that the electrical potential in absence of the charged particle is zero on the plane $x = 0$. Now, when a constant external electric field, E_∞, is applied, the potentials on the planes PA and QB can be written as

$$\psi = -x E_\infty \quad \text{at PA and QB}$$

where x is the axial coordinate of these planes. Using the above procedure, the boundary conditions on the segments PA and QB are specified as Dirichlet conditions in terms of the electric potentials. On the boundary AB, the imposed electric field is specified as a boundary condition for the electric displacement. In other words,

$$\mathbf{D} = \epsilon \mathbf{E}_\infty = \epsilon E_\infty \mathbf{i}_x \quad \text{on AB}$$

For the Navier–Stokes and continuity equations, the appropriate far field boundary condition is that the overall fluid stress (both pressure and viscous stresses) vanishes. Accordingly, the total fluid stress on the line segment AB is expressed using

$$\bar{\bar{\sigma}} \cdot \mathbf{n} = \left\{ -p\bar{\bar{I}} + \mu\left[\nabla \mathbf{u} + (\nabla \mathbf{u})^T\right]\right\} \cdot \mathbf{n} = 0 \tag{14.63}$$

where p is the hydrostatic pressure. On the segments PA and QB, we assign a zero pressure condition. Finally, for the Nernst–Planck equations, the ion concentrations in the far field (on segments PA, AB, and BQ) are given by their bulk values

$$n_i = n_{i\infty} \tag{14.64}$$

which should satisfy the bulk electroneutrality condition, $\sum z_i n_{i\infty} = 0$.

It is important to note that the far field boundary conditions should strictly apply at an infinite distance from the particle. To ensure that application of these boundary conditions on the finite computational domain does not introduce appreciable errors in the calculations, one needs to ensure that the radius of the cylindrical domain b is significantly large compared to the particle radius. In all calculations, we consider an outer domain that has a radius equal to 12 to 20 times the radius of the particle. It is also ensured that the boundaries AP and BQ are located at a sufficiently large distance from the particle. Typically, the distances of these boundaries from the particle center were also set at 12 to 20 times the particle radius. Using such a large domain ensures that the fluid velocity at the outer boundaries of the domain are not significantly affected by the particle. One should recall from our discussion of hydrodynamic interactions in Chapter 6 that the hydrodynamic effects of a particle on the Stokes flow in the surrounding fluid extends to very large distances. Consequently, the numerical solutions obtained over a finite domain using the boundary condition given by Eq. (14.63) will never be "exact". However, noting that all numerical solutions are approximate to within some preset tolerance, it is sufficient to place the outer boundary at a distance where the perturbation of the fluid velocity due to the particle is below the preset numerical tolerance. In the present calculations, setting $b = 12a_p - 20a_p$ ensures that the hydrodynamic perturbations of the far field velocity due to the particle are adequately suppressed below the numerical accuracy of the finite element program. One can compare the size of the domain used in this problem with the requirement imposed in the solution of the Poisson–Boltzmann equation (Section 14.2), where we used $b = 5a_p$ for the location of the outer boundary. Clearly, the computational domain in the present problem is larger due to the long-range nature of hydrodynamic interactions. If one uses a domain that is sufficiently large to attenuate the hydrodynamic interactions, it is imperative that the electrical boundary conditions will also be valid at this large distance, particularly when the parameter κa_p, where κ is the inverse Debye length, is large.

The problem is solved for a symmetric $(z : z)$ electrolyte. Additionally the ionic diffusivities are assumed to be identical for every species as was done in the case of electroosmotic flow discussed in the previous section. All lengths are scaled with respect to the particle radius, a_p. Thus, the scaled radial and axial coordinates are represented as $\bar{r} = r/a_p$ and $\bar{x} = x/a_p$, respectively. This scaling is different from our use of the Debye length, κ^{-1}, as the length scale in the previous problems discussed in this chapter. The governing equations are non-dimensionalized using the scaled variables described in Table 14.7. The electric field and velocity scalings used here are identical to those used in Chapter 9 in context of the O'Brien and White (1978) model of electrophoretic mobility.

The non-dimensional equations are written using the scaled variables as follows. Poisson equation:

$$\bar{\nabla}^2\Psi = -\frac{1}{2}(\kappa a_p)^2(N_p - N_n) \tag{14.65}$$

Navier–Stokes equation:

$$\bar{\rho}\mathbf{u}^* \cdot \bar{\nabla}\mathbf{u}^* = -\bar{\nabla}P + \bar{\nabla}^2\mathbf{u}^* + \frac{1}{2}(\kappa a_p)^2\beta^*(N_p - N_n) \tag{14.66}$$

TABLE 14.7. Scaled Variables Used for Non-Dimensionalizing the Governing Equations for Electrophoresis.

Dimensionless Variable	Expression
Electric potential, Ψ	$\frac{ze\psi}{k_B T}$
Particle radius, κa_p	$a_p\sqrt{\frac{2z^2e^2 n_{i\infty}}{\epsilon k_B T}}$
Ion concentration, N_i	$\frac{n_i}{n_{i\infty}}$
Gradient operator, $\bar{\nabla}$	$a_p\nabla$
Density, $\bar{\rho}$	$\frac{\epsilon}{\mu^2}\left(\frac{k_B T}{ze}\right)^2 \rho$
Pressure, P	$\frac{1}{\epsilon}\left(\frac{a_p ze}{k_B T}\right)^2 p$
Diffusivity, $\bar{D}_i$	$\frac{\mu D_i}{\epsilon}\left(\frac{ze}{k_B T}\right)^2$
Electric field, β^*	$\left(\frac{a_p ze}{k_B T}\right)\mathbf{E}$
Particle charge, Q_p^*	$\left(\frac{ze}{\epsilon k_B T a_p}\right) Q_p$
Velocity, $\mathbf{u}^*$	$\frac{\mu a_p}{\epsilon}\left(\frac{ze}{k_B T}\right)^2 \mathbf{u}$

Nernst–Planck equation:

$$\bar{\nabla}\cdot[\mathbf{u}^* N_i - \bar{D}_i\bar{\nabla}N_i - \text{sign}(z_i)\bar{D}_i N_i\bar{\nabla}\Psi] = 0 \tag{14.67}$$

For the present problem involving a binary electrolyte, we deal with two Nernst–Planck equations, one for the positively charged ion, denoted by subscript p, and a second for the negatively charged ion, represented by subscript n. For each type of ionic species, the scaled Nernst–Planck equation should incorporate the appropriate sign of the ionic charge in the migration flux term. The parameter β^* in Eq. (14.66) represents the scaled electric field,

$$\beta^* = \left(\frac{a_p ze}{k_B T}\right)\mathbf{E} \tag{14.68}$$

where **E** is the local electric field.

One final point needs to be clarified regarding the calculation of the electrophoretic mobility using the above equations. This pertains to a statement made at the outset regarding how the equations are written in a particle fixed reference frame. One should note that to a stationary observer, a particle bearing a total charge Q_p, under the influence of an externally imposed electric field of magnitude E_∞, will undergo an electrophoretic motion even when there are no ions present in the fluid surrounding it. The steady state particle velocity in this case will evolve from a balance of the electrical force, $Q_p E_\infty$, and the Stokes drag force, $6\pi \mu a_p u_p$. In this so-called Hückel limit ($\kappa a_p = 0$), the particle velocity is given by

$$u_p = \frac{Q_p E_\infty}{6\pi \mu a_p}$$

However, one can observe from Eq. (14.66) that the fluid velocity becomes zero as $\kappa a_p \to 0$, since there is no electrical body force acting on the fluid in this case. This is because, the present set of equations written in a particle fixed reference frame only provides the electroosmotic flow of the electrolyte surrounding the particle. In other words, as $\kappa a_p \to 0$, the electrophoretic mobility of the particle predicted by the present model will approach zero, much in the manner in which the cell models for concentrated suspensions predict a zero mobility for finite volume fractions of particles as $\kappa a_p \to 0$. Thus, one should only employ the present model for finite values of the parameter κa_p. In this study, we will restrict our attention to the calculation of the mobility for $\kappa a_p > 1$.

In the following, we briefly discuss numerical models obtained by simplification of the above governing equations that represent Henry's (1931) approach, as well as the linear perturbation analysis presented in Chapter 9.

14.4.1.1 *Formulation of Henry's Problem* The formulation described above represents the general set of governing equations for electrophoresis, without any assumption of linear superposition of the external electric field on the electric field generated by the charged particle. In Chapter 9, when discussing the linear perturbation theory, we noted that the methodology involves two steps, namely, evaluation of the electric potential distribution, ψ^{eq}, around a charged particle, followed by imposing the external field, E_∞, and adding the local potential perturbation, $\delta\psi$, due to this field on ψ^{eq}. One might implement this methodology by separately solving a Laplace equation governing the potential distribution due to the applied electric field (as in Henry's approach). In the following, we present this procedure, which numerically emulates Henry's technique. Recall that Henry's approach involved solving the Poisson–Boltzmann equation for the potential distribution due to the charged particle, ψ, the Laplace equation for the potential distribution due to the externally imposed electric field, ϕ, and the Navier–Stokes equation. This approach does not consider convective relaxation effects, and hence, assumes that the ion distributions are not affected by the fluid flow. Consequently, one does not need to explicitly solve the Nernst–Planck equations.

The numerical implementation of Henry's approach involves solving the following equations, written in scaled form.
Poisson–Boltzmann equation:

$$\bar{\nabla}^2 \Psi^{eq} = (\kappa a_p)^2 \sinh(\Psi^{eq}) \tag{14.69}$$

where $\Psi^{eq} = ze\psi^{eq}/k_B T$ is the equilibrium potential distribution around the charged particle.
Laplace equation:

$$\bar{\nabla}^2 \Phi = 0 \tag{14.70}$$

where $\Phi = ze\phi/k_B T$ represents the potential distribution due to the applied electric field.
Navier–Stokes equation:

$$\bar{\rho}\mathbf{u}^* \cdot \bar{\nabla}\mathbf{u}^* = -\bar{\nabla}P + \bar{\nabla}^2\mathbf{u}^* + (\kappa a_p)^2 \sinh(\Psi^{eq})\bar{\nabla}(\Psi^{eq} + \Phi) \tag{14.71}$$

Here, the electric field used in the Navier–Stokes equation is given in terms of the total potential gradient, $\bar{\nabla}(\Psi^{eq} + \Phi)$.

The boundary conditions for the above set of equations are written in a slightly different manner. For the Poisson–Boltzmann equation, we set the electric potential on the particle surface (boundary CDE) to $\Psi^{eq} = \Psi_p = ze\zeta_p/k_B T$. At the outer boundaries of the domain (PA, AB, BQ), we set $\Psi^{eq} = 0$. For the Laplace equation, as in Henry's approach, the charge density is set to zero at the particle surface (CDE). On the outer boundaries PA and QB, the electric potential, Φ is given by

$$\Phi = -\bar{x}\beta^*_\infty \quad \text{at PA and QB} \tag{14.72}$$

where β^*_∞ is the scaled applied electric field. On the boundary AB, the applied electric field is specified as the electric displacement boundary condition for the Laplace equation. The boundary conditions for the Navier–Stokes equation are similar to the general formulation discussed earlier.

The above formulation based on Henry's approximation is considerably simpler than the general procedure for solving the problem. However, the formulation completely ignores any redistribution of ion concentrations around the particle due to the fluid convection.

14.4.1.2 Formulation of the Perturbation Problem This formulation implements the equations described in Section 9.3.1, and is similar to the set of governing equations solved by Wiersema *et al.* (1966) and O'Brien and White (1978). The pertinent variables are written as sums of the equilibrium value plus the perturbation in presence of the applied field. In scaled form, these variables are

$$
\begin{aligned}
\text{Potential:} &\quad \Psi^{eq} + \delta\Psi \\
\text{Velocity:} &\quad 0 + \mathbf{u}^* \\
\text{Ion concentration:} &\quad N_i^{eq} + \delta N_i \\
\text{Pressure:} &\quad P^{eq} + \delta P
\end{aligned}
$$

For the equilibrium situation, we solve Eqs. (9.43) to (9.45), which yields the Poisson–Boltzmann equation as in Eq. (14.69). Solution of this problem provides the equilibrium parameters, namely, Ψ^{eq} and the scaled ion concentrations, N_i^{eq}. Following this, the linear perturbation equations, namely, Eqs. (9.49) to (9.52) are solved. In scaled form, these equations are given by
Poisson equation:

$$\bar{\nabla}^2 \delta\Psi = -\frac{1}{2}(\kappa a_p)^2(\delta N_p - \delta N_n) \tag{14.73}$$

Navier–Stokes equation:

$$\mathbf{u}^* \cdot \bar{\nabla}\mathbf{u}^* = -\bar{\nabla}\delta P + \bar{\nabla}^2\mathbf{u}^* - \frac{1}{2}(\kappa a_p)^2\left[(N_p^{eq} - N_n^{eq})\bar{\nabla}\delta\Psi + (\delta N_p - \delta N_n)\bar{\nabla}\Psi^{eq}\right] \tag{14.74}$$

Nernst–Planck equation:

$$\bar{\nabla} \cdot \left\{\mathbf{u}^* N_i^{eq} - \bar{D}_i \bar{\nabla}\delta N_i - \text{sign}(z_i)\bar{D}_i\left[N_i^{eq}\bar{\nabla}\delta\Psi + \delta N_i \bar{\nabla}\Psi^{eq}\right]\right\} = 0 \tag{14.75}$$

where $N_i = N_p, N_n$ for the positive and negative ionic species, respectively.

The boundary conditions for this perturbation problem are given by Eqs. (9.53) and (9.54). A solution of these equations will provide the fluid velocity field around the particle, from which the electrophoretic mobility can be determined. The perturbation formulation, unlike the Henry formulation, accounts for the convective relaxation of the electric double layer. However, it is applicable when the magnitude of the externally imposed electric field is much smaller than the field due to the charged particle.

14.4.2 Mesh Generation and Numerical Solution

As with all the other problems discussed in this chapter, the computational domain of Figure 14.20(b) was discretized using a non-uniform mesh consisting of quadratic triangular elements. A typical mesh used in the numerical calculations is shown in Figure 14.21. The nature of the physical problem imposes serious demands on the mesh. On one hand, to ensure that the far field fluid velocity is not affected by the particle, and that the boundary conditions at the outer periphery of the domain are sufficiently accurate, one needs to consider a large computational domain. Secondly, since the electric potential and ion concentrations decay rapidly near the particle, particularly for larger values of the parameter κa_p, one needs to have a sufficiently

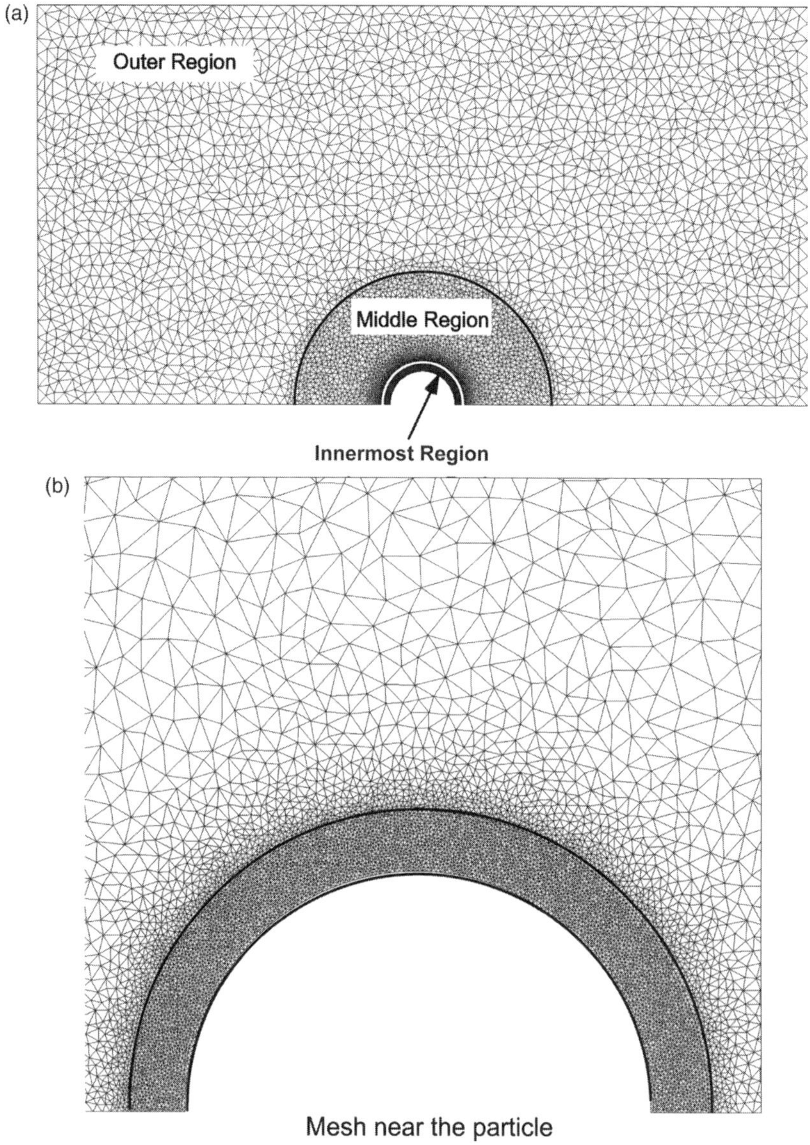

Figure 14.21. Finite element mesh used for modeling electrophoretic mobility of a charged spherical particle. (a) Discretization of the computational domain into three regions, namely, outer, middle, and innermost. (b) The refined mesh around the particle (innermost region).

refined mesh near the particle surface. Consequently, finite element calculations for electrophoretic mobility can become highly computation intensive.

The discretization of the computational domain is performed by selecting three regions surrounding the particle as shown in Figure 14.21(a). The thickness of the innermost spherical region surrounding the particle is approximately $2\kappa^{-1}$, and it contains a highly refined mesh. The decay in the electric potential and ion concentrations are generally confined within this domain. The middle region has an outer radius of approximately $6a_p$. The fluid velocity virtually attains the far field values at this outer radius. This intermediate shell has a slightly coarser mesh compared to the innermost region. Finally, the outermost region has the coarsest mesh. The flow in this outer region is generally uniform, and no electrical effects due to the particle are perceptible in this outer region. In this context, one may even solve the Navier–Stokes equation in this outer region without explicitly considering the electrical body forces. Furthermore, the solution of the Laplace equation may suffice in this region, since the electrolyte solution is electroneutral. The division of the computational domain into these three sub-domains ensures that the highly non-linear Poisson and Nernst–Planck equations are solved accurately near the particle surface. The interior boundaries demarcating the innermost and middle, as well as, the middle and outer regions are "continuous", which implies that for every equation, the corresponding continuity condition applies at these internal boundaries.

The mesh in these calculations generally consist of approximately 20,000 elements. The maximum element size in the innermost domain is selected such that the decay in the electric potential and ion concentrations around the particle is captured accurately. This refined region is depicted in Figure 14.21(b). The maximum element size, as well as the thickness of this spherical shell domain is dependent on the value of the parameter κa_p. As κa_p increases, the domain shrinks in thickness.

The numerical solution of the governing equations is performed to obtain the steady state far field velocity of the fluid under an applied electric field. The parameters and their ranges used in the simulations are shown in Table 14.8. The solution methodology involves an initialization step, which requires solving the Poisson and Nernst–Planck equations simultaneously to obtain the ion concentration and electric potential distributions in absence of any externally imposed electric field. For this initialization step, a constant surface potential condition is used on the particle. The particle surface charge density calculated from this initial solution is then substituted as a boundary condition to the Poisson equation for subsequent computations. The Poisson and Nernst–Planck equations are next solved with the constant surface charge density condition on the particle and the imposed external field at the outer boundaries. Using the volumetric charge density and the electric field distribution obtained from the above solution, the Navier–Stokes and continuity equations are solved to obtain the fluid velocity field. Once the fluid velocity field is calculated, it is substituted back in the Poisson and Nernst–Planck equations to evaluate the updated electric potential and ion concentrations. The sequential solution of the Poisson, Nernst–Planck, and Navier–Stokes equations in the manner described above is continued until the solutions (velocity, electric potential, and concentrations of ions) converge to within a specified tolerance.

Upon convergence of the pertinent variables to within the preset tolerance, the scaled far-field fluid axial velocity (evaluated at any location of the outer boundary), v_∞^* becomes equal and opposite to the scaled electrophoretic velocity, u_p^*, of the particle (*i.e.*, $v_\infty^* = -u_p^*$). The particle electrophoretic mobility is thus directly determined from the scaled far field fluid velocity.

TABLE 14.8. Parameter Values Employed in the Model for Calculation of Electrophoretic Mobility.

Parameter	Value/Range
Solvent permittivity, ϵ	6.954×10^{-10} C^2/Nm^2
Scaled particle surface potential, $\Psi_p = ze\zeta_p/k_BT$	-1 to -7
Particle radius	10^{-7} m
Scaled particle radius, κa_p	1 to 30
Ion valence, $\lvert z_i \rvert$	1
Ion diffusivity, D	1×10^{-9} m^2/s
Temperature	298 K
Fluid density, ρ	1000 kg/m^3
Fluid viscosity, μ	0.001 N.s/m^2
Applied axial electric field, E_∞	10^4 V/m

The electrophoretic mobility, or the particle velocity per unit applied electric field, was written in Chapter 9 in context of Henry's solution as

$$\eta = \frac{u_p}{E_\infty} = \frac{2}{3}\frac{\epsilon\zeta_p}{\mu} f(\kappa a_p) \tag{14.76}$$

where $f(\kappa a_p)$ is Henry's function, given by Eq. (9.85). In terms of the scaled velocity, $u_p^* = (\mu a_p/\epsilon)(ze/k_BT)^2 u_p$, and the scaled electric field, $\beta_\infty^* = a_p ze E_\infty/(k_BT)$, one can express the mobility as

$$\eta^* = \frac{3u_p^*}{2\beta_\infty^*} = \frac{3\mu}{2\epsilon}\left(\frac{ze}{k_BT}\right)\frac{u_p}{E_\infty} = \frac{3\mu}{2\epsilon}\left(\frac{ze}{k_BT}\right)\eta = \frac{ze\zeta_p}{k_BT} f(\kappa a_p) \tag{14.77}$$

We will use the scaled notations of Eq. (14.77) to represent the numerical results pertaining to Henry's function and the mobilities calculated by O'Brien and White (1978).

The numerical procedures applied to solve the simplified models pertaining to the finite element implementations of Henry's approach and the perturbation approach are briefly discussed here. In both cases, the Poisson–Boltzmann equation is first solved to determine the equilibrium electric potential and ion concentration distributions. Following this, in Henry's approach, the Laplace equation is solved using the applied electric field as a boundary condition. The total electric potential is obtained by adding the equilibrium potential and the potential from the Laplace equation. Following this, the Navier–Stokes equation is solved to determine the velocity distribution. No iterations are required in this solution. In the perturbation analysis based formulation, the perturbed Poisson and Nernst–Planck equations are first solved using the equilibrium potential and ion concentration distributions. Following this, the perturbed Navier–Stokes equation is solved. Generally a single iteration is sufficient for the solution of these perturbed equations, since the perturbed-Nernst–Planck equations are linear. However, as we use the Navier–Stokes instead of the Stokes equation, a slight non-linearity exists that can be resolved through a couple of iterations. In subsequent discussions, we will refer to these numerical solutions as Henry's numerical solution and numerical solution of perturbation equations, respectively. All three types of models were solved using the same finite element mesh, same sets of model parameters, and same numerical precision.

The requirement of a highly refined mesh to capture the rapid decay of the electric potential near the particle surface, particularly for large surface potentials and κa_p imposes restrictions on the range of parameter values that can be explored using the numerical model described here. In particular, the procedure delineated here imposes serious demands on the number of elements for large values of κa_p, and accordingly, most of the simulations are confined to a small range $1 < \kappa a_p < 30$.

14.4.3 Representative Simulation Results

We now present a comparison of the electrophoretic mobility predictions from the numerical solution with the results of Henry (1931) and O'Brien and White (1978) for a constant surface potential particle[3] undergoing electrophoresis in presence of an external electric field. The purpose of these comparisons is to simultaneously validate the numerical results, and show how the analytical results of Henry (1931), although approximate, provides a remarkably good prediction of the electrophoretic mobility as long as the assumptions made in the analytical approach remain valid.

Figure 14.22 shows a comparison of the function $f(\kappa a_p)$ obtained numerically (symbols), and from Eq. (9.85) (line). The numerical simulations were conducted for a particle with a scaled surface potential of $\Psi_p = -1$. A range of scaled particle radii ($1 \leq \kappa a_p \leq 30$) was studied numerically. The numerical simulations were conducted using two approaches, namely, by direct solution of complete set of governing equations (Poisson, Nernst–Planck, and Navier–Stokes), as well as the set of simplified equations numerically emulating the methodology of Henry (1931). As seen from Figure 14.22, the numerical predictions of Henry's function obtained from the simplified numerical implementation of Henry's methodology (open circles) are virtually identical to the analytical values predicted using Eq. (9.85). The solid symbols, representing the results of the complete numerical simulation, on the other hand, provide a slightly lower mobility. The methodology adopted by Henry (1931) to arrive at the electrophoretic mobility was quite different from the methodology employed in the complete numerical simulations. First, Henry used the linearized Poisson–Boltzmann equation in his perturbation analysis. Secondly, he assumed an additional Laplace equation superimposed on the Poisson–Boltzmann equation to incorporate the effect of the external field. The complete numerical procedure does not use any of these assumptions, and yet, the two results are within 3% of each other. Furthermore,

[3]Although we use the terminology constant surface potential, it should be recalled that we are actually keeping the particle surface charge density fixed during the electrophoretic motion in presence of the applied field. The constant surface potential simply refers to the fact that we initially determine the particle surface charge density (in absence of any external field) for each value of κa_p assuming a constant surface potential. The constant surface potential also refers to the fact that the particle has a constant surface potential prior to imposing the external electric field.

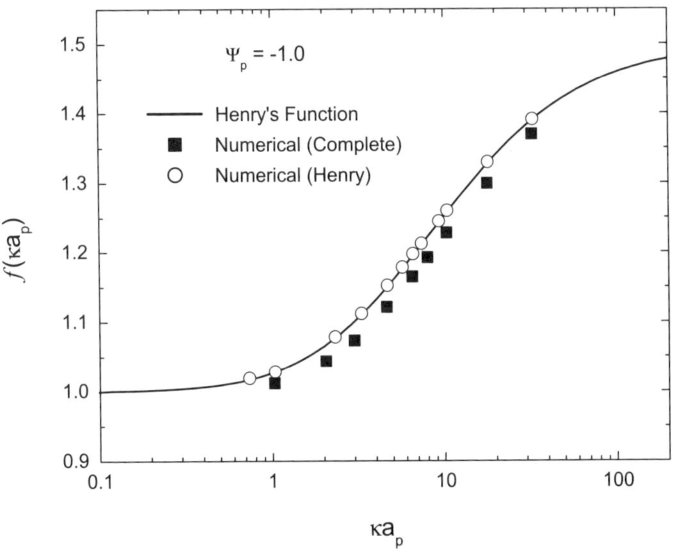

Figure 14.22. Comparison of Henry's function for electrophoretic mobility, Eq. (9.85) (line), with the corresponding predictions obtained from the finite element simulations (symbols) for a scaled particle surface potential of $\Psi_p = -1$. Open circles represent the numerical implementation of Henry's formulation. Solid squares represent the full numerical solution of the governing equations.

when one numerically implements the simplifications of the general governing equations that emulates Henry's approximations, one obtains a result that is identical to his analytic expression. This comparison provides us confidence regarding the accuracy and validity of the numerical result for electrophoretic mobility. Furthermore, the comparison underscores the fact that although Henry's analytical result is based on completely ignoring the convective relaxation effects, it is remarkably accurate, as long as the particle surface potentials are kept small.

That said, however, the deviation between Henry's result (line) and the complete numerical solution (solid symbols) in Figure 14.22 is real, and should not be solely construed as a computational error in the numerical calculations. Even for scaled particle surface potentials of $\Psi_p = -1$, one observes a slight effect of convective relaxation as the imposed electric field becomes large. This reduces the particle electrophoretic mobility slightly. Such a deviation was observed even by Booth (1950) when the particle surface charge density was large. A discussion of this deviation is given by Saville (1977) (see Figure 3 in Saville, 1977).

We now compare the predictions of the scaled electrophoretic mobility, η^*, obtained from the numerical calculations with the corresponding predictions of O'Brien and White, (1978) for larger particle surface potentials $|\Psi_p| > 1$. The numerical calculations of O'Brien and White (1978) pertain to solution of the perturbation equations of Section 9.3.1. Their results incorporate the convective relaxation effects, and hence, provide a more accurate prediction of the electrophoretic mobility at higher particle surface potentials compared to Henry's function. The comparison of the finite

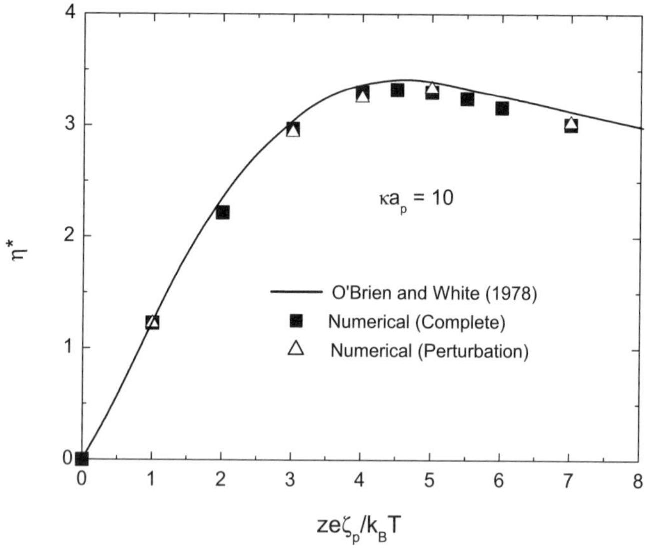

Figure 14.23. Variation of the scaled particle electrophoretic mobility, $\eta^* = 3U^*/2\beta^*$, with scaled particle surface potential at $\kappa a_p = 10$. Line: Digitized data from O'Brien and White (1978). Solid symbols: Complete numerical solution of the governing equations. Open triangles: Numerical solution of the perturbation equations.

element results with those of O'Brien and White (1978) is shown in Figure 14.23. The comparisons are done for $\kappa a_p = 10$. The numerical results were determined using values of Ψ_p between -1 to -7. The convergence of the numerical results become increasingly difficult at higher values of Ψ_p. The solid and open symbols in Figure 14.23 represent the complete numerical simulation and the numerical solution of the perturbation equations, respectively. It is evident that the numerical results obtained using the two approaches are virtually identical. These numerical results are also in good accordance with the results of O'Brien and White (1978) (line). Once again, although the numerical techniques used in the two approaches, and the governing equations solved are quite different, since the same physical problem is solved, the results should be identical. The comparison simply shows that the rigorous numerical solution of the governing equations indeed depicts the same behavior as predicted by the perturbation approach.

We will restrict our presentation of results from the numerical model to the above two comparisons, since these already illustrate the applicability of direct numerical simulations to electrokinetic phenomena associated with particle electrophoresis. One can, however, obtain substantial information from these simulations regarding the detailed maps of the velocity, electric potential, and ion concentration fields. These can lead to analysis of forces acting on the particle, the total volumetric charge in the electrolyte solution surrounding the particle, the current density distribution, and similar quantities. Although we restricted the application of the general simulation to small applied fields, one can increase the applied electric field and observe the onset of several nonlinear effects on electrophoresis as well.

The above comparisons lead to the conclusion that direct numerical solutions, despite their unique approximations arising from consideration of a finite computational domain, provide sufficiently accurate predictions of electrophoretic mobility. Refinement in discretization schemes, improvement in the solution methodology, and enhanced computational resources can lead to further refinement of these numerical results. However, the "cost" of these direct numerical calculations *vis a vis* the approximate analytical and numerical solutions should be borne in mind when considering these simulations. As a rough estimate of the computational cost, let us consider the computational time required by the three models of electrophoresis discussed in this section. The first approach, which involves the solution of the complete set of governing equations, typically requires 22 minutes on a given computer[4] to converge. The numerical implementation of the perturbation equations using the same number of elements as the complete solution requires less than 3 minutes to converge. Finally, the numerical implementation of Henry's approach requires only a few seconds to provide the solution. In this context, using a complete numerical solution is perhaps an overkill unless an extremely large electric field is applied (such that the linear perturbation approach becomes invalid). The general numerical solution may also be warranted when different particle shapes are considered, or asymmetric electrolytes with different ionic mobilities are employed.

The mathematical model for electrophoretic mobility developed in this section can be modified slightly to predict the particle mobility in a cylindrical capillary. Hsu *et al.* (2004) numerically solved the problem when the particle translates along the axis of a charged cylindrical capillary. The presence of the charged capillary wall imposes two modifications in the model. First, the no-slip condition applies at the capillary wall, modifying the fluid velocity profile, and secondly, the charged capillary wall modifies the electrolyte concentration distribution in the computational domain, thereby modifying the electric field around the particle. In this case, the problem cannot be solved readily using a particle fixed reference frame. Instead, one starts with a guess value for the particle velocity, u_p, and solves the steady state equations using this velocity. From the solution, one then calculates the electrical and hydrodynamic forces on the particle. Noting that for steady state translation of the particle, these two forces will be equal and opposite, one can develop an iterative solution scheme where different guess values of the particle velocity are used until the force balance equation is satisfied. Using this methodology, Hsu *et al.* (2004) studied the modification of the particle electrophoretic velocity due to the presence of the charged capillary confinement. They observed that the direction of the particle motion can be reversed by using an appropriately charged cylindrical capillary confinement.

14.5 CONCLUDING REMARKS

Finite element analysis of three problems relevant to electrokinetic transport phenomena was presented in this chapter. The key message from these exercises is that

[4]An Intel Pentium IV 2.8 GHz PC with 1GB RAM.

the analytical approaches discussed in the previous chapters of this book often work with remarkable precision when applied judiciously to simple geometries. Numerical solution procedures, however, become indispensable when considering more complex geometries, asymmetric electrolytes with different ionic mobilities, and multi-component electrolyte solutions. In particular, when considering electrokinetic phenomena in narrow confined domains, application of numerical techniques becomes mandatory to assess the influence of the confinement on an electrokinetic phenomenon. In this context, numerical simulations become necessary in modelling microfluidic systems, where most of the phenomena like electroosmotic flow, electrophoresis, and colloidal interactions between two charged particles are modified due to the presence of the confining walls of the microfluidic channels.

The numerical simulations presented in this chapter may not be regarded by some as "elegant" approaches for treating the simple problems that they were applied to. These were rigorous implementations of the exact governing equations without any attempt at simplifying these equations. However, once these detailed numerical simulation methods are developed and validated for simple systems, they can easily be applied to more complicated systems where the governing equations cannot be readily simplified.

14.6 NOMENCLATURE

a, a_p	particle radius, m
b	radius of capillary, m
c_∞	bulk electrolyte molar concentration, mol/L
D	ion diffusion coefficient, m^2/s
$\mathbf{E}$	electric field, Vm^{-1}
$\mathbf{E}_\infty$	externally imposed electric field, Vm^{-1}
E_x	component of electric field along x direction, Vm^{-1}
e	magnitude of electronic charge, C
$\mathbf{F}$	force, N
F_x	component of force along axial direction, N
f_x	scaled force component along axial direction
$\mathbf{g}$	gravitational acceleration, m/s^2
h	separation distance between two colloidal particles, m
$\overline{\overline{I}}$	unit tensor
$I_0(-)$	modified Bessel function of the first kind and zero order
$I_1(-)$	modified Bessel function of the first kind and first order
I	total current, A
$\mathbf{j}_i^{**}$	ionic flux based on number concentration, $m^{-2}s^{-1}$
k_B	Boltzmann constant, J/K
L	length of a capillary microchannel, m
$N_p,\ N_n$	scaled concentration of positive and negative ions
n_∞	bulk electrolyte number concentration, m^{-3}
n_i	number concentration of ith ionic species, m^{-3}

$\mathbf{n}$	unit surface normal
n_r, n_z	components of unit normal along r and z coordinates
P	non-dimensional pressure
p	pressure, Pa
p_x	axial pressure gradient $(-\partial p/\partial x)$, Pa/m
q_p, q_c	surface charge density, Cm^{-2}
R	source term in a general partial differential equation
r	radial coordinate, m
$\bar{r}$	scaled radial coordinate, κr or r/a_p
T	absolute temperature, K
$\bar{\bar{T}}$	total stress tensor (hydrostatic and Maxwell stress), $\mathrm{N/m^2}$
$\bar{\bar{T}}_e$	Maxwell stress tensor, $\mathrm{N/m^2}$
t	time, s
$\mathbf{U}$	non-dimensional velocity vector
U_i	degrees of freedom in a finite element formulation
$\mathbf{u}$	fluid velocity vector, ms^{-1}
$\mathbf{u}^*$	scaled velocity vector
u, v	components of fluid velocity vector along r and x directions, respectively, m/s
u_p	particle electrophoretic velocity, m/s
u_p^*	scaled particle electrophoretic velocity
v_∞^*	scaled far-field axial fluid velocity
x	axial coordinate, m
$\bar{x}$	scaled axial coordinate, κx or x/a_p
z	valence of ion

Greek Symbols

β^*	scaled electric field
β_∞^*	scaled applied electric field
Γ	spatial gradient of dependent variable
∇	gradient operator, m^{-1}
η	electrophoretic mobility, $\mathrm{m^2V^{-1}s^{-1}}$
η^*	scaled electrophoretic mobility, $3u_p^*/2\beta_\infty^*$
ξ_i	basis function used in finite element technique
Φ	scaled electric potential (for external field)
ϕ	dependent variable of a partial differential equation electric potential (externally imposed), V
$\partial\Omega$	computational domain boundary
ϵ	dielectric permittivity, $\mathrm{C^2N^{-1}m^{-2}}$
κ	inverse Debye length, m^{-1}
λ	ratio of capillary to particle radius, b/a
μ	fluid viscosity, Pa.s
Ψ	scaled electric potential, $ze\psi/k_BT$
Ψ_p	scaled particle surface potential, $ze\zeta_p/k_BT$

ψ	electric potential, V
ρ	fluid material density, kgm^{-3}
ρ_f	volumetric free charge density, Cm^{-3}
σ_p, σ_c	scaled surface charge density
τ	dimensionless time
ζ_p	zeta potential on a particle surface, V
Ω	computational domain

14.7 PROBLEMS

14.1. The Poisson–Boltzmann equation was solved numerically in Section 14.2 for a symmetric electrolyte. Generalize the model for an asymmetric electrolyte $(z_1 : z_2)$.

14.2. Develop a numerical model for electrokinetic flow of an electrolyte through a slit microchannel. To render the model two-dimensional, assume the channel width to be much larger than the channel height. Solve the general electrochemical transport equations numerically, and compare the results with the analytical results of Chapter 8 for a slit-microchannel.

14.3. During the treatment of electrophoresis of a spherical particle using the perturbation analysis (Section 14.4.1), the perturbation equation for the ion concentration fields was obtained as Eq. (14.75). In this equation, the dependent variable is the perturbed ion concentration δN_i. Note that in this equation, all the quantities with superscript "eq" are known functions. Now, suppose you have a computer program that solves the steady-state convection diffusion equation with a source/sink term, expressed as

$$-\nabla \cdot [-D\nabla c + \mathbf{u}c] = R$$

where D is the diffusion coefficient, c is the dependent variable (concentration), $\mathbf{u}$ is the fluid velocity vector, and R is the source/sink term. Rewrite Eq. (14.75) to recast it in the form of the above convection diffusion equation, where $c \equiv \delta N_i$. In this case, what will be the functions $\mathbf{u}$ and R?

14.8 REFERENCES

Anderson, J. L., and Malone, D. M., Mechanism of osmotic flow in porous membranes, *Biophys J.*, **14**, 957–982, (1974).

Anderson, J. L., Colloid transport by interfacial forces, *Ann. Rev. Fluid Mech.*, **21**, 61–99, (1989).

Basu, S., and Sharma, M. M., An improved space-charge model for flow through charged microporous membranes, *J. Membrane Sci.*, **124**, 77–91, (1997).

Bhattacharjee, S., Chen, J. C., and Elimelech, M., Coupled model of concentration polarization and pore transport in crossflow nanofiltration, *AIChE J.*, **47**, 2733–2745, (2001).

Booth, F., The cataphoresis of spherical, solid nonconducting particles in a symmetrical electrolyte, *Proc. Roy. Soc. Lond. Ser. A*, **203**, 514–533, (1950).

Bowen, W. R., and Sharif, A. O., Long-range electrostatic attraction between like-charge spheres in a charged pore, *Nature*, **393**, 663–665, (1998).

Bowen, W. R., and Sharif, A. O., Long-range electrostatic attraction between like-charge spheres in a charged pore (vol. 393, p. 663, 1998), *Nature*, **402**, 841–841, (1999).

Burgreen, D., and Nakache, F. R., Electrokinetic flow in ultrafine capillary slits, *J. Phys. Chem.*, **68**, 1084, (1964).

Carnie, S. L., Chan, D. Y. C., and Stankovich, J., Computation of forces between spherical colloidal particles – nonlinear Poisson–Boltzmann theory, *J. Colloid Interface Sci.*, **165**, 116–128, (1994).

Cwirko, E. H., and Carbonell, R. G., Transport of electrolytes in charged pores – analysis using the method of spatial averaging, *J. Colloid Interface Sci.*, **129**, 513–531, (1989).

Daiguji, H., Yang, P., and Majumdar, A., Ion transport in nanofluidic channels, *Nano Lett.*, **4**, 137–142, (2004).

Das, P. K., Bhattacharjee, S., and Moussa, W., Electrostatic double layer force between two spherical particles in a straight cylindrical capillary: finite element analysis, *Langmuir*, **19**, 4162–4172, (2003).

Das, P. K., and Bhattacharjee, S., Electrostatic double layer interaction between spherical particles inside a rough capillary, *J. Colloid Interface Sci.*, **273**, 278–290, (2004).

Das, P. K., and Bhattacharjee, S., Finite element estimation of electrostatic double layer interaction between colloidal particles inside a rough cylindrical capillary: effect of charging behavior, *Colloids Surf. A*, **256**, 91–103, (2005).

Deen, W. M., Hindered transport of large molecules in liquid-filled pores, *AIChE J.*, **33**, 1409–1425, (1987).

Glendinning, A. B., and Russel, W. B., The electrostatic repulsion between charged spheres from exact solutions to the linearized Poisson–Boltzmann equation, *J. Colloid Interface Sci.*, **93**, 95–104, (1983).

Grant, M. L., and Saville, D. A., Electrostatic interactions between a nonuniformly charged sphere and a charged surface, *J. Colloid Interface Sci.*, **171**, 35–45, (1995).

Gray, J. J., Chiang, B., and Bonnecaze, R. T., Colloidal particles – origin of anomalous multibody interactions, *Nature*, **402**, 750, (1999).

Gross, R. J., and Osterle, J. F., Membrane transport characteristic of ultrafine capillaries, *J. Chem. Phys.*, **49**, 228, (1968).

Hall, M. S., Starov, V. M., and Lloyd, D. R., Reverse osmosis of multicomponent electrolyte solutions. 1. Theoretical development, *J. Membrane Sci.*, **128**, 23–37, (1997).

Helmholtz, H. V., Studien uber elctrische grenschichten, *Ann. der Physik und Chimie*, **7**, 337–387, (1879).

Henry, D. C., The cataphoresis of suspended particles, Part 1. The equation of cataphoresis, *Proc. Roy. Soc. Lond.*, **133A**, 106–129, (1931).

Hogg, R. I., Healy, T. W., and Fuerstenau, D. W., Mutual coagulation of colloidal dispersions, *Trans. Faraday Soc.*, **62**, 1638–1651, (1966).

Hsu, J.-P., and Hung, S.-H., Electrophoresis of a charge-regulated spheroid along the axis of an uncharged cylindrical pore, *J. Colloid Interface Sci.*, **264**, 121–127, (2003).

Hsu, J.-P., Ku, M.-H., and Kao, C.-Y., Electrophoresis of a spherical particle along the axis of a cylindrical pore: effect of electroosmotic flow, *J. Colloid Interface Sci.*, **276**, 248–254, (2004).

Hunter, R. J., *Zeta Potential in Colloid Science*, Academic Press, London, (1981).

Jacazio, G., Probstein, R. F., Sonin, A. A., and Yung, D., Electrokinetic salt rejection in hyperfiltration through porous materials: Theory and experiments, *J. Phys. Chem.*, **76**, 4015–4023, (1972).

Keh, H. J., and Chen, S. B., Particle interactions in electrophoresis. 1. Motion of 2 spheres along their line of centers, *J. Colloid Interface Sci.*, **130**, 542–555, (1989a).

Keh, H. J., and Chen, S. B., Particle interactions in electrophoresis. 2. Motion of 2 spheres normal to their line of centers, *J. Colloid Interface Sci.*, **130**, 556–567, (1989b).

Keh, H. J., and Tseng, H. C., Transient electrokinetic flow in fine capillaries, *J. Colloid Interface Sci.*, **242**, 450–459, (2001).

Larsen A. E., and Grier, D. G., Like-charge attractions in metastable colloidal crystallites, *Nature*, **385**, 230–233, (1997).

Levine, S., Marriott, J. R., Neale, G., and Epstein, N., Theory of electrokinetic flow in fine cylindrical capillaries at high zeta-potentials, *J. Colloid Interface Sci.*, **52**, 136–149, (1975).

Lewis, A. F., and Myers, R. R., The efficiency of streaming potential generation, *J. Phys. Chem.*, **64**, 1338–1339, (1960).

Li, D. Q., *Electrokinetics in Microfluidics*, Elsevier, Amsterdam, (2004).

Mansouri, A., Scheuerman, C., Bhattacharjee, S., Kwok, D. Y., and Kostiuk, L. W., Transient streaming potential in a finite length capillary, *J. Colloid Interface Sci.*, **292**, 567–580, (2005).

Morrison, F. A., Transient electrophoresis of a dielectric sphere, *J. Colloid Interface Sci.*, **29**, 687, (1969).

O'Brien, R. W., and White, L. R., Electrophoretic mobility of a spherical colloidal particle, *J. Chem. Soc. Faraday Trans.*, II, **74**, 1607–1626, (1978).

Osterle, J. F., Unified treatment of thermodynamics of steady-state energy conversion, *Appl. Sci. Res. A*, **12**, 425, (1964).

Rice, C. L., and Whitehead, R., Electrokinetic flow in a narrow cylindrical capillary, *J. Phys. Chem.*, **69** , 4017–4024, (1965).

Sader, J. E., and Chan, D. Y. C., Electrical double-layer interaction between charged particles near surfaces and in confined geometries, *J. Colloid Interface Sci.*, **218**, 423–432, (1999).

Sader, J. E., and Chan, D. Y. C., Long-range electrostatic attractions between identically charged particles in confined geometries and the Poisson–Boltzmann theory, *Langmuir*, **16**, 324–331, (2000).

Sasidhar, V., and Ruckenstein, E., Electrolyte osmosis through capillaries, *J. Colloid Interface Sci.*, **82**, 439–457, (1981).

Sasidhar, V., and Ruckenstein, E., Anomalous effects during electrolyte osmosis across charged porous membranes, *J. Colloid Interface Sci.*, **85**, 332–362, (1982).

Saville, D. A., Electrokinetic effects with small particles, *Ann. Rev. Fluid Mech.*, **9**, 321–337, (1977).

Shugai, A. A., Carnie, S. L., Chan, D. Y. C., and Anderson, J. L., Electrophoretic motion of two spherical particles with thick double layers, *J. Colloid Interface Sci.*, **191**, 357–371, (1997).

Shugai, A. A., and Carnie, S. L., Electrophoretic motion of a spherical particle with a thick double layer in bounded flows, *J. Colloid Interface Sci.*, **213**, 298–315, (1999).

Smit, J. A. M., Reverse-osmosis in charged membranes – analytical predictions from the space-charge model, *J. Colloid Interface Sci.*, **132**, 413–424, (1989).

Smith III, F. G., and Deen, W. M., Electrostatic double-layer interactions for spherical colloids in cylindrical pores, *J. Colloid Interface Sci.*, **78**, 444–465, (1980).

Smoluchowski, M. von, Contribution a la theorie de l'endosmose electrique et de quelques phenomenes correlatifs, *Bull. International de l'Academie des Sciences de Cracovie*, **8**, 182–200, (1903).

Stankovich, J., and Carnie, S. L., Electrical double layer interaction between dissimilar spherical colloidal particles and between a sphere and a plate: Nonlinear Poisson–Boltzmann theory, *Langmuir*, **12**, 1453–1461, (1996).

Teubner, M., The motion of charged colloidal particles in electric fields, *J. Chem. Phys.*, **76**, 5564–5573, (1982).

Yang, R. J., Fu, L. M., and Hwang, C. C., Electroosmotic entry flow in a microchannel, *J. Colloid Interface Sci.*, **244**, 173–179, (2001).

Warszynski, P., and Adamczyk, Z., Calculations of double-layer electrostatic interactions for the sphere/plane geometry, *J. Colloid Interface Sci.*, **187**, 283–295, (1997).

Wiersema, P. H., Loeb, A. L., and Overbeek, J. Th. G., Calculation of the electrophoretic mobility of a spherical colloid particle, *J. Colloid Interface Sci.*, **22**, 78–99, (1966).

Ye, C., and Li, D. Q., Electrophoretic motion of two spherical particles in a rectangular microchannel, *Microfluid. Nanofluid.*, **1**, 52–61, (2004).

Zienkiewicz, O. C., and Taylor, R. L., *The Finite Element Method*, 5th ed., McGraw-Hill, New York, (1989).

CHAPTER 15

ELECTROKINETIC APPLICATIONS

15.1 INTRODUCTION

We shall consider in this chapter some selected applications of electrokinetic transport phenomena to illustrate the influence of electric double layer and other colloidal forces on the transport of ions, charged particles, and liquids. Electrokinetic transport phenomena play an important role in many industrial processes, waste water treatment, and biological and physiological functions. Examples of selected applications will be presented spanning mechanisms of salt rejection by reverse osmosis membranes, iontophoretic drug delivery, electrokinetic management of hazardous wastes, prediction of transport properties of colloidal dispersions, application of electrokinetic and colloidal phenomena in liberation of crude petroleum from mined oil (or tar) sands, and application of electrokinetics in microfluidic systems.

15.2 ELECTROKINETIC SALT REJECTION IN POROUS MEDIA AND MEMBRANES

Clays, various porous solids, and membranes have the property of partially rejecting salt when a saline solution filters through. In other words, the salt concentration in the filtrate is lower than that in the saline solution upstream of the porous medium or membrane. In order to conserve salt mass balance, the porous solid acts as a filter by not allowing the salt to pass through. Analysis of salt transport and rejection

Electrokinetic and Colloid Transport Phenomena, by Jacob H. Masliyah and Subir Bhattacharjee
Copyright © 2006 John Wiley & Sons, Inc.

through porous media or membranes can be performed by assuming the medium to comprise of a bundle of circular cylindrical capillaries. The physical mechanism of salt rejection in a charged circular cylindrical capillary can be modeled using electrokinetic principles discussed in this book. The surface charge on the pore walls gives rise to a to a electric potential field which extends to a distance comparable to the Debye length into the capillary pore. Salt rejection analysis for a single idealized circular cylindrical pore can be extended to the actual porous medium by relating the porosity and thickness of the medium to the capillary dimensions. It should be recognized that salt rejection in a capillary is a reverse osmosis phenomenon where flow of a liquid occurs from high to low salt concentrations. Such a flow occurs under the influence of an imposed pressure gradient. The term hyperfiltration is also used for the salt rejection phenomenon to be discussed in this section.

Jacazio *et al.* (1972) give an excellent account of the salt rejection phenomenon in a circular cylindrical capillary. Their analysis is presented here to demonstrate the application of electrokinetic principles. As in the case of electroosmosis, the total electric potential $\Phi(r, x)$ is assumed to be a linear combination of the induced potential $\phi(x)$ and the electric double layer potential $\psi(r, x)$:

$$\Phi(r, x) = \phi(x) + \psi(r, x) \tag{15.1}$$

where r and x represent the radial and axial directions, respectively. Boltzmann equilibrium is assumed, giving

$$c_{\pm}(r, x) = f_{\pm}(x) \exp[\mp\bar{\psi}(r, x)] \tag{15.2}$$

where $\bar{\psi} = z\mathcal{F}\psi/R_gT$ and $f_{\pm}(x)$ is a function related to the ionic concentration.

For the case where the inverse Debye length is very small compared to the capillary radius, the potential in the central core of the capillary depends only on the axial direction, x, leading to

$$f_{\pm}(x) = c(x) \tag{15.3}$$

and Eq. (15.2) becomes

$$c_{\pm}(r, x) = c(x) \exp[\mp\bar{\psi}(r, x)] \tag{15.4}$$

The Poisson equation in cylindrical coordinates is

$$\frac{1}{r}\frac{\partial}{\partial r}\left(r\frac{\partial \Phi}{\partial r}\right) + \frac{\partial^2 \Phi}{\partial x^2} = -\frac{\rho_f}{\epsilon} \tag{15.5}$$

leading to

$$\lambda^2 \frac{1}{R}\frac{\partial}{\partial R}\left(R\frac{\partial \bar{\psi}}{\partial R}\right) = \sinh(\bar{\psi}) \tag{15.6}$$

where

$$\frac{\partial^2 \bar{\psi}}{\partial x^2} \ll \frac{\partial^2 \bar{\psi}}{\partial r^2} \tag{15.7}$$

$$\frac{\partial^2 \phi}{\partial x^2} = 0 \tag{15.8}$$

$$R = \frac{r}{a} \tag{15.9}$$

$$\lambda = \frac{\lambda_D}{a} \tag{15.10}$$

and

$$\lambda_D = \left(\frac{\epsilon R_g T}{2z^2 \mathcal{F}^2 c(x)} \right)^{1/2} \tag{15.11}$$

Equation (15.6) can be solved subject to the boundary conditions

$$\bar{\psi}(r, x) = \bar{\psi}_w \quad \text{at } R = 1 \quad \text{(capillary wall)} \tag{15.12}$$

and

$$\frac{\partial \bar{\psi}}{\partial R} = 0 \quad \text{at } R = 0 \quad \text{(axis of symmetry)} \tag{15.13}$$

It should be noted that the Debye length λ_D is a function of the axial position. The solution of Eq. (15.6) reflects the variation of λ with axial position. For a given axial position, the solution of Eq. (15.6) provides the radial distribution of the potential $\bar{\psi}(r, x)$.

Jacazio *et al.* (1972) assumed that the fluid flow within the capillary is given by the Poiseuille relation as

$$u_x(r) = 2V(1 - R^2) \tag{15.14}$$

where V is the average fluid velocity and $u_x(r)$ is the axial fluid velocity at radial position r.

Using the fact that at any axial location, the total current is zero in steady state process, Jacazio *et al.* (1972) arrived at the axial ionic concentration equation

$$\left(\frac{H}{Pe} \right) \frac{d\bar{c}}{d\bar{x}} - \bar{c} = -\bar{c}_2 G \tag{15.15}$$

where

$\bar{x} = x/L$, dimensional axial distance

$Pe = VL/D_\infty$, flow Peclet number

$\bar{c} = c(x)/c_1$, dimensionless axial concentration

c_1 is the feed salt concentration and L is the capillary length. The exit salt concentration c_2 is given by

$$c_2 = \frac{J_s}{(2\pi a^2 V)} \tag{15.16}$$

Here, J_s (mol/s) is the total flux of the dissociated salt molecules. The salt rejection coefficient is defined as

$$R_r = 1 - \frac{c_2}{c_1} \tag{15.17}$$

R_r is equal to unity for complete salt rejection by the capillary tube and zero for no rejection.

The coefficients H and G of Eq. (15.15) are functions of radial position, $\bar{\psi}$ and λ. The initial condition for Eq. (15.15) is $\bar{c} = 1$ at $\bar{x} = 0$. The evaluation of R_r is obtained by simultaneous solution of Eq. (15.6) and (15.15). Solution details are provided by Jacazio *et al.* (1972). It is clear from Eq. (15.15) that the salt rejection is a function of surface charge, Peclet number and Debye length. Usually, the Debye length at the capillary inlet is specified.

The salt rejection coefficient for a cylindrical pore as a function of the Peclet number for different values of the wall potential is shown in Figure 15.1. Here, λ_1 represents the value of λ evaluated using the bulk molar feed concentration, c_1. Clearly, the salt rejection increases with an increasing Peclet number, and becomes constant as $Pe \rightarrow \infty$. At small, Pe, diffusion is dominant with the result that there is a lower rejection. At large values of Pe, the rejection increases for higher surface potentials on the pore wall.

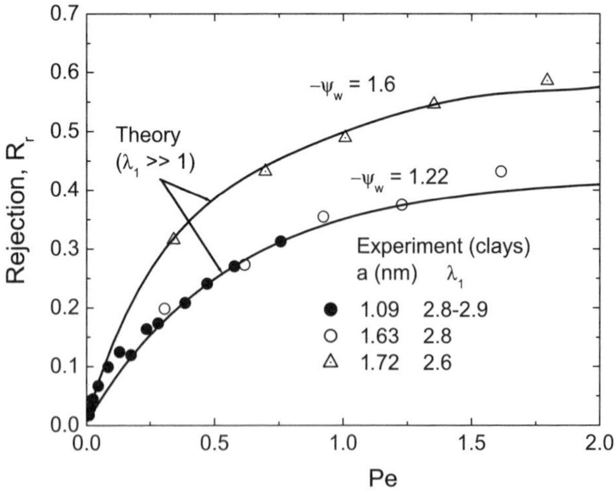

Figure 15.1. Salt rejection coefficient of a cylindrical pore with a constant surface potential for a large Debye length as a function of the Peclet number, and comparison with experiments on compacted clays (Jacazio *et al.*, 1972).

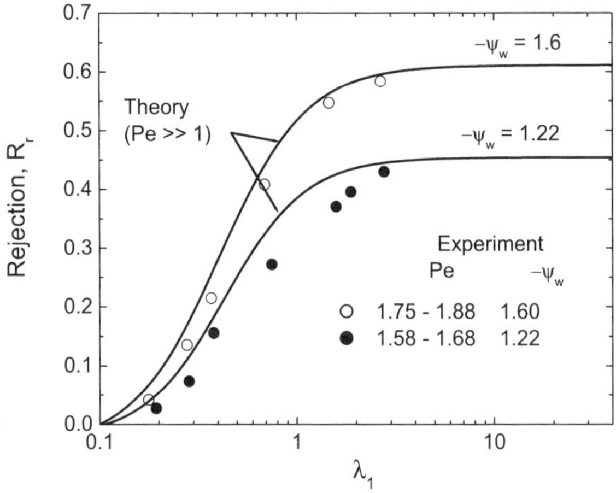

Figure 15.2. Salt rejection coefficient, R_r, of a cylindrical capillary with constant surface potential for a large Peclet number as a function of the dimensionless Debye length, λ_1, and comparison with experiments on compacted clays (Jacazio *et al.*, 1972).

For $Pe \gg 1$, Figure 15.2 shows the variation of salt rejection with the inlet Debye length λ_1. The salt rejection coefficient approaches zero as the Debye length approaches zero. As salt rejection takes place entirely within the Debye sheath, a small λ_1 would indicate that nearly the entire pore has $\bar{\psi} = 0$ and the electric double layer is confined to an area close to the capillary surface. Consequently, no salt rejection is expected to occur. For larger λ_1, the electric double layer extends to the entire capillary pore and rejection is enhanced. Clearly, Figure 15.2 indicates that higher salt rejection is expected for a saline feed having quite low salt concentrations.

The experimental results with clays and cellophane membranes agree quite well with the capillary model put forward by Jacazio *et al.* as shown in Figures 15.1 and 15.2. Although a highly simplified model was used for prediction of salt rejection, it is evident from the example that electrokinetics offers an excellent means to understand the experimental results on salt rejection.

15.3 ELECTROOSMOTIC CONTROL OF HAZARDOUS WASTES

Electroosmosis entails migration of a liquid through a porous charged medium under the influence of an applied electric field. Electroosmosis has been used for the dewatering and consolidation of soils, mine tailings, and waste sludges.

As fluid flow occurs due to an applied electric field, appropriate placement of electrodes would direct the fluid flow in a controlled manner. Consequently, it is possible to direct the flow of specially injected fluids or *in-situ* fluids in a fashion that would divert groundwater from a spill site, direct a chemical sealant toward a waste site, or dewater the region surrounding a hazardous waste area. The various modes

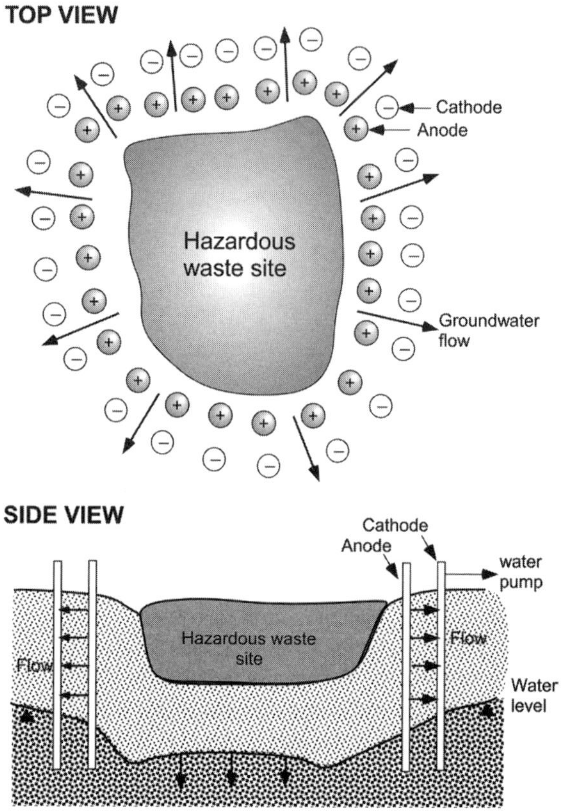

Figure 15.3. Soil dewatering close to a contaminated site (Renaud and Probstein, 1987).

of electroosmotic control of hazardous wastes are given below. They are detailed by Renaud and Probstein (1987).

A case of water drainage close to a hazardous waste site is shown in Figure 15.3. Here, the groundwater level is lowered under the waste site by implanting positive electrodes (anodes) around the waste area. The cathodes are located further away from the waste site. For negatively charged porous media, the salt bearing groundwater flows from the anode to the cathode, thus lowering the groundwater level in the immediate vicinity of the waste site.

Figure 15.4 shows that chemical sealants injected at the anode can be directed away from the hazardous waste and thus can isolate the waste site from the groundwater flow.

Renaud and Probstein (1987) discussed in detail the case where the groundwater flow toward a contaminated waste site is controlled and directed away from the contaminated area. This control is achieved by creating an adverse pressure gradient surrounding the waste site by the appropriate placements of the electrodes as shown

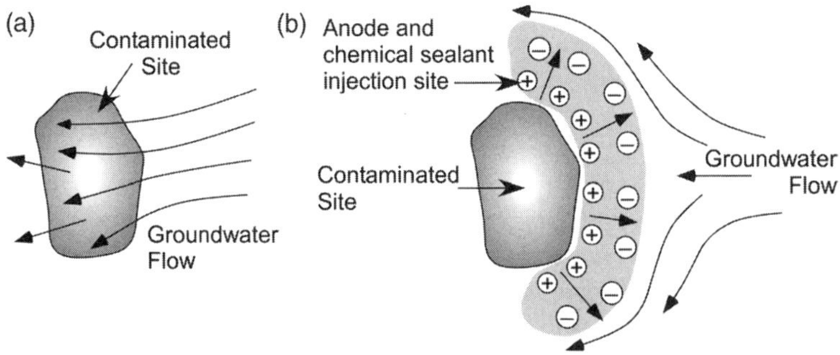

Figure 15.4. Chemical sealant controlled injection: (a) groundwater flow prior to sealant injection, and (b) groundwater flow after sealant injection (Renaud and Probstein, 1987).

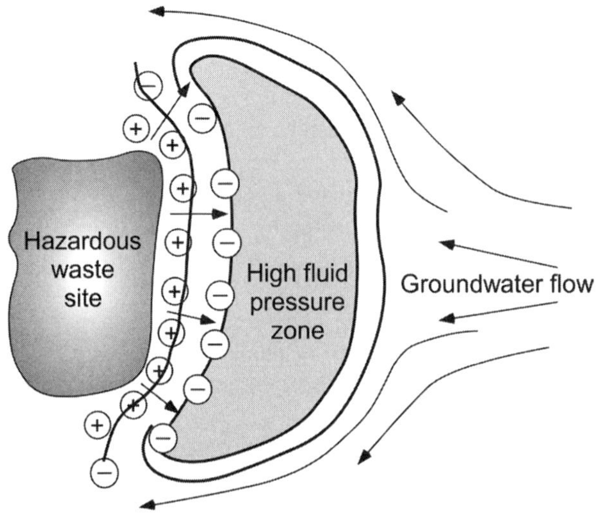

Figure 15.5. Contours of constant pressure near a waste site due to the placement of electrodes (adapted from Renaud and Probstein, 1987).

in Figure 15.5. It was shown that a large adverse pressure gradient can be easily maintained through the use of the applied electrical potential.

15.4 IONTOPHORETIC DELIVERY OF DRUGS

Transdermal (skin) delivery of drugs has gained increasing importance in recent years since this route of drug administration bypasses gastrointestinal degradation and hepatic (liver) metabolism. Iontophoresis, a process which causes an increase in the migration of ionic species into the skin or tissue under a gradient of electrical potential, is used to enhance the penetration of charge molecules (Liu *et al.*, 1988). Thus, iontophoresis involves the transfer of ions or charged molecules (*e.g.*, insulin, pilocarpine) into the body by an electric field. Ions with a positive charge are driven into the skin at the anode and those with negative charges, at the cathode (Banga and Chien, 1988). Reviews discussing iontophoresis are given by Harris (1967) and Banga and Chien (1988). To a large extent, iontophoresis is similar to electrophoresis where with iontophoresis one deals with the transport of ions rather than colloidal particles.

Skin manifests a large impedance to charged molecules which are transmitted through the skin under an applied electric field. The stratum corneum (outer skin layer) is the least conductive layer of the skin. Skin also has hair follicles and sweat ducts that can provide a possible pathway for the migration and diffusion of ions across the skin (Chien, 1982). Under the influence of an electric field, ionic species can penetrate the skin *via* the hair follicles and sweat ducts, which are referred to as "shunt" (Siddiqui, *et al.*, 1985, 1987).

Liu *et al.* (1988) discussed the electrical properties of the stratum corneum. They indicated that the stratum corneum has two important properties that influence iontophoresis. First, the stratum corneum can be polarized by a direct electric field (dc). Secondly, its impedance changes with the frequency of the applied field, *e.g.*, an electric field alternating between zero and a positive value with a given waveform. Consequently, the skin can be represented by an equivalent electrical analogue as shown in Figure 15.6.

The stratum corneum is represented by a combination of resistive and a capacitative component which is a function of the pulse frequency. Yamamoto and Yamamoto (1976, 1978) found that the impedance of human skin decreases with the increase in the pulse frequency. The viable skin is represented by pure resistance R_{vs}, which does not change with the pulse frequency.

From the above characteristics of the stratum corneum, it becomes clear that when a direct electric field is utilized to enhance the penetration of charged molecules, electrochemical polarization becomes established in the skin. The induces polarization operates against the applied electrical field and reduces the current density. As pointed

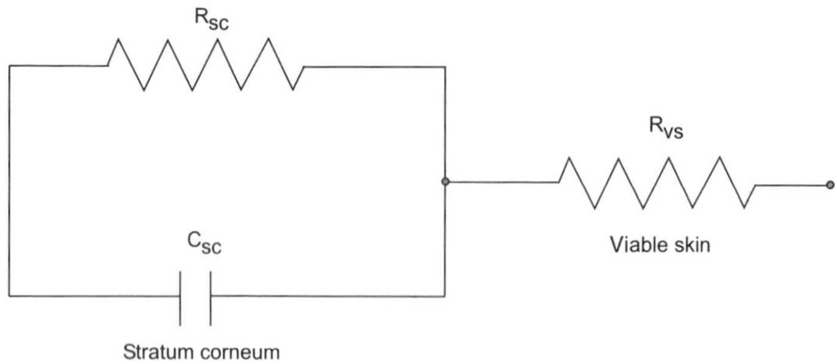

Figure 15.6. Equivalent circuit of skin impedance where R_{vs} and R_{sc} are the resistors for the viable skin and stratum corneum, respectively. C_{sc} is the capacitance for the stratum corneum (Liu *et al.*, 1988).

out by Liu *et al.* (1988), the polarization of the stratum corneum is similar to the charging of a capacitor C_{sc} as shown in Figure 15.6 with an initial current i. The input current decays exponentially across the stratum corneum when a constant dc voltage is applied. As the current is due to ionic movement in the skin, it is clear that the movement of the ionic species through the skin is expected to decay when a dc voltage is applied.

To avoid polarization of the stratum corneum, a pulse dc voltage is normally used in the application of iontophoresis. This pulse mode is a dc voltage that periodically alternates between "on" and "off" for the applied voltage. When the voltage is "on", the charged molecules penetrate the skin while the stratum corneum is being polarized. During the "off" period, no ionic penetration takes place and the stratum corneum becomes depolarized. The manner by which the "on" and "off" cycles are administered controls the rate of the ionic species penetration into the skin.

Iontophoretic delivery of insulin to diabetic rats was investigated by Liu *et al.* (1988). The effectiveness of insulin delivery was monitored by measurement of the blood glucose. The lowering of the blood glucose was indicative of insulin delivery. Figure 15.7 shows the variation of the blood glucose level with time due to application of simple dc voltage and pulse dc voltage. The experiments were conducted on diabetic hairless rats. The dc voltage mode gave a slight initial change in blood glucose level (BGL); however, little change in BGL occurred after four hours. This is due to the polarization of the stratum corneum. However, pulsed dc voltage shows excellent penetration of the insulin into the rat's blood stream.

An electrokinetic model has been proposed by Schwendeman *et al.* (1992) to simulate iontophoresis.

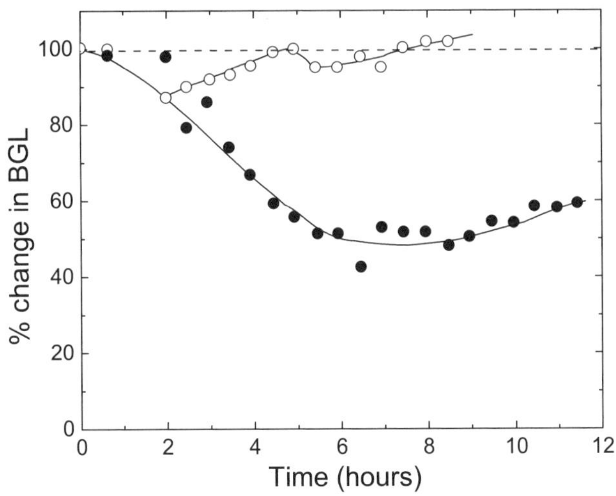

Figure 15.7. Effect of delivery mode on blood glucose levels (BGL) in diabetic hairless rats treated with transdermal periodic iontophoretic system at 2 mA (0.33 mA/cm^2) for 40 min: ○, simple dc; •, pulse dc (2000 Hz) (Liu *et al.*, 1988).

15.5 FLOTATION OF OIL DROPLETS AND FINE PARTICLES

Rising air bubbles have been used in the mineral processing industries to float valuable minerals from solid suspensions. As a bubble rises within the solids suspension, the fine solid particles are separated by adhering to the bubble surface. Such a separation technique has been shown to be quite effective in removing oil droplets from oil wastewaters (Hung, 1978; Van Ham *et al.*, 1983; Pal and Masliyah, 1990).

In order to analyze the deposition process between an air bubble and an oil droplet or a solid particle, it becomes necessary to measure the zeta potentials on the air bubbles, solid particles, and oil droplets. To this end, Okada and Akagi (1987) developed an apparatus to measure the zeta potential of air bubbles. Figure 15.8 shows a schematic diagram of the experimental setup of Okada and Akagi (1987). The apparatus is of the microelectrophoresis type and the zeta potential is determined by measuring the electrophoretic velocity of the bubbles. The measuring system consists of a microscope, an electrophoresis cell, and a video camera attached to a recorder and a monitor. The air bubbles were generated by releasing the pressure of water containing dissolved air. The bubble size was 20–40 μm.

The experimental results for the air bubble zeta potential are shown in Figure 15.9 for the case of an aqueous cationic surfactant cetyl trimethylammonium bromide (CTAB) solution (5×10^{-5} mol/L) containing 0.5 vol % ethanol. Clearly, the zeta potential of the air bubbles is a strong function of the ionic strength of the solution.

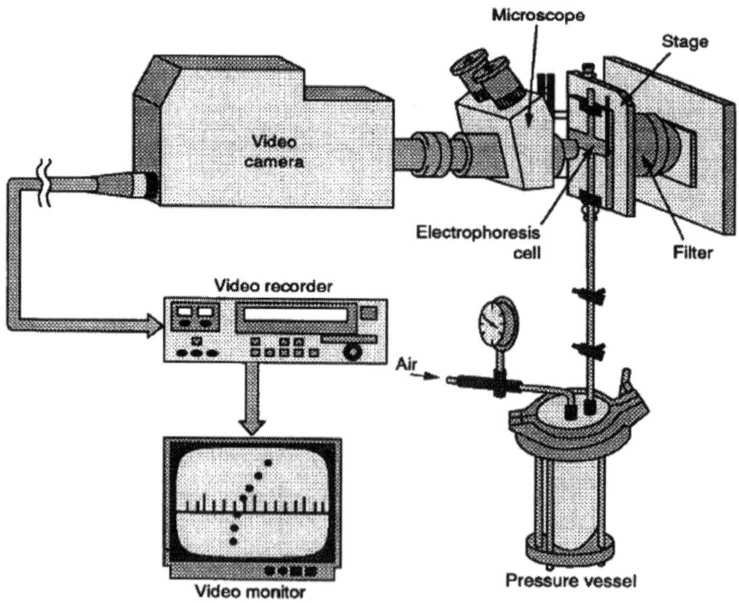

Figure 15.8. Schematic diagram of experimental apparatus used to measure zeta potential of air bubbles (Okada and Akagi, 1987).

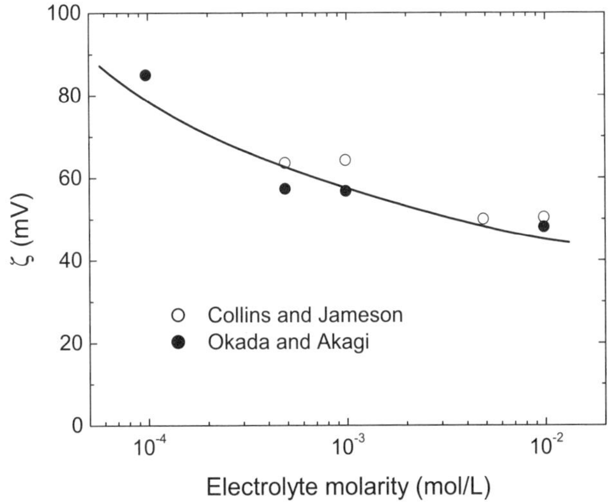

Figure 15.9. Effect of concentration of Na_2SO_4 electrolyte on the zeta potential of an air bubble at 25°C (Okada and Akagi, 1987).

This is similar for the case of the zeta potential dependence on ionic strength for liquids and solid surfaces.

Okada *et al.* (1988) carried out further zeta potential measurements on heavy oil ($a_p = 1.4 \times 10^{-6}$ m) and air bubbles in an anionic surfactant, sodium dodecyl sulfate (SDS) (1×10^{-4} mol/L) in the presence of different electrolytes (Figure 15.10). The

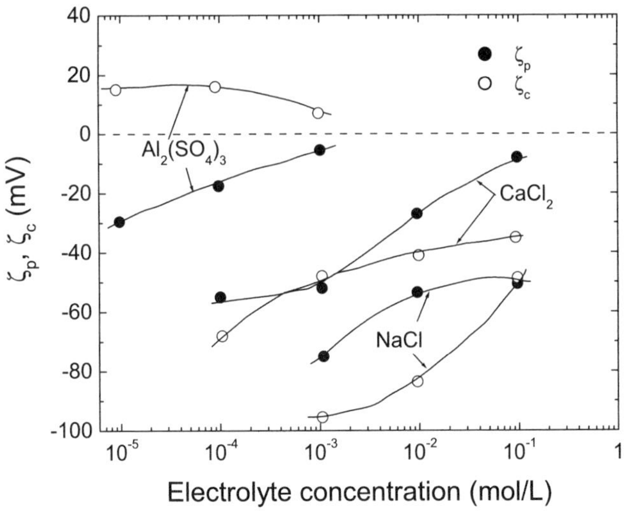

Figure 15.10. Variation of the zeta potential of air bubbles, ζ_c, and oil droplets, ζ_p, with electrolyte molarity (Okada *et al.*, 1988).

zeta potentials are both a function of electrolyte type and concentration. Yoon and Yordan (1986) gave extensive measurements of bubble zeta potentials and the effect of various surfactants. Okada *et al.* (1990a,b) carried out flotation experiments for oil droplets. They measured the flotation efficiency η_T where

$$\eta_T = 1 - \frac{c_f}{c_i} \tag{15.18}$$

with c_i and c_f being the initial and final oil droplet concentration, respectively, in the flotation experiments.

Okada *et al.* (1990a,b) also carried out a trajectory analysis similar to that discussed in Chapter 13. For the case of flotation of oil droplets, they found that maximum floatability was achieved when the value of the dimensionless adhesion number $m = 4\pi\epsilon\zeta_p\zeta_c/\kappa a$ was in the range of $m < 1.0$. Their dimensionless adhesion number is related to the electrostatic repulsion number N_ζ, which was defined in Eq. (13.110). Figure 15.11 shows a comparison between the calculated collection efficiency using Eq. (13.108) and the experimentally measured values. The particles employed are polystyrene latex spheres with $a_p = 1.47 \times 10^{-6}$ m. Clearly, the zeta potentials of the particles and air bubbles play a crucial role in the collection efficiency. Fair agreement is evident between the theory and experiment.

From the above, the application of the DLVO theory coupled with the hydrodynamic interaction can be effectively used to study particle flotation in a systematic manner that can be very useful in optimizing solids and oil droplets collection in an industrial process.

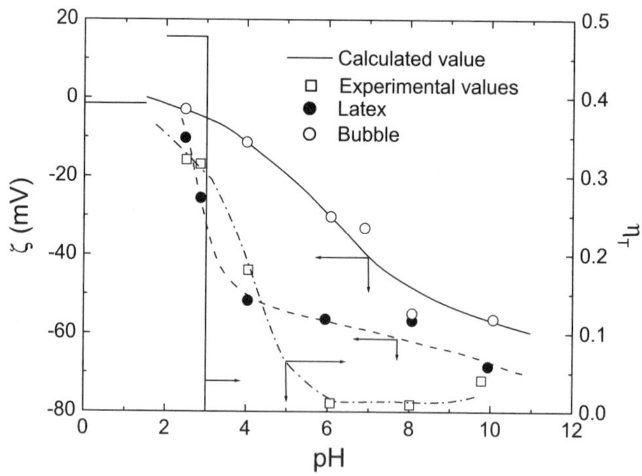

Figure 15.11. Comparison of calculated and experimental values of η_T, and effect of the zeta potentials of the bubble and particles on η_T in polyoxyethylene 23-lauryl ether solution (1×10^{-5} mol/L) with change of pH. The diameters of the bubbles and particles were 29.7 and 2. 95 μm, respectively (Okada *et al.*, 1990b).

15.6 RHEOLOGY OF COLLOIDAL SUSPENSIONS

The rheology of a colloidal suspension is a very important property that influences the usage of the system in an industrial environment. In many situations, it is possible to alter the rheological behavior of a colloidal suspension by changing the electrolyte concentration of the continuous phase or by anchoring long-chain polymers to the surface of the colloidal particles in the suspension.

15.6.1 Historical Background

When a fluid is subjected to a simple shear, the shear stress τ required to produce a shear rate (strain rate) $\dot{\gamma}$ is given by

$$\tau = \mu(\dot{\gamma})\dot{\gamma} \tag{15.19}$$

The coefficient $\mu(\dot{\gamma})$ is known as the apparent viscosity of the fluid. When $\mu(\dot{\gamma})$ is independent of the shear rate, one can write

$$\tau = \mu\dot{\gamma} \tag{15.20}$$

where μ is a constant. Fluids obeying the formulation of Eq. (15.20) are known as Newtonian fluids. Water and simple organic liquids are Newtonian fluids.

Kreiger and Eguiluz (1976) gave a tabulation of the dimensionless groups that control the rheology of a colloidal system under the influence of Brownian, viscous, and electrostatic forces. In the limiting case of negligible inertial effects, and at steady state, the relative viscosity μ_r of a colloidal system is given by

$$\mu_r = \frac{\mu_s}{\mu_c} = f(\alpha_p, \tau_r, q_r, \kappa a_p) \tag{15.21}$$

where, μ_s and μ_c are viscosities of the colloidal suspension and the continuous phase, respectively. Here α_p is the colloidal particle volume fraction in the suspension. The dimensionless groups are τ_r, q_r, and κa_p, where

$$\tau_r = \frac{\tau a_p^3}{k_B T} = \frac{\dot{\gamma}\mu_c a_p^3}{k_B T} \quad \text{reduced shear stress}$$

$$q_r = \frac{n_\infty e}{Nq} \quad \text{charge ratio}$$

and

$$\kappa a_p = \left[\frac{2z^2 e^2 n_\infty}{\epsilon k_B T}\right]^{1/2} a_p \quad \text{dimensionless inverse Debye length}$$

It is possible to define a particle Peclet number, Pe' as

$$Pe' = \frac{\dot{\gamma} a_p^2}{D_\infty} \tag{15.22}$$

Making use of the Stokes–Einstein relationship for the particle diffusion coefficient, D_∞:

$$D_\infty = \frac{k_B T}{6\pi \mu_c a_p}$$

the Peclet number becomes

$$Pe' = \frac{6\pi \dot{\gamma} \mu_c a_p^3}{k_B T} = 6\pi \tau_r \tag{15.23}$$

Rescaling the particle Peclet number, Pe', one can write $Pe = \tau_r$. In some literature, the particle Peclet number Pe' is used in lieu of the reduced shear τ_r.

In the absence of electrostatic forces, the relative suspension viscosity becomes

$$\mu_r = f(\alpha_p, Pe) \tag{15.24}$$

Here the viscosity of a suspension is affected only by the viscous forces, the Brownian motion, and the excluded volume of the particles. In such circumstances, the system is referred to as a "hard sphere" model. In the limit of infinite dilution, *i.e.*, $\alpha_p \to 0$, the relative viscosity of a colloidal system becomes a function of the volume fraction only.

15.6.2 Hard Sphere Model

This is the case where only the viscous and Brownian motion forces are present. The relative viscosity is expected to depend only on the particle volume fraction and the particle Peclet number, Pe. At a fixed volume fraction α_p, it is expected that rheological data for different particle sizes and for continuous phase viscosities (μ_c) collapse together when the relative suspension viscosity, μ_r, is plotted against Pe. This is indeed the case as is shown in Figure 15.12. The plot is an S-shaped curve characteristic of uncharged particles. For given continuous medium viscosity, temperature, and particle size, the plot of Figure 15.12 indicates that the relative viscosity of a colloidal suspension is a function of the shear rate $\dot{\gamma}$ and hence, by definition, exhibits non-Newtonian behavior. In general, due to Brownian diffusion, the relative viscosity of a suspension of colloidal particles can exhibit non-Newtonian characteristics. At low particle Peclet numbers, the Brownian motion dominates. However, at high particle Peclet numbers, the shearing force becomes dominant.

In both limiting cases, the colloidal suspension exhibits Newtonian characteristics. deKruif *et al.* (1986) gave the limiting values for the suspension viscosities as

$$\frac{\mu_0}{\mu_c} = \left(1 - \frac{\alpha_p}{0.63}\right)^{-2} \tag{15.25}$$

and

$$\frac{\mu_\infty}{\mu_c} = \left(1 - \frac{\alpha_p}{0.71}\right)^{-2} \tag{15.26}$$

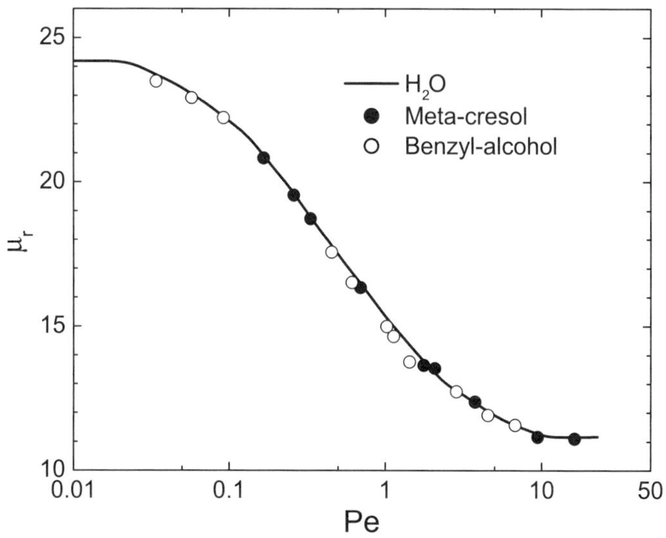

Figure 15.12. Relative viscosities *versus* Peclet number for a monodisperse suspension of polystyrene spheres of various sizes in different media with a fixed $\alpha_p = 0.5$ (Krieger, 1972).

where μ_0 and μ_∞ are the limiting values of the relative viscosities at $Pe \to 0$ and $Pe \to \infty$, respectively. For a given colloidal system, μ_0 and μ_∞ correspond to $\dot{\gamma} \to 0$ (low shear rate) and $\dot{\gamma} \to \infty$ (high shear rate), respectively. The S-shaped curve can be cast as (Krieger, 1972)

$$\mu_r = \mu_0 + \frac{(\mu_0 - \mu_\infty)}{1 + 2.32Pe} \tag{15.27}$$

In the limit of infinite dilution, Einstein's result for the relative viscosity is given as

$$\mu_r = 1 + [\eta]\alpha_p + \mathcal{O}(\alpha_p^2) \tag{15.28}$$

where $[\eta]$ is the intrinsic viscosity and has a value of $5/2$ for the case of hard spheres. In the limit of $\alpha_p \to 0$, there should be no dependence of the relative viscosity on the flow Peclet number.

Batchelor (1977) extended Einstein's relationship for $Pe \ll 1$ to give

$$\mu_r = 1 + \frac{5}{2}\alpha_p + 6.2\alpha_p^2 + \mathcal{O}(\alpha_p^3) \tag{15.29}$$

Figure 15.13 shows plots for Eqs. (15.25) to (15.29). It can be observed that the Einstein and Batchelor equations are valid only for $\alpha_p \leq 0.05$.

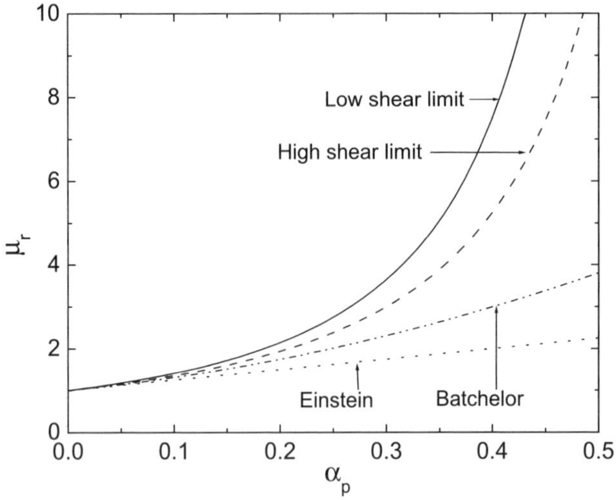

Figure 15.13. Variation of the relative viscosity of a colloidal suspension with the particle volume fraction. Hard sphere model.

15.6.3 Electroviscous Effects

Three distinct effects of electrical charge on colloidal suspension rheology have been identified (Watterson and White, 1981). The first is known as the primary electroviscous effect. It arises from the distortion of the diffuse part of the electric double layer due to shear. The second is known as the secondary electroviscous effect. It arises from interparticle interactions, which modify the particle trajectories and give rise to an increase in the effective particle excluded volume. The third effect is known as the tertiary electroviscous effect. It is due to the expansion and contraction of stabilizing polymer chains on the particle surface due to changes in the electrolyte concentrations (Hirtzel and Rajagopalan, 1985; Stein, 1985).

For dilute systems, the first electroviscous effect is more significant than the second effect as it influences the coefficient of $\mathcal{O}(\alpha_p)$ in Eq. (15.29). The second electroviscous effect influences the coefficient of $\mathcal{O}(\alpha_p^2)$ of Eq. (15.29). Consequently, the second electroviscous effect becomes significant for moderately concentrated suspensions (Russel, 1980).

The primary electroviscous effect was first identified by Smoluchowski as being due to an enhanced energy dissipation rate caused by the interaction of the diffuse part of the electric double layer with the flow field. For a dilute system, in the presence of an electric double layer, the coefficient of α_p becomes

$$[\eta] = \frac{5}{2}[1 + E] \tag{15.30}$$

where $[1 + E]$ is the augmentation factor due to the presence of the electric double layer. Watterson and White (1981) gave a plot of $[1 + E]$ as influenced by the dimensionless zeta potential $(e\zeta/k_BT)$ of the particles and inverse dimensionless Debye

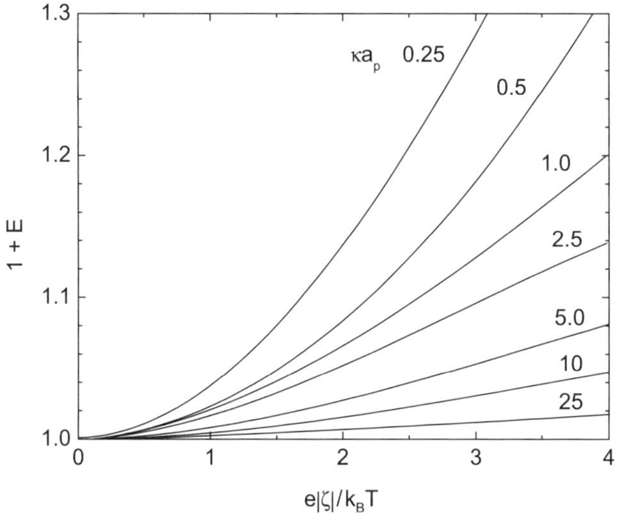

Figure 15.14. The primary electroviscous effect as a function of the electrokinetic potential in a (1 : 1) electrolyte solution (Watterson and White, 1981).

length, κa_p. Their plot is shown in Figure 15.14. For $\kappa a_p > 10$, *i.e.*, a thin electric double layer, the term $(1 + E)$ does not exceed 1.05 for $(e\zeta/k_B T)$ as high as 4 indicating little change in $[\eta]$ due to the first electroviscous effect. However, when the electric double layer is thick, *i.e.*, $\kappa a_p \ll 1$, the term $(1 + E)$ becomes appreciably greater than unity and the primary electroviscous effect becomes significant.

For non-dilute colloidal suspensions, the secondary electroviscous effect becomes significant where the probability of particle-particle interaction increases. Although in non-dilute colloidal systems both the primary and secondary electroviscous effects are present, the secondary effect becomes dominant at increasing particle concentrations. It is for this reason that one assumes that deviation from the hard sphere model at high concentrations is due to secondary electroviscous effects.

Figure 15.15 shows variations of the relative viscosity of polystyrene latex with the continuous phase counterion concentration at different particle Peclet numbers for $\alpha_p = 0.509$. At high Pe, say $Pe = 100$, the relative viscosity is not sensitive to the variation of the electrolyte concentration. Here, the viscous forces are dominant. However, at lower Pe values, there is a sharp increase in the relative viscosity of the colloidal suspension with decreasing electrolyte concentration where κa_p is smaller and the effect of the electric double layer extends further away from the particle surface. For all values of Pe, there is a distinct minimum in the μ_r corresponding to an electrolyte concentration c_{min} that is independent of the Peclet number. Electrolyte concentrations greater than c_{min} correspond to the flocculation of the colloidal system where the repulsive force becomes weaker with the addition of electrolytes.

Variation of the relative viscosity with the flow Peclet number is shown in Figure 15.16 for polystyrene latex particles, $\alpha_p = 0.4$, at different HCl concentrations ranging from 0 to 0.1 M. All curves tend to approach a common horizontal asymptote

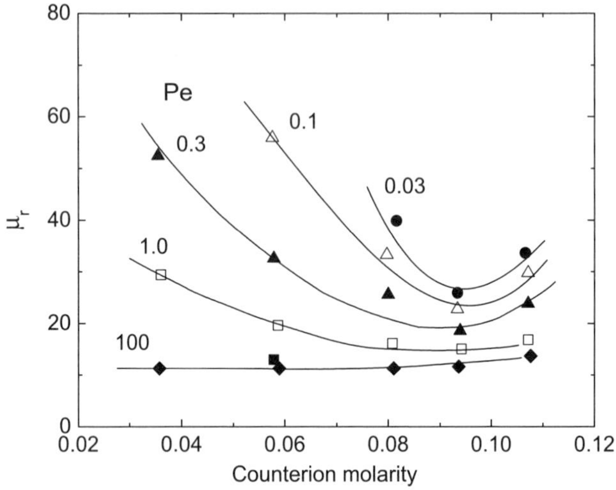

Figure 15.15. Effect of electrolyte molarity on the relative viscosity of monodisperse polystyrene latex, $a_p = 96$ nm, $\alpha_p = 0.509$, pH 7.0 with 21% anionic and 79% anionic surfactants (Krieger, 1972).

with an increasing Pe where the electroviscous forces are negligible in comparison to the viscous forces. According to Eq. (15.26), the asymptotic value is 5.25. At low values of Pe (or low shear stress), the viscosity of the deionized latex climbs as though approaching a vertical asymptote which is indicative of a yield stress. The apparent yield stress decreases with the addition of an electrolyte. At high electrolyte

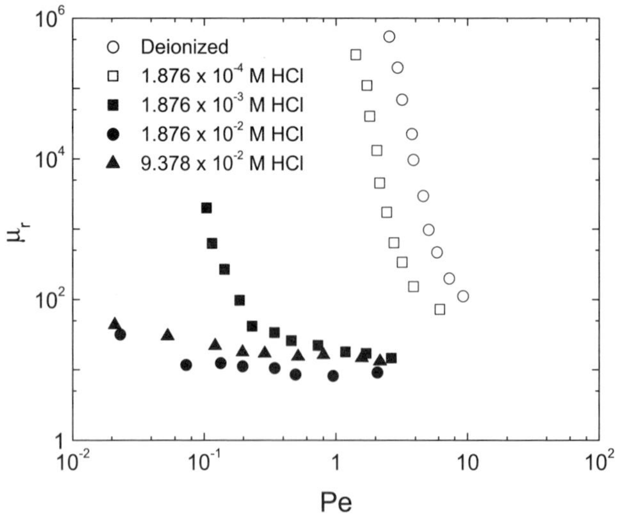

Figure 15.16. Variation of the relative viscosity of polystyrene latex ($a_p = 110$ nm) at $\alpha_p = 0.4$ for various HCl concentrations (Krieger and Eguiluz, 1976).

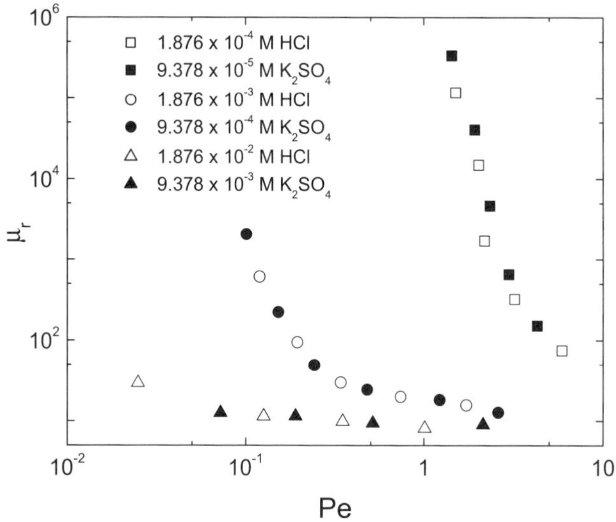

Figure 15.17. Effect on the relative viscosity due to different electrolytes at the same inverse Debye length, κ (Krieger and Eguiluz, 1976).

concentrations, there is no indication of a vertical asymptote, and there is no possibility of a yield stress being present (Krieger and Eguiluz, 1976). The plot of Figure 15.16 suggests that the secondary electroviscous effects are extremely important in modifying the rheological behavior of a colloidal suspension where the relative viscosity at a given Pe value can vary by several orders of magnitude.

A similar plot for polystyrene latex at $\alpha_p = 0.4$ is shown in Figure 15.17 where additions of K_2SO_4 and HCl are made. The Debye length has the same value when the concentration of HCl is double that of K_2SO_4. Figure 15.17 indicates that for given κa_p, the data collapse together onto one curve. The variation of the particle surface charge at the various HCl and K_2SO_4 concentrations was not reported.

The rheological behavior of latex particles sterically stabilized by a polymer has been investigated by Willey and Macosko (1978) and Liang *et al.* (1992). A plot of the relative viscosity of sterically stabilized monodisperse suspensions is given in Figure 15.18. For a given colloidal suspension, the relative viscosity increased sharply as the particle volume fraction approached its maximum packing value, a case similar to the hard sphere model. The relative viscosity data could be correlated using Dougherty and Krieger type equation (Krieger, 1972) that was derived for a hard sphere model. The relative viscosity is given by

$$\mu_r = \left(1 - \frac{\alpha_p}{\alpha_m}\right)^{-[\eta]\alpha_m} \tag{15.31}$$

where α_m is the maximum packing volume fraction of the particles. It should be recognized that due to the attached polymeric chains at the colloidal particle surface, the volume fraction α_p is simply a nominal value. The effective volume fraction of

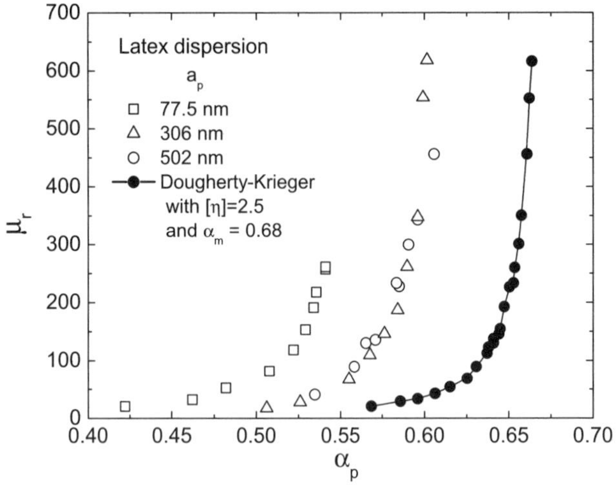

Figure 15.18. Relative viscosity *versus* volume fraction of polystyrene latex dispersions for various particle sizes (Liang *et al.*, 1992).

the dispersed phase is given by

$$\alpha_{eff} = \alpha_p \left[1 + \left(\frac{\delta}{a_p} \right) \right]^3 \tag{15.32}$$

where δ is the polymeric chain length, and is a function of the volume fraction α_p and shear rate $\dot{\gamma}$.

The relative viscosity data at $\alpha_p = 0.3$ of Willey and Macosko (1978) are plotted against Pe in Figure 15.19 for a different PVC plastisol particle size and solvent quality. The collapse of the data onto a single curve would suggest that the PVC plastisols behave as an ideal colloidal suspension of rigid spheres subject to Brownian diffusion and they are little affected by the repulsive forces. The limiting relative viscosity value at $Pe \to 0$ is about 300. Comparison with the prediction from Eq. (15.25) indicates that a much greater effective volume fraction is operative than the nominal value of $\alpha_p = 0.2$ (Russel, 1980).

The rheology of a colloidal system is clearly a strong function of the prevailing colloidal forces. By altering the relative magnitude of these forces, it is possible to obtain the desired rheological characteristic. It is for this reason, among others, that colloidal suspensions have found many applications in paints, dyestuffs, and pharmaceutical and pesticidal formulations.

15.7 BITUMEN EXTRACTION FROM OIL SANDS

We will explore in this section the application of electrokinetic phenomena to understand and appreciate a major industrial energy process provider. The theoretical

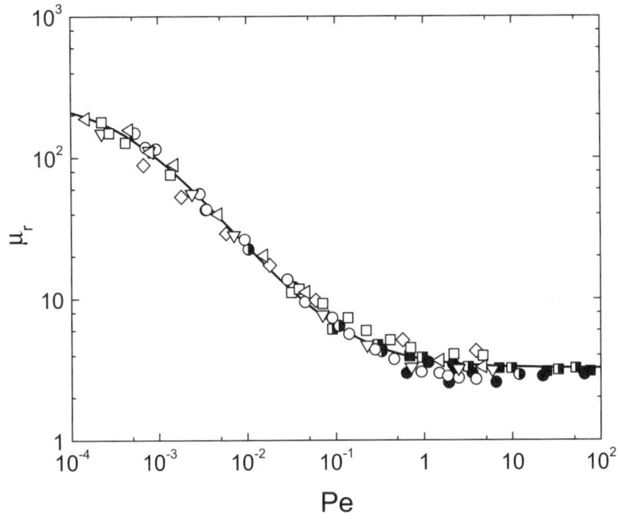

Figure 15.19. Relative viscosity of sterically stabilized PVC plastisols at $\alpha_p = 0.2$ for different particle sizes and solvents (Willey and Macosko, 1978).

analysis presented in earlier chapters will be used to shed some understanding on bitumen extraction from oil sands.

Oil sands are also known as tar sands and bituminous sands. They are unconsolidated sand deposits that are impregnated with high molar mass viscous petroleum, normally referred to as bitumen. Bitumen within the oil sands ore can be thought of as being a very viscous oil embedded within a sand matrix. At room temperature, bitumen is like cold molasses having very high viscosity and it is difficult to make it flow under gravity. Bitumen must be treated in upgraders to convert it into "synthetic crude oil" before it can be fractionated by refineries to produce gasoline, heating oils, and diesel fuels.

Oil sands are found throughout the world, usually in the same geographical location as conventional petroleum. The world's two largest sources of bitumen are in Canada and in Venezuela. Canada's bitumen resources are located almost entirely within the province of Alberta. Alberta's oil sand deposits are grouped on the basis of geology, geography and bitumen content, (NEB-EMA, 2004) and they are found in three locations: The Athabasca, Peace River and Cold Lake regions. The bitumen deposits in these three areas are found in sedimentary formations of sand and carbonate that collectively cover six million hectares. The largest deposit in the world is in the Athabasca area in the northeast part of the province of Alberta, Canada (Camp, 1976; Oil & Gas, 2004).

Based on current data, the Alberta Energy and Utilities Board (AEUB) estimates that the ultimate bitumen volume, a value that represents bitumen volume expected to be found by the time all exploration has ceased to be 400×10^9 m^3 (2.5×10^{12} barrels). Of the ultimate bitumen volume-in-place, 50×10^9 m^3 (315×10^9 barrels) are estimated to be recoverable using current and anticipated new technologies.

With current technologies, it is estimated that about 5.6×10^9 m^3, which are at depths of less than 75m, are amenable to surface mining. At higher buried depths, about 22.7×10^9 m^3 of the oil sands are amenable to underground type mining and *in-situ* (in-place) production (NEB-EMA, 2004; Oil & Gas, 2004). In 2004, the total Canadian daily bitumen production from open pit operations was about 625,000 barrels/day.

The oil sands deposits are composed of mainly quartz sand, clays, connate water and associated salts and bitumen. The clays are predominately kaolinite. The bitumen content varies from 7 to 15% by weight. The mineral solids are about 80–85% by weight and the connate water is 3–5%, by weight. The mineral solids particle size varies from less than 0.1 to 300 microns. The density of bitumen is about 1050 kg/m^3 with a viscosity of about 1000 Pas at room temperature.

As stated by Czarnecki *et al.* (2005), there is a general belief, as yet to be proven, that because for the most part, the mineral sand grains in Athabasca oil sands are hydrophilic, they are also "water-wet". In other words, the bitumen is not in direct contact with the sand grains, but instead separated from the sand grains by a nano-thick water film. The presence of this water film is assumed to be the major difference between Alberta oil sands and Utah tar sands that are considered to be "oil wet". Subsequently, bitumen from "water-wet" oil sands can be recovered using a water based extraction process where bitumen can be easily liberated from a hydrophilic or a "water-wet" surface. On the other hand, bitumen from hydrophobic or "oil-wet" sands would not be easily recovered in a water-based extraction process as it would be more difficult to dislodge bitumen that is directly attached to the sand grains. Due to the nature of the mineral solids, recovery of bitumen is achieved my mixing warm water with mined oil sand ore to liberate the bitumen with its subsequent flotation in gravity settlers. The processing temperature can vary from 35°C to 75°C.

Conceptually, bitumen recovery from oil sands using water-based extraction processes as applied to open pit mining, involves the following steps (Masliyah *et al.*, 2004):

(i) *Bitumen Liberation from Sand Grains*: To recover bitumen from the oil sand ore, the bitumen needs first to be liberated from the sand grains. This step is controlled by the process temperature, mechanical agitation and interfacial properties. The dislodging of the bitumen from the sand grains is the first step in the bitumen recovery process.

(ii) *Bitumen Aeration*: As bitumen density is very close to that of water, in order to recover the liberated bitumen in gravity settlers, it is necessary to aerate the liberated bitumen by having air attach to it.

It is well known in the oil sands industry that oil sand ores containing high mineral fines (fines are defined as mineral solids less than 44 microns in size) processed in a slurry containing high concentrations of divalent ions give a low bitumen recovery (Masliyah *et al.*, 2004). The divalent ions are calcium and magnesium. Conversely, oil sand ores containing low mineral fines processed in a slurry containing low concentrations of divalent ions, if not aged, would lead to a high bitumen recovery. The

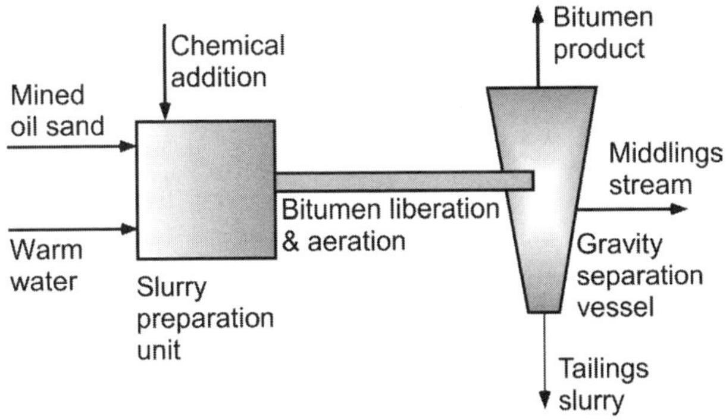

Figure 15.20. Schematic process chart for bitumen extraction from oil sands.

hypothesis is that the presence of high mineral fines coupled with high concentration of divalent ions leads to the coating of the bitumen droplets by the mineral fines, *i.e.*, hetero-coagulation. As these mineral fines are in general hydrophilic, fines-coated bitumen droplets become less hydrophobic and would not efficiently attach to hydrophobic air bubbles. As bitumen has similar density as water, the bitumen-air attachment process is vital to reduce the bitumen density for its recovery. A poor bitumen-air attachment would lead to a low bitumen recovery.

A schematic bitumen recovery from oil sands is shown in Figure 15.20. From the afore-mentioned bitumen recovery steps, it becomes essential to provide the environment for efficient bitumen liberation from the sand grains and bitumen-air attachment. Any factor that adversely affects these two steps would give a lower bitumen recovery.

In this Section, zeta potential distribution measurements and colloidal force measurements will be used to show that (i) repulsion and attraction forces between bitumen and silica surfaces can lead to poor bitumen liberation from a silica surface, and (ii) undesirable mineral fines when coupled with high divalent ion concentration would lead to surface coating of solid mineral fines on a bitumen surface.

15.7.1 Zeta Potential of Oil Sand Components

The origin of the surface charge depends on the surface chemistry and the electrolyte solution. For the case of bitumen-water systems, it is assumed that the natural surfactants present in the bitumen are responsible for the origin of the bitumen-water interface charge. The negative charge that is usually present at the bitumen-water interface can be explained by the dissociation of the carboxyl groups belonging to the surfactants that are naturally present in bitumen (Takamura and Chow, 1985).

$$\mathrm{RCOOH} + \mathrm{OH}^- \rightleftharpoons \mathrm{RCOO}^- + \mathrm{H_2O} \tag{15.33}$$

$$\mathrm{RCOOH} \rightleftharpoons \mathrm{RCOO}^- + \mathrm{H}^+ \tag{15.34}$$

The presence of NaOH would modify the dissociation state and can subsequently alter the surface charge. One would expect that, with the addition of NaOH, the above reaction, Eq. (15.33), proceeds more to the right hand side, whereby the bitumen

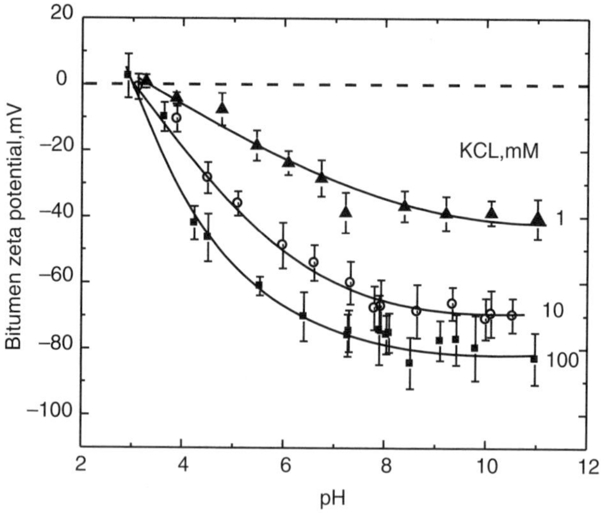

Figure 15.21. Variation of bitumen zeta potential at room temperature with bulk pH at different KCl concentrations.

surface becomes more negatively charged. Industrial bitumen extraction processes utilize water with a pH of about 8.5. Figure 15.21 shows the variation of zeta potential of bitumen at room temperature as a function of potassium chloride concentration. Two observations can be made. For a given KCl molarity, the zeta potential becomes more negative at high pH and that the zeta potential of bitumen, for a given, pH, is less negative at higher KCl molarity.

When a multivalent cation is present in the bulk solution (*e.g.*, Ca^{2+}), ion-binding between the carboxyl group present in bitumen and the cations takes place (Takamura and Chow, 1985; Takamura *et al.*, 1986).

$$\mathrm{RCOO^- + Ca^{2+} \rightleftharpoons RCOOCa^+} \tag{15.35}$$

Here, one would expect an increase in the surface potential (less negative) with the addition of, say, $CaCl_2$. Figure 15.22 shows the bitumen zeta potential variation with pH at various calcium chloride concentrations.

The surface charge of silica mineral-water interface can also be explained using the surface groups ionization concept of Healy and White (1978). The silica surface becomes either positively or negatively charged depending on the equilibrium associated with the following surface ionization reactions (Takamura and Isaacs, 1989)

$$\text{At acidic conditions:}\quad \mathrm{-SiOH + H^+ \rightleftharpoons -SiOH_2^+} \tag{15.36}$$

$$\text{At alkaline conditions:}\quad \mathrm{-SiOH + OH^- \rightleftharpoons -SiO^- + H_2O} \tag{15.37}$$

Figure 15.23 shows the variations of silica zeta potential with pH for different $CaCl_2$ concentrations.

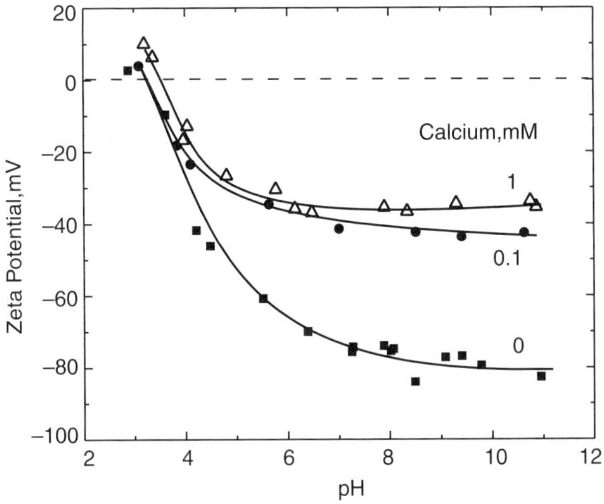

Figure 15.22. Variation of bitumen zeta potential at room temperature with bulk pH for different $CaCl_2$ concentrations. The background KCl concentration is 1 mM (Liu *et al.*, 2003).

In the case of clays, the crystal charge is due to substitution of aluminum for silicon in the tetrahedral layer with a consequent imbalance of negative charge. For example, Al^{3+} may replace Si^{4+} at the surface of the clay, producing a negative surface charge. In this case, a point of zero surface charge, PZC, can be reached by reducing the pH. Here, the added H^+ ions combine with the negative charges on the surface to form OH groups. Such a charge origin is referred to as isomorphous substitution.

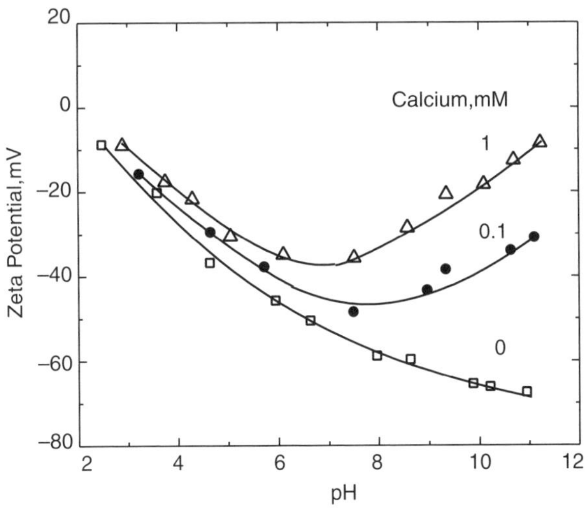

Figure 15.23. Variation of silica zeta potential with pH: Effect of added calcium ion on silica zeta potential at room temperature. KCl at 1 mM is the background electrolyte concentration (Liu *et al.*, 2003).

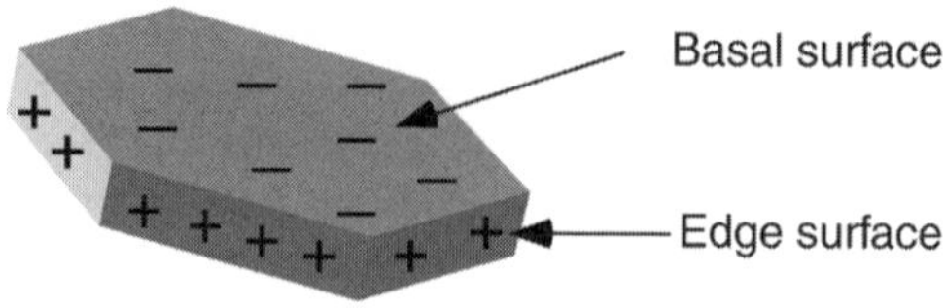

Figure 15.24. A kaolinite clay crystal showing surface charges at a pH of, say 5.

In some clays, *e.g.*, kaolinite, when a platelet is broken, the exposed edges contain AlOH groups which take up H^+ ions to give a positively charged edge at a low pH. The edge positive charge decreases to zero as the pH is raised to about 7. The positive edge surface charge may coexist with the negatively charged basal surfaces leading to special properties. In this case, there will be no single point of zero charge, PZC, but each surface (edge and basal) has its own value. Figure 15.24 shows a typical kaolinite platelet with a negative charge at the basal surface and positive charge at the edge. The positive charge is eliminated at a pH above 7 (Hunter, 2001; Everett, 1988).

The variation of zeta potential of kaolinite clays with pH at different calcium chloride molarity is shown in Figure 15.25. At low pH values, the zeta potential is positive and it is increasingly negative at higher pH values. For a given pH, it is clear that the presence of the divalent calcium ions has a very large effect on the clay zeta potential. The zeta potential becomes closer to zero at higher calcium chloride concentrations.

From the zeta potential plots for the bitumen and mineral solids, it becomes clear that bitumen processability would be affected by the process water pH and electrolyte type and concentration.

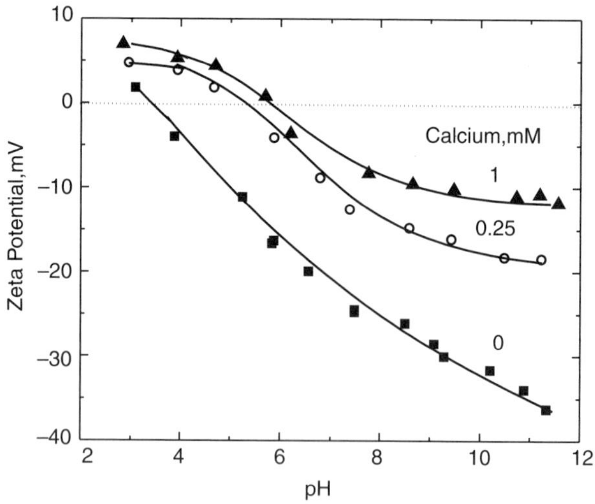

Figure 15.25. Variation of the zeta potential of kaolinite clays with bulk pH for different calcium ion concentrations, (Liu *et al.*, 2004a).

15.7.2 Zeta Potential Distribution Measurements Technique

From the zeta potential measurements for both bitumen and silica, it comes clear that presence of high calcium concentrations would decrease the repulsion force between silica and bitumen surface with the possibility of a net attractive force between them.

Electrokinetic properties of bitumen and silica have been investigated extensively by many researchers, using conventional electrophoretic zeta meter. In the early measurements, only the average electrophoretic mobility or zeta potential for a single component system was reported. In the case of a binary mixture system, the measured average zeta potentials of the two components often either over or under-estimate the electrokinetic behavior of the system, thereby giving rise to misguided information regarding to the interpretation of interactions between the two components. With the capability of measuring electrophoretic mobility or zeta potential distributions, it is possible to identify the attraction of one component on the other in a binary suspension system. The concept for this application is described as follows.

When zeta potential distributions are measured separately using emulsified bitumen droplets and silica particles, each has its own unique zeta potential distribution centered at ζ_B and ζ_S, respectively. For illustrative purposes, these two distributions are overlaid schematically in Figure 15.26(a). When bitumen and silica are mixed together under the same physicochemical conditions, the measured zeta potential

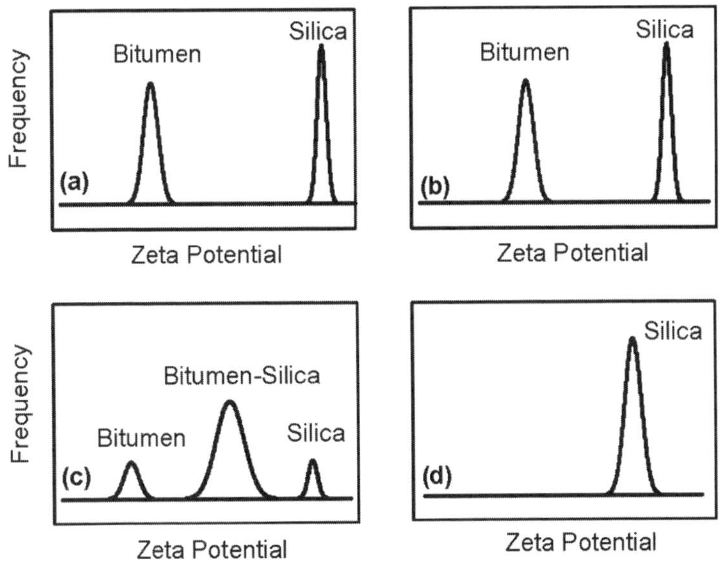

Figure 15.26. Schematic zeta potential distributions for a binary particulate component system made up of bitumen and silica: (a) zeta potential distribution of bitumen and silica measured individually; (b) bitumen and silica mixture showing no attraction; (c) partial covering of bitumen by silica and/or partial covering of silica by bitumen; and (d) strong attraction between bitumen and silica where the bitumen droplets are fully covered by the silica particles (Adapted from Liu *et al.*, 2002).

distribution of the resultant mixture can be interpreted in terms of the interactions between the two components. If the bitumen and silica do not interact with each other, a bimodal zeta potential distribution with two peaks centered at ζ_B and ζ_S as shown in Figure 15.26(b) is anticipated. Due to hydrodynamic interaction of moving particles at different electrophoretic mobilities, a slight shift of ζ_B and ζ_S towards each other may be observed. In the case where there is a weak attraction between the bitumen and silica particles, the measured zeta potential distribution would show several distribution peaks whose positions depend on the bitumen to silica ratio, Figure 15.26(c). When the silica particles completely cover the bitumen droplets, one peak would be observed whose location would be similar that of individual silica particles, Figure 15.26(d).

The use of zeta potential distribution measurements will be illustrated later in this Section.

15.7.3 Atomic Force Microscope Technique

The classical DLVO (Derjaguin–Landau–Verwey–Overbeek) theory, which considers the sum of van der Waals and electric double layer interactions, is often used to predict colloidal particle interactions. Calculation of the electrostatic double layer and van der Waals interactions is relatively straightforward for well defined systems with known system parameters such as surface potential of the interacting colloidal particles and Hamaker constant. However, in complex systems, the classical DLVO theory might not accurately predict the colloidal interactions and an extended DLVO theory has to be used, which would include repulsive hydration force for hydrophilic surfaces, attractive hydrophobic force for hydrophobic surfaces, repulsive steric force and attractive bridge force for polymer bearing surfaces.

The theory for describing these additional non-DLVO forces is less well developed. Subsequently, these forces are often inferred to from the deviation of the experimentally measured colloidal forces from those predicted by the classical DLVO theory. As the DLVO theory cannot be depended upon for a complex system, direct force measurement is needed. To that end, direct force measurement with atomic force microscope, AFM is a most direct method for the measurement of the interaction forces between bitumen, silica and clays.

The atomic force microscope, AFM, consists of a piezoelectric translation stage, a cantilever tip, a laser beam system, a split photodiode and a fluid cell. The working principle of AFM for colloidal force measurement can be found in the literature (Ducker *et al.*, 1992). In bitumen extraction investigations, the probe particles are model spherical silica, and pseudo-spherical fine solids of about $5 \sim 10\,\mu m$ in diameter. A probe particle was glued with a two-component epoxy onto the tip of a short, wide beam AFM cantilever under an optical microscope. Bitumen substrates were prepared by spin-coating bitumen on silica wafers (Liu *et al.*, 2003; Liu, 2004; Liu *et al.*, 2004a,b). The spin-coated bitumen substrate was glued onto a magnetic plate that was mounted on the piezo-electric translation stage as shown in Figure 15.27. The cantilever substrate with a probe particle was mounted in a fluid cell. When the piezo-stage brings the bitumen substrate to approach or retreat from the probe particle in

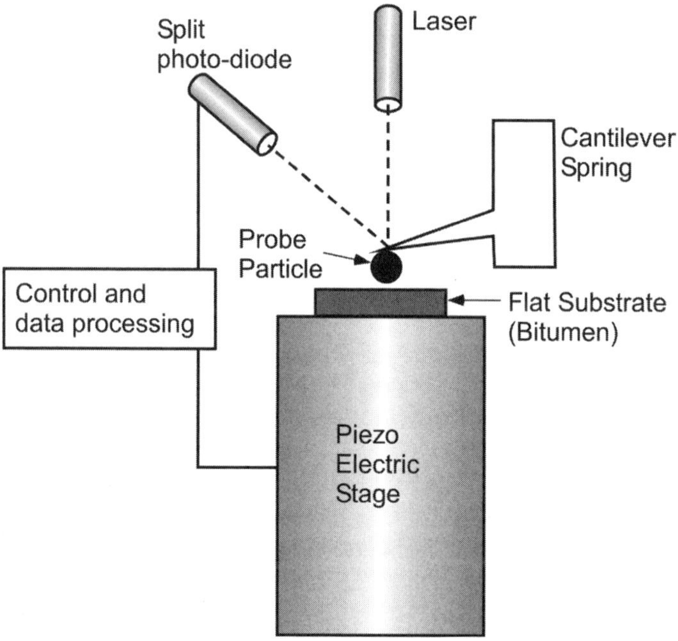

Figure 15.27. Schematic view of an atomic force microscope (AFM) showing the particle probe, the sample surface and the measuring photo diode.

the vertical direction, the force between the two surfaces causes the cantilever spring to deflect upward or downward, depending on the nature of the force between them. The deflection of the cantilever spring is detected by the position-sensitive laser beam that is focused on the upper surface of the spring cantilever, and reflected to the split photodiode through a mirror. From the displacement of the piezo-stage and the deflection of the spring cantilever, the long-range force and adhesive force (pull-off force) between the substrate and the probe particle can be obtained. Figures 15.27 to 15.29 show some illustrations of the AFM configuration.

Adhesive force data can be collected under given loading force. A typical set of raw data of probing surface forces is shown in Figure 15.29. The repulsive force barrier and adhesive force are of great concern since they control the coagulation behavior between two substrates.

15.7.4 Electrokinetic Phenomena in Bitumen Recovery from Oil Sands

15.7.4.1 Bitumen-Silica Interaction: Bitumen Liberation from Sand Grains

Zeta Potential Measurements: To demonstrate the interaction between colloidal bitumen droplets and silica particles, the zeta potential distributions of bitumen emulsion and silica suspension, individually and in a mixture, were measured

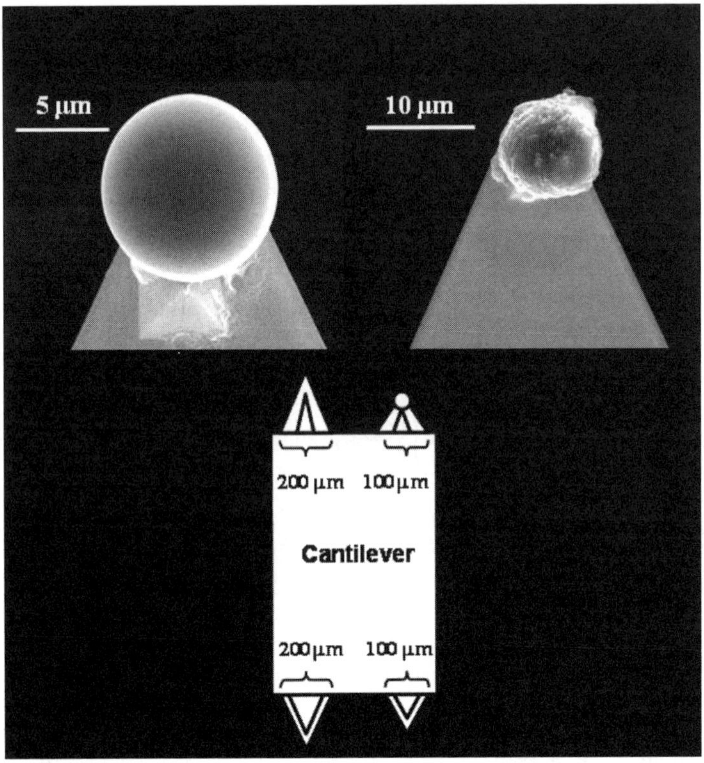

Figure 15.28. AFM schematic showing (a) a model silica probe particle; (b) clay particle; and (c) a cantilever.

and the results are interpreted in terms of colloidal interactions. For illustrative purposes, only the results with and without calcium addition at solution pH of 10.5 are presented here. Let us first discuss the case of absence of calcium. As shown in the overlaid histogram of Figure 15.30(a), the zeta potential distributions measured with bitumen droplets or silica particles alone in the electrolyte solution without calcium addition, are centered at −82 and −67 mV, respectively. A similar zeta potential distribution histogram is obtained for the mixture of the two species as shown in Figure 15.30(b). The presence of two distinct distribution peaks at −82 and −69 mV indicates that bitumen droplets and silica particles in the mixture are non-coagulative, *i.e.*, they are present separately as individual particles.

Now let us include calcium in the water where 1 mM calcium chloride, in the form of calcium chloride, is added. As shown in Figure 15.31(a) the zeta potential distribution histograms for bitumen droplets and silica particles measured separately were found to be much less negative. When zeta potential distribution was measured with the mixture of the emulsified bitumen and silica particles, only a single broad distribution peak located in between the two original peaks is observed [Figure 15.31(b)]. The disappearance of the original zeta potential distribution peaks of silica and bitumen and the appearance of a new broad distribution peak in

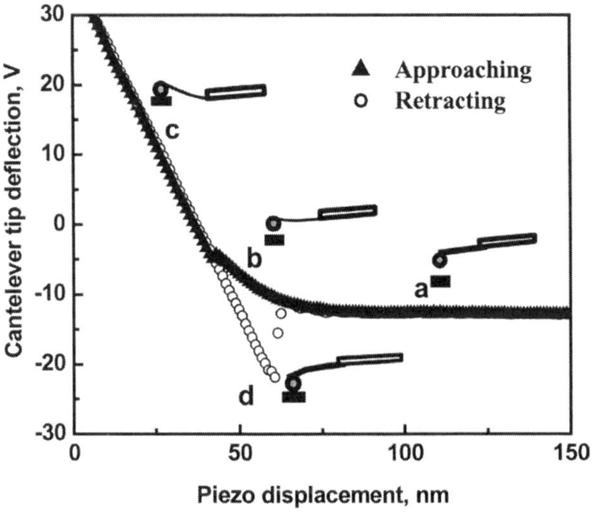

Figure 15.29. AFM force measurement modes: Curves a to b up to c, are in repulsion mode. Contact of a flat surface with a particle probe at point c. Curves c and d are in adhesion measuring mode. At point d, separation between the surface and probe takes place. At this location, the adhesion force is measured.

between suggest that the silica particles and bitumen droplets are hetero-coagulated to form composite aggregates.

The zeta potential distribution measurements of Figures 15.30 and 15.31 clearly show that the presence of calcium ions has the impact of adhering bitumen and

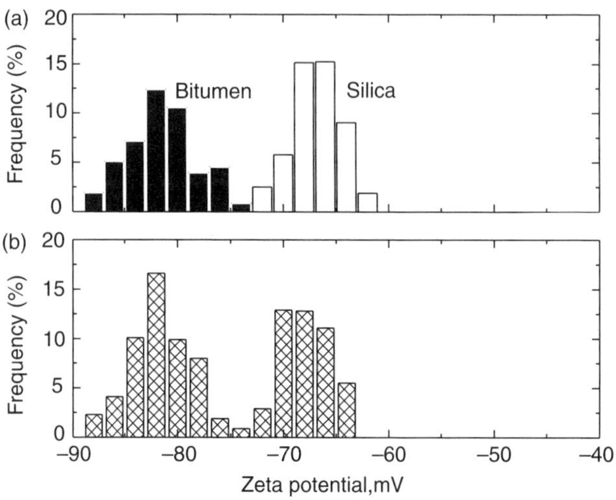

Figure 15.30. Zeta potential distribution of bitumen emulsion and silica suspension in 1 mM KCl solution at pH 10.5 without calcium, measured (a) separately and (b) as a mixture (Liu *et al.*, 2003).

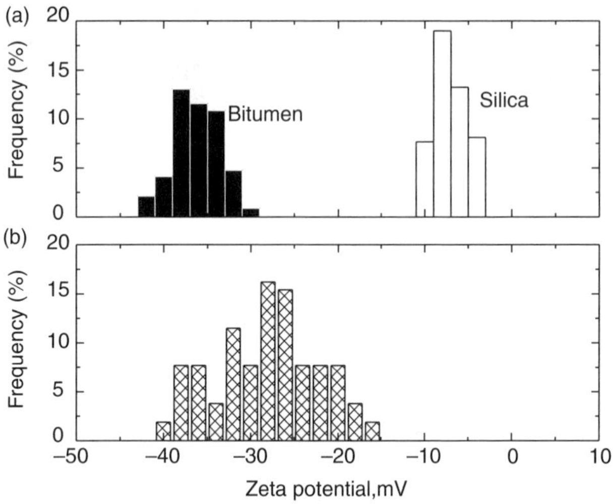

Figure 15.31. Zeta potential distribution of bitumen emulsion and silica suspension in 1 mM KCl solution at pH 10.5 with 1 mM calcium ion addition, measured (a) separately and (b) as a mixture (Liu *et al.*, 2003).

silica together, where hetero-coagulation takes place, a situation that would hinder the liberation of bitumen from the silica sand grains.

Atomic force microscope measurements: To take into account all the prevailing colloidal forces between a bitumen surface and a silica probe, atomic force microscope measurements are conducted and they are shown in Figure 15.32 as a function of pH in the absence of calcium and in Figure 15.33 in the presence of calcium at a pH of 10.5. The solution pH is a critical operating parameter in bitumen recovery, and in most cases, the controlling parameter for surface charge. In a 1mM KCl solution, the effect of pH on the interaction forces between the bitumen surface and silica particle is shown in Figure 15.32. The measured long-range force profiles are monotonically repulsive. The repulsion force increases with increasing pH. At pH 3.5, a very weak long-range repulsive force is observed. As shown by the solid lines of Figure 15.32, at separations greater than about 2–3 nm, all the measured force profiles can be reasonably well fitted with the classical DLVO theory. The good fit shown in Figure 15.32 by the solid lines suggests that the long-range repulsive forces are predominantly from the electrostatic double layer interactions. The fitted Debye decay length (κ^{-1}) in the range of 8.5–10 nm agrees well with that calculated for a 1mM KCl solution used in the experiment, further confirming electrostatic nature of the long-range repulsive force. It should be noted that the observed repulsive force at separation distances less than 2–3 nm is inconsistent with the attractive force regime as predicted by the DLVO theory, suggesting the presence of an additional repulsive force. Although the exact reason for this contradiction is not clear, considering a 2–3 nm range, this additional repulsive force would appear to originate from brush-like surfaces or small protrusions at the bitumen/water interface, resulting in a steric type of repulsion.

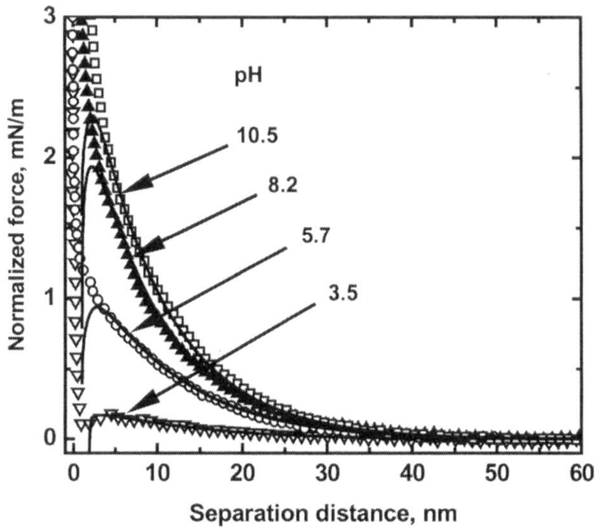

Figure 15.32. Interaction forces between bitumen and silica as a function of separation distance in 1 mM KCl solution at different solution pH. Solid lines represent the DLVO fitting using $A_{BWS} = 3.3 \times 10^{-21}$ J with best-fitted decay length and Stern potential of: down triangles, pH 3.5: $\kappa^{-1} = 9.4$ nm, $\psi_B = -20$ mV, $\psi_S = -25$ mV; circles, pH 5.7: $\kappa^{-1} = 9.4$ nm, $\psi_B = -58$ mV, $\psi_S = -48$ mV; up triangles, pH 8.2: $\kappa^{-1} = 9.1$ nm, $\psi_B = -76$ mV, $\psi_S = -59$ mV; squares, pH 10.5: $\kappa^{-1} = 8.6$ nm, $\psi_B = -83$ mV, $\psi_S = -64$ mV. Force normalization is made by dividing the force by the probe particle radius (Liu *et al.*, 2003).

A dramatic effect of calcium addition on both long-range colloidal force was observed at pH 10.5, as shown in Figure 15.33. The long-range colloidal forces changed progressively from repulsive to attractive with increasing calcium ion addition to 1 mM. Again, the long-range forces can be well fitted by the DLVO theory, indicating that the change of interaction force profiles can be simply attributed to the diminished electrical double layer forces by the specific adsorption of calcium ions.

To illustrate the electrokinetic findings as illustrated by the zeta potential distributions of bitumen and silica and AFM direct measurements, the recession of an initially bitumen disc patch on a silica glass slide immersed in water is shown in Figure 15.34. The recession rate is fast at low calcium concentrations and fairly slow at higher concentrations. Moreover, the contact angle is fairly small at low calcium concentrations and it is more favorable for bitumen detachment from a silica surface.

15.7.4.2 Bitumen-Fine Solids Interaction: Bitumen-Fine Solids Hetero-Coagulation

Zeta Potential Measurements

Fine Solids from Good Processing Ore: After performing bitumen extraction using a good processing ore, the mineral fine solids are collected and used for

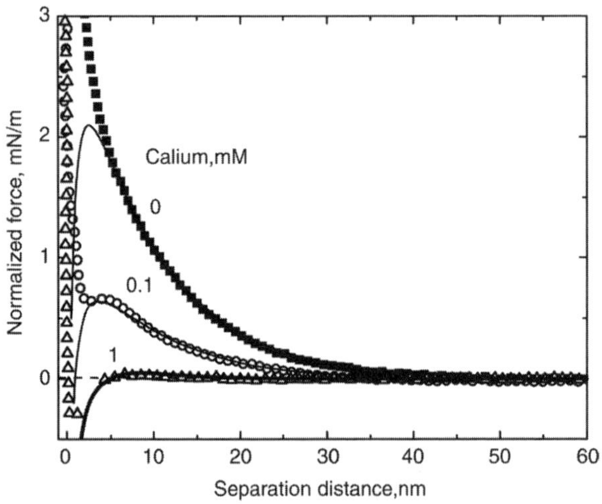

Figure 15.33. Interaction of normalized forces between bitumen and silica as a function of separation distance in solution containing 1 mM KCl and different concentrations of calcium ions at pH 10.5. Solid lines represent the DLVO fitting using $A_{BWS} = 3.3 \times 10^{-21}$ J with best-fitted decay length and Stern potential of: squares, 0 mM $CaCl_2$: $\kappa^{-1} = 9.1$ nm, $\psi_B = -83$ mV, $\psi_S = -64$ mV; circles , 0.1 mM $CaCl_2$: $\kappa^{-1} = 7.8$ nm, $\psi_B = -45$ mV, $\psi_S = -35$ mV; up triangles, 1 mM $CaCl_2$: $\kappa^{-1} = 4.5$ nm, $\psi_B = -38$ mV, $\psi_S = -8$ mV. Force normalization is made by dividing the force by the probe particle radius (Liu *et al.*, 2003).

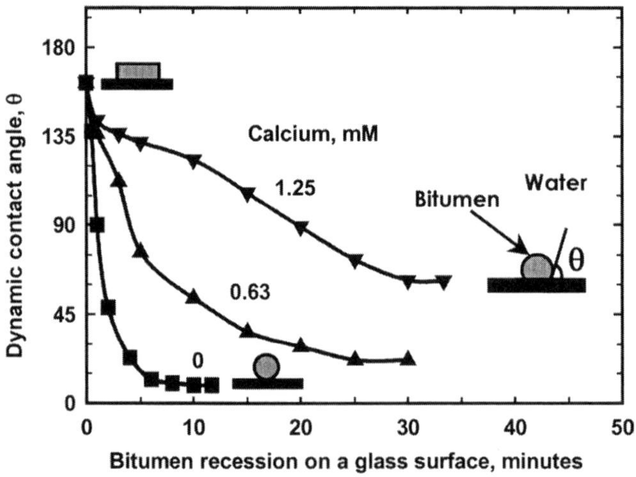

Figure 15.34. Dynamic contact angle of bitumen on a glass slide at pH = 9 and 22°C: Effect of calcium addition (Masliyah *et al.*, 2004).

zeta potential distribution measurements together with emulsified bitumen. The water used in these tests was that obtained from the extraction process itself. The results in Figure 15.35a show two distinct distribution peaks at −71 and −45 mV, corresponding to bitumen and fine solids, respectively. For the mixture of

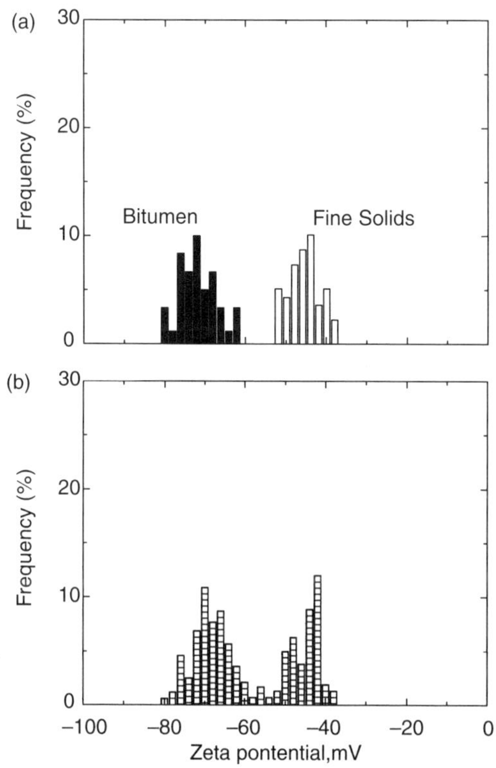

Figure 15.35. Good processing ore: Zeta potential distributions in its corresponding extraction process water of (a) individual emulsified bitumen and fine solids suspension and (b) their mixture (Liu *et al.*, 2005).

the bitumen and fine solids, a bimodal distribution histogram as shown in Figure 15.35b is observed. The peak values of the zeta potential distribution histogram correspond to those for the bitumen and fine solids, respectively, thereby illustrating a negligible attraction between the two components. The results confirm the negligible attraction between the fine solids and bitumen droplets as observed in industrial practice while dealing with a good processing ore.

Fine Solids from Poor Processing Ore: The zeta potential distributions of bitumen droplets and mineral fines from poor processing ores are also measured separately in the corresponding process water from the extraction test itself. The results of Figure 15.36(a) show that the distribution for individual emulsified bitumen and mineral fine solids is presented by two peaks at −50 and −28 mV. Compared with Figure 15.35(a) for the case of a good processing ore, a reduction of zeta potential values for both bitumen and fine solids is observed as anticipated from the compression of electric double layer by the presence of calcium and magnesium ions in the extraction water from a poor processing ore. The zeta potential distribution of

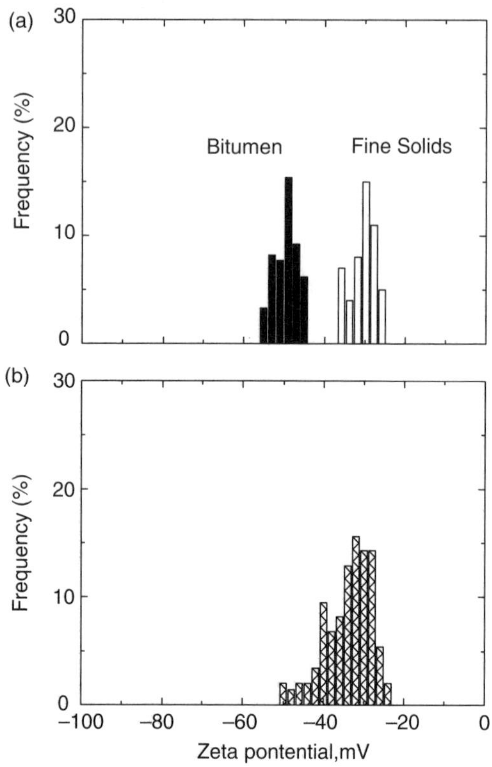

Figure 15.36. Poor processing ore: Zeta potential distributions in corresponding extraction process water of (a) individual emulsified bitumen and fine solids suspension and (b) their mixture (Liu *et al.*, 2005).

Figure 15.36(b) for the mixture of the bitumen and fine solids shows a distribution peak at a zeta potential value corresponding to that for the fine solids, with a small tail spreading towards the bitumen distribution peak. Comparing the results for the fine solids from the good processing ore as shown in Figure 15.35(b), the results here suggest a stronger attachment between the bitumen and mineral fine solids derived from a poor processing ore in its corresponding extraction process water.

Atomic Force Microscope Measurements

The surface force profiles measured with bitumen-fine solids pairs in their corresponding extraction process water are shown in Figure 15.37. Once again the mineral fine solids are obtained from the individual oil sand ore. The interaction force between bitumen and fine solids from a good processing ore is strongly repulsive. However, the interaction force between the bitumen and fine solids from a poor processing ore is attractive at a separation less than 10nm.

The adhesion force between bitumen and fine solids pairs in their corresponding extraction process water is shown in Figure 15.38. A stronger adhesion force

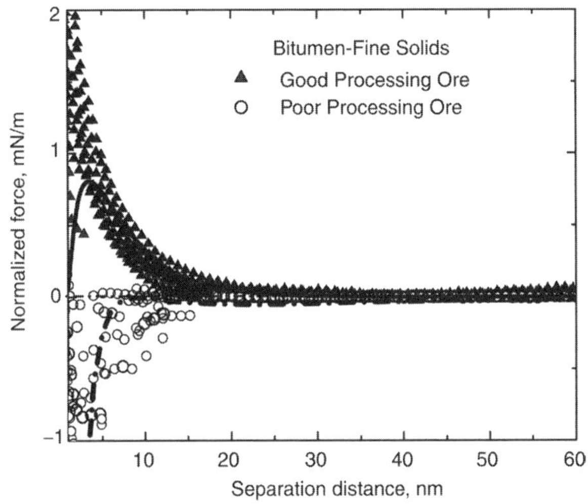

Figure 15.37. Normalized interaction forces between bitumen and fine solids collected from the oil sand ores as a function of separation distance in their corresponding extraction process water. Force normalization is made by dividing the force by the probe particle radius (Liu *et al.*, 2004b, 2005). Solid circles: Bitumen-fine solids from good processing ore. Open circles: Bitumen-fine solids from poor processing ore. Solid line represents the classical DLVO fitting using $A_{BWF} = 6.5 \times 10^{-21}$ J with best-fitted decay length and Stern potential being: $\kappa^{-1} =$ 4.6 nm, $\psi_B = -60$ mV, $\psi_F = -30$ mV. Dotted line represents extended DLVO fitting using $A_{BWF} = 6.5 \times 10^{-21}$ J with best-fitted decay length, Stern potential, $\kappa^{-1} = 4.3$ nm, $\psi_B = -42$ mV, $\psi_F = -28$ mV with added hydrophobic force.

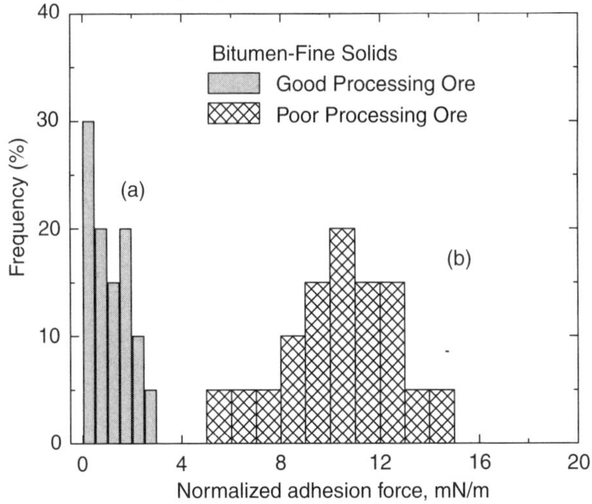

Figure 15.38. Distribution of normalized adhesion forces between bitumen and fine solids in their corresponding extraction process water. (a) Bitumen-fine solids collected from good processing ore and (b) Bitumen-fine solids collected from a poor processing ore. Force normalization is made by dividing the force by the probe particle radius (Liu *et al.*, 2004b, 2005).

between bitumen and fine solids is observed from the poor processing ore than between bitumen and fine solids from a good processing ore.

The measured strong repulsive force and weak adhesion force between the bitumen and mineral fine solids, Figures 15.37 and 15.38 together with the zeta potential distributions, Figure 15.35, for the good processing ore account for the negligible slime coating of the fine solids on bitumen during bitumen extraction. The findings, from electrokinetic considerations, suggest a high bitumen flotation rate from good processing ores. A situation that is often the case in practice.

The measured non-contact attractive force and strong adhesion force, Figures 15.37 and 15.38, together with the zeta potential distributions of Figure 15.36, explain the observed strong attachment between the fine solids and bitumen. The results imply a stronger coagulation of fine solids from a poor processing ore with bitumen during bitumen extraction. Such a hetero-coagulation leads to harmful slime coating. Therefore, a low bitumen recovery and poor bitumen product quality are anticipated for poor processing ores. Indeed, this is a situation experienced in an industrial oil sands operation.

The electrokinetic experimental results from both the zeta potential distribution and atomic force microscope measurements shed good understanding on the industrial bitumen recovery process.

15.8 MICROFLUIDIC AND NANOFLUIDIC APPLICATIONS

Conventional complex chemical analysis, generally conducted in laboratories using large equipment, require large volumes of analytes, considerable human resources, and are time consuming. During the past decade, tremendous progress has been made toward development of technologies that pertain to miniaturization of such complex analytical procedures into extremely small microchips. Such chips are commonly referred to as microfluidic chips, lab-on-chips, or micro total analysis systems (μTAS). One can consider such chips as a subset of micro-electromechanical systems (MEMS) (Gad-el-Hak, 2002). Microfluidic chips are comprised of microchannels ranging in dimension from a few microns to less than a millimeter. The handling of fluids in such narrow channels falls under the purview of the subject of microfluidics. Microfluidic chips can be fabricated on a variety of materials, including glass, polymers, silicon, *etc*., employing surface micromachining procedures such as photolithography, glass etching using hydrofluoric acid (HF), molding, and embossing. Using a synergistic combination of pressure, electrical fields, interfacial and material properties of different elements in these chips, as well as, utilizing the electromechanical response of different materials to electromagnetic impulses (such as piezoelectric and dielectric polarization effects), one can perform an astounding array of fluid manipulation in these devices. The devices combine a wide range of functions including sample loading, pumping, screening, separation, mixing, and diagnostics (measurement of properties). The small volumes (typically nanoliters to microliters) of fluid required for the analysis, portability, and the rapid nature of the assays make microfluidics based chemical analysis a highly attractive alternative to conventional chemical analysis.

With the remarkable global interest in nanotechnology during the past decade, there has been a significant resurgence of interest in nanofluidics or fluid flow in and around nanoscale objects.[1] It is, however, important to note that nanofluidics existed, albeit implicitly, within the scope of many disciplines over the last century. Literature in diverse disciplines, such as biological sciences, catalysis, colloid transport and electrochemistry, membrane science, physics of fluids in confined media, and soil science, to name a few, provide a significant knowledge base in this area. The modern thrust in nanofluidics research has sometimes overlooked this enormous knowledge base, and reinvented ideas that have been existent for over a century. Common examples cited in this context (Eijkel and van den Berg, 2005) are the rediscovery of Donnan exclusion (Pu *et al.*, 2004) and ion rejection through narrow capillaries (Daiguji *et al.*, 2004) after nearly a century of their original incorporation in our knowledge base. Nevertheless, the promise of nanofluidics lies in the recent ability to fabricate nanoscale structures and channels with unprecedented geometrical precision, which can potentially lead to novel techniques for separation, single molecule detection, and other biological and chemical assays.

In context of the continuum theory described in this book, the fluid flow behavior in microfluidic and nanofluidic systems can be addressed in light of the same physical models. The demarcations imposed between microfluidics and nanofluidics often stem from the somewhat misleading notion that the flow mechanisms are very different in these domains. However, from the point of view of classical electrohydrodynamic theories, the fluid mechanical and electrokinetic transport behavior observed over the length scales spanning nano- and microfluidics is tractable using the same theoretical principles. The only length scale at which such continuum mechanics based theories can break down is typically below a couple of nanometers, when the solvent cannot be treated as a continuum, and quantum effects might become important. Above this length scale, however, it is generally believed that continuum approaches are adequate in resolving the flow behavior.

The history of modern microfluidics dates back to the work of Manz *et al.* (1990), who introduced the concept that a hands-off, self-contained chemical analysis system can be constructed on a miniaturized chip. This was followed by successful demonstration of chip-based fast separation of fluorescent dyes (Manz *et al.*, 1992; Jacobson *et al.*, 1994) and fluorescent labelled amino acids by capillary electrophoresis (CE) (Harrison *et al.*, 1993). Capillary electrophoresis utilizes the different electrophoretic mobilities of various molecular entities (such as proteins) suspended in a carrier fluid to separate them employing an applied electric field. Confining the carrier fluid into narrow microchannels also causes an electroosmotic flow in presence of the applied electric field. This electroosmotic flow transports the fluid along the capillary microchannel, thus obviating the requirement of any additional pumping devices.

[1]Loosely defined, systems with dimensions ranging from 1 to 1000 nanometers.

However, one can additionally utilize pressure gradients to precisely control the flow behavior in microfluidic networks.

Following these early developments, there has been an explosive growth in the research and development on microfluidic devices. Many aspects of these developments have been summarized in several excellent reviews and books on the subject (Reyes *et al.*, 2002; Auroux *et al.*, 2002; Beebe *et al.*, 2002; Verpoorte and de Rooij, 2003; Erickson and Li, 2004; Mogensen *et al.*, 2004; Cooper *et al.*, 2004; Darhuber and Troian, 2005; Li, 2004). These reviews elaborately document different aspects of microfluidics, including microfabrication techniques, fundamental aspects of microfluidic flow control, detection and diagnostic systems (such as optical and electrochemical detection techniques), and application in different areas of research such as proteomics, biochemical assays, microfluidic cooling of microchips, *etc*. A critical aspect in development of microfluidic devices is precise control of fluid transport through the system (Yoshida, 2005). Microfluidic pumping devices constitute the heart of microfluidic chips. Typically, in highly miniaturized microchannels, enormous pressure differences need to be set up to cause a pressure driven flow of a liquid. Consequently, it is more common in such systems to employ electroosmotic flow. Embedding appropriately designed electrodes within a microchannel network can lead to suitable manipulation of the liquid through these channels with very little power requirement. Microchannels also afford a high surface to volume ratio, allowing enhanced heat and mass transfer. The flows are generally in the creeping flow regime ($Re \ll 1$).

As the size of the system is reduced from microscale to the nanoscale, the size of the confining domain becomes similar to the confined entities, necessitating modification of the boundary conditions of, or the rigor needed in solving, the governing transport equations. As an illustration, consider the flow of an electrolyte solution through a capillary microchannel. For this type of flow, the parameter κa, which represents the ratio of the capillary radius, a, to the thickness of the electric double layer, κ^{-1}, is of great significance. For a large capillary radius, the parameter κa will have a large value. Most earlier models of electrokinetic flow were developed using the assumption $\kappa a \gg 1$. Such an assumption led to considerable simplification of the governing electrochemical transport equations, eventually yielding simple analytical expressions for the electrokinetic velocity of the electrolyte solution in the capillary (see Chapter 8). These simple analytical expressions (for instance, the Helmholtz–Smoluchowski expression) are remarkably accurate for flows in microfluidic channels with radii of the order of 10 to 100 microns. However, when one deals with channel radii of < 1 micron, the parameter κa may no longer be considered large. Consequently, one cannot apply the simple analytical solutions obtained using the assumption of $\kappa a \gg 1$. Hence, in nanofluidic applications, the governing electrochemical transport equations are either solved with an alternate set of assumptions (for instance, $\kappa a \ll 1$) to obtain another limiting analytical expression for nanoscale capillaries, or with an alternate set of boundary conditions that incorporate the influence of the narrow capillary dimensions in the mathematical model with greater accuracy. Such alterations in the solution methodology do not constitute an alteration of the underlying physics of the transport process. This was demonstrated in the numerical calculations of Chapter 14, where the same numerical model based on a set of fundamental governing equations for electrochemical transport worked with great accuracy over the entire range of large and small κa.

In this section, we will briefly discuss two applications, which emphasize a slightly different aspect of microfluidics. Commonly, the overarching focus of microfluidics is on miniaturization, leading to applications that are essentially confined to the microscale. While this is definitely a niche, there are few microfluidic applications that directly contribute to the macroscale industrial and analytical operations by enhancing or facilitating them. In these applications, the driving force toward miniaturization was to achieve a technical breakthrough in conventional large scale technologies. In this section, we will briefly discuss two applications that demonstrate how miniaturization can potentially lead to a significant advancement in how we measure properties of large-scale surfaces, or enhance large-scale separation processes.

15.8.1 Measurement of Zeta Potential of Macroscopic Surfaces

Measurement of zeta (ζ) potential of a solid surface immersed in an electrolyte solution is of immense practical significance. The measurement of zeta potential provides insight into the electrical properties of the surface, and how it might affect different processes of industrial relevance. For instance, surface properties of commercial polymeric membranes used for water treatment are routinely probed through measurement of zeta potential to predict their propensity of fouling. In industrial applications dealing with surface coating, it is of paramount importance to have an adequate knowledge of the electrical properties of the surfaces to be coated to ensure proper adhesion of the coating chemicals. Finally, to prevent biofouling on surfaces immersed in an aqueous environment over a long period (such as a ship's hull), one needs to assess and modify the properties of these surfaces. In all these applications, it is often necessary to measure a representative zeta potential of a very large surface.

Virtually any of the electrokinetic transport phenomena discussed in this book can be utilized to provide a measure of the ζ potential of surfaces. These techniques include streaming potential measurement, electroosmotic flow, electrophoresis, and sedimentation potential. Streaming potential measurement in appropriately designed channels have become a commonly used technique for characterizing electrical properties of planar (sheet like) surfaces. There are commercially available instruments for conducting such measurements. Typically, in these devices, one deals with a slit-channel geometry, where two facing walls of the channel are prepared by cutting appropriately sized samples of the material to be tested. These samples fit into a fixed-dimension rectangular channel cell. Cutting solid samples to fit them in the cell constitutes a destructive testing procedure. In this context, development of a streaming potential measurement tool that can conduct *in-situ* non-destructive measurements of zeta potential of large planar surfaces can be construed as a significant improvement.

Walker *et al.* (2002) conducted a study on the viability of such an electrokinetic cell, dubbed the asymmetric clamping cell, which was originally developed by Anton Paar (Graz, Austria), and is currently marketed by Brookhaven Instruments Corporation. While these authors never explicitly claimed this cell to be a microfluidic system in their work, the sub-millimeter channel dimensions used in their study, and the direct applicability of the technique to macroscale surface characterization make it an interesting case study on application of microfluidics to address an important measurement in many large-scale industrial systems.

The fundamental concept behind the measurements using the asymmetric clamping cell is to have a probe cell that can be pressed against planar substrates of varied dimensions, followed by taking a quick and facile measurement of the streaming potential. The probe cell consists of a non-conducting grooved poly-methylmethacrylate (PMMA) plate containing several parallel sub-millimeter rectangular channels fabricated using micromachining techniques. The channels are approximately 1 mm wide, 140 μm deep, and 20 mm long. When a planar surface of a test material is pressed against this grooved PMMA manifold, a number of rectangular parallel microchannels are formed. The test material forms one wall of each channel, while the other three walls are formed by the PMMA spacer. The term *asymmetric clamping cell* stems from the fact that the two opposing walls of the channel develop different zeta potentials when brought in contact with an electrolyte solution. Figure 15.39 schematically depicts the construction of the clamping cell by pressing together a grooved PMMA manifold against a flat substrate, resulting in formation of several parallel microchannels.

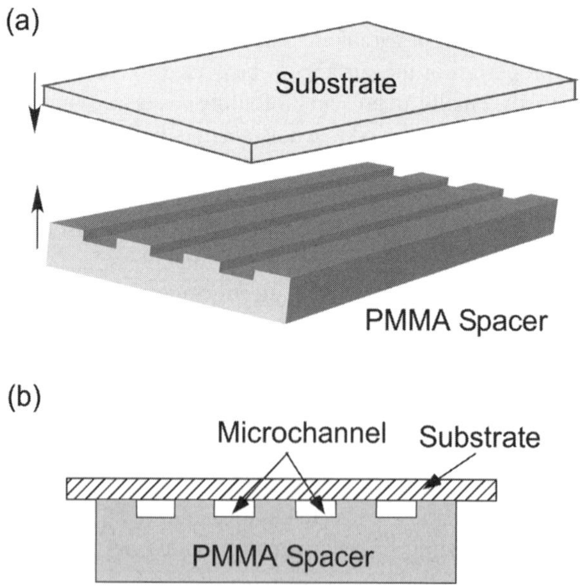

Figure 15.39. Schematic representation of (a) a grooved PMMA (poly- methylmethacrylate) spacer and the planar substrate used to construct the asymmetric clamping cell. The assembled structure is shown in (b), depicting the parallel rectangular microchannels formed by pressing the PMMA spacer against the substrate. The top surface of each channel represents the properties of the substrate, while the other surfaces represent the properties of PMMA.

The analysis of Walker *et al.* (2002) for the asymmetric clamping cell involved re-deriving the Helmholtz–Smoluchowski formulation for the streaming potential through a two-dimensional slit-microchannel geometry. The two opposing walls of the microchannel have different zeta potentials: one wall represents the zeta potential of PMMA, ζ_{PMMA}, while the other represents the zeta potential of the test substrate, ζ_S. For aqueous electrolyte solutions with monovalent electrolyte concentrations in the range of 1 mM to 100 mM, and a channel half-height, h, of 70 μm, the parameter κh is large ($\kappa h \gg 1$), allowing the use of the assumptions inherent in the Helmholtz–Smoluchowski analysis. In this section, we present the generalized derivation of the results obtained by Walker *et al.* (2002).

The geometry of the slit-microchannel used for the analysis is shown in Figure 15.40. The dimensions of the channel are such that $h \ll W \ll L$ (Figure 15.40a). Under these conditions, we can assume that the two side walls (edges) of the channel have negligible influence on the electrokinetic flow, and use a two-dimensional slit channel geometry for the analysis as shown in Figure 15.40(b). Furthermore, assuming a long channel, $L \gg h$, we can essentially employ the analysis developed for a slit microchannel geometry in Chapter 8. The sole modification of the analysis in this section stems from the use of different zeta potentials on the two opposing channel walls.

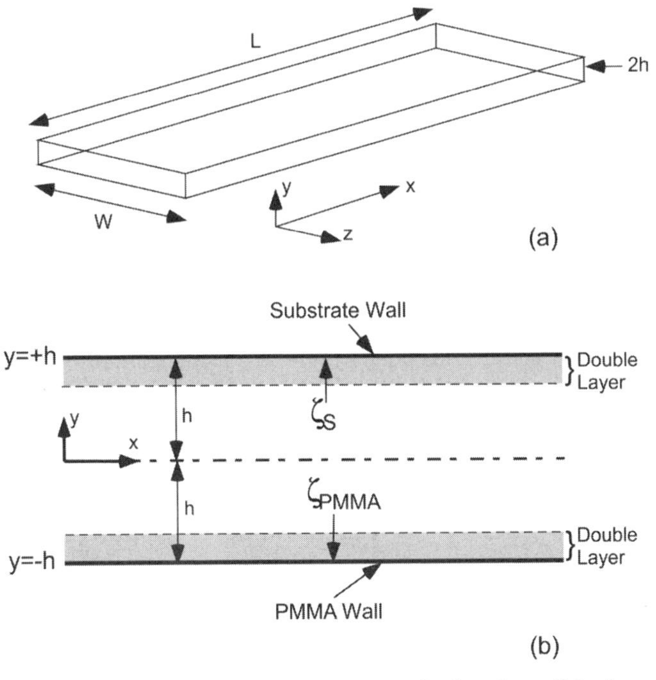

Figure 15.40. Geometric representation of the asymmetric clamping cell for the mathematical analysis of streaming potential.

Under the assumptions stated above, the electric potential distribution in the slit-microchannel can be represented by the Poisson–Boltzmann equation, linearized assuming the Debye-Hückel limit (low potentials)

$$\frac{d^2\psi}{dy^2} = \kappa^2\psi \tag{15.38}$$

where κ is the inverse Debye length. The linearized Poisson–Boltzmann equation can be solved employing the boundary conditions at the two walls of the slit-microchannel, given by

$$\psi = \zeta_S \quad \text{at } y = +h \tag{15.39}$$

$$\psi = \zeta_{PMMA} \quad \text{at } y = -h \tag{15.40}$$

The solution of Eq. (15.38) subject to the above boundary conditions provides the electric potential distribution as

$$\psi = \frac{\zeta_S + \zeta_{PMMA}}{2\cosh(\kappa h)}\cosh(\kappa y) + \frac{\zeta_S - \zeta_{PMMA}}{2\sinh(\kappa h)}\sinh(\kappa y) \tag{15.41}$$

The axial momentum balance equation for the electrolyte solution in the slit-microchannel is given by

$$\mu\frac{d^2u_x}{dy^2} = -p_x - \rho_f E_x \tag{15.42}$$

where μ is the fluid viscosity, u_x is the axial velocity of the fluid, $p_x = -\partial p/\partial x$ represents the axial pressure gradient, ρ_f is the volumetric free charge density, given by

$$\rho_f = -\epsilon\kappa^2\psi \tag{15.43}$$

and E_x is the local electric field component in the axial direction.

Applying no slip boundary conditions ($u_x = 0$) at the two stationary walls of the slit-microchannel, one can integrate Eq. (15.42) to obtain an expression for the axial velocity profile

$$u_x(y) \equiv u_x = \frac{p_x h^2}{2\mu}\left(1 - \frac{y^2}{h^2}\right) + \frac{\epsilon E_x}{\mu}\left\{\frac{\zeta_S + \zeta_{PMMA}}{2}\left[\frac{\cosh(\kappa y)}{\cosh(\kappa h)} - 1\right] + \frac{\zeta_S - \zeta_{PMMA}}{2}\left[\frac{\sinh(\kappa y)}{\sinh(\kappa h)} - \frac{y}{h}\right]\right\} \tag{15.44}$$

It may be noted that when both the walls have the same zeta potential, the last term in Eq. (15.44) vanishes, yielding the velocity profile derived in Eq. (8.30). Furthermore, if the electric field in axial direction is vanishingly small, Eq. (15.44) simplifies to the parabolic pressure-driven flow profile

$$u_x = \frac{p_x h^2}{2\mu}\left(1 - \frac{y^2}{h^2}\right) \tag{15.45}$$

Using the expressions for the electric potential and velocity distributions, one can now determine the convective transport current or streaming current per unit channel width along the axial direction, I_t (see Eq. 8.61). For the present geometry, the streaming current is expressed as

$$I_t = -\epsilon \int_{-h}^{h} u_x \frac{d^2\psi}{dy^2} dy = \epsilon \int_{-h}^{h} \frac{du_x}{dy}\frac{d\psi}{dy} dy \tag{15.46}$$

where the final expression was obtained employing integration by parts (with $u_x = 0$ at $y = \pm h$). Substituting the velocity profile, Eq. (15.44), and the potential distribution, Eq. (15.41), in Eq. (15.46) one can obtain the streaming current as

$$\begin{aligned} I_t = & -\frac{2\epsilon p_x h}{\mu}\left(\frac{\zeta_S + \zeta_{PMMA}}{2}\right)\left[1 - \frac{\tanh(\kappa h)}{\kappa h}\right] \\ & + \frac{\epsilon^2 \kappa^2 h E_x}{\mu}\left\{\left(\frac{\zeta_S + \zeta_{PMMA}}{2}\right)^2\left[\frac{\tanh(\kappa h)}{\kappa h} - \frac{1}{\cosh^2(\kappa h)}\right]\right. \\ & \left. + \left(\frac{\zeta_S - \zeta_{PMMA}}{2}\right)^2\left[\frac{1}{\sinh^2(\kappa h)} + \frac{\coth(\kappa h)}{\kappa h} - \frac{2}{(\kappa h)^2}\right]\right\} \end{aligned} \tag{15.47}$$

It may be noted that when the zeta potential of the two walls become identical, *i.e.*, $\zeta_S = \zeta_{PMMA}$, Eq. (15.47) becomes identical to Eq. (8.67). In the limiting case of $\kappa h \to \infty$, Eq. (15.47) takes the form

$$I_t = -\frac{2\epsilon p_x h}{\mu}\left(\frac{\zeta_S + \zeta_{PMMA}}{2}\right) \tag{15.48}$$

The conduction current per unit channel width along the axial direction can be expressed as (see Eq. 8.54)

$$I_c = \frac{2e^2 z^2 D n_\infty E_x}{k_B T}\int_{-h}^{h}\left[1 + \frac{1}{2}\left(\frac{ze\psi}{k_B T}\right)^2\right] dy = \sigma^\infty E_x \int_{-h}^{h}\left[1 + \frac{1}{2}\left(\frac{ze\psi}{k_B T}\right)^2\right] dy \tag{15.49}$$

where σ^∞ is the bulk electrolyte solution conductivity. Substitution of the electric potential distribution, Eq. (15.41), in Eq. (15.49) followed by evaluation of the resulting integral yields

$$I_c = 2\sigma^\infty E_x h \left\{ 1 + \frac{1}{4}\left[\left(\frac{\Psi_S + \Psi_{PMMA}}{2}\right)^2 \left(\frac{\tanh(\kappa h)}{\kappa h} + \frac{1}{\cosh^2(\kappa h)}\right) + \left(\frac{\Psi_S - \Psi_{PMMA}}{2}\right)^2 \left(\frac{\coth(\kappa h)}{\kappa h} - \frac{1}{\sinh^2(\kappa h)}\right) \right] \right\} \quad (15.50)$$

where $\Psi = ze\zeta/k_B T$. As in the case of streaming current, we note that when $\zeta_S = \zeta_{PMMA}$, the conduction current given by Eq. (15.50) becomes identical to the result of Eqs. (8.57) and (8.58). One can further simplify Eq. (15.50) in the Helmholtz–Smoluchowski limit ($\kappa h \to \infty$), yielding

$$I_c = 2\sigma^\infty E_x h \quad (15.51)$$

Equations (15.48) and (15.51) provide the streaming and conduction currents per unit channel width in the Helmholtz–Smoluchowski limit ($\kappa h \to \infty$). One can calculate the total current through a channel of width W by adding these two contributions

$$I = W(I_t + I_c) = -\frac{\epsilon p_x h W}{\mu}(\zeta_S + \zeta_{PMMA}) + 2\sigma^\infty E_x h W \quad (15.52)$$

Setting the total current to zero in Eq. (15.52) yields the required condition for the streaming potential flow, given by

$$\frac{E_x}{p_x} = \frac{\epsilon}{\mu\sigma^\infty}\frac{(\zeta_S + \zeta_{PMMA})}{2} = \frac{\epsilon}{\mu\sigma^\infty}\zeta_{Avg.} \quad (15.53)$$

where

$$\zeta_{Avg.} = \frac{(\zeta_S + \zeta_{PMMA})}{2} \quad (15.54)$$

With a constant axial electric field ($E_x = \Delta V/L$) and a linear axial pressure gradient ($p_x = -\partial p/\partial x = \Delta p/L$), where ΔV and Δp are the electric potential and pressure differences over a length, L, of the slit-microchannel, respectively, one can rewrite Eq. (15.53) as

$$\frac{\Delta V}{\Delta P} = \frac{\epsilon}{\mu\sigma^\infty}\zeta_{Avg.} \quad (15.55)$$

It is thus observed that the streaming potential in the Helmholtz–Smoluchowski limit evolves from a simple arithmetic average of the zeta potentials of the two surfaces. Consequently, if the zeta potential of one of the channel walls is known, the corresponding potential on the other wall can be easily determined from the average zeta potential obtained through the streaming potential measurement.

Equation (15.55) was employed in the analysis of streaming potential by Walker *et al.* (2002). The average zeta potential, $\zeta_{Avg.}$, was evaluated from the slope of the streaming potential *vs.* applied pressure difference plots corresponding to a fixed

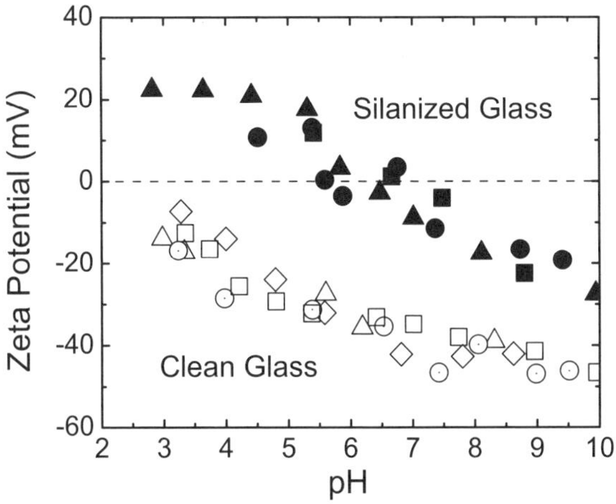

Figure 15.41. Variation of the zeta potential of clean and silanized glass surfaces with electrolyte solution pH measured using the asymmetric clamping cell. The electrolyte contains 0.01 M KCl in all measurements. Experimental data taken from Walker *et al.* (2002).

electrolyte concentration and pH. Using a known value of the PMMA zeta potential under the same experimental conditions, the unknown zeta potential of the substrate, ζ_S, was then evaluated using Eq. (15.54). Figure 15.41 depicts representative results from the analysis of Walker *et al.* (2002), where the zeta potential of clean quartz glass and a silanized sample of glass coated with aminoethyl-aminomethyl-phenethyl-trimethoxysilane (AAPhenol) were measured in an aqueous solution of 0.01 M KCl over a wide range of pH values. It is evident from the figure that the silanized glass has a distinctly different and more positive surface compared to the native glass. The point of zero charge (corresponding to the pH where the zeta potential is zero) was found to be shifted from about 2 for quartz glass to about 6.5 for the silanized glass. These results clearly indicate the ability of the asymmetric clamping cell to provide reliable estimates of the zeta potential of differently charged substrates using facile measurements of the streaming potential. The striking feature of these measurements is that the small self-contained electrokinetic cell with the grooved polymeric spacer was simply clamped to large pieces of the substrates without having to "process" the surfaces to be tested.

One can apply the principles of the asymmetric clamping cell to construct even narrower channels, thus reducing the overall size of the device. However, one should note that the pressure gradients required to drive the fluid through narrower microchannels will increase considerably in such situations. As well, the general forms of the governing equations applicable for smaller channel heights (small κh) need to be used in such situations.

15.8.2 AC Electrokinetics: Application in Membrane Filtration

In this book, we solely focussed on electrokinetic phenomena in a direct current (DC) electric field. The application of alternating current (AC) fields to cause electrokinetic effects is an equally well-developed subject, with far reaching applications in microfluidics and biology (Pohl, 1978; Jones, 1995; Hughes, 2003). The alternating electric field, as the term *alternating* suggests, arises from periodic (oscillatory) variations in the magnitude and direction of the electric field with time. When a polarizable multicomponent fluid is subjected to such an alternating electric field, one observes additional electrokinetic phenomena that are dependent on the frequency of the AC field. Some of the key features of electrokinetic phenomena observed in an AC field are:

1. The dielectric permittivities of materials subjected to an AC field become a complex function of the frequency of the imposed field. The complex dielectric permittivity, ϵ^*, of a material is expressed as

$$\epsilon^* = \epsilon - i\frac{\sigma}{\omega} \tag{15.56}$$

where ϵ is the real part of the permittivity, $i = \sqrt{-1}$ is the imaginary number, σ is the electric conductivity of the material, and ω is the frequency of the AC signal.[2]

2. A charged particle placed in a spatially uniform AC electric field will experience no Coulomb force, since the time average of the electric field will be zero. In other words, averaged over time, the net Coulomb force is given by

$$\langle \mathbf{F}_p \rangle = Q_p < \mathbf{E} >= 0 \tag{15.57}$$

where $\langle \cdots \rangle$ denotes a time averaged quantity, $\mathbf{F}_p$ is the Coulomb force, Q_p is the charge on the particle, and $\mathbf{E}$ is the electric field. This is an important concept in AC electrokinetics, since at high frequencies, say $\omega \to 1$ MHz (megaHertz), even an electrolyte solution with a substantial salt concentration will become virtually non-conducting. Furthermore, there will be no electrode polarization (formation of electric double layers) in presence of such high frequency AC fields.

3. A particle with a complex dielectric permittivity, ϵ_p^*, placed in a continuous medium with a complex dielectric permittivity, ϵ_m^*, will experience a dielectric polarization force in a nonuniform AC field, which is proportional to $\nabla(\mathbf{E} \cdot \mathbf{E})$. This force is called the *dielectrophoretic* force (Pohl, 1978; Jones, 1995). An approximate expression for the time averaged dielectrophoretic force, $< \mathbf{F}_{DEP} >$, on a spherical particle of radius a_p is

$$\langle \mathbf{F}_{DEP} \rangle = 2\pi a_p^3 \epsilon_m Re[f_{CM}] \nabla(\langle \mathbf{E} \cdot \mathbf{E} \rangle) \tag{15.58}$$

where ϵ_m is the real part of the permittivity of the suspending medium and $Re[f_{CM}]$ is the real part of a complex number, f_{CM}, known as the Clausius–Mossotti factor. The Clausius–Mossotti factor gives the frequency dependence

[2]When we refer to "AC signal", we imply an applied voltage that varies with time between a positive and negative peak value, usually sinusoidally.

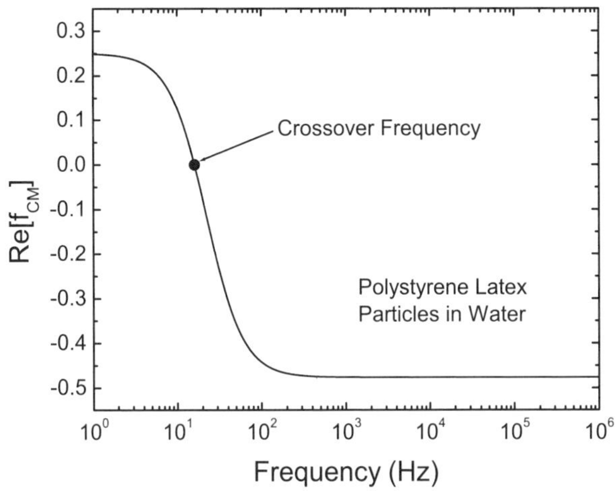

Figure 15.42. Variation of the real part of the Clausius–Mossotti factor with frequency of the AC signal for polystyrene latex particles in water. The crossover frequency demarcates the two regimes of dielectrophoresis. At lower frequencies, the system will exhibit positive dielectrophoresis, while at higher frequencies, the same system will exhibit negative dielectrophoresis.

of the polarization effect, and is expressed by

$$f_{CM} = \frac{\epsilon_p^* - \epsilon_m^*}{\epsilon_p^* + 2\epsilon_m^*} \tag{15.59}$$

The real part of the Clausius–Mossotti factor changes with the frequency of the AC signal, and varies between the limits of $+1$ to -0.5. The frequency dependent real part of the Clausius–Mossotti factor for polystyrene latex particles suspended in water is depicted in Figure 15.42. When the factor is positive, the dielectrophoretic force acts along the direction of the positive gradient of the electric field intensity (toward regions of high field intensities). In contrast, when the factor is negative, the force acts along the negative gradient of the field intensity (toward regions of low field intensities). In Eq. (15.58) the term $\langle \mathbf{E} \cdot \mathbf{E} \rangle$ represents the time averaged electric field intensity. Note that in an AC system, although the time averaged electric field, $\langle \mathbf{E} \rangle$ is zero, the corresponding time averaged field intensity, $\langle \mathbf{E} \cdot \mathbf{E} \rangle =< E^2 >$ is a finite quantity. Migration of particles along the positive electric field intensity gradient is termed as positive dielectrophoresis, while the opposite phenomenon is referred to as negative dielectrophoresis (Hughes, 2003).

It is important to note that the dielectrophoretic forces are felt only when a spatial gradient of the electric field intensity exists. These forces are absent when the electric field is spatially uniform. Another important aspect of dielectrophoretic forces is that one requires an enormous field (generally in the range of 10^6 V/m) to generate appreciable forces. In macroscopic systems, when

> the electrodes are large, one requires very large voltages to obtain measurable dielectrophoretic effects. However, if the electrode dimensions and spacings are decreased to the micrometer range, one can easily generate local electric fields of the order of 10^6 V/m by applying potentials of 1 to 10 V. Thus miniaturization is an important aspect of dielectrophoretic manipulation.

AC electrokinetic techniques based on dielectrophoresis have been utilized for manipulation, separation, and characterization of colloidal scale entities for several decades, although their application received heightened attention during the past decade (Green and Morgan, 1997; Hughes, 2000; Gascoyne and Vykoukal, 2002). The application of dielectrophoresis is particularly prevalent in the field of biology, where these forces are used to trap and sort different types of cells, and fractionate components of a cellular mixture (Pethig and Markx, 1997). Application of dielectrophoresis in flocculation of colloidal entities is also known (Eow *et al.*, 2001).

In this section, we will briefly describe a novel concept of combining dielectrophoresis with tangential flow membrane filtration, which can be used as a convenient mechanism for mitigation of particulate fouling of membranes. Once again, the rationale for describing such an application is to demonstrate that by judiciously applying a microscale technique, one might affect an advancement in a conventional large-scale technology.

Membrane filtration processes such as microfiltration and ultrafiltration operate in commercial scale applications in a continuous mode. In these applications, the feed suspension is circulated under high pressure through a membrane module. The porous wall of the membrane module is made of polymeric or ceramic membranes, which selectively allows the solvent to permeate due to the transmembrane pressure difference, and retain the solutes. In most applications involving microfiltration, the feed suspension contains colloidal particles or macromolecules. The retained solutes (colloidal particles) accumulate near the membrane surface, and eventually block the membrane pores. This phenomenon is termed as particulate fouling of the membrane. Fouling results in reduction of the solvent permeation rate through the membrane, as well as, over a long term operation, causes irreversible damage to the membrane, necessitating its replacement. Numerous techniques for mitigation of membrane fouling have been devised, although no viable technique for *in situ* fouling prevention is currently available (Molla and Bhattacharjee, 2005).

In this context, Molla *et al.* (2005) and Molla and Bhattacharjee (2005) proposed the use of dielectrophoresis on a parallel electrode array placed over a membrane as a possible mechanism for abatement of membrane fouling. The underlying concept of their approach is to embed a microfabricated parallel electrode array on a membrane. A photograph of a sample microfabricated parallel electrode array on a glass substrate is shown in Figure 15.43. Each parallel electrode in this array is about 50 μm wide, with a gap of 50 μm between consecutive electrodes. The overall array consists of 200 electrodes. The electrodes are arranged as intercalated combs, with consecutive electrodes connected to a different busbar. The two busbars are powered by a 180° phase shifted AC power supply with an adjustable root mean square (RMS) voltage and frequency setting. The 180° phase shift implies that at a given

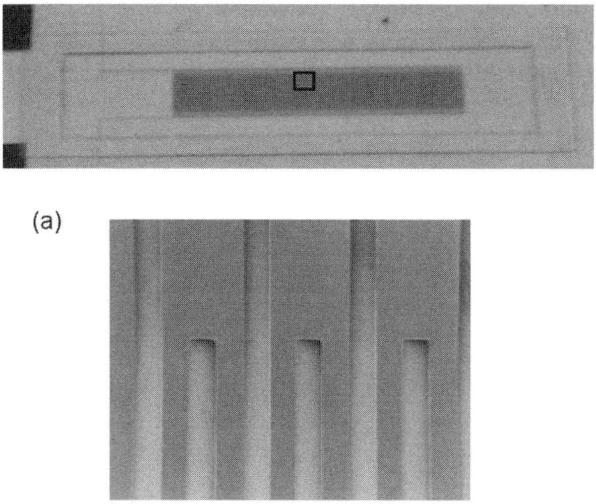

(a)

(b)

Figure 15.43. Photograph of a microfabricated gold electrode array on a glass substrate. (a) The full array with 200 electrodes. (b) Magnified view of a few electrodes within the rectangle in part (a) showing the intercalated comb structure.

instant, two consecutive electrodes will have their surface potentials set to $+V$ and $-V$. Application of such a potential to the parallel electrode array ensures that we have stationary wave dielectrophoresis in the system. In other words, the electric field intensity gradients, $\nabla(\mathbf{E} \cdot \mathbf{E})$, do not move spatially with time in this case.

Molla and Bhattacharjee (2005) conducted numerical simulations of colloidal particle trajectories near a membrane during tangential flow filtration in presence of dielectrophoretic forces. The two dimensional rectangular channel geometry used in their simulations is depicted in Figure 15.44. They numerically evaluated the electric potential distribution, the electric field intensities, and the time averaged dielectrophoretic forces acting on individual colloidal particles at different distances from the membrane surface. They used the finite element technique to solve the Laplace equation for the electric potential distribution in the fluid. The boundary conditions at the electrode surfaces were set to constant surface potential, although the potential was varied sinusoidally with time emulating the AC signal. From the potential and electric field distributions, the electric field intensity gradient components parallel and perpendicular to the membrane surface (along the x and y directions in Figure 15.44, respectively) were determined. These were substituted in Eq. (15.58) to calculate the components of the time averaged dielectrophoretic force on a particle along these directions.

In their study, Molla and Bhattacharjee (2005) also obtained a numerical solution of the Navier-Stokes equation for the fluid flow in presence of an intermittently permeable wall channel. It should be noted that placing impermeable electrodes covering part of the membrane decreases the membrane permeability, and the product water flux. Consequently, one needs to appropriately factor in this reduction of flux to study

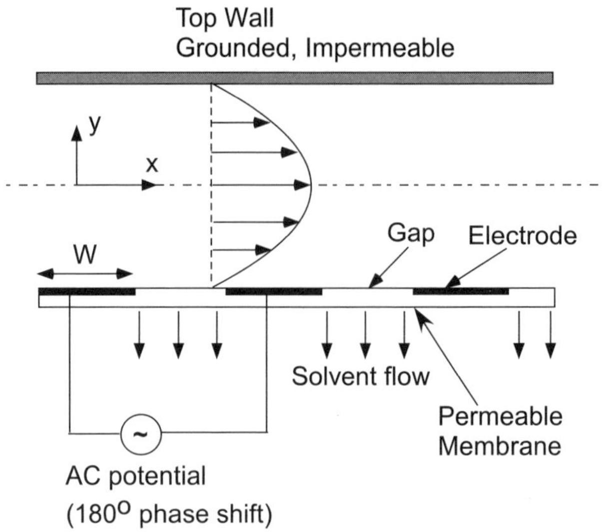

Figure 15.44. Schematic diagram of the geometry used for modeling the dielectrophoretic crossflow membrane filtration system. The tangential velocity in the channel is represented through the parabolic distribution. The gap between the electrodes is assumed to be the same as the electrode width. All dimensions are scaled with respect to the electrode width, W.

the benefits of using dielectrophoretic membrane filtration as a mechanism for reducing fouling. This necessitates a detailed accounting of the influence of intermittently permeable membrane on the solvent flux and the particle hydrodynamics.

Utilizing the dielectrophoretic forces and the fluid velocity field in a particle trajectory analysis as outlined in Chapter 13, Molla and Bhattacharjee (2005) obtained representative particle trajectories in presence of attractive and repulsive dielectrophoretic forces. They studied a system of polystyrene latex particles in water undergoing steady-state crossflow filtration in a two-dimensional rectangular channel. In all simulations, the normal fluid velocity at the membrane surface (termed suction velocity) was set to 10 μm/s. The average tangential velocity in the channel was 1 mm/s. Figure 15.45 depicts some of the representative particle trajectories obtained under different conditions. The trajectory labeled "uniform suction" represents the motion of the particle in absence of any dielectrophoresis on an unmodified membrane. Since there are no electrodes blocking the membrane surface, this situation corresponds to a standard particle trajectory in conventional membrane filtration. Notably, since the particle trajectory terminates at the membrane surface, this condition implies that the particle will deposit on the membrane and foul it. When one places the electrodes on the membrane, thus making it intermittently permeable, one obtains the trajectory labelled "intermittent suction". It is evident from these two trajectories that partially blocking the membrane surface slows down the approach of the particle toward the membrane, thus slightly reducing the fouling tendency.

The two trajectories labelled positive and negative dielectrophoresis in Figure 15.45 were obtained by applying the dielectrophoretic force. The positive

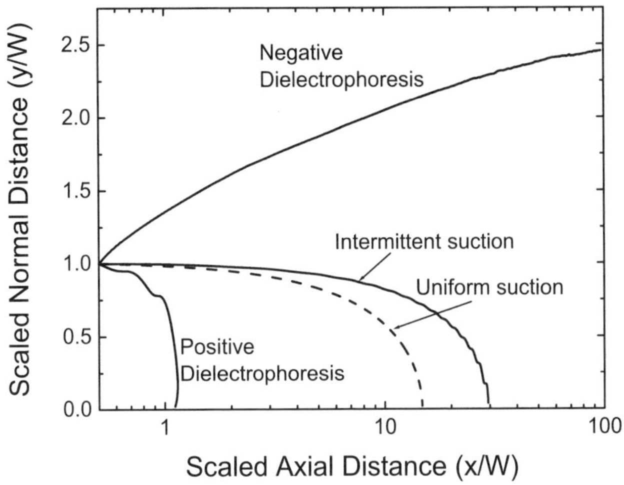

Figure 15.45. Comparison of four trajectories of a 1 μm radius polystyrene latex particle obtained under different conditions. All trajectories were calculated assuming the same fluid velocities. The average tangential and normal permeation velocity components were 10^{-3} m/s and 10^{-6} m/s, respectively. The initial position of the particle was fixed at a height of 50 μm from the electrode surface. All lengths are scaled with respect to the electrode height $W = 50$ μm. (Adapted from Molla and Bhattacharjee, 2005.)

dielectrophoresis occurs in the system at a low frequency of the applied AC signal, while negative dielectrophoresis occurs at a high frequency. This is evident from the frequency dependence of the Clausius–Mossotti function depicted in Figure 15.42. The remarkably large magnitude and range of the dielectrophoretic forces is evident from Figure 15.45, since a particle at a distance of 50 μm from the electrode surface can evidently feel this force, and migrate toward or away from the membrane under the influence of this force. The magnitude and range of the dielectrophoretic forces are much larger compared to those of the electric double layer and van der Waals forces.

It is evident from the negative dielectrophoretic trajectory that one can use these forces to repel the particles from the membrane, levitating them further away from the membrane. Such a force is independent of the presence of charge on the colloidal particles. The technique can thus serve as a unique mechanism for *in-situ* prevention of membrane fouling without having to stop the filtration operation for cleaning. In presence of dielectrophoretic levitation, since the fluid immediately adjacent to the membrane will be essentially devoid of particles, there will be no reduction in permeate flux. In fact, the membrane will operate as if it is "filtering" pure solvent, thus maximizing the solvent flux through its pores.

The above example of a microfluidic application based on AC electrokinetics to a conventional large-scale separation once again demonstrates the immense potential of microscale phenomena in providing significant breakthroughs in conventional industrial applications. The example also illustrates that the principles underlying AC electrokinetics are essentially embedded in the general framework of electrostatics and electrokinetics dealt with in this book. The interesting fact about the underlying electrostatics in both AC and DC electrokinetics is that the same Poisson and Laplace equations can be used over a wide range of frequencies. Dynamic effects, such as electromagnetic coupling become important only when extremely high frequencies are used ($\omega > 10$ MHz). Thus, an in depth knowledge of electrostatics, and some additional insight regarding frequency dependence of dielectric polarizability and permittivity of materials allow one to take the theories of DC electrokinetics to the realm of AC electrokinetics.

15.9 NOMENCLATURE

a	capillary tube radius, m
a_p	particle radius, m
BGL	blood glucose level
$c(x)$	ionic concentration at an axial distance x along the capillary axis, mol/m^3
c_1	molar feed salt concentration, mol/m^3
c_2	molar outlet salt concentration, mol/m^3
$\bar{c}$	dimensionless ionic concentration
c_i	initial oil droplet concentration
c_f	final oil droplet concentration
$c_{\pm}(r, x)$	ionic concentration, mol/m^3
C_{sc}	electric capacitance of the stratum corneum, F
D_{∞}	diffusion coefficient at infinite dilution, m^2/s
e	elementary charge, C
E_x	axial component of electric field, V/m
$\mathbf{E}$	electric field vector, V/m
f_{CM}	Clausius–Mossotti factor
$\mathcal{F}$	Faraday constant, C/mol
$\mathbf{F}_p$	Coulomb force on a charged particle, N
$\mathbf{F}_{DEP}$	dielectrophoretic force, N
h	half-height of slit-microchannel, m
i	imaginary number, $\sqrt{-1}$
J_s	total flux of dissociated salt, mol/s
k_B	Boltzmann constant, J/K
L	capillary tube length, m
n	particle concentration m^{-3}
p_x	axial pressure gradient in slit microchannel, Pa/m
Pe	rescaled particle Peclet number
Pe'	particle Peclet number
Q_p, q	charge on a particle, C
r	radial coordinate, m
R	scaled radial coordinate
Re	Reynolds number

R_g	universal gas constant, J mol^{-1} K^{-1}
R_r	salt rejection coefficient of capillary
R_{sc}	electrical resistance of the stratum corneum, Ω
R_{vs}	electrical resistance of viable skin, Ω
T	absolute temperature, K
$u_x(r)$	local axial velocity, m/s
V	average fluid velocity, m/s
x	axial coordinate, m
z	valency

Greek Symbols

α_m	maximum packing volume fraction of particles
α_p	volume fraction of dispersed phase
$\dot{\gamma}$	shear rate, s^{-1}
ΔV	streaming potential, V
ϵ	dielectric permittivity of medium, C/(Vm)
$\epsilon^*, \epsilon_m^*, \epsilon_p^*$	complex dielectric permittivities
η_T	flotation efficiency
$[\eta]$	intrinsic viscosity
κ	inverse Debye length, m^{-1}
λ	dimensionless Debye length parameter
λ_1	dimensionless Debye length at capillary inlet
λ_D	Debye length parameter, m^{-1}
μ	fluid viscosity, Pa.s
μ_0	limiting suspension viscosity at $\dot{\gamma} \to 0$
μ_∞	limiting suspension viscosity at $\dot{\gamma} \to \infty$
μ_c	continuous phase viscosity, Pa.s
μ_r	relative viscosity of a colloidal suspension
μ_s	suspension viscosity, Pa.s
$\phi(x)$	induced potential, V
$\Phi(r, x)$	total potential, V
σ	conductivity of material, S/m
σ^∞	bulk electrolyte solution conductivity, S/m
$\zeta_{Avg.}$	average zeta potential, V
ζ_{PMMA}	zeta potential of PMMA, V
ζ_S	zeta potential of test substrate, V
$\psi(r, x)$	potential due to electric double layer, V
$\bar{\psi}(r, x)$	dimensionless potential due to electric double layer
$\bar{\psi}_w$	dimensionless capillary wall potential
Ψ_S, Ψ_{PMMA}	scaled zeta potentials, $(ze\zeta/k_BT)$
ρ_f	free charge density, C/m^3
τ	shear stress, N/m^2
τ_r	reduced shear stress
ω	frequency of an AC voltage (signal), s^{-1} or Hz

15.10 REFERENCES

Auroux, P.-A., Iossifidis, D., Reyes, D. R., and Manz, A., Micro total analysis systems. 2. Analytical standard operations and applications, *Anal. Chem.*, **74**, 2637–2652, (2002).

Banga, A. K., and Chien, Y. W., Iontophoretic delivery of drugs: Fundamentals, developments, and biomedical applications, *J. Controlled Release*, **7**, 1–24, (1988).

Basu, S., Nandakumar, K., Lawrence, S., and Masliyah, J. H., Effect of Calcium Ion and Montmorillonite Clay on Bitumen Displacement by Water on a Glass Surface, *Fuel*, **83**, 17–22, (2004).

Batchelor, G. K., The effect of Brownian motion on the bulk stress in a suspension of spherical particles, *J. Fluid Mech.*, **83**, 97–117, (1977).

Beebe, D. J., Mensing, G. A., and Walker, G. M., Physics and applications of microfluidics in biology, *Annu. Rev. Biomed. Eng.*, **4**, 261–286, (2002).

Camp, F. W., The Tar Sands of Alberta, Canada, Cameron Engineers Inc., Colorado, U.S.A., (1976).

Chien, Y. W., *Novel Drug Delivery Systems*, Marcel Dekker, New York, (1982).

Cooper, J. W., Wang, Y., and Lee, C. S., Recent advances in capillary separations for proteomics, *Electrophoresis*, **25**, 3913–3928, (2004).

Czarnecki, J., Radoev, B., Schramm, L. L., and Slavchev, R., On the nature of Athabasca oil sands, *Adv. Colloid Interface Science*, **114–115**, 53–60, (2005).

Daiguji, H., Yang, P., and Majumdar, A., Ion transport in nanofluidic channels, *Nano Lett.*, **4**, 137–142, (2004).

Darhuber, A. A., and Troian, S. M., Principles of microfluidic actuation by modulation of surface stresses, *Ann. Rev. Fluid Mech.*, **37**, 425–455, (2005).

deKruif, C. G., van Iersel, E. M. F., Vry, A., and Russel, W. B., Hard sphere colloidal dispersions: Viscosity as a function of shear rate and volume fraction, *J. Chem. Phys.*, **83**, 4717–4725, (1986).

Ducker, W. A., Senden, T. J., and Pashley, R. M., Measurement of forces in liquids using a force microscope, *Langmuir*, **8**, 1831–1836, (1992).

Eijkel, J. C. T., and van den Berg, A., Nanofluidics: what is it and what can we expect from it?, *Microfluidics Nanofluidics*, **1**, 249–267, (2005).

Eow, J. S., Ghadiri, M., Sharif, A. O., and Williams, T. J., Electrostatic enhancement of coalescence of water droplets in oil: A review of the current understanding, *Chem. Eng. Sci.*, **84**, 173–219, (2001).

Erickson, D., and Li, D. Q., Integrated microfluidic devices, *Anal. Chim. Acta*, **507**, 11–26, (2004).

Everett, D. H., *Basic Principles of Colloid Science*, Royal Society of Chemistry, London, (1988).

Gad-el-Hak, M., (Ed.), *The MEMS Handbook*, CRC Press, Boca Raton, (2002).

Gascoyne, P. R. C., and Vykoukal, J., Particle separation by dielectrophoresis, *Electrophoresis*, **23**, 1973–1983, (2002).

Green, N. G., and Morgan, H., Dielectrophoretic separation of nano-particles, *J. Phys. D-Appl. Phys.*, **30**, L41–L44, (1997).

Harris, R., Iontophoresis, in *Therapeutic Electricity and Ultraviolet Radiation*, Licht, S. (Ed.), Waverly Press, Baltimore, MD, (1967).

Harrison, D. J., Fluri, K., Seiler, K., Fan, Z. H., Effenhauser, C. S., and Manz, A., Micromachining a miniaturized capillary electrophoresis-based chemical analysis system on a chip, *Science*, **261**, 895–897, (1993).

Healy, T. W., and White, L. R., Ionizable surface group models of aqueous interfaces, *Adv. Colloid Interface Sci.*, **9**, 303–345, (1978).

Hirtzel, C. S., and Rajagopalan, R., *Colloidal Phenomena: Advanced Topics*, Noyes Pub. N.J., (1985).

Hughes, M. P., AC electrokinetics: Applications for nanotechnology, *Nanotechnology*, **11**, 124–132, (2000).

Hughes, M. P., *Nanoelectromechanics in Engineering and Biology*, CRC Press, Boca Raton, (2003).

Hung, Y. T., Evaluation of flotation and sedimentation in treating oil refinery wastes, *Proc. Nat. Acad. Sci.*, **30**, 51–63, (1978).

Hunter, R. J., *Foundations of Colloid Science*, 2nd ed., Oxford Press, (2001).

Jacazio, G., Probstein, R. F., Sonin, A. A., and Yung, D., Electrokinetic salt rejection in hyperfiltration through porous materials: Theory and experiments, *J. Phys. Chem.*, **76**, 4015–4023, (1972).

Jacobson, S. C., Hergenröder, R., Koutny, L. B., and Ramsey, J. M., High speed separations on a microchip, *Anal. Chem.*, **66**, 1114–1118, (1994).

Jones, T. B., *Electromechanics of Particles*, Cambridge University Press, Cambridge, (1995).

Krieger, I. M., Rheology of monodisperse latices, *Adv. Colloid Interface Sci.*, **3**, 111–136, (1972).

Krieger, I. M., and Eguiluz, M., The second electroviscous effect in polymer latices, *Trans. Soc. Rheol.*, **20**, 29–45, (1976).

Li, D. Q., *Electrokinetics in Microfluidics*, Elsevier, Amsterdam, (2004).

Liang, W., Tadros, Th. F., and Luckham, P. F., Rheological studies on concentrated polystyrene latex sterically stabilized by poly(ethylene oxide) chains, *J. Colloid Interface Sci.*, **153**, 131–139, (1992).

Liu, J.-C., Sun, Y., Siddiqui, O., Chien, Y. W., Shi, W. M., and Li, J., Blood glucose control in diabetic rats by transdermal iontophoretic delivery of insulin, *Int. J. Pharm.*, **44**, 197–204, (1988).

Liu, J., Zhou, Z., Xu, Z., and Masliyah, J. H., Bitumen-clay interactions in aqueous media studied by zeta potential distribution measurement, *J. Colloid Interface Sci.*, **252**, 409–418, (2002).

Liu, J., Xu, Z., and Masliyah, J. H., Studies on bitumen-silica interaction in aqueous solutions by atomic force microscopy, *Langmuir*, **19**, 3911–3920, (2003).

Liu, J., *Role of colloidal interactions between oil sand components in bitumen recovery from oil sands*, Ph.D. Thesis, Chemical and Materials Engineering, University of Alberta, (2004).

Liu, J., Xu, Z., and Masliyah, J. H., Role of fine clays in bitumen extraction from oil sands, *AIChE J.*, **50**, 1917–1927, (2004a).

Liu, J., Xu, Z., and Masliyah, J. H., Interaction between bitumen and fines in oil sands extraction system: Implication to bitumen recovery, *Can. J. Chem. Eng.*, Special Issue in Oil Sands, **82**, 655–666, (2004b).

Liu, J., Xu, Z., and Masliyah, J. H., Interaction forces in bitumen extraction from oil sands, *J. Colloid Interface Sci.*, **287**, 507–520, (2005).

Manz, A., Graber, N., Widner, H. M., Miniaturized total chemical analysis systems: A novel concept for chemical sensing, *Sens. Actuators*, **B1**, 244–248, (1990).

Manz, A., Harrison, D. J., Verpoorte, E. M. J., Fettinger, J. C., Paulus, A., Ludi, H., and Widmer, H. M., Planar chips technology for miniaturization and integration of separation techniques into monitoring systems – Capillary electrophoresis on a chip, *J. Chromatogr.*, **593**, 253–258, (1992).

Masliyah, J. H. Zhou, Z., Xu, Z., Czarnecki J., and Hamza, H., Understanding water-based bitumen extraction from athabasca oil sands, *Can. J. Chem. Eng.*, Special Issue in Oil Sands, **82**, 628–654, (2004).

Mogensen, K. B., Klank, H., and Kutter, J. P., Recent developments in detection for microfluidic systems, *Electrophoresis*, **25**, 3498–3512, (2004).

Molla, S. H., and Bhattacharjee, S., Prevention of colloidal membrane fouling employing dielectrophoretic forces on a parallel electrode array, *J. Membrane Sci.*, **255**, 187–199, (2005).

Molla, S. H., Masliyah, J. H., and Bhattacharjee, S., Simulations of a dielectrophoretic membrane filtration process for removal of water droplets from water-in-oil emulsions, *J. Colloid Interface Sci.*, **287**, 338–350, (2005). Corrigendum to "Simulations of a dielectrophoretic membrance filtration process for removal of water droplets from water-in-oil emulsions (vol 287, pg 338, 2005)" *J. Colloid Interface Sci.*, **290**, 303, (2005).

NEB-EMA, National Energy Board (Canadian), An Energy Market Assessment: Canada's Oil Sands: Opportunities and Challenges to 2015, May (2004).

Okada, K., and Akagi, Y., Method and apparatus to measure the ζ-potential of bubbles, *J. Chem. Eng. Jpn.*, **20**, 11–15, (1987).

Okada, K., Akagi, Y., Kogure, M., and Yoshioka, N., Effect of zeta potentials of oil droplets and bubbles on flotation of oil-in-water mixtures, *Can. J. Chem. Eng.*, **66**, 276–281, (1988).

Okada, K., Akagi, Y., Kogure, M., and Yoshioka, N., Effect of surface charges of bubbles and fine particles on air flotation process, *Can. J. Chem. Eng.*, **68**, 393–399, (1990a).

Okada, K., Akagi, Y., Kogure, M., and Yoshioka, N., Analysis of particle trajectories of small particles in flotation when the particles and bubbles are both charged, *Can. J. Chem. Eng.*, **68**, 614–621, (1990b).

Pal, R., and Masliyah, J. H., Oil recovery from oil in water emulsions using a flotation column, *Can. J. Chem. Eng.*, **68**, 959–967, (1990).

Pethig, R., and Markx, G. H., Applications of dielectrophoresis in biotechnology, *Trends Biotechnol.*, **15**, 426–432, (1997).

Pohl, H. A., *Dielectrophoresis*, Cambridge University Press, Cambridge, (1978).

Pu, Q. S., Yun, J. S., Temkin, H., Liu, S. R., Ion-enrichment and ion-depletion effect of nanochannel structures, *Nano Lett.*, **4**, 1099–1103, (2004).

Renaud, P. C., and Probstein, R. F., Electroosmotic control of hazardous wastes, *PCH Physicochemical Hydrodynamics*, **9**, 345–360, (1987).

Reyes, D. R., Iossifidis, D., Auroux, P.-A., and Manz, A., Micro total analysis systems. 1. Introduction, theory, and technology, *Anal. Chem.*, **74**, 2623–2636, (2002).

Russel, W. B., Review of the role of colloidal forces in the rheology of suspensions, *J. Rheol.*, **24**, 287–317, (1980).

Schwendeman, S. P., Amidon, G. L., Meyerhoff, M. E., and Levy, R. J., Modulated drug release using iontophoresis through heterogeneous cation exchange membranes: Membrane preparation and influence of resin cross linkage, *Macromolecules*, **25**, 2531–2540, (1992).

Siddiqui, O., Roberts, M. S., and Polack, A. E., The effect of iontophoresis and vehicle pH on the in-vitro permeation on lignocaine through human stratum corneum, *J. Pharm. Pharmacol.*, **37**, 732–735, (1985).

Siddiqui, O., Sun, Y., Liu, J. C., and Chien, Y. W., Facilitated transdermal transport of insulin, *J. Pharm. Sci.*, **76**, 341–345, (1987).

Stein, H. N., Rheological behavior of suspensions, in *Encyclopedia of Fluid Mechanics*, Cheremisinoff, N.P. (Ed.), **5**, 3–48, Gulf, Pub. Co. (1985).

Takamura, K., and Chow, R. S., The electric properties of the bitumen/water interface II. Application of the ionizable surface-group model, *Colloids Surf.*, **15**, 35–48, (1985).

Takamura, K., Chow, R. S., and Tse, D. L., The prediction of electrophoretic mobilities and the coagulation behavior of bitumen-in-water emulsions in aqueous NaCl and $CaCl_2$ solutions using ionizable surface-group model, Proc. Symposium on Flocculation in Biotechnology and Separation Systems, San Francisco, CA, July, (1986).

Takamura, K., and Isaacs, E. E., Interfacial Properties, Chapter 5 in *AOSTRA Technical Handbook on Oil Sands, Bitumens and Heavy Oils*, Hepler, L.E., and Hsi, C., (Eds.), AOSTRA Technical Publication Series #6, Alberta (1989).

Van Ham, N. J. M., Behie, L. A., and Svrcek, W. Y., The effect of air distribution on the induced air flotation of fine oil in water emulsions, *Can. J. Chem. Eng.*, **61**, 541–547, (1983).

Verpoorte, E., and de Rooij, N. F., Microfluidics meets MEMS, *Proc. IEEE*, **91**, 930–953, (2003).

Walker, S. L., Bhattacharjee, S., Hoek, E. M. V., and Elimelech, M., A novel asymmetric clamping cell for measuring streaming potential of flat surfaces, *Langmuir*, **18**, 2193–2198, (2002).

Watterson, I. G., and White, L. R., Primary electroviscous effect in suspensions of charged spherical particles, *J. Chem. Soc. Faraday Trans. II*, **77**, 1115–1128, (1981).

Willey, S. J., and Macosko, C. W., Steady shear rheological behavior of PVC plastisols, *J. Rheol.*, **22**, 525–545, (1978).

Worldwide Look at Reserves and Production, *Oil & Gas J.*, Tulsa, Dec. 20, **102**, 47, 22, (2004).

Yamamoto, T., and Yamamoto, Y., Electrical properties of the epidermal stratum corneum, *Med. Biol. Eng. Comp.*, **14**, 151–158, (1976).

Yamamoto, Y., and Yamamoto, T., Dispersion and correlation of the parameters for skin impedance, *Med. Biol. Eng. Comp.*, **16**, 592–594, (1978).

Yoon, R.-H., and Yordan, J. L., Zeta-potential measurements on microbubbles generated using various surfactants, *J. Colloid Interface Sci.*, **113**, 430–438, (1986).

Yoshida, H., The wide variety of possible applications of micro-thermofluid control, *Microfluidics Nanofluidics*, **1**, 289–300, (2005).

INDEX

Electrokinetic and Colloid Transport Phenomena, by Jacob H. Masliyah and Subir Bhattacharjee
Copyright © 2006 John Wiley & Sons, Inc.